Astrophysics, Astronomy and Space Sciences in the History of the Max Planck Society

History of Modern Science

VOLUME 4

The titles published in this series are listed at *brill.com/hims*

Astrophysics, Astronomy and Space Sciences in the History of the Max Planck Society

By

Luisa Bonolis
Juan-Andres Leon

BRILL

LEIDEN | BOSTON

This book was funded by the Research Program History of the Max Planck Society (GMPG).

Cover illustration: Image of the protoplanetary disc surrounding the star HL Tauri taken by the Atacama Large Millimeter/submillimeter Array and showing footprints of planets forming in the gas patches within the system. Credit: ALMA (ESO/NAOJ/NRAO). CC BY 2.0.

The Library of Congress Cataloging-in-Publication Data is available online at https://catalog.loc.gov
LC record available at https://lccn.loc.gov/2022044793

Typeface for the Latin, Greek, and Cyrillic scripts: "Brill". See and download: brill.com/brill-typeface.

ISSN 2352-7145
ISBN 978-90-04-44975-6 (hardback)
ISBN 978-90-04-52913-7 (e-book)

This book is printed on acid-free paper and produced in a sustainable manner.

Contents

Preface

This book was written as part of a research project in the context of a major research program on the history of the Max Planck Society (MPG) directed by Jürgen Kocka, Carsten Reinhardt, and myself, with Florian Schmaltz as project manager. It covers the history of the MPG from its beginnings in 1948 up to 2002 and brings together the perspectives of intellectual, institutional, and contemporary history. The project was generously funded by the MPG and involved a massive effort to digitize historical documents that could be analyzed using innovative digital humanities tools. Consequently, the co-evolution of research clusters and institutional structures of more than 100 institutes could be closely followed for more than half a century. This undertaking enabled new insights into the dynamic history of one of Germany's leading research organizations and also into the interaction between science and society in the second half of the twentieth century. The results of the project will be published in a larger synthesis volume, as well as in research papers and monographic studies.

This book by Luisa Bonolis and Juan-Andres Leon is one such study and embodies the overall aims of the project on the history of the MPG in an exemplary way. It reveals a fascinating history of powerful scientific subjects, of ambitious efforts to found institutions, but also of conflicts at all levels—from the institutional to the political. Beyond a careful examination of historical documents, the book also owes much to interviews and interactions with some of the key players in this history. Taking into account these personal views and yet still maintaining a sober, objective, and sometimes critical perspective is one of the main challenges in tackling an investigation of such recent history. In my view, Luisa Bonolis and Juan-Andres Leon have mastered this challenge in an exemplary way. Their book will be a standard reference for the history of the astro- and space sciences in the MPG for many years to come.

Today astronomy, astrophysics, and the space sciences play a leading role in the society and its practitioners are among the foremost global players in their fields. This standing is the culmination of a process initiated in the aftermath of the Second World War, from rather modest beginnings. What were the origins of this extraordinary development? How did the MPG deal with the fact that the center of gravity of physics had shifted to the U.S.? What enabled it to nevertheless take advantage of the opportunities offered by the Space Age? Which opportunities were missed and why? Did the MPG follow a strategy of expansion or did this happen more by accident? Who took the critical decisions and on what basis? How did female researchers fare in a male-dominated environment? How did instrumental innovations contribute and who were the technicians and engineers who made them possible? When and how did the MPG develop from a predominantly German body into an increasingly international organization? What were the major obstacles and conflicts associated with the reorientation of the astro-institutes toward the most dynamic topics of 21st-century astrophysics? What role did science diplomacy, the making of Europe, and the situating of observatories in politically problematic environments play in this success story? These are just some of the questions addressed by the authors in this fascinating, thoroughly researched, and comprehensive account.

Jürgen Renn

Foreword

In this impressive 700-plus page book, the two experienced and well-known science historians Luisa Bonolis and Juan-Andres Leon present the results of their five-year study of the evolution of the astro-sciences in the Max Planck Society (MPG), from a very small subfield of nuclear physics in the pre-war Kaiser Wilhelm Society, to one of the larger and more highly visible activities of the MPG in the physical sciences five decades later. Bonolis and Leon present the reader with a rich historical narrative and analysis built from a detailed study of the archival sources, as well as from many discussions with individuals who shaped this story. Starting with the difficult situation post World War II, where research activities in nuclear/particle physics, radioactivity, and rockets were forbidden by allied regulations, the leading physicists in the newly formed MPG around Werner Heisenberg, Walter Bothe, Carl Friedrich von Weizsäcker, and Wolfgang Gentner had to find new directions for their work. Bonolis and Leon describe how in a first phase, two centers in cosmic ray studies and plasma physics, including theory, began to form in Göttingen/Munich and Heidelberg around Heisenberg and Gentner, the nuclei of the new astro-science activities.

In the second phase, and with the help of the next generation of leaders, including Ludwig Biermann, Reimar Lüst, Joachim Trümper, Peter Mezger, and Rudolf Kippenhahn, activities broadened. Following the 1957 Sputnik "shock" space research began, and observational astronomy across the entire electromagnetic spectrum was taken up in the 1960s and 1970s in a "cluster" of new research groups and institutes. Benefitting from the "German *Wirtschaftswunder*" and the overall growth of the MPG during this period, these new activities broadened into many emerging research areas: the solar system, the interstellar medium, star formation, cosmochemistry, and stellar evolution in galaxies including our own, as well as high-energy phenomena in neutron stars, supernovae, black holes, and finally the evolution of the universe on the largest scales. Bonolis and Leon tell the fascinating story of the growth, rivalry, and collaboration between the strong-minded, powerful directors building this new landscape. The advantages of the MPG structure soon paid off with well-funded, independent institutes and ambitious long-term research programs. In the experimental areas this opened up many new avenues, with substantial technical support groups allowing the new institutes to tackle complex and sometimes risky projects.

In the next decades the astro-cluster grew still further, becoming increasingly connected with the European activities at the European Laboratory for

Particle Physics (CERN), the European Southern Observatory (ESO), and the European Space Agency (ESA), as well as in the United States. This growth happened despite the fact that the overall budget of the MPG had begun to stagnate. At its peak in the 1980s, a remarkable 12 to 25%[1] of all MPG resources (including extra space research funds from the Ministry for Research and Technology, and from several foundations) went to the astro-cluster. At the turn of the century, with the third and fourth generations of directors at the wheel, this development of the astro-cluster had led to a world-class set of enterprises, in several cases at the forefront of global competition, achievements often owing to the specific "advantages" of the MPG funding and operating model. In their analysis of this "success story," Bonolis and Leon also touch on some of the downsides and disappointments, such as major project failures, the tensions between MPIs and local universities, the low proportion of female leaders, and the danger that the MPG "Harnack principle" can generate a dangerous power imbalance between directors and the scientists working with them.

The book finishes with a detailed study of the MPG leadership in the emerging new fields of astro-particle physics (neutrinos, TeV-γ-astronomy) and gravitational wave astronomy.

I found this detailed account highly stimulating and interesting. The early history of the cluster happened before my time, and for me a number of the threads were quite unexpected, and in many ways clarifying for the cluster's later history. Many of the more recent developments of this history I have participated in myself and I largely agree with the authors' analysis. Clearly this book succeeds in its goal of telling a history of a (large) subfield within a major basic research organization. It is not an overall history of astronomy and astrophysics in the last 50 years.[2] At the global level a similar expansion took place, driven by the same opportunities in technology, space technology, and international collaboration. Considering just the last decade, work in astronomy and astrophysics (including astro-particle physics) has garnered a number of Nobel prizes in physics, demonstrating that it is currently one of the most active fields of the physical sciences. But in comparison to this wider story,

---

1 The exact value depends on which MPG activities are counted as belonging to the astro-cluster, whether or not one only counts "core" institutes fully dedicated to astro/space-research, and how one treats external contributions from outside the MPG core budget.

2 A nice recent summary of such a broader history can be found in Malcolm Longair's "100 Years of Astronomy, Astrophysics and Cosmology: A celebration of the centenary of the IAU," in *Under One Sky: The IAU Centenary Symposium Proceedings IAU Symposium No. 349*, eds. C. Sterken, J. Hearnshaw and D. Valls-Gabaud (Cambridge University Press, 2019).

the MPG astro-history is unique in that the beginnings were so modest; Germany before World War II was not a formidable force in astronomy, and until recently the development of astronomy and astrophysics at German universities has not been as dynamic.

It is impossible to predict the future. I would expect that the comparative advantages of the MPG model in the astro-sciences will continue for some time, and there is still much to learn and discover. Yet astronomy and astrophysics projects have been increasing in cost and duration, both on the ground and in space, as well as in numerical simulations. This may push the field's central activities outside of the natural windows of advantage for a national, basic research organization based on individual excellence.

Reinhard Genzel
Max Planck Institute for Extraterrestrial Physics

Acknowledgments

This book builds on the work carried out within the Research Program History of the Max Planck Society (GMPG), which also generously funded its production and open access availability. For their contribution in one way or another to this volume we wish to thank our many colleagues from the GMPG team with whom we discussed a wide number of topics and whose views and suggestions were invaluable at all stages of writing. Many—but by no means all—of their intellectual contributions are reflected in the references. For a more definitive view of the intellectual connections between this book and the GMPG project, see the main research output of the project: *Die Max-Planck-Gesellschaft: Wissenschafts- und Zeitgeschichte 1945–2005*, vol. 1. Jürgen Kocka, Carsten Reinhard, Jürgen Renn und Florian Schmaltz (eds.), forthcoming in 2023 with Vandenhoeck & Ruprecht.

We are extremely indebted to all those who read the working drafts of particular chapters at crucial moments of their development. In particular we thank Angela Creager, Alessandro De Angelis, Reinhard Genzel, Dieter Hoffmann, Werner Hoffmann, Till Kirsten, Roberto Lalli, Adele La Rana, Paolo Lipari, Carsten Reinhardt, Jürgen Renn, Florian Schmaltz, Christian Spiering, and Joachim Trümper.

In the publication phase, we benefited from the invaluable editorial support of Lindy Divarci and Birgit Kolboske. We are also grateful to Jill Denton for her meticulous copyediting of the English language.

We would like to acknowledge the generous assistance we received from Florian Spillert during our research in the Archives of the Max Planck Society in Berlin and are particularly indebted to Susanne Uebele whose unstinting support gave us the opportunity to include the photographic materials presented in the first chapter. The Director of the Archives, Kristina Starkloff, consistently supported and facilitated the consultation of archival materials. We also thank Anita Hollier at the CERN Archives in Geneva and Antonella Cotugno at the Archives of the Department of Physics at Sapienza University of Rome for their kindness in supporting our archival research.

Special thanks go to the library staff at the Max Planck Institute for the History of Science for their patient guidance and help in obtaining rare materials. We appreciate the technical support provided by the Max Planck Institute for the History of Science for our recording of the oral history interviews. Finally, we thank the Max Planck Society's GWDG branch of our project for providing the digital infrastructure and staffing through which we, as all in the GMPG project, were able to access digitized archival material and novel analytical tools.

We are deeply grateful to the two anonymous reviewers for their insightful feedback on an earlier version of the manuscript.

Last but not least, our deepest gratitude goes to all the researchers who generously provided interviews, conversations, and personal communications that were crucial and enlightening in understanding the developments, dynamics, and intricacies of the complex and multi-layered reality of the Max Planck Society. They are acknowledged individually in the list of interviewees and footnotes. This research has made extensive use of the SAO/NASA's Astrophysics Data System.

Luisa Bonolis and Juan-Andres Leon
May 31, 2022

Oral History Interviews

Appenzeller, Immo: interview by Juan-Andres Leon, Heidelberg, August, 2016

Baars, Jacob W. M.: interview by Luisa Bonolis and Juan-Andres Leon, Bonn, February 5–7, 2018. DA GMPG, BC 601050

Begemann, Friedrich: interview by Ursula B. Marvin in 2001, Rome. Published in Oral histories in meteoritics and planetary science: VIII. Friedrich Begemann. Meteoritics and Planetary Science 37: B69–B77 (2002), https://doi.org/10.1111/j.1945-5100.2002.tb00905.x

Biermann, Ludwig Franz Benedikt: interview by Owen Gingerich, June 23 and July 7, 1978. Transcript. Niels Bohr Library & Archives, American Institute of Physics, College Park, MD USA

Biermann, Ludwig Franz Benedikt: interview by Martin Harwit, February 16, 1984. Niels Bohr Library & Archives, American Institute of Physics, College Park, MD USA

Biermann, Ludwig Franz Benedikt: interview by Woodruff T. Sullivan III, September 15, 1978, National Radio Astronomy Observatory Archives, https://www.nrao.edu/archives/items/show/896

Bondi, Hermann: interview by David DeVorkin, March 20, 1978, Niels Bohr Library & Archives, American Institute of Physics, College Park, MD USA, https://www.aip.org/history-programs/niels-bohr-library/oral-histories/4519

Cowling, Thomas: interview by David DeVorkin, March 22, 1978, Niels Bohr Library & Archives, American Institute of Physics, College Park, MD USA, https://www.aip.org/history-programs/niels-bohr-library/oral-histories/4563

Drever, Ronald W. P.: interview by Shirley K. Cohen, January-June 1997. Transcript, California Institute of Technology Archives, Oral History Project, http://resolver.caltech.edu/CaltechOH:OH_Drever_R

Falcke, Heino: interview by Juan-Andres Leon, Roberto Lalli and Luisa Bonolis, Berlin, August 23, 2019

Feustel-Büechl, Jörg: interview by Helmut Trischler on 09.04.2010. Historical Archives of the European Union. Oral History of Europe in Space Collection, https://archives.eui.eu/en/oral_history/INT065

Forward, Robert L.: interview by Harry Collins available at http://sites.cardiff.ac.uk/harrycollins/webquote/

Gentner, Wolfgang: interview by Charles Weiner, November 15, 1971. Transcript, Niels Bohr Library & Archives, American Institute of Physics, College Park, MD USA, https://www.aip.org/history-programs/niels-bohr-library/oral-histories/5080

Giazotto, Adalberto: interview by Luisa Bonolis. In Carlo Bemporad and Luisa Bonolis: *Storie di Uomini e quarks. La fisica sperimentale a Pisa e lo sviluppo della Sezione dell'Istituto Nazionale di Fisica Nucleare (1960–2010)*. Società Italiana di Fisica (2012) 175–190

Gottstein, Klaus: interview by Luisa Bonolis and Juan-Andres Leon, Munich, September 7, 2017. DA GMPG, BC 601006

Hachenberg, Otto: interview by Woodruff T. Sullivan III, Bonn, February 22, 1973. Papers of Woodruff T. Sullivan III: Tapes Series, National Radio Astronomy Observatory Archives, https://www.nrao.edu/archives/items/show/14922

Haerendel, Gerhard: interview by Helmuth Trischler and Matthias Knopp, April 9, 2010. Transcript, Historical Archives of the European Union. Oral History of Europe in Space Collection, https://archives.eui.eu/en/oral_history/INT066

Hoerner, Sebastian von: interview by Woodruff T. Sullivan III February 23, 1977. Papers of Woodruff T. Sullivan III: Tapes Series, National Radio Astronomy Observatory Archives, https://www.nrao.edu/archives/items/show/15272

Hoerner, Sebastian von: interview by Woodruff T. Sullivan III, Montreal, August 20, 1979. Papers of Woodruff T. Sullivan III, Tapes Series, National Radio Astronomy Observatory Archives, https://www.nrao.edu/archives/items/show/15273

Hofmann, Werner: interview by Luisa Bonolis and Juan-Andres Leon, Heidelberg, November 10–11, 2017. DA GMPG, BC 601010

Kafka, Peter: interview by Michael Langer, March 27, 1999. Live-Gespräch-Sendung "Zwischentöne" Deutschlandfunk, http://www.gegen-den-untergang.de/zwischentoene1999.html

Kafka, Peter: interview by Harry Collins, http://sites.cardiff.ac.uk/harrycollins/webquote/

Keller, Horst-Uwe: interview by Helmuth Trischler and Matthias Knopp, June 10, 2010. Transcript, Oral History of Europe in Space Collection, https://archives.eui.eu/en/oral_history/INT078

Kellermann, Kenneth: interview by Woodruff T. Sullivan III, Tapes Series, National Radio Astronomy Observatory Archives, https://www.nrao.edu/archives/items/show/14994

Kippenhahn, Rudolf: interview by Owen Gingerich, June 18, 1978. Transcript, Niels Bohr Library & Archives, American Institute of Physics, College Park, MD USA, https://www.aip.org/history-programs/niels-bohr-library/oral-histories/5091

Kirsten, Till: interview by Luisa Bonolis and Juan-Andres Leon, Heidelberg, October 24–25, 2017. DA GMPG, BC 601051

Kowarski, Lew: interview by Charles Weiner, 8 sessions, March 20, 1969–November 20, 1971. Transcripts, Niels Bohr Library & Archives, American Institute of Physics, College Park, MD USA, https://www.aip.org/history-programs/niels-bohr-library/oral-histories/4717-1

Krabbe, Alfred, Thomas Keilig, Christian Fischer, Dörte Mehlert, and Zaheer Ali: interview by Markus Völter, 2015. Omega Tau, http://omegataupodcast.net/189-sofia-part-1-basics/

Longair, Malcolm: interview by Robert W. Smith, June 14, 1984. Space Telescope History Project, National Air and Space Museum, Smithsonian Institution, https://sova.si.edu/record/NASM.1999.0035?s=0&n=10&t=C&q=oral+history+interview+with+Longair&i=0

Lüst, Reimar: interview by Hans von Storch and Klaus Hasselmann, December 2, 2000. Transcript, Niels Bohr Library & Archives, American Institute of Physics, College Park, MD USA, https://www.aip.org/history-programs/niels-bohr-library/oral-histories/33761

Lüst, Reimar: interview by Horst Kant and Jürgen Renn, Hamburg, May 18, 2010. DA GMPG Archive, ID 601068

Lüst, Reimar: interview by Helmut Trischler on 08.09.2010. Historical Archives of the European Union. Oral History of Europe in Space Collection, https://archives.eui.eu/en/oral_history/INT070

Marshall, L.: interview by Alan Mitchell, June 4, 1978. Transcript. Graduate Research Center of the Southwest, Center for Advanced Studies (SCAS) Collection, Special Collections and Archives Division, The University Archives, University of Texas at Dallas, Box 3, Folder 23

Menten, Karl: interview by Luisa Bonolis and Juan-Andres Leon, Bonn, February 5–8, 2018. DA GMPG, BC 601052

Mezger, Peter G.: interview by Woodruff T. Sullivan, Bonn, November 22, 1973. Transcript. Papers of Woodruff T. Sullivan III: Tapes Series, National Radio Astronomy Observatory Archives, https://www.nrao.edu/archives/items/show/15066

Mirzoyan, Razmik: interview by Juan-Andres Leon, Munich, August 13–14, 2018. GMPG DA, BC 601021

Mirzoyan, Razmik: interview by Eleni Chatzichristou, March 11, 2015. Transcript. http://www.appec.org/roadmap

Morrison, Philip: interview by Owen Gingerich, February 22, 2003, Cambridge MA, Session I. Transcript, Niels Bohr Library & Archives, American Institute of Physics, College Park, MD USA, https://www.aip.org/history-programs/niels-bohr-library/oral-histories/30591-1

Pinkau, Klaus: interview by Helmuth Trischler, March 9, 2010. Transcript, Historical Archives of the European Union. Oral History of Europe in Space Collection, https://archives.eui.eu/en/oral_history/INT072

Pirani, Felix: interview by Dean Rickles, June 23, 2011. Transcript, Niels Bohr Library & Archives, American Institute of Physics, College Park, MD USA, https://www.aip.org/history-programs/niels-bohr-library/oral-histories/34463

Priester, Wolfgang: interview by Woodruff T. Sullivan III August 30, 1976. Papers of Woodruff T. Sullivan III: Tapes Series, National Radio Astronomy Observatory Archives, https://www.nrao.edu/archives/items/show/15130

Prix, Reinhard, Christoph Affeldt, Carsten Aulbert: interviews by Markus Völter, Hannover, June 2016. *Omega Tau*, 2016. http://omegataupodcast.net/223-relativitat-und-gravitationswellen/

Rauck, Horst: interview by Helmut Trischler, June 19, 2010. Historical Archives of the European Union. Oral History of Europe in Space Collection, https://archives.eui.eu/en/oral_history/INT073

Schunck, Hermann: interview by Adele La Rana, May 14, 2019

Schwarzschild, Martin: interview by Spencer Weart, July 30, 1975. Transcript, Niels Bohr Library & Archives, American Institute of Physics, College Park, MF USA, https://www.aip.org/history-programs/niels-bohr-library/oral-histories/28321

Schwarzschild, Martin: interview by David DeVorkin and Spencer Weart, December 16, 1977, Session III. Transcript, AIP, https://www.aip.org/history-programs/niels-bohr-library/oral-histories/4870-3

Schwarzschild, Martin: interview by William Aspray, November 18, 1986, Transcript, Charles Babbage Institute, https://conservancy.umn.edu/handle/11299/107629

Spitzer, Lyman: interview by Joan Bromberg, March 15, 1978. Transcript, Niels Bohr Library & Archives, American Institute of Physics, College Park, MF USA, https://www.aip.org/history-programs/niels-bohr-library/oral-histories/4900

Townes, Charles Hard: interview by Suzanne B. Riess, 1991–1992. Transcript, Selected oral histories from the UC Berkeley Oral History Center, Online Archive of California, https://oac.cdlib.org/ark:/13030/kt3199n627/?brand=oac4

Trümper, Joachim: interview by Helmuth Trischler and Matthias Knopp, Munich, March 18, 2010. Transcript, Historical Archives of the European Union. Oral History of Europe in Space Collection, https://archives.eui.eu/en/oral_history/INT076

Trümper, Joachim: interview by Luisa Bonolis and Juan-Andres Leon, Munich, August, 7–8, 2017. DA GMPG, BC 601036

Unsöld, Albrecht: interview by Owen Gingerich, June 6, 1978. Transcript, Niels Bohr Library & Archives, American Institute of Physics, College Park, MD USA, https://www.aip.org/history-programs/niels-bohr-library/oral-histories/4924

Völk, Heinrich: interview by Luisa Bonolis and Juan-Andres Leon, Heidelberg, October 9–10, 2017. DA GMPG, BC 601037

Wänke, Heinrich: interview by Ursula B. Marvin in 2002. Published in Oral histories in meteoritics and planetary science: IX. Heinrich Wänke. Meteoritics and Planetary Science 37: B79–B88 (2002), https://onlinelibrary.wiley.com/doi/abs/10.1111/j.1945-5100.2002.tb00906.x

Weiss, Reiner: interview by Shirley K. Cohen, May 10, 2000. Transcript. California Institute of Technology Archives, Oral History Project, https://resolver.caltech.edu/CaltechOH:OH_Weiss_R

Weizsäcker, Carl Friedrich, Freiherr von: interview by Karl Hufbauer, April 18, 1978. Transcript, Niels Bohr Library & Archives, American Institute of Physics, College Park, MD USA, www.aip.org/history-programs/niels-bohr-library/oral-histories/4948

Wheeler, John Archibald: interview by Kenneth W. Ford, Session IX, March 4, 1994, https://www.aip.org/history-programs/niels-bohr-library/oral-histories/5908-9, and Session XII, 28 March 1994, https://www.aip.org/history-programs/niels-bohr-library/oral-histories/5908-12 Niels Bohr Library & Archives, American Institute of Physics, College Park, MD USA

Illustrations

Unless otherwise specified, all the photographs are reproduced with kind permission of the Archives of the Max Planck Society, Berlin-Dahlem. Chapter 1, relating to the foundational era, is the only chapter with illustrations.

Acronyms and Abbreviations

4LGSF	Laser Guide Star Facility
ABM	Anti-Ballistic Missile
ABMA	Army Ballistic Missile Agency
ABRIXAS	A BRoadband Imaging X-ray All-sky Survey
AEC	Atomic Energy Commission
AERE	Atomic Energy Research Establishment
AGN	Active Galactic Nuclei
AIGO	Australian International Gravitational Observatory
AIP	American Institute of Physics, Niels Bohr Library & Archives
AIROBICC	Air shower Observation By angle Integrating Cherenkov Counters
ALMA	Acatama Large Millimeter/submillimeter Array
AMBER	Astronomical Multi BEam Recombiner
AMPG	Archiv der Max-Planck-Gesellschaft
AMPTE	Active Magnetospheric Particle Tracer Explorer
AMS	Alpha Magnetic Spectrometer
AOMC	Army Ordinance Missile Command
APEX	Atacama Pathfinder Experiment
ARI	Astronomisches Rechen-Institut
ARPA	Advanced Research Projects Agency
ASPERA	AStroParticle European Research Area
ATCA	Australia Telescope Compact Array
AVA	Aerodynamische Versuchsanstalt
AVIDAC	Argonne Version of the Institute's Digital Automatic Computer
BC	Barcode
BESSY	Berlin Electron Storage Ring Society for Synchrotron Radiation
BKG	Bundesamt für Kartographie und Geodäsie
BMAt	Bundesministerium für Atomfragen
BMBF	Bundesministerium für Bildung und Forschung (Federal Ministry for Education and Research)
BMBW	Bundesministerium für Bildung und Wissenschaft (Federal Ministry for Education and Science)
BMFT	Bundesministerium für Forschung und Technologie (Federal Ministry for Research and Technology)
BNL	Brookhaven National Laboratory
BNSC	British National Space Centre
BOREXINO	BORon solar neutrino EXperiment
Caltech	California Institute of Technology

CANGAROO	Collaboration of Australia and Nippon for Gamma-Ray Observation in the Outback
CASA-MIA	Chicago Air Shower Array—Michigan muon Array
CCD	Charge-Coupled Device
CDU	Christian Democratic Union
CERN	Conseil Européen pour la Recherche Nucléaire
CGRO	Compton Gamma Ray Observatory
CMB	Cosmic Microwave Background
CNES	Centre National d'Études Spatiales
CNRS	Centre National de la Recherche Scientifique
COBE	Cosmic Background Explorer
COMSAT	Communications Satellite Corporation
CONICA	COudé Near Infrared CAmera
COPERS	Commission Préparatoire Européenne de Recherches Spatiales
COSPAR	Committee on Space Research
CPTS	Chemisch-Physikalisch-Technische Sektion
CRT	Cosmic Ray Tracking
CRIRES	Cryogenic high-resolution InfraRed Echelle Spectrograph
CSIRO	Commonwealth Scientific and Industrial Organization
CSNSM	Centre de Sciences Nucléaires et de Sciences de la Matière
CSU	Christian Social Union
CTA	Cherenkov Telescope Array
DA GMPG	Digital Archive of the Research Program History of the Max Planck Society
DARPA	Defense Advanced Research Projects Agency
DDR	Deutsche Demokratische Republik
DESY	Deutsches Elektronen-Synchrotron
DFG	Deutsche Forschungsgemeinschaft (German Research Foundation)
DFL	Deutsche Forschungsanstalt für Luftfahrt (German Research Institute for Aviation)
DFVLR	Deutsche Forschungs-und Versuchsanstalt für Luft- und Raumfahrt (German Test and Research Institute for Aviation and Space Flight)
DIRBE	Diffuse Infrared Background Experiment
DLR	Deutsches Zentrum für Luft- und Raumfahrt (German Aerospace Center)
DMR	Differential Microwave Radiometer
DOE	Department of Energy
DPG	Deutsche Physikalische Gesellschaft
DVL	Deutsche Versuchsanstalt für Luftfahrt (German Test Institute for Aviation)

EAS	Extensive Air Showers
ECFA	European Committee for Future Accelerators
EGRET	Energetic Gamma Ray Experiment Telescope
EHTC	Event Horizon Telescope Collaboration
EISCAT	European Incoherent Scatter Scientific Association
ELDO	European Launcher Development Organisation
ELT	Extremely Large Telescope
EPIC	European Photo Imaging Camera
ERC	European Research Council
ERIC	European Research Infrastructure Consortium
ERIS	Enhanced Resolution Imager and Spectrograph
ERNO	Entwicklungsring Nord (Northern development circle)
eROSITA	extended ROentgen Survey with an Imaging Telescope Array
ESA	European Space Agency
ESO	European Southern Observatory
ESO-ELT	ESO Extremely Large Telescope
ESO-VLT	ESO Very Large Telescope
ESRIN	European Space Research Institute
ESRO	European Space Research Organisation
EURATOM	European Atomic Energy Community
EXOSAT	European X-Ray Observatory Satellite
FIAT	Field Information Agency Technical
FIRAS	Far Infrared Absolute Spectrophotometer
FORS	FOcal Reducing Imager and Spectrograph
GALLEX	Gallium Experiment
GAMM	Gesellschaft für Angewandte Mathematik und Mechanik
GDR	German Democratic Republic
GENIUS	GErmanium in liquid NItrogen Underground Setup
GERDA	Germanium Detector Array
GeV	Gigaelectronvolt
GIRL	German Infrared Laboratory
GLAST	Gamma-ray Large Area Space Telescope
GMPG	Geschichte der Max-Planck-Gesellschaft
GNO	Gallium Neutrino Observatory
GRB	Gamma Ray Burst
GREAT	German Receiver for Astronomy at Terahertz Frequencies
GUT	Grand Unified Theory
GVMPG	Generalverwaltung der Max Planck Gesellschaft
HAEU	Historical Archives of the European Union
HDMS	Heidelberg Dark Matter Search

HEAO	High Energy Astronomy Observatories
HEGRA	High Energy Gamma-Ray Astronomy
HEOS	Highly Eccentric Orbit Satellite
H.E.S.S.	High Energy Stereoscopic System
HEXE	High Energy X-Ray Experiment
HHT	Heinrich Hertz Submillimeter Telescope
IACT	Imaging Atmospheric Cherenkov Telescopes
IAEA	International Atomic Energy Agency
IAS	Institute for Advanced Study
IAU	International Astronomical Union
IBM	International Business Machines Corporation
ICBM	InterContinental Ballistic Missiles
ICRC	International Cosmic Ray Conference
ICSU	International Council of Scientific Unions
IGN	Instituto Geográfico Nacional
IGY	International Geophysical Year
IKI	Space Research Institute Moscow
ILL	Institut Laue-Langevin
INFN	Istituto Nazionale di Fisica Nucleare
INSTN	Institut National des Sciences et Techniques Nucléaires
INTEGRAL	INTErnational Gamma-Ray Astrophysics Laboratory
IPP	Institute for Plasma Physics
IRAM	Institut de Radioastronomie Millimétrique
IRAS	Infrared Astronomical Satellite
ISO	Infrared Space Observatory
ISS	International Space Station
ITER	International Thermonuclear Experimental Reactor
JADE	Japan Deutschland England detector
JINR	Joint Institute for Nuclear Research
JIVE	Joint Institute for VLBI
JPL	Jet Propulsion Laboratory
JWST	James Webb Space Telescope
KAGRA	Kamioka Gravitational Wave Detector
KAMIOKANDE	Kamioka Nucleon Decay Experiment
KeV	Kiloelectronvolt
KMOS	K-band Multi Object Spectrograph
LAS	Large Astronomical Satellite
LBT	Large Binocular Telescope
LEP	Large Electron-Positron Collider
LETG	Low Energy Transmission Grating

LHC	Large Hadron Collider
LIGO	Laser Interferometer Gravitational Wave Observatory
LINC-NIRVANA	LBT Interferometric Camera—Near-IR/Visible Adaptive Interferometer for Astronomy
LISA	Laser Interferometer Space Antenna
LLNL	Lawrence Livermore National Laboratory
LMU	Ludwig-Maximilians-Universität
LNGS	Gran Sasso National Laboratories
LOFAR	Low Frequency Array
LST	Large-Sized Telescope
LUCI	LBT Near Infrared Spectroscopic Utility with Camera and Integral Field Unit for Extragalactic Research
MACAO	Multi-Application Curvature Adaptive Optics
MACRO	Monopole, Astrophysics and Cosmic Ray Observatory
MAGIC	Major Atmospheric Gamma-ray Imaging Cherenkov
MAN	Maschinenfabrik Augsburg-Nürnberg
MANIAC	Mathematical Analyzer Numerical Integrator and Computer
MATISSE	Multi-AperTure mid-Infrared SpectroScopic Experiment
MBB	Messerschmitt-Bölkow-Blohm
MeV	Megaelectronvolt
MIDAS	Missile Defense Satellite System
MIDI	Mid-Infrared Interferometric instrument
MIMOSA	Milano-Monaco-Saclay collaboration
MIT	Massachusetts Institute of Technology
MORABA	Mobile RAketenBasis (Mobile Rocket Base)
MPA	Max-Planck-Institut für Astrophysik
MPAE	Max-Planck-Institut für Aeronomie
MPE	Max-Planck-Institut für Extraterrestrische Physik
MPG	Max-Planck-Gesellschaft (Max Planck Society)
MPI	Max-Planck-Institut
MPIA	Max-Planck-Institut für Astronomie
MPIB	Max-Planck-Institut für Biochemie
MPIC	Max-Planck-Institut für Chemie
MPIfR	Max-Planck-Institut für Radioastronomie
MPIK	Max-Planck-Institut für Kernphysik
MPIP	Max-Planck-Institut für Physik
MPIPA	Max-Planck-Institut für Physik und Astrophysik
MUSE	Multi-Unit Spectroscopic Explorer
NACO	NAOS—CONICA
NAOS	Nasmyth Adaptive Optics System

NASA	National Aeronautics and Space Administration
NATO	North Atlantic Treaty Organization
NIVR	Netherlands Agency for Aerospace Programmes
NOAO	National Optical Astronomy Observatory
NOBEYAMA	National Astronomical Observatory of Japan
NOEMA	NOrthern Extended Millimeter Array
NPT	Non Proliferation Treaty
NRAO	National Radio Astronomy Observatory
NRAO/AUI	National Radio Astronomy Observatory/Associated Universities Inc.
NSF	National Science Foundation
NTT	New Technology Telescope
OPAL	Omni-Purpose Apparatus at LEP
OTRAG	Orbital Transport und Raketen (Orbital Transport and Rockets)
PEP	Positron-Electron Project
PETRA	Positron Electron Tandem Ring Accelerator
PSAC	President's Scientific Advisory Committee
ROSAT	ROentgen SATellite
SAGE	Soviet-American Gallium Experiment
SERC	Science and Engineering Research Council
Sgr A*	Sagittarius A*
SHARP	System for High Angular Resolution Pictures
SINFONI	Spectrograph for INtegral Field Observations in the Near Infrared
SKA	Square Kilometre Array
SLAC	Stanford Linear Accelerator Center
SOFIA	Stratospheric Observatory for Infrared Astronomy
SOUSY	SOunding SYstem for atmospheric structure and dynamics
SPHERE	Spectro-Polarimetric High-contrast Exo-Planet REsearch
SPIFFI	SPectrometer for Infrared FaintField Imaging
SPIM	Service de prévision ionosphérique de la Marine
SPS	Super Proton Synchrotron
SRC	Science Research Council
SRG	Spectrum-Roentgen-Gamma
SSC	Superconducting Super Collider
SWS	Short Wavelength Spectrometer
TDRSS	Tracking and Data Relay Satellite System
TERLS	Thumba Equatorial Rocket Launching Station
TeV	Teraelectronvolt
TEXUS	Technologische EXperimente Unter Schwerlosigkeit (Technological microgravity experiments)
TPC	Time Projection Chamber

TUM	Technische Universität München
UHE	Ultra High Energy
UHECR	Ultra High Energy Cosmic Rays
UKSA	United Kingdom Space Agency
UNIVAC	UNIVersal Automatic Computer
USSR	Union of Soviet Socialist Republics
VERITAS	Very Energetic Radiation Imaging Telescope Array System
VHE	Very High Energy
VLA	Very Large Array
VLBA	Very Long Baseline Array
VLBI	Very-Long-Baseline Interferometry
VLTI	Very Large Telescope Interferometer
XMM-Newton	X-Ray Multi-Mirror Mission-Newton Space Telescope
WRK	West German Rectors' Conference
ZETA	Zero Energy Thermonuclear Assembly
ZIAP	Zentralinstitut für Astrophysik

Introduction

A Cosmic Explosion

The 20th century saw one of the most radical transformations in our worldview, brought about by remarkable advances in astronomy, astrophysics, and outer space exploration, which we refer to collectively as cosmic research.[1] Within one century, our place in the Universe was augmented, from a static picture in what was thought to be 'the Galaxy,' to a dynamic perspective with a distinct origin and evolution, made possible by a synergy between theoretical advancements in physics and the development of a wide array of new instruments and methodologies for observing the environments beyond our home planet.

While the theoretical seeds for much of this understanding date from the early years of the century, it was largely after World War II that new observational techniques and the ability to access outer space made it possible to turn hypothetical approaches into a very established picture of the Universe with countless galaxies and other exotic bodies, which have also evolved and existed in unique periods of its 15-billion-year existence. Scientists were able to ascertain how stars are formed, 'work,' evolve, and die, based on the new understandings of nuclear and subnuclear physics and on general relativity, and then were able to infer the existence of these processes (sometimes even see them happen) from observations with different kinds of 'telescopes.' Radio telescopes, in particular, were the first to reveal that the Universe to a large part consists of ionized matter—plasma—whose constituent particles exhibit an exceedingly complicated collective behavior. The opening of the plasma universe has permanently changed our picture of the cosmos, from a 'void' to a rich, energetic environment of particles and electromagnetic fields. Concurrently, the development of an entire framework of plasma physics to

---

1 For comprehensive surveys of the scientific developments in Astronomy and Astrophysics, see: Malcolm S. Longair: *The Cosmic Century. A History of Astrophysics and Cosmology*. Cambridge: Cambridge University Press 2006. Malcolm Longair: 100 Years of Astronomy, Astrophysics and Cosmology. A Celebration of the Centenary of the IAU. In: C. Sterken, J. Hearnshaw, and D. Valls-Gabaud (eds.): *Under One Sky: The IAU Centenary Symposium Proceedings IAU Symposium*. 349. Cambridge: Cambridge University Press 2019, 3–24. doi:10.1017/S1743921319000097. Martin Harwit: *Cosmic Discovery. The Search, Scope, and Heritage of Astronomy*. Cambridge, MA: Cambridge University Press 2019. Martin Harwit: *Cosmic Messengers. The Limits of Astronomy in an Unruly Universe*. Cambridge: Cambridge University Press 2021. For the space sciences, see: Johan A. M. Bleeker, Johannes Geiss, and Martin C. E. Huber (eds.): *The Century of Space Science*. Dordrecht: Kluwer Academic Publishers 2001.

describe the complex thermonuclear processes in stars and their vicinity, with the spread of high-energy radiation and particles that accompanies them, was applied to the experimental reproduction of some of these processes, turning nuclear fusion into the promise of an imaginable energy source on Earth.

The origin of the chemical elements in the early Universe and in the center of stars was also explained, and through this understanding, scientists managed to follow the evolution of galaxies over time and determine the age of our planet and its solar system. Scientists understood the planets' evolution and interaction with the Sun, and even 'visited' them with interplanetary probes, before it was confirmed, in the last decade of the century, that our solar system and its planets are typical of what we can expect in similar stars. Moreover, many of the techniques and concepts necessary for understanding our fragile biosphere were first developed in the quest for the infinite beyond, before we 'turned them around' to understand and help shape our own fragile planetary existence.

A multitude of exotic cosmic bodies was discovered and then understood thanks to the observation of incoming radiation beyond visible light, the traditional optical 'window,' in which all telescopes operated before the mid-20th century. The entire electromagnetic spectrum became available, incorporating the long waves of radio and infrared, and the short waves in the ultraviolet, X-ray, and gamma-ray domains. Observation at each of these wavelengths originated in separate traditions in astronomy, experimental physics, and engineering, which only gradually combined over several generations, in the latter part of the century, while all these observations were increasingly interwoven with theoretical astrophysics.

Since many of these new 'windows' are accessible only from very high, remote locations on Earth, or outside Earth's atmosphere altogether, the 'new' astronomies led to the most significant scientific use of spaceflight, with the creation of satellite-based observatories. And even on the ground, the need for observations from increasingly remote locations fostered an organization of observational infrastructure and international division of scientific labor akin to that pioneered by space-based astronomy and experimental high-energy physics.

Eventually, a magnificent convergence of observations in new wavelengths, combined with new theoretical insight from physicists originating in many different traditions, led to the confirmation both of the Big Bang as the origin of the Universe, and the existence of exotic entities such as neutron stars and black holes.

Futhermore, in the last decades of the 20th century, a multitude of researchers initially apprenticed in experimental particle physics applied their expertise to develop creative new means of observing the Universe, intertwining detection, theory, and computational simulation in order to overcome the problem of feeble, exotic signals and particles hidden in a sea of 'noise.'

Most significantly for the current cutting edge of scientific research, cosmic research during the second half of the 20th century turned from being an arena of speculation on, and extrapolation of understandings of physics seen first experimentally on Earth, into a real-time laboratory that has meanwhile become the source of the further comprehension and completion of physical theories beyond our experimental means. Cosmic rays and multiple other 'astroparticles' such as neutrinos, as well as gravitational waves, are our sole means of access to phenomena crucial for the further development of fundamental physics. At the same time, these entities are also 'messengers' via which we augment our understanding of the complex, irreducible 'ecosystem' that is the Universe.[2]

Right at the time of writing this, all these developments came together with the spectacular attainment of 'multi-messenger astronomy': in 2017, the collision of two neutron stars—entities and processes that had been the realm of theoretical astrophysics for half a century, and were predicted to be the origin of the majority of heavy chemical elements—was observed in real time around the world (save for a time delay of over one hundred million years), by telescopes for all wavelengths, as well as by astroparticle 'telescopes' triggered by the detection of the event by brand-new gravitational wave observatories. The latter were the result of a century of developments in an often, marginalized branch of theoretical astrophysics, which served to perfect a meticulous observational technique guided by theoretical and simulational projections, an enterprise that detected gravitational waves only four decades after its inception, to finally join the mainstream of astrophysical research in the last few years. Completely independently, also in 2017, an extremely energetic neutrino was traced with the aid of gamma-ray astronomy to its origin near a violent, supermassive black hole at the center of a galaxy four billion light years away. This 'blazar,' one-third of the age of the Universe away from us, became the first identified source of high energy cosmic rays, providing answers to a century-old foundational mystery of modern physics. That same year, millimeter-wavelength observatories around the world pointed simultaneously towards the center of our galaxy as well as to M87, and via the technique of very long baseline interferometry yielded the first 'image' of the

---

2 Simon D. M. White: Fundamentalist Physics: Why Dark Energy Is Bad for Astronomy. *Reports on Progress in Physics* 70/6 (2007), 883–897. doi:10.1088/0034-4885/70/6/R01.

shadow of a black hole. And deeper, multi-decade observations of the center of our galaxy have first observed and then confirmed the effects of Einstein's general theory of relativity under the extreme conditions in the vicinity of 'our' supermassive black hole, sitting at the center of the Milky Way.

Along the way, such advances have also deeply transformed cosmology, from a largely speculative science into an observational science that uses telescopes and detectors to study the structure, evolution, and origin of the Universe, and further blurs the lines between its sibling sciences, astronomy and astrophysics.

The starting point of our book is the reminder that in all the developments described above, the institutes, observatories, scientists, and instrumentation of the Max Planck Society (Max Planck Gesellschaft, MPG) have played significant, often pivotal roles. And it need hardly be said that such scientific developments were closely intertwined with the most important technological and sociopolitical changes of the 20th century, spanning the end of World War II and the entire Cold War era, immersed in a march towards globalization that was particularly resolute in the research fields treated here.

What This Book Is Not

This book does not attempt to offer an exhaustive survey of all the research conducted in astronomy, astrophysics, and space science in the Max Planck Society in the 20th century. Instead, its aim is to show scientific developments in their sociopolitical context and pinpoint the ways in which research in the Max Planck Society was organized to adapt to the expanding global constellation of specialties in this wide field. This adaptive process began under postwar Allied constraints on the reemergence of a national research system in West Germany, yet at a time when leadership in science was key to the country's resurgence within Europe and worldwide in the Cold War era. The Max Planck Society was successor to the Kaiser Wilhelm Society (Kaiser Wilhelm Gesellschaft, KWG) which was founded in the later imperial era and became Germany's foremost research institution in the course of the Weimar and Nazi eras, while adapting to the changing sociopolitical circumstances. Throughout this early history, astronomy and astrophysics had a modest footprint in the research focus of its scientists. Their expansion occurred rather during the postwar era, both through the reorientation of existing research traditions towards cosmic questions, and the incorporation of completely new fields and researchers. As we will see, cosmic research in West Germany came to be heavily dominated by the Max Planck Society within a single generation. The participating sciences played a prominent role in the modernist imaginary, and Max Planck scientists benefited from this, latching on first to the 'nuclear age' and then, after Sputnik, to the 'space age.' It was in this postwar era that cosmic research, too, began showcasing the newfound credo of the Max Planck Society: to foster 'fundamental science'—knowledge for its own sake. Our book illustrates, however, how all these activities were deeply intertwined with the ideological and technological needs of the Cold War era. Key to the success of the different scientific traditions treated in this book was their participation in scientific networks with colleagues in the Allied countries (United States, Britain, and France), who were often immersed in research relevant to military purposes, such as nuclear fusion, nuclear bomb testing, long-distance communication, radar, rockets, ballistic missile tracking, spy satellites, and even space-based weapons. The growing prestige of cosmic research after Sputnik even allowed for a previously inconceivable expansion of the Max Planck Society into the existing discipline of observational astronomy, leading to a series of ground- and space-based observatories in all the observational wavelengths, and novel messengers such as neutrinos and gravitational waves.

The end of the Cold War then consolidated the transition to internationalization that was already underway in scientific research around the world. The last decades of the century saw a shift away from the construction of national research infrastructures, towards a globally intertwined ecosystem of scientific research projects based in multinational mega-infrastructural platforms. Links to the military-industrial complex of the Western Alliance became subtler, while a more market-oriented logic now justified its generous public patronage. This was further catalyzed by German reunification in 1990, which provoked a reshuffle both of the research landscape in the former GDR, under conditions of austerity, and of the extant institutes in the West, and risked ending in a zero-sum game. The research institutes in the West had to mobilize their international prestige and their influence in Germany to assure their immediate survival. Nonetheless, cosmic research in the Max Planck Society has maintained astonishing continuities in the first two decades of the 21st century, which we analyze as scientific traditions dating all the way back to the early postwar years and beyond, with roots in the golden age of German science of the interwar era. The names and activities may have changed, yet an unmistakable lineage can still be traced in the epistemic and technological cosmic research practices of the Max Planck Society, as well as in those of its regional, national, and international allies.

How This Book Fits in the Wider Historiography of the Max Planck Society

Cosmic research is an excellent example of how this interplay of scientific developments and larger sociopolitical forces was ably channeled by the Max Planck Society, in its rise to dominance over the scientific landscape in Germany. The Society, typically West German in its constitution as a private and independent society (like a charity or club), despite its vast dependence on public funding, is an exemplar of the organizations emerging in the postwar era, in constant overlap and competition with one another at the federal and regional levels and across the sphere of influence of various government ministries. In this book we will see, too, how, within the coordinated capitalist system of West Germany and its allies, 'fundamental research' was repeatedly put to ideological use in interactions between the military-industrial complex and predominantly state-funded scientists.

The historical role of the Max Planck Society in the second half of the 20th century is the central inquiry of the Research Program on the History of the Max Planck Society (GMPG), under the auspices of which this book has been written. For this reason, aspects of the Society that are common to all its institutes and scientific activities are not treated in detail in this book. Concurrent with this publication is the ongoing edition of a *Synthesis Volume*, a collective work of over one thousand pages, which will ultimately bring together the particulars of the various individuals and teams in the project in a unified whole. Readers of our book on the history of astronomy, astrophysics, and space research, no matter if German or from elsewhere in the world, will inevitably have follow-up questions regarding more general aspects of the Max Planck Society, the German research system, or even the workings of the German state and society. Whatever they fail to find explored in depth here is treated much more extensively in the *Synthesis Volume*. We have decided to deal in our book primarily with those aspects of the Max Planck Society which relate specifically to cosmic research and are not addressed elsewhere in the GMPG program. We focus on exemplary topics and episodes that are particularly useful for understanding cosmic research itself, and which also best explain the key role of cosmic research in shaping the wider sociopolitical history of Germany during the period in question.

Accordingly, we do not dwell exhaustively on aspects treated elsewhere, such as the antagonistic relationship between the Max Planck Society, universities, and other state-funded research institutes and organizations. We mention them in passing, but only insofar as is necessary to clarify the context, and especially if the developments in our field of research had a significant impact on the broader society. Chapter 1, for example, contains analyses of the importance of the nuclear age, while Chapter 2 delves equally deeply into the effects of the space age, for both these eras are uniquely powerful strands in cosmic research.

Likewise, one other aspect that permeates everything but cannot be treated encyclopedically, here, is the role of international relations and of the internationalization of cosmic research in the history of the Max Planck Society. Readers will identify an implicit narrative thread in this regard, as the international 'embeddedness' of Max Planck scientists is one of the defining features of the research we describe in this book. Moreover, Chapter 4 deals explicitly with the period around the last three decades of the century, when internationalization and then globalization became the leading sociopolitical trend. But still, in this book we cannot encompass the entire variety of international connections and collaborations related to the Society's cosmic research, or the intricate diplomatic processes that accompanied them. A book on the foreign relations of the Max Planck Society has been written by Carola Sachse,[3] while the *Synthesis Volume* and several other publications likewise more emphatically turn the spotlight on international themes. In consequence, we recommend that readers look to other publications in the Research Program for a more comprehensive treatment of anything related to foreign relations addressed too briefly here.

The *Synthesis Volume*, the volume on the foreign relations of the MPG, and this one on the history of astronomy, astrophysics, and space research were written concurrently and in full awareness of the respective authors' specialized areas of interest; readers who tackle all three will doubtless easily identify the common threads.

Clusters and Clustering: An Analytical Framework for Research Organizations

The internal structure of the Max Planck Society reflects a multiplicity of tensions. An early masterstroke by the figures behind the relaunch of the politically tainted Kaiser Wilhelm Society was to carve out an immaculate role for the new Max Planck Society as the institution in charge of so-called 'fundamental research' in West Germany, under a remarkably self-organized leadership. But as we will see throughout the book, this self-government by no means implied internal harmony nor a realm impervious to politics. We will show how the internal leadership structures of the Society—through

3 Carola Sachse: *Wissenschaft und Diplomatie. Die Max-Planck-Gesellschaft im Feld der internationalen Politik* (1945–2000). Göttingen: Vandenhoeck & Ruprecht 2023.

those senior scientists and institute directors are recommended by commissions and appointed by the MPG Senate as decision-making 'Scientific Members'—refracted the political, economic, and regional interests that negotiated West German scientific policy. But above and beyond simply navigating and embodying these 'external' pressures, Scientific Members also fiercely defended their unique scientific research traditions. It is within this interplay of scientific practices and sociopolitical forces that the Max Planck Society shaped its role in postwar Germany, seeking to manage and channel internal plurality, while maintaining a common front in the face of external threats and competitors. The cosmic sciences are the exemplary case of how this modus operandi helped the Society gain practically a monopoly on research in a wide-ranging scientific landscape, within one generation. And the central feature of the Max Planck Society that explains this success, is what our Research Program calls 'clustering.'[4]

Cosmic phenomena have been an object of study in many institutes and outposts of the Max Planck Society. While each of them is nominally independent, we are interested in how this mandated autonomy underlies what is, in practice, a much more deeply interrelated system of researchers and scientific traditions.

The 'cluster' hypothesis is one of the central analytical frameworks of the Research Program. It will be detailed in the *Synthesis Volume* published at the end of the project how this clustering explains the strength of the Max Planck Society in many scientific fields, and our work on cosmic research is one of the main pillars supporting this argument. The present volume lays this out in detail: the cluster of astronomy, astrophysics, and the space sciences (cosmic research, in short) corresponds to the group of institutes and researchers that, in coordinating its activities within the Max Planck Society around the study of outer space, gained a position from which it could ably exploit the sociopolitical interest in this topic during the second half of the 20th century.

The categorization of other clusters in the Society is ongoing work that will not be addressed in our book, with one exception: we will allude frequently to two clusters closely related to the cosmic sciences. Firstly, we will show how, especially before the launch of Sputnik, cosmic research existed within a worldview oriented towards the nuclear age, and was hence implicitly also part of a 'nuclear' cluster. Secondly, we examine how, from the 1970s on, a growing sociopolitical interest in environmental research led to the formation of a new cluster, which initiated an 'Earth system' and, perhaps, a wider

4 In addition to the *Synthesis Volume*, this analytical framework will appear in a separate journal publication.

'environmental' cluster, rooted in expertise and methodologies inherited from the nuclear and cosmic sciences. Both the 'nuclear age' and the 'environmental turn' will be addressed in detail in other GMPG publications; they feature in this book only as far as is necessary to pinpoint their interrelationship with cosmic research and the space age.

To be clear, 'cluster' is our Research Program's own analytical entity, and not one ever claimed or pursued by the subjects of our research. We identify as clusters those informal arrangements in the Max Planck Society that encompass institutes, departments, scientists, expert commissions, administrative bodies, and governance instances. While each of these represent different and often, diverging scientific and political standpoints, they also act in coordinated ways, such that the reach of the Max Planck Society over wide scientific realms is maximized (and was maximized, historically).

The concept of cluster and clustering emerges from empirical observations and is useful also as an organizing category within the very wide array of scientific activities pursued in the Society. In the case of the cosmic sciences, this analytical framework is mobilized to analyze the dynamics of the research and interaction of Scientific Members of the Max Planck Society across different institutes. While there exists a substantial body of work dealing with the activities of *individual* Max Planck Institutes conducting cosmic research—from historical monographs to oral histories, to related documentation—the (collective, personal, official, ad hoc, or incidental) *interaction of the various institutes* is generally granted only a casual mention in the existing historiography.

Cosmic research presents us with one of the most successful clusters of the Max Planck Society. This kind of activity was almost entirely absent from the prewar Kaiser Wilhelm Society, yet from the 1950s to the 1980s, it rose quickly to prominence as one of the main fields of research, distributed among a dozen institutes. As we will see in detail, by the late 1980s Max Planck Institutes were considered to have practically a national monopoly on these sciences; and likewise since the end of the Cold War—while shifting their focus away from national scientific programs and infrastructures towards an emphasis on specialized roles within international scientific collaborations—they maintained their relative standing among comparable fields of research conducted by the Max Planck Society, as well as their dominant position among German research organizations in their own cosmic sciences field. Throughout the entire period in question, their participation in international collaboration networks has been exemplary, increasingly shifting the emphasis towards becoming a key player within a globalized scientific research ecosystem.

For all that, clusters and clustering will often feature in this book only implicitly, as we follow the dynamics of competition, collaboration, complementarity, and coordination between the various research traditions, to align them with wider sociopolitical goals. A much more explicit and analytical treatment of the same processes features in the *Synthesis Volume*, which takes cosmic research as an exemplar in its chapter devoted to clusters. The text there presents a more sociological analysis of what is presented in this present book, tracing scientific traditions through the major sociopolitical eras of the nuclear age, the space age, and end-of-century globalization.

Scientific Traditions: A Leading Thread

In following scientific developments, our scale of analysis focuses primarily on what we call *scientific traditions*. These traditions operate at a more intimate level than scientific disciplines (as defined by formal educational programs and degrees), and often overlap among disparate fields. Ultimately, scientific traditions draw their strength from the kind of scientific expertise that exists at the leading edge of research, namely activity based on interpersonal and informal dynamics—which may include long apprenticeship, orality, shifting and, possibly, temporary forms of knowledge and, ultimately, that which becomes 'embodied' or 'tacit knowledge'—hence, elements of scientific expertise that not even practitioners themselves may formulate explicitly, and which generally do not appear in the formal scientific literature. These aspects are, rather, learned through long exposure to, and participation in a group, and may be intermingled with deep personal friendships or even kinship of the intellectual, political, and blood varieties.

In order to acquaint ourselves with all these subtle aspects in the field of cosmic research, we often needed to look beyond the published literature and draw on personal accounts, oral histories, and interviews. In the few cases in which it was possible, we complemented these by reference to the available relevant historical, anthropological, and sociological literature.

In many scientific fields, these traditions are manifest in the importance that those who uphold them grant to academic genealogies as well as geographical trajectories. In the case of this book, one even sees the prevalence of powerful patriarchs and founders, especially among the generation that established careers already before the war. And while hierarchical structures feature to some degree in any place where science is practiced, in Germany in particular they can be shown to have played an essential role in the evolution of scientific research institutions, where autonomy and longevity around the top of the pyramid have been forged in deliberate opposition to dependency and precarity at the wider base.[5] We often use deliberately loaded terminology such as 'family' to highlight the ways in which the scientists in our period of inquiry were bound not only by common scientific traditions and political and economic dependencies, but also by deep loyalties, emotional bonds, and camaraderie; and this arguably resulted in far sharper boundaries than our fluid 21st-century professional relationships and networks.

Furthermore, the term 'patriarch' mentioned above highlights the importance of examining gender dynamics in the history of the Max Planck Society. Unfortunately, however, in the scientific traditions and period under study here, almost all of the intellectual leaders and decision makers were male. In consequence, women are conspicuous by their absence from this book. This reflects a lamentable reality of postwar West Germany, namely that the early Federal Republic perpetuated the regressive legal and professional status afforded women under the Third Reich, a status itself exacerbated by the dire lack of opportunities for women scientists in Imperial Germany, in contrast to the situation in other European countries and the United States throughout the 19th and 20th centuries. Then as now, women scientists' status in scientific research was generally unfavorable, if mitigated somewhat by access to teacher training for women, which served a few of them as a springboard to research positions. In smaller disciplines such as astronomy, however, the situation was much more precarious: whereas in countries like the United States, women were valued in astronomy (albeit often in subservient positions and data analysis), young male astronomers in Germany were in oversupply from the late 19th century on, and accordingly took priority, also in the subservient positions, men's need for employment being used explicitly as a reason to exclude women. Astronomy was a small, separate discipline led by rather socially conservative figures, and there was little need for schoolteachers with astronomical expertise. German astronomy was thus even less open to women than other natural sciences. And while the Weimar Republic saw improvements in women's legal status and more progressive attitudes on women's rights among some (male) scientific decision makers, their already uneven impact was further limited by the poor economic circumstances.[6]

---

5 Vita S. Peacock: *We, the Max Planck Society. A Study of Hierarchy in Germany*. London: Doctoral thesis, University College London 2014.

6 See: Annette Vogt: Astronominnen in Berlin und Potsdam. In: Wolfgang R. Dick, and Klaus Fritze (eds.): *300 Jahre Astronomie in Berlin und Potsdam. Eine Sammlung von Aufsätzen aus Anlaß des Gründungsjubiläums der Berliner Sternwarte*. Frankfurt am Main: Thun 2000, 121–141. During the postwar era, early work with computers provided a unique window of access for women to scientific research worldwide, and some mathematicians reached quite

These circumstances shaped men's careers, too. The institutional authoritarianism dating back to the 19th century compounded the effects of the first half of the 20th century. Economic hardship further exacerbated the existing close-knit patriarchalism, and this in turn facilitated the compartmentalization of research during World Wars I and II, in an environment marked by political surveillance and persecution, but also by compromise and willing participation: before 1945, deep, lifetime loyalties at the personal level were forged in small close circles, while distrust and competitiveness were projected onto those excluded from them. This had profound effects on the way the wartime generation conducted scientific debates, even decades after 1945.[7]

On top of all this historical baggage, the Max Planck Society itself clearly must take responsibility for having reinvigorated this hierarchical patriarchalism during its postwar constitution; for the young successor perpetuated the

---

senior positions—especially if they were unmarried—as in the case of Eleanore Trefftz at the Max Planck Institute for Physics and Astrophysics. But the gender balance in computing worsened in subsequent decades. See Jennifer S. Light: When Computers Were Women. *Technology and Culture* 40/3 (1999), 455–483. http://www.jstor.org/stable/25147356. Last accessed 7/3/2018. Another case better illustrates the limitations of forging a career as a woman in the early years of the Federal Republic: Rhea Lüst (née Kulka) was a promising astronomer in postwar Göttingen, who then married one of the central figures in this book, Reimar Lüst. Her scientific career after marriage diminished to informal contracts for part-time work in the research group of Ludwig Biermann. While she made important contributions to the analysis of comet tails, her trajectory, in which family life ultimately took priority over research, was that of a 'first lady' of an influential scientist. Reimar Lüst, and Paul Nolte: *Der Wissenschaftsmacher. Reimar Lüst im Gespräch mit Paul Nolte*. München: C.H. Beck 2008. p. 89–91. It is only in the last decade of the century (a generation later than in the United States) that the proportion of women in senior roles at the Max Planck Society began to improve. For a general study of this problematic see Birgit Kolboske: *Hierarchien. Das Unbehagen der Geschlechter mit dem Harnack-Prinzip. Arbeits- und Lebenswelten von Frauen in der Max-Planck-Gesellschaft, 1948–1998*. Dissertation. Leipzig 2021. See also Birgit Kolboske: Hierarchies. Lotta Support, Little Science? Scientists and Secretaries in the Max Planck Society. In: Ulla Weber (ed.): *Fundamental Questions. Gender Dimensions in Max Planck Research Projects*. Baden-Baden: Nomos 2021, 105–134. doi:10.5771/9783748924869.

7 Mark Walker: *German National Socialism and the Quest for Nuclear Power 1939–1949*. Cambridge: Cambridge University Press 1989, 95. For example, Walker brings up the rivalries between Werner Heisenberg and Kurt Diebner during the wartime nuclear project to make the following point: "But as is often the case with personal feuds among scientists, the personal conflict between Heisenberg and Diebner almost always assumed a professional guise. Diebner would be attacked as a mediocre physicist, or Heisenberg's circle would be accused of performing second-rate experiments. In fact, the researchers on both sides were capable scientists doing the best to make the nuclear project a success, even though each of the two factions was quite ambitious and believed sincerely that it was better suited to conduct the uranium machine experiments."

key mantra of the Kaiser Wilhelm Society, the so-called Harnack principle. This stated that every institute should be 'built around' a stellar scientist, and it was at this level of leadership that the so fiercely defended, constitutionally enshrined 'freedom of research' actually applied. The stellar scientists' dozens if not hundreds of researchers and employees essentially remained subalterns, even after the 1960s, when the universities and other organizations made some advances towards flattening hierarchies and extending participation less exclusively. Reforms in the MPG during the 1960s towards 'collegiate leadership' by a group of directors sharing a single institute did limit individual power, somewhat; but the deep abyss between directors and the people below them endures to this day and, moreover, continues to be defended as the key distinguishing feature and competitive advantage of the MPG in the global scientific ecosystem.

Until the 1970s, when the last wartime generation went into retirement, deep personal loyalties also continued to shape the interaction of Max Planck scientists. It is a key part of this study to show how these patriarchal features of the early postwar period gave way to a new form of organization from the 1960s on. By then, the number of researchers in the Max Planck Society had exploded, while their individual ability to grasp large areas of science had diminished; expertise and power was increasingly distributed among medium-sized collectives, which were, however, still strongly based on specific research traditions and personal links. This book will show how this demographic change and the political transformations it afforded further advanced the cluster behavior of the cosmic sciences in the Society.

Strong scientific traditions extended beyond academic researchers. In Germany in particular, they were also crucially sustained by an array of technical and instrumental expertise embodied by teams of engineers and technicians, who sometimes kept a lower public profile than the academically trained scientists. This low profile of technicians in the MPG reflected the entrenched division of labor in Germany along vocational or academic lines, respectively the distinct tracks laid for this in the country's system of secondary education. Yet it will be demonstrated repeatedly throughout this book, how, even in the case of 'theoretical' scientific traditions, the weighty but understated protagonism of technicians and their workshops was often actually the key comparative advantage of Max Planck Institutes, particularly with respect to researchers based in universities in Germany and elsewhere.

Scientific traditions were also already deeply intertwined with a field of political, economic, and social forces. We are not proposing a divide between internal scientific work and its external 'influences.' Rather, at the level of scientific traditions, these 'external' sociopolitical elements are embedded in an unformulated, informal way, and are ritually de-emphasized, while the main identifying trait projected outwards in the wider world is scientific expertise. At the same time, as we will see, successful scientific traditions are those able to 'jump' and make the most of wider sociopolitical trends, obtaining therein a prominent role by leveraging their established expertise. In the case of the cosmic sciences, this is very clearly what happened, firstly, by linking cosmic research to nuclear questions, and later on, by benefiting, in far greater scope and scale, from the space age inaugurated by Sputnik.

The main traditions that will be followed in this book, roughly in their order of appearance, include theoretical plasma astrophysics, experimental cosmochemistry, cosmic ray research, radio astronomy, optical astronomy, high-energy astrophysics, space-based astronomy, and relativistic astrophysics. We often continue to name their original identities to emphasize long continuities, even though, as we will see, successful traditions mutate considerably over the decades. For example, experimental cosmochemistry had a prehistory in 1930s nuclear physics and cosmic ray research, and much later ended up becoming one of the foundations of what is now called astroparticle physics. Along the way, this tradition seeded swaths of communities in widely varied areas, ranging from Mössbauer spectroscopy, planetary exploration and the Earth system sciences, to neutrino astronomy.

Ultimately, scientific traditions are not necessarily entrenched in a fixed scientific object (such as cosmic entities and phenomena), but rather, are strongly dependent on specific practices and expertise developed over a relatively long time and through intense interaction within scientific collectives. Several times in this book we will see how successful traditions were able to quickly redefine their object of study and sociopolitical roles, while maintaining strong continuity in their actual concrete research practices. Usually, however, such reconfigurations were fostered either by opportunities to grow the scale and scope of their research in the global arena, as with Sputnik, or by crises that forced their hand, leading some core practices to be abandoned in favor of others, in line with shifts in the global organization of scientific production such as occurred, for example, at the end of the Cold War.

Existing Histories and Printed Sources

This book spans a period of more than half a century as well as tens of Max Planck Institutes and other research entities in Germany, Europe, and around the world, while seeking to connect scientific traditions and developments with contemporary sociopolitical forces.

Our core period of analysis is from 1945 to the end of the century, but whenever necessary, we reach back to the prewar history of the actors and scientific traditions involved. Moreover, due to the very long timescales involved in cosmic research projects, we often follow developments beyond the official cut-off date, occasionally extending even to the present, if the core of the process had been already on course before the end of the century. Due to data privacy restrictions, however, we can only access archival material dating from before 2004.

For such large scales of analysis, from the personal to the geopolitical, it would in any case be insufficient to work solely with archival material. Moreover, we would often be duplicating high-quality historical work that has been published already in widely different forms. Instead, whenever possible, we mobilize an existing corpus of institute-centered histories and autobiographies written by scientists. These are complemented by interviews, either conducted by ourselves or sourced from previous historical projects. All these materials, produced over the past half century, are treated by us as primary and secondary sources. While varying widely in style and content, they generally contain nuanced, first-person understandings not just of the sciences, but also of the political and economic forces that influenced the course of scientific research. Only very rarely, however, do these kinds of source contain wider analyses of the Max Planck Society as a whole and the participants' historical role, such as are likely to be of interest to non-specialist historians of modern Germany. It is our role to translate the specific scientific developments described by these actors into the wider patterns that we seek to elucidate for the purposes of this book.

There have been some remarkable precursors to, and inspirations for, the current book, which from different points of view provided premonitions of the collective entity that we call the cosmic research cluster. Two Max Planck scientists, Joachim Trümper[8] and Dietrich Lemke,[9] have independently written on the development of astronomy, astrophysics, and space science in the Max Planck Society, identifying the main institutes, research traditions, and even major sociopolitical drivers of the research conducted in the MPG in the second half of the 20th century. We are much indebted to these two scientists for the initial impetus to explore this cluster of institutes, and their conceptual maps remain largely valid. In many ways, this current book seeks to flesh out

8 Joachim Trümper: Astronomy, Astrophysics and Cosmology in the Max Planck Society. In: André Heck (ed.): *Organizations and Strategies in Astronomy*. Dordrecht: Springer Netherlands 2004, 169–187.

9 Dietrich Lemke, and Astronomische Gesellschaft (eds.): *Die Astronomische Gesellschaft 1863–2013. Bilder und Geschichten aus 150 Jahren*. Heidelberg: Astronomische Gesellschaft 2013.

the political and institutional forces that underly the scientific traditions and institute genealogies described in their work.

Then there is the work of historian Ulf von Rauchhaupt, who in the course of several works provided an extensive analysis of an exemplary branch of space research which paved the way to the founding of the Institute for Extraterrestrial Physics. Rauchhaupt's research on this Institute and its first director Reimar Lüst, a central figure in this book, can be seen as a higher-resolution study among dozens of cases that will be mentioned in the coming chapters. From early on, Rauchhaupt's work on the nascent 'space age' (discussed in more detail in Chapter 2), influenced our understanding of the foundational years of space research in Germany and the political dynamics surrounding them, which we subsequently analyzed beyond space exploration, to encompass the entire cosmic sciences.

Michael Seiler's work on solar astronomy and ionospheric research, although focused primarily on World War II, provides ample research and analysis on the impact that such work had in subsequent decades, and is a fantastic exemplar through which to trace the ways in which German scientists' wartime experience, also of the occupation of foreign countries, shaped their postwar careers and their acceptance (or not) by their counterparts in other countries and in specific organizations in Germany, including the Max Planck Society. Besides being generally important examples of these features, this book also specifically lays the foundations for a history of the Max Planck Institute for Aeronomy.[10]

Extending the scope beyond the Max Planck Society, the works of Cathryn Carson[11] and Mark Walker[12] set the stage for the postwar configuration of physical research and the heavy baggage it brought from research projects (especially nuclear ones), organized during the war. Carson's research on Werner Heisenberg's postwar career in particular provides many clues, focusing on one historical figure and the social world around him; a vignette that we draw on, in sketching a collective history with dozens of actors. Mark Walker's work provides detail on many precursors who shaped nuclear and cosmic research in postwar West Germany, and provides the clearest insight into the way that

10 Michael P. Seiler: *Kommandosache "Sonnengott". Geschichte der deutschen Sonnenforschung im Dritten Reich und unter alliierter Besatzung*. Frankfurt am Main: Verlag Harri Deutsch 2007.

11 Cathryn Carson: *Heisenberg in the Atomic Age. Science and the Public Sphere*. Cambridge: Cambridge University Press 2010.

12 Walker, *The Quest for Nuclear Power*, 1989.

personal and scientific rivalries were negotiated by the generation that dominated the first decades of the Max Planck Society, including figures such as Werner Heisenberg and Wolfgang Gentner. Gentner himself was the subject of a commemorative volume to which a wide spectrum of historians and scientists contributed.[13]

Moving forward to the core period of our research, the work by Hohn and Schimank[14] provides a magnificent historical framework for positioning the Max Planck Society within the West German research system, and furthermore, it includes many historical episodes related to the nuclear and cosmic sciences directly relevant to this study. This work is complemented very well by the work of Osietzki and Eckert on the political and economic forces behind the development of German postwar research in the physical sciences,[15] which again features many of the main actors and episodes whose involvement or entanglement in nuclear research helped determine the future of the Max Planck Society and the consolidation of a cosmic sciences cluster after Sputnik. A bit broader in scope are the works of Reinke,[16] Weyer,[17] and Trischler[18] on West German space policy and its related sociopolitical ecosystems (treated in more detail in Chapter 2), which feature the protagonism of the Max Planck Society, particularly concerning the Institutes for Extraterrestrial Physics and Aeronomy. Finally, a comprehensive review of the West German scientific

---

13 Dieter Hoffmann, and Ulrich Schmidt-Rohr (eds.): *Wolfgang Gentner. Festschrift zum 100. Geburtstag*. Berlin: Springer 2006. See also: V. Soergel et al.: *Wolfgang Gentner 1906–1980*. Geneva: CERN 1982.

14 Hans-Willy Hohn, and Uwe Schimank: *Konflikte und Gleichgewichte im Forschungssystem. Akteurkonstellationen und Entwicklungspfade in der staatlich finanzierten außeruniversitären Forschung*. Frankfurt am Main: Campus Verlag 1990.

15 Michael Eckert, and Maria Osietzki: *Wissenschaft für Macht und Markt. Kernforschung und Mikroelektronik in der Bundesrepublik Deutschland*. München: Beck 1989. Maria Osietzki: *Wissenschaftsorganisation und Restauration. Der Aufbau ausseruniversitärer Forschungseinrichtungen und die Gründung des westdeutschen Staates*. Köln: Böhlau 1984.

16 Niklas Reinke: *The History of German Space Policy. Ideas, Influences, and Interdependence 1923–2002*. Translated by Barry Smerin, and Barbara Wilson. Paris: Beauchesne 2007.

17 Johannes Weyer: *Akteurstrategien und strukturelle Eigendynamiken. Raumfahrt in Westdeutschland 1945–1965*. Göttingen: Schwartz 1993. Johannes Weyer: Die Raumfahrtpolitik des Bundesforschungsministeriums. In: Peter Weingart, and Niels C. Taubert (eds.): *Das Wissensministerium. Ein halbes Jahrhundert Forschungs- und Bildungspolitik in Deutschland*. Weilerswist: Velbrück Wissenschaft 2006, 64–91. Johannes Weyer (ed.): *Technische Visionen–politische Kompromisse. Geschichte und Perspektiven der deutschen Raumfahrt*. Berlin: Edition Sigma 1993.

18 Helmuth Trischler: *Luft-und Raumfahrtforschung in Deutschland 1900–1970. Politische Geschichte einer Wissenschaft*. Frankfurt am Main: Campus 1992.

landscape conducted by the journal *Nature* in early 1982 provides brilliant insight from an outsider's perspective.[19]

Similarly, there are the classic works on the history of the Institute for Plasma Physics (IPP) by Susan Boenke,[20]which extend far beyond the scientific developments to paint a detailed picture of the particularities of so-called *Großforschung* (large-scale research) in Germany. These histories constitute another exemplar for writing about the synergies between research traditions and wider sociopolitical forces. Furthermore, this bibliography on the IPP is directly relevant to the history of the cosmic sciences because, as we will see throughout this book, plasma physics was a scientific tradition in many institutes and the IPP itself, and while initially embedded in a more 'nuclear' rather than 'space age' framework, it played a critical role in shaping, consolidating, and increasing the influence of cosmic research in the Max Planck Society.

Some of the cited works originated in a more personal perspective, while still providing an analysis of the sociopolitical forces that accompanied these scientific lives. The aforementioned works by the Max Planck scientist Dietrich Lemke, though more focused on scientific developments, paint a very comprehensive picture of ground-based and space-based astronomy in Germany, in which the Max Planck Society features prominently. The Max Planck radio astronomer Richard Wielebinski has also published widely as a historian on specific topics of the history of his scientific discipline and the Institute for Radio Astronomy.[21] Reimar Lüst, a former president of the Max Planck Society and a central figure in the constitution of a cosmic sciences cluster, produced an extended corpus of writing (cited here often) on the matters under discussion throughout this book, while his retrospective reflections on them feature in a major autobiographical volume based on a series of interviews with Paul Nolte.[22] Just published from Gerhard Haerendel, Lüst's successor at the department conducting space-based experiments, is an autobiography which sheds additional light on many episodes, yet which we had access to only after completing the present manuscript.[23]

---

19 John Maddox: Science in West Germany. Discovery and Disappointment. *Nature* 297/5864 (1982), 261–280. doi:10.1038/297261a0. Other articles highlighting further specific aspects related to the Max Planck Society will be mentioned in the proper contexts.

20 Susan Boenke: *Entstehung und Entwicklung des Max-Planck-Instituts für Plasmaphysik 1955–1971*. Frankfurt am Main: Campus Verlag 1991.

21 See multiple citiations in the bibliography and respective episodes.

22 Lüst, and Nolte, *Der Wissenschaftsmacher*, 2008.

23 Gerhard Haerendel: *My Life in Space Exploration*. Cham: Springer 2022. doi:10.1007/978-3-031-10286-8.

These are just some key examples of the dozens of institutional histories and (auto)biographical works related to Max Planck Institutes and associate organizations, which are listed in the bibliography, along with details of various oral histories and interviews conducted by us or sourced from other historical research.

As for international connections, finally, this study is informed by the large-scale institutional histories of both CERN[24] and ESA,[25] and several smaller-scale works on ESO.[26] The presence of these international scientific organizations has been pivotal to the development of the Max Planck Society, as one of the central questions in science policy in Germany since 1945 has been how to balance research conducted at the regional, national, or European scale.

Archival Research and Unpublished Sources

The largest component of extensive examination was the archival material related to the governing commissions of the Max Planck Society, specifically the decision-making processes of the CPT (chemistry, physics, and technology) section. As was described earlier, we consider these commissions the points of encounter of different interest groups and stakeholders embodied by the appointed participants. Generally, the selection of commission members sheds substantial light on the interests and power relations underlying a decision. The reports of these commissions to the Scientific Council and the Senate (AMPG, II. Abt., Rep. 62) contain the different points of view discussed before a decision was made. Another important record group are the so-called *Institutsbetreuerakten* (also II. Abt., Rep. 62) documenting information on many affairs concerning the development of the Max Planck Institutes that the *Institutsbetreuer* (custodians or 'mentors' of the institutes) dealt with at the interface of the MPG general administration and the MPI directors.

24 Armin Hermann et al.: *History of CERN. Launching the European Organization for Nuclear Research*. Vol. 1. Amsterdam: North-Holland 1987. Armin Hermann et al.: *History of CERN. Building and Running the Laboratory, 1954–1965*. Vol. 2. Amsterdam: North-Holland 1990. John Krige (ed.): *History of CERN*. Vol. 3. Amsterdam: North Holland 1996.

25 John Krige, and Arturo Russo: *A History of the European Space Agency 1958–1987. The Story of ESRO and ELDO, 1958–1973*. Vol. 1. Noordwijk: European Space Agency 2000. John Krige, Arturo Russo, and Lorenza Sebesta: *A History of the European Space Agency 1958–1987. The Story of ESA, 1973 to 1987*. Vol. 2. European Space Agency 2000.

26 To date, all historical research on the European Southern Observatory has been done by former directors and officials of the organization: Adriaan Blaauw: *ESO's Early History. The European Southern Observatory from Concept to Reality*. ESO 1991. Lodewijk Woltjer: *Europe's Quest for the Universe. ESO and the VLT, ESA and Other Projects*. Les Ulis: EDP Sciences 2006. Claus Madsen: *The Jewel on the Mountaintop. The European Southern Observatory through Fifty Years*. ESO 2012.

Familiarity with the actors, the interests, and the historical period is necessary to interpret the full significance of certain statements and decisions, identify what was left unsaid, and distinguish between ritualized bureaucratic maneuvering and the 'deeper' issues it may hide. We often provide relevant citations from these sources in the footnotes so that interested readers can grasp their subtleties without interruption to the narrative flow of the book.

Even which types of commissions are convened[27] highlights the rising stakes in policymaking: many of them begin with the task of finding a director for a new institute or department, or a successor; but the proceedings, if not straightforward, may escalate into a reassessment of the whole scientific significance and relevance of a research group, or even of an entire scientific field and how, if at all, the Max Planck Society should tackle it. All of the major Society-wide decisions regarding the subjects of our research are contained in such commissions' reports; and we refer to the composition of the commissions themselves, to further elucidate the relevance to the process in question of their various resolutions. One of the most salient manifestations of the clustering of institutes is that specific groups of scientists wielded control over the Society's policymaking bodies, as these commissions well illustrate. Occasionally, as with the Presidential Commission instituted in the early 1990s (see Chapter 4), such deliberations had the power to redefine the future of the entire cosmic research cluster.

Another crucial source for deeper insight into specific episodes is the scientists' papers held by the Archives of the Max Planck Society, among them those of the former MPG presidents Adolf Butenandt (III. Abt., Rep. 84), Otto Hahn (III. Abt., Rep. 14), and Reimar Lüst (III. Abt. Rep. 145). Even more important, given their level of detail, are the papers of influential scientific organizers in the field, those of Ludwig Biermann (III. Abt., ZA 1), Wolfgang Gentner (III. Abt., Rep. 68A), and Werner Heisenberg (III. Abt., Rep. 93), for example; or of more recent scientists such as Heinrich Völk (III. Abt. ZA 166).

Quantitative financial data was examined early in our research and continued to inform the narrative that is presented in the book. It was decided, however, not to detail these finances within the chapters themselves, where the interaction of sociopolitical dynamics with scientific developments was to remain the primary focus. Rather, at any mention of a particularly relevant financial event, we refer the reader to the Financial Appendix at the end of the

27 Mainly *Berufung*—[appointment], *Nachfolge*—[succession], *Zukunft*—[future], *Stamm*—[general direction or 'regular'], and *Präsidenten*—[presidential: convened by the MPG President].

book. The Appendix thus constitutes a complementary analysis of the cosmic research cluster, based on financial planning, as per the so-called Budget Plans (*Haushaltspläne*: II. Abt. Rep 69).

The Archives of the Max Planck Society provided the majority of the primary sources used in this book: in addition to the specific collections of papers mentioned above, we were able to draw on the growing body of searchable, digitized materials made possible by the GMPG. We also occasionally turned to external archives, and to relevant archival material that we had collected during previous research projects.

The reader should bear in mind that the aforementioned commissions lasted for years and that their deliberations therefore constitute a microcosm of the Max Planck Society: the material available on every single case would sustain a more detailed historical research article. In fact, several GMPG projects will provide deeper insight into these matters via single case studies of the commissions' business, and the research group's *Synthesis Volume* will include not only a detailed explanation of the procedural mechanisms of these decision-making bodies, but also reflections on their circumvention by Scientific Members, as historically, there has been a very marked preference within the Max Planck Society for informal arrangements and intrigue.

Oral Histories, Interviews, and Workshops

Many historical figures featured in this study had already been interviewed in the context of ongoing oral history projects and specific historical projects, the transcripts of which are increasingly available for external use. We have made extensive use of oral histories preserved at the Niels Bohr Library & Archives of the American Institute of Physics,[28] and of interviews conducted either within the framework of the History of the European Space Agency,[29] or by Woodruff Sullivan in the context of his book on the early history of Radio Astronomy, the latter now held by the Archives of the National Radio Astronomical Observatory/Associated Universities Inc.[30] In addition to these large depositories, we

28 Oral History Interviews, Niels Bohr Library & Archives, American Institute of Physics, College Park, MD USA (from now on AIP Archives), https://www.aip.org/history-programs/niels-bohr-library/oral-histories. Last accessed 4/18/2021.

29 Historical Archives of the European Union, European Space Agency historical Archives, Oral History of Europe in Space (from now on ESA Archives), https://archives.eui.eu/en/oral_history/#ESA. Last accessed 4/18/2021.

30 National Radio Astronomy Observatory /Associated Universities Inc. Archives (from now on NRAO/AUI Archives), https://www.nrao.edu/archives/. See listing of radio astronomers inverviewed by Sullivan at https://www.nrao.edu/archives/sullivan-individuals-listing. Last accessed 4/18/21.

drew on the Oral History Collection of the California Institute of Technology Archives,[31] and a multitude of interviews published in journals.

Our research program organized three workshops between 2015 and 2016, which brought together historians and participant scientists in an effort to elucidate and map out the path ahead that ultimately led to this book (and, over the last four years, also to additional interviews with workshop participants). A full list of these sources can be found at the end of the book.

Finally, another rich historical source for us was the great number of publications produced at the interface of (auto)biography, interviews, and archival compilation by prominent figures or the Max Planck Institutes themselves.

Scientific Content and Publications

One of the ambitions of this study is to integrate global scientific developments in a wide array of disciplines and traditions in the sociopolitical environment in which they originated. The most formidable challenge is how to talk *enough* about the scientific content without alienating what we hope will mainly be a 'general interest' readership, i.e., people who are not overly familiar with these sciences. The key aspect to bear in mind is that this study does not aim to be encyclopedic in its treatment of scientific developments in the Max Planck Society. Scientific developments appear when they are relevant to our dual analytical focus, namely the evolution of longstanding scientific traditions and how these have variously clustered over time. Accordingly, our overriding emphasis is on those traditions we believe best explain the trajectory of the cosmic sciences in the Max Planck Society; and much research that we have not mentioned explicitly still falls in one way or another into these traditions. We decided not to focus much on fields that do not rank among the Max Planck Society's strengths, even if they were notable 'missed opportunities.' Addressing these would have been very difficult, moreover, given that our methodologies of choice were generally qualitative and selective. The ongoing GMPG research program will address some of the most significant among them, using sophisticated network analysis of a vast corpus of data.[32]

We aim to mention scientific content in the main text only inasmuch as it is necessary to follow the central narrative. Further details can be found in the footnotes and the scientific literature cited. Moreover, the latter should be understood as a starting point for further reading, not as a comprehensive treatment of the scientific issue in question.

---

31 Caltech Archives, Oral History Collection (from now on Caltech Archives): https://archives.caltech.edu/collections/oral-histories.html. Last accessed 4/18/2021.

32 See forthcoming work by Roberto Lalli and Dirk Wintergrün.

At the end of the study (Chapter 5), the reader will have the opportunity to 'zoom in' on three cases studies that illustrate the emergence in the Max Planck Society of new scientific fields, to which the interplay of research traditions and the clustering process are central. The reader can imagine that many other scientific developments in the cosmic sciences could have been treated in similar depth.

In our analysis of scientific developments throughout this book, we aim to be inclusive of a broad readership; yet, inevitably, some specialized terminology does appear without an immediate explanation. We try to restrict this to cases where the interested reader will easily find the relevant guidance elsewhere, or where ignorance of 'what this means' does not prevent them from following the argument. If instead we think our particular analytical perspective and guidance require the readers to understand the scientific content, we go into the necessary depth ourselves. Whenever possible, this is done within the main text; and sometimes further detail is provided in the footnotes and the references they contain. Please bear in mind that our descriptions are not general learning tools, but rather seek to provide the non-physicist reader with enough background information to follow the main storylines

Story Outline and Chapters

This book traces scientific traditions in the cosmic sciences and how they have clustered over time in the Max Planck Society, in a roughly chronological fashion, punctuated by the major underlying historical processes: the 'nuclear age,' the 'space age,' the absorption of observational astronomy, and scientific globalization. Within these processes and chapters, we identify and address various parallel developments in different Max Planck Institutes and, crucially, how researchers from these institutes interacted with one another. The final chapter then allows us to 'zoom in,' to follow in detail three emerging fields that span all these developments and best illustrate the long-term constitution of scientific traditions, and their interplay both with one another and with wider epistemic pathways and sociopolitical forces. The book is then capped, not by a conclusion, but by a Financial Appendix that presents a complementary perspective on all that has been described qualitatively throughout the chapters, through the quantitative lens of the individual institutes' finances, including the costs of the cluster and their aggregate behaviors, all in relation to the MPG as a whole.

Chapter 1 of the book focuses on astrophysics in the first postwar decade, when it was embedded in the promises of the 'nuclear age' and, in West Germany, in the hardship of reconstruction and Allied restrictions. A key element at the time was the ambiguity of 'nuclear' physics, which tacitly implied the potential of nuclear energy and weapons, while also connoting a disparate array of scientific subjects in addition to the atomic nucleus, such as fusion and plasmas, cosmic rays, and subatomic particles. This ambiguous overlap enabled astrophysicists to obtain financial and political support for any research signaling future expansion into the relevant applied fields, while yet respecting postwar prohibitions. The backdrop to this support was essentially the regionally focused competition between the Allies, which played out in the various occupation zones and nascent states of West Germany. Scientists rooted in diverse scientific traditions and political orientations adapted to the different regional interests and the priorities of the Allied occupiers. It was the strength acquired in this decade, partly as a result of competition among these different factions, that assured the Max Planck Society a good headstart when the space age took center stage after Sputnik.

In Section 1, we describe how a community of scientists converged in Göttingen in the aftermath of World War II to become part of what was to become—at the initiative of the famous physicist Werner Heisenberg—the Max Planck Institute for Physics and Astrophysics, the primary hub and powerhouse of the Max Planck Society. This chapter focuses on its trajectory during the early nuclear age, until the move to Munich in 1958. Göttingen was the birthplace of the research tradition rooted in the theoretical plasma astrophysics led by the astrophysicist Ludwig Biermann and his disciples, who were able to make contact and collaborate closely with scientists working on nuclear fusion in the United States, thanks to their pioneering work in plasma astrophysics, supported by the first fully-modern digital computers made in West Germany, right at their institute.

In Section 2, we introduce the counterweight to the community described in the previous section. This was a research tradition based on experimental nuclear physics, making use of particle detectors and accelerators. The precursors of this tradition were Walther Bothe, one of Germany's most prominent experimental physicists, and his disciple Wolfgang Gentner, who emerged as a central political figure and played a key role in the scientific Europeanization of West Germany. Gentner and his colleagues pursued a path to scientific excellence in the first decade of postwar scarcity and research restrictions: they built large infrastructures at CERN, while conducting fundamental nuclear research locally by entering the field of cosmochemistry, in which mineral samples and meteorites were analyzed to gain insight into fundamental physical processes. This tradition spanned a growing network centered on the Max Planck Institute for Nuclear Physics in Heidelberg and the Max Planck Institute for Chemistry in Mainz, including allies in nearby universities in the host cities and other locations such as Freiburg and Bern.

Weaker traditions were part of the early history of the Max Planck Society, and Section 3 describes the reasons for their weakness as well as the contingencies on which becoming a Max Planck Institute depended during this early era. The eventual outcome was the patchwork Max Planck Institute for Aeronomy in Lindau, Lower Saxony, which would continue to be a somewhat problematic scenario, and one reason why coordinated action from the different power centers in the Max Planck Society was repeatedly necessary. The key figures in this section are Erich Regener, Germany's top cosmic ray researcher using balloons; Walter Dieminger, head of the wartime radio disturbance forecasting network which surveilled the ionosphere; and Karl-Otto Kiepenheuer, leading astronomer in the wartime project to predict radio disturbances with solar observatories. Their fates inside and outside of the Max Planck Society illustrate the patterns and contingencies of membership in the organization in its first decade.

Following the introduction in the previous sections to the field and early players, Section 4 is the first to focus explicitly on the interaction of the different research traditions and centers of power. Alignment with regional power bases is emphasized, as this was a key factor in the early decades of the Federal Republic of Germany. These regional and political rivalries played out in the scientific world particularly after the end of Allied restrictions in 1955, when a race began for leadership in the development of nuclear power. The cosmic sciences, which up until 1955 were a path toward excellence and global connections under precarious circumstances, faded (temporarily) into the background, as the development of large-scale projects such as nuclear reactors, thermonuclear fusion facilities, and particle accelerators turned into a battlefield, with both Heisenberg and Gentner among the key protagonists. The section shows how disunity and rivalries in the context of nuclear ambitions would set a precedent for failure not to be repeated later in the cosmic sciences.

Chapter 2 follows the enormous expansion of the space sciences around the world after the launch of Sputnik, as well as the uniquely constrained West German response; and it focuses on how the Max Planck Society maneuvered itself into a role of predominance in the space sciences, under these circumstances. Thanks to its strong scientific traditions and political backers, the Max Planck Society was singularly well placed to take advantage of the rising interest in the study and conquest of outer space: while guaranteeing a concerted emphasis on 'fundamental research' and international collaboration, it mobilized existing projects in plasma physics, cosmochemistry, and balloon-based cosmic rays, and joined in diverse space activities with the United States and various European countries. This entry into the space age paved the way to the Society's subsequent expansion into astronomy (the subject of the next chapter), and also allowed the scientific traditions of the early postwar era to diversify: dependency on 'nuclear' sociopolitical interests and funding was now succeeded by a focus on astrophysical subjects proper. As we will see in subsequent chapters, this reorientation ultimately became one of the vehicles propelling these longstanding traditions towards the most effervescent topics of 21st-century astrophysics.

In Section 1, we describe the transition from the predominantly 'nuclear' period up to 1957 to the nascent space age. This took place under the Allied constraints on military technologies, which seriously hindered the West Germans' construction of a fully national (as in sovereign) space launch capability. Within only a few months of the launch of the Soviet satellite, the status of disciplines such as astronomy and astrophysics changed dramatically, as they now became integrated into the Cold War apparatus, just as experimental physics had been in 1945. Key players in this radical shift were those scientists around the world who had preexisting strengths and interests in the cosmic sciences, but had formulated their research in terms of 'nuclear' topics during the postwar years. Space exploration initiatives in the United States, Soviet Union, France, Britain, and other European countries would now become the model for the German MPI scientists described in the previous chapter, and, eventually, their collaboration counterparts, too.

In Section 2 we focus more narrowly on the Max Planck Society. Thanks to preexisting expertise as well as the global connections forged during the first postwar decade, scientists at the Max Planck Institutes versed in all the traditions described in Chapter 1 were ideally placed to jump on the space age bandwagon. Each of these traditions used its particular expertise to position itself within international collaborations. Munich theoretical astrophysicists pivoted toward space-based plasma experiments. Southwestern cosmochemists, with their sample analysis expertise, participated in Apollo missions. The Max Planck Institute for Aeronomy in Lindau had the greatest stake in this new era, as its work had been the closest to what would later become space probes and satellites. Here, we foreshadow the problematic direct competition between Lindau and Munich, which will unfold fully in subsequent chapters. The figure of Reimar Lüst emerges in Munich as someone originally from the plasma astrophysics tradition who then transitions to 'space,' collaborating on French and American rocket-based projects and serving as a delegate to the international bodies that created institutions such as ESRO and its successor ESA. Lüst then went on to become President of the Max Planck Society in 1973.

Chapter 3 describes the arrival of astronomy in the Max Planck Society. Until the 1960s, observational astronomy was not considered a field of interest by the Max Planck Society, whose astrophysical pioneers were strongly oriented toward topics intersecting with the nuclear age. In West Germany, astronomy retained an aura of antiquatedness, and was based largely in observatories dating from previous centuries and still the purview of individual federal states. This changed radically after Sputnik, when astronomy underwent a revival around the world. Even before 1957, an astronomical revolution had been spearheaded by radio astronomy. This was the case also in Germany, where radar pioneers had built the first radio telescopes and forged an international reputation during the first postwar decade. The Max Planck Society, in its moment of most radical expansion, now absorbed these scientists and turned their projects into national infrastructures. This model was then repeated, with the absorption of the most promising observatory project in the traditional optically visible wavelengths, while, simultaneously, there was a major drive toward space-based astronomy in wavelengths inaccessible from the ground. In all these fields, the Max Planck Society grew by attracting external experts who, in addition to their flagship projects, continued to expand into adjacent wavelengths in subsequent decades, at their respective institutes. This absorption of astronomy led to a significant shift within the Max Planck Society itself, an institution where astrophysics had hitherto been dominated by theoretical plasma physicists in Munich, and by experimental nuclear and particle physicists in Heidelberg. The growth of astronomy and its corresponding political influence led to a major reconfiguration of the disciplinary focus of several Max Planck Institutes in the 1970s, and this also signaled a transition from the space sciences of the early post-Sputnik era to the more differentiated astronomy, astrophysics, and planetary sciences of the coming decades.

The most significant transformation resulting from Sputnik in the Max Planck Society was the incorporation of observational astronomy as a research field, and Section 1 of Chapter 3 deals with the incorporation of ground-based astronomy. Up until 1957, there was a strong incipient research tradition in radio astronomy outside of the Society. As with the other strong traditions, this one had a powerful political base, in North Rhine-Westphalia in the context of radar development, which reemerged as a dual-use technology after 1955. In the drive to expansion in astronomy, and taking advantage of regional rivalries, the Max Planck Society subsequently also absorbed the fledging project of what would become the Max Planck Institute for Astronomy, namely the construction of Germany's national optical telescopes, one in each hemisphere. After these two starters, the strategy and narrative of opening new wavelength windows became central to the Society's expansion, first internally, at the

Institute for Radio Astronomy, and soon through additional directorships at many other institutes.

Section 2 follows with the incorporation of satellite-based astronomy. By the mid-1960s, there were initial attempts, internationally, to base astronomical observatories directly in outer space, a decades-old dream, as many wavelengths are blocked by the atmosphere even at mountain altitudes. This section follows the transition of the Max Planck Institute for Extraterrestrial Physics, from an early nuclear era focused on near-Earth space plasma experiments to the institution's increasing dedication to space astronomy. As with the ground-based astronomers, Max Planck leaders invited external pioneers in the field to become directors and participated in several revolutionary astronomical satellites in the gamma-ray and X-ray domains. Space-based astronomy was embedded in European collaboration as well as in competition with the United States. These satellite observatories then guaranteed further German—and hence Max Planck scientists'—participation in all the major missions in these fields, in Europe, the United States, and the Soviet Union. High-energy space-based astronomers differed significantly from their ground-based colleagues, having come from a tradition of experimental particle physics, and their appointment further shifted the center of gravity away from the plasma astrophysicists of previous decades.

Finally, Section 3 deals with the political transformations brought about by the growth of astronomy in the Max Planck Society. The major coordination process that strengthened the monopoly of the cosmic sciences in the Max Planck Society was related to generational renewal and the shifting emphasis of scientific research. The initial 'space science' generation had focused on plasma physics problems, first theoretically and then experimentally. By the late 1960s, however, the future lay in space-based astronomy. In parallel, the longstanding factional rivalry between the two strongholds in Munich and Heidelberg peaked around the election of the next president of the Max Planck Society in 1973, but when Reimar Lüst was elected, he worked towards reconciliation. This increased the circulation of scientists among the cosmic Max Planck Institutes, as new directors were appointed, facilitating the division of scientific labor among them. Extraterrestrial Physics specialized further in space-based astronomy; space plasmas was concentrated in Lindau, and the institute there also moved into planetary exploration, together with the Mainz institute. Cosmochemistry in Heldelberg increasingly shifted towards pioneering work in what is now called astroparticle physics. Several plasma physicists actually became theoretical astrophysicists and inaugurated new lines of research in Heidelberg. Moreover, the enormous Institute for Plasma Physics was readmitted to the Max Planck Society and its infrastructure and institutional support were mobilized to the benefit of many institutes conducting cosmic research.

Chapter 4 concentrates on the transformations brought about by globalization and the end of the Cold War. International collaboration was always key to Max Planck leadership; in the first postwar decade, astrophysicists and cosmochemists were frequent guests within much larger projects based in Allied countries. During the post-Sputnik boom, one of the objectives of the vast expansion was to be able to mobilize national strengths to obtain a stronger voice in international collaborations. The chapter takes up this process of internationalization as it matures from the 1970s on, to become the main mode of research in the Max Planck Society, which it is still to this day. Unexpectedly, this was thanks not so much to the large German-owned infrastructures but, rather, to the weight of longstanding scientific and technical traditions, which brought to the global table theoretical insights, innovative experimentation, and superior instrument-making capacities. The end of the Cold War and German reunification further accelerated the Max Planck Society's transition toward this 21st-century mode of scientific production. Reform in the 1990s coincided with these geopolitical shifts, as well as with the retirement of many of those leading Max Planck Institute directors who had led the wavelength expansion in the previous 30 years. Their successors de-emphasized the construction and ownership of observatories, focusing instead on scientific research within large collaborations, secure in the knowledge that their institutes' instrumental expertise would provide political leverage and a comparative advantage over their partners. Political pressures to relocate institutes to the former East Germany, or even to close them down, were successfully turned into opportunities for expansion, and ultimately, even the one most seriously under threat from these reforms, Biermann's original (theoretical) Institute for Astrophysics, found a reinvigorated mission within the cluster of Max Planck Institutes dedicated to cosmic research, as well as in the, by then, global powerhouse of Garching.

Section 1 focuses on internationalization. The giant telescopes and satellites of the 1960s were national projects, and several ended up becoming major disappointments, while by the 1970s, the parallel track of Europeanization began to bear fruit. Institutions such as the European Southern Observatory (ESO) and the Institute for Millimeter Astronomy (IRAM, founded by France and Germany) paved the way, and the Max Planck Society aimed to maximize its influence within such organizations. In parallel, from different starting points, all the observational institutes converged technologically on infrared astronomy, blurring wavelength as a demarcation between institutes and leading to intense inter-institute collaboration in the 1980s and '90s. As the large telescopes and astronomical satellites came to be built predominantly as international collaborations, and to operate as infrastructures, Max Planck Institutes reoriented much of their work, towards scientific publication on the one hand, and to instrument development on the other; and this afforded them privileged access within the new mode, namely the division of scientific labor. In this context, Max Planck Institutes innovated in many instrumental techniques, such as adaptive optics and interferometry, taking advantage of technical traditions and many decades of participation in collaborations that initiated and benefited from these novel techniques.

Section 2 then shows the changes that coincided with the end of the Cold War, a turning point which shifted the relative position of power of the cosmic sciences, globally. But in Germany in particular, this was further magnified by the challenges brought about by German reunification. Even before the fall of the Berlin Wall, the German scientific community had recommended a reshuffle to revert the excessively dominant position of the Max Planck Society. The rapid and unexpected reunification of the country then tested these plans to the limit, intensifying regional demands and financial pressures on the Society. Yet, despite the succession crises at several institutes, closures were averted and instead there was an expansion eastward. Amid these financial and regional pressures, however, projects such as a planned gravitational wave interferometer had to be scaled down.

Section 3 centers on the most difficult rescue of the decade, namely that of the Institute for Astrophysics (MPA) dating back to Ludwig Biermann's arrival in Göttingen in the late 1940s, and by then the most veteran among the entire cosmic research institute cluster. While already facing doubts about the contemporary significance of 'theoretical astrophysics,' the institute now had to confront the rising predominance of observational astronomy in all wavelength domains within the ensemble of institutes conducting larger-scale cosmic research in the Society. German reunification increased pressure to relocate or close down Max Planck Institutes, while a local institutional crisis and independent institutes in the Garching area made the small theoretical institute particularly vulnerable. The resulting institutional debates reached beyond this particular institute, however, to question how cosmic research was to be conducted in the Max Planck Society overall, and specifically to ask what the function of a theoretical institute within this constellation should be in the 21st century. The solutions devised to save the institute further strengthened both the Society's 'clustering' approach to cosmic research and its international connections, and helped propel appointments and reforms at other institutes in crisis, leading them into the new century.

A deeper, complementary view of the history so far is then presented in Chapter 5, which breaks with the bird's-eye perspective to engage in depth with three episodes that best highlight the intense interrelationship of long-standing scientific traditions in the Max Planck Society and global leadership in scientific fields. All three case studies have in common their roots in traditions that date from before Sputnik and that benefited from unique features of the Max Planck system, such as interdisciplinarity, embeddedness in international collaboration, and strong theoretical, experimental, and instrumental expertise. These all facilitated the rise of astroparticle physics and multi-messenger astronomy in Europe, in contrast to the difficulties experienced by their American counterparts, and this smoothed the path of their early participation in the entirely new field of gravitational wave astronomy. But the growing scale of scientific infrastructures and shifts in conditions at the end of the Cold War also heralded the constraints that Max Planck scientists would face in the 21st century, given that their scientific and technological achievements are meanwhile interwoven with vast multinational research organizations, where successes are not easily accredited.

Section 1 lays down the comparative analytical framework for the case studies.

Section 2 focuses on the quest for solar neutrinos and the related puzzles raised by the nature of this elusive particle. The first newly emerging field benefited directly from the research tradition of cosmochemistry in southwestern Germany, introduced in Chapter 1. Through experimental techniques of mass spectroscopy and small sample radiochemistry, scientists from Freiburg and, later, the Max Planck Institute for Nuclear Physics in Heidelberg were able to collaborate with Brookhaven National Laboratory, where they met Ray Davis, father of solar neutrino detection and the 'solar neutrino deficit' paradox. Researchers from Heidelberg, led by Till Kirsten, improved the instrumentation and, in the 1970s, were even able to overtake the Americans by setting up the GALLEX collaboration, the next-generation experiment in which Germans, Italians, the French, and Israelis worked together with indirect support from the USA and the Soviet Union. Two decades after its conception, in the early 1990s, came the experimental results from GALLEX, which were part of the 'Decade of the Neutrino' that culminated in Nobel Prizes for the founders of the field. This leadership guaranteed a subsequent foothold in neutrino research, even as it evolved away from cosmochemistry toward the electronic detection methods which have now become a central aspect of neutrino-based multi-messenger astronomy.

Section 3 then follows on with gravitational wave research. This second emerging field was the result of the research tradition in theoretical astrophysics in Göttingen and Munich. The 1960s saw an explosion of interest in the new field of relativistic astrophysics, boosted by the unveiling of the violent universe by radio astronomy and by spectacular astrophysical discoveries. Then came the decisive push through the pioneering experiments of Joseph Weber, who claimed to have detected gravitational waves (1969). Munich scientists quickly entered the field with a three-branched approach: experimental detection, statistical analysis of the results, and a deep theoretical footing in general relativity with the appointment of the renowned relativist Jürgen Ehlers. This initial strength then allowed them to shift toward the new method of laser interferometry, taking advantage of expertise at the nearby Max Planck Institute for Plasma Physics. In the 1970s and 1980s, this effort was led by an itinerant group of experts circulating through institutes in the Munich area, facilitating the transition from resonant bars towards laser interferometry and its innovation at increasingly large scales, eventually finding a dedicated site in Hannover, in the early 1990s. Resistance from the worldwide astronomical community and financial constraints resulting from German reunification then forced the Europeanization of the project and, ultimately, the scaling down of the proposed experiment to pilot scale. The German approach never developed into a fully-scaled detector, emphasizing instead the need to perfect experimental systems and build excellence in technology and instrumental innovation. In parallel, Jürgen Ehlers founded an institute for gravitational physics in Potsdam, and soon both branches were unified as the Albert Einstein Institute of the Max Planck Society, one of the central contributors to the detection of gravitational waves in 2015.

Finally, Section 4 tells the story of ground-based gamma-ray astronomy. This third and final emerging field is the complex result of the evolution, throughout the entire 20th century, of the question about the origin and nature of cosmic rays. Until the 1960s, cosmic ray particles were one of the key research areas in experimental physics, part of all the research traditions mentioned in Chapters 1–3. From the late 1950s onward, however, ground-based cosmic ray research declined, as most of its stellar researchers moved toward particle accelerators or jumped on the Sputnik bandwagon to became space scientists. In the following three decades, cosmic rays were studied at less prestigious institutions, in Kiel, for example, which nonetheless obtained results in the early 1980s that attracted worldwide attention. A new generation of accelerator-based particle physicists from both the Max Planck Institute for Physics and the Max Planck Institute for Nuclear Physics then began collaborating with Kiel, which was crucially also joined by a community of Armenians from the Yerevan Physics Institute, who had pioneered the innovative, stereoscopic Imaging Atmospheric Cherenkov Technique (IACT). This technique turned out to be its most promising feature, finalizing the tradition's leap towards ground-based gamma-ray astronomy. Armenian success with Cherenkov telescopes, increasingly supported by Max Planck scientists, sparked competition between two Max Planck Institutes in Munich and Heidelberg, to become world leaders in what promised to become an entirely new form of ground-based astronomy, thereby absorbing the Armenian scientists. Max Planck Institutes then built the most successful telescopes of the subsequent generation, MAGIC and H.E.S.S., while competing both with each other and with other global players. Thanks to their complementary double presence in the field, the two Max Planck Institutes won the race towards ground-based, gamma-ray telescopes, leading to the global Cherenkov Telescope Array (CTA) collaboration with over 100 telescopes, which the Americans then entered as junior partners.

The book concludes with the Financial Appendix. Quantitative financial data was examined early in our research and continued to inform the narrative that is presented in the book. It was decided, however, not to detail these finances within the chapters themselves, where the interaction of sociopolitical dynamics with scientific developments was to remain the primary focus. Rather, at any mention of a particularly relevant financial event, we refer the reader to the Financial Appendix. The Appendix thus constitutes a complementary analysis of the cosmic research cluster, based on financial planning and money flow that, as the reader will see, significantly mirrored or even propelled the events described throughout the book; and on occasion, this offers additional insight that no other historical source could have afforded.

CHAPTER 1

Nuclear Age (1945–1957): Reconstruction under Regional Fragmentation

The focus of this first chapter is astrophysics in the first decade following World War II, when the discipline was embedded in the promises of the 'nuclear age' and, in West Germany, the hardship of reconstruction and Allied restrictions. A key element at the time was the ambiguity of the term 'nuclear' physics, which tacitly implied the potential of nuclear energy and weapons, while also connoting a disparate array of scientific subjects, in addition to the atomic nucleus, such as fusion and plasmas, cosmic rays, and subatomic particles. This ambiguous overlap enabled astrophysicists to obtain financial and political support for any research signaling future expansion into the relevant applied fields, while yet respecting postwar prohibitions. The backdrop to this support was, in essence, the regionally focused competition between the Allies, which played out in the various occupation zones and nascent states of West Germany. Scientists rooted in diverse scientific traditions and political orientations adapted to the different regional interests and the priorities of the Allied occupiers. It was the strength acquired in this decade, partly as a result of competition among these different factions, that assured the Max Planck Society a good headstart after Sputnik, when the 'space age' took center stage.

1 Postwar Scientific Traditions in Göttingen

A community of scientists converged in Göttingen, in the aftermath of World War II, to become part of what was to become—at the initiative of the famous physicist Werner Heisenberg—the Max Planck Institute for Physics and Astrophysics, the primary hub and powerhouse of the young Max Planck Society. This chapter focuses on its trajectory during the early post-war nuclear age, until the move to Munich in 1958. Göttingen was the birthplace of the research tradition rooted in the theoretical plasma astrophysics led by the astrophysicist Ludwig Biermann and his disciples, who were able to make contact and collaborate closely with scientists working on nuclear fusion in the United States.

Wartime Legacy and Reconstruction via Theoretical Physics

In the immediate postwar years, Germany was facing devastation, poverty, trauma resulting from the enormous loss of lives, and the collapse of economic and political organization, with the country divided into four occupation zones. Since the final days of World War II, the Allies had taken the most useful German researchers permanently out of the country, particularly those who were far ahead of their counterparts in the Allies' own countries, as was the case with rocket experts.[1] The Western Allies also competed to host, in their respective occupation zone, all the other, more dispensable scientists, including those who had contributed to the comparatively modest efforts of the German wartime nuclear program.[2] Between spring and summer 1945, ten German nuclear scientists were captured as part of the Allied Alsos Mission—the science intelligence unit whose chief focus was the German nuclear project—and interned at Farm Hall, a country estate near Cambridge, UK.[3]

In order to determine how close Nazi Germany had been to constructing a nuclear weapon, their conversations were secretly recorded.[4] Werner Heisenberg, Otto Hahn, Max von Laue, and other German nuclear scientists were

1 Tom Bower: *The Paperclip Conspiracy. The Hunt for the Nazi Scientists*. Boston: Little, Brown 1987. Annie Jacobsen: *Operation Paperclip. The Secret Intelligence Program to Bring Nazi Scientists to America*. New York, NY: Little, Brown and Company 2014. Linda Hunt: *Secret Agenda. The United States Government, Nazi Scientists, and Project Paperclip, 1945–1990*. New York: St. Martin's Press 1991.

2 On the German nuclear work during the war, see Mark Walker: *German National Socialism and the Quest for Nuclear Power 1939–1949*. Cambridge: Cambridge University Press 1989. Mark Walker: *Die Uranmaschine. Mythos und Wirklichkeit der deutschen Atombombe*. Berlin: Siedler Verlag 1990. Mark Walker: *Nazi Science. Myth, Truth, and the German Atomic Bomb*. New York, NY: Plenum Press 1995. Ruth Lewin Sime: The Politics of Forgetting. Otto Hahn and the German Nuclear-Fission Project in World War II. *Physics in Perspective* 14/1 (2012), 59–94. doi:10.1007/s00016-011-0065-6. Mark Walker: Physics, History, and the German Atomic Bomb. *Berichte Zur Wissenschaftsgeschichte* 40/3 (2017), 1–18. doi:10.1002/bewi.201701817. David Cassidy: *Farm Hall and the German Atomic Project of World War II. A Dramatic History*. Cham, Switzerland: Springer 2017.

3 Samuel A. Goudsmit: *Alsos*. Woodbury, NY: AIP Press 1996. Boris T. Pash: *The Alsos Mission*. New York, NY: Award House 1969. Leo J. Mahoney: *A History of the War Department Scientific Intelligence Mission (ALSOS) 1943–1945*. Dissertation/ PhD Thesis. Ann Arbor, MI: University of Michigan 1981. Mary A. McPartland: *The Farm Hall Scientists. The United States, Britain, and Germany in the New Atomic Age, 1945–46*. Dissertation. Washington, DC: The George Washington University 2013. https://search.proquest.com/pqdtglobal/docview/1436978271/abstract/DDC811FAFC464047PQ/2. Last accessed 4/1/2019.

4 Charles Frank (ed.): *Operation Epsilon. The Farm Hall Transcripts*. London: Institute of Physics Publishing 1993. On the Farm Hall internment, see Jeremy Bernstein: *Hitler's Uranium Club. The Secret Recordings at Farm Hall*. Woodbury, NY: American Institute of Physics

ILLUSTRATION 1 The Farm Hall house, near Cambridge, where ten German physicists (Erich Bagge, Kurt Diebner, Walther Gerlach, Otto Hahn, Paul Harteck, Werner Heisenberg, Horst Korsching, Max von Laue, Carl Friedrich von Weizsäcker, and Karl Wirtz) were detained from July 1945 to early January 1946.

finally released from their internment at Farm Hall, and brought back to Germany on January 3, 1946.[5]

1996. For the current state of research on this diaspora, see Matthias Judt, and Burghard Ciesla (eds.): *Technology Transfer out of Germany after 1945*. Amsterdam: Harwood Academic Publishers 1996. Burghard Ciesla: Das "Project Paperclip". Deutsche Naturwissenschaftler und Techniker in den USA (1946 bis 1952). In: Jürgen Kocka (ed.): *Historische DDR-Forschung. Aufsätze und Studien*. Berlin: Akademie Verlag 1993, 287–301. Burghard Ciesla, and Helmuth Trischler: Legitimation through Use. Rockets and Aeronautical Research in the Third Reich and the USA. In: Mark Walker (ed.): *Science and Ideology. A Comparative History*. London: Routledge 2003, 156–185. Michael J. Neufeld: The Nazi Aerospace Exodus. Towards a Global, Transnational History. *History and Technology* 28/1 (2012), 49–67. Monique Laney: *German Rocketeers in the Heart of Dixie. Making Sense of the Nazi Past During the Civil Rights Era*. New Haven, CT: Yale University Press 2015.

5 Besides Hahn, Heisenberg, and von Laue, the group included von Weizsäcker (theoretical nuclear physicist), Erich Bagge (expert in isotope separation), Kurt Diebner (a leader of nuclear research in the German Army Weapons Bureau), Walther Gerlach (chief adminis-

ILLUSTRATION 2 Left to right: Werner Heisenberg, Max von Laue, and Otto Hahn in Göttingen, soon after their return to Germany on January 3, 1946.

Upon arrival in the small village of Alswede, Heisenberg immediately wrote to his wife Elisabeth:

> My dear Li! This is the first evening back in Germany since the end of the war. This long time of captivity seemed to us only bearable through the scientific work. How it's going to be here, we do not know yet. The purpose of our being here is as follows: The highest authorities have decided that we all should in the future have our workplaces in the British occupation zone.[6]

In the British zone, the city of Göttingen had survived World War II without major damage. The George Augusta University, one of the most prestigious in

---

trator of nuclear research, 1944–45), Paul Harteck (heavy water as a neutron moderator for reactor design, neutron physics), Horst Korsching (isotope separation), Karl Wirtz (heavy water and isotope separation). For details on the detainees, see Appendix D in Bernstein, *Hitler's Uranium Club*, 1996. In the meantime, Otto Hahn had been awarded the Nobel Prize in Chemistry 1944 for the discovery of nuclear fission, but he received the Prize one year later, in 1945. He was still in Farm Hall when the announcement was made, and learned of the award when reading the *Daily Telegraph*. Bernstein, *Hitler's Uranium Club*, 1996, 282–283.

6 Werner Heisenberg, and Elisabeth Heisenberg: *My Dear Li. Correspondence, 1937–1946*. Edited by Anna Maria Hirsch-Heisenberg. New Haven, CT: Yale University Press 2016.

ILLUSTRATION 3 Left to right: Otto Hahn, Karl-Friedrich Bonhoeffer, Werner and Elisabeth Heisenberg, and Werner Hoppenstedt. Göttingen, March 8, 1949

Germany, was the first university in the country to resume teaching already in September 1945. On January 12, Hahn and Heisenberg were accompanied to Göttingen, where they found Max Planck, who had arrived there as a refugee, seeking shelter with relatives. From 1930 to 1937, Planck had been President of the Kaiser Wilhelm Society (KWG), the extra-universitary research organization founded in 1911 to conduct specialized research in its own institutes, predominantly in the natural sciences. Thanks to its outstanding scientific achievements, the Society had quickly established itself, nationally and internationally; but now, because of its involvement in Hitler's regime and armament research, the Allies were urging that it be dissolved.[7]

7 Bernhard vom Brocke: Die Kaiser-Wilhelm-Gesellschaft im Kaiserreich. Vorgeschichte, Gründung und Entwicklung bis zum Ausbruch des Ersten Weltkriegs. In: Rudolf Vierhaus, and Bernhard vom Brocke (eds.): *Forschung im Spannungsfeld von Politik und Gesellschaft. Geschichte und Struktur der Kaiser-Wilhelm-/Max-Planck-Gesellschaft.* Stuttgart: Deutsche Verlags-Anstalt 1990, 17–162. Bernhard vom Brocke: Die Kaiser-Wilhelm-Gesellschaft in der Weimarer Republik. Ausbau zu einer gesamtdeutschen Forschungsorganisation (1918–1933). In: Rudolf Vierhaus, and Bernhard vom Brocke (eds.): *Forschung im Spannungsfeld von Politik und Gesellschaft. Geschichte und Struktur der Kaiser-Wilhelm-/Max-Planck-Gesellschaft.*

ILLUSTRATION 4 The Max Planck Institute for Physics in Göttingen on Bunsenstrasse, housed since the summer of 1946 in Building No. 10 of the Aerodynamics Research Institute (AVA), which formerly contained a cooling tunnel.

Once back in Alswede, in another letter to Elisabeth, on January 14, 1946, Heisenberg sketched his first impressions of Göttingen:[8]

> I just spent three days in Göttingen together with an unusually nice British officer and have deliberated on the future of my institute. There are many indications that we all will come to Göttingen not too far down the line. They have huge empty institute rooms there, so that the external givens are not bad. Difficulty: proximity of the Russians and lack of housing…

Stuttgart: Deutsche Verlags-Anstalt 1990, 197–355. Helmuth Albrecht, and Armin Hermann: Die Kaiser-Wilhelm-Gesellschaft im Dritten Reich (1933–1945). In: Rudolf Vierhaus, and Bernhard vom Brocke (eds.): *Forschung im Spannungsfeld von Politik und Gesellschaft. Geschichte und Struktur der Kaiser-Wilhelm-/Max-Planck-Gesellschaft*. Stuttgart: Deutsche Verlags-Anstalt 1990, 356–406.

8 Werner Heisenberg to Elisabeth, January 14, 1946. Heisenberg, and Heisenberg, *My Dear Li*, 2016.

ILLUSTRATION 5 The library of the Max Planck Institute for Physics in Göttingen

During the first postwar decade, much of the science that had been conducted during the war was not permitted under the terms of the occupation.[9] This was the case with nuclear fission research, in particular; but in Göttingen, it also included the aerodynamic research conducted at what had been the Kaiser Wilhelm Institute for Fluid Dynamics (*Kaiser-Wilhelm-Institut für*

9 Control Council and Coordinating Committee of the Allied Control Authority: Enactments and Approved Papers of the Control Council and Coordinating Committee. Allied Control Authority, Germany (1945–1948). 9 Volumes. *Military Legal Resources. Federal Research Division. Library of Congress*. https://www.loc.gov/rr/frd/Military_Law/enactments-home.html. Last accessed 10/30/2018. The first lines of the Control Council Law No. 25 are related to the Control of Scientific Research, as specified in volume III, on pages 103–105: "In order to prohibit for military purposes scientific research and its practical application, to control them in other fields in which they may create a war potential, and to direct them along peaceful lines, the Control Council enacts as follows [...]." The meaning of "fundamental scientific research" and "applied scientific research" are specified on p. 105. See "Applied nuclear physics" in the list of Prohibited Applied Scientific Research on p. 108. On the law No. 25 and the Allied research control see Manfred Heinemann: Überwachung und »Inventur« der deutschen Forschung. Das Kontrollratsgesetz Nr. 25 und die alliierte Forschungskontrolle im Bereich der Kaiser-Wilhelm-/Max-Planck-Gesellschaft (KWG/MPG) 1945–1955. In: Lothar Mertens (ed.): *Politischer Systemumbruch als irreversibler Faktor von Modernisierung in der Wissenschaft?* Berlin: Duncker & Humblot 2001, 167–199.

Strömungsforschung)[10] and its associated Aerodynamics Research Institute (*Aerodynamische Versuchsanstalt*, AVA), with its famous wind tunnels.[11] The tunnels themselves had been dismantled and shipped to the UK and in fact it was these now vacant buildings that housed both the newly established Max Planck Society and Heisenberg's Institute for Physics, in their early years.[12]

Heisenberg and his wife were not able to reunite as a family until some months after his arrival in Germany from internment at Farm Hall. Having long had no news of her husband, Elisabeth had assumed throughout their enforced separation that he was "in the more fortunate America" and on January 18, she asked him whether there had been "any choice at all" to avoid that "encroaching misery."[13] Heisenberg answered her question on January 25:[14]

> You ask whether we had any choice about staying in Germany or going to America. I do not believe that they wanted our entire group over there, but Hahn and I were asked semiofficially: Goudsmith [*sic*] asked me right at the first 'interrogation' in Heidelberg whether I wanted to go to America, and Blackett, in England, reiterated the question later on. I had already pondered it very thoroughly before and arrived at the following position: it is completely clear to me that in the next decades America

10 K. Oswatitsch, and K. Wieghardt: Ludwig Prandtl and His Kaiser-Wilhelm-Institut. *Annual Review of Fluid Mechanics* 19/1 (1987), 1–25. doi:10.1146/annurev.fl.19.010187.000245.

11 Florian Schmaltz: Aeronautical Research Under Nazi Occupation in Paris. The Aerodynamische Versuchsanstalt Göttingen and the Mobilisation of Resources, French Scientists and Engineers, 1940–1944. In: Claudine Fontanon, and Irina Gouzévich (eds.): *Les Ingénieurs Civils et La Circulation Des Savoirs En XIXe–XXe Siècles*. Paris: Garnier im Erscheinen. Florian Schmaltz: Luftfahrtforschung auf Expansionskurs. Die Aerodynamische Versuchsanstalt in den besetzten Gebieten. In: Sören Flachowsky, Rüdiger Hachtmann, and Florian Schmaltz (eds.): *Ressourcenmobilisierung. Wissenschaftspolitik und Forschungspraxis im NS-Herrschaftssystem*. Göttingen: Wallstein Verlag 2016, 326–382.

12 Aerodynamische Versuchsanstalt Göttingen: *Die Aerodynamische Versuchsanstalt Göttingen von 1945 bis 1969*. Göttingen: Max-Planck-Gesellschaft 1969. Albert Betz: Aus der Geschichte der Aerodynamischen Versuchsanstalt Göttingen. In: Generalverwaltung der Max-Planck-Gesellschaft (ed.): *Jahrbuch 1957 der Max-Planck-Gesellschaft zur Förderung der Wissenschaften e.V.* Göttingen 1957, 40–59. See also the dedicated section in Eckart Henning, and Marion Kazemi: *Handbuch zur Institutsgeschichte der Kaiser-Wilhelm-/Max-Planck-Gesellschaft zur Förderung der Wissenschaften 1911–2011. Daten und Quellen*. Vol. 1. Berlin: Archiv der Max-Planck-Gesellschaft 2016, 27–45.

13 Elisabeth Heisenberg to Werner, January 18, 1946. Heisenberg, and Heisenberg, *My Dear Li*, 2016.

14 Werner Heisenberg to Elisabeth, January 25, 1946. Heisenberg, and Heisenberg, *My Dear Li*, 2016.

> will be the center of scientific life, and that working conditions for me in Germany will be much worse than over there. Exactly because of this, on the other hand, I am not needed there as much: many excellent, competent physicists are there. Here, however, it matters a great deal that an intellectual life should again become viable. Since 1933 it has been clear to me that here a terrible tragedy for Germany was in progress, only I could not have imagined the extent and the ending; and I stayed here at the time so that I might also be here afterward and help. This was exactly what I also told my American friends in the summer of 1939, and the best among them could understand it; this intention remains firm and will not be betrayed.[15]

Heisenberg was of course well aware that the letter would be screened by the Allies. But still, the sense of what he wrote in the last sentence was later confirmed by Edoardo Amaldi, a member of Enrico Fermi's group in Rome, better known as 'the Via Panisperna boys.'[16] Like Heisenberg, Amaldi was deeply frustrated by the passage of leadership in physics from Europe to the United States. Fermi had left Italy in fall 1938, partly because the fascist government had rejected his repeated funding requests to build an institute for nuclear physics equipped with modern research tools, but in particular, following promulgation of the racial laws which menaced his Jewish wife Laura Capon and their two children. Fermi's team had disbanded, but Amaldi, instead of trying, like many Italian physicists, to move to the US, chose to remain in Italy, tackling the catastrophic situation and becoming one of the main promoters of the postwar reconstruction of physics in Italy as well as in Europe (he later was a key figure in the birth of CERN and the European Space Agency).[17]

As Amaldi recalled, he had been influenced in this choice also by Heisenberg's underlying reasons for remaining in Germany and which Heisenberg had explained to his colleagues immediately before the outbreak of the war,

---

15 Heisenberg himself described later his long conversation with Enrico Fermi during which the latter tried to convince him to emigrate to the US and start a new scientific life. Werner Heisenberg: *Physics and Beyond. Encounters and Conversations*. New York: Harper & Row Publishers 1971, 169–172.

16 Gerald Holton: Striking Gold in Science: Fermi's Group and the Recapture of Italy's Place in Physics. *Minerva* 12/2 (1974), 159–198. https://www.jstor.org/stable/41820198. Last accessed 1/8/2020.

17 Carlo Rubbia: *Edoardo Amaldi: Scientific Statesman*. Vol. 91–09. Geneva: CERN 1991. doi:10.5170/CERN-1991-009.

ILLUSTRATION 6 Werner Heisenberg (right) at CERN in 1960 with Edoardo Amaldi and Giuseppe Fidecaro (left). Fidecaro was one of the first physicists to work at the Geneva international laboratory (© CERN ARCHIVES).

during discussions on the ominous situation in Europe that took place in the US in the summer of 1939:[18]

> It was Sunday afternoon [...] the Fermis had invited several colleagues and young physicists for a small welcome reception for Werner Heisenberg coming, if I well remember, from Berkeley and directed to Germany. The only—and central—topic of conversation was political events in Europe, where the situation appeared to be growing increasingly grim.

18 Edoardo Amaldi: 1939, Perché scelsi l'America. *La Repubblica* (12/6/1996). https://ricerca.repubblica.it/repubblica/archivio/repubblica/1996/12/06/1939-perche-scelsi-america.html?ref=search. Last accessed 1/20/2022. Edoardo Amaldi: *Da Via Panisperna all'America. I fisici italiani e la seconda guerra mondiale*. Edited by Giovanni Battimelli, and Michelangelo De Maria. Roma: Editori Riuniti 1997, 72.

> I remember that S. Goudsmit asked Heisenberg what he thought and whether he would consider opportunities to leave Germany and move to the United States. Heisenberg said that he absolutely intended to return to Germany. A discussion followed [...]. I do not remember all of Heisenberg's considerations, except one point that has remained impressed on my mind [...] He associated the decision to emigrate to the United States with the aspiration to work in the peace and quiet so indispensable to intense scientific work, and the decision to remain in his own country with the desire *to preserve a certain form of culture and keep it alive* [...] This reasoning remained impressed on my mind and certainly had an influence on the decisions that I took six or seven years later..." [our translation, emphasis added].[19]

The Italian scientific community, traditionally strong in cosmic ray and nuclear physics, had been decimated by the racist policies implemented by Mussolini's fascist government, and so deprived of some of its most influential and prestigious members, such as Enrico Fermi and Bruno Rossi, the fathers of modern physics in Italy. After the stagnant phase and isolation of wartime, Amaldi, in collaboration with Gilberto Bernardini, initiated an intensive program for the revival of physics in Italy[20] and, in parallel, began promoting an international strategy to relaunch physical sciences in Europe.[21]

Interestingly, a similar 'protective' attitude toward national realities can be observed also in Frédéric Joliot, who, together with his wife Irène Curie, had

19 Relationships with the Italian community went back to the early 1920s, when Heisenberg and Fermi met in Göttingen at Max Born's Institute for Theoretical Physics. In 1948 Amaldi invited Heisenberg to visit the University of Rome and to collaborate (Heisenberg's Papers, AMPG, III. Abt., Rep. 93, No. 18).

20 Edoardo Amaldi: The Years of Reconstruction. In: Giovanni Battimelli, and Giovanni Paoloni (eds.): *20th Century Physics. Essays and Recollections. A Selection of Historical Writings by Edoardo Amaldi*. Singapore: World Scientific 1998, 263–294.

21 The postwar history and the successful rebuilding of Europe after WWII has been widely discussed in several books, some of which have specifically focused on the relaunching of physics. See, for example, Michelangelo De Maria, Mario Grilli, and Fabio Sebastiani (eds.): *The Restructuring of Physical Sciences in Europe and the United States, 1945–1960. Proceedings of the International Conference*. Singapore: World Scientific 1989. John Krige: *American Hegemony and the Postwar Reconstruction of Science in Europe*. Cambridge, MA: MIT Press 2006. Martin Kohlrausch, and Helmuth Trischler: *Building Europe on Expertise. Innovators, Organizers, Networkers*. Basingstoke: Palgrave Macmillan 2014. Naomi Oreskes, and John Krige (eds.): *Science and Technology in the Global Cold War*. Cambridge, MA: MIT Press 2014. Roberto Lalli: Crafting Europe from CERN to Dubna: Physics as Diplomacy in the Foundation of the European Physical Society. *Centaurus* 63/1 (2021), 103–131. doi:https://doi.org/10.1111/1600-0498.12304.

ILLUSTRATION 7 Left to right: Werner Heisenberg, Enrico Fermi, Louis Leprince-Ringuet, and Bruno Rossi during the Summer School on "Nuclear Physics and Cosmic Rays" held in Varenna, Lake Como from July 26 to August 2, 1954. Heisenberg lectured on the structure of atomic nuclei, Fermi reported on the production of pions in nucleon-nucleon collisions at the Cosmotron, while Rossi lectured on both the origin of cosmic rays and "fundamental particles." The school's program thus illustrates both the ongoing transition in particle physics from cosmic rays to accelerators and the remaining ambiguity between the nuclear and subnuclear realm. Fermi was already very ill and passed away a few months later in Chicago; the Varenna Summer School was subsequently renamed after him.

been awarded the Nobel Prize in Chemistry 1935, for the discovery of artificial radioactivity. During the war, he was approached in secret more than once by Allied organizations who offered him the chance to leave France—if only temporarily—to work with colleagues abroad. He always refused, and used quite similar arguments, saying "he wanted to ensure the survival of French nuclear physics and the education of the next generation of scientists."[22]

22 Spencer R. Weart: Scientists in Power: France and the Origins of Nuclear Energy, 1900–1950. *Bulletin of the Atomic Scientists* 35/3 (1979), 41–50, 44. doi:10.1080/00963402.1979.11458599.

ILLUSTRATION 8 Otto Hahn and Frédéric Joliot-Curie in Lindau, 1958, at the 8th Nobel Laureate Meeting dedicated to chemistry. Joliot-Curie died only a few weeks later.

In a more subliminal way, Patrick Blackett, who was very active in promoting the relaunch of physics in Europe, used the occasion of the Nobel Prize ceremony in 1948 to emphasize at the beginning of the Banquet speech that he liked to think of the Prize not only as a recognition of his own scientific work, but "as a tribute to the vital school of European Experimental Physics" in which he had been trained. He also added that the fact that all four Nobel Prizes that year and so many others in previous years had been awarded to Europeans was

> surely a striking tribute to the astonishing vitality in the Arts and Sciences of our irrepressible, colorful, turbulent but war-scarred Continent of Europe.[23]

Blackett had visited the physicists interned at Farm Hall in September 1945, and had long conversations on the future of German science with Heisenberg, who was his old friend, and who considered Blackett "a sensible man with whom one can get down to brass tacks."[24]

Well aware that the center of gravity of physics had now shifted to the US, these physicists were now each in their own way fighting a battle for the reconstruction of Europe, at both the national and international level.

The British authorities did not support the idea that the Kaiser Wilhelm Society should be dissolved. The integrity of scientists such as Max Planck, Otto Hahn, and Max von Laue had never been in doubt. They represented the 'crystallization nucleus' around which it became possible to initiate the reconstruction of fundamental scientific activity in Germany.[25]

Under the supervision of the British occupation authorities, and with the help of two British officials of the Intelligence Division, Bertie Blount and Ronald G. J. Fraser, the latter a physicist as well as Scientific Advisor to the British Military Government, Heisenberg was set the task of refounding the Kaiser Wilhelm Institute for Physics that he had led as acting head from 1942 on, in Berlin-Dahlem, when the secret German nuclear program was still headquartered there.[26] But the postwar period could be no simple continuation of

---

23 Patrick M. S. Blackett, speech at the Nobel Banquet in Stockholm, December 10, 1948. https://www.nobelprize.org/prizes/physics/1948/blackett/speech/. Last accessed 5/17/2020.

24 Bernstein, *Hitler's Uranium Club*, 1996, 215–230.

25 Bagge's diary of October 9, 1945. Erich Bagge, Kurt Diebner, and Kenneth Jay: *Von der Uranspaltung bis Calder Hall.* Edited by Ernesto Grassi. Hamburg: Rowohlt 1957, 61. On the role of the British authorities in the foundation of the Max Planck Society see Peter Alter: Die Kaiser-Wilhelm-Gesellschaft in den deutsch-britischen Wissenschaftsbeziehungen. In: Rudolf Vierhaus, and Bernhard vom Brocke (eds.): *Forschung im Spannungsfeld von Politik und Gesellschaft. Geschichte und Struktur der Kaiser-Wilhelm-/Max-Planck-Gesellschaft.* Stuttgart: Deutsche Verlags-Anstalt 1990, 726–746, 743–746.

26 On the history of the KWG institute in Berlin and Heisenberg's research program during the war, see: Horst Kant: Albert Einstein, Max von Laue, Peter Debye und das Kaiser-Wilhelm-Institut für Physik in Berlin (1917–1939). *Die Kaiser-Wilhelm-/Max-Planck-Gesellschaft und ihre Institute: Studien zu ihrer Geschichte: Das Harnack-Prinzip,* 1996, 227–243. https://pure.mpg.de/pubman/faces/ViewItemOverviewPage.jsp?itemId=item_2276913_1. Last accessed 1/17/2020. Helmut Rechenberg: Werner Heisenberg und das Forschungsprogramm des Kaiser-Wilhelm-Instituts für Physik (1940–1948). In:

ILLUSTRATION 9 Max Planck (left) and Max von Laue (right) in Göttingen, 1947

the Kaiser Wilhelm Society era. Heisenberg had arrived at this institute during the war, when it was still located in Berlin, with the task of contributing to (and later coordinating) the *Uranverein* (Uranium Club), Nazi Germany's wartime effort to explore the military potential of the recent discovery—at the Kaiser

Bernhard Vom Brocke, and Hubert Laitko (eds.): *Die Kaiser-Wilhelm-/Max-Planck-Gesellschaft und ihre Institute. Studien zu ihrer Geschichte. Das Harnack-Prinzip*. Berlin: Walter de Gruyter 1996, 245–262. Helmut Rechenberg: Werner Heisenberg und das Kaiser-Wilhelm—(Max-Planck-)Institut für Physik. *Physikalische Blätter* 37/12 (1981), 357–364. doi:https://doi.org/10.1002/phbl.19810371206. Eckart Henning, and Marion Kazemi: *Handbuch zur Institutsgeschichte der Kaiser-Wilhelm-/Max-Planck-Gesellschaft zur Förderung der Wissenschaften 1911–2011. Daten und Quellen*. Vol. 2. Berlin: Archiv der Max-Planck-Gesellschaft 2016, 1177–2016.

ILLUSTRATION 10 The Kaiser Wilhelm Institute for Physics in Berlin-Dahlem, with the laboratory for low-temperature physics visible on the right. Photo taken at the institute's inauguration in 1938.

Wilhelm Institute for Chemistry in Berlin-Dahlem—of the phenomenon of nuclear fission.[27]

Like many other Berlin institutes, Heisenberg's staff and equipment had relocated to the southwest of Germany in the final years of the war, and then reassembled in peacetime in Göttingen, with the support of the British occupation forces. This meant proximity to an important university; Heisenberg was appointed to a lecturer position there, and it is no surprise to learn that the first postwar meeting of the *Deutsche Physikalische Gesellschaft* (German

27 For an outline of Heisenberg's research work during the 1940s, see: Rechenberg, Werner Heisenberg, 1996, 245–262. Mark Walker: *Eine Waffenschmiede? Kernwaffen- und Reaktorforschung am Kaiser-Wilhelm-Institut für Physik*. Ergebnisse. Vorabdrucke aus dem Forschungsprogramm »Geschichte der Kaiser-Wilhelm-Gesellschaft im Nationalsozialismus«, 26. Berlin: Forschungsprogramm »Geschichte der Kaiser-Wilhelm-Gesellschaft im Nationalsozialismus«. Max-Planck-Institut für Wissenschaftsgeschichte 2005. Werner Heisenberg: Über die Arbeiten zur technischen Ausnutzung der Atomkernenergie in Deutschland. *Die Naturwissenschaften* 33/11 (1946), 325–329. doi:10.1007/BF00842932. The canonical work on the German nuclear fission program is: Monika Renneberg, and Mark Walker: *Science, Technology, and National Socialism*. Cambridge: Cambridge University Press 2003. See also Walker, *Eine Waffenschmiede?*, 2005. Walker, *Nazi Science*, 1995.

ILLUSTRATION 11 The Cockcroft-Walton accelerator in the tower building of the Kaiser Wilhelm Institute for Physics in Berlin-Dahlem, 1938

Physical Society, DPG) was held in Göttingen, in October 1946.[28] Four Nobel Prize laureates, Max Planck, Otto Hahn, and Max von Laue among them, were

28 Ernst Brüche: Physiker-Tagung in Göttingen. Vorträge von der Göttinger Tagung. *Physikalische Blätter* 3/9 (1947), 317–325. doi:10.1002/phbl.19470030909. See also Friedrich

crucial actors in the revival of the Kaiser Wilhelm Society. With the support of Max Planck, who was unanimously regarded as an outstanding scientist with an impeccable international reputation, Hahn's efforts to relaunch activities in the British zone succeeded in gaining British approval. The Kaiser Wilhelm Society was eventually refounded in Göttingen on February 26, 1948, as a successor organization but under the new name Max Planck Society, in honor of the recently deceased founding figure and renowned trailblazer in modern physics, who, despite his opposition to the Nazi dictatorship, had remained in Germany throughout the war.[29]

Heisenberg's Institute for Physics in Göttingen was initially manned largely by scientists who had been isolated at Farm Hall while their nuclear expertise was under assessment.[30] However, once settled in the new headquarters, they found themselves lacking most of their wartime experimental equipment, which had not yet been allowed to leave Hechingen in southwest Germany.

A. Paneth: Scientific Research in the British Zone of Germany. *Nature* 161/4084 (1948), 191–192. doi:10.1038/161191a0.

29 Manfred Heinemann: Der Wiederaufbau der Kaiser-Wilhelm-Gesellschaft und die Neugründungen der Max-Planck-Gesellschaft (1945–1949). In: Rudolf Vierhaus, and Bernhard vom Brocke (eds.): *Forschung im Spannungsfeld von Politik und Gesellschaft. Geschichte und Struktur der Kaiser-Wilhelm-/Max-Planck-Gesellschaft*. Stuttgart: Deutsche Verlags-Anstalt 1990, 407–470. See also Otto Gerhard Oexle: *The British Roots of the Max-Planck-Gesellschaft*. Translated by Jane Rafferty. London: German Historical Institute 1995. Otto Gerhard Oexle: Wie in Göttingen die Max-Planck-Gesellschaft entstand. In: Max-Planck-Gesellschaft zur Förderung der Wissenschaften e.V. (ed.): *Max-Planck-Gesellschaft Jahrbuch 1994*. Göttingen: Vandenhoeck & Ruprecht 1994, 43–60. Ruth Lewin Sime: *Otto Hahn und die Max-Planck-Gesellschaft. Zwischen Vergangenheit und Erinnerung*. Edited by Carola Sachse. Vol. 14. Berlin: Max-Planck-Gesellschaft zur Förderung der Wissenschaft 2004. Jürgen Renn, Horst Kant, and Birgit Kolboske: Stationen der Kaiser-Wilhelm-/Max-Planck-Gesellschaft. In: Jürgen Renn, Birgit Kolboske, and Dieter Hoffmann (eds.): *»Dem Anwenden muss das Erkennen vorausgehen«. Auf dem Weg zu einer Geschichte der Kaiser-Wilhelm-/Max-Planck-Gesellschaft*. 2nd ed. Berlin: epubli 2015, 5–120. For an in-depth analysis of the origins of the Max Planck Society see Jaromír Balcar: *Die Ursprünge der Max-Planck-Gesellschaft. Wiedergründung—Umgründung—Neugründung*. Berlin: GMPG-Preprint 2019.

30 A document dated June 1946 lists the staff of the new Kaiser Wilhelm Institute for Physics, founded on June 1, 1946, under the leadership of Werner Heisenberg: Prof. Dr. Otto Hahn (President of the Kaiser-Wilhelm-Gesellschaft), Prof. Dr. Max von Laue (Deputy Director), Prof. Dr. C. F. Von Weizsäcker (head of Department), Dr. Karl Wirtz (Head of Department), Dr. Horst Korsching (assistant), Dr. Erich Bagge (assistant), und Frl. Dr. Elisabeth Rall (Librarian), Frl. Helene Gleitz (administrative secretary), AMPG, II. Abt., Rep. 66, No. 3047. All the aforementioned scientists of the staff had been detained at Farm Hall, that is, seven of the ten captured by the Alsos Mission; the remaining three were: Kurt Diebner, Walther Gerlach and Paul Harteck.

ILLUSTRATION 12 Max Planck congratulates Otto Hahn for having been awarded the 1944 Nobel Prize for Chemistry "for his discovery of the fission of heavy nuclei." Max von Laue, Adolf Windaus, and Werner Heisenberg are visible in the background. The four Nobel Laureates were crucial actors in the revival of the Kaiser Wilhelm Society.

This was where the Kaiser Wilhelm Institute of Physics in Berlin-Dahlem had relocated during the war, to be safe from Allied bombing raids; and it was meanwhile in the French occupation zone. At the original institute in Dahlem, all the books, research devices, and equipment had been seized by the Russians, while all the materials related to work on the uranium pile had been taken to the United States by the Alsos Mission.[31]

---

31 Heisenberg was extremely disappointed by such circumstances. In an anonymous memorandum reporting a meeting between Heisenberg and Otto Hahn (October 4, 1947), the current situation is clearly referred to: "He [Heisenberg] points out that the work possibilities in Göttingen are so far very limited and that the Institute's most important large facilities, namely the uranium pile, the low-temperature laboratory, a part of the high-voltage system and the library, had been lost by the end of the war. For this reason, he attaches great importance to the fact that he will also be able to fully dispose of the part of the Institute that remains in Hechingen as soon as the political situation again permits it" [our translation]. AMPG, II. Abt., Rep. 66, No. 3047. David Cassidy: Controlling German Science, I: U.S. and Allied Forces in Germany, 1945–1947. *Historical Studies in the Physical and Biological Sciences* 24/2 (1994), 197–235.

Internationally and in Germany, after 1945, nuclear research branched out in three different directions. The first one was straight nuclear physics, that is, the study of nuclear structure, nuclear energy levels, and nuclear reactions. The second branch of postwar nuclear research was nuclear engineering, specifically, the application of nuclear energy in nuclear power plants, nuclear weapons, medical equipment, and other settings. The third line of development in nuclear physics was represented by the study of interactions at the nuclear and subnuclear levels. Such investigations could now be carried out either through the study of cosmic radiation or through the study of nuclear processes artificially generated by accelerating machines, which, by the late 1940s, began to be competitive with cosmic rays that hitherto—since the early 1930s—had been the sole source of high-energy particles. In the early 1950s Germany was in possession of two cyclotrons, a couple of betatrons, and some electrostatic accelerators, all machines with energies in the range of a few MeV.[32] The possibility of building higher-energy accelerators, like the synchrotrons, in combination with the, owing to wartime progress, outstanding role and prestige of nuclear physics as a research field, had become a powerful trigger for the construction of a new generation of machines producing artificial beams of particles of great intensity, which would permit investigation of nuclear and subnuclear processes. However, due to the prohibition on constructing high-energy accelerators bigger than 100 MeV, in force until the mid-1950s, no really big project of this sort could be implemented in Germany, initially. In any case, in 1946—and until the early 1950s—the experimental data on new elementary particles still derived almost entirely from cosmic radiation.

The nuclear energy program had been one of the 'three pillars' of Heisenberg's wartime research program, along with research on cosmic rays and, of course, on the continuation of his personal work on the theoretical and math-

doi:10.2307/27757723. David Cassidy: Controlling German Science, II: Bizonal Occupation and the Struggle over West German Science Policy, 1946–1949. *Historical Studies in the Physical and Biological Sciences* 26/2 (1996), 197–239. doi:10.2307/27757762. Helmut Rechenberg: Gentner und Heisenberg. Partner bei der Erneuerung der Kernphysik- und Elementarteilchenforschung im Nachkriegsdeutschland (1946–1958). In: Dieter Hoffmann, and Ulrich Schmidt-Rohr (eds.): *Wolfgang Gentner. Festschrift zum 100. Geburtstag*. Berlin: Springer 2006, 63–94.

32 Ulrich Schmidt-Rohr: *Die deutschen Teilchenbeschleuniger von den 30er Jahren bis zum Ende des Jahrhunderts*. Heidelberg: Max-Planck-Institut für Kernphysik Heidelberg 2001.

ematical foundations of quantum field theory.[33] Since the very start of the 1930s, when cosmic ray research became a branch of modern physics, Heisenberg, like other theoreticians, had closely followed cosmic ray experiments in parallel with his investigations of nuclear structure, in particular early studies on the interaction of cosmic particles with matter-producing effects like the multiple production of secondary particles.[34] In the years 1941–42, this research field was the topic of seminars at the Berlin Kaiser Wilhelm Institute for Physics, which had become Germany's main fission research laboratory, with Heisenberg as its acting head from July 1942, after having been theoretical advisor on experiments since 1940. There, he brought from Leipzig his main research activity: investigations of cosmic ray physics and elementary particles.[35] The need to find an adequate theory to explain effects induced by high-energy cosmic rays in the Earth's atmosphere or in lead sheets, like the showers of newly created particles and photons, led Heisenberg to formulate his theory of the scattering matrix (S-matrix), an approach to describing interactions in elementary particles theory solely in terms of directly observable quantities, which he laid out in a series of papers between 1942 and 1944.[36]

---

33 Alexander S. Blum: *Heisenberg's 1958 Weltformel and the Roots of Post-Empirical Physics.* Cham, Switzerland: Springer Nature 2019, 6. doi:10.1007/978-3-030-20645-1. Alexander S. Blum: The State Is Not Abolished, It Withers Away. How Quantum Field Theory Became a Theory of Scattering. *Studies in History and Philosophy of Modern Physics*, 2017, 33. doi:10.1016/j.shpsb.2017.01.004. Helmut Rechenberg: The Early S-Matrix Theory and Its Propagation (1942–1952). In: Laurie M. Brown, Max Dresden, and Lillian Hoddeson (eds.): *Pions to Quarks. Particle Physics in the 1950s.* Cambridge, MA: Cambridge University Press 1989, 551–578.

34 Werner Heisenberg: Theoretische Überlegungen zur Höhenstrahlung. *Annalen der Physik* 405/4 (1932), 430–452. doi:10.1002/andp.19324050404. Werner Heisenberg: Über die durch Ultrastrahlung hervorgerufenen Zertrümmerungsprozesse. *Naturwissenschaften* 20/21 (1932), 365–366. doi:10.1007/BF01504936. See Section 12.4 in Helmut Rechenberg: *Werner Heisenberg—Die Sprache der Atome. Leben und Wirken—Eine wissenschaftliche Biographie. Die "Fröhliche Wissenschaft" (Jugend bis Nobelpreis).* Berlin: Springer-Verlag 2009.

35 Werner Heisenberg (ed.): *Vorträge über Kosmische Strahlung.* Berlin: Springer-Verlag 1943. See also Werner Heisenberg: Das Kaiser-Wilhelm-Institut für Physik. Geschichte eines Instituts. In: Generalverwaltung der Max-Planck Gesellschaft zur Förderung der Wissenschaften e.V. (ed.): *Jahrbuch 1971 der Max-Planck-Gesellschaft zur Förderung der Wissenschaften e.V.* Göttingen 1971, 45–89.

36 Rechenberg, The Early S-Matrix Theory, 1989, 551–578. For a discussion on Heisenberg's theory of the scattering matrix see also Reinhard Oehme's comment in Werner Heisenberg: *Collected Works. Series A/2: Original Scientific Papers.* Edited by Walter Blum, Hans-Peter Dürr, and Helmut Rechenberg. Vol. 2. Berlin: Springer-Verlag 1985, 605–610. doi:10.1007/978-3-642-70078-1.

During their postwar internment at Farm Hall, Heisenberg and his colleagues had discussed future research, as we know from the secretly recorded conversations. They were aware that it would be impossible for them to pursue research on nuclear physics, and this was indeed the case, owing to the Legislation of the Allied Control Council, which prevented any kind of nuclear research in Germany for (as it turned out) a whole decade after the war. That left only work on cosmic rays. However, in conversations with his colleagues during their detention, Heisenberg clearly expressed the opinion that it would not be worth their while to work in cosmic ray research "with a few shabby proportional counters," on subjects which had "already been exhausted by the Americans"; for surely, they would not be allowed to freely use the airplanes or balloons necessary to investigate the high atmosphere and, in particular, to detect the actual primary cosmic rays. Accordingly, they would be "permanently put on ice," and able only to "do physics on the Romanian or Bulgarian scale."[37] At Farm Hall, Heisenberg had imagined "doing nuclear physics and cosmic ray work in greater style in peace time" ("I do not want to do petty physics. Either, I want to do proper physics or none at all");[38] but instead, in the early postwar years, research at the Max Planck Society was conducted under Allied prohibitions on carrying out 'nuclear' research and in precarious circumstances, too, subject to a level of scarcity that in many experimental fields quickly ruled out any attempts to keep up.[39] And so, in the broader framework of reconstruction of West Germany's scientific research, mainly only experiments with cosmic rays could be conducted. Often with excellent results, despite the very simple and low-cost instruments, as had been the case in Italy, too, whose small scientific community had been nearly destroyed by the fascist racial laws, and then severely damaged by the war and the subsequent lack of resources.[40] During the war, these experiments had kept alive a research tradition which was to flourish again over the next half century, when subnuclear physics could recommence with balloons and accelerators,

37 Frank, *Operation Epsilon*, 1993, 202. These conversations took place September 14–15, 1945.

38 Frank, *Operation Epsilon*, 1993, 203.

39 Dieter Hoffmann (ed.): *Physik im Nachkriegsdeutschland*. Frankfurt am Main: Harri Deutsch 2002. Ulrich Herbert, and Axel Schildt (eds.): *Kriegsende in Europa. Vom Beginn des deutschen Machtzerfalls bis zur Stabilisierung der Nachkriegsordnung 1944–1948*. 1st ed. Essen: Klartext 1998.

40 Amaldi, The Years of Reconstruction, 1998, 263–294. For an overview of the history of cosmic ray research in Italy from the 1930s to the 1950s, see Giulio Peruzzi, and Sofia Talas: The Italian Contributions to Cosmic-Ray Physics from Bruno Rossi to the G-Stack. A New Window into the Inexhaustible Wealth of Nature. *La Rivista Del Nuovo Cimento* 30/5 (2007), 197–257. doi:10.1393/ncr/i2007-10020-0.

once the resources for such expensive experimental research became available.

In general, research on cosmic ray physics during the 1940s could be divided into two broad categories, according to the two different types of questions researchers were then trying to answer: What are the constituents of the local cosmic radiation? What is the origin of the primary cosmic radiation and what are its effects on Earth? The latter query became the main task of cosmic ray physicists when the nuclear/particle aspect of their work was taken over by scientists using accelerators.

Postwar Fundamental 'Nuclear' Research: Practices and Semantics

Some scientific background is worth discussing at this point, to understand the linguistic complexities and ambiguities of the term 'nuclear' as used in the postwar decade, and around which physicists maneuvered to maximize their support and autonomy in this initial phase. By 1945, eight particles were known: the electron, the positron, the proton, the neutron, the photon, the (still hypothetical) neutrino, and the cosmic ray meson, the so-called positive and negative mesotron—the very penetrating component of local cosmic rays, so termed because of its mass, intermediate between that of electron and proton—which was wrongly thought at the time to be Hideki Yukawa's meson, the predicted field quantum associated with the extraordinary attractive forces binding together the neutrons and protons in nuclei.[41] Such a question was definitively clarified in 1947 by Cecil F. Powell's group in Bristol, UK. Newly developed nuclear emulsions enabled Powell's group to detect the π-meson in cosmic radiation, which was identified as Yukawa's meson. The mesotron of cosmic rays, which they now termed the μ-meson, was actually recognized to be the product of the π-meson's decay (accompanied by an electron and a neutrino).[42]

---

41 These predictions were blown to bits by a crucial experiment carried out in Rome during the war by Marcello Conversi, Ettore Pancini and Oreste Piccioni providing the first demonstration that the mesotron of cosmic rays, was almost completely unreactive in a nuclear sense, and thus was not behaving as it should, if it were the meson predicted by Yukawa as the mediator of nuclear forces. M. Conversi, E. Pancini, and O. Piccioni: On the Disintegration of Negative Mesons. *Physical Review* 71/3 (1947), 209–210. doi:10.1103/PhysRev.71.209. As stressed by Louis Alvarez in his Nobel lecture, such experiment was marking the beginning of "modern particle physics." L. W. Alvarez, Recent developments in particle physics, Nobel Lecture, December 11, 1968, https://www.nobelprize.org/prizes/physics/1968/alvarez/lecture/. Last accessed 4/28/2020.

42 Cesare M. G. Lattes et al.: Processes Involving Charged Mesons. *Nature* 159/4047 (1947), 694–697. doi:10.1038/159694a0. Powell's connections to the Max Planck Society are fur-

At the time, the entire research relating to questions of the basic structure of matter and the laws behind it, i.e., nuclear forces, mesons, field theory, etc., belonged to 'nuclear physics,' the notion of which, in the 1940s, was still much broader and quite different from its present-day meaning.[43] A central task occupying physicists since the 1930s had been to establish the law describing the nuclear force. As was now clear, the π-meson was responsible for mediating the strong interaction between particles forming the atomic nucleus. For this reason, these studies could throw light on the nature of such forces. Understanding how cosmic ray mesons were produced, as well as measuring their main properties (such as mass, charge, spin, and lifetime), and studying their interaction with matter, became a major area of investigation in the second half of the 1940s. The new challenge was to make mesons in the laboratory and study them in number and in detail—which was an impossible undertaking as long as they could be sourced solely from cosmic rays. However, the energy required still excluded Europe from such a possibility. Not even the UK, where the construction of new-generation accelerators had been planned since fall 1945,[44] had access to 'homemade mesons' in the early postwar years. The outstanding role of nuclear physics as a research field, established during the war, became a powerful trigger for the construction of new accelerating machines. But in the late 1940s, what we now call high-energy physics with accelerators had not yet emerged as a field distinct from nuclear research and many of the machines planned in the UK, for example, were intended as tools for 'applied nuclear physics.'[45]

Since the start of the 1930s, both theoretical and experimental physicists had tried to learn the secrets of nuclear structure by bombarding nuclei with α-

ther detailed in Chapters 3 and 5. All these achievements made 1947 a high point in the history of cosmic rays and elementary particles, recognized by the Nobel Prize to Patrick Blackett, Hideki Yukawa and Cecil Powell between 1948 and 1950.

43 For a comprehensive account of the historical development of experimental and theoretical nuclear physics up to the 1950s see Bernard Fernandez, and Georges Ripka: *Unravelling the Mystery of the Atomic Nucleus. A Sixty Year Journey 1896–1956*. New York: Springer 2013.

44 Ulrike Mersits: From Cosmic-Ray and Nuclear Physics to High-Energy Physics. In: Armin Hermann et al. (eds.): *History of CERN. Launching the European Organization for Nuclear Research*. Amsterdam: North-Holland 1987, 3–52.

45 On the connection between nuclear, particle, and cosmic ray physics still in the late 1940s, see Erwin Schopper: Janossy: Cosmic Rays and Nuclear Physics/Powell und Occhialini: Nuclear Physics in Photographs. *Physikalische Blätter* 4/10 (1948), 449–450. doi:10.1002/phbl.19480041012.

particles or neutrons, or studying high-energy cosmic ray interactions.[46] Many aspects of nuclear and particle physics, which were still part of cosmic ray research, included the use of cosmic rays as a source of high-energy particles as well as analysis of the primary cosmic radiation and its interaction with nuclei in the atmosphere.[47] On the other hand, the study of showers of secondary particles generated as a result of such interactions provided information on nuclear and subnuclear processes, which in turn were of general interest to theoretical physicists, who, in this time of transition, were struggling to find a theory for both the strong and weak interactions which could be observed, for example, in the cascade decay processes involving the π- and μ-meson: the *strongly* interacting one, the pion, was produced primarily at high altitude in showers generated by primary cosmic ray interactions with nuclei in the high

46 The earliest evidence for such interactions had come from the observation of the so-called 'cosmic ray stars' discovered in photographic plates in the 1930s. These 'stars' were groups of heavily ionizing particles which were thought to arise from the disruptions of nuclei. On energy grounds, it had been found that these stars could not be due to a radioactive contamination of the plates, but must be produced by cosmic rays. When showers in matter and in the atmosphere were discovered, it was assumed that they were the result of nuclear collisions by cosmic rays, in which nuclei were disrupted, but later it was shown that showers were cascade phenomena, involving alternate radiation processes by electrons and pair production by photons. Thus, at the end of the 1930s, the direct evidence for the nuclear interactions of cosmic rays was rather scarce and, in any case, it concerned only events of comparatively low energy, such as the 'stars' in photographic plates. On the other hand, the indirect evidence for high-energy nuclear interactions had become quite compelling. It came from the very presence of what where still called mesotrons in the local radiation: being unstable they could not come from outer space as part of the primary radiation, but must be produced locally from nuclear interactions, as they were too heavy to be produced by electromagnetic interactions in such significant numbers. Once the occurrence of nuclear interactions in cosmic rays was firmly established during the 1940s, it was taken as an indication that the primary cosmic radiation itself consists of nuclear-active particles. This conclusion confirmed the results obtained between 1940 and 1941 by Schein and coworkers according to which primary cosmic rays should be, at least for the most part, protons: Marcel Schein, William P. Jesse, and E. O. Wollan: The Nature of the Primary Cosmic Radiation and the Origin of the Mesotron. *Physical Review* 59/7 (1941), 615–615. doi:10.1103/PhysRev.59.615. For a detailed review on nuclear interactions of cosmic rays, see Chapter 10 in Bruno Rossi: *Cosmic Rays*. New York, NY: McGraw-Hill 1964.

47 In 1948, it was definitely established that primary cosmic rays are mainly protons and, to a lesser extent, bare heavier nuclei, solving a problem which had preoccupied scientists since the discovery of cosmic rays during the balloon flight of Victor Hess in 1912. Experiments with balloons carrying nuclear emulsion plates up to nearly 29 kilometers led to the conclusion that some recorded tracks were due to bare atomic nuclei heavier than protons. Phyllis Freier et al.: Evidence for Heavy Nuclei in the Primary Cosmic Radiation. *Physical Review* 74/2 (1948), 213–217. doi:10.1103/PhysRev.74.213.

atmosphere, and it decayed rapidly into the *weakly* interacting muon. Moreover, new particles had been observed since 1946 in photographic plates and cloud chambers, the so-called V-particles, whose strange forked tracks testified to a variety of decay schemes and raised hopes that higher energies would reveal an entire zoo of new particles. This new subnuclear world began to be intensively explored by cosmic ray physicists on the top of high mountains, in airplanes, and through high-altitude balloon flights.

In this early postwar scenario, which was turning out to exceed by far all that Heisenberg had imagined when in Allied custody in Britain, his institute in Göttingen was starting scientific activities, well aware that the dream of returning German physics to its prewar international position was a long way from being realized. In beginning to build up experimental physics, Heisenberg could in any case pursue only cosmic ray research, which, moreover, had no conceivable military purpose or application, and could be tackled with the limited budget and infrastructure available at that time. Following the arrival in 1950 of Martin Deutschmann, an expert in cloud chambers, from Freiburg, Germany,[48] and of Peter Meyer, from Göttingen University, experimental work at Heisenberg's institute came to focus mainly on cosmic ray physics, also with the aid of Geiger-Müller counters, a fundamental tool in nuclear and particle physics.

At the same time, a cosmic ray group led by Martin Teucher, a former student of Fritz Houtermans at the University of Göttingen (more on him later in the chapter), was working with the new nuclear emulsions coming from England, which were exposed at high altitudes—also using balloons—to investigate 'nuclear disintegrations' produced by high-energy cosmic rays.[49]

48 Martin Deutschmann's dissertation (1944, Berlin) investigating cosmic ray showers with a big cloud chamber designed by Hans Geiger, was published in part in 1947, as a result from work carried out at the Physics Institute of the Technical University in Berlin under the direction of Geiger and with the support of Otto Haxel and Friedrich Bopp. Martin Deutschmann: Untersuchung Der Kosmischen Strahlenschauer Mit Hilfe Einer Großen Wilson-Kammer. *Zeitschrift Für Naturforschung A* 2/2 (1947), 61–69. doi:10.1515/zna-1947-0201.

49 See, for example, M. Teucher: Die Absorption der Nukleonenkomponente der kosmischen Strahlung in Luft zwischen Seehöhe und 4000 m. *Zeitschrift für Naturforschung A* 7/1 (1952), 61–63. doi:10.1515/zna-1952-0111. This work was part of a special issue of the *Zeitschrift für Naturforschung* dedicated to Heisenberg on the occasion of his 50th birthday. A description of cosmic ray studies going on at the institute can be found in Tätigkeitsbericht der Kaiser-Wilhelm-Gesellschaft und der Max-Planck-Gesellschaft Für die Zeit vom 1. 1. 1946 bis 31. 3. 1951. Chemisch-Physikalisch-Technische Sektion. Berichte aus den Einzelnen Instituten. *Die Naturwissenschaften* 38/16 (1951), 365–372. doi:10.1007/BF00637817.

ILLUSTRATION 13 Göttingen at the beginning of the 1950s. Klaus Gottstein and Werner Heisenberg, on the left Fritz Houtermans. The working group for the study of nuclear and elementary particle processes in the exposure of photographic plates to cosmic rays was formed by Gottstein, Teucher, and Houtermans.

They were one of the first groups—probably even the very first—to work in Germany with the new visual techniques.

In 1952, when Teucher followed Fritz Houtermans to Bern (see Section 2), Klaus Gottstein became leader of the experimental group, which continued to explore the high-energy nuclear processes generated by cosmic rays,[50] also in connection with Heisenberg's theoretical models for such events.[51]

---

50 See, for example, Klaus Gottstein, and Martin Teucher: Zur Mesonenerzeugung beim Zusammenstoß energiereicher Nukleonen. *Zeitschrift für Naturforschung A* 8/2–3 (1953), 120–126. doi:10.1515/zna-1953-2-303. Klaus Gottstein: Zur Aufspaltung der schweren Kerne in der kosmischen Strahlung. *Naturwissenschaften* 40/3 (1953), 104–105. doi:10.1007/BF00597050. Klaus Gottstein: On the Fragmentations of Heavy Cosmic Ray Nuclei. *Il Nuovo Cimento (1943–1954)* 11/2 (1954), 377–380. doi:10.1007/BF02781100. Klaus Gottstein et al.: Heavy Unstable Particles in Nuclear Emulsions. *Il Nuovo Cimento (1955–1965)* 4/2 (1956), 440–444. doi:10.1007/BF02747915.

51 After the war, a new volume on cosmic rays with contributions by Heisenberg's group was published in the early 1950s: Werner Heisenberg (ed.): *Kosmische Strahlung. Vorträge gehalten im Max-Planck-Institut für Physik Göttingen.* 2nd ed. Berlin: Springer 1953. The volume clearly shows how cosmic rays, which had been regarded primarily as a domain

ILLUSTRATION 14 Inflation of the balloons that would lift the weight of photographic plates in order to study the nuclear reactions generated by cosmic rays in the upper atmosphere, April 17, 1952.

Heisenberg's 'cosmic ray program' was still focused on cosmic ray showers and multiparticle production processes in strong interactions, which he had been investigating since the second half of the 1930s. Now he began to study the multiple production of mesons, whose main properties had been identified in 1947 by Powell's group in the UK, and whose significance could now provide hints on the as yet unexplored field of strong interaction in elementary particle physics.[52] The Institute for Physics was

of physics, were now becoming of growing relevance for astronomers and astrophysicists. For a review and an outline of the field in the early 1950s, see Stefan Temesváry: Vorträge gehalten im Max-Planck-Institut für Physik. *Mitteilungen der Astronomischen Gesellschaft* 5 (1954), 99–101. http://adsbit.harvard.edu/full/seri/MitAG/0005//0000100.000.html. Last accessed 5/24/2020.

52 Werner Heisenberg: Die Erzeugung von Mesonen in Vielfachprozessen. *Il Nuovo Cimento (1943–1954)* 6/3 (1949), 493–497. doi:10.1007/BF02822044. Werner Heisenberg: Production of Meson Showers. *Nature* 164/4158 (1949), 65–66. doi:10.1038/164065c0. Werner Heisenberg: Über die Entstehung von Mesonen in Vielfachprozessen. *Zeitschrift für Physik* 126/6 (1949), 569–582. doi:10.1007/BF01330108. Werner Heisenberg: Bemerkungen zur Theorie der Vielfacherzeugung von Mesonen. *Naturwissenschaften* 39 (1952), 69–69. doi:10.1007/BF00596818. Werner Heisenberg: Mesonenerzeugung als Stoßwellenproblem. *Zeitschrift*

ILLUSTRATION 15 Göttingen, October 1951. The Federal President, Theodor Heuss, during a visit to the Max Planck Institute for Physics, sitting at a microscope and observing a photographic plate with Martin Teucher; behind them stands Ms. Baumbach.

involved between 1952 and 1954 in a European collaboration for the launch of nuclear emulsions assembled in stacks and flown to high altitude by bal-

für Physik A Hadrons and nuclei 133/1 (1952), 65–79. doi:10.1007/BF01948683. Werner Heisenberg: The Production of Mesons in Very High Energy Collisions. *Il Nuovo Cimento Series 10* 2/1 (1955), 96–103. doi:10.1007/BF02746079. Heisenberg's work on meson showers and multiparticle production between 1949–52 was commented by R. Hagedorn in Werner Heisenberg: *Collected Works. Series A/3: Original Scientific Papers*. Edited by Walter Blum, Hans-Peter Dürr, and Helmut Rechenberg. Vol. 3. Berlin: Springer-Verlag 1985, 75–85. doi:10.1007/978-3-642-70078-1.

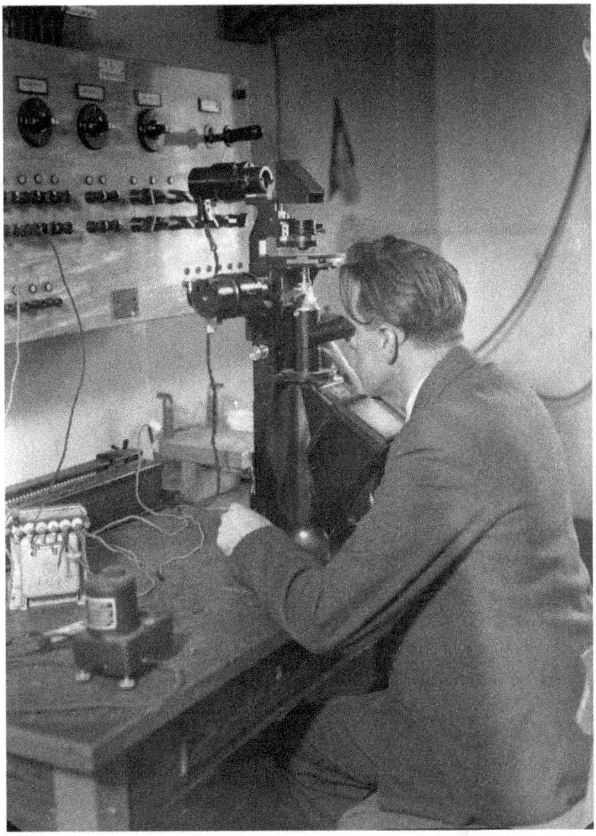

ILLUSTRATION 16 Fritz Houtermans observing a photographic plate through a microscope in April 1948.

loons in the Mediterranean area.[53] This cooperation, promoted by C. F. Powell of the University of Bristol, was instrumental in the process of recon-

53 Klaus Gottstein: interview by Luisa Bonolis and Juan-Andres Leon, Munich, September 7, 2017. DA GMPG, BC 601006. See related archival material in Klaus Gottstein papers, Correspondence during the 1950s, particularly with Marcello Ceccarelli (1952–67), AMPG, III. Abt., ZA 58, No. O 143. After his *Habilitation* in 1960, Gottstein became a Scientific Member of the institute (meeting minutes of the Chemistry, Physics and Technology Section of the Scientific Council—from now on CPTS meeting minutes—of 06.06.1961, 11.03.1961, AMPG, II. Abt., Rep. 62, No. 1737, 1738. The displayed date format for all cited archival documents follows the form dd.mm.yyyy). See also Werner Heisenberg: Kosmische Strahlungen und Atomphysik. *Jahrbuch 1951 der Max-Planck-Gesellschaft zur Förderung der Wissenschaften e.V.* Göttingen: Max-Planck-Gesellschaft zur Förderung der Wissenschaften 1951, 229–263. Wolfgang Gentner: Einige Rückblicke auf die Anfänge der 50 jährigen Forschung über die kosmische Strahlung. *Die Naturwissenschaften* 50/8 (1963), 317–318. doi:10.1007/BF00645924.

ILLUSTRATION 17 The working group for the study of nuclear processes in photographic plates on a lunch break, March 1952. Left to right: Ms. Ahrens, Juan Roederer, Christa Schriel, Alfred Gierer, Xula Vigon, Klaus Gottstein, and Ms. Baumbach

structing physics in Europe and in preparing later joint research activities at CERN.[54]

Even after the advent of the first powerful accelerators in the early 1950s, cosmic ray physicists were still able to discover new elementary particles. But during the 1950s, modern particle physics—then still called *high-energy nuclear physics*—branched out from cosmic rays and nuclear physics into an autonomous field, with accelerators as its primary research tool.[55] In the second half of the 1950s, nuclear emulsions began to be exposed to beams of

54 Cristina Olivotto: The G-Stack Collaboration (1954): An Experiment of Transition. *Historical Studies in the Natural Sciences* 39/1 (2009), 63–103. doi:10.1525/hsns.2009.39.1.63.

55 For the birth and evolution of cosmic rays and particle physics between 1930 up to the end of the 1950s see M. Laurie Brown, and Lillian Hoddeson (eds.): *The Birth of Particle Physics. Based on a Fermilab Symposium*. Cambridge: Cambridge University Press 1983. Laurie M. Brown, Max Dresden, and Lillian Hoddeson (eds.): *Pions to Quarks. Particle Physics in the 1950s*. Cambridge: Cambridge University Press 1989. The 'invasion' of accelerators was explicitly mentioned in 1953, during the seminal international cosmic ray conference held at Bagnères-de-Bigorre, at the foot of the Pyrenees mountains, dividing

ILLUSTRATION 18 Institute's excursion in "the old eagle," (Heisenberg's old car). Left to right: Juan Roederer, H.M. Mayer, ?, ?, Klaus Gottstein

ILLUSTRATION 19 Women scanning nuclear emulsions with microscopes, March 12, 1954. From left to right: Bischoff, Ahrens, Koebe, Behm, Baumbach, Arndt, and Pätzold

ILLUSTRATION 20 Max Planck Institute for Physics, spring 1954. Left to right: Klaus Gottstein, Baumbach, Schriel, Ahrens, Lindenberger, Bette, and H.-M. Mayer

ILLUSTRATION 21 Klaus Gottstein on the ship used for the recovery of photographic plates, from which balloons were launched in the Mediterranean Sea.

ILLUSTRATION 22 Naples, early 1950s. Departure of two darex balloons, in the background is the Göttingen group's hut.

particles produced by accelerators and Heisenberg's team began to use new detecting techniques, such as bubble chambers, to study processes involving fundamental interactions using artificial high-energy particles.[56] In 1958, however, at the time of their relocation to the new seat in Munich, the era of cosmic rays came to an end at what had, tellingly, just been renamed the 'Institute for Physics and Astrophysics.' The newly institutionalized interest in the latter term came from an entirely different direction.

France from Spain, where the famous French high-altitude Pic du Midi Observatory was located. James W. Cronin: The 1953 Cosmic Ray Conference at Bagnères de Bigorre. The Birth of Sub Atomic Physics. *The European Physical Journal H* 36/2 (2011), 183–201, 197. doi:10.1140/epjh/e2011-20014-4.

56 N. N. Biswas et al.: Decay Modes and Mean Life of Scattered K^+-mesons. *Il Nuovo Cimento (1955–1965)* 4/3 (1956), 631–636. doi:10.1007/BF02745387. Klaus Gottstein: Die Blasenkammer und ihre Anwendung in der Physik der Elementarteilchen. *Die Naturwissenschaften* 46/3 (1959), 97–102. doi:10.1007/BF00638309. Gottstein's group at the Institute for Physics pioneered the technique in Germany with materials that Gottstein himself had brought back from Berkeley in 1957, when he interrupted his collaboration with Louis Alvarez group (Gottstein to Gentner, February 23, 1970, AMPG, II. Abt., Rep. 62, No. 437, Fol. 397).

ILLUSTRATION 23 International collaboration in Sardinia (Italy). Filling of darex balloons, June-August 1953.

Plasma physics as both astrophysics and a path toward nuclear fusion

Heisenberg was one of the revolutionary founders of quantum mechanics in the interwar period, and hence a living legend in theoretical physics.[57] The fact that he remained in Germany during World War II and actively contributed to Nazi efforts in nuclear research had distanced him from the network of theoretical physicists with whom he had made his name, such as Niels Bohr, and after the war he was no longer a leading figure in theoretical physics. In the early postwar period, also because the pursuit of experimental nuclear and large-scale cosmic ray physics was prohibited at his institute, Heisenberg turned to "foundational pursuits in high theory."[58] But in the early 1950s, his personal scientific interests in a unified field theory were considered by mainstream theoretical physicists to be more of a niche endeavor, and they had little impact, even though some of his original insights contributed to later developments in elementary particle theories.[59] Heisenberg continued to be a household name, giving frequent lectures on popular scientific subjects, or highlighting the role of scientists in contemporary society; yet in his own fields of expertise, he moved within a politically and socially much less valued aspect of physics.[60] Dating from his university years in Munich,[61] Heisenberg had always had a problematic relationship with experimental physicists and the old rivalry of these fields in the 1920s was exacerbated in the 1930s by attacks

57 Jagdish Mehra, and Helmut Rechenberg: *The Historical Development of Quantum Theory*. Vol. 4. New York, NY: Springer Verlag 1982. Rechenberg, *Werner Heisenberg*, 2009.

58 Blum, *Heisenberg's 1958 Weltformel*, 2019, 60. See also Chapter 5 of Cathryn Carson: *Particle Physics and Cultural Politics. Werner Heisenberg and the Shaping of a Role for the Physicist in Postwar West Germany*. Dissertation. Cambridge, MA: Harvard University 1995. ProQuest.

59 Konrad Bleuler: Werner Heisenberg's Ideas on Particle Physics in the Light of Recent Achievements. *Zeitschrift Für Naturforschung A* 45/9 und 10 (2014), 1051–1058. doi:10.1515/zna-1990-9-1001.

60 Cathryn Carson: *Heisenberg in the Atomic Age. Science and the Public Sphere*. Cambridge: Cambridge University Press 2010. Rechenberg, Werner Heisenberg, 1996, 245–262. Ulrich Schmidt-Rohr: *Die deutschen kernphysikalischen Laboratorium*. Heidelberg: Max-Planck-Institut für Kernphysik Heidelberg 2005. Schmidt-Rohr, *Teilchenbeschleuniger*, 2001. For Heisenberg's later opinions on modern experimental physics, see Werner Heisenberg: *Encounters with Einstein and Other Essays on People, Places, and Particles*. Princeton, NJ: Princeton University Press 1989.

61 On the well-known episode of Heisenberg's near failure at his doctoral oral exam with Wilhelm Wien, in charge of the physics laboratory, see David C. Cassidy: *Uncertainty. The Life and Science of Werner Heisenberg*. New York: W. H. Freeman 1992, 149–154.

by the adherents of what was known then as *Deutsche Physik*.[62] In the postwar era, Heisenberg continued to publicly depreciate the role of experimental physics in Germany.[63] Yet he was also increasingly estranged from the mainstream of theoretical physics; by the 1960s, his skepticism of new concepts like quarks alienated him even from the experimental teams at his own institutes.[64]

Heisenberg and von Weizsäcker's strategy of putting renewed emphasis on theoretical studies—which were cheaper than experiments—turned into a promising path, quickly reconnecting them with scientific research at the international level. But in the end, it was a very different kind of theoretical physics that afforded Heisenberg's institute a significant foothold in postwar physics, namely, plasma astrophysics.

Plasma, a state of matter consisting of free charged particles and atomic nuclei in complex interaction, makes up almost 100 percent of the visible universe, including the interplanetary, interstellar, and intergalactic medium. All the stars—as well as our Sun—are about 100 percent plasma. This was determined theoretically back in the 1920s and 1930s, as part of one of the key physical discoveries of the 20th century, namely, the process of nuclear fusion, which explained, once and for all, the age-old question of the source of the Sun's energy. Inside a star in hydrostatic equilibrium, the inward force of gravity is balanced by the outward force of gas pressure, and energy produced in the stellar core through thermonuclear reactions is transported to the surface by radiation and convection mechanisms. The fundamental equations governing the structure of a star in radiative equilibrium had been established

62 How the 'Deutsche Physik' episode was intertwined with the rivalry between experimental and theoretical physics was best explored by Reinald Schröder: Die "schöne deutsche Physik" von Gustav Hertz und der "weiße Jude" Heisenberg. Johannes Starks ideologischer Antisemitismus. In: Helmuth Albrecht (ed.): *Naturwissenschaft und Technik in der Geschichte. 25 Jahre Lehrstuhl für Geschichte der Naturwissenschaften und Technik am Historischen Institut der Universität Stuttgart*. Stuttgart: Verlag für Geschichte der Naturwissenschaften und Technik 1993. For a wider discussion see also Mark Walker: National Socialism and German Physics. *Journal of Contemporary History* 24/1 (1989), 63–89. https://www.jstor.org/stable/260700. Last accessed 1/27/2020.

63 As recalled, for example, by Heinz Maier-Leibnitz: "The theoretical physicist Werner Heisenberg has repeatedly been carried away by statements such as: 'I know of no experimental physicist who has done anything of importance in Germany in the last forty years'" [our translation]. Anna-Lydia Edingshaus: *Heinz Maier-Leibnitz. Ein halbes Jahrhundert experimentelle Physik*. München: Piper 1986, 122.

64 See Chapter 3, Section 3. Distancing himself from contemporary particle physics in the 1960s had a profound effect on the last years of his directorship in Munich and on his succession.

already by the mid-1920s,[65] but the problem of the source of solar energy remained one of the major unsolved scientific puzzles until the late 1930s, when German physicists made a remarkable contribution to solving it. Hans Bethe (who had relocated to England in 1933, and then to the United States in 1935) made use of the recent understandings of subatomic physics to show that the source of the Sun's energy was in the process of nuclear fusion. With the extreme conditions occurring at the core of the stars, the lightest atoms in the universe, such as hydrogen, helium, lithium, and oxygen, are fused through a series of chain reactions to make more massive nuclei and release, in the course of this process, an incredible amount of energy in the form of light and heat.[66]

Nuclear fusion in fact predated the scientific interest in nuclear fission, the process behind the atomic bomb, by several years. Carl Friedrich von Weizsäcker, who was later to become one of the Deputy Directors of Heisenberg's postwar Max Planck Institute,[67] had made his early scientific career in stellar physics and remained in Germany after the Nazis came to power. In 1938, von Weizsäcker, in parallel to Bethe, developed one of the definitive descriptions of the nuclear processes by which the Sun produces its energy.[68]

However, fusion was not pursued seriously as a source of energy until after the end of the war. Scientists working on the Manhattan Project had pointed to the potential of nuclear fusion to produce a bomb of unlimited explosive power, what would subsequently become the hydrogen bomb.[69] At the time,

---

65 Arthur S. Eddington: *The Internal Constitution of the Stars*. Cambridge: Cambridge University Press 1926.

66 Hans Bethe: Energy Production in Stars. *Physical Review* 55/1 (1939), 103–103. doi:10.1103/PhysRev.55.103. Hans Bethe: Energy Production in Stars. *Physical Review* 55/5 (1939), 434–456. doi:10.1103/PhysRev.55.434.

67 Von Weizsäcker, who was leading the Theoretical Department at the institute, as well as Karl Wirtz, head of the experimental division, became Scientific Members in 1950 (see CPTS meeting minutes of 14.06.1950, AMPG, II. Abt., Rep. 62, No. 1724). Biermann was appointed Scientific Member in 1951.

68 Carl F. von Weizsäcker: Über Elementumwandlungen im Inneren der Sterne. I. *Physikalische Zeitschrift* 38/6 (1937), 176–191. Carl F. von Weizsäcker: Über Elementumwandlungen im Inneren der Sterne. II. *Physikalische Zeitschrift* 39/17/18 (1938), 633–646.

69 It was understood at a very early stage that while practical considerations limited the maximum explosive power of fission bombs, thermonuclear weapons could be made as big and powerful as available resources would allow, so the energy release of even the first ones was expected to be hundreds of times that of bombs such as the one used in Hiroshima and Nagasaki. On US nuclear projects see Richard Rhodes: *The Making of the Atomic Bomb*. 2nd ed. Simon and Schuster: London 1988. Bruce Cameron Reed: *The History and Science of the Manhattan Project*. Berlin: Springer 2014. doi:10.1007/978-3

however, nuclear fusion lacked the key aspect that made fission technologically viable: the possibility of a self-sustaining chain reaction.[70] More than the splitting of uranium itself, what inspired the Manhattan Project as well as the German *Uranverein*[71] was the observation that the uranium-235 isotopes, after splitting upon absorbing a neutron, release not only a vast amount of energy, but also additional neutrons that are able in turn to split further uranium nuclei nearby and, perhaps, thus sustain a domino-like fission chain reaction.[72] It is this chain reaction that can be used as a source of energy in nuclear reactors and as an explosive system in nuclear weapons. Establishing how to channel such a chain reaction to cause an explosion was the main wartime objective, and controlling this reaction at a slow rate to produce energy (as well as more fissile materials) was the working principle behind nuclear fission reactors.

Nuclear fusion does not provide such an easy pathway, and it is here that plasma physics enters the picture. In the United States, parallel to the secret developments in nuclear fission during the war, a small section of the Manhattan Project dealt with how to create nuclear fusion explosions, which were known to release much more energy. By the end of the war, this was believed

---

-642-40297-5. Lillian Hoddeson et al. (eds.): *Critical Assembly. A Technical History of Los Alamos during the Oppenheimer Years, 1943–1945*. Cambridge: Cambridge University Press 1993. Richard Rhodes: *Dark Sun. The Making of the Hydrogen Bomb*. New York, NY: Simon & Schuster 2005.

70 The first human-made self-sustaining nuclear chain reaction was achieved by Enrico Fermi in December 1942, about three years after Otto Hahn and Fritz Straßmann's publication announcing that the element barium was a product of the bombardment of uranium with neutrons, a discovery providing evidence to identify the previously unknown phenomenon of the splitting of uranium, as immediately recognized by Lise Meitner and Otto Frisch who named the new process *nuclear fission*. Otto Hahn, and Fritz Straßmann: Über den Nachweis und das Verhalten der bei der Bestrahlung des Urans mittels Neutronen entstehenden Erdalkalimetalle. *Die Naturwissenschaften* 27/1 (1939), 11–15. doi:10.1007/BF01488241. Lise Meitner, and Otto Frisch: Disintegration of Uranium by Neutrons. A New Type of Nuclear Reaction. *Nature* 143/3615 (1939), 239–240. doi:10.1038/143239a0. For a detailed reconstruction of Fermi's work leading to the first prototype of nuclear reactor see Carlo Bernardini, and Luisa Bonolis (eds.): *Enrico Fermi. His Work and Legacy*. Berlin: Springer-Verlag 2004. doi:10.1007/978-3-662-01160-7.

71 Walker, Physics, 2017, 1–18.

72 Hans von Halban, Frédéric Joliot, and Lew Kowarski: Liberation of Neutrons in the Nuclear Explosion of Uranium. *Nature* 143/3620 (1939), 470–471. doi:10.1038/143470a0. Hans von Halban, Frédéric Joliot, and Lew Kowarski: Number of Neutrons Liberated in the Nuclear Fission of Uranium. *Nature* 143/3625 (1939), 680–680. doi:10.1038/143680a0. Herbert L. Anderson, Enrico Fermi, and Leo Szilard: Neutron Production and Absorption in Uranium. *Physical Review* 56/3 (1939), 284–286. doi:10.1103/PhysRev.56.284.

to be possible, if the necessary extreme conditions were created with the aid of a nuclear fission device. Working this out, however, was considered to be the most difficult theoretical endeavor of the Manhattan Project, and became one of the tasks of the program's theoretical division, led by stellar astrophysicists such as Hans Bethe and Edward Teller, and a younger generation trained in nuclear fusion on the basis of the existing knowledge in stellar astrophysics.[73] In addition to theoreticians capable of providing brilliant unconventional insights, wartime work on thermonuclear processes also required new theoretical methods, and this was one of the areas with the greatest need for calculating machines, one of the major drivers of the development of modern computers.[74]

In the postwar era, while the basic subatomic processes occurring in nuclear fusion were well documented, the problem was the highly complex behavior of multiple particles in the extreme conditions necessary to sustain such reactions. Plasma's unique properties make its mathematical treatment extremely difficult. In practice, much of the analysis of its behavior has to be conducted by means of two concurrent strategies: firstly, using theoretical insight and mathematical methods to simplify the problems to limited tractable cases and, secondly, feeding these principles to calculating machines and computers to trace their evolution. These were some of the earliest forms of computer simulation. These techniques were mostly sought in the postwar era for their application to thermonuclear weapons, as well as the possible creation of controlled fusion processes in a reactor; but the theoretical insights, methods, and use of calculating machines for this field had originated in stellar astrophysics and studies of cosmic plasmas, as a sequel to the work on astrophysics that began in the 1930s.[75]

---

73 The definitive book on the hydrogen bomb project is: Rhodes, *Dark Sun*, 2005.

74 Peter Galison: *Image and Logic. A Material Culture of Microphysics*. Chicago: The University of Chicago Press 1997. In addition to complex plasma physics, the other major driver of early computing was the simulation of random processes, or Monte Carlo simulations, which were later to play an important instrumental role at the interface between theory and experiment in nuclear and particle physics. Nicholas Metropolis: The Beginning of the Monte Carlo Method. *Los Alamos Science* 15/Special Issue (1987), 125–130. https://library.lanl.gov/cgi-bin/getfile?00326866.pdf. Last accessed 7/19/2020.

75 The astrophysicist Martin Schwarzschild remembered how astrophysical problems were used to test computers also used for thermonuclear research. He himself explored the interiors of stars by means of numerical models, problems for which analytical solutions were not known and that could thus become a test of the value of computers to scientific research, providing the possibility of modeling phenomena which could not be directly investigated through laboratory experiments. Martin Schwarzschild:

It was within this dual-purpose potential that the long scientific tradition in plasma physics—and plasma astrophysics—at the Max Planck Society first began. This was to remain one of its leading scientific fields up to the present day.

Ludwig Biermann's Tradition of Plasma Astrophysics

The roots of the establishment of astrophysics as a research field within the Max Planck Society lie in the expertise developed during the war, as well as in the closer relationship between physics and astrophysics that began to be forged in the 1920s and 1930s, when the problem of the interior of the stars and the problem of stellar energy became a common ground of interest and discussion. Arthur Eddington's major monograph *The Internal Constitution of the Stars*, published in 1926, concluded and summarized the results obtained over the previous two decades.[76] The fundamental equations governing the structure of a star in radiative equilibrium had been established, but the fundamental problem of which nuclear processes keep the Sun shining had still to be solved. The discovery of the neutron in 1932[77] and the evolving knowledge on nuclear matter and nuclear reactions provided sufficient conceptual and

---

interview by David DeVorkin and Spencer Weart, December 16, 1977, Session III. Transcript, Niels Bohr Library & Archives, American Institute of Physics, College Park, MD USA, https://www.aip.org/history-programs/niels-bohr-library/oral-histories/28321. Last accessed 7/30/2019. The astrophysicist Louis Henyey, for example, spent the year 195–52 at Princeton University, where he was involved in the classified defense work on Project Matterhorn, the US top secret project to control thermonuclear reactions. Henyey also developed links with the Livermore Radiation Laboratory (now LLNL), which had the UNIVAC, probably the world's most powerful computational facility during the 1950s. There, in collaboration with scientists at the laboratory, he was able to develop specific numerical methods for the automatic solution of the equations of stellar evolution, also applicable to a wide range of physical conditions and phases in the lifetime of a star. The Henyey method became the standard tool for the theory of stellar interior. Louis G. Henyey: The Evolution of Stars Near the Main Sequence. *Publications of the Astronomical Society of the Pacific* 68/405 (1956), 503–504. doi:10.1086/126986. Louis G. Henyey et al.: A Method for Automatic Computation of Stellar Evolution. *Astrophysical Journal* 129 (1959), 628–636. doi:10.1086/146661. As we will see, in the early 1960s, an upgraded version of this method became the starting point for computer simulations of stellar structure and evolution carried out at Biermann's Institute for Astrophysics.

76 Eddington, *Internal Constitution*, 1926.

77 James Chadwick: Possible Existence of a Neutron. *Nature* 129/3252 (1932), 312. doi:10.1038/129312a0. Valery Nesvizhevsky, and Jacques Villain: The Discovery of the Neutron and Its Consequences (1930–1940). *Comptes Rendus Physique* 18/9 (2017), 592–600. doi:10.1016/j.crhy.2017.11.001. Edoardo Amaldi: From the Discovery of the Neutron to the Discovery of Nuclear Fission. *Physics Reports* 111/1 (1984), 1–331. doi:10.1016/0370-1573(84)90214-X.

ILLUSTRATION 24 The wreath-laying ceremony on the hundredth birthday of Max Planck, Göttingen, April 23, 1958. From left to right: Otto Hahn, Ludwig Biermann, Werner Heisenberg, and Ernst Telschow

theoretical foundations for the decisive work of Hans Bethe, Charles Critchfield, and Carl von Weizsäcker toward the end of the 1930s.[78] In his second article on the problem of energy production in stars, von Weizsäcker also proposed as origin of the universe a cosmic explosion from a superdense compressed nuclear state. Given its strong nuclear physics content, it later provided key inspiration for George Gamow's Big Bang cosmology, published at the end of the war.[79] During the war, when he participated in the nuclear project led by Heisenberg, von Weizsäcker had also formulated a theory on the creation of a planetary system around a star as a possible final stage in the formation of the star itself, which had aroused great interest and inspired others

78 Weizsäcker, Elementumwandlungen I, 1937, 176–191. Weizsäcker, Elementumwandlungen II, 1938, 633–646. Bethe, Energy Production, 1939, 434–456.

79 George Gamow: Expanding Universe and the Origin of Elements. *Physical Review* 70/7–8 (1946), 572–573. doi:10.1103/PhysRev.70.572.2.

to expand upon his work.[80] During their internment at Farm Hall, he clearly stated that he had no interest in continuing to work in nuclear physics: "What I would like to do would be to lecture on physics at some German University and to study cosmology and philosophy";[81] and thus he decided to redirect his research activity toward astrophysics, a field he had been deeply interested in since the mid-1930s, when he first arrived at the Institute for Physics in Berlin-Dahlem, still led by Peter Debye.[82] Heisenberg and von Weizsäcker worked together on turbulence, in particular as applied to the problem of galaxy formation.[83]

In early 1946, when Heisenberg and his colleagues were given the opportunity to relaunch the Kaiser Wilhelm Institute for Physics in Göttingen, neither Karl Wirtz (a nuclear physicist leading the Experimental Department of the new institute) nor von Weizsäcker (head of the Theoretical Department), were allowed to conduct any kind of 'nuclear' research, as emphasized above.[84]

---

80 Von Weizsäcker had pointed out that turbulent gas motions must be present in all gaseous systems in free space and that the shape of the spiral nebulae must most probably be determined by these turbulent effects, and applied all these ideas to formulate his theory. Carl F. von Weizsäcker: Über die Entstehung des Planetensystems. *Zeitschrift für Astrophysik* 22 (1943), 319–355. http://adsabs.harvard.edu/abs/1943ZA.....22..319W. Last accessed 10/30/2018.

81 Frank, *Operation Epsilon*, 1993, 111. This conversation on his future took place between von Weizsäcker and Hahn on August 11, 1945, after the announcement that the United States had dropped two nuclear weapons on Hiroshima and Nagasaki, and the declaration of the surrender of Japan. On von Weizsäcker's involvement in the German nuclear project, see Mark Walker: "Mit der Bombe leben"—Carl Friedrich von Weizsäckers Weg von der Physik zur Bombe. In: Klaus Hentschel, and Dieter Hoffmann (eds.): *Carl Friedrich von Weizsäcker. Physik–Philosophie–Friedensforschung. Leopoldina-Symposium vom 20. bis 22. Juni 2012 in Halle (Saale)*. Stuttgart: Wissenschaftliche Verlagsgesellschaft 2014, 343–356.

82 Carl F. von Weizsäcker: interview by Karl Hufbauer, April 18, 1978. Transcript, AIP, www.aip.org/history-programs/niels-bohr-library/oral-histories/4948. Last accessed 5/8/2019. Helmut Rechenberg: Vom Atomkern Zum Kosmischen Wirbel. *Physik Journal* 1/6 (2002), 59–61. https://www.pro-physik.de/restricted-files/114961. Last accessed 4/29/2020.

83 Werner Heisenberg: On the Theory of Statistical and Isotropic Turbulence. *Proceedings of the Royal Society of London* 195/1042 (1948), 402–406. www.jstor.org/stable/98337. Last accessed 10/30/2018. Werner Heisenberg: Zur statistischen Theorie der Turbulenz. *Zeitschrift für Physik* 124/7 (1948), 628–657. doi:10.1007/BF01668889. Werner Heisenberg, and Carl F. von Weizsäcker: Die Gestalt der Spiralnebel. *Zeitschrift für Physik* 125/4–6 (1948), 290–292. For a discussion on this group of articles see annotation by S. Chandrasekhar in Werner Heisenberg: *Collected Works. Series A/1: Original Scientific Papers*. Edited by Walter Blum, Hans-Peter Dürr, and Helmut Rechenberg. Vol. 1. Berlin: Springer-Verlag 1985, 19–24. doi:10.1007/978-3-642-61659-4.

84 As Biermann recalled later, experimental nuclear physics at the time meant mainly cosmic ray physics (exploiting a natural source of high-energy particles to probe the nuclear

ILLUSTRATION 25 Carl Friedrich von Weizsäcker (left) and Karl Wirtz (right) during the General Meeting of the Max Planck Society in Munich in 1951

However, as we will see in a moment, the necessity of reorganizing and redirecting the activities of the group for numerical computations, formerly of the meanwhile disbanded Aerodynamic Research Institute of the Kaiser Wilhelm Institute for Fluid Dynamics, was to prove to be a stroke of luck.[85]

---

and subnuclear realm), because it was the only form of experimentation with 'elementary particles' that was allowed. Ludwig Franz Benedikt Biermann: interview by Martin Harwit, March 16, 1984. Transcript, AIP. Both von Weizsäcker and Wirtz officially became Scientific Members of the Institute for Physics in June 1950 (CPTS meeting minutes of 14.06.1950, AMPG, II. Abt., Rep. 62, No. 1724).

85 Aerodynamische Versuchsanstalt Göttingen, *Die Aerodynamische Versuchsanstalt*, 1969.

Indeed, von Weizsäcker's astrophysical turn during the war had a decisive influence on the relaunch of research activities at Heisenberg's institute in Göttingen. However, the key transformative role in opening up a novel research perspective fell to the astrophysicist Ludwig Biermann, who established and developed astrophysics as a brand-new research field within the Kaiser Wilhelm Society/ Max Planck Society. Biermann had obtained his doctoral title from Göttingen University in 1932, with a dissertation on the convection zones in the interior of stars, and was one of the first 'native' plasma astrophysicists, a recognized expert in the physics of stellar atmospheres.[86] When he first began his career in the early 1930s, research in astrophysics was moving on, from a basic understanding of nuclear fusion as the energy source of stars to the more complex problem of how this energy propagated outward from the center of stars among a turbulent plasma. Before 1939, Biermann's work was mainly on stellar interior structure and convection. His early insight was to be the first to apply the concept of mixing length formulated by his mentor Ludwig Prandtl (who led in Göttingen the aforementioned Kaiser Wilhelm Institute for Fluid Mechanics, or *Strömungsforschung*, and had founded the meanwhile disbanded Aerodynamic Research Institute)[87] to this extreme astrophysical scenario, calculating the transport of energy by convection.[88]

---

86 Ludwig Biermann: Neuere Fortschritte der Theorie des inneren Aufbaues und der Entwicklung der Sterne. In: Ferdinand Trendelenburg (ed.): *Ergebnisse der Exakten Naturwissenschaften*. Berlin: Springer-Verlag 1945, 1–49. Ludwig Biermann, and Peter Wellmann: Physik der Sternatmosphären. In: Paul ten Bruggencate (ed.): *Astronomie, Astrophysik und Kosmogonie*. Wiesbaden: Dieterich'sche Verlagsbuchhandlung 1948, 119–159. Ludwig Biermann: Der innere Aufbau der Sterne. In: Paul Bruggencate (ed.): *Astronomie, Astrophysik und Kosmogonie*. Wiesbaden: Dieterich'sche Verlagsbuchhandlung 1948, 161–179.

87 Michael Eckert: *Ludwig Prandtl. Strömungsforscher und Wissenschaftsmanager. Ein unverstellter Blick auf sein Leben*. Berlin: Springer 2017. Eberhard Bodenschatz, and Michael Eckert: Prandtl and the Göttingen School. In: Peter A. Davidson et al. (eds.): *A Voyage Through Turbulence*. Cambridge: Cambridge University Press 2011, 40–100.

88 Rudolf Kippenhahn: Ludwig Biermann und die Theorie der Konvektionszonen in Sternen. In: Max-Planck Gesellschaft (ed.): *Ludwig Biermann. 1907–1986*. München 1988, 11–23. Biermann's pioneering ideas on the problem of convection in stars were appreciated by Arthur Eddington, as recalled by the well-known astrophysicist Thomas Cowling, with whom a strong relationship was established during Biermann's stay in the UK in the first half of the 1930s, and which continued after the war (see correspondence in AMPG, III. Abt., ZA 1, No. 1). Thomas Cowling: interview by David DeVorkin, March 22, 1978. AIP, https://www.aip.org/history-programs/niels-bohr-library/oral-histories/4563. Last accessed 1/31/2022. For an outline of Biermann's seminal work on the convective stellar model, laying the premise for the understanding of phenomena such as sunspots and the solar corona, see Thomas G. Cowling, and Leon Mestel: Obituary—Biermann,

ILLUSTRATION 26 Ludwig Prandtl in the wind tunnel at the Aerodynamics Research Institute, Göttingen, 1940

Biermann was also strongly influenced by the astronomer Hans Kienle, with whom he had obtained his doctoral title, and who was Professor of Astrophysics and Astronomy at Göttingen University and Director of the local observatory up to 1939, when he became Director of the Astrophysical Observatory in Potsdam, near Berlin.[89]

Biermann belonged to a generation of astrophysicists, mainly born in the first decade of the century, who became acquainted with modern physics as

---

L.F.B. *Quarterly Journal of the Royal Astronomical Society* 27/4 (1986), 698–700. http://adsabs.harvard.edu/abs/1986QJRAS..27..698C. Last accessed 1/13/2020. See also Eleonore Trefftz: In memoriam. Nachruf Professor Ludwig Biermann. *Journal of Geophysics* 60/3 (1986), 204–206. http://resolver.sub.uni-goettingen.de/purl?PPN1015067948_0060. Last accessed 2/9/2021.

89 Otto Heckmann: Nachrufe. Hans Kienle. *Mitteilungen der Astronomischen Gesellschaft* 38 (1976), 9–10. https://ui.adsabs.harvard.edu/#abs/1976MitAG..38....9H. Last accessed 10/30/2018.

students, and increasingly shared their research topics with physicists during the 1930s.[90] Theoretical astrophysicists like the Norwegian Svein Rosseland and the German Albrecht Unsöld, who had been trained in physics respectively by Niels Bohr and Arnold Sommerfeld, the fathers of quantum mechanics, were interested in applying physics to a wide range of cosmic phenomena. As new branches of physics grew, they found application in astrophysics. At the same time, astrophysics came to be seen by physicists as a highly fertile ground for research. By then, nuclear astrophysics had already developed into a research field attracting physicists with knowledge of theoretical nuclear physics, who were in particular drawn to study the constitution and energy source of the Sun and stars, as was described above. This was the intellectual framework within which a physicist like von Weizsäcker and other nuclear physicists, such as Fritz Houtermans, Hans Bethe, George Gamow, and Edward Teller, determined nuclear processes in stars, based upon the transformation of hydrogen into heavier elements through nuclear fusion. By the early 1940s, when astrophysics and physics were merging, wartime physicists introduced new standards of practice, which led to new forms of the organization of, and approaches to, research.

Von Weizsäcker and Biermann knew each other well, since they had both been students in Göttingen, taking part, among other things, in the astronomy seminar led by Hans Kienle. During their studies, Biermann introduced von Weizsäcker to Arthur Eddington's classic theory on the interior of stars; so, in a sense, Biermann was at the root of von Weizsäcker's interest in stellar physics. Later, when Biermann arrived in Berlin in 1937, as theoretician at the Berlin-Babelsberg Observatory in Potsdam, together with von Weizsäcker and Siegfried Flügge—the latter likewise interested in nuclear processes in stars—the three of them organized seminars on astrophysical topics at the Kaiser Wilhelm Institute for Physics in Berlin-Dahlem led by Debye.[91] It was then

90 These issues are widely examined in Luisa Bonolis: Stellar Structure and Compact Objects before 1940. Towards Relativistic Astrophysics. *The European Physical Journal H* 42/2 (2017), 311–393. doi:10.1140/epjh/e2017-80014-4. See also Martin Schwarzschild's opinion on the relevance of Biermann's research on stellar interiors at an international level ("He [L. Biermann] played quite a role…") Martin Schwarzschild: interview by Spencer Weart, July 30, 1975. Transcript, AIP, https://www.aip.org/history-programs/niels-bohr-library/oral-histories/28321. Last accessed 5/8/2019.

91 In this regard, see von Weizsäcker to Biermann, March 29, 1972, and Biermann to von Weizsäcker, April 5, 1972, AMPG, III. Abt., ZA 1, No. 21.

that Biermann first came into contact with plasma physics.[92] In particular, around 1938, in Berlin, he met Robert Rompe, an expert in plasma physics and co-author, with Max Steenbeck, of a review article which,[93] according to Biermann, "for many people was the beginning of the use of plasma physics in other fields."[94] On the other hand, Biermann was von Weizsäcker's main contact in Berlin "in matters of astrophysics" and they regularly met for long discussions on problems related to stellar interiors, on which Biermann had published several articles since the early 1930s.[95]

In those years, Biermann also interacted with the astronomer Karl Wurm, who was then at work at the Babelsberg Observatory in Potsdam, and one of

92 Carl F. von Weizsäcker: interview by Karl Hufbauer, April 18, 1978. Transcript, AIP, www.aip.org/history-programs/niels-bohr-library/oral-histories/4948. Last accessed 5/8/2019. Ludwig Franz Benedikt Biermann: interviews by Owen Gingerich, June 23 and July 7, 1978. Transcript, AIP. See also Richard Wielebinski: Ludwig Franz Benedikt Biermann: The Doyen of German Post-War Astrophysics. *Journal of Astronomical History and Heritage* 18/3 (2015), 277–284. http://www.narit.or.th/en/files/2015JAHHvol18/2015JAHH...18.277W.pdf. Last accessed 10/30/2018. Reimar Lüst: Ludwig Biermann. 13.3.1907-12.1.1986. *Berichte und Mitteilungen der Max-Planck-Gesellschaft*. München 1986, 78–81.

93 Robert Rompe, and Max Steenbeck: Der Plasmazustand der Gase. In: Ferdinand Trendelenburg (ed.): *Ergebnisse der Exakten Naturwissenschaften*. Berlin: Springer 1939, 257–376. doi:10.1007/BFb0112028.

94 Ludwig Biermann: interview by Woodruff T. Sullivan III, September 15, 1978, National Radio Astronomy Observatory Archives, https://www.nrao.edu/archives/items/show/896. Last accessed 1/31/2022.

95 Carl F. von Weizsäcker: interview by Karl Hufbauer, April 18, 1978. Transcript, AIP, www.aip.org/history-programs/niels-bohr-library/oral-histories/4948. Last accessed 5/8/2019. Already in 1931, Biermann had tackled the brand-new problem of white dwarfs, whose unusual high density was a challenging puzzle for astrophysicists that had been explained thanks to the new quantum statistics. The dense core of these compact stars was a new physical system (a so-called degenerate gas), of interest both for physics *and* astrophysics. At that time, both Lev Landau and Subramahnian Chandrasekhar were in fact proposing that such stars must have a maximum mass, beyond which there would be no more hydrostatic equilibrium between gravitation self-attraction and ionized gas pressure. And it is not surprising that astronomers reacted with skepticism to the limiting mass proposal coming from outside the astronomical profession. In his article on the internal constitution of stars with degenerate cores, submitted in December 1931, while he was still in Göttingen, Biermann thanked Chandrasekhar for several "valuable communications." Ludwig Biermann: Untersuchungen über den inneren Aufbau der Sterne, III. Über Sternmodelle mit entartetem Kern. *Veroeffentlichungen der Universitaets-Sternwarte zu Goettingen* 2 (1931), 163–171, 166. This early interest in the internal constitution of compact stars later became instrumental in focusing Biermann's interest on much denser objects like neutron stars, which require general relativity for their description. For the evolution of research on such compact astrophysical objects during the 1920s and '30s see Bonolis, Stellar Structure, 2017, 311–393.

the first to introduce methods of quantitative analysis to the study of spectral lines of molecular compounds in stellar atmospheres and in comets at different distances from the Sun. Wurm alerted Biermann to problems in the interpretation of observations of cometary tails, and cometary phenomena subsequently became a major interest for Biermann, one he continued to develop throughout his whole scientific career and which features repeatedly throughout this book.[96]

At that time, Biermann also established a personal relationship with Heisenberg, who since 1939 had been working closely with the groups focusing on the German nuclear project, both at his institute in Leipzig and in Berlin-Dahlem. He would later become head of the Kaiser Wilhelm Institute, once Debye decided to resign and move to the US. In 1941, Biermann accompanied von Weizsäcker and Heisenberg during their visit to Copenhagen, on the occasion of a week-long conference concerned, primarily, with the composition of the atmosphere of stars; organized by the newly founded Danish seat of the German Scientific Institute (*Deutsches Wissenschaftliches Institut*), it took place from 11 to 24 September. This visit—in particular the related private conversations between Heisenberg and Bohr—became the subject of a most lively controversy that culminated in the theatrical play *Copenhagen* by the British playwright and novelist Michael Frayn.[97]

96 Wielebinski, Ludwig Franz Benedikt Biermann, 2015, 277–284, 277.

97 On this visit, see also Mark Walker: Physics and Propaganda: Werner Heisenberg's Foreign Lectures under National Socialism. *Historical Studies in the Physical and Biological Sciences* 22/2 (1992), 339–389. doi:10.2307/27757685. In March 1941, von Weizsäcker himself had given lectures in Copenhagen, including Bohr's Institute (see correspondence between Bohr and Weizsacker in Niels Bohr Archive, Niels Bohr Scientific Correspondence, Folder 368). Like Biermann, Kienle and Unsöld, who also took part in the event, were renowned experts in the theme of the conference, which was also a central research focus for Bengt Strømgren, who had just succeeded his father Svante Elis as Director of the Royal Copenhagen Observatory, and was heartily invited to participate with a lecture (von Weizsäcker to Strømgren, August 15, 1941, Niels Bohr Archive, Folder 368, BSC-WEIC-410815tb). Strømgren's hypothesis that hydrogen is the main constituent of a star (and not the heavier elements, as was generally assumed in the late 1920s) and his model according to which helium is the second most abundant element in the Sun, paved the way for investigations by theoretical physicists and in particular to Bethe's and von Weizsäcker's 1938 theories on stellar energy production through the conversion of hydrogen into helium in nuclear reactions. On August 15, von Weizsäcker informed Bohr of their arrival and told him that he would talk about transformation of elements in stars, while Heisenberg would lecture on cosmic rays (Niels Bohr Archive, Folder 368, BSC-WEIC-410815ta). Since April 1940 Denmark had been occupied by Hitler, and thus von Weizsäcker specified that they would be glad if Danish scientists ("as many as possible") would attend; on the other hand, he added, Bohr personally "should not be forced to come," and in case

From 1935, the astronomer and solar astrophysicist Paul ten Bruggencate, who had been Kienle's assistant in Göttingen during the 1920s and meanwhile held a leading position at the Berlin-Babelsberg Observatory, participated in these seminars, too; in particular, he was in charge of the solar telescope which, up until the arrival of the Nazis, had been called the Einstein Tower; then in 1941 ten Bruggencate moved to Göttingen.[98] Also in Potsdam was the Astronomical Calculation Institute (*Astronomisches Recheninstitut*), the site of Biermann's first serious encounters with astronomical computing. Here, he worked on the application of astronomical navigation techniques to the calculation of navigational tables for airplanes and submarines. This involved handling large amounts of numerical data every day, so that calculations for the production of these mathematical tables was an early extensive application of mechanical computers. Apart from playing a leading role in the preparation of tables for astronomical navigation for the Luftwaffe,[99] during the 1940s, Biermann further developed his longstanding interest in atomic physics; and from 1941 he used the calculating machines he had at his disposal for interpreting observations of stellar spectra and for examining oscillator strengths, that is, quantities expressing the probability of absorption or emission of electromagnetic radiation in atomic transitions. These were among the first such computed data, on which Biermann published pioneering articles.[100]

---

they would be free to make visits and meet privately. We are very grateful to Peter Biermann for having drawn our attention to Ludwig Biermann's presence in Copenhagen in September 1941.

98 Hans Kienle: Paul ten Bruggencate. 24.2.1901–14.9.1961. *Die Naturwissenschaften* 49/4 (1962), 73–74. doi:10.1007/BF00622019. F.W. Jäger: Nachruf auf P. ten Bruggencate. *Mitteilungen der Astronomischen Gesellschaft* 15 (1962), 21–23. http://adsabs.harvard.edu/abs/1962MitAG..15...21J. Last accessed 10/30/2018. For more background on the Einstein Tower itself, see Klaus Hentschel: *The Einstein Tower. An Intertexture of Dynamic Construction, Relativity Theory, and Astronomy*. California: Stanford University Press 1997.

99 BArch, No. R 26-III/8.

100 Biermann's activities as a theoretical astrophysicist at the Babelsberg Observatory were related to the structure of stars, theory of stellar atmospheres, novae, atoms and ions, stellar spectra, and the probability of absorption or emission of electromagnetic radiation in atomic transitions. He recognized the need for quantitative data on opacity (the ability of solar material to absorb radiation) and abundances of chemical elements for a proper modeling of the solar interior and atmosphere and computed oscillator strengths for intermediate mass ions such as sodium, potassium, magnesium, silicon, and aluminum. Ludwig Biermann: Die Oszillatorenstärken einiger Linien in den Spektren des Na I, K I und Mg II. *Zeitschrift für Astrophysik* 22 (1943), 157–164. http://adsabs.harvard.edu/abs/1943ZA.....22..157B. Last accessed 10/30/2018. Ludwig Biermann: Normierte Wellenfunktionen verschiedener Zustände des Leuchtelektrons und Oszillatorenstärken der Übergänge zwischen ihnen für Na I, K I, Mg II, Si II

Thus, personal ties and common interests, rooted in the growing interaction between physicists and astrophysicists, characterized the relationship between Biermann, von Weizsäcker, and ten Bruggencate from the time of their youth, and grew stronger during the Berlin years.[101] Ten Bruggencate, who was on the board of the journals *Die Naturwissenschaften* and *Zeitschrift für Astrophysik*, after the war also became editor of the FIAT review volume dedicated to astronomical and astrophysical sciences and cosmology.[102] Biermann himself and von Weizsäcker contributed to one of these volumes.[103]

At the time, von Weizsäcker continued to be deeply interested in the interplay between turbulence and rotation and its prominent role in a protoplanetary nebula in which long-lived vortices can capture a large quantity of solid particles and initiate the formation of planets.[104] This common cultural background was the nucleus of what subsequently became the Institute for Astrophysics within the Max Planck Society.

Plasma Astrophysics with Calculating Machines in Postwar Göttingen

Before World War II, astrophysics and astronomy had been research fields pursued at several German universities, but there was no dedicated institute for

---

und Al I. *Veröffentlichungen der Universitäts-Sternwarte zu Göttingen* 5 (1947), 245–248. http://adsabs.harvard.edu/abs/1947VeGoe...5..245B. Last accessed 10/30/2018. For Biermann's scientific work and a list of his papers during the 1930s and early 1940s, see BArch, No. R 4901/13259, R 4901/25961, R 73/10317, R 9361-II/76118, R 9361-VIII/2480232.

101 Since November 1945, Biermann and ten Bruggencate had corresponded about a work on the theory of the solar corona that was published in 1947: Ludwig Biermann, and Paul ten Bruggencate: Über die Ursachen der hohen Temperatur der Sonnenkorona nebst einer Bemerkung über das Nachthimmellicht. *Veröffentlichungen der Universitäts-Sternwarte zu Göttingen* 5 (1947), 223–228. Bruggencate, Paul ten: *Astronomie, Astrophysik und Kosmogonie.* Wiesbaden: Dieterich'sche Verlagsbuchhandlung 1948. Carl F. von Weizsäcker: Zur Kosmogonie. *Zeitschrift für Astrophysik* 24/1/2 (1948), 181–206. http://adsabs.harvard.edu/abs/1948ZA.....24..181V. Last accessed 10/30/2018.

102 The FIAT (Field Information Agency Technical) series of reports on the status of German science in various disciplines was published after the war to provide information on scientific research which had been cut off from scientific publications by Nazi control. On these topics, see also the related scientific correspondence between Biermann and ten Bruggencate in Ludwig Biermann's papers, AMPG, III. Abt., ZA 1, No. 1.

103 Biermann, and Wellmann, Physik der Sternatmosphären, 1948, 119–159. Biermann, Der innere Aufbau, 1948, 161–179. Carl F. von Weizsäcker: Kosmogonie. In: Paul Bruggencate (ed.): *Astronomie, Astrophysik und Kosmogonie.* Wiesbaden: Dieterich'sche Verlagsbuchhandlung 1948, 413–426.

104 Carl F. von Weizsäcker: Die Rotation kosmischer Gasmassen. *Zeitschrift für Naturforschung A* 3/8–11 (1948), 524–539. https://ui.adsabs.harvard.edu/#abs/1948ZNatA...3..524W. Last accessed 11/2/2017.

them within the Kaiser Wilhelm Society.[105] Heisenberg and von Weizsäcker's common interest in astrophysical topics, cosmic rays, and solar physics in general, collided with a fortunate coincidence in postwar Göttingen: the necessity of reorienting activities of the computing group (*Rechengruppe*) from the disbanded Aerodynamic Research Institute, a precious resource that was not to be wasted.

Like several astronomers from Potsdam and Babelsberg, Biermann had moved after the war to the Hamburg area, to the Bergedorf Observatory directed since 1941 by Otto Heckmann,[106] which at the time was perhaps the principal observatory in Germany.[107] In November 1946, Biermann received a letter from ten Bruggencate, who had been appointed to the Chair of Astronomy in Göttingen. Head of the university's observatory since 1941, as well as President of the Academy of Sciences, he had also built a solar observatory there.[108] In his letter, ten Bruggencate asked Biermann to train the computing group at the Aerodynamic Research Institute, so as to launch new activity in

105 Joachim Trümper: Astronomy, Astrophysics and Cosmology in the Max Planck Society. In: André Heck (ed.): *Organizations and Strategies in Astronomy*. Dordrecht: Springer Netherlands 2004, 169–187, 171–174. For the early history of astronomy and astrophysics in Germany and in the KWG/Kaiser Wilhelm Society, see also Dietrich Lemke, and Astronomische Gesellschaft (eds.): *Die Astronomische Gesellschaft 1863–2013. Bilder und Geschichten aus 150 Jahren*. Heidelberg: Astronomische Gesellschaft 2013. For an overview of postwar expansion, see Rolf-Peter Kudritzki, and Reinhold Häfner: German Astronomy. Edited by Paul Murdin. *Encyclopedia of Astronomy and Astrophysics*, 2001, 1–3. doi:10.1888/0333750888/2933.

106 Heckmann had worked in Göttingen since 1927 and had developed there as an astrophysicist when physics and mathematics were still flourishing, before the 1933 catastrophe brought about the loss of its excellence following the decimation of the world-class mathematics and physics faculties due to the forced departure of Jews and 'political undesirables.' He was especially interested in cosmology, which he continued to pursue in Hamburg, where he was in constant contact with Pascual Jordan, whose research program on general relativity made Hamburg one of the centers of the revival of the field in the 1950s. Hans-Heinrich Voigt: Nachruf auf Otto Heckmann. *Mitteilungen der Astronomischen Gesellschaft* 60 (1983), 9–12. https://ui.adsabs.harvard.edu/#abs/1983MitAG..60....9V. Last accessed 10/30/2018. At Bergedorf, Bierman also found Wurm, who had moved there from Babelsberg in 1941.

107 Gerard P. Kuiper: German Astronomy during the War. *Popular Astronomy* 54 (1946), 263–286, 266–268. http://adsabs.harvard.edu/abs/1946PA.....54..263K. Last accessed 9/17/2018. According to Kuiper's review, also containing a list of publications related to the war period, some research activities conducted at Bergedorf Observatory were especially appreciated abroad.

108 During the war, in order to pursue his interest in solar observations based on his previous work at the Einstein Tower in Potsdam, ten Bruggencate had built a solar telescope with the aid of the military, because of its interest in forecasts of short-wave communication conditions. As we will see in later chapters, in this ambition, ten Bruggencate was

ILLUSTRATION 27 Biermann sitting at his desk in Göttingen, April 1948

numerical computations on stellar spectra.[109] Considering Biermann's expertise, such a request was far from surprising. At the time, Biermann could never have imagined that such a letter would completely change his life and, too, the future of astrophysics at the Kaiser Wilhelm Society soon to be reborn as the Max Planck Society.

The Kaiser Wilhelm Society and the Göttingen Academy of Sciences came to an agreement on their joint funding of the computing group previously

---

competing with the solar astrophysicist Karl-Otto Kiepenheuer, who had similar, albeit more grandiose, aims. See pp. 78–79 in Michael P. Seiler: *Kommandosache "Sonnengott". Geschichte der deutschen Sonnenforschung im Dritten Reich und unter alliierter Besatzung*. Frankfurt am Main: Verlag Harri Deutsch 2007.

109 Letter from ten Bruggencate to Biermann, November 18, 1946, related to a conversation with Jacob Sommer about using the old 'AVA *Rechengruppe*' for calculations of astrophysical interest (AMPG, III. Abt., ZA 1, No. 1). See also the letter from Biermann to Sommer, dated January 21, 1947, announcing his arrival in Göttingen from Hamburg on the 27th, and his stay for a week (AMPG, II. Abt., Rep. 66, No. 3058). It is also relevant to recall that, immediately after the war, before moving from Babelsberg Observatory to the University Observatory in Hamburg, Biermann had been a temporary member of the Mathematical Institute of Göttingen University (see Biermann's Curriculum Vitae in Heisenberg's papers, AMPG, III. Abt., Rep. 93, No. 64).

ILLUSTRATION 28 Carl Friedrich von Weizsäcker at his desk in Göttingen, April 1948

led by Hans. G. Küssner, which was now to be led by Biermann and work mainly on topics of astrophysical interest.[110] Ten Bruggencate was clearly keen to have an efficient group focusing on astronomical numerical computations, to support his large projects. Against this backdrop, a more ambitious idea grew around the reorganization of the AVA *Rechengruppe* and, already in April 1947, Heisenberg officially invited Biermann to become a member of the Institute for Physics and take charge of the computing group, now incorporated into what was still called the Kaiser Wilhelm Society.[111] A special Depart-

110 AMPG, II. Abt., Rep. 66, No. 3047, fol. 501–506.

111 Following a meeting of the commission for the future of the *Rechengruppe* held on April 24, and after consulting with Hahn, the President of the Kaiser Wilhelm Society, Heisenberg asked Biermann to accept a position as head of a department at the KWG Institute for Physics and as leader of the computing group which would be attached as a department to the institute. The commission had decided that the group should primarily deal with astrophysical problems that Biermann would propose and also on other possible tasks that might arise. The Academy of Sciences would reserve the right to have young scientists work as guests in the group (Heisenberg to Biermann, April 25, 1947, AMPG, III. Abt., Rep. 93, No. 1687). Heisenberg arranged for an apartment to be assigned to Biermann's family (Heisenberg to Biermann, April 30, 1947, AMPG, III. Abt., ZA 1, No. 2). An invitation for a meeting of the "Kommission für das Recheninstitut der Kaiser-Wilhelm-

ment for Astrophysics led by Biermann was officially established on July 1, 1947.[112]

In 1948, Biermann met Heinz Billing from the AVA, an excellent physicist and skilled instrument builder, who was already developing the magnetic drum memory, the first magnetic storage system for computers, at what was then the Society's Institute for Scientific Instruments (*Institut für Instrumentenkunde*).[113]

This pioneering device immediately attracted Biermann's attention and he became enthusiastic about its potential applications.[114] No commercial computers were available at the time, so he seized this opportunity to build in-house electronic computers, which would greatly reduce the calculation time for astrophysical computations. From 1948 to 1949, Billing struggled to complete the first prototype of his machine, but was secretly considering migrating to Argentina. Simultaneously, he was invited to go to Australia instead, and ended up working in Sydney for the Commonwealth Scientific and Industrial Organization (CSIRO). Ever since Billing announced his imminent move to

---

Gesellschaft," which was scheduled for April 24, 1947, had been sent by Heisenberg to the members of the commission: Becker, ten Bruggencate, Eucken, Telschow, von Weizsäcker, Sommer.

112 Rhea Lüst, and Rudolf Kippenhahn: Max-Planck-Institut für Physik und Astrophysik. Institut für Astrophysik München. *Max-Planck-Gesellschaft. Berichte und Mitteilungen* 1/77 (1977), 1–64.

113 Billing had studied mathematics and physics in Göttingen and received his PhD in Munich under the direction of Walter Gerlach, in 1938. After graduating from university, he began to work at AVA and in fall 1946 he moved to the Institute for Scientific Instruments of the Kaiser Wilhelm Society, which was also housed on the site of the AVA. There, in the aftermath of WWII, he developed the magnetic drum memory. J.A.N. Lee: Heinz Billing. Edited by J.A.N. Lee, and IEEE Computer Society. *Computer Pioneers*. IEEE Computer Society 2012, 102–106. http://history.computer.org/pioneers/billing.html. Last accessed 10/30/2018. See also Heinz Billing: *Meine Lebenserinnerungen*. Garching: Selbstverlag 1994.

114 Ludwig Franz Benedikt Biermann: interview by Martin Harwit, February 16, 1984. Transcript, AIP. See also AMPG, II. Abt., Rep. 66, No. 1787. In December 1947, Billing had started work on an electronic adding machine with his magnetic drum storage, and in the spring of 1948, Biermann visited his laboratory, accompanied by Heisenberg, Wirtz, and Bagge. In the following July, Billing presented his plans at the annual meeting of the Society for Applied Mathematics and Mechanics (Gesellschaft für Angewandte Mathematik und Mechanik, GAMM). As an example of a practical application, Billing chose the numerical solution of the Schrödinger differential equations—a problem first tackled with desk calculators by Biermann's computing group—and only later found out that the so elegant method he used derived in fact from Biermann. Heinz Billing: Ludwig Biermann und die Rechenmaschinen. In: Max-Planck Gesellschaft (ed.): *Ludwig Biermann. 1907–1986*. München 1988, 51–62, 51–52.

ILLUSTRATION 29 Laboratory of the Max Planck Institut für Instrumentenkunde in 1948, where Heinz Billing invented the first magnetic storage system for computers. The institute was founded in 1946 to develop new scientific apparatuses, which would enable other institutes to start new projects or improve their ongoing research with more effective devices.

Australia, Biermann had been trying to convince Heisenberg to have him at the Institute for Physics. He insisted that there was a very good chance his work would be funded and, moreover, it was not on the list of prohibited research topics.[115] Therefore, Heinz Billing, one of the Germans with the most experience of calculating machines, was brought back to Göttingen in 1950, where he built West Germany's first electronic computers for the Max Planck Insti-

115 Ludwig Franz Benedikt Biermann: interview by Martin Harwit, February 16, 1984. Transcript, AIP. Biermann also mentioned that the computer pioneer Konrad Zuse used many of Billing's findings for the computing machines he built in his company and so a percentage of his earnings ("thousands of DM [Deutschmarks] per year") went to the Institute: "This enabled us, when we later went into (nuclear) fusion and questions about patents came up, that we—Schlüter and I—could pay all our patent attorneys' bills without additional cost to the Max Planck Society, just with income that we had from there" [our translation]. See Billing's biography: Lee, Heinz Billing, 2012, 102–106. See also Martin Schwarzschild: interview by David DeVorkin and Spencer Weart, December 16, 1977, Session III. Transcript, AIP, https://www.aip.org/history-programs/niels-bohr-library/oral-histories/4870-3. Last accessed 19/2/2019.

ILLUSTRATION 30 Heinz Billing standing near the computing machine, G1

tute for Physics.[116] The G1 was completed in the fall of 1952 and Billing later developed and built the calculating machine G2 and, subsequently, the fully automatic G3.[117]

These tools, which allowed for numeric simulation and data modeling, played a crucial role in the institute's research activities and in building expertise in a brand-new field. Apart from computations related to atomic and nuclear theories, as well as to the non-linear shock waves especially relevant in astrophysical realms, they also calculated hundreds of paths of the charged cosmic ray particles from outer space interacting with the Earth's magnetic field, according to Carl Størmer's theory—an issue in which Biermann was par-

116 In 1950 Billing formed a working group that later became the *Numerische Rechenmaschinen* Department. He became a Scientific Member in 1961.

117 Several modified copies of the G1, called G1a, were made. The later model G2, ready in 1954, was ten times faster. See Ludwig Biermann's note on the early machines in Generalverwaltung der Max-Planck-Gesellschaft (ed.): *Mitteilungen aus der Max-Planck-Gesellschaft. Heft* 2. Göttingen 1952, 16–19. Ludwig Biermann, and Heinz Billing: Der Stand der Entwicklung und Ausnutzung der elektronischen Rechenanlagen in Göttingen. *Mitteilungen aus der Max-Planck-Gesellschaft zur Förderung der Wissenschaften* Heft 1 (1954), 35–38.

ILLUSTRATION 31 Heinz Billing and Hermann Oehlmann at the G2 computer in 1954

ILLUSTRATION 32 Heinz Billing near the computing machine, G3, beginning operations. At the console is Arno Carlsberg. Göttingen, 1960.

ILLUSTRATION 33 Heinz Billing and Ludwig Biermann in 1972 at the shutdown of the G3 machine

ticularly interested—and later simulated processes in plasmas.[118] In the days before "the US supercomputer centers were up to speed," Biermann's Institute for Astrophysics became "a 'mecca' for theorists who wanted to do computational work," and this contributed significantly to establishing its leadership at the global level.[119] In 1955, Billing was invited to spend six months at the prestigious Institute for Advanced Study in Princeton, where John von Neumann had built the IAS computer, which was meant to be mainly used for experiments in computations and made available to researchers for various purposes.[120]

118 Ludwig Biermann: Elektronische Rechenmaschinen und physikalische Forschung. In: Generalverwaltung der Max-Planck-Gesellschaft (ed.): *Jahrbuch 1956 der Max-Planck-Gesellschaft zur Förderung der Wissenschaften e.V.* Göttingen 1956, 19–33. A round of calculating the orbit of a particle took three to four hours with the G1, while doing the same task by hand with the aid only of mechanical machines would have taken a week of full-time work.

119 We owe this remark to the astrophysicist Alastair G.W. Cameron, who pioneered computational astrophysics while working at the Chalk River atomic energy laboratory in Canada in the 1950s, and became one of the founders of postwar nuclear astrophysics. Eliot Marshall: Astrophysics Institute at Risk. *Science* 257/5070 (1992), 606–606. doi:10.1126/science.257.5070.606-a.

120 On his travels abroad during the 1950s, see Billing, *Lebenserinnerungen*, 1994, 102–131.

ILLUSTRATION 34 Apparatus for simulating the trajectories of cosmic-ray particles in the Earth's magnetic field, built at the Max Planck Institute for Physics in Göttingen in the very early 1950s. More or less all the particles come from the same direction of the Sun (which is to be imagined at the right of the observer). As a result of their electric charge, they take quite complicated paths. Observing the device, from left to right, are: the Federal President, Theodor Heuss, during a visit at the Max Planck Institute for Physics, Ludwig Biermann, Otto Hahn, Werner Heisenberg, and Karl Wirtz, November 1951. Right: device for studying the motions of charged particles in magnetic fields, January 1954

ILLUSTRATION 35 Workshop of the Max Planck Institute for Physics directed by Heisenberg, March 1954

In parallel to his traditional research on quantum theoretical problems in astrophysics, stellar plasma physics, and solar questions, Biermann developed a more general interest in the behavior of astrophysical plasmas and cosmic rays during this period, when Hannes Alfvén's theory of magnetohydrodynamic waves in plasmas was used by Enrico Fermi to formulate a mechanism for the acceleration of charged cosmic ray particles by interstellar magnetic fields embedded in diluted plasma clouds acting as 'magnetic mirrors.'[121] From then on, many authors explored the potential of magnetic fields to trap and accelerate cosmic rays, and later their importance for the deceleration of relativistic electrons was recognized, too. During this period, also thanks to the advent of radio astronomy, interstellar space came to be seen as the site of more complex phenomena in which magnetic fields and turbulent gas motions play a significant part. A remarkable relationship was established in the very early 1950s between the radio waves emission and high-energy electrons moving in galactic magnetic fields. Magnetic fields accelerate charged particles in circular motions making them spiral around the field lines, a motion which generates radio waves. Thus, the energy of the radiating electrons is due not to thermal motions—associated with particles' thermal energy—but to a *non-thermal* radiation process, which had been called 'synchrotron radiation,' as it was first observed to be emitted by electrons accelerated in a synchrotron. This phenomenon gave scientists a chance to indirectly

121 The existence of magnetohydrodynamic waves in plasmas proposed by Alfvén in 1941 was generally considered nonsense, until it became fully supported by Enrico Fermi in his most cited article of 1949, arising from discussions with Alfvén and Chandrasekhar in Chicago in the late 1940s. Enrico Fermi: On the Origin of the Cosmic Radiation. *Physical Review* 75/8 (1949), 1169–1174. doi:10.1103/PhysRev.75.1169. Apparently, Biermann had a copy of this article in mimeographed form before publication, as mentioned in the correspondence between Biermann and Kiepenheuer of April-May 1949 (AMPG, III. Abt., ZA 1, No. 2). On Alfvén's discovery of the magnetohydrodinamic waves see Alexander J. B. Russell: 75th Anniversary of 'Existence of Electromagnetic–Hydrodynamic Waves'. *Solar Physics* 293/5 (2018), 83. doi:10.1007/s11207-018-1296-3. Alfvén's waves propagating into the corona, the aura of plasma surrounding the Sun, became one of the clues to explain the puzzle of its incredibly high temperature despite being farther from the solar core, a question which Biermann himself had tackled, proposing that the chromosphere—a deeper layer of the Sun's atmosphere lying under the corona—could be heated by dissipation processes in shock waves thus transporting energy towards the corona. Ludwig Biermann: Über die Ursache der chromosphärischen Turbulenz und des UV-Exzesses der Sonnenstrahlung. *Zeitschrift fur Astrophysik* 25 (1948), 161–177. https://ui.adsabs.harvard.edu/#abs/1948ZA.....25..161B. Last accessed 2/19/2019.

detect cosmic plasmas and led to a growing understanding of the Universe as consisting largely of magnetized plasma.[122]

The existence of diffuse synchrotron radiation from plasmas in our Galaxy, as proposed by Kiepenheuer (see Section 3 of this chapter), could not but attract the attention of such a solar plasma specialist as Biermann.[123] In the late 1940s, his research agenda was already focused on topics such as radio waves from the Sun, or the connection between cosmic rays and interstellar magnetic fields, which were becoming the subject of 'cosmical electrodynamics' and which he—and his collaborator Arnulf Schlüter—were presenting at meetings.[124] By 1953, Biermann was considered such an authority on the problem of cosmic rays that Arnold B. Lovell, an influential British cosmic ray physicist, who later developed the 76 m Jodrell Bank radio telescope, asked

122 Hannes Alfvén was probably the first to suggest that radio waves are emitted by charged particles moving in galactic magnetic fields thus connecting radio emission with cosmic particles. Hannes Alfvén, and Nicolai Herlofson: Cosmic Radiation and Radio Stars. *Physical Review* 78/5 (1950), 616–616. doi:10.1103/PhysRev.78.616.

123 In parallel with Alfvén and Nicolai Herlofson, Kiepenheuer proposed that galactic radio background was synchrotron radiation. Karl-Otto Kiepenheuer: Cosmic Rays as the Source of General Galactic Radio Emission. *Physical Review* 79/4 (1950), 738–739. doi:10.1103/PhysRev.79.738. On Biermann's early interest in this regard, see his correspondence during the period 1946–1951, particularly with Kiepenheuer and Unsöld (AMPG, III. Abt., ZA 1, No. 1, 2) as well as a cutting dated May 12, 1949 on "Signale aus dem Weltenraum" [signals from space] (AMPG, II. Abt., Rep. 66, No. 3058).

124 Arnulf Schlüter: Zur Theorie der Kurzwellenstrahlung der Sonne. *Die Naturwissenschaften* 35/5 (1948), 154–155. doi:10.1007/BF00631599. Arnulf Schlüter, and Gerd Burkhardt: Ausbreitung und Ausstrahlung radiofrequenter Wellen in der Sonnenkorona. *Zeitschrift für Astrophysik* 26 (1949), 295–304. http://adsabs.harvard.edu/abs/1949ZA.....26..295B. Last accessed 10/30/2018. Ludwig Biermann, and Arnulf Schlüter: Interstellare Magnetfelder. *Zeitschrift für Naturforschung A* 5/5 (1950), 237–251. doi:10.1515/zna-1950-0501. Arnulf Schlüter: Dynamik des Plasmas I. Grundgleichungen, Plasma in gekreuzten Feldern. *Zeitschrift für Naturforschung A* 5/2 (1950), 72–78. doi:10.1515/zna-1950-0202. Arnulf Schlüter: Solare Ultrastrahlung und Erdmagnetfeld. *Zeitschrift für Naturforschung A* 6/11 (1951), 613–618. doi:10.1515/zna-1951-1108. Ludwig Biermann, and Arnulf Schlüter: Cosmic Radiation and Cosmic Magnetic Fields. II. Origin of Cosmic Magnetic Fields. *Physical Review* 82/6 (1951), 863–868. doi:10.1103/PhysRev.82.863. Ludwig Biermann: Origin and Propagation of Cosmic Rays. *Annual Review of Nuclear Science* 2 (1953), 235–364. doi:10.1146/annurev.ns.02.120153.002003. See also Biermann's contribution on the origin of cosmic rays in Heisenberg, *Kosmische Strahlung*, 1953. Schlüter had studied Physics in Bonn and after his PhD in theoretical physics he moved to Göttingen with Biermann in 1948. Uwe Schumacher: Arnulf Schlüter. 22. August 1922–24. Juni 2011. *Jahresbericht der Max-Planck-Gesellschaft. Beilage Personalien*, 2012, 29–30.

ILLUSTRATION 36 Ludwig Biermann at the blackboard during a seminar at the Max Planck Institute for Physics, Göttingen, 1956

Biermann to write a book on the subject.[125] As can be inferred from Biermann's correspondence, the interest in radio, ultraviolet, and X-ray emission from the Sun continued to attract his group's attention over the years. In particular, Biermann and his collaborators were fully aware of the great potential of radio astronomy, also as a natural link to their interest in cosmic plasmas and magnetic fields.[126]

It later became clear that cosmic X-rays and gamma rays, too, can derive from high-energy magnetized plasmas and, as we will see, all this background

125 Lovell to Biermann, 13.08.1953, AMPG, III. Abt., ZA 1, No. 34. The following document in the folder is the program of a conference on radio astronomy held at the Jodrell Bank experimental station the previous July.

126 See, for example, an article on the non-thermal origin of the radio-frequency emission from the solar chromosphere and corona by Biermann and Reimar Lüst, mentioning work on the same issues by Arnulf Schlüter and Lüst himself. Biermann, Ludwig, and Reimar Lüst: Remarks on the Energy of the Non-Thermal Radio-Frequency Emission. In: Hendrik Christoffel Van de Hulst (ed.): *Radio Astronomy, Proceedings from 4th IAU Symposium. International Astronomical Union*. Radio astronomy. Cambridge: Cambridge University Press 1957, 354–355. http://adsabs.harvard.edu/abs/1957IAUS....4..354B. Last accessed 6/25/2020.

was an important premise for plans developed in the early 1960s (see Chapter 2), at what would become the Max Planck Institute for Extraterrestrial Physics (MPE), led by Biermann's collaborator Reimar Lüst. The same scientific background also motivated Biermann and Lüst in subsequent years to support the founding, in 1966, of a dedicated Institute for Radio Astronomy (MPIfR) (see Chapter 3).[127]

This interdisciplinary background, combined with competence in making numerical calculations with computing machines, lay the groundwork and set the stage for the future research agenda of Biermann's department at the Institute for Physics.[128] Its research areas were astrophysical plasmas, cosmic magnetic fields, and cosmic rays (calculations of paths of charged particles in the Earth's magnetic field, with the G1 computer), comet tails, structure and evolution of stars, and use of computers.[129] At the same time, as has been clearly shown in several of its members' publications, this pioneering activity gave the younger generation a chance to develop skills unique at that time, which made them more likely to be headhunted by scientific institutions abroad, notably in the United States, where people able to conduct relevant research for immediate application were much in demand.[130]

---

127 From then on, Biermann was a member of the Scientific Advisory Board (*Fachbeirat*) of the Institute for Radio Astronomy and in 1975, at the time of Biermann's retirement, Lüst, as President of the Max Planck Society, acknowledged his valuable work on the board and especially his precious assistance in the founding phase, during which many problems had had to be solved before the large Effelsberg telescope was commissioned (Lüst to Biermann, June 30, 1975, AMPG, III. Abt., ZA 1, No. 87). On the founding period of the Institute for Radio Astronomy see AMPG, III. Abt., Rep. 145, No. 292, 862.

128 In 1950, the Departments for Astrophysics and Theoretical Physics led by Biermann and von Weizsäcker, respectively, employed Arnulf Schlüter and Eleonore Trefftz as assistants, Reimar Lüst, I. Lucas, E. v. Roka, Peter Stumpff (the son of the famous astronomer Karl Stumpff), Stefan Temesváry, and Sebastian von Hoerner as scientific collaborators, and four members of the computing group. Ludwig Biermann, and Carl Friedrich von Weizsäcker: Jahresberichte deutscher Sternwarten für 1950. Göttingen. Max-Planck-Institut für Physik (Abteilungen für Astrophysik und für theoretische Physik). *Mitteilungen der Astronomischen Gesellschaft* 2 (1950), 41–43. http://adsabs.harvard.edu/abs/1950MitAG...2...30. Last accessed 10/30/2018.

129 Arnulf Schlüter: Biermanns Göttinger Schule der Kosmischen Elektrodynamik. In: Max-Planck-Gesellschaft zur Förderung der Wissenschaften (ed.): *Ludwig Biermann. 1907–1986*. München 1988, 24–34. Lüst, and Kippenhahn, Max-Planck-Institut für Physik und Astrophysik, 1977, 1–64.

130 See Billing's personal recollections and historical work on early computers, Billing, Ludwig Biermann, 1988, 51–62. Billing, *Lebenserinnerungen*, 1994. Ulf Hashagen, and Raul Rojas (eds.): *The First Computers. History and Architectures*. Cambridge, MA: MIT Press 2000. Lee, Heinz Billing, 2012, 102–106. Heinz Billing: Schnelle Rechenmaschinenspeicher

Bringing Together Cosmic Rays, Experimental Particle Physics, and Plasma Physics

Around the mid-1940s, the term 'cosmic-ray physics' was related to two basic questions that researchers were trying to answer: a) What are the constituents of cosmic radiation? b) What is the origin of cosmic radiation and what are its effects on Earth? The second question became the main subject of cosmic ray research, which shifted gradually to the problems related to astrophysics, once the 'particle aspect' was taken over by physicists using accelerators, after 1948, when π-mesons began to be produced artificially. Later on, with the more powerful machines put into operation in the course of the 1950s, it became possible to study even particles such as kaons and hyperons, which had been discovered in high-energy collisions of cosmic rays in the stratosphere.

Biermann's interest took hold in a period in which cosmic rays were briefly at the crossroads of all the major physical sciences: astronomy and high-energy astrophysics, nuclear physics, plasma physics, and elementary particle physics. Indeed, it was the quest for an understanding of the nature and behavior of cosmic rays and the challenges facing researchers that gave rise to new scientific disciplines, technologies, and astrophysical concepts. Particle and high-energy physics, as well as magnetic fields and plasmas of astrophysical origin, are typical research fields born of cosmic ray research.[131] Up until the mid-1950s, cosmic rays, as a source of high-energy particles, were a substantial experimental field of Heisenberg's Institute for Physics, providing a forum for nuclear and subnuclear phenomena to be explored—also from a theoretical point of view—at a time when these questions could not be pursued with particle accelerators in Europe, where powerful machines were still lacking, let alone with nuclear reactors, because of the restrictions imposed by the Allies

---

und ihre Geschwindigkeits- und Kapazitätsgrenzen. In: Generalverwaltung der Max-Planck Gesellschaft zur Förderung der Wissenschaften e.V. (ed.): *Jahrbuch 1962 der Max-Planck-Gesellschaft zur Förderung der Wissenschaften e.V.* Göttingen 1962, 52–79. Billing became a Scientific Member in 1961 (CPTS meeting minutes of 19.01.1961, 06.06.1961, AMPG, II. Abt., Rep. 62, No. 1736, 1737). Related archival material can be found in Biermann's papers: AMPG, III. Abt., ZA 1, No. 99. For financial data of the Billing's group of computing machines from the early 1950s up to 1967, see AMPG, II. Abt., Rep. 66, No. 3218.

131 John Simpson: The Cosmic Radiation. In: Johan A. M. Bleeker, Johannes Geiss, and Martin C. E. Huber (eds.): *The Century of Space Science*. Dordrecht: Kluwer Academic Publishers 2001, 117–151. For a historical review of the developments during the 100 years since the discovery of cosmic rays, a summary of the current research and future perspectives see Jonathan F. Ormes (ed.): *Centenary Symposium 2012: Discovery of Cosmic Rays, 26–28 June 2012, Denver, Colorado, USA. AIP Conference Proceedings*. Vol. 1516. New York, NY: American Institute of Physics 2013.

on Germany after WWII. As recalled by Reimar Lüst, who had participated in the scientific life of Heisenberg's institute since the late 1940s:

> The topic that united the whole institute at the internal colloquia was cosmic radiation. For two years, this subject was treated at each of the colloquia, which took place weekly on Saturday mornings. Nearly every member of the scientific staff was expected to contribute, and also to prepare a manuscript for the second edition of the book on cosmic radiation, edited by Heisenberg and published in 1953 by Springer-Verlag [...] The observation of cosmic radiation was part of the experimental program of the institute. The radiation was detected with the help of photographic plates that were carried by balloons at great heights. The expeditions to follow the balloons, either by car in, among others, Heisenberg's Mercedes, or in Italian warships were always a special attraction to Heisenberg, since they reminded him of his *Wandervogel* time at the beginning of the 1920s. This was a wonderful time in Göttingen for all of us, enormously productive scientifically, but also characterized by a very close personal living and working environment. Shortly before his death Heisenberg said, "That time in Göttingen—it was the happiest time of my life."[132]

However, by the early 1950s, the process leading to a bifurcation in cosmic ray research was clearly manifest. On the one hand there was its role in nuclear and particle physics: accelerators such as the Cosmotron at Brookhaven and the Bevatron at Berkeley were producing at the time intense 'homemade' meson beams, even if at lower energies than energetic cosmic ray particles from outer space. The discovery of the antiproton at the Bevatron in 1955, resulting in the Nobel Prize in Physics 1959 for Emilio Segrè and Owen Chamberlain, definitely marked the transition from a style of research based on cosmic rays to the authoritative evidence provided by the more systematic studies that became possible with accelerators.[133] On the other hand, there was a growing interest in the 'cosmic' nature of the cosmic ray particles and their

132 Reimar Lüst: Heisenberg and the Scientist's Responsibility. In: Gerd Buschhorn, and Julius Wess (eds.): *Fundamental Physics—Heisenberg and Beyond. Werner Heisenberg Centennial Symposium "Developments in Modern Physics."* Berlin: Springer 2004, 15–24, 19.

133 Owen Chamberlain et al.: Example of an Antiproton-Nucleon Annihilation. *Physical Review* 102/3 (1956), 921–923. doi:10.1103/PhysRev.102.921. See also Luisa Bonolis: Emilio Gino Segrè. Research Profile. *Lindau Nobel Mediatheque*, 4/11/2018. http://www.mediatheque.lindau-nobel.org/research-profile/laureate-segr. Last accessed 4/11/2018.

origin. Theories of the acceleration of cosmic rays provided the earliest interpretations for the signals detected by the first generations of radio telescopes, which were little more than repurposed wartime radars.[134] Toward the end of the 1950s, the development of new detectors and techniques allowed a more detailed analysis of extensive atmospheric showers of secondary particles generated by the interaction of very-high-energy cosmic rays with atmospheric nuclei. Such high energies, which were at the time—and still are—definitely far removed from what can be achieved using accelerators, brought about novel questions on the astrophysical sources and acceleration mechanisms of the primary particles.[135] Cosmic ray research continued to perform a vital role, but increasingly in relation to problems of an astrophysical nature, together with their interaction with galactic matter, as well as galactic and interplanetary magnetic fields and plasma clouds, which were a main research subject both for Biermann and his collaborator Schlüter.[136] Biermann had explored the connection between plasmas and magnetic fields in stars and in interstellar space already in 1950, suggesting a mechanism for generating magnetic fields in plasmas.[137] But at the time, all these problems gained a special status,

---

134 For more details, see the account of Kiepenheuer's work later in this chapter, as well as the early history of radio astronomy in the Max Planck Society, in Chapter 3.

135 John Linsley, Livio Scarsi, and Bruno Rossi: Extremely Energetic Cosmic-Ray Event. *Physical Review Letters* 6/9 (1961), 485–487. doi:10.1103/PhysRevLett.6.485.

136 Biermann, and Schlüter, Origin of Cosmic Magnetic Fields, 1951, 863–868. Biermann, and Schlüter, Interstellare Magnetfelder, 1950, 237–251. Schlüter, Solare Ultrastrahlung, 1951, 613–618. Schlüter, Dynamik I, 1950, 72–78. Arnulf Schlüter: Dynamik des Plasmas II. Plasma mit Neutralgas. *Zeitschrift für Naturforschung A* 6/2 (1951), 73–78. doi:10.1515/zna-1951-0202. Schlüter, Solare Ultrastrahlung, 1951, 613–618. Schlüter, Solare Ultrastrahlung, 1951, 613–618. Ludwig Biermann: Entstehung von Magnetfeldern in bewegten Plasmen. *Annalen der Physik* 445/8 (1952), 413–417. doi:10.1002/andp.19524450802. See also Schlüter's review article of Alfvén's *Cosmical Electrodynamics*, discussing magnetic fields in cosmic physics and plasma-related phenomena, which remained for years a classic in the foundations of plasma physics and plasmas in space: Arnulf Schlüter: Besprechungen. Cosmical Electrodynamics von H. Alfvén. Oxford. Clarendon Press, 1950. *Zeitschrift für Naturforschung A* 6/1 (1951), 55–56. doi:10.1515/zna-1951-0111.

137 Plasma is an ionized gas, where spatial separation of positive and negative charges can create electric currents (electrons, which have a smaller mass, tend to be accelerated much more than the ions for given conditions) which, in turn, can lead to growing magnetic fields due to a mechanism later named 'Biermann Battery,' first proposed by Biermann as a mechanism for the thermal generation of stellar magnetic fields. Ludwig Biermann: Über den Ursprung der Magnetfelder auf Sternen und im interstellaren Raum. *Zeitschrift für Naturforschung A* 5/2 (1950), 65–71. doi:10.1515/zna-1950-0201. This article was the result of work carried out by Biermann during the war (see footnote 1 in the article) and contains an Appendix written by Schlüter. See also Biermann, Entstehung von Magnetfeldern, 1952, 413–417.

being a subject of interest both for astrophysics *and* plasma physics, which since the early 1950s was being studied in (still secret) laboratories dedicated to the development of thermonuclear fusion devices.[138]

Yet even those entities esoterically named 'cosmic rays' by Robert Millikan in the 1920s were deeply embedded in dual-use potential in the first post-war decade: we will see in Section 2 of this chapter how research in the early 1950s on cosmic ray interactions in the atmosphere was also methodologically linked to nuclear bomb tests.

Plasma Astrophysics in the Nuclear Age

Under the guidance of von Weizsäcker and Biermann, mathematicians such as Arnulf Schlüter and astronomers such as Sebastian von Hoerner (a key protagonist in Chapter 3)[139] were recruited from the nearby University of Göttingen, which was to become one of the main feeders of the Max Planck Institute for Physics.[140] Von Weizsäcker's influence can be seen also in articles by Eleonore

138 An example of the link between concepts familiar to cosmic ray physicists and research on laboratory plasmas is the reflection of charged particles spiraling along magnetic field lines as they move into regions of increasing density of the lines in what is known as the 'mirror machine,' a type of magnetic confinement device based on the magnetic trap, one of the earliest approaches to fusion power, along with the stellarator and the Z-pinch machines. This phenomenon is typical of charged particles spiraling along the geomagnetic field, which are repeatedly reflected back and forth along geomagnetic field lines by opposite mirror points in the two hemispheres, such as those that constitute the Van Allen radiation belts and the ring current carried by charged particles trapped in a planet's magnetosphere.

139 Von Hoerner had received his PhD under von Weizsäcker, working with him in cosmic hydrodynamics, on turbulence and shock fronts in astrophysical plasmas: "Actually, he is the reason I came to astrophysics. I wanted to be his student no matter which field." Sebastian von Hoerner: interview by Woodruff T. Sullivan III, Montreal, August 20, 1979. Papers of Woodruff T. Sullivan III., Tapes Series, National Radio Astronomy Observatory Archives, https://www.nrao.edu/archives/Sullivan/sullivan_transcript_vonhoerner_1977.shtml. Last accessed 7/22/2020. In 1951 and in 1956 von Hoerner got a Fulbright fellowship, which he spent at Mt. Wilson and Mt. Palomar observatories in California. Richard Wielebinski: Sebastian von Hoerner. *Mitteilungen der Astronomischen Gesellschaft* 86 (2003), 9–10. http://adsabs.harvard.edu/abs/2003MitAG..86....9. Last accessed 10/30/2018.

140 The group around Biermann grew during the 1950s from three scientific assistants to seven by 1955, and to nine by 1960. The computing group (three members plus a mathematician) remained constant during this period (AMPG, II. Abt. Rep. 66, No. 3214). On scientific interaction between Biermann and von Weizsäcker and research activities going on at the Institute for Astrophysics (involving their young collaborators Arnulf Schlüter, Sebastian von Hoerner, Eleanore Trefftz, Stephan Temesváry) see, for example, letters exchanged between the two in fall 1949 (AMPG, III. Abt., ZA 1, No. 4). Temesváry

Trefftz,[141] who had a PhD in theoretical physics from Dresden Technical University, and became a member of Biermann's Department for Astrophysics in 1948, beginning to work on calculations made with computing machines on atomic and molecular spectral lines that were especially interesting for astrophysical purposes.[142]

Reimar Lüst himself, a future President of the Max Planck Society and a central figure in this book, arrived in Göttingen from Frankfurt in 1949, after gaining his diploma in physics. During his very first visit to the institute, after a conversation with von Weizsäcker, he was invited to take part in a seminar:

> The lecturer was Arnulf Schlüter, presenting his first work on plasma physics which later became important for my whole scientific work; von Weizsäcker accepted me as a doctoral student. At first, he wanted to give me a theme regarding the general theory of relativity, but the experts said this was too difficult. I was therefore provided with another problem which I found more interesting, namely the question "What had slowed down the Sun's rotation? How had the angular momentum been transported?" For the Sun rotates relatively slowly in our planetary system,

---

had received his PhD in Heidelberg in astronomy and after the war had joined Karl-Otto Kiepenheuer's group at the Schauinsland Observatory, which was engaged in solar observations. In 1949 Biermann had offered him a stipend at the Max Planck Institute for Physics in Göttingen. Trefftz, Eleonore: Obituary—Temesvary, S. *Quarterly Journal of the Royal Astronomical Society* 27/1 (1986), 129–130. http://adsabs.harvard.edu/abs/1986QJRAS..27..129T. Last accessed 1/13/2020.

141 Eleonore Trefftz: Zur Entwicklung einer rotierenden Gasmasse. *Zeitschrift für Naturforschung A* 7/1 (1952), 99–103. doi:10.1515/zna-1952-0119.

142 Eleonore Trefftz had studied physics and mathematics at Dresden Technical University and Leipzig University, gaining a PhD in theoretical physics in 1945. She had excellent teachers, such as the mathematician Bartel L. van der Waerden and the physicist Friedrich Hund, who supervised her early research work. Eleonore Trefftz: Zur Statistik der Mischkristalle und Ferromagnetica. *Zeitschrift für Physik* 127/4 (1950), 371–380. doi:10.1007/BF01329834. She was an assistant in Dresden before moving to Biermann's Department in 1948, where she began work on wave functions and transition probabilities in atoms of astrophysical interest. Ludwig Biermann, and Eleonore Trefftz: Wellenfunktionen und Übergangswahrscheinlichkeiten der Leuchtelektronen des Atoms Mg I. I. Teil. *Zeitschrift für Astrophysik* 26 (1949), 213–239. http://adsabs.harvard.edu/abs/1949ZA.....26..213B. Last accessed 11/2/2017. For many years Trefftz led the Department for Quantum Physics at the Institute for Astrophysics, also extending into quantum chemistry, which was important for the physics of cometary nuclei. See Trefftz's own review article on related research activities in Eleonore Trefftz: Zur Berechnung der Eigenschaften von Atomen und Molekülen. *Mitteilungen aus der Max-Planck-Gesellschaft* 5/73 (1973), 311–320.

ILLUSTRATION 37 Göttingen, early 1950s, from left: Ms. Kugel, Reimar Lüst, Ms. Schulten, Stefan Temesváry (with a bicycle), Eleonore Trefftz (Courtesy of Milian Trefftz)

> while most of the total angular momentum of the solar system resides in Jupiter. So, my task was to calculate, using hydrodynamical equations, whether such an angular momentum transfer was actually feasible in a gas disk.[143]

In his doctoral thesis Lüst was the first to make practical use of von Weizsäcker's hydrodynamical equations that the latter had formulated in the aforementioned work on the origin of the planetary system, published in 1948.[144]

---

143 Reimar Lüst: interview by Hans von Storch and Klaus Hasselmann, December 2, 2000. Transcript, AIP, https://www.aip.org/history-programs/niels-bohr-library/oral-histories/33761. Last accessed 5/8/2019. See also Reimar Lüst, and Paul Nolte: *Der Wissenschaftsmacher. Reimar Lüst im Gespräch mit Paul Nolte*. München: C.H. Beck 2008.

144 Reimar Lüst: Die Entwicklung einer um einen Zentralkörper rotierenden Gasmasse. I. Lösungen der hydrodynamischen Gleichungen mit turbulenter Reibung. *Zeitschrift für Naturforschung A* 7/1 (1952), 87–98. doi:10.1515/zna-1952-0118. This work appeared in a special issue of the journal *Zeitschrift für Naturforschung*, dedicated to Heisenberg's 50th birthday. A very similar topic was tackled by Eleonore Trefftz on the same occasion: Trefftz, Zur Entwicklung, 1952, 99–103.

Throughout the early 1950s, while still subject to harsh economic conditions and Allied restrictions on any research with potential direct military application, Biermann and his colleagues became key participants in the worldwide scientific community of stellar astrophysicists. In the early postwar years, Biermann continued his astrophysical work on the Sun, extending his interest to the behavior of charged particles once they left the Sun itself, looking at the collective behavior of charged particles in the solar corona and space beyond, including Earth–Sun interaction. Back in 1948, Biermann and Erich Bagge, who had been studying nuclear processes in cosmic rays in Heisenberg's group since the early 1940s,[145] wrote about the origin of solar cosmic rays (Bagge is a central 'shadow figure' in this book: see Chapters 1, 3, and 5).[146]

This article marked the start of Biermann's longstanding and articulated interest in the emission of charged particles from the Sun, which would later provide the scientific platform for launching space activities within the future Max Planck Institute for Extraterrestrial Physics.

---

145 Bagge had been Heisenberg's student in Leipzig for his doctorate and *Habilitation* ("Beiträge zur Theorie der schweren Atomkerne und Kernzertrümmerungen", 1938; "Schwere Teilchen in der kosmischen Strahlung", 1941). Erich Bagge: Beiträge zur Theorie der schweren Atomkerne. I. Zur Frage des Neutronenüberschusses in den schweren Atomkernen. *Annalen der Physik* 425/4 (1938), 359–388. doi:10.1002/andp.19384250406. Erich Bagge: Kernzertrümmerungen und schwere Teilchen in der kosmischen Strahlung. *Naturwissenschaften* 29/21 (1941), 318–318. doi:10.1007/BF01479547. He was later Heisenberg's collaborator in the nuclear war project and moved to Hamburg in 1948, now a professor, continuing to cultivate cosmic ray research in parallel with nuclear physics. As we will see in later chapters, two of his students, Klaus Pinkau and Joachim Trümper, would typically move from cosmic ray physics to cosmic ray astronomy, opening the new windows of gamma and X-ray astronomy at the Institute for Extraterrestrial Physics. For a comment on Heisenberg's group of articles of the 1930s on cosmic ray phenomena and related theoretical problems, see Bagge's annotation in Heisenberg, *Heisenberg. Collected Works A/2*, 1985, Vol. 2, 239–249.

146 Erich Bagge, and Ludwig Biermann: Die Erzeugung von Ultrastrahlung auf der Sonne. *Die Naturwissenschaften* 35/4 (1948), 120–121. doi:10.1007/BF00626776. Erich Bagge, and Ludwig Biermann: Über die Entstehung der solaren Komponente der Ultrastrahlung. *Zeitschrift für Naturforschung A* 4/4 (1949), 303–315. doi:10.1515/zna-1949-0410. They thanked Alfred Ehmert (Regener's collaborator) for showing them his results before publication. Alfred Ehmert: Ultrastrahlung von der Sonne. *Zeitschrift für Naturforschung A* 3/5 (1948), 264–285. doi:10.1515/zna-1948-0504. Erich Bagge: Die Sonne und die Fixsterne als Quellen kosmischer Strahlung. *Il Nuovo Cimento* 6/3 (1949), 327–329. doi:10.1007/BF02822006. See also Brüche, Physiker-Tagung in Göttingen, 1947, 317–325. The problem of solar influence on the flux of cosmic rays was also investigated within Biermann's group. E.G. v. Roka: Sonnenaktivität und kosmische Strahlung. *Zeitschrift für Naturforschung A* 6/3 (1950), 117–122. doi:10.1515/zna-1951-0301. Of the same period, see also Ludwig Biermann, Otto Haxel, and Arnulf Schlüter: Neutrale Ultrastrahlung von der Sonne. *Zeitschrift für Naturforschung A* 6/1 (1951), 47–48. doi:10.1515/zna-1951-0107.

ILLUSTRATION 38 Erich Bagge working on cosmic rays in Göttingen, end of the 1940s.

Clearly advantageous, during the late 1940s and 1950s, was that Biermann's theoretical approaches to plasma physics could be checked in the light of existing astronomical data. And thanks to his theoretical insights, well-known observations would sometimes lead to brilliant new discoveries. This was most prominently the case with his hypothesis of 1951, that the Sun emits a constant large flow of particles as a plasma embedding the solar magnetic field.[147] Until the early 1950s, it was believed that the space occupied by the solar system was a vacuum, that corpuscular streams occasionally emitted by the Sun disturbed the geomagnetic field, and that aurorae were thus produced by fast charged particles from the Sun speeding along magnetic field lines and interacting with atoms in the upper atmosphere, while plunging into the Earth at the magnetic

147 Ludwig Biermann: Kometenschweife und solare Korpuskularstrahlung. *Zeitschrift für Astrophysik* 29 (1951), 274–286. https://ui.adsabs.harvard.edu/#abs/1951ZA.....29..274B. Last accessed 10/30/2018. Ludwig Biermann: Über den Schweif des Kometen Halley im Jahre 1910. *Zeitschrift für Naturforschung A* 7/1 (1952), 127–136. doi:10.1515/zna-1952-0122. Ludwig Biermann: Solar Corpuscular Radiation and the Interplanetary Gas. *The Observatory* 77 (1957), 109–110. https://ui.adsabs.harvard.edu/abs/1957Obs....77..109B/abstract. Last accessed 8/14/2020.

poles. In the late 1930s, the Swedish scientist Hannes Alfvén had proposed that plasmas usually pervade the interplanetary and interstellar space and can carry electric currents capable of generating galactic magnetic fields.[148] The notions that the streams of plasma flowing from our star transport the solar magnetic field and that magnetized plasmas fill the whole Universe were only gradually accepted—or even disputed by scientists in space physics—but they were attentively considered by Ludwig Biermann. Already in the early 1950s, through studying the radial distribution of comet tails in space, based on material he obtained from his national and international network of colleagues, Biermann concluded that there must be a *continuous* emission of solar plasma throughout the interplanetary medium, and not simply intermittent bursts.[149] This process would predict the shape of a comet's tail: the interaction between the ionized gases in a comet's tail and the stream of solar particles distorts magnetic field lines, giving rise to a comet's 'magnetotail,' which points outward, away from the Sun.[150] Biermann's kinematic view was contradicted by Sidney Chapman's theory of a solar corona in static equilibrium, consisting of electrons and protons and extending beyond the Earth's orbit.[151] At that time, as of September 1955, Reimar Lüst spent an entire year in the United States working with John Simpson, a well-known cosmic ray physicist based at the Fermi Institute for Nuclear Studies in Chicago, who had developed the detector for neutrons produced by incoming cosmic rays.[152] When Lüst was invited by Simpson,

---

148 R. S. Pease, and S. Lindqvist: Hannes Olof Gösta Alfvén. 30 May 1908–2 April 1995. *Biographical Memoirs of Fellows of the Royal Society* 44 (1998), 3–19. doi:10.1098/rsbm.1998.0001.

149 See Biermann's correspondence on this topic from 1946 (AMPG, III. Abt., Rep. ZA 1, No. 1).

150 Ludwig Biermann: Physical Processes in Comet Tails and Their Relation to Solar Activity. In: P. Swings (ed.): *La Physique Des Comètes. Communications Présentées Au Quatrième Colloque International d'Astrophysique, Tenu à Liège Les 19, 20 et 21 Septembre 1952. Mémoires de La Société Royale Des Sciences de Liège. Quatrième Série*. Liège: Institut d'Astrophysique de l'Université de Liège 1953, 251–262. Biermann, Solar Corpuscular Radiation, 1957, 109–110.

151 Sydney Chapman, and Harold Zirin: Notes on the Solar Corona and the Terrestrial Ionosphere. *Smithsonian Contributions to Astrophysics* 2 (1957), 1–14. http://adsabs.harvard.edu/abs/1957SCoA....2....1C. Last accessed 2/2/2020. About some of Chapman's views on magnetospheric physics see Syun Akasofu: Chapman and Alfvén. A Rigorous Mathematical Physicist Versus an Inspirational Experimental Physicist. *EOS Transactions* 84 (2003), 269–274. doi:10.1029/2003EO290002.

152 Ludwig Biermann, and Carl F. Weizsäcker: Jahresberichte deutscher Sternwarten und Institute für 1955, Göttingen, Max-Planck-Institut für Physik. *Mitteilungen der Astronomischen Gesellschaft* 7 (1956), 106–108, 106. http://articles.adsabs.harvard.edu/pdf/1956MitAG...7..106. Last accessed 10/30/2018.

> [...] he was already known for his work on particle propagation in the geomagnetic field—a most important tool for the analysis of solar-flare nuclei. He joined in the work at Chicago on charged-particle trajectory calculations and the important questions of the acceleration mechanisms and magnetohydrodynamics that must underlie the startling solar-flare phenomena.[153]

Studies on the particles coming from the giant solar flare of 1956 provided evidence of magnetic fields in space and of the plasma in which the field is embedded.[154] In this regard, Lüst had the occasion to talk about Biermann's ideas with Eugene Parker, who was working in Chicago on his theory of the solar corona and its production of the interplanetary medium, and later Biermann himself discussed the problem with Parker, during a stay at the Chicago Institute.[155] Both Biermann's and Chapman's conclusions were firmly based on well-established observations and basic theoretical inferences and thus Parker suggested that the solar atmosphere must expand continually into space, filling the whole solar system and generating a high-velocity radial flow with speeds of hundreds of kilometers per second. Parker incorporated Biermann's "stream of particles flowing out from the Sun at high speeds" and Sydney Chapman's static solar atmosphere extending beyond the Earth into a theory in which the solar corona is continuously expanding.[156] In the early 1960s, Parker's—and thus Biermann's—theory was confirmed first by the Russian rockets and later, with more dedicated detectors, by US space probes.[157] This

---

153 Lüst was thus considered "an invaluable collaborator," since he had extensive experience in calculating charged particle trajectories in the geomagnetic field with early electronic computers. Peter Meyer, and John A. Simpson: Reminiscences of Solar Flares and the Chicago Years of Reimar Lüst. *Topics in Plasma-, Astro- and Space Physics. A Volume Dedicated to Reimar Lüst on the Occasion of His 60th Birthday*. München: Max-Planck-Institut für Physik und Astrophysik, Institut Extraterrestrische Physik 1983, 117–134, 121.

154 Reimar Lüst, and John A. Simpson: Initial Stages in the Propagation of Cosmic Rays Produced by Solar Flares. *Physical Review* 108/6 (1957), 1563–1576. doi:10.1103/PhysRev.105.1827.

155 See correspondence with Parker in AMPG, III. Abt., ZA 1, No. 42.

156 Eugene Parker: Dynamics of the Interplanetary Gas and Magnetic Fields. *Astrophysical Journal* 128 (1958), 664–676. doi:10.1086/146579. Eugene N. Parker: Coronal Expansion and Solar Corpuscular Radiation. In: C. C. Chang, and S. S. Huang (eds.): *Proceedings of the Plasma Space Science Symposium. Held at the Catholic University of America Washington, D.C.*, June 11–14, 1963. New York, NY: Springer 1965, 99–114.

157 K.I. Gringauz et al.: Results of Observations of Charged Particles Observed Out to R = 100,000 Km, with the Aid of Charged-Particle Traps on Soviet Space Rockets. *Soviet Astronomy* 4 (1961), 680–695. https://ui.adsabs.harvard.edu/#abs/1961SvA.....4..680G. Last

stream of particles later became known as the solar wind, and would be one of the first important confirmations of Biermann's theories obtained with the use of space-based probes in the early years of spaceflight.[158] Such predictive successes brought Biermann and his group at the Institute for Astrophysics notable scientific prestige in the earliest days of the nascent space age. As recalled by Thomas Cowling and Leon Mestel, Biermann's interest in cosmical electrodynamics "made his group one of the foremost of those working to elucidate the properties of magnetized plasmas."[159]

After six months in Chicago, Lüst went to Princeton University Observatory because he "also wanted to learn from Martin Schwarzschild," a great astrophysicist, the son of Karl Schwarzschild.[160] Later, he was invited to spend a year in Chicago by the German mathematician Richard Courant, who had been dismissed from his position as Director of the Mathematical Institute in Göttingen in 1933:

---

accessed 10/30/2018. H.S. Bridge et al.: Direct Observations of the Interplanetary Plasma. *Journal of the Physical Society of Japan* 17/Supplement A-II (1962), 553–559. C.W. Snyder, M. Neugebauer, and U.R. Rao: The Solar Wind Velocity and Its Correlation with Cosmic-Ray Variations and with Solar and Geomagnetic Activity. *Journal of Geophysical Research* 68/24 (1963), 6361–6370. doi:10.1029/JZ068i024p06361.

158 Ludwig Biermann: On the History of the Solar Wind Concept. In: Wilfried Schröder, and International Association of Geomagnetism and Aeronomy (eds.): *Historical Events and People in Geosciences. Selected Papers from the Symposia of the Interdivisional Commission on History of IAGA during the IUGG General Assembly, Held in Hamburg, 1983*. Frankfurt am Main: Peter Lang 1985, 39–47.

159 Cowling, and Mestel, Biermann, 1986, 698–700, 699. On such topics, Biermann was also lecturing at Göttingen University (see 150-page typescript "Kosmische Elektrodynamik" dated summer semester 1954 in Lüst's papers, AMPG, III. Abt., Rep. 145, No. 1163). Biermann's early interest on the interaction between plasma tails and the solar wind developed into a longstanding engagement in the physics of comets that eventually led in the involvement of MPA in preparation and data analysis for the Giotto mission in the 1980s. ESA's first deep-space mission, launched in March 1986 to encounter and study the Halley's Comet, provided the first pictures ever of a cometary nucleus, also confirming theoretical work done by Biermann's group in the early 1960s.

160 Both Simpson and, in particular, Schwarzschild influenced Lüst: "I spent half a year working with him. He was an especially open, forthcoming person. The most remarkable aspect was that Schwarzschild as well as his co-director Spitzer, who played a major role in fusion, were Jews. Nevertheless, they accepted me, a German. Those were the two persons I learnt a lot of new things from, who influenced me in their way of doing physics. In Chicago, I had adopted the habit there of always leaving the door to my office open, and I introduced that in Garching later: to always keep the doors open." Reimar Lüst: interview by Hans von Storch and Klaus Hasselmann, December 2, 2000. Transcript, AIP, https://www.aip.org/history-programs/niels-bohr-library/oral-histories/33761. Last accessed 5/8/2019.

> Courant and Friedrich had written a book on shock waves. I did not realize until later why Courant and Friedrich dealt with shock waves. It was connected with the development of the atomic bomb during the war. In 1953, I had written a paper on hydromagnetic shock waves which was published in *Zeitschrift für Naturforschung* [A Journal of Physical Sciences].[161] It was the first paper ever on this problem. Courant and Friedrich had seen it and therefore invited me and asked whether I would like to work at the Courant Institute for a year.[162]

In the first postwar decade, Biermann's group work on space plasmas was thus primarily admired abroad for what it said about behaviors that played a role not only in astrophysics, but also in thermonuclear processes applicable to fusion bombs and reactors: in fact, many design features of early thermonuclear reactors were based on direct analogies with phenomena first studied in astrophysical contexts.[163] In the United States, also thanks to his experience with the G1 computer built by Billing, Lüst had an opportunity to work with big electronic computers, such as the AVIDAC at the Argonne National Laboratory, and to calculate the orbits of cosmic ray particles starting in the vicinity of the Sun and passing through the geomagnetic field.[164] Early computers, such

161 Reimar Lüst: Magneto-hydrodynamische Stoßwellen in einem Plasma unendlicher Leitfähigkeit. *Zeitschrift für Naturforschung A* 8/5 (1953), 277–284. http://zfn.mpdl.mpg.de/data/Reihe_A/8/ZNA-1953-8a-0277.pdf. Last accessed 11/2/2017.

162 Reimar Lüst: interview by Hans von Storch and Klaus Hasselmann, December 2, 2000. Transcript, AIP, https://www.aip.org/history-programs/niels-bohr-library/oral-histories/33761. Last accessed 5/8/2019. During his stay at Princeton in 1955, Lüst had also worked on plasma shock waves using the powerful electronic computer available there. Preliminary calculations had been done previously by Schlüter, and Martin Schwarzschild was thanked for suggesting writing the article and for the discussions. Reimar Lüst, and M. Scholer: Kompressionswellen in Einer Isothermen Atmosphäre Mit Vertikalem Magnetfeld. *Zeitschrift Für Naturforschung A* 21/7 (1966), 1098–1106. doi:10.1515/zna-1966-0734.

163 Gary J. Weisel: Properties and Phenomena. Basic Plasma Physics and Fusion Research in Postwar America. *Physics in Perspective* 10/4 (2008), 396–437. doi:10.1007/s00016-007-0371-1. Gary J. Weisel: The Plasma Archipelago. Plasma Physics in the 1960s. *Physics in Perspective* 19/3 (2017), 183–226. doi:10.1007/s00016-017-0205-8. A clear sign of how Biermann was held in esteem by the international scientific community can be found in a letter to him penned at the Yerkes Observatory in the US by the astrophysicist Subrahmanyan Chandrasekhar, who would be awarded the Nobel Prize in Physics 1983 for his studies on the structure and evolution of the stars: "I need not say that if I accept [an invitation by ten Bruggencate to spend a year in Göttingen] it would be very largely because of you and the Max-Planck-Institut being at Göttingen." Chandrasekhar to Biermann, 05.08.1956, AMPG, III. Abt., ZA 1, No. 41.

164 Reimar Lüst: Impact Zones for Solar Cosmic-Ray Particles. *Physical Review* 105/6 (1957), 1827–1839. doi:10.1103/PhysRev.105.1827.

as the MANIAC, had been developed in the US primarily for the purpose of making the calculations required to build the hydrogen bomb. Physicists such as John Wheeler in the US and Yakov Zeldovich in the USSR, who had each worked on nuclear weapon projects, easily spotted in the early 1950s that the physics of stars—particularly very dense stars—and the physics of a nuclear explosion have much in common. Already in the 1950s, Stirling A. Colgate and Montgomery H. Johnson—working at the Livermore Laboratory, in California, where Teller had his general headquarters for the development of the H-bomb project—had conducted precise and extensive simulations to investigate the outcomes of an H-bomb explosion. Computer calculations using a hydrodynamic code which had been modified to include gravitation found that an H-bomb is quite similar to a supernova explosion, and the material spalled from the surface is the source of cosmic radiation.[165] At the same time, well-known astrophysical problems were being used to test computers used for thermonuclear research, as recalled by Martin Schwarzschild:

> Von Neumann was very interested to have a problem which was non-linear and sufficiently complicated to really need the whole power of his machine, but where lots of hand computations for checks were available; and therefore the stellar evolution work, which I think von Neumann also considered interesting in itself, though not all that deeply—he thought that that was an excellent one. So, actually, next to the official major program, the meteorological dynamics for which the machine officially was funded, stellar evolution, with its implicit thermonuclear inquiries, got the biggest share of time.[166]

---

165 Their attempt to understand the mechanism of a supernova explosion was later followed by systematic studies in which Colgate, in collaboration with Richard White, created models of collapsing stars by combining equations of state of superdense matter with software used to design bombs. Their work showing that stars really could undergo an ongoing and endless catastrophic collapse, and also confirming the enormous release of neutrinos into space, was eventually published in 1966. Stirling A. Colgate, and Richard H. White: The Hydrodynamic Behavior of Supernovae Explosions. *Astrophysical Journal* 143 (1966), 626–681. doi:10.1086/148549.

166 Martin Schwarzschild: interview by David DeVorkin and Spencer Weart, December 16, 1977, Session III. Transcript, AIP, https://www.aip.org/history-programs/niels-bohr-library/oral-histories/28321. Last accessed 5/8/2019. Schwarzschild also remembered that Wheeler was joined for one year by the theoretical astronomer Louis Henyey, who spent the year 1951–52 at Princeton University, where he was involved in the classified defense work on Project Matterhorn, the USA's top-secret project to control thermonuclear reactions. Martin Schwarzschild: interview by William Aspray, Princeton, November

Through all such developments, in this decade, space plasma physicists such as Biermann, working largely with theoretical tools aiming to interpret known observations, changed scientists' understanding of outer space: instead of the endless void, there now emerged a more dynamic picture full of particles and fields in complex interaction, which might well prove to be a permanently fertile area of research.[167] Such a connection to a new scientific network would have been quite advantageous even if this had been an isolated 'pure' scientific niche. But research on thermonuclear fusion still had its scientific roots in astrophysics, in particular in the subfield plasma physics, including magnetohydrodynamics, the study of its magnetic and electric behavior, which in turn was directly connected to the problem of the magnetic confinement of hot plasmas in fusion reactors. And so, one of the achievements that won the Göttingen astrophysicists worldwide fame was their results in plasma physics theory, which had in fact been previously developed, but kept secret, by researchers in the United States and the Soviet Union. American interest in what Biermann's team was doing led to an altogether different scale of cooperation, since the skill set related to this scientific field transferred directly to knowledge useful for fusion reactors and thermonuclear weapons. In fact, the majority of American partners in the scientific conversation on stellar astrophysics were involved at the time also in classified research. Most notably, Lyman Spitzer, the key international contact for plasma researchers at the Max Planck Institute, worked as an astrophysicist at Princeton, and simultaneously ran the secret thermonuclear reactor program based there, the focus of which was the stellarator design of fusion reactors. An expert in star formation and plasma physics, Spitzer had devised a new concept to confine a plasma for long periods, and was the founding director of the Project Matterhorn, which early code name covered the secret fusion research at Princeton, a pioneering program in thermonuclear research. Through the early involvement of German researchers there, the stellarator design would become one

18, 1986. Transcript. N. J. Charles Babbage Institute. Retrieved from the University of Minnesota Digital Conservancy, http://hdl.handle.net/11299/107629. Last accessed 7/7/2019. In turn, Henyey realized that what he had learned at Princeton from von Neumann was extremely useful for the stellar interior and developed what came to be known as the aforementioned 'Henyey method,' which became the standard tool for the theory of stellar interior. Henyey et al., Automatic Computation, 1959, 628–636.

167 Ludwig Biermann: Plasmaphysik im Kosmos und für die Fusion. In: Max-Planck Gesellschaft (ed.): *Ludwig Biermann. 1907–1986.* München 1988, 63–76. Reimar Lüst: Terrestrische und extraterrestrische Plasmen. *Die Naturwissenschaften* 62/6 (1975), 255–263. doi:10.1007/BF00608951.

of the key specialties of the Max Planck Society in later decades.[168] At Princeton, access to the computing machine at the Institute for Advanced Study had been offered by the mathematician John von Neumann, who built it, to the astrophysicist Martin Schwartzschild, for the purpose of complicated calculations on gravitational contraction and the evolution of very dense and massive stars; a testing ground, namely, to establish its potential.[169] All this clearly explains why the Göttingen astrophysicists, already well-known correspondents in the field, were invited in significant numbers to the United States. Yet, while brought in nominally for collaboration on astrophysical problems, their hosts used their guests also to assess the state of research in plasma-related fields in West Germany, at times also recruiting the most brilliant among them.[170] Many of the major figures who would later become a Max Planck Institute director (Biermann, Schlüter), a scientific member (Eleonore Trefftz), or even president (Reimar Lüst), visited the main fusion-related research sites in the United States during the 1950s.[171] Biermann, in particular, was often in

168 Biermann and Spitzer discussed the German stellarator project when they met in Geneva during the second 'Atoms for Peace' conference, in September 1958 (Spitzer to Biermann, 05.02.1959, AMPG, III. Abt., ZA 1, No. 42).

169 Martin Schwarzschild: interview by David DeVorkin and Spencer Weart, December 16, 1977, Session III. Transcript, AIP, https://www.aip.org/history-programs/niels-bohr-library/oral-histories/4870-3). Last accessed 5/8/2019.

170 See Lyman Spitzer's recollections of the close connection between work on fusion and astrophysical problems, which afforded opportunities to invite well-known astrophysicists with expertise in astrophysical plasmas: "We had a number of people here as visiting lecturers during this period. Alfvén came for a while, and Ludwig Biermann from the Max Planck Institute, and Arnulf Schuelter [sic.]. They didn't know what was going on out at Project Matterhorn (as the Plasma Lab was called until declassification occurred in 1958). We couldn't tell them. They would give lectures on various problems, on the relations between plasma and magnetic fields, and we would sit and take notes, and then rush them out to Matterhorn." Lyman Spitzer: interview by Joan Bromberg, March 15, 1978. Transcript, AIP, https://www.aip.org/history-programs/niels-bohr-library/oral-histories/4900. Last accessed 5/8/2019. Conversely, as recalled by Lüst, "I was at Princeton Observatory for six months in 1956. The director, Lyman Spitzer, was the head of fusion research at Princeton, but I didn't hear a thing about it; it was strictly classified information, although every now and then he would invite me to have a chat. I had just done work on magnetohydrodynamic shock waves that was relevant to it, but he never revealed himself, and then when I returned to Göttingen in December of 1956, the first question from Heisenberg was: 'Did you hear about nuclear fusion at Princeton?' I say, 'Nothing; not a word about it' [laughs]" [our translation]. Reimar Lüst: interview by Horst Kant and Jürgen Renn, Hamburg, May 18, 2010 (DA GMPG, ID 601068).

171 Trefftz visited Ohio University as early as 1951 and later moved to the Institute for Advanced Study (IAS) in Princeton, where the IAS machine, one of the first large-scale computers, was beginning to work. A very similar machine, the MANIAC, had been devel-

Princeton, and in other US centers, notably the California Institute of Technology in Pasadena, and later, Boulder, Colorado, as well as Washington D. C. In practice, Biermann visited the United States every year from 1954 on, establishing a multitude of personal contacts.[172] As he later remarked, this also allowed him "to see his own institute constantly from outside [...]."[173] From the early 1950s, Heisenberg himself attracted many visitors and young theoreticians from abroad, who helped establish new interaction channels with physicists outside Germany.[174]

From Theoretical Astrophysics to Experimental Plasma Physics

It was not true that astrophysics in Göttingen was 'just' a cover for prohibited nuclear research. But it was not an entirely innocent endeavor, either, as was indeed the case with several other research fields where scientific excellence thrived in relative obscurity, including other areas of astrophysics, such as topics related to gravitational cosmology. Instead, as we will see throughout this volume, plasma astrophysics fits with the overall trajectory of many research institutions in Germany in the first postwar decade, seeking scientific excellence within the constraints of economic scarcity and prohibitions, while

---

oped by Nicholas Metropolis at Los Alamos laboratories. The main use of these early computers was, of course, weapon simulations, particularly thermonuclear weapons. Trefftz's trip marked the start of a strong relationship with US scientists, which became more intense around the mid-1950s, when Trefftz and other members of Biermann's group visited the UK and the US. From 1958, Trefftz led the Department of Quantum Mechanics for many years before finally becoming a Scientific Member of the Max Planck Society in 1971 (CPTS meeting minutes of 09.02.1971, 23.06.1971, 22.10.1971, AMPG, II. Abt., Rep. 62, No. 1761, 1762, 1763). For an outline of the work of the group, see Trefftz, Eigenschaften von Atomen und Molekülen, 1973, 311–320.

172 Biermann was invited to give talks on astrophysical plasmas and magnetohydrodynamics at the California Institute of Technology in Pasadena, at Haverford College in Pennsylvania, and at Princeton University. See the annual reports and especially Biermann's papers for correspondence relating to his interaction with foreign scientific centers, travel, and his collaborators' longer sojourns abroad: AMPG, III. Abt., ZA 1, No. 28 (Correspondence with Foreign Countries 1946–1961), No. 33 (Correspondence with Belgium, Holland/the Netherlands, and Sweden), No. 34 (Correspondence England 1946–59), No. 41 (Correspondence USA 1956–1960, A–K), No. 42 (Correspondence USA 1956–1960, L–Z). See also his course notes "Astrophysical Theory of Stellar Electromagnetism and Plasma Physics," Second Term 1954–1955, English typescript preserved in Lüst's papers, AMPG, III. Abt., Rep. 145, No. 1203.

173 Ludwig Franz Benedikt Biermann: interview by Martin Harwit, March 16, 1984. Transcript, AIP.

174 See short note on the growing number of contacts with physicists all over the world in Generalverwaltung der Max-Planck-Gesellschaft (ed.): *Mitteilungen aus der Max-Planck-Gesellschaft. Heft 3*. Göttingen 1952, 31–32.

preparing the field for the opportunities for expansion that would arise once restrictions were lifted and the economic situation permitted a larger scale of involvement; in the case of Göttingen, it was the prospect of experimental programs in fields in which theoretical expertise had been built up during the first postwar decade.

Theoretical plasma physics turned out to be the field in which the most interesting, promising science could be conducted in the first postwar decade, and this provided a foothold that helped sustain the scientific reputation of the Max Planck Society for the rest of the 20th century. Furthermore, as will be seen later, space plasma astrophysics also created a close-knit community of researchers (that went on to hold disproportionate power within the Society), and even fostered a mode of governance that coordinated the work of several independent Max Planck Institutes. In the immediate postwar years, during the 1940s and early 1950s, space plasmas had been a secondary pursuit at Heisenberg's institute, at a time when its director harbored more explicit intentions in the nuclear realm.[175] The reestablishment of the institute, like that of the Max Planck Society itself, had been possible due to the recently acquired prestige of nuclear fission. Otto Hahn, the first President, was one of the discoverers of nuclear fission and had just been awarded the Nobel Prize 1944 for this work. Heisenberg, awarded Nobel Prize 1932 for his contributions to quantum theory, had coordinated Germany's research into nuclear fission in the last years of the war.[176] Then, during the first postwar decade, Heisenberg aspired to become the person who would lead West Germany toward the peaceful uses of nuclear energy, lobbying extensively with the Allied powers and the new federal government to be in charge of this mission.[177]

---

175 See, for example, Schlüter, Dynamik II, 1951, 73–78. Biermann, Entstehung von Magnetfeldern, 1952, 413–417. Lüst, Magneto-hydrodynamische Stoßwellen, 1953, 277–284.

176 Mark Walker has stressed how the change in the Nazi regime's attitude to Heisenberg (who had been attacked in 1937 by National Socialist scientists promoting the *Deutsche Physik* and was generally considered politically unreliable) was due to the rehabilitation of modern physics and the great interest in nuclear power, which improved his position to the point that in June 1942 he was appointed Director of the Kaiser Wilhelm Institute for Physics in Dahlem as well as to a professorship at the University of Berlin. Walker, Physics and Propaganda, 1992, 339–389, 372.

177 Michael Eckert: Heisenberg and the Beginnings of Nuclear Energy in the FRG. In: Michelangelo De Maria, Mario Grilli, and Fabio Sebastiani (eds.): *Proceedings of the International Conference on The Restructuring of Physical Sciences in Europe and the United States 1945–1960. Università "La Sapienza", Rome, Italy, 19–23 September 1988*. Singapore: World Scientific 1989, 247–256. The underwhelming outcome of Heisenberg's ambitions in nuclear fission is described further in Chapter 3.

ILLUSTRATION 39 Model of nuclear reactor. Max Planck Institute for Physics, July 16, 1956

In fact, research related to nuclear fission continued even during the 'prohibition era,' when, for example, researchers from Heisenberg's institute headed by Karl Wirtz collaborated with physicists in Franco's Spain to circumvent these limitations, while providing the technical expertise for beginning a nuclear program in the then isolated dictatorship.[178] Unfortunately for Heisenberg, his ambitions for the peaceful uses of nuclear fission were frustrated once the restrictions were lifted in the mid-1950s, as we detail later in this chapter. The wartime nuclear program had left an array of experts in nuclear fission, who now competed to be the standard-bearers of the peaceful atom, while in the field of fission reactors, in particular, others gained the

178 Albert Presas i Puig: Science on the Periphery. The Spanish Reception of Nuclear Energy. An Attempt at Modernity? *Minerva* 43/2 (2005), 197–218. doi:10.1007/s11024-005-2332-7.

upper hand, to the point of Heisenberg announcing in 1956 that he would leave the field altogether.[179]

In the mid-1950s, with the end of the Allied occupation of West Germany, these political failures in nuclear fission further propelled the shift toward nuclear fusion at Heisenberg's institute, as did the publication of previously secret scientific findings related to peaceful uses of fusion research. In April of 1956, the Soviet nuclear physicist Igor V. Kurchatov visited the United Kingdom with a Soviet delegation led by Nikita S. Khrushchev and including the Premier, Nikolai A. Bulganin. Kurchatov surprised Western scientists with a lecture at AERE (Atomic Energy Research Establishment), Great Britain's famous nuclear research center at Harwell, offering deep insights into the problems of controlled thermonuclear fusion and unveiling research conducted in the Soviet Union.[180] This event and, in particular, early triumphal results of the ZETA machine (Zero Energy Thermonuclear Assembly) at Harwell—an early experiment in fusion power research published at the time (but shown by later analysis to be a blunder)—prompted France and Italy to enter the field. But these countries were starting from scratch, whereas Ludwig Biermann's group in Göttingen, as we have already detailed, had a significant lead, because plasmas had been its daily bread since the very beginning.[181]

Only a few months after the Harwell event, in November 1956, a plasma physics research group led by Arnulf Schlüter was established at the Max

---

179 Joachim Radkau: *Aufstieg und Krise der deutschen Atomwirtschaft 1945–1975. Verdrängte Alternativen in der Kerntechnik und der Ursprung der nuklearen Kontroverse.* Reinbek: Rowohlt 1983. The large-scale experimental reactor that he had lobbied to have built in Munich as part of the planned relocation of his institutes to Bavaria, was instead built in Karlsruhe near the French border; and even Bavaria's first small experimental reactor, known as the 'Atomic Egg' was built by one of his competitors. On the wartime competition among German scientists in this field, see Walker, *The Quest for Nuclear Power*, 1989. For the way this rivalry unfolded in nuclear fission in the 1950s, see Michael Eckert, and Maria Osietzki: *Wissenschaft für Macht und Markt. Kernforschung und Mikroelektronik in der Bundesrepublik Deutschland.* München: Beck 1989. This episode will be revisited in Section 3.

180 Sir John Cockcroft: British Research in Controlled Thermonuclear Fusion. Kurchatov Memorial Article for Atomnaya Energiya. *Journal of Nuclear Energy. Part C, Plasma Physics, Accelerators, Thermonuclear Research* 5/6 (1963), 388–391. doi:10.1088/0368-3281/5/6/311.

181 Correspondence shows how, from the late 1950s, Biermann's group supported early activities led in Rome by Edoardo Amaldi and Enrico Persico (see 'Edoardo Amaldi Archives,' Physics Department, Sapienza University of Rome, Edoardo Amaldi papers, Box 198, Folder 1, and Enrico Persico's papers, Box 1, Folder 266). On the birth of fusion research in Italy in the late 1950s, see Luisa Bonolis: *Il sogno di Prometeo. Dagli anni della ricostruzione alla nascita delle ricerche sulla fusione nucleare in Italia*. Labirinti 2022.

Planck Institute in Göttingen, with the objective of studying the theory of plasmas, shock waves, and thermonuclear processes. Later, the Institute for Plasma Physics was founded, and Schlüter became a Scientific Member of the Institute for Astrophysics.[182]

In early 1957, the United States and the United Kingdom began to declassify fusion research for peaceful uses. The Third International Conference on Ionization Phenomena in Gases, held in Venice in June 1957, was the

> first regular International Conference, where thermonuclear fusion was an official part of the proceedings. And indeed, there were contributions on fusion not only from nearby continental Europe, but also from the United Kingdom, the United States and the Union of Socialist Soviet Republics.[183]

This meeting prepared the ground for the Second Conference on the Peaceful Uses of Atomic Energy (better known as the 'Atoms for Peace' conference) held in Geneva in 1958, where the US even organized a major exhibition on its fusion research.[184] On that occasion, more than a hundred papers on plasma

---

182 See CPTS meeting minutes of 02.06.1959, AMPG, II. Abt., Rep. 62, No. 1734. A list of research fields at the Institutes for Physics (plasma physics, elementary particles experimental, theoretical nuclear physics, field theory, construction of cloud chambers, detection of neutrons) and Astrophysics (electronic computing machines, atomic quantum theory, theoretical astrophysics, theoretical plasma and fusion physics, logic of computing, methods of numerical analysis) show that, in 1959, both had groups working on theoretical and experimental plasma physics (AMPG, II. Abt., Rep. 66, No. 3069). See also contributions by Biermann and Schlüter on the fundamentals of plasma physics and on research work conducted by the group in Generalverwaltung der Max-Planck-Gesellschaft (ed.): *Mitteilungen aus der Max-Planck-Gesellschaft zur Förderung der Wissenschaften. Heft 2/1957*. Göttingen 1957, 66–73. Generalverwaltung der Max-Planck-Gesellschaft (ed.): *Mitteilungen aus der Max-Planck-Gesellschaft zur Förderung der Wissenschaften. Heft 3/1957*. Göttingen 1957, 146–163. See also related folder in Biermann's papers (AMPG, III. Abt., ZA 1, No. 82, 83, 84, 85). For an overview of research work done by members of Biermann's group during the 1950s, creating the premises and expertise for the birth of the Institute for Plasma Physics, see Reinhard Breuer, and Uwe Schumacher: *Max-Planck-Institut für Plasmaphysik. Garching bei München*. Edited by Max-Planck-Gesellschaft. München: Max-Planck-Gesellschaft 1982, 10–12.

183 Arnulf Schlüter: Fusion at Venice, Remembered 32 Years Later. *Plasma Physics and Controlled Fusion* 31/10 (1989), 1725–1726, 1725. http://stacks.iop.org/0741-3335/31/i=10/a=318. Last accessed 10/30/2018.

184 On the promotion "of the benign atom as an instrument of American foreign policy and hegemonic ambitions" in the early years of the Cold War, see John Krige: Atoms for Peace, Scientific Internationalism, and Scientific Intelligence. *Osiris* 21/1 (2006), 161–181.

physics and controlled thermonuclear fusion were submitted.[185] This second conference effectively marked the unveiling of fusion research for peaceful uses and Biermann, too, participated, with a paper on "Recent Work on Controlled Thermonuclear Fusion in Germany (Federal Republic)." The only contribution representing West Germany, it was published in the proceedings, in the chapter "Possibility of Controlled Fusion." What became clear at the conference was that the secrecy had led to an overlap of work and findings in the different research projects all over the world. In this regard, stressed Biermann,

> I should like to say that I share very much the satisfaction that has been expressed by earlier speakers that now the period of duplication and non-communication has apparently come to an end and that international cooperation gives better promise for the future of physics.[186]

During the conference, thanks to the unveiling of research for the achievement of controlled nuclear fusion, Biermann was able to discuss with Lyman Spitzer their respective plans to build their own stellarator, including the possibility of an exchange between members of the Institute for Astrophysics and Spitzer's collaborators at the Project Matterhorn.[187] Immediately after the conference,

---

doi:10.1086/507140. Indeed, in early 1957, Biermann raised the problem of discussing the general rules of conduct in the matter of the communication of their research results, also in view of the conference. See his memorandum on the international exchange of scientific knowledge and experience on nuclear fusion within EURATOM and other countries, dated March 1957 and sent to Wilhelm Grau (Director, Bundesministerium für Atomfragen), AMPG, III. Abt., Rep. 83, No. 98.

185 United Nations (ed.): *Proceedings of the Second United Nations International Conference on the Peaceful Uses of Atomic Energy. Held in Geneva 1 September–13 September 1958.* Geneva: United Nations Publication 1958.

186 Ludwig Biermann: Recent Work on Controlled Thermonuclear Fusion in Germany (Federal Republic). *Proceedings of the Second United Nations International Conference on the Peaceful Uses of Atomic Energy. Held in Geneva 1 September–13 September 1958. Vol. 31. Theoretical and Experimental Aspects of Controlled Nuclear Fusion.* Geneva: United Nations 1958, 21–26, 25. http://www-naweb.iaea.org/napc/physics/2ndgenconf/data/Proceedings%201958/NG900088.pdf. Last accessed 10/30/2018. See also related research papers of the group listed within the references.

187 See a letter of 05.02.1959 from Spitzer to Biermann: "Dear Ludwig […] As I remember you had asked whether we could send you a copy of our report outlining what was planned for our model C […] Since that time the necessary arrangements have been made and all our reports are now completely unclassified. Accordingly, I am sending you, under separate cover, a copy of two of our reports, PM-S-14 and PM-S-29. The first of these discusses what a full-scale reactor might look like […] The second report, outlining our Model C Stellarator, gives a fairly detailed description of the problems involved in designing

John B. Adams, Director of the Proton Synchrotron division at CERN, and later Director of the Culham Fusion Laboratory in the UK, promoted the formation of the CERN Study Group on Fusion Problems, in order to coordinate research and prevent duplication of effort by exchanging information and discussing programs undertaken in the various laboratories. Seven European nations (as well as members of the US Atomic Energy Commission and EURATOM) cooperated with the Study Group (Belgium, Denmark, France, Italy, Netherlands, Norway, Sweden, Switzerland, the UK, the US, and, of course, West Germany). Of the five German members, three came from the Max Planck Society: Biermann, Schlüter, and Gerhard von Gierke.[188]

By the late 1950s, the scientific capital that had been accumulated over the long postwar decade in theoretical plasma astrophysics was expected to finally be channeled toward the construction and operation of a large-scale experimental thermonuclear facility to match those in the leading countries, such as the United States, Soviet Union, United Kingdom, and France. Until the last few months of 1957, the vast majority of work in astrophysics had been pointing in this direction.[189] In Göttingen, astrophysics before 1958 existed in the shadow of the atomic mushroom. But it was in these years that a clear scientific tradition emerged, a team of researchers closely collaborating in a research field, sharing a theoretical background, methodological skill set, and significant connections with the leading researchers in the United States, a country most of them would visit at a key moment in their careers. This is one of the two core scientific traditions in the Max Planck Society in the cosmic sciences which dated from the early postwar period and extended successfully all the way to the present day.

---

this large facility" (AMPG, III. Abt., ZA 1, No. 42). Over the following months, the two men continued to discuss the stellarator scheme. In the same folder, see also a letter of 19.02.1959 from Biermann to Spitzer, about the possibility of an exchange program (i.e., of each institution hosting the other's staff).

188 Adams, John Bertram. 1959. European fusion research: John Bertram Adams: *European Fusion Research. Report of the CERN Study Group on Fusion Problems.* CERN Yellow Reports. Monographs, CERN-59-16. Geneva: CERN 1959. http://cds.cern.ch/record/214328. Last accessed 10/30/2018. See related folder in Biermann's papers (AMPG, III. Abt., ZA 1, No. 55).

189 Susan Boenke: *Entstehung und Entwicklung des Max-Planck-Instituts für Plasmaphysik 1955–1971*. Frankfurt am Main: Campus Verlag 1991. Ludwig Biermann: Relations between Plasma Physics and Astrophysics. *Reviews of Modern Physics* 32/4 (1960), 1008–1011. doi:10.1103/RevModPhys.32.1008.

2 Postwar Research Traditions in Southwest Germany

We introduce the counterweight to the community described in the previous section. This was a research tradition based on experimental nuclear physics, making use of particle detectors and accelerators. The precursors of this tradition were Walther Bothe, one of Germany's most prominent experimental physicists, and his disciple Wolfgang Gentner, who emerged as a central political figure and played a key role in the scientific Europeanization of West Germany. Gentner and his colleagues pursued a path to scientific excellence in the first decade of postwar scarcity and research restrictions: they built large infrastructures at CERN, while conducting fundamental nuclear research locally by entering the field of cosmochemistry, in which mineral samples and meteorites are analyzed to gain insight into fundamental physical processes. This tradition spanned a growing network centered on the Max Planck Institute for Nuclear Physics in Heidelberg and the Max Planck Institute for Chemistry in Mainz, including allies in nearby universities in the host cities and other locations such as Freiburg and Bern.

Walther Bothe, Wolfgang Gentner, and Experimental Nuclear Physics

The second major scientific tradition in the Max Planck Society in the cosmic sciences emerged in the southwestern part of Germany, occupied by France. This tradition was strongly tied to the figure of Wolfgang Gentner and, as will be shown later, developed into the most significant counterweight to Heisenberg's institutes in Göttingen and, later, Munich. Astrophysics was a relatively modest element in the complex relationship between Heisenberg and Gentner, which played out primarily in rivalries in the 'nuclear' realm. In the cosmic realm, the very different scientific traditions led to a relationship of complementarity, allowing the two to expand relatively undisturbed by each other, in contrast to what would be their conflictive overlap in nuclear and particle physics. The combination of rivalry in nuclear physics and complementary growth in astrophysics is illustrative of how scientific traditions, political ambitions, and regional allegiances reinforced one another in the first three decades of the Max Planck Society.

The southwestern scientific tradition of the Max Planck Society was based on a prominent lineage of 20th-century experimental physics, which dated back to Walther Bothe, whose outstanding skills in both theoretical and experimental physics were deeply rooted in his formation as a doctoral student of Max Planck and as assistant to Hans Geiger, during the first quarter of the 20th century, a time of major revolutionary developments in physics; and they also owed much to the influence of Albert Einstein. During the birth of quantum

ILLUSTRATION 40 Walther Bothe and Hans Geiger sitting in a cafe in the 1920s.

mechanics, between 1923 and 1926, Bothe made a major contribution to elucidating the particle-wave duality of light in a series of elegant and laborious experiments in which "the interplay between experimental and theoretical ideas" played an essential role.[190] This crucial test, confirming the existence of light quanta and establishing the validity of conservation principles in elementary processes which had been called into question by the Bohr-Kramers-Slater theory,[191] was based on the novel coincidence method devised by Bothe and Geiger when researching the simultaneous appearance of two different signals in two separate detectors. With the invention of electronic circuits, the coincidence technique achieved its full potential, becoming one of the basic tools for the study of nuclear reactions and in cosmic ray physics.

In 1925, after working about ten years with Hans Geiger, Bothe became his successor as Director of the Laboratory for Radioactivity at the Imperial Physical Technical Institute (*Physikalisch-Technische Reichsanstalt*) in Berlin-

190 Dieter Fick, and Horst Kant: Walther Bothe's Contributions to the Understanding of the Wave-Particle Duality of Light. *Studies in History and Philosophy of Science Part B: Studies in History and Philosophy of Modern Physics* 40/4 (2009), 395–405. doi:10.1016/j.shpsb.2009.08.005.

191 Helge Kragh: Bohr—Kramers—Slater Theory. In: Daniel Greenberger, Klaus Hentschel, and Friedel Weinert (eds.): *Compendium of Quantum Physics*. Berlin: Springer 2009, 62–64. doi:10.1007/978-3-540-70626-7_19.

Charlottenburg, and began to study the transformation of light elements by bombardment with alpha rays, a work which would provide hints as to the existence of an unusual penetrating radiation emitted by beryllium, which was very soon identified by Chadwick as the neutron hypothesized and long searched for by Ernest Rutherford.[192]

In 1929, Bothe had introduced the coincidence method also into cosmic ray research, in a pioneering study conducted in collaboration with the astronomer Werner Kolhörster. The existence of a radiation coming from above, constantly bombarding Earth from outer space, later named cosmic rays by Robert Millikan, had been verified by Victor Hess and Domenico Pacini between 1911 and 1912.[193] However, whereas the first decades of cosmic ray science and radioactivity research had depended on rudimentary tools such as electroscopes and ionization chambers, by the late 1920s, the Geiger-Müller counter, a meanwhile ubiquitous tool that detected the passage of individual particles and emitted an electric signal as output, became the emblematic marker of the start of cosmic ray research as a branch of experimental physics.[194] Bothe and Kolhörster revolutionized the field: by aligning two such detectors in sequence and combining their output they provided evidence of the corpuscular nature of cosmic rays, which at the time were instead generally considered 'ultra-gamma rays,' because of their incredible penetrating power, far exceeding that of rays from any known radioactive substance.[195] The coincidence method, which Bruno Rossi soon turned into electronic signals, remained the basis of modern subatomic particle detection for the rest of the 20th century.[196] This technique opened the door to the sophisticated

192 Chadwick, Possible Existence, 1932, 312. In the first lines of this article Chadwick acknowledged Bothe's contribution in providing a decisive experimental insight.

193 Victor Hess: Über Beobachtungen der durchdringenden Strahlung bei sieben Freiballonfahrten. *Physikalische Zeitschrift* 13 (1913), 1084–1091. For the contemporary discovery of cosmic rays by the Italian scientist Domenico Pacini see Per Carlson, and Alessandro De Angelis: Nationalism and Internationalism in Science. The Case of the Discovery of Cosmic Rays. *The European Physical Journal H* 35/4 (2011), 309–329. doi:10.1140/epjh/e2011-10033-6.

194 Hans Geiger, and Walther Müller: Das Elektronenzählrohr. *Physikalische Zeitschrift* 29 (1928), 839–841.

195 Walther Bothe, and Werner Kolhörster: Die Natur der Höhenstrahlung. *Naturwissenschaften* 17/17 (1929), 271–273. doi:10.1007/BF01507590. According to Millikan, cosmic rays—a mixture of high-energy photons—were born of the energy released during the synthesis of heavier elements from primordial hydrogen spread throughout the universe.

196 Walther Bothe: Coincidence Method. *Science* 122/3175 (1955), 861–863. http://www.jstor.org/stable/1749457. Last accessed 10/30/2018. Georg Pfotzer: Early Evolution of Coincidence Counting a Fundamental Method in Cosmic Ray Physics. In: Yataro Sekido, and

statistical analysis that became predominant in cosmic ray studies, as well as in nuclear and particle physics.[197]

In 1932, Bothe was appointed Director of the Physical and Radiological Institute at the University of Heidelberg, as successor to Philipp Lenard, yet was driven out of office the following year by supporters of the *Deutsche Physik*. Through Max Planck himself, then President of the Kaiser Wilhelm Society, Bothe was offered a position heading the Institute for Physics at the nearby Kaiser Wilhelm Institute for Medical Research.[198] This was nominally a department dedicated to medical physics, which at the time was raising interest in the therapeutic use of radiation and radioactive substances. In practice, however, Bothe continued the research line that he had followed at the university, focusing on nuclear physics—still a novel field of inquiry restricted to a worldwide yet small community—as well as further developments in cosmic ray research, at a time when the two fields continued to be closely connected.[199] As will be seen throughout this chapter, this German tradition of electronic detection techniques was to prove a valuable scientific asset, during the 1930s, for it could lead to opportunities for visits and various forms of collaboration

Harry Elliot (eds.): *Early History of Cosmic Ray Studies. Personal Reminiscences with Old Photographs*. Dordrecht: Springer Netherlands 1985, 39–44. doi:10.1007/978-94-009-5434-2_5. The coincidence method by combination of three counters arranged in a triangular form with lead boards piled up between the counters was further devised by Rossi to study the secondary processes produced by incident cosmic ray particles. Luisa Bonolis: Walther Bothe and Bruno Rossi. The Birth and Development of Coincidence Methods in Cosmic-Ray Physics. *American Journal of Physics* 79/11 (2011), 1133–1150. doi:10.1119/1.3619808. Luisa Bonolis: International Scientific Cooperation During the 1930s. Bruno Rossi and the Development of the Status of Cosmic Rays into a Branch of Physics. *Annals of Science* 71/3 (2014), 355–409. doi:10.1080/00033790.2013.827074.

197 Galison, *Image and Logic*, 1997, 6. For his invention of a new detecting method and for the resulting discoveries, Bothe shared the Nobel Prize for Physics 1954 with Max Born. Bothe's award was the first Nobel Prize for the young Max Planck Society.

198 The best account of Bothe's trajectory in the Max Planck Society was written by his disciple, the future Max Planck director Ulrich Schmidt-Rohr: *Erinnerungen an die Vorgeschichte und die Gründerjahre des Max-Planck-Instituts für Kernphysik*. Heidelberg: Selbstverlag 1996. See also biographical contributions written by other disciples and colleagues of Bothe's after his death: Wolfgang Gentner: Nachruf für Walter Bothe. *Zeitschrift für Naturforschung A* 12/2 (1957), 175–176. doi:10.1515/zna-1957-0213. Rudolf von Fleischmann: Walter Bothe und sein Beitrag zur Atomkernforschung. *Die Naturwissenschaften* 44/17 (1957), 457–460. doi:10.1007/BF00640879. Lise Meitner: Prof. Walter Bothe. *Nature* 179/4561 (1957), 654–655. doi:10.1038/179654a0. Heinz Maier-Leibnitz: W. Bothe, Experimental Nuclear Physicist. *Science* 126/3267 (1957), 246–247. doi:10.1126/science.126.3267.246. Maier-Leibnitz, W. Bothe, 1957, 246–247.

199 See, for example, Walther Bothe, and Siegfried Flügge (eds.): *Nuclear Physics, Cosmic Rays*. Vol. 1. Wiesbaden: Dieterich'sche Verlagsbuchhandlung 1948.

ILLUSTRATION 41 From left: Walther Bothe, Otto Haxel, and Hans Kopfermann in the late 1930s. All had been members of the 'Uranium Club', the German nuclear project.

with scientists from abroad. It was difficult for researchers in other countries to duplicate these techniques at a distance on the basis of published descriptions alone. Moreover, this gap in 'tacit knowledge' kept the door open for the community of experimental physicists around Bothe, even though resources at this time were greatly reduced because of the Great Depression and the increasing political isolation as a result of being based in Nazi Germany.

Wolfgang Gentner was the key beneficiary of this incipient tradition. Having completed his doctoral research in his native Frankfurt, he was invited to continue his work with the Joliot-Curies in Paris in 1933.[200] Gentner's for-

200 For an outline of Gentner's activities up to 1943, see his personal record, where, as in Bothe's case, his main research field is nuclear physics (BArch, No. R26-III/8). On Wolfgang Gentner's recollections of his residence in Paris and his collaboration with Frédéric Joliot, see Wolfgang Gentner: interview by Charles Weiner, November 15, 1971. Transcript, AIP, https://www.aip.org/history-programs/niels-bohr-library/oral-histories/5080. Last accessed 5/8/2019. See also: Bonolis, Walter Bothe and Bruno Rossi, 2011, 1133–1150. Dieter Hoffmann, and Ulrich Schmidt-Rohr: Wolfgang Gentner. Ein Physiker als Naturalist. In: Dieter Hoffmann, and Ulrich Schmidt-Rohr (eds.): *Wolfgang Gentner. Festschrift zum 100. Geburtstag*. Berlin: Springer 2006, 1–60.

mative years in Paris would prove pivotal to his dominant role in postwar German experimental physics. He closely befriended the scientific circles around Frédéric Joliot and his wife Irène Curie, and became a cultural Francophile, acquiring tastes, habits, and a political affinity for the country that he proudly maintained throughout his life.[201] While in Paris, Gentner contributed his expertise in particle detectors and, notably, Geiger counters made by him were used to verify the Joliot-Curies' Nobel Prize-winning discovery of artificial radioactivity. It was in Paris that he first became acquainted with accelerators and, more generally, with research on the atomic nucleus, through discussions with Joliot.[202] Knowledge of the central core of the atom was very limited at the time, and bombarding nuclei with neutrons was only just beginning to be explored.[203] In 1935 Gentner returned, "full of enthusiasm for nuclear physics,"[204] and found his way into Bothe's Kaiser Wilhelm Institute, where it was his and Bothe's intention to apply this expertise to building one of Germany's first accelerators, a Van de Graaff generator, actually the very first of this kind in the country. It was ready by 1936, and they used it for nuclear research.[205] This was a time before nuclear fission and fusion, when 'nuclear' research had already become established as an important mainstream field in physics but was not yet seen to hold any major societal promise. Nuclear research in the 1930s focused on the processes of radioactivity and the transmutation of elements based on the number of particles in their nucleus; and,

201 Dieter Hoffmann, and Ulrich Schmidt-Rohr (eds.): *Wolfgang Gentner. Festschrift zum 100. Geburtstag*. Berlin: Springer 2006. Pages.

202 See the report on Gentner's activity at the Curies' chemistry laboratory, dated July 4, 1935, and signed by the director Andreé Debierne (Musée Curie, Archives, Paris, Box 20).

203 Amaldi, Discovery of the Neutron, 1984, 1–331. Nesvizhevsky, and Villain, The Discovery of the Neutron, 2017, 592–600.

204 Victor F. Weisskopf: Wolfgang Gentner—ein Forscherleben in unserer Zeit. In: Generalverwaltung der Max-Planck-Gesellschaft (ed.): *Max-Planck-Gesellschaft. Berichte und Mitteilungen. Gedenkfeier Wolfgang Gentner*. München 1981, 23–27, 24.

205 This electrostatic accelerator, invented in 1929 by the American physicist Robert Van de Graaff, used a moving belt to accumulate electric charge on a metal globe, thus creating very high electric potentials. Later it was used as injector for high-energy accelerators. As recalled by Gentner, this was the first Van de Graaff machine in Heidelberg, but they did not need special support to build it: "[…] this was a very cheap machine. We could build this machine with our own resources, in our own workshop […] We got about… the first machines about for 600 thousand volts, and the second was about for one million." Gentner had no experience, but he had used high tensions while working on his PhD dissertation and was able to build the first and the second Van de Graaff more or less alone (Wolfgang Gentner: interview by Charles Weiner, November 15, 1971. Transcript, AIP, https://www.aip.org/history-programs/niels-bohr-library/oral-histories/5080. Last accessed 5/8/2019).

in that decade, the nucleus was also the source of new theoretical insights into elementary particles such as neutrons, positrons, or neutrinos.

Going beyond nuclear studies of light elements, Bothe wanted to explore with his group the much more complex reactions in elements of higher atomic weights, which however required higher-energy bombarding particles. Up to the 1930s, such beams had been obtained from natural or even artificial radioactive sources, after the discovery made by the Joliot-Curies in Paris and by Enrico Fermi in Rome.[206] Experimental nuclear physics research was the crucible of the charged particle accelerator, which from the early 1930s onward became the most important tool.[207] In 1937, Gentner, together with Bothe, made his first major scientific contribution with research on the nuclear photo-effect, whereby electromagnetic radiation of the right energy can induce processes inside the atomic nucleus.[208] Gentner had already worked on nuclear photo-effects in Paris, using gamma rays from radioactive sources, but now, by bombarding lithium with protons accelerated with the Van de Graaff, it was possible to get gamma rays of much higher energy and in a specific wavelength, with which photo-effects were created in all elements, not only in the light ones such as deuterium and beryllium.[209]

---

206 Francesco Guerra, Matteo Leone, and Nadia Robotti: The Discovery of Artificial Radioactivity. *Physics in Perspective* 14/1 (2012), 33–58. doi:10.1007/s00016-011-0064-7. Francesco Guerra, and Nadia Robotti: *The Lost Notebook of Enrico Fermi. The True Story of the Discovery of Neutron-Induced Radioactivity*. Cham: Springer Verlag 2018. doi:10.1007/978-3-319-69254-8.

207 The era of accelerator-based experimental nuclear physics began in 1932—the so-called *annus mirabilis* of nuclear physics—when Chadwick announced the existence of the neutron, Carl Anderson identified the positron, and John Cockcroft and Ernest Walton were able to use their new electrostatic accelerator to perform the first artificial disintegration of an atomic nucleus and the first artificial transmutation of one element (lithium) into another (helium); in parallel, Ernest Lawrence's first cyclotron had just gone into operation in Berkeley.

208 Gentner gained his post-doctoral teaching qualification (*Habilitation*) from the University of Frankfurt in 1937 with a dissertation entitled "Experimentelle Untersuchungen zur Absorption, Streuung und Sekundärstrahlung harter gamma-Strahlen." Hoffmann, and Schmidt-Rohr, Wolfgang Gentner, 2006, 1–60, 11.

209 Walther Bothe, and Wolfgang Gentner: Atomumwandlungen durch γ-Strahlen. *Zeitschrift für Physik* 106/3–4 (1937), 236–248. doi:10.1007/BF01340320. Heinz A. Staab: 50 Jahre Kaiser-Wilhelm/Max-Planck-Institut für Medizinische Forschung Heidelberg. In: Universitäts—Gesellschaft (ed.): *Heidelberger Jahrbücher*. Berlin: Springer 1980, 47–70. Wolfgang U. Eckart: Max-Planck-Institut für medizinische Forschung Heidelberg. In: Reinhard Rürup, and Peter Gruss (eds.): *Denkorte. Max-Planck-Gesellschaft und Kaiser-Wilhelm-Gesellschaft. Brüche und Kontinuitäten 1911–2011*. Dresden: Sandstein 2010, 174–183. Hermann Weber: Max-Planck-Institut für medizinische Forschung in Heidelberg.

ILLUSTRATION 42 The Van de Graaff electrostatic generator at the Kaiser Wilhelm Institute for Medical Research in Heidelberg, 1936. This device, successfully employed by Bothe and Gentner for their experiments, accumulates electric charge, thereby creating very high potentials that can accelerate charged particles to high speeds, which in turn can be used to produce a variety of nuclear reactions. Such accelerators played a key role in the development of nuclear physics during the 1930s.

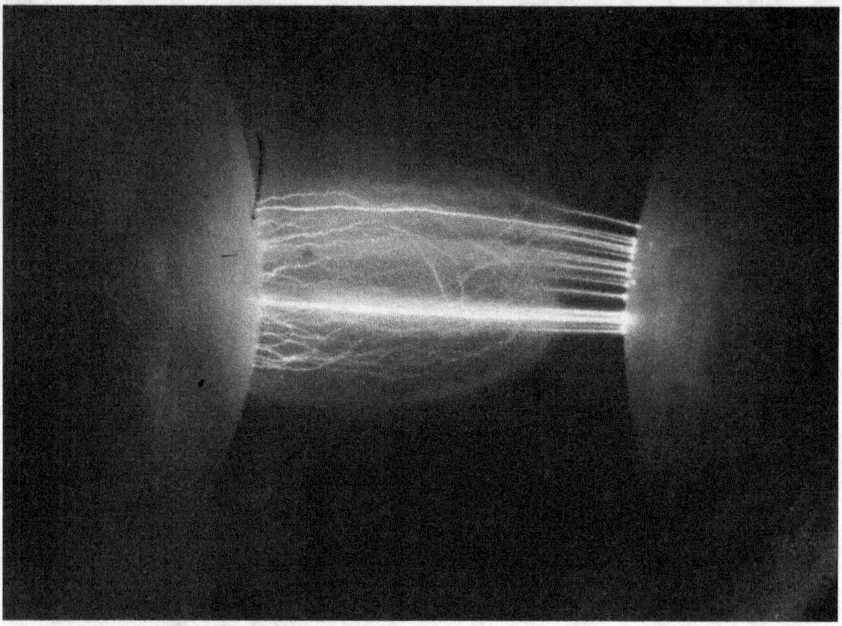

ILLUSTRATION 43 Van de Graaff accelerator: flashover due to electric discharge of high current made through the air between the spheres at very high electric potential.

The method used to produce this effect, was described by Niels Bohr as "beautiful."[210] As Gentner himself recalled,

> nobody at that time had used gamma rays to induce a process related to the nuclear realm: […] people said, "Oh, that's a very difficult thing, to use gamma rays." […] We were the first to use gamma rays.[211]

Throughout this period, Bothe and Gentner attempted, but failed, to obtain funding to build a much more powerful machine, a cyclotron, far better suited

---

In: Generalverwaltung der Max-Planck-Gesellschaft zur Förderung der Wissenschaften e.V. (ed.): *Jahrbuch 1961 der Max-Planck-Gesellschaft zur Förderung der Wissenschaften e. V. Teil II*. Göttingen 1962, 535–556.

210 Niels Bohr: Nuclear Photo-Effects (Letter to the Editor). *Nature* 141/3564 (1938), 326–327. doi:10.1038/141326a0.

211 Wolfgang Gentner: interview by Charles Weiner, November 15, 1971. Transcript, AIP, https://www.aip.org/history-programs/niels-bohr-library/oral-histories/5080. Last accessed 10/1/2022.

ILLUSTRATION 44 Wolfgang Gentner (left) and Walther Bothe (right) in Paris with Frédéric Joliot-Curie in 1937 during the physicists' meeting organized on the occasion of the World Fair. Behind Joliot-Curie is Bruno Pontecorvo, who fled Paris in June 1940 when the German troops invaded the city.

as a working tool of nuclear physics, as it made it possible to produce nuclear collisions, and even useful quantities of new isotopes, or to conduct fundamental research.[212] During the 1930s, the dream of building a cyclotron was cultivated in every European laboratory for nuclear physics. Gentner was thus most interested in deepening his knowledge of cyclotron accelerators, a sub-

212 Schmidt-Rohr, *Teilchenbeschleuniger*, 2001. Correspondence between Bothe and the Deutsche Forschunsgemeinschaft (1931–43) shows that he began to ask for funds in spring 1938, shortly after having completed the Van de Graaff, with which they had successfully carried out the aforementioned work on nuclear photo-effects (BArch, No. R 73/10419). As part of German scientific work during the war, the Heidelberg cyclotron was described by Gentner in a volume of the Field Information Agency's technical reviews, a series of reports on the status of German science in various disciplines (FIAT Review of German Science 1939–1946) published after the war: Wolfgang Gentner: Das Heidelberger Zyklotron. *Nuclear Physics and Cosmic Rays*. Vol. 2, 1948. See also Horst Kant: Von der Lichttherapie zum Zyklotron. Das Institut für Physik im Heidelberger Kaiser-Wilhelm-Institut für medizinische Forschung bis 1945. *Dahlemer Archivgespräche* 13 (2008), 49–92. http://opac.ifz-muenchen.de/webOPACClient.ifzsis/start.do?Login=woifz&Query=10="BV040926399". Last accessed 10/30/2018.

ject in which the United States was far ahead of Europe at the time. Their work with the photo-effect, and interest in cyclotrons, gave Gentner the opportunity to tour the United States in 1938, making himself known in person to another scientific community that would prove crucial for his postwar trajectory. As in France, people in the US were interested in his tacit knowledge of experimental physics, which included detectors, accelerators, and his recent experimental work on the nuclear photo-effect. Gentner spent the period 1938–39 in Berkeley, where Ernest O. Lawrence had built the first cyclotron in 1931, and was then building a new, more powerful machine.[213] Gentner was at Berkeley when Lawrence received a telegram from Washington with news of the discovery of fission and they were able to immediately use the cyclotron to produce the fission of uranium and see its effects in an ionization chamber.[214] By that time, Lawrence's 60-inch cyclotron was capable of delivering 20 MeV protons, twice the energy of the most energetic alpha particles emitted by radioactive sources.[215]

The kind of hands-on experimental physics represented by Gentner was perfectly suited for making very close personal connections with his scientific peers. While theoretical physics was sustained by a well-developed 'republic of letters' and common cultural background,[216] scientific work as in the case of Gentner required long hours in the lab, in direct personal contact with collaborators, in a culture centered around a fascination with building one's own

213 See Gentner's CV, AMPG, III Abt. Rep 68 A, Nachlass Wolfgang Gentner, No. 138. See also Gentner's review article on accelerators as tools for nuclear physics written after his US tour: Wolfgang Gentner: Die Erzeugung schneller Ionenstrahlen für Kernreaktionen. In: Ferdinand Trendelenburg (ed.): *Ergebnisse der Exakten Naturwissenschaften*. Berlin: Verlag von Julius Springer 1940, 107–169. Wolfgang Gentner: Mitteilungen aus der Kernphysik. Das neue 1,5 Meter-Zyklotron in Berkeley (Calif.). *Die Naturwissenschaften* 28/25 (1940), 394–396. doi:10.1007/BF01479460. See also Maria Osietzki: The Ideology of Early Particle Accelerators: An Association between Knowledge and Power. In: Monika Renneberg, and Mark Walker (eds.): *Science, Technology and National Socialism*. Cambridge: Cambridge University Press 1994, 255–270.

214 Wolfgang Gentner: interview by Charles Weiner, November 15, 1971. Transcript, AIP, https://www.aip.org/history-programs/niels-bohr-library/oral-histories/5080. Last accessed 5/8/2019.

215 For these achievements, Lawrence was awarded the Nobel Prize in 1939.

216 See, for example, David Kaiser: *Drawing Theories Apart. The Dispersion of Feynman Diagrams in Postwar Physics*. Chicago, IL: University of Chicago Press 2005. David Kaiser: Bringing the Human Actors Back on Stage. The Personal Context of the Einstein-Bohr Debate. *The British Journal for the History of Science* 27/2 (1994), 129–152. http://www.jstor.org/stable/4027432. Last accessed 10/30/2018.

ILLUSTRATION 45 Wolfgang Gentner with Peter H. Jensen (on his right) and Arnold Flammersteld (left) during WWI, evaluating measurements of the energy of fission neutrons, whose pulses were recorded on photographic paper strips via oscillographs. The institute's Van de Graaff provided neutrons through the bombardment of beryllium with accelerated nuclei of deuterium.

instruments.[217] However, Gentner himself was considered by experimental physicists to have a particularly good nose for the theoretical implications of his work, which translated into an uncanny ability to formulate new research directions and research programs that linked the existing skills and resources in experimental physics with the latest theoretical issues.[218]

---

217 Wolfgang Gentner: interview by Charles Weiner, November 15, 1971. Transcript, AIP, https://www.aip.org/history-programs/niels-bohr-library/oral-histories/5080. Last accessed 5/8/2019.

218 Paul Kienle: Festveranstaltung anlässlich des 100. Geburtstags von Prof. Heinz Maier-Leibnitz. "Rückblick eines Zeitzeugens" (1952–2000). Meinem verehrten Lehrer, Kollegen und Freund zum 100ten Geburtstag gewidmet. Physics Department of the Technical University of Munich, 3/28/2011. https://www.frm2.tum.de/fileadmin/woobnv/www/_migrated_content_uploads/Wie_Kernphysik_in_Muenchen_einzog.pdf. Last accessed 10/30/2018.

Wolfgang Gentner During the Shaping of the French Occupation Zone

A year after the outbreak of World War II, the Germans occupied Paris, and Frédéric Joliot's laboratories at the Collège de France were put at the service of German science.[219] In particular, the full-fledged cyclotron recently completed by Joliot was in the basement and could not be moved from its location. Joliot decided not to sabotage or dismantle it, but to use the machine instead to leverage better terms and conditions from the occupying forces. Walther Bothe, who was building his own cyclotron in Heidelberg at the time, agreed it was best to leave the accelerator in Paris—it was very difficult to move it—and therefore sent Wolfgang Gentner, who had worked there in the early 1930s, to liaise between the French scientists, occupation forces, and the German physics community working on the uranium project.

Gentner's wartime service would become legendary and later earn him the highest French distinction, the Legion of Honor: in Paris, he protected Joliot and his colleagues, occasionally saving them from arrest. More controversial is the degree to which this protection resulted in a more effective collaboration on fulfilling the needs of Bothe's experimental program, now deeply related to the German nuclear project.[220] But contemporary witnesses tell how Bothe's direct attempts to use the cyclotron for analytical purposes related to the uranium project were sabotaged with the passive acquiescence of Gentner, who is said to have also turned a blind eye to resistance activities organized in the laboratory. As the war intensified, Gentner was accused of aiding the French, and was relieved of his duties and ordered to return to Germany. Shortly afterwards, Joliot went underground.[221]

Gentner's wartime residence in Paris meant that at the end of the war, when Joliot was appointed head of the new French Atomic Energy Commission, Gentner was Joliot's closest German scientific advisor as they worked together to shape the scientific future of the French occupation zone.[222]

---

219 Gabriele Metzler: Wissenschaft im Krieg. Frédéric Joliot-Curie und die deutschen Besatzer am Collège de France. In: Stefan Martens, and Maurice Vaïsse (eds.): *Frankreich und Deutschland im Krieg (November 1942–Herbst 1944). Okkupation, Kollaboration, Résistance*. Bonn: Bouvier 2000, 685–700.

220 von Fleischmann, Walther Bothe, 1957, 457–460.

221 Michel Pinault: *Frédéric Joliot-Curie*. Paris: Editions Odile Jacob 2000. The entire Section III deals with Joliot's wartime experience. This particular episode is recounted on pages 198–201.

222 For discussions about the reorganization of the French zone, see the document "Auszug aus der Niederschrift über die Sitzung des Wissenschaftlichen Rates vom 21.7.49," AMPG, II. Abt., Rep. 66, No. 3047.

ILLUSTRATION 46 Kaiser Wilhelm Institute for Medical Research, Heidelberg, 1943. Transportation of the cyclotron magnet for Bothe's Institute for Physics

ILLUSTRATION 47 Kaiser Wilhelm Institute for Medical Research, Heidelberg. The cyclotron magnet ready for installation, 1943. Russian prisoners of war were employed for transport and installation operations.

ILLUSTRATION 48 Kaiser Wilhelm Institute for Medical Research, Bothe's Institute for Physics, Heidelberg, 1943. The cyclotron with its vacuum chamber

Joliot directly helped reestablish the former Kaiser Wilhelm Institute for Chemistry in Mainz, now affiliated with the Max Planck Society,[223] and made sure that Gentner was given a professorship at the University of Freiburg near the French border, which would be his German base for the next decade. Gentner also remained affiliated with Bothe's laboratory at the Institute for Medical Research in Heidelberg (in the American zone), as an External Scientific Member.[224] Within the Max Planck Society, Gentner also tried to persuade

223 Otto Hahn, A. Flammersfeld, and W. Kroebel: Persönliche Erinnerungen an Frédéric Joliot. *Physikalische Blätter* 14/11 (1958), 510–514. doi:10.1002/phbl.19580141106.

224 On Gentner's appointment as an External Scientific Member of Bothe's Institute for Physics at the Max Planck Institute for Medical Research, see CPTS meeting minutes of 14.06.1950, AMPG, II. Abt., Rep. 6, No. 1724. At the same meeting, two of Bothe's wartime collaborators, the nuclear physicists Heinz Maier-Leibnitz and Rudolf Fleischmann, (the latter later a leading figure in establishing and building nuclear research reactors), were appointed Internal and External Scientific Members, respectively. When the former was later appointed Chair for Technical Physics in Munich, he became an External Scientific Member (CPTS meeting minutes of 19.05.1953, AMPG, II. Abt., Rep. 62, No. 1727). As described later in this volume, Maier-Leibnitz became a leading figure in establishing and building scientific centers around nuclear research reactors.

ILLUSTRATION 49 Kaiser Wilhelm Institute for Medical Research, Bothe's Institute for Physics, Heidelberg. The cyclotron vacuum chamber where particle trajectories are bent by the magnetic field and repeatedly accelerated by an electric field.

his former mentor Friedrich Dessauer, meanwhile a politician in exile and professor in Freiburg (Switzerland), to return to Frankfurt and renew his involvement in the Kaiser Wilhelm Institute for Biophysics which he had founded in 1921.[225]

Most importantly, when, in the early 1950s, the first conversations took place about founding a joint European Organization for Nuclear Research, CERN, Gentner was one of the German representatives, along with Hahn and Heisenberg.[226] In 1955, on the basis of his experience with Joliot in Paris and with Bothe in Heidelberg, he was appointed Director of CERN's SC Division,

225 Dessauer had left Germany in 1934, at first for Istanbul; but Gentner later mediated via Bothe to help secure him the position in Switzerland in 1936. Michael Habersack: *Friedrich Dessauer (1881–1963). Eine politische Biographie des Frankfurter Biophysikers und Reichstagsabgeordneten.* Paderborn: Ferdinand Schöningh 2011, 403, 449.

226 About Germany's role and the negociations concerning the setting-up of CERN see Armin Hermann: Germany's Part in the Setting-up of CERN. In: Armin Hermann et al. (eds.): *History of CERN. Launching the European Organization for Nuclear Research.* Amsterdam: North-Holland 1987, 383–429.

ILLUSTRATION 50 Kaiser Wilhelm Institute for Medical Research, Bothe's Institute for Physics, Heidelberg, 1940–1941. Graphite sphere for measuring the absorption cross-section of neutrons in carbon, expressing the probability of a particular kind of interaction between an incident neutron and a target nucleus, a key concept in nuclear and particle physics. After the cyclotron was able to go into operation in February 1942, it became by far the most powerful neutron source available to German nuclear physics. Up until the end of the 1950s, Bothe's Institute for Physics had the only cyclotron available in Germany at the time.

ILLUSTRATION 51 From left to right: Wolfgang Gentner, with Werner Heisenberg and Alexander Hocker (both German delegates to the Conference on the Establishment of the European laboratory) in 1955, during the meeting held on June 11 to sign the agreement between the Council of CERN and the Swiss Federal Council defining the legal status of the organization in Switzerland. On October 1, Gentner took up his appointment as director of the 600-MeV synchro-cyclotron division, and was responsible for its construction and for directing research using the accelerator, which produced its first 600 MeV proton beam on August 1, 1957. This medium-energy machine was primarily intended to bridge the gap until the introduction of the 28 GeV proton synchrotron, which began operation at the end of 1959.

becoming leader of the group tasked to design and build the 600 MeV Synchro-Cyclotron, the organization's first particle accelerator.[227]

Accelerators at CERN, Cosmochemistry in Germany

As mentioned earlier in this chapter, 'nuclear' research was prohibited by the terms of the occupation. For Gentner, this had consequences very different from those in Göttingen at the time. In the French occupation zone, with his very good contacts to Joliot, and taking on a new role as part of

227 John Bertram Adams: Wolfgang Gentner and CERN. In: Dieter Hoffmann, and Ulrich Schmidt-Rohr (eds.): *Wolfgang Gentner. Festschrift zum 100. Geburtstag*. Berlin: Springer 2006, 139–145.

the pan-European postwar collaboration, Gentner adopted a division of scientific labor, between his activities for the international organization, and those that he carried out while based in Germany: in the first postwar decade, Gentner provided his accelerator expertise to CERN, while focusing in Germany on research in astrophysics-related activities, making use of an instrument closely derived from accelerators, the mass spectrometer. These instruments work on the same principles as circular accelerators (where charged particles are deflected by magnetic fields), but on a smaller scale, and are optimized not for accelerating charged particles toward collisions but instead for separating them according to the mass of their nucleus. Isotopes of different elements, having the same number of protons but different numbers of neutrons, differ in mass but not in chemical behavior, and thus this method, bridging the gap between nuclear chemistry and nuclear physics, is one of the most accurate for determining chemical composition. But while large cyclotrons and mass spectrometers had also specific uses in the production of atomic weapons, Gentner specialized in the mass spectrometry of very small samples. In entering mass spectrometry in the early 1950s, Gentner was joining a longstanding tradition of expertise in mass spectrometry at the Kaiser Wilhelm Institute for Chemistry, dating back to the Dahlem days under Joseph Mattauch, one of the leading experts in the field, whose department was reestablished in Mainz after the war.[228]

228 Mattauch had first succeeded Lise Meitner as Head of the Physics Department of the Kaiser Wilhelm Institute for Chemistry, and then Hahn as Director in 1947. Heinrich Hintenberger: Josef Mattauch. 21.11.1895-10.8.1976. *Berichte und Mitteilungen Max-Planck-Gesellschaft* Nachrufe, Jahresbericht 1976, Jahresrechnung 1975 (1977), 19–21. About Hahn and Mattauch's plans, from the 1930s on, for large-scale equipment such as mass spectrographs and accelerators for nuclear research, see Burghard Weiss: The "Minerva" Project. The Accelerator Laboratory at the Kaiser Wilhelm Institute / Max Planck Institute for Chemistry. Continuity in Fundamental Research. In: Monika Renneberg, and Mark Walker (eds.): *Science, Technology and National Socialism*. Cambridge: Cambridge University Press 2003, 271–290. On the history of the Institute, see Josef Mattauch: Max-Planck-Institut für Chemie (Otto-Hahn-Institut) in Mainz. In: Generalverwaltung der Max-Planck-Gesellschaft zur Förderung der Wissenschaften e.V. (ed.): *Jahrbuch 1961 der Max-Planck-Gesellschaft zur Förderung der Wissenschaften e.V. Teil II*. Göttingen 1962, 215–224. Carsten Reinhard, and Horst Kant: *100 Jahre Kaiser-Wilhelm-/Max-Planck-Institut für Chemie (Otto-Hahn-Institut). Facetten seiner Geschichte*. Vol. 22. Berlin 2012. On the occasion of Mattauch's retirement in 1965, Wilhelm Walcher gave a talk on the history and development of mass spectroscopy. Präsidialbüro der Max-Planck-Gesellschaft (ed.): *Mitteilungen aus der Max-Planck-Gesellschaft zur Förderung der Wissenschaften. Heft 2/1966*. München 1966, 89–111.

ILLUSTRATION 52 Kaiser Wilhelm Institute for Chemistry, Mainz. Mass spectrograph. This instrument operates on the principle that moving ions can be deflected by electric and magnetic fields. This enables various investigations to take place, including the identification of isotopes of chemical elements and determination of their precise masses and abundances, the dating of geological samples, the analysis of small amounts of impurities in organic and inorganic chemicals, and even the analysis of chemical and isotopic constituents in unknown materials, such as lunar samples.

Shortly afterwards, in 1953, when Fritz Strassmann moved to the nearby University of Mainz, the Institute for Chemistry gained another figure of international renown in the form of the exiled, Austrian-born chemist Friedrich Paneth, who founded the Cosmochemistry Department, opening up a new area of cross-disciplinary research in Mainz.[229] While himself a radiochemist, Paneth had also pioneered the use of mass spectrometry for the purpose of

229 Strassmann was immediately appointed an External Scientific Member and different candidates were discussed for his succession (CPTS meeting minutes of 19.05.1953, AMPG, II. Abt., Rep. 62, No. 1727); but in November, Josef Mattauch presented the proposal to call

ILLUSTRATION 53 Josef Mattauch at the mass spectrograph for precise measurement of the masses of atomic nuclei, Max Planck Institute for Chemistry, Mainz, 1956

answering astrophysical questions, during his time as professor in Durham, through the 1930s and 1940s.[230] After the war, as a consequence of improvements both in vacuum technology and electronics due to the needs of nuclear energy projects, all analytic techniques benefited from the development of electronic measuring instruments; mass-spectrometers, for instance, became far more reliable machines than the prewar delicate devices, which had

Paneth to the Institute for Chemistry as Scientific Member and Director. (The commission was made up of Bothe, Hahn, Kuhn, and Mattauch. See further communication by Karl F. Bonhoeffer, President of the CPT Section, dated November 10, 1953, attached to the minutes of the aforementioned meeting of 19.05.1953). The proposal was accepted during the Senate meeting of January 29, 1954 (AMPG, II. Abt., Rep. 60, No. 17).

230 On Paneth's scientific contributions, see: L.T. Aldrich et al. (eds.): *Cosmological and Geological Implications of Isotope Ratio Variations. Proceedings of an Informal Conference, Massachusetts Institute of Technology, June 13–15, 1957*. Washington, D.C.: National Academy of Sciences, National Research Council 1958. Paneth himself was an old guard radiochemist, but he was the first to suggest using mass spectrometry to examine helium samples to find Helium-3.

ILLUSTRATION 54 From left: Josef Mattauch, Otto Hahn, and Hans D. Jensen, Eltville am Rhein, October 27, 1954

needed constant expert attention. All this had an enormous influence on the growth of geochemistry, isotope geology, and geochronology. Paneth was one of the pioneers of the field known as cosmochemistry, as he himself liked to call it.[231] His plan in Mainz was thus to work on new cosmochemical methods based on both radiochemistry and mass spectrometry, in collaboration with the institute's experimental expert in the field, Heinrich Hintenberger, who had begun working at the institute under Mattauch, in 1949. In Paneth's department, they researched the production of cosmogenic radioisotopes resulting from the interaction of high-energy cosmic rays with meteorites, and conducted age dating by detecting extremely small quantities of the noble gases helium and neon generated on Earth in iron meteorites and metallic iron.[232]

231 See Josef Mattauch's obituary in Generalverwaltung der Max-Planck-Gesellschaft (ed.): *Mitteilungen aus der Max-Planck-Gesellschaft zur Förderung der Wissenschaften. Heft 6/1958*. Göttingen 1958, 313–319.

232 On this, see, for example, Heinrich Wänke, and Heinrich Hintenberger: Notizen. Helium und Neon als Reaktionsprodukte der Höhenstrahlung in Eisenmeteoriten. *Zeitschrift für Naturforschung A* 13/10 (1958), 895–897. doi:10.1515/zna-1958-1017. Heinrich Wänke, one of

ILLUSTRATION 55 Friedrich Paneth with apparatus for the microanalysis of noble gases, Max Planck Institute for Chemistry, Mainz, June 6, 1956

In the late 1940s and early 1950s, nuclear physics underwent a significant transformation based in part on the incredible development of the field during the war and the advent of high-energy accelerators, but also following a deep evolution of the theoretical tools required for research on the nucleus as a complex physical system. At that time, geochemistry, with its deep interactions with astrophysics and nuclear physics, was at the origin of the shell nuclear model, a highly successful scheme describing the way protons and neutrons are arranged inside a nucleus, which between 1948 and 1949 was

---

Paneth's student and collaborators since their time in Durham, continued with meteorite research after becoming Director of the Cosmochemistry Department in 1967. He studied the early history of the solar system, rocks from the Apollo-11 expedition to the Moon, and because of his deep interest in the geochemistry of Mars, the Institute was involved also in the analysis of Martian soils and rocks with a special spectrometer launched on board the rover of the Mars Pathfinder mission of 1997.

ILLUSTRATION 56 Van de Graaff accelerator, Max Planck Institute for Chemistry, Mainz, June 16, 1953

proposed independently by Maria Goeppert and by Hans D. Jensen in collaboration with Otto Haxel and Hans Suess.[233] More generally, as emphasized by Helge Kragh, "geochemists supplied astrophysicists, cosmologists, and nuclear physicists with important data that could not be obtained otherwise."[234]

233 The empirical regularities in the data related to the isotopic-abundance distribution of elements in rocks and meteorites established by geo- and cosmochemists, as well as the knowledge on the nuclear cross-sections accumulated during the war, proved crucial in the process that led toward such improvement of theoretical knowledge on the atomic nucleus. Luisa Bonolis: J. Hans D. Jensen. Research Profile. *Lindau Nobel Laureate Meetings*, 2014. https://www.mediatheque.lindau-nobel.org/research-profile/laureate-jensen. Last accessed 9/9/2018.

234 Helge Kragh: An Unlikely Connection: Geochemistry and Nuclear Structure. *Physics in Perspective* 2/4 (2000), 381–397. doi:10.1007/s000160050051. On the interaction between

ILLUSTRATION 57 Van de Graaff generator for 3 up to 5 million volts. Max Planck Institute for Chemistry, Mainz, June 6, 1956

nuclear chemistry and geochemistry and the role of interdisciplinary investigations in the origin of the nuclear shell model, see also Karen E. Johnson: From Natural History to the Nuclear Shell Model: Chemical Thinking in the Work of Mayer, Haxel, Jensen, and Suess. *Physics in Perspective* 6/3 (2004), 295–309. doi:10.1007/s00016-003-0203-x.

ILLUSTRATION 58 1.5 million-volt Cockcroft-Walton accelerator, Max Planck Institute for Chemistry, Mainz, June 6, 1956

Pioneering work by Paneth and his colleagues on the cosmogenic production of helium in meteorites stimulated Wolfgang Gentner's interest in the

ILLUSTRATION 59 Max Planck Institute for Chemistry, Mainz. Lower end of the Cockcroft-Walton cascade generator with deflecting magnet, June 6, 1956.

problem.[235] Parallel to his involvement with the founding and early research

235 Friedrich A. Paneth, P. Reasbeck, and K.I. Mayne: Helium 3 Content and Age of Meteorites. *Geochimica et Cosmochimica Acta* 2/5 (1952), 300–303. doi:10.1016/0016-7037(52)

activities of CERN,[236] Gentner used his excellent research skills to examine the radioactive decay of potassium into argon in order to be able to date rocks and meteorites. Gases in meteorites can be of primordial, radiogenic, or cosmogenic origin. Nuclides formed by nuclear reactions induced by high-energy cosmic rays are more common in meteorites than on Earth, where the atmosphere and the geomagnetic field screen out most cosmic rays. Around the mid 1950s, while his group in Freiburg was working on cosmic rays on the nearby Schauinsland mountain, and on low-energy nuclear reactions with their small Van de Graaff generator, Gentner and his colleague Josef Zähringer explored the presence of argon and helium as products of nuclear reactions in meteorites.[237]

In 1958, upon the death of his mentor Walther Bothe (who had finally received a Nobel Prize, in 1954), Gentner was the obvious successor and ended up founding the entirely new Institute for Nuclear Physics.[238] He was thus

---

90013-6. Friedrich A. Paneth, P. Reasbeck, and K.I. Mayne: Production by Cosmic Rays of Helium-3 in Meteorites. *Nature* 172/4370 (1953), 200–201. doi:10.1038/172200a0. On Gentner's contribution to cosmochemistry and geochronology, see Till Kirsten's essay in Gentner's centennial volume. Till A. Kirsten: Gentner und die Kosmochemie. Hobby oder Symbiose? In: Dieter Hoffmann, and Ulrich Schmidt-Rohr (eds.): *Wolfgang Gentner. Festschrift zum 100. Geburtstag*. Berlin: Springer 2006, 177–208. According to Kirsten, Paneth often visited Gentner in Freiburg and discussed measurement techniques. See Josef Zähringer: Isotope Chronology of Meteorites. *Annual Review of Astronomy and Astrophysics* 2 (1964), 121–148. doi:10.1146/annurev.aa.02.090164.001005.

236 His review of mesons physics written in 1959, when the powerful CERN Proton Synchrotron first went into operation—becoming for a brief period the world's highest energy particle accelerator—shows Gentner's deep knowledge of elementary particle physics from the dual perspective of cosmic ray research and high-energy physics with accelerators.

237 Wolfgang Gentner, and Josef Zähringer: Argon- und Heliumbestimmungen in Eisenmeteoriten. *Zeitschrift für Naturforschung A* 10/6 (1955), 498–499. doi:10.1515/zna-1955-0610. Wolfgang Gentner, and Josef Zähringer: Argon und Helium als Kernreaktionsprodukte in Meteoriten. *Geochimica et Cosmochimica Acta* 11/1–2 (1957), 60–71. Wolfgang Gentner, Hugo Fechtig, and G. Kistner: Edelgase und ihre Isotopenverschiebung im Eisenmeteorit Treysa. *Zeitschrift Naturforschung Teil A* 13/7 (1958), 569–570b. doi:10.1515/zna-1958-0719. Wolfgang Gentner, and Joseph Zähringer: Kalium-Argon-Alter einiger Tektite. *Zeitschrift Naturforschung Teil A* 14/7 (1959), 686–687. doi:10.1515/zna-1959-0716.

238 Bothe's succession and Gentner's appointment as Director of the newly founded Max Planck Institute for Nuclear Physics were briefly discussed during the meeting of the CPT session of June 26, 1957 (CPTS meeting minutes of 26.06.1957, AMPG, II. Abt., Rep. 62, No. 1731). The decision of the commission in charge of examining the call was accepted without comment by the members participating in the meeting. Heisenberg declared himself in favor of Gentner's appointment in a message sent to Karl Ziegler, Director of the Max-Planck-Institut für Kohlenforschung (MPI for Coal Research) and President

definitely entering into the field of cosmochemistry in coincidence with the inauguration of the space age.[239] At that time, the study of extraterrestrial materials was opening up a new research perspective which could provide novel insights into the origin and timing of the birth of our solar system and even into the workings of the Sun itself. These discoveries brought nuclear physicists and cosmic ray physicists into the field, and they realized that meteorites are a kind of "poor man's space probe," containing a wealth of information concerning the constancy in time and space of cosmic radiation and conditions in space that date back millions or even billions of years. Even members of the astrophysical community would eventually be attracted by the chance to obtain information on solar abundances from the study of solar wind gases that were implanted over time in meteorites or lunar soil.[240]

Joseph Zähringer, who made his initial contributions to cosmochemistry, together with Gentner, while obtaining his doctoral degree, then spent time in Brookhaven in the mid-1950s. The Brookhaven choice was no accident. Gentner had established contacts with US scientists from 1938 onward, and now the Cosmotron, which was one of the most powerful accelerators in the world, allowed studies on the interaction of very energetic protons with heavy nuclei. Such processes shed light on the cosmogenic production of different nuclear species in meteorites. At Brookhaven, Zähringer met Oliver Schaeffer, an experienced radiochemist, with whom he established a very fruitful collaboration.[241]

---

of the *Wissenschaftlicher Rat* (Scientific Council) on April 11, 1957: "After returning to Göttingen, I heard that the name Gentner was frequently discussed in connection with the matter of who would succeed Mr. Bothe. I can only agree with this proposal. Mr. Gentner is an excellent experimental physicist and has also proved this excellence in his work at CERN in Geneva. So, if Bothe's Institute is to maintain its previous line of work, and this seems to be the general opinion, then Gentner would certainly be the right successor" [our translation] (AMPG, III. Abt., ZA 1, No. 74).

239 Ulrich Schmidt-Rohr: *Die Aufbaujahre des Max-Planck-Instituts für Kernphysik.* Heidelberg: Max-Planck-Institut für Kernphysik 1998. Schmidt-Rohr, *Erinnerungen*, 1996.

240 Friedrich Begemann: Edelgase in Meteoriten. In: Generalverwaltung der Max-Planck-Gesellschaft (ed.): *Jahrbuch der Max-Planck-Gesellschaft zur Förderung der Wissenschaften e.V. 1972*. Göttingen 1972, 59–82. Friedrich Begemann: Noble Gases and Meteorites. *Meteoritics & Planetary Science* 31/2 (1996), 171–176. doi:10.1111/j.1945-5100.1996.tb02012.x. See also Ursula B. Marvin: Oral Histories in Meteoritics and Planetary Science. VIII. Friedrich Begemann. *Meteoritics & Planetary Science* 37/S12 (2002), B69–B77. doi:10.1111/j.1945-5100.2002.tb00905.x.

241 Oliver Schaeffer, and Josef Zähringer: High-Sensitivity Mass Spectrometric Measurement of Stable Helium and Argon Isotopes Produced by High-Energy Protons in Iron. *Physical Review* 113/2 (1959), 674–678. doi:10.1103/PhysRev.113.674. Oliver Schaeffer, and Josef

ILLUSTRATION 60
Josef Zähringer

ILLUSTRATION 61
Oliver Adam Schaeffer

Zähringer: Helium- und Argon-Erzeugung in Eisentargets durch energiereiche Protonen. *Zeitschrift für Naturforschung A* 13/4 (1958), 346–347. doi:10.1515/zna-1958-0413. Schaeffer was appointed external member at the institute in 1972 (CPTS meeting minutes of 22.04.1972, 20.06.1972, AMPG, II. Abt., Rep. 62, No. 1765, 1766). The relationship with him became instrumental for later work on the lunar samples. Till A. Kirsten: Oliver Adam Schaeffer. 20.02.1919-11.11.1981. *Berichte und Mitteilungen Max-Planck-Gesellschaft. Jahresbericht 1981 und Jahresrechnung 1980, Nachrufe.* München 1982, 33–36.

This national laboratory in the United States, a consortium of East Coast American universities on which CERN would be modeled,[242] became one of the main places for Gentner's scientific collaborations throughout the 1950s. In similarity to how theoretical plasma astrophysics in Göttingen was a foot in the door of American institutions such as Princeton, MIT, and Chicago, cosmochemistry provided Gentner with a point of entry into American experimental physics. Southwestern Germany's cosmochemists were to maintain global leadership in the field for the next half century, and this expertise in turn opened up new scientific fields to them; some of these fields were conceptually close, such as geochemistry, or solar system research, but others, such as neutrino physics, had a wider reach in experimental physics.

However, as in the case of the Göttingen scientists invited to the United States, there were strings attached: Cold War interests facilitated the integration of scientists from southwest Germany into American research projects. The radiochemical and mass spectrometric analysis of very small samples, the area of expertise of cosmochemists such as Gentner and his disciples, as well as his allies in Mainz, Heidelberg, Freiburg, and Bern, was needed at the time also to analyze the radioactive products created in the atmosphere by nuclear weapons explosions. Prominent researchers in the field included Gentner's friends Hans Jensen and Hans Suess at the University of Heidelberg.[243] The latter migrated to the United States in the 1950s to contribute to this research, while also paving the way for modern atmospheric science.

Interest in cosmic rays themselves at the time resulted from research on the effect of nuclear weapons explosions: it was vital to determine the extent to which cosmic rays in interaction with the atmosphere create new isotopes, which is to say, the latter's 'natural' occurrence, in order then to establish how many isotopes derive from nuclear weapons explosions. This research also led to another central methodological contribution by German scientists in the southwest: age determination, the analysis of rare isotopes to determine the age of a substance. This had begun with the cosmochemistry of meteorites but, by the 1950s, it was also extending to exotic new areas such as the carbon-14 dating of biological remnants to determine the time that had passed since

242 Accelerator Tradition Is Thriving at Brookhaven. *CERN Courier* 42/2 (2002), 24–27. https://cds.cern.ch/record/1733305. Last accessed 3/17/2021.

243 On relations between the Max Planck Institute and Heidelberg University after 1945, see Hans Kopfermann: Zur Geschichte der Heidelberger Physik seit 1945. In: Universitäts-Gesellschaft (ed.): *Heidelberger Jahrbücher*. Berlin: Springer 1960, 159–164. doi:10.1007/978-3-642-45950-4_9.

their death, finally providing a definitive, calibrated scale for geological and even historical dating.[244]

Collaboration in cosmochemistry was a means for German experimentalists to make themselves known through excellence in a specific field, opening the door to wider collaboration in other areas, including the mainstream of experimental physics: accelerator-based particle research. But ultimately, what made Gentner outshine others in the cosmochemistry field was his ability, together with his closest collaborators, to go beyond the questions linked to meteorite samples, like those first raised by Paneth, which related mostly to the composition and age of the solar system, toward much more profound, abstract, and fundamental questions at the heart of particle physics and cosmology, concerning the nature and behavior of neutrinos. In the mid-1960s, his disciple Till Kirsten and colleagues in Brookhaven used meteorite samples which had been bombarded by cosmic rays for billions of years to prove the existence of nuclear double-beta decay, a rare subatomic process which had been predicted theoretically decades earlier.[245] The fame that ensued, still grounded mostly in radiochemical methods and mass spectrometry, was a direct path to their continued participation in neutrino detection experiments in the following decades, which will be described in more detail in Chapter 5.

Southwestern Europeanism and the International Division of Scientific Labor

Gentner's successes in the 1950s and 1960s exemplified his approach to the regional, national, and international division of scientific labor. Through his cultural affinity and excellent relationships with the French, particularly Joliot, but also the cosmic ray physicist Pierre Auger, one of the founding figures of CERN, Gentner was the closest German ally of CERN, and his vision was

---

244 Collision of secondary neutrons produced by cosmic rays striking the atmosphere produce ^{14}C which combines with oxygen to form radioactive $^{14}CO_2$. Willard F. Libby: Atmospheric Helium Three and Radiocarbon from Cosmic Radiation. *Physical Review* 69/11–12 (1946), 671–672. doi:10.1103/PhysRev.69.671.2. While plants and living organisms are continuously renewing their content in carbon-12, which remains stable in the atmosphere, radioactive carbon-14 in dead animals, humans, and other samples, decays into nitrogen-14 over time at a predictable rate, providing a clock for the technique of radioactive dating. Willard F. Libby: Radiocarbon Dating. Nobel Lecture, 12/12/1960. https://www.nobelprize.org/nobel_prizes/chemistry/laureates/1960/libby-lecture.pdf. Last accessed 10/30/2018.

245 Till A. Kirsten et al.: Experimental Evidence for the Double-Beta Decay of Te^{130}. *Physical Review Letters* 20/23 (1968), 1300–1303. doi:10.1103/PhysRevLett.20.1300.

to organize German experimental particle physics around a hierarchy, with the laboratory in Geneva at the top of a system in which universities and other research institutions, such as the Max Planck Institutes, would participate equally.[246] In the late 1950s, when Gentner returned to Heidelberg in the newly founded Max Planck Institute for Nuclear Physics, he established from the outset a division of cosmochemistry that further strengthened the regional expertise in the field.[247] The new institute also had large experimental divisions centered around medium-sized tandem Van de Graaff accelerators, which did work that was complementary, not competing with CERN, and were accessible to researchers on a national scale.[248]

Gentner's interest in establishing a strong local accelerator division is testified by his parallel proposal to appoint as Scientific Member Ulrich Schmidt-Rohr, Bothe's former student and collaborator, who had participated in the building of a second cyclotron in Heidelberg in the 1950s.[249] In 1965, when Gentner proposed a collegial directorship, Schmidt-Rohr became one of the Directors of the Max Planck Institute for Nuclear Physics.[250] Such structure was soon further strengthened on the theoretical side by the appointment of

246 Adams, Gentner and CERN, 2006, 139–145. About Gentner's involvement with CERN, and attendant connections with German accelerator center DESY, see AMPG, III. Abt., Rep. 68 A, No. 51, 55, 56, 57, 58, 85, 86, 87. See also Hermann, Germany's Part, 1987, 383–429.

247 Josef Zähringer, Gentner's student and collaborator in Freiburg, was in charge of the Cosmochemistry Department. Gentner himself pursued cosmochemistry as a scientific activity and, in the early 1960s, as part of a wide-scale organizational strategy, he proposed Zähringer as a Scientific member and Head of the Department (see CPTS meeting minutes of 09.06.1964, 03.12.1964, 05.03.1965, AMPG, II. Abt., Rep. 62, No. 1743, 1744, 1745).

248 The Tandem accelerator was mentioned by Adolf Butenandt in his presidential address of June 1961 (held on the occasion of the 50th anniversary of the foundation of the Kaiser Wilhelm Society) as an excellent example of how an expensive large-scale equipment financed by the Max Planck Society could also be made available for university research and teaching, thus strengthening cooperative relationships with the academic world. Adolf Butenandt: *Das Werk eines Lebens. Wissenschaftspolitische Aufsätze, Ansprachen und Reden*. Vol. 2. Göttingen: Vandenhoeck & Ruprecht 1981, 27.

249 See CPTS meeting minutes of 19.01.1961, AMPG, II. Abt., Rep. 62, No. 1736. During the period 1963–65, as director of the newly founded Nuclear Research Center in Jülich, Schmidt-Rohr was involved in the building of an isochronous cyclotron. He also led a research group working with a similar cyclotron already running in Karlsruhe, where in particular he examined the shell model of the nuclear structure, which had always been one of his main research fields.

250 Such a proposal, discussed between 1964 and 1965, involved directors Gentner, Mayer-Kuckuk, Schmidt-Rohr, and Zähringer (CPTS meeting minutes of 03.12.1964 and 05.03.1965, AMPG, II. Abt., Rep. 62, No. 1744, 1745).

ILLUSTRATION 62 Inauguration of the experimental hall containing the tandem accelerator, Max Planck Institute for Nuclear Physics, Heidelberg, 1962. From left: Wolfgang Gentner, Otto Hahn, Siegfried Balke (then Federal Minister for Nuclear Energy), Adolf Butenandt (then President of the Max Planck Society), Werner Heisenberg. The tandem configuration of the Van de Graaff generator achieves a two-step acceleration of particles, thus providing a beam with twice the energy for the cost of one electrostatic generator.

Hans-A. Weidenmüller as a new Scientific Member, a theoretician whose experience of working in strong relationship with experimentalists had been built during the previous years in the United States.[251]

Weidenmüller was also meant to continue the strong tradition of collaboration with eminent nuclear physicists at the Heidelberg University, notably Hans Jensen, who had been the driving force behind rebuilding physics research in Heidelberg. In Heidelberg after the war, Jensen, a former member of the Uranium Club, had pulled in Hans Kopfermann and Otto Haxel to

251 CPTS meeting minutes of 07.04.1967, 07.06.1967, AMPG, II. Abt., Rep. 62, No. 1749, 1750. On this, see Hans-Arwed Weidenmüller: Nuclear Physics in Heidelberg in the Years 1950 to 1980. Personal Recollections. *European Physical Journal H* 40/3 (2015), 279–299. doi:10.1140/epjh/e2015-60019-4.

ILLUSTRATION 63 Brigitte Huck at the control panel of the tandem accelerator. Max Planck Institute for Nuclear Physics, Heidelberg

collaborate with him on developing the shell model for which Jensen shared the Nobel Prize in Physics 1963 with Maria Goeppert Mayer.[252]

Similarly to Heidelberg, the Max Planck Institute for Chemistry in Mainz also operated its own medium-sized accelerator for nuclear physics research in conjunction with the nearby university.[253] Over the following decades, this expertise with medium-sized accelerators in southwestern Germany was used to extend alliances outside of Europe, when, for example, one of the early instances of collaboration with the Weizmann Institute of Science in Israel

252 This network of influential nuclear physicists was instrumental in supporting Gentner's project for the foundation of a dedicated Max Planck Institute for Nuclear Physics. Kopfermann's collaborator Peter Brix later became a Scientific Member of the institute (CPTS meeting minutes of 12.11.1970, 09.02.1971, AMPG, II. Abt., Rep. 62, No. 1760, 1761). On Brix, see Hans-Arwed Weidenmüller: Peter Brix 25.10.1918-21.01.2007. *Jahresbericht der Max-Planck-Gesellschaft* Beilage Personalien (2008), 14–15. See also Wolfgang Gentner: Max-Planck-Institut für Kernphysik in Heidelberg. In: Generalverwaltung der Max-Planck-Gesellschaft zur Förderung der Wissenschaften e.V. (ed.): *Jahrbuch 1961 der Max-Planck-Gesellschaft zur Förderung der Wissenschaften e.V. Teil II*. Göttingen 1962, 486–491.

253 The Mainz accelerator had an interesting parallel story stemming back to the war years, when the institute was still in Berlin. See Weiss, The "Minerva" Project, 2003, 271–290.

ILLUSTRATION 64 Maria Goeppert Mayer and Hans D. Jensen in Heidelberg, in 1957.

was based on work with a similar device there, housed inside that institute's most iconic tower building.[254] The Heidelberg Institute for Nuclear Physics was part of a global network using these medium-sized devices, which also became the workhorses of American research universities, while the larger circular accelerators were increasingly centralized in only a few places around the world. The scientific work conducted with medium-sized linear accelerators was largely about the atomic nucleus, but it also had some uses in cosmochemistry itself, in order to produce nuclear reactions whose final products could be compared with those found on meteorites. Ultimately, Gentner extended the use of his accelerators to even more exotic purposes, such as accelerating macroscopic dust particles, for the experimental research on

254 Dietmar K. Nickel, and Ulrich Schmidt-Rohr: Wolfgang Gentner und die Begründung der deutsch-israelischen Wissenschaftsbeziehungen. In: Dieter Hoffmann, and Ulrich Schmidt-Rohr (eds.): *Wolfgang Gentner. Festschrift zum 100. Geburtstag*. Berlin: Springer 2006, 147–170. Dietmar K. Nickel: *Brückenpfeiler. Vierzig Jahre wissenschaftliche Zusammenarbeit zwischen Deutschland und Israel*. München: Minerva Stiftung 1998. On Gentner's involvement and role in the collaboration between Germany and the Weizman Institute of Science, see Thomas Steinhauser, Hanoch Gutfreund, and Jürgen Renn: *A Special Relationship. Turning Points in the History of German-Israeli Scientific Cooperation*. Berlin: GMPG-Preprint 2017.

micrometeorites. Then this initially exotic expertise on 'dusts' became generalized to a wide research field that, decades later, guaranteed the Heidelberg and Mainz institutes participation in interplanetary probes with an emphasis on comets and interplanetary dust analysis.[255]

Finally, as will be discussed in detail later, Gentner maintained a closely knit territorial network extending outwards from what had been the French occupation zone (containing Mainz and Freiburg), but which, by the late 1950s, also included his stronghold, the Max Planck Institute for Nuclear Research in Heidelberg, the University of Heidelberg (Hans Jensen), as well as the University of Bern, to which Fritz Houtermans, founder of the Bern tradition of cosmochemistry, had been appointed in 1952, after a tortuous trajectory of persecution by both Nazis and Communists.[256] And of course, there was CERN itself in Geneva. Married to a Swiss citizen, Gentner benefited from an extended homeland that spanned France, Switzerland, and southwest Germany. In the spirit of such European dimension, he was involved during the 1960s in the founding process of the European Physical Society by the Italian physicist Gilberto Bernardini, his colleague and good friend since the years of the construction of the CERN Synchro-Cyclotron.[257]

---

255 Hugo Fechtig, a former student of Gentner's at the University of Freiburg, applied the potassium-argon dating method to meteorites. He worked at the Max Planck Institute for Nuclear Physics from 1958 on, became a Scientific Member in 1974, and was Managing Director from 1979 to 1981. Together with Eberhard Grün, Fechtig studied the chemical composition of micrometeorites and interplanetary dust, a field that was continued with space probes (Helios and Giotto) and was extended to comets. Hugo Fechtig et al. (eds.): *Interplanetary Dust.* Berlin: Springer 2001. Hans Balsiger, Hugo Fechtig, and Johannes Geiss: A Close Look at Halley's Comet. *Scientific American* 259/3 (1988), 96–103. http://www.jstor.org/stable/24989232. Last accessed 10/30/2018. On the early development of dust accelerators, see Edingshaus, *Heinz Maier-Leibnitz*, 1986.

256 On Houtermans's career, see Edoardo Amaldi: *The Adventurous Life of Friedrich Georg Houtermans, Physicist (1903–1966).* Edited by Saverio Braccini, Antonio Ereditato, and Paola Scampoli. Berlin: Springer-Verlag 2012. Iosif B. Khriplovich: The Eventful Life of Fritz Houtermans. *Physics Today* 45/7 (1992), 29–37. doi:10.1063/1.881313. Konrad Landrock: Friedrich Georg Houtermans (1903–1966)—Ein bedeutender Physiker des 20. Jahrhunderts. *Naturwissenschaftliche Rundschau* 56/4 (2003), 187–199. http://www.naturwissenschaftliche-rundschau.de/navigation/dokumente/Beitrag-Landrock-4-2003.pdf. Last accessed 10/30/2018. Viktor J. Frenkel: *Professor Friedrich Houtermans. Arbeit, Leben, Schicksal. Biographie eines Physikers des zwanzigsten Jahrhunderts.* Edited by Dieter Hoffmann. Preprint 414. Max-Planck-Institut für Wissenschaftsgeschichte 2011. https://www.mpiwg-berlin.mpg.de/Preprints/P414.PDF. Last accessed 10/30/2018.

257 See related documents in AMPG, III. Abt., Rep. 68 A, No. 105, 106, 107, 108. On Gilberto Bernardini and Gentner's involvement in the foundation of the European Physical Society, see Lalli, Crafting Europe from CERN to Dubna, 2021, 103–131.

Throughout his period as a leader in this research tradition, Gentner was not shy about his cultural and national allegiances, chain-smoking French cigarettes, driving a Citroën DS—a cultural icon notable for its innovative, futuristic design—and displaying in his office a copy of the portrait of Frédéric Joliot-Curie drawn by Picasso.[258]

3 The Orphan Scenario: Regener, Kiepenheuer, and Dieminger

Weaker traditions were part of the early history of the Max Planck Society, and this section describes the reasons for their weakness as well as the contingencies on which becoming a Max Planck Institute depended during this early era. The eventual outcome was the patchwork Max Planck Institute for Aeronomy in Lindau, Lower Saxony, which would continue to be a somewhat problematic scenario, and one reason why coordinated action by the different power centers in the Max Planck Society was repeatedly necessary. The key figures in this section are Erich Regener, Germany's top cosmic ray researcher using balloons; Walter Dieminger, head of the wartime radio disturbance forecasting network which surveilled the ionosphere; and Karl-Otto Kiepenheuer, leading astronomer in the wartime project to predict radio disturbances with solar observatories. Their fates inside and outside the Max Planck Society illustrate the patterns and contingencies of membership in the organization in its first decade.

Preconditions of Success in the Early Max Planck Society

As we have seen, a combination of three factors led to the emergence of powerful factions within the Max Planck Society:

- A deeply rooted, globally recognized scientific tradition that adapted to survive the austere postwar era through a temporary reorientation of efforts toward theoretical and smaller-scale experimental programs, maintaining the Society's presence in the global scientific community and preparing it to move on to larger experimental and technological programs, once West Germany's political and economic renaissance permitted this. At the head of these traditions stood larger-than-life personalities such as Heisenberg and Gentner.
- The alignment of these preexisting scientific traditions, with a regional power base, often accompanied by a specific political orientation. The most

258 Hoffmann, and Schmidt-Rohr, *Wolfgang Gentner*, 2006, 248.

obvious case here is Gentner's, who effectively united universities and Max Planck Institutes in southwestern Germany in his initial role as favored intermediary with the French occupiers. The case of Heisenberg's family of institutes shows how scientific traditions were not inextricably linked to a single location, as the 'seat' of a scientific community could move in search of better conditions. Over the course of the 1950s, Göttingen and Lower Saxony became too small a playground for Heisenberg's ambitions, and his move to Munich in 1958, as well as that of the Max Planck general administration and presidency, signified one of the major shifts of power from the north to the south.

- Alignment of a preexisting scientific tradition with a politically and socially dominant contemporary cause, in this case, the 'nuclear age.' Regardless of the actual internal interests of researchers, the growth in scale and influence within Heisenberg's and Gentner's factions can only be explained by their association with the 'nuclear,' particularly in the 1950s. This came with a series of ritual behaviors that were occurring also in other countries and in the nascent international collaborations. These underlined the importance of the autonomy of scientists and, in West Germany, carried a certain weight justified by a particular interpretation of the recent past. Beyond Heisenberg's or Gentner's institutes and their allies, the Max Planck Society itself, through its president Otto Hahn, derived its early legitimacy from nuclear science.

The perfect counterexample to this tripartite alignment of conditions is constituted by what would have been a third faction, in fact better described as the 'orphan' scenario of astrophysics and space science in the Max Planck Society. These were a series of institutes and scientists, some of which eventually comprised the Max Planck Institute for Aeronomy in Lindau near Göttingen.

The institute was the result of the forced merger of two very different entities.

Erich Regener's Tradition of Airborne Probes

On the one hand, there was the scientific tradition of Erich Regener, one of the top German scientists in the first half of the century, known for his experimental skills and ingenuity in the design of instruments.[259] Following in the steps of the pioneering experimentalist Victor Hess, in the late 1920s and early

259 Hans-Karl Paetzold, Georg Pfotzer, and Erwin Schopper: Erich Regener als Wegbereiter der extraterrestrischen Physik. *Zur Geschichte der Geophysik. Festschrift zur 50jährigen Wiederkehr der Gründung der Deutschen Geophysikalischen Gesellschaft*. Berlin Hei-

ILLUSTRATION 65 Werner Heisenberg and Ludwig Biermann on August 21, 1956, at the ceremony for laying the foundation stone of the new headquarters of the new Max Planck Institute for Physics and Astrophysics in Munich

1930s, Regener became one of the world leaders in cosmic ray research, carrying out experiments on the absorption of cosmic rays in Lake Constance and

delberg: Springer 1974, 167–188. Hans-Karl Paetzold: Erich Regener. A Pioneer of Geophysical Research. In: Wilfried Schröder, and International Association of Geomagnetism and Aeronomy (eds.): *Historical Events and People in Geosciences. Selected Papers from the Symposia of the Interdivisional Commission on History of IAGA during the IUGG General Assembly, Held in Hamburg, 1983*. Frankfurt am Main: Peter Lang 1985, 59–63. Walther Bothe: Erich Regener 70 Jahre. *Zeitschrift für Naturforschung A* 6/11 (1951), 564–567. doi:10.1515/zna-1951-1101. Alfred Ehmert, and Erwin Schopper: In Memoriam Erich Regener. *Die Naturwissenschaften* 43/4 (1956), 69–71. doi:10.1007/BF00631846. Georg Pfotzer: On Erich Regener's Cosmic Ray Work in Stuttgart and Related Subjects. In: Yataro Sekido, and Harry Elliot (eds.): *Early History of Cosmic Ray Studies: Personal Reminiscences with Old Photographs*. Dordrecht: Springer Netherlands 1985, 75–89. doi:10.1007/978-94-009-5434-2_8. See also Bothe and Hahn's contributions in the special issue of the *Zeitschrift für Naturforschung* published on the occasion of Erich Regener's 70th birthday: Bothe, Erich Regener 70 Jahre, 1951, 564–567. Otto Hahn: Erich Regener und die Max-Planck-Gesellschaft. *Zeitschrift für Naturforschung A* 6/11 (1951), 568–569. doi:10.1515/zna-1951-1102.

ILLUSTRATION 66 Inauguration ceremony of the new Max Planck Institute for Physics and Astrophysics in Munich, May 9, 1960. From left: the Secretary of State, Fritz Staudinger, Otto Hahn, Ludwig Biermann, and Werner Heisenberg

in the high atmosphere, up to the stratosphere.[260] According to Bruno Rossi, himself one of the fathers of modern cosmic ray studies since the beginning of the 1930s, and a protagonist of later developments:

> In the late 1920s and early 1930s the technique of self-recording electroscopes, carried by balloons into the highest layers of the atmosphere or sunk to great depths under water, was brought to an unprecedented degree of perfection by the German physicist Erich Regener and his group. To these scientists we owe some of the most accurate measurements ever made of cosmic ray ionization as a function of altitude and depth.[261]

260 Erich Regener: Spectrum of Cosmic Rays. *Nature* 127/3198 (1931), 233–234. doi:10.1038/127233b0. Erich Regener: Intensity of Cosmic Radiation in the High Atmosphere. *Nature* 130/3279 (1932), 364–364. doi:10.1038/130364a0. Erich Regener: Messung der Ultrastrahlung in der Stratosphäre. *Naturwissenschaften* 20/38 (1932), 695–699. doi:10.1007/BF01494465.

261 Rossi, *Cosmic Rays*, 1964, 9–10.

Regener's scientific prestige was therefore associated with the delivery vehicles and the design of automated airborne instrumentation that could be small, light, and resistant to extreme conditions. Regener also gained worldwide renown for his stratospheric research and, in particular, for his work on the ozone layer, which he started in the 1930s.[262]

Regener was professor at the Technical University of Stuttgart until 1937, when he was forced to leave by rivals, because of the Jewish background of his wife Viktoria Mintschin and for having signed the "Heisenberg-Wien-Geiger Memorandum," a *Denkschrift* protesting National Socialists' attacks in the press on physics (especially theoretical physics). Similarly to Bothe, who had found a new home at the Max Planck Institute for Medical Research in Heidelberg, Regener negotiated with the Kaiser Wilhelm Society for a largely self-funded Research Laboratory for the Physics of the Stratosphere, in Friedrichshafen on Lake Constance, to continue his work.[263] In the mid-1930s, in collaboration with his student Georg Pfotzer, Regener discovered

---

262 In 1906, Regener had been one of the first to study the decomposition of ozone under the action of ultraviolet light. Erich Regener: Über die chemische Wirkung kurzwelliger Strahlung auf gasförmige Körper. *Annalen der Physik* 325/10 (1906), 1033–1046. doi:10.1002/andp.19063251008. Once it had been firmly established at the beginning of the 20th century that ozone was responsible for the absorption of ultraviolet radiation from the Sun—as well as for its location in the stratosphere—Chapman's classic articles proposing the first model of the distribution of ozone as a function of altitude in the atmosphere and explaining the existence of an ozone layer launched the modern era of atmospheric ozone research. Sidney Chapman: On Ozone and Atomic Oxygen in the Upper Atmosphere. *The London, Edinburgh, and Dublin Philosophical Magazine and Journal of Science* 10/64 (1930), 369–383. https://doi.org/10.1080/14786443009461588. Last accessed 3/1/2018. Sidney Chapman: A Theory of Upper-Atmospheric Ozone. *Memoirs of the Royal Meterological Society* 3/26 (1930), 103–125. In 1934, while deeply involved in his investigations of cosmic radiation, Regener measured with his son Victor the solar ultraviolet absorption with a stratospheric balloon, showing that the ozone maximum is located near 25 km. Erich Regener, and Victor H. Regener: Ultra-Violet Solar Spectrum and Ozone in the Stratosphere. *Nature* 134 (1934), 380. doi:10.1038/134380a0. Erich Regener: Das atmosphärische Ozon. In: Field Information Agencies Technical, and Office of Military Government for Germany (eds.): *Geophysics. Part II*. Wiesbaden: Dieterich'sche Verlagsbuchhandlung 1948, 297–307.

263 On his involvement with the Kaiser Wilhelm Society, see Carl Freytag: »Bürogenerale« und »Frontsoldaten« der Wissenschaft. Atmosphärenforschung in der Kaiser-Wilhelm-Gesellschaft während des Nationalsozialismus. In: Helmut Maier (ed.): *Gemeinschaftsforschung, Bevollmächtigte und der Wissenstransfer. Die Rolle der Kaiser-Wilhelm-Gesellschaft im System kriegsrelevanter Forschung des Nationalsozialismus*. Göttingen: Wallstein Verlag 2007, 215–267, 238–246.

ILLUSTRATION 67 From left: Georg Pfotzer and Erich Regener around 1950 in Weissenau

the altitude at which the production of ionization in the atmosphere reaches a maximum.[264]

Regener was also one of the first to estimate the energy density of cosmic rays, an estimate that was used in 1933 by astronomers Walter Baade and Fritz Zwicky (who had both moved to the United States on a Rockefeller Fellowship) to propose that they might originate in supernova explosions.[265]

---

264 Per Carlson, and Alan A. Watson: Erich Regener and the Ionisation Maximum of the Atmosphere. *History of Geo- and Space Sciences* 5 (2014), 175–182. doi:10.5194/hgss-5-175-2014. The authors also argue that Regener's name is less recognized by present-day cosmic ray physicists than it should be, largely because of his forced early retirement.

265 Walter Baade, and Fritz Zwicky: Cosmic Rays from Super-Novae. *Proceedings of the National Academy of Sciences* 20/5 (1934), 254–259. doi:10.1073/pnas.20.5.259.

In 1939, together with his colleague Alfred Ehmert, Regener studied at different altitudes extensive air showers generated by the collision in the atmosphere of very-high-energy particles, in other words, the 'primary' cosmic radiation.[266] This strand of research directed the attention of scientists toward the astrophysical sources of cosmic rays and the mechanisms of their acceleration to such high energies. Ehmert then explored the possibility of a connection between the solar activity and the observed variation of cosmic rays, which he discovered in 1942.[267]

All this increased interest in the Sun–Earth interaction and, later, also in the interaction of cosmic rays with galactic magnetic fields. Throughout the 1940s, Regener was part of the scientific conversation with astrophysicists such as Karl-Otto Kiepenheuer (detailed below), Albrecht Unsöld, Ludwig Biermann, and, in particular, Hannes Alfvén (treated in Section 1 of this chapter; later awarded a Nobel Prize in Physics for his theoretical work leading to applications in different areas of plasma physics), that marked the transition from a basic interest in individual solar particles to the general field of space plasmas coming from the Sun and in cosmic sources.[268] Their brilliant theoretical insights notwithstanding, Regener's and Kiepenheuer's traditions were eminently experimental, and when finding themselves having to deal with complex plasma physics questions, they sought input from theoretical astrophysicists such as Biermann or Alfvén.

Similar to Bothe, although he had been persecuted by ideological Nazi sympathizers at his university, Regener's new research aligned him well with the interests of the more pragmatic Nazi rulers, and, in particular, he was supported by Hermann Göring's air force.

Parallel to his relevant activity in cosmic ray research, in collaboration with Kiepenheuer, Regener developed instruments flown in sounding balloons which were able to measure the ultraviolet radiation of the Sun at a height of more than 30 km (ionization at altitudes where the primary cosmic radiation is interacting with the atmosphere).[269] During the war, Regener, invited

266 Erich Regener, and Alfred Ehmert: Über die Schauer der kosmischen Ultrastrahlung in der Stratosphäre. *Zeitschrift für Physik* 111/7–8 (1939), 501–507. doi:10.1007/BF01329511. All these results were achieved with balloon-based research. Georg Pfotzer: History of the Use of Balloons in Scientific Experiments. *Space Science Reviews* 13/2 (1972), 199–242. doi:10.1007/BF00175313.

267 Ehmert, Ultrastrahlung, 1948, 264–285.

268 Hannes Alfvén: *Cosmical Electrodynamics*. Oxford: Oxford University Press 1950.

269 See Erich Regener: Über die Temperatur der höchsten Atmosphärenschichten. *Die Naturwissenschaften* 29/32–33 (1941), 479–484. doi:10.1007/BF01485940. Michael Globig: Mit

ILLUSTRATION 68 Walther Bothe and Erich Regener in 1955, in front of the Max Planck Institute for Medical Research in Heidelberg

by Wernher von Braun to Peenemünde, drew on his earlier work with balloon instruments for Kiepenheuer[270] to help with the instrumentation for the V-2 rockets and, by the end of the war, had even famously designed and built the world's first extraterrestrial scientific payload, a device for detecting ultraviolet radiation, intended to be carried aloft by one of the V-2s.[271] After a successful test flight, the project was cancelled in September 1944 because of war priorities, since the rockets were being used as long-range missiles against

der Tonne in die Atmosphäre. *Max Planck Forschung* 4 (2006), 56–57. https://www.mpg.de/971205/S003_Rueckblende_056_057.pdf. Last accessed 10/30/2018.

270 Wolfgang Mattig: Nachrufe. Karl-Otto Kiepenheuer. *Mitteilungen der Astronomischen Gesellschaft* 38 (1976), 11–13. https://ui.adsabs.harvard.edu/#abs/1976MitAG..38...11M. Last accessed 10/30/2018. On Kiepenheuer's involvement in wartime activities with V-2 rockets, see Cornelis de Jager: Early Solar Space Research. In: Johan A. M. Bleeker, Johannes Geiss, and Martin C. E. Huber (eds.): *The Century of Space Science*. Dordrecht: Kluwer Academic Publishers 2001, 203–223. See also Karl-Otto Kiepenheuer: Solar-terrestrische Erscheinungen. In: Paul ten Bruggencate (ed.): *Astronomie, Astrophysik und Kosmogonie*. Wiesbaden: Dieterich'sche Verlagsbuchhandlung 1948, 229–284.

271 Globig, Mit der Tonne in die Atmosphäre, 2006, 56–57.

ILLUSTRATION 69 Rare image of the fully equipped 'Regener Tonne,' the first scientific rocket payload designed to reach the upper part of the atmosphere. This image was found by chance by the engineer Gerhard H. R. Reisig, who had been in charge of the instrumentation and measurement parts of the instrument. The project, developed in the early 1940s, was under strict secrecy, so no pictures or sketches of the device could be officially shown.

the United Kingdom.[272] After the war, it was rumored that the *Regener-Tonne* (Regener barrel) had been confiscated by the Allies, although it has never resurfaced to this day.

In any case, as part of the Alsos Mission, the Allies, in a task force that included astronomers Gerard Kuiper and Fritz Zwicky, learned of Regener's rocket-borne scientific plans and brought back instruments to kick-start (or rather, continue) this type of work with captured rockets.[273] When the parts and pieces of German V-2 arrived in New Mexico, the US suddenly had the

272 On Regener's scientific contributions during the war, see, for example, David H. DeVorkin: *Science With A Vengeance. How the Military Created the US Space Sciences After World War II*. New York, NY: Springer-Verlag 1992, 3–4. For an assessment of his relationship with the Third Reich, see Freytag, »Bürogenerale« und »Frontsoldaten«, 2007, 215–267, 231–264.

273 DeVorkin, *Science With A Vengeance*, 1992, 4.

ILLUSTRATION 70 Preparation for the launch of an A-4 rocket, 1942: the control section of the world's first large rocket to travel in space, which was 14 meters high and weighed 12.9 tons with full tanks, is open; above there is space for a payload of one ton. The A4 missile (A stands for Aggregate), a series of ballistic rockets developed by Wernher von Braun's group in Peenemünde, was later named the V2, Vergeltungswaffe (weapon of retaliation).

capability to launch heavy payloads to altitudes well exceeding 150 kilometers and thus also to fly scientific instruments. In 1946, British and American scientists mounted their own instrumentation on captured V-2 rockets made available for research and obtained the first photograph of the solar ultraviolet spectrum, which had been one of Regener's main objectives.[274] The application of sounding rockets to solar physics, high-atmosphere research (aeronomy), and astronomy was successfully continued in the United States by Herbert Friedman, a pioneer in the space sciences, who in 1949 detected solar X-ray and ultraviolet radiation with a V-2 rocket.[275] These early rocket research projects opened a gap with European scientists which would be filled only with the advent of the space age.

When the Max Planck Society was founded, Regener had the powerful position of Vice-President, and continued with his research site in Weissenau, southwestern Germany, to where he had moved in the final years of the war, to escape Allied air raids.[276] There, his institute even took control of the equipment that Heisenberg's Kaiser Wilhelm Institute for Physics had relocated to southwestern Germany to build its nuclear reactor. After the war, the French occupying forces prohibited the return of this equipment to the British zone and, in a deal brokered by Gentner and Heisenberg, the instrumentation and personnel remained under Regener's stewardship for several years.[277] Regener had soon applied to reopen the aerodynamics section of his KWG institute in support of his high-altitude cosmic ray research, which was of the utmost interest and importance to all British-zone physicists.[278] He had been a forerunner in extraterrestrial research, and having "weathered the devastation of the Hitler regime and of the War," as remarked by Patrick Blackett, afterwards did much to rebuild the great tradition of German science.[279]

---

274 Richard Tousey et al.: The Solar Ultraviolet Spectrum from a V-2 Rocket. *The Astronomical Journal* 52/1162 (1947), 158–159. doi:10.1086/106028.

275 Herbert Gursky: Herbert Friedman. *Physics Today* 54/3 (2001), 94. doi:10.1063/1.1366078. Herbert Friedman: From the Ionosphere to High Energy Astronomy. A Personal Experience. In: Johan A. M. Bleeker, Johannes Geiss, and Martin C. E. Huber (eds.): *The Century of Space Science*. Dordrecht: Kluwer Academic Publishers 2001, 277–286. DeVorkin, *Science With A Vengeance*, 1992, 237–241.

276 Hahn, Erich Regener, 1951, 568–569.

277 Rechenberg, Gentner und Heisenberg, 2006, 63–94, 70–71.

278 Cassidy, Controlling German Science, II, 1996, 197–239, 218.

279 P. M. S. Blackett: Prof. E. Regener. 4469. *Nature* 175/4469 (1955), 1107–1108. doi:10.1038/1751107a0.

ILLUSTRATION 71 Research Center for Stratospheric Physics, Weissenau, early post-war years. Preparing the launch of a Regener's balloon tandem. The telescope was used to track the balloon tandem so that the payload could potentially be recovered intact before it touched the ground.

Then, however, something happened that was to resurface as a consistently grave problem in the Max Planck Society: the death of a prominent figure. Regener, who was born in 1881, passed away in 1955, too early for the Weissenau site to have become a full-fledged Max Planck Institute with long-term stability.[280] The future of the institute in Weissenau did not look rosy: it was obvious to people like Heisenberg that no one of equivalent stature could ever be found to replace Regener.[281] Instead it was decided to move the entire institute to Lindau, near Göttingen,[282] and merge it with an entirely different scientific tradition under a nominally

280 Ehmert, and Schopper, In Memoriam Erich Regener, 1956, 69–71.

281 Memorandum of March 7, 1955, by unknown author, "Conversation with Heisenberg," AMPG, II. Abt., Rep. 66, No. 3047.

282 Erhard Keppler: *Der Weg zum Max Planck Institut für Aeronomie. Von Regener bis Axford—eine persönliche Rückschau.* Katlenburg-Lindau: Copernicus 2003. Julius Bartels, Walter Dieminger, and Alfred Ehmert: Max-Planck-Institut für Aeronomie in Lindau. In: Generalverwaltung der Max-Planck-Gesellschaft zur Förderung der Wissenschaften e.V. (ed.): *Jahrbuch 1961 der Max-Planck-Gesellschaft zur Förderung der Wissenschaften e.V. Teil II.* Göttingen 1962, 16–45. Max-Planck-Gesellschaft (ed.): *Max-Planck-Gesellschaft. Berichte*

ILLUSTRATION 72 Research Center for Stratospheric Physics, Weissenau. Balloon ascent in early 1951

common interest in atmospheric research. This was the group led by Walter Dieminger.

Walther Dieminger's Research on Radio Disturbances in the Ionosphere

Dieminger's group was one among a series of research teams who found themselves within the Max Planck Society somewhat begrudgingly. The group dated from the wartime project to predict radio communication disturbances caused by changes in the ionosphere, that part of the upper atmosphere consisting of charged particles which permits radio communication at great distances. German researchers had been forerunners in radio-based ionosphere research since the early 20th century, with pioneers such as Jonathan Zenneck and his disciple Johannes Plendl, the latter having in fact coined the term *ionosphere* in the early 1930s.[283] Plendl had then applied his expertise in radio signals to the development of the navigation systems that directed the Luftwaffe's

---

und Mitteilungen 4/81. Max-Planck-Institut für Aeronomie Katlenburg-Lindau. München 1981.

283 Georg Schmucker: *Jonathan Zenneck 1871–1959. Eine technisch-wissenschaftliche Biographie.* Universität Stuttgart 1999.

ILLUSTRATION 73 Research Center for Stratospheric Physics, Weissenau. Observation of the balloon tandem flight

nighttime bombardments of Britain. During the war, he became plenipotentiary (*Bevollmächtigter*) in the field of high frequency research and, among other programs, Dieminger's network was included in this and, at its peak, it had about a dozen stations distributed throughout occupied Europe, from Norway to Sicily, from Ukraine to France.[284] This service was most useful for communications with the submarine fleet and it even coordinated communications between Europe and the African forces under Erwin Rommel. The headquarters where Dieminger was stationed shifted during the war, from

284 The best source on both Dieminger and Kiepenheuer is Seiler, *Kommandosache*, 2007. See also, Walter Dieminger: Radio Communication and Solar-Terrestrial Physics. In: Wilfried Schröder, and International Association of Geomagnetism and Aeronomy (eds.): *Historical Events and People in Geosciences. Selected Papers from the Symposia of the Interdivisional Commission on History of IAGA during the IUGG General Assembly, Held in Hamburg, 1983*. Frankfurt am Main: Peter Lang 1985, 11–27.

Reichlin in the Berlin area, to the outskirts of Vienna, and then to Upper Austria. Its activities consisted in sending signals of varying frequencies to the ionosphere and registering their echoes and their reception by the network stations, then issuing recommendations for which frequencies to use at different times and locations for the next two weeks, which was known as a radio-weather (*Funkwetter*) prediction. After the war, it turned out that this prediction network was much better organized than any Allied counterpart, making Dieminger an attractive cooperation partner for the occupiers.[285]

Karl-Otto Kiepenheuer's Solar Disturbances Prediction Network

Dieminger's radio interference surveillance network based on direct ionospheric research had an astronomical counterpart directed by Plendl. This also happened to be the largest German wartime application of astronomy: a project aiming to predict radio interference but based on solar observations. It was already known at the time that solar activity was the cause of the disturbances in the ionosphere, but there had been no large-scale attempts to understand these phenomena to the point of making predictions. The communication needs of the war now provided an opportunity to conduct this research. This effort was directed by the solar astronomer Karl-Otto Kiepenheuer, who had recruited many observatories in Germany and throughout the occupied and neutral territories of Europe, sometimes even requisitioning instrumentation to be redistributed among the various stations. Most of these observatories also erected antennae for Dieminger's network.[286]

Similar to Gentner, Kiepenheuer would subsequently be praised by scientists in the occupied countries for having defended their autonomy and interests against the German occupation forces, and Kiepenheuer was even accused of disloyalty to Germany in the final years of the war. His personnel requests rescued many astronomers from having to serve on the frontlines,

---

285 Karl Rawer: *Meine Kinder umkreisen die Erde. Der Bericht eines Satellitenforschers.* Freiburg im Breisgau: Herder 1986. Rawer was Dieminger's deputy in this program and this book details the entire history of wartime ionosphere research and its postwar fate.

286 Seiler, *Kommandosache*, 2007. Pages 48 and 88 include historical diagrams of these networks. This book contains a comprehensive account of the development of solar physics in Germany during the war, in particularly emphasizing solar-terrestrial relations (and the related connection between solar activity and ionospheric disturbances), which were investigated through a large network of solar observation stations by Kiepenheuer and Hans Plendl (the latter responsible at the Air Force Research Center of the Luftwaffe in Rechlin for the monitoring of solar activity and its influence on long-range radio communications).

and he occasionally even used the network of observatories in occupied countries to provide an escape route for persecuted scientists; in the later years of the war, he was suspected of preparing a similar exit for himself.[287] What is more problematic is the fact that, as in the case of Gentner, such a 'disloyal' figure could much more effectively coordinate research with scientists in occupied countries. Ultimately, however, unlike Dieminger's radio-based service, Kiepenheuer's attempts to forecast disturbances based on direct solar observations provided little useful information during the war, and he had relatively little military value for the Allies. From the end of the war to the early 1950s, Kiepenheuer had a key ally in the Dutch-American astronomer Gerard Kuiper, who had himself authored the Allied report on German activities in astronomy during the war, and who had singled out Kiepenheuer among all the German astronomers as not having been a Nazi collaborator. This was despite Kiepenheuer's leading role in the network of observatories spanning German-occupied Europe.[288]

Due to this background, Kiepenheuer, like Gentner, found a new home in the postwar years in Freiburg, within the French zone, where in the final years of the war he had already established a new solar research institute (now the Kiepenheuer Institute for Solar Physics), which to this day coordinates German solar astronomy and its participation in European and international collaborative projects. Kiepenheuer's work in the postwar years led to the founding, decades later, of the world's largest ground-based solar observatory, the Teide Observatory on Tenerife. All this was done independently of the Max Planck Society.

Distributing the Spoils of Ionospheric Research among the Allies

In Dieminger's case, to their own surprise, the Germans were in fact far more advanced than the Allies in this type of ionospheric observation, and as in other war-relevant fields, the months after the unconditional surrender became a race for the scientific spoils of the war.[289] An inspection team guided by Dieminger's deputy Karl Rawer departed from the French zone for Dieminger's headquarters in the American zone of Austria, inspecting several of the network's outposts along the way. At Dieminger's headquarters,

287 All these events in Kiepenheuer's life are central to Seiler, *Kommandosache*, 2007.

288 Seiler, *Kommandosache*, 2007, 224–225. Kuiper was in turn accused by German astronomers of using technology developed in Germany to obtain his pioneering infrared spectra of Mars (p. 219).

289 The Allied 'race' toward Dieminger at the end of the war is best described in Rawer, *Meine Kinder umkreisen die Erde*, 1986.

a decision was made among the Allies to transfer the entirety of Dieminger's installations to the British zone, while the French would return to their zone with Rawer and his colleagues. The Americans overruled this plan, however. The British decided, nonetheless, to quickly 'steal' Dieminger's installations in a clandestine operation with 70 trucks crossing through the American sector, and ultimately reinstalled them near the Harz mountains in their German occupation zone, putting them back into operation in 1948.[290] The French established their own counterpart near Freiburg; directed by Rawer as a forecasting service owned by the French navy, the *Service de prévision ionosphérique de la Marine* (SPIM) operated for the next decade. This ionospheric research network expanded globally under the French, to French colonial possessions in Africa and even as far as Vietnam.[291]

The very different fates of Kiepenheuer, Rawer, and Dieminger illustrate how strongly the development of the young Max Planck Society depended on the predecessor institutes' location in different occupation zones and on how much military value a research team was perceived to have. In the British zone, there was a strong emphasis on founding the Max Planck Society and on quickly expanding its footprint even beyond that of the prewar Kaiser Wilhelm Institutes. This expansion was pushed through against the wishes of the major figures in the new Society in the British zone, such as Heisenberg or Hahn, and included institutions like the Göttingen Central Workshop or Institute for Scientific Instruments, as well as Dieminger's station, initially not a full-fledged Max Planck Institute, but one of the institutes 'under the tutelage' of the Max Planck Society. Dieminger himself recognized that the MPG leadership expected to be able to dispose of the institute once British pressure had subsided.[292] After all, since the institute had been largely a forecasting service in a military context, it did not fit with the Society's newly self-proclaimed mission: to support fundamental science.

In the French zone the complete opposite happened. First of all, there was no equivalent pressure from the French to bring the research institutes in their zone into the Max Planck Society, as they initially fostered a network

290 Rawer, *Meine Kinder umkreisen die Erde*, 1986, 72–74. For more details on the early history of the Institute in Lindau, see Peter Czechowsky, and Rüdiger Rüster (eds.): *60 Jahre Forschung in Lindau. 1946–2006. Vom Fraunhofer-Institut zum Max-Planck-Institut für Sonnensystemforschung. Eine Sammlung von Erinnerungen*. Katlenburg-Lindau: Copernicus 2007, 24–32.

291 Rawer, *Meine Kinder umkreisen die Erde*, 1986, 92–98.

292 Hans-Willy Hohn, and Uwe Schimank: *Konflikte und Gleichgewichte im Forschungssystem. Akteurkonstellationen und Entwicklungspfade in der staatlich finanzierten außeruniversitären Forschung*. Frankfurt am Main: Campus Verlag 1990. 95–96.

of independent institutes closely allied with the nearby universities. In the early postwar years, there was even deliberate resistance to the widespread incorporation of institutes into the Max Planck Society, although this was primarily organized not by the French themselves but by scientists in their zone (the *Tübinger Herren*, headed by biochemist Adolf Butenandt), who objected to what they regarded as a too indiscriminate expansion of the Max Planck Society in the British zone.[293] Eventually, institutes in the French zone did enter the Max Planck Society, but most of these were former Kaiser Wilhelm Institutes, including the one headed by Butenandt, which had relocated from Berlin to Tübingen. More applied institutes such as Rawer's ionospheric forecasting service continued instead under the stewardship of the French until the mid-1950s, when it was handed over to the *Bundespost* (German Federal Post Office). In a manner typical of the French zone, Rawer also established an institute at the University of Freiburg that eventually participated in rocket-based ionospheric research with French rockets launched from Algeria, and later even led some satellites with the Americans. But Rawer remained outside the Max Planck Society throughout his entire career.[294]

Other, even more 'fundamental' institutes did not make it into the Max Planck Society, however. Kiepenheuer counted on the full political support of the French administrators, as well as the astronomical community of Western Europe, fostering the kind of independent organization epitomized, ultimately, by the Kiepenheuer Institute itself, namely, one open to close collaboratation with other similar institutes within the French zone. For example, Kiepenheuer's solar observatory on the Schauinsland mountain above Freiburg was used in the early postwar period by Wolfgang Gentner, to install a cosmic ray laboratory. Kiepenheuer's work fit the definition of leading-edge, fundamental science, in this case observational solar astronomy, and he also made major contributions to explaining how radio emissions from galaxies—one of the first findings of radio astronomy—could be created by the phenomenon of synchrotron radiation, produced by cosmic rays propagating in interstellar magnetic fields.[295]

293 Jeffrey Lewis: Kalter Krieg in der Max-Planck-Gesellschaft. Göttingen und Tübingen—eine Vereinigung mit Hindernissen, 1948–1949. In: Wolfgang Schieder, and Achim Trunk (eds.): *Adolf Butenandt und die Kaiser-Wilhelm-Gesellschaft. Wissenschaft, Industrie und Politik im »Dritten Reich«*. Göttingen: Wallstein Verlag 2004, 403–443. See also Eckart Henning, and Marion Kazemi: *Chronik der Kaiser-Wilhelm-/Max-Planck-Gesellschaft zur Förderung der Wissenschaften 1911–2011. Daten und Quellen*. Berlin: Duncker & Humblot 2011, 276–316.

294 Rawer, *Meine Kinder umkreisen die Erde*, 1986.

295 Kiepenheuer, Cosmic Rays, 1950, 738–739.

Even when the French did begin integrating their zone's 'civilian' institutes into the Max Planck Society in the early 1950s, Kiepenheuer's institute remained exempt. Between the scientists working for Kiepenheuer and for Göttingen (for the Max Planck Society and the university, respectively, in the latter case), a political antipathy had developed over their differing interpretation of their wartime activities, and this isolated Kiepenheuer from many of his former German colleagues, while he remained closer to those in the formerly Nazi-occupied countries.[296] Kiepenheuer's closest ally within the Max Planck Society would have been Erich Regener, with whom he had collaborated in the past and whose son, Victor, was meanwhile his research colleague. But, as mentioned above, Regener passed away in 1955.

An equally important factor was that the Max Planck Society's most prominent physicists were invested primarily in scientific research that intersected with nuclear questions, and observational astronomy generally fell outside of their area of interest, at least until after 1957. Even the kind of solar astrophysics pursued by Biermann, who, as we have explained, gained through this field a foothold in the crucial 'nuclear' realm of plasma physics, saw little need to 'own' solar observatories, and, when necessary, simply interacted with established observatories, first in Germany and, increasingly, also in the United States.[297] Kiepenheuer's solar astronomy institute is just one example of how, before 1957, 'space' was not a field of primary interest for the leading Max Planck Society scientists, except when it was directly related to 'nuclear' questions. None of the preexisting astronomical observatories in the Western occupation zone was absorbed by the Max Planck Society, in contrast to developments in the East, where the Astrophysical Observatory in Potsdam was very

---

296 Seiler, *Kommandosache*, 2007. Astronomers who had worked for the wartime solar astronomy project coordinated by Kiepenheuer were dismayed to find themselves categorized by the Allied evaluators headed by Gerard Kuiper according to specific levels of allegiance to National Socialism, while the only senior German astronomer of the group to be absolved was Kiepenheuer himself, the main source of information for Kuiper (p. 225). This strained his relationship with his former colleagues in the postwar era. A few years later, after Kiepenheuer visited Kuiper in Chicago, their relationship, too, was not without conflict. Kuiper subsequently spread the word that Kiepenheuer was a mediocre scientist, as reported by Reimar Lüst during the Roundtable "*Astronomy and Astrophysics in the History of the Max Planck Society* with a special focus on the *Changes in the 'cluster' of astronomy and astrophysics within the MPG*," Max Planck Institute for the History of Science, October 21, 2016. *Research Program History of the Max Planck Society. Report 2014–2017*. Edited by Florian Schmaltz et al. 2014–2017. Berlin 2017, 108–109. Such circumstances most probably further damaged Kiepenheuer's career opportunities in Germany.

297 Conversation with Reimar Lüst (Roundtable *Astronomy and Astrophysics in the History of the Max Planck Society*).

quickly brought under the aegis of the Soviet-style counterpart to the MPG, the Academy of Sciences.[298]

The Patchwork Max Planck Institute for Aeronomy

The final move in this history of 'orphans' came with the appointment of a successor to Regener, Professor Julius Bartels of the University of Göttingen, a prominent figure in magnetospheric research since the 1930s.[299] During the war, Bartels had headed geophysical research at the Potsdam observatory, which worked closely with Kiepenheuer and Dieminger's radio forecasting teams, and his experience with British and American initiatives in this area up to 1941 informed the German project.[300] Like many other scientists from the Berlin area, he had moved to Göttingen at the end of the war and continued to be formidable, connected internationally through his extensive prewar experience.

Bartels accepted the directorship of the relocated institute on the condition that he could retain his professorship in Göttingen, strengthening the case for reestablishing Regener's group in the area.[301] The decision was made to create a new institute, loosely defined by high-atmosphere research or 'Aeronomy,' with two independent sub-institutes: Regener's former Institute for Physics of the Stratosphere, now under Bartels, dedicated to stratosphere research with high-altitude probes; and Dieminger's group, which had been incorporated in the Kaiser Wilhelm Society in 1946, and renamed Max Planck Institute for Ionospheric Research in 1948, now continuing its ionosphere observations based on the wartime technique of bounced radio signals.[302] The ensuing

298 See, for example, Wolfgang R. Dick (ed.): *300 Jahre Astronomie in Berlin und Potsdam. Eine Sammlung von Aufsätzen aus Anlaß des Gründungsjubiläums der Berliner Sternwarte.* Thun: Deutsch 2000.

299 Julius Bartels: *Geophysics. Part I.* Wiesbaden: Dieterich'sche Verlagsbuchhandlung 1948. Julius Bartels: *Geophysics. Part II.* Wiesbaden: Dieterich'sche Verlagsbuchhandlung 1948. Julius Bartels: Erdmagnetismus II. Zeitliche Variationen, Beziehungen zur Sonnenphysik, zum Polarlicht, zur Ionosphäre. *Geophysics Part I.* Wiesbaden: Dieterich'sche Verlagsbuchhandlung 1948, 39–91. Keppler, *Max Planck Institut für Aeronomie*, 2003.

300 Seiler, *Kommandosache*, 2007, 11.

301 In 1954, Regener himself had suggested Bartels as his successor. Bartels, Dieminger, and Ehmert, Max-Planck-Institut für Aeronomie, 1962, 16–45, 21. Julius Bartels: Zur Vorgeschichte der Weltraumforschung. *Die Naturwissenschaften* 49/14 (1962), 313–323. doi:10.1007/BF00602195.

302 The committee that appointed Bartels (CPTS meeting minutes of 13.06.1955, AMPG, II. Abt., Rep. 62, No. 1729) was composed of Walter Dieminger, Otto Hahn, Carl F. von Weizsäcker, Walter Tollmien, Friedrich Paneth, Boris Rajewski, Werner Heisenberg, and Erwin Schopper. See also Walter Dieminger: Ionosphäre. *Geophysics. Part I.* Wiesbaden: Dieterich'sche Verlagsbuchhandlung 1948, 93–163.

ILLUSTRATION 74
Julius Bartels, May 11, 1956

years, up to the mid-1950s, saw an uptick in research in these areas, culminating in the International Geophysical Year (1957–58), behind which Bartels, among others, was a driving force.[303] Overall, it appeared that this merger of orphans with Bartels might be promising. At the same time, however, the weaknesses were already evident: half of the institute, Dieminger's, fell into the category of 'legacy' research dating from the Nazi era, and the trend over the next two decades was to gradually move away from this.[304]

Moreover, the move of Regener's team to Göttingen occurred just when Lower Saxony was losing its scientific influence, owing to the relocation of the Max Planck Society's headquarters and presidency, as well as Heisenberg's institutes, from Göttingen to Munich. For some years, the institute's survival was facilitated by the 'personal union' of Bartels with the University of Göttingen, but as we shall see below, counting on the brilliance of one individual carries the risk of problems when that person is no longer around; and in the

303 Karl-Heinz Glaßmeier, Manfred Siebert, and Emilio Herrero-Bervera: Bartels, Julius (1899–1964). Edited by David Gubbins. *Encyclopedia of Geomagnetism and Paleomagnetism.* Dordrecht: Springer 2007, 42–42, 42. https://doi.org/10.1007/978-1-4020-4423-6_15. Last accessed 11/3/2018. In relation to Bartel's involvement in the preparation of the IGY, see documents related to the period 1957–58 in AMPG, II. Abt., Rep. 36, No. 2.5, 2.6.

304 This general tendency to dispose of/close down antiquated, politically problematic institutes began during the Butenandt presidency, and will be a major theme in the chapter on governance and centralization of the forthcoming *Synthesis Volume* on the history of the Max Planck Society by the Research Program "Geschichte der Max-Planck-Gesellschaft", GMPG ("History of the Max Planck Society").

history of the Institute for Aeronomy, this happened at least three times, first with Regener (1954), then with Bartels (1964),[305] and finally, with the retirement of Ian Axford at the end of the century.

Regener's institute in Weissenau had in fact produced a line of brilliant researchers, such as Ehmert himself (who was later its director, from 1965 to 1971) or the rising star Erhard Keppler (see Chapters 2 and 3). But we can see how, isolated from the university context, having forged their careers entirely within the institute, these researchers had a legitimacy problem vis-à-vis the scientific community, including other Max Planck Institutes. Over the next few decades, this was expressed as the Institute for Aeronomy being categorized as the home of excellent instrument builders, but with few people with a general scientific and, specifically, a theoretical insight,[306] which put it in contrast, in particular, with the theorist-dominated Institute for Extraterrestrial Physics—an offshoot of Heisenberg and Biermann's Institute for Physics and Astrophysics—which conducted very similar research. As we will see in Chapters 2 and 3, unlike the case in Munich, aeronomy insiders such as Keppler could not be appointed full-fledged scientific members and permanent directors of their institute as the successors to Regener or, later, Bartels. The definitive solution to this problem came with a major reorganization of the Max Planck Society, which is the subject at the end of Chapter 3.

We will continue to revisit this permanent crisis of the Institute for Aeronomy, and to examine how, within this crisis, its researchers still sometimes found paths to scientific excellence, pushing against a bias, largely originating in Munich, that underestimated instrumental expertise in the Max Planck Society in favor of theorists. As can be illustrated by its early days, the weakness of the Institute for Aeronomy resulted from several factors:

- Identity drift: having emerged from a haphazard collection of orphan organizations, the institute could never pinpoint an area of clear, unique scientific leadership in a field of research deemed crucial. In its early years, up until 1957, its high-atmosphere research—since unconnected with nuclear questions—was not considered glamorous. Furthermore, as we will see

305 After Bartel's death, a collegial directorship was established at the Institute for Stratospheric Physics with Ehmert and Pfotzer as directors, both having been Regener's collaborators since before the war (CPTS meeting minutes of 09.06.1964, 03.12.1964, 05.03.1965, AMPG, II. Abt., Rep. 62, No. 1743, 1744, 1745). The two Institutes for Ionospheric and Stratospheric Physics had merged into what was named the Institute for Aeronomy, but a long difficult period began which reached its peak around the early 1970s, as will be discussed later.

306 Heinrich Völk: interview by Luisa Bonolis and Juan-Andres Leon, Heidelberg, October 9–10, 2017. DA GMPG, BC 601037.

ILLUSTRATION 75 Weissenau, early 1950s. From right: Alfred Ehmert, Erich Regener, and an unknown person in Weissenau, at the Research Center for Stratospheric Physics from which the Max Planck Institute for Stratospheric Physics originated in 1952. The study of cosmic rays was Ehmert's main research focus.

next, after 1957, when the high atmosphere and outer space became key subjects of scientific study, the institute had no distinctive scientific tradition to differentiate it from what could be done at other sites, so quickly ended up in competition with Munich, and at a disadvantage. Over the next decades, this lack of a competititve edge led to its confinement to the narrower fields of research deliberately left behind by Munich and which were also, given their growing focus on solar research and planetary science, increasingly far removed from the stated mission of 'Aeronomy' or high-atmosphere research. To make matters worse, sociopolitical interest in environmental matters during the 1970s led both to the reorientation of research in several Max Planck Institutes and the creation of new ones, so that at least two others, the Institute for Chemistry in Mainz and the Institute for Meteorology in Hamburg, ended up conducting more atmospheric research than the institute named after this field. The Institute for Aeronomy remained focused on a form of high-atmosphere research rooted in plasma physics questions.

ILLUSTRATION 76 Max Planck Institute for Aeronomy, Katlenburg-Lindau: preparation of a balloon ascent in the hangar, on the left in a white lab coat is Georg Pfotzer. May 1965

ILLUSTRATION 77 Max Planck Institute for Aeronomy, Katlenburg-Lindau: preparation of a balloon ascent in the hangar, far left: Georg Pfotzer, May 1965

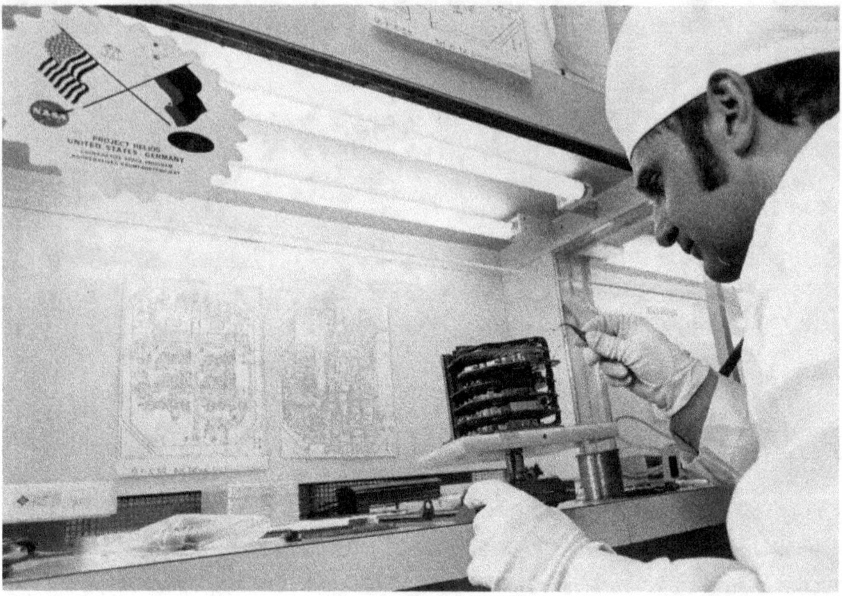

ILLUSTRATION 78 Max Planck Institute for Aeronomy, Lindau/Harz, assembly of parts of Experiment 8 of the Helios solar probe, around 1975

– Regional weakness: as explained earlier, the Institute for Aeronomy had suffered, since its establishment in Lindau, from a problem systemic to those Max Planck Institutes in the weaker federal states. Bavaria, North Rhine-Westphalia, and Baden-Württemberg had the means to aim for national leadership by attracting as much research and industry as possible. The smaller or weaker federal states could not compete with them, however, and their Max Planck Institutes (and universities) consequently experienced a relative decline.[307]

307 In the case of Göttingen, the major decline in physics and mathematics began with the Nazification of the University in 1933. The postwar Max Planck Institute temporarily revitalized the disciplines somewhat, but the general feeling in the second half of the century was of a loss of leadership, exacerbated then by the move of Heisenberg's institute to Munich. On the struggles of Max Planck Institutes in the weaker federal states, see Hohn, and Schimank, *Konflikte*, 1990. Chapter 4 deals specifically with the Max Planck Society, where these regional disparities were a major problem.

ILLUSTRATION 79 General Meeting of the Max Planck Society in Saarbrücken, 1959. From left: Otto Hahn, Ludwig Biermann, and Walter Dieminger

4 Regional Alliances and Rivalries

Following the introduction in the previous sections to the field and early players, this is the first section to focus explicitly on the interaction of the different research traditions and centers of power. Alignment with regional power bases is emphasized, as this was a key factor in the early decades of the Federal Republic of Germany. These regional and political rivalries played out in the scientific world particularly after the end of Allied restrictions in 1955, when a race began for leadership in the development of nuclear power. The cosmic sciences, which up until 1955 were a path toward excellence and global connections under precarious circumstances, faded (temporarily) into the background, as the development of large-scale projects such as nuclear reactors, thermonuclear fusion facilities, and particle accelerators turned into a battlefield, with both Heisenberg and Gentner among the key protagonists. The section shows how disunity and rivalries in the context of nuclear ambitions would set a precedent for failure not to be repeated later in the cosmic sciences.

Regional Alignment of the First Postwar Generation: Southwestern Germany versus Bavaria

Given the presence of very powerful scientists and scientific traditions, as in the case both of Heisenberg and Gentner, what marked the first three decades of the postwar era (roughly up until the mid-1970s), was the persistence of almost independent spheres of influence, at times stronger than the Max Planck Society itself. During this era, the general administration and decision-making processes of the Society became the interface between these separate spheres of influence. Most of the time, the different factions respected one another's territories, as in the appointment of Scientific Members within their institutes, but occasionally—when there was an overlap of interests, for instance—the Max Planck Society could be one of the arenas where such conflicts were settled. This went hand in hand with the gradual consolidation of political power, whereby the scientific traditions of the interwar era now became full-fledged factions which, in addition to sharing a common research background and certain aims, were aligned with wider social, political, and economic forces.

The most important source of political and economic power for these factions dated back to the division of Germany into separate zones of Allied occupation. As described above, Gentner's postwar role was inseparable from his affinity to French scientific and political circles, and initially coincided with the French occupation zone. After the normalization of West Germany in the mid-1950s, this sphere of influence extended further, to include roughly the quadrant south of Frankfurt am Main and west of Stuttgart. Within this zone, the Max Planck Institutes ensured that they maintained a close, cordial relationship with one another as well as with the nearby universities, and as allies they had the southwestern states of Baden-Württemberg and Rhineland-Palatinate, as well as the local governments of their host cities.[308] In addition, as mentioned earlier, this southwestern 'confederation' of universities

308 At the end of the 1970s, the general problem of the relationship between the Max Planck Society and the universities was specifically examined and widely discussed during a meeting of the Scientific Council where it was remarked that while the Max Planck

and Max Planck Institutes was complemented by institutions in neighboring France and Switzerland.

The alignment of Heisenberg's institute with the state of Bavaria would come half a decade later, and in a similar manner to Gentner's faction. Heisenberg, himself a Bavarian, negotiated his institute's move to Munich during the 1950s.[309] Originally, the main motive for this move was Heisenberg's ambition to be the father of nuclear fission in Germany. An experimental heavy water fission reactor was to be constructed near Munich with this purpose in mind. This episode, however, turned into the most monumental failure of Heisenberg's postwar career. The federal government, under pressure from a constellation of competing regional interests, managed to shift the location of the research reactor to Karlsruhe in the southwest of Germany.[310] While the part of Heisenberg's institute led by Karl Wirtz, which was focusing on nuclear fission, went on to lead this project, Heisenberg himself terminated his involvement with nuclear fission altogether, but still completed the move

Society was a privileged place for research, "qualification" was still completely in the hands of the universities, so the Society should also be given the opportunity for collaboration. On this, see, in particular, a long report by Heinz A. Staab, attached as an appendix to the CPTS meeting minutes of 08.03.1977, AMPG, II. Abt., Rep. 62, No. 1780.

309 According to Biermann, "That was actually an idea of Otto Hahn's, not Heisenberg's, to do it like that [...] the Institute practically had to leave Göttingen because it grew and grew and there was simply too little space in Göttingen (the Aerodynamic Experimental Institute, in whose space we were working, received permission and wanted to resume its work). At that particular time there was also apparently no money to be had for it [the Institute] in Lower Saxony, while at the same time we, Heisenberg and I, could both here in Munich and in Karlsruhe (they were in competition so to speak) simply wish for whatever we wanted. In this transitional period there were three department heads in Heisenberg's institute. They were Weizsäcker, Wirtz, and I. Wirtz was responsible for experimental nuclear physics, that was at that time above all cosmic ray physics, because it was the only form of experimentation with elementary particles that was allowed. But that was loosening up just at this time; it became possible to work on nuclear physics and to develop reactors. That was above all the background in Karlsruhe. It was decided to found a large institute in Karlsruhe that was more focused on nuclear technology. Heisenberg himself tended more towards Munich because he could ask for whatever he wanted there. Weizsäcker had begun around this time, around '52 or '53 perhaps, to be again more interested in philosophy [...] so Weizsäcker went to Hamburg, Wirtz went to Karlsruhe to the institute that I just spoke about, and at the same time to the technical university there, only I remained with Heisenberg" [our translation]. Ludwig Franz Benedikt Biermann: interview by Martin Harwit, February 16, 1984. Transcript, AIP.

310 When Wirtz became Director of the Institute of Neutron Physics and Reactor Technology at the Karlsruhe Center for Nuclear Research, in 1957, he also became an External Scientific Member of the Institute for Physics (CPTS meeting minutes of 26.06.1957, AMPG, II. Abt., Rep. 62, No. 1731).

to Munich in 1958, after which the institute was renamed Institute for Physics and Astrophysics, with Biermann's sub-institute becoming an institute in its own right, even if under this common umbrella.[311]

But while relocation to Munich was a highly favorable move, without which the success of Heisenberg's institute and its successors would be unthinkable, it came at a price, namely the ensuing conflictual relationship with universities: the University of Göttingen and the nearby Max Planck Institute for Physics had completely rebuilt themselves, and there was an intellectual and political affinity between the physicists at both. The move to Bavaria left the Göttingen university physicists in a weaker position, and at the same time, as we will see, made the universities in Munich uneasy about how the competition with Heisenberg's institutes would affect their own growth and influence. This unease was exacerbated by competition from the two different factions leading Bavaria toward the nuclear age: by the mid-1950s, the reach of Wolfgang Gentner's alliances had already extended to Munich, through the appointment there of his close collaborator Heinz Maier-Leibnitz, who in

311 This decision had been taken as early as 1956 (CPTS meeting minutes of 11.06.1956, AMPG, II. Abt., Rep. 62, No. 1731). A historical outline of the Max Planck Institute for Physics up to 1960 was written by Heisenberg and Biermann in connection with the inauguration of the new location in Munich: Werner Heisenberg: Max-Planck-Institut für Physik und Astrophysik in München. In: Generalverwaltung der Max-Planck-Gesellschaft (ed.): *Jahrbuch 1961 der Max-Planck-Gesellschaft zur Förderung der Wissenschaften e.V. Teil II.* Göttingen 1962, 632–643. Ludwig Biermann: 50 Jahre physikalische Grundlagenforschung in der Kaiser-Wilhelm und der Max-Planck-Gesellschaft. *Die Naturwissenschaften* 48/1 (1961), 2–10. doi:10.1007/BF00600935. See also Horst Kant: Das Max-Planck-Institut für Physik. Berlin–München. In: Peter Gruss, and Reinhard Rürup (eds.): *Denkorte. Max-Planck-Gesellschaft und Kaiser-Wilhelm-Gesellschaft. Brüche und Kontinuitäten 1911–2011.* Dresden: Sandstein Verlag 2010, 316–323. On the move to Munich, see Joachim Radkau, and Lothar Hahn: *Aufstieg Und Fall Der Deutschen Atomwirtschaft.* München: Oekom 2013. Radkau, *Aufstieg*, 1983. Cathryn Carson: Heisenberg als Wissenschaftsorganisator. In: Christian Kleint, Helmut Rechenberg, and Gerald Wiemers (eds.): *Werner Heisenberg, 1901–1976. Beiträge, Berichte, Briefe. Festschrift zu seinem 100. Geburtstag.* Leipzig: Verlag der Sächsischen Akademie der Wissenschaften zu Leipzig 2005, 214–222. Carson, *Heisenberg in the Atomic Age*, 2010. See also Carl F. von Weizsäcker: Werner Heisenberg 5.12.1901-1.2.1976. In: Generalverwaltung der Max-Planck-Gesellschaft (ed.): *Berichte und Mitteilungen Max-Planck-Gesellschaft. Nachrichten, Personalien, Würdigungen 1.4.1975-1.4.1976.* München: Generalverwaltung der Max-Planck Gesellschaft zur Förderung der Wissenschaften e.V. 1976, 5–11. Cathryn Carson: Nuclear Energy Development in Postwar West Germany. Struggles over Cooperation in the Federal Republic's First Reactor Station. *History and Technology* 18/3 (2002), 233–270. doi:10.1080/0734151022000020166.

turn mentored Rudolf Mössbauer,[312] a future giant in Bavarian physics, whom Maier-Leibnitz had sent to complete his studies with Bothe in Heidelberg in the mid-1950s. Maier-Leibnitz, who had been recalled by Bothe and Gentner from the Eastern Front during the war, maintained the deepest admiration for Gentner, in personal, intellectual, and moral terms, all his life long.

Moreover, this closeness to Gentner increasingly meant a carefully distanced relationship with the institutes originating from Heisenberg's relocation to Munich,[313] a stance later also inherited by Rudolf Mössbauer upon his appointment as professor there in 1965. This uneasy relationship expressed itself physically in the colonization of the new scientific city of Garching, north of Munich, which was to be shared between the Technical University of Munich (TUM) and Heisenberg's growing array of institutes. The central feature of the city (visible in its coat of arms), is the 'Atomic Egg,' the first nuclear reactor built in Germany. This reactor reiterated Heisenberg's defeat in the forced relocation of 'his' reactor project to Karlsruhe, since in this case it was Maier-Leibnitz who was behind the construction of the small, American-made 'Atomic Egg,' which became the source of his expertise in neutron physics, the basis of a lifetime career in the field.[314] In building the 'Atomic Egg,' Maier-Leibniz came to control one of the few reactors in Germany specialized in neutron generation, which propelled the remainder of his career in this direction. Thanks to this expertise and his connections both with Gentner's field and the French scientific establishment, he was later appointed the first Director of the Institut Laue-Langevin (ILL) in Grenoble, whose central feature was a much more powerful neutron source. Years later, his disciple Mössbauer succeeded Maier-Leibniz as Director of this French-German collaboration.[315]

312 Georg Michael Kalvius, and Paul Kienle (eds.): *The Rudolf Mössbauer Story. His Scientific Work and Its Impact on Science and History*. Berlin: Springer 2012. Further biographical details can be found in the semi-autobiographical book about his mentor Maier-Leibnitz: Anne-Lydia Edingshaus: *Heinz Maier-Leibnitz. Eine halbes Jahrhundert experimentelle Physik*. München: Piper & Co. Verlag 1986.

313 Edingshaus, *Heinz Maier-Leibnitz*, 1986. For instance, the author tells how, while academic dinner parties at the Meier-Leibniz home were usually light-hearted affairs, also attended by the family's children, on the one occasion Heisenberg was invited, the atmosphere was tense and no children were present (pp. 142–43). This 'territorial' rivalry was also exacerbated by Heisenberg's dismissive stance on experimental physicists (see pp. 101, 122, 142–43).

314 Edingshaus, *Heinz Maier-Leibnitz*, 1986.

315 This institute in Grenoble was named after Paul Langevin, close collaborator (and lover) of Marie Curie and mentor of Joliot. Langevin's grandson Michel later became a disciple of Joliot, whose daughter, Hélène, he then married. One of Wolfgang Gentner's most daring moves during World War II had been to help free the young Langevin from arrest

ILLUSTRATION 80 Rudolf Mössbauer and B. Schimmer at the Max Planck Institute for Medical Research, Heidelberg, around 1955–1958. After finishing his master's thesis in Munich in 1955, Rudolf Mössbauer started his doctorate thesis in Heidelberg, which he completed in 1958. While working on his thesis, Mössbauer discovered the nuclear phenomenon now known as the 'Mössbauer effect' which opened up the possibility of many applications for precision experiments and became a powerful research tool in many diverse areas of natural sciences and technology. For his discovery, R. Mössbauer received the Nobel Prize for Physics in 1961, when he was only 32 years old.

As a result of this overlap of interests, there was continuous rivalry between Heisenberg's institute and the Technical University of Munich, instead of

---

by the German occupying forces. See: Pinault, *Frédéric Joliot-Curie*, 2000. See also Metzler, Wissenschaft im Krieg, 2000, 685–700.

ILLUSTRATION 81 Mössbauer's apparatus at the Max Planck Institute for Medical Research, Heidelberg, around 1958. The discovery of the nuclear process that bears his name at the Institute of Physics of the MPI for Medical Research in 1957 took place during the founding phase of the new Max Planck Institute for Nuclear Physics promoted by Wolfgang Gentner.

the easy integration experienced by the Max Planck Institutes and neighboring universities in southwestern Germany (Heidelberg, Mainz, Freiburg). This meant that extending the size and influence of Heisenberg's institute was more likely to occur through internal growth than through interaction with people at the TUM. Relationships with the University of Munich (LMU) were more cordial and Max Planck directors of this first generation held honorary positions and taught there.

The relationship was somewhat one-directional, however, since the Max Planck directors teaching at the university identified the best students there and brought them to their institutes, while, in contrast to southwestern Germany, here, in the early years, there was little flow in the opposite direction. Consequently, whereas in the southwest an easy circulation of scientists and projects between universities and Max Planck Institutes developed (extending as far as to the Technical University in Munich), those working at the Max Planck Institute for Physics and Astrophysics, up until to the mid-1970s, largely cooperated with their immediate colleagues and with scientists at geographically more distant institutions.

'Cell Division' in Munich

This intellectual inbreeding in Munich is best exemplified by what became the growth strategy already known at the time as 'cell division.'[316] The single

---

316 For use of the terminology, see Trümper, Astronomy, Astrophysics and Cosmology, 2004, 169–187. The Munich 'cell division' dated from the initial period of Max Planck Institutes with a single strong director, in this case, Heisenberg. By 1958, at the time of the relocation to Munich of Heisenberg's Institute for Physics, Biermann's sub-Institute for Astrophysics comprised four departments, which continued the work of Biermann's former Astrophysics Department at the parent Institute for Physics: Theoretical Astrophysics (led by Reimar Lüst), Quantum Physics applied to atoms and molecules of astrophysical interest, in particular to comets (led by Eleonore Trefftz). These activities were especially connected with research conducted at the Department of Numerical Computing Machines (led by Heinz Billing), where codes were developed also for the automatic evaluation of data taken in bubble chambers. The fourth Department, for Theoretical Plasma Physics (and fusion research), was led by Arnulf Schlüter. See the scheme for such research fields both at the Institute for Physics and Astrophysics in the special number of the Max Planck Society Bulletin dedicated to the inauguration of the new building in Munich, Generalverwaltung der Max-Planck-Gesellschaft (ed.): *Mitteilungen aus der Max-Planck-Gesellschaft zur Förderung der Wissenschaften. Heft 6/1960*. Göttingen 1960, 334. As research areas grew, Heisenberg appointed new Scientific Members with their own sub-institutes. In 1961 Lüst was appointed Director of the new Department for Extraterrestrial Physics, which was transformed into a sub-institute in May 1963. Schlüter's Department became the root of the Institute for Plasma Physics, founded in 1960.

ILLUSTRATION 82 Reimar Lüst with Ludwig Biermann in 1963, at the time of the foundation of the Max Planck Institute for Extraterrestrial Physics

Institute for Physics headed by Heisenberg in Göttingen quickly became an umbrella organization for the many independent subunits linked to it. The first was, in 1958, the Institute for Astrophysics headed by Biermann, who had always presided over the growth of the astrophysics part of the Max Planck Institute for Physics, which culminated in the establishment, in 1960, of the Institute for Plasma Physics (IPP), led by Arnulf Schlüter, and in 1963, of the Institute for Extraterrestrial Physics, led by Reimar Lüst, which will both be treated in detail later. Both directors were former associates of Biermann and represented his manifold scientific interests.

In subsequent decades, these were followed by the Institute for Quantum Optics (1981), the second site of the Institute for Plasma Physics, in Greifswald (1994), and the Institute for Gravitational Physics, in Potsdam (1995), which was named after Albert Einstein, and later included the Hanover branch focusing on gravitational wave detection, previously part of the Institute of Quantum Optics. There were also two humanistic offshoots led by the veterans of Heisenberg's Institute for Physics in Göttingen, Carl Friedrich von Weizsäcker and Klaus Gottstein. In 1970, von Weizsäcker was the Founding Director of the Max Planck Institute for the Research of Living Conditions in

ILLUSTRATION 83
Arnulf Schlüter

the Modern World, located in Starnberg near Munich.[317] A decade later, in 1984, a special arrangement was made for Klaus Gottstein in the form of a separate 'research station' for Peace Studies, after a long career in experimental particle physics.[318]

It was only from the 1990s onward that many of these cells became fully independent Max Planck Institutes, as we describe further in Chapter 4. From the 1950s to the 1980s, all of them except the Institute for Plasma Physics, with its offshoot Quantum Optics, and von Weizsäcker's humanistic institute were formally sub-institutes of the Max Planck Institute for Physics and

---

317 See documents on the foundation of the Institute in BArch, No. B 196/7168. For more on von Weizsäcker's humanistic career and trajectory inside the Max Planck Society/MPG, see Horst Kant, and Jürgen Renn: Eine utopische Episode. Carl Friedrich von Weizsäcker in den Netzwerken der Max-Planck-Gesellschaft. In: Klaus Hentschel, and Dieter Hoffmann (eds.): *Carl Friedrich von Weizsäcker. Physik–Philosophie–Friedensforschung*. Stuttgart: Wissenschaftliche Verlagsgesellschaft 2014, 213–242.

318 He had become scientific advisor to German diplomatic circles and a colleague of von Weizsäcker's in Starnberg in the 1970s, while on leave from his duties as a Scientific Member at Heisenberg's institute, after the latter's retirement. From 1980 to 1984, he had been given his own working group at the Institute for Physics (Arbeitsgruppe Gottstein im Institut für Physik) while a leading representative at the Pugwash Conferences. On Gottstein, see Carola Sachse: *Die Max-Planck-Gesellschaft und die Pugwash Conference on Science and World Affairs (1955–1984)*. Preprint 479. Berlin: Max Planck Institute for the History of Science 2016.

Astrophysics, as it was called at the time, although they managed their own budgets.[319] The parent institute in Munich-Freimann functioned largely as a buffer that kept many decisions at the Munich level which could otherwise have exposed the sub-institutes to the wider politics of the Max Planck Society. The sub-institutes themselves were generally located elsewhere, primarily in Garching, but later also in Hanover and Potsdam. Their formal existence as a single Max Planck Institute was necessary, given the hostility within the Society concerning the growing power and influence of the Bavarian capital and Heisenberg's family of institutes born of 'cell division' at Heisenberg's Institute for Physics.[320]

Political Orientation and Regional Interests

In addition to differences in scientific traditions and geographical allegiances between the southwest of Germany and Bavaria, there was also the more explicit political orientation of the main figures, and the effect this had on the readmission of German scientists to the European and international scientific communities in the postwar era. As discussed earlier, Gentner's position vis-à-vis the French occupiers and scientific community was particularly distinctive and set much of the political tone for the southwestern community of university and Max Planck researchers. In 1953, the appointment of Friedrich Paneth to Mainz was not only a key move to establish a cosmochemistry tradition in the region, but also of symbolic importance, in that it was one of the very few

319 The Max Planck Society's budgets are a good measure of whether an institute is financially independent, as in the case of the family of institutes in Munich and Garching. The institutes became fully independent following major reforms in 1991 (see Chapter 4).

320 Heisenberg's own practice of appointing new Scientific Members with their own sub-institutes went largely uncontested by external parties in the early years. In the 1960s, however, when rivals such as Gentner in Heidelberg attempted to do the same, Heisenberg protested such 'insider' moves. The most notable case was that of Anselm Citron's appointment as head of an accelerator group in Heidelberg and the attendant de facto creation of a full-fledged sub-institute, which were vetoed on account of the evident mentor-disciple relationship between Gentner and Citron. The underlying issue was competition for funding and influence in nuclear and particle physics. See Hoffmann, and Schmidt-Rohr, Wolfgang Gentner, 2006, 1–60, 44. These impasses were among the initial causes of the institutes' gradual transition to a collegiate form of leadership from the mid-1960s onward. After this period, new sub-institutes became highly exceptional, the system favoring instead either departments with their own Scientific Members, or completely new Max Planck Institutes. It was only after 1991 that the 'cosmic' institutes in Garching became full-fledged Max Planck Institutes independent of the Max Planck Institute for Physics, still in Munich-Freimann. In contrast, spin-offs from the Institute for Plasma Physics, such as the Institute for Quantum Optics, officially became independent earlier.

cases of a German exiled due to his Jewish heritage returning to the country.[321] Houtermans in Bern also had a past as a political exile, and even after his imprisonment in the Soviet Union, remained ideologically left-wing.

Furthermore, with Gentner and his collaborators' involvement in CERN came the first ever opportunity for a relationship between West Germany and Israel, in which scientific collaboration and political rapprochement went hand in hand, many years before formal political relations could be established.[322] In the scientific relationship between Israel and Germany, Gentner was most influential owing to his authority in the Minerva Foundation, which coordinated the exchange between Israelis and Germans, carefully vetting the right people to collaborate with the Israelis. Gentner was also pivotal in the first nuclear binational project with France, in Grenoble, which made available a nuclear reactor with a high flux of neutrons, something that, because of its dual potential, could not easily have been set up in Germany. This was the origin of the aforementioned Institute Laue-Langevin, which was headed first by Heinz Maier-Leibnitz (1967–72), and then by Rudolf Mössbauer (1972–77).[323] Finally, as mentioned above, there was a sustained close relationship with universities in Switzerland, particularly with the University of Bern. Overall, while, as with any institution in Germany, relations were cultivated with the United States, scientists in the southwest of the country had much easier access, especially in the early decades, to European collaborations.

---

321 Paneth was especially recommended as Strassmann's successor by Lise Meitner, a choice also favored by Otto Hahn: Reinhard, and Kant, *100 Jahre Kaiser-Wilhelm-/Max-Planck-Institut für Chemie (Otto-Hahn-Institut)*, 2012, Vol. 22, 113. See also Mattauch, Max-Planck-Institut für Chemie, 1962, 215–224. Michael Schüring: *Minervas verstoßene Kinder. Vertriebene Wissenschaftler und die Vergangenheitspolitik der Max-Planck-Gesellschaft*. Göttingen: Wallstein Verlag 2006. Otto Hahn: Friedrich A. Paneth. *Zeitschrift für Elektrochemie. Berichte der Bunsengesellschaft für physikalische Chemie* 61/9 (1957), 1121–1122. doi:10.1002/bbpc.19570610902.

322 Steinhauser, Gutfreund, and Renn, *A Special Relationship*, 2017.

323 David L. Worcester, Antonio Faraone, and Giuseppe Zaccai: The Summer of 1954 and Paths to the Institut Laue-Langevin. *Neutron News* 28/3 (2017), 15–19. doi:10.1080/10448632.2017.1342480. Bernard Jacrot: *Des neutrons pour la science. Histoire de l'Institut Laue-Langevin, une coopération internationale particulièrement réussie*. Les Ulis: EDP sciences 2006. See English translation (*Neutrons for Science. The story of the first forty years of the Institut Max von Laue-Paul Langevin, Grenoble 1967–2007, a successful European Cooperation*) at https://neutronsources.org/media/jacrot_history_of_the_ill_s.pdf. Last accessed 6/16/2020. See also Eberhard Moll: The Franco-German High Flux Reactor and Its Facilities for Nuclear Research. In: János Erő, and J. Szűcs (eds.): *Nuclear Structure Study with Neutrons. Proceedings of the International Conference on Nuclear Structure Study with Neutrons. Budapest, 31 July–5 August 1972*. Boston, MA: Springer 1974, 313–326. doi:10.1007/978-1-4613-4499-5_14.

The political orientation of Heisenberg and scientists at his institutes, by contrast, needed to be more delicately calibrated and remain low key. During the war, Heisenberg and his collaborators had managed to maintain some independence from ideological Nazism, while contributing scientifically to the war effort; but as they worked deep inside Germany, they had not had the same opportunities to connect and ally themselves with scientists in the occupied countries,[324] as had Gentner or the solar astrophysicist Karl-Otto Kiepenheuer (as discussed in the previous section), who, paradoxically, were leading participants in the occupation yet were later seen to have proven their moral character by saving scientists from arrest and identifying with their cause; ultimately, they had been accused by the German secret services of colluding with the local scientists.[325]

Heisenberg and his colleagues were subject to significant isolation after the war, particularly in European circles, and this further propelled their inclination to establish a separate domain and reach out resolutely to the United States. Within West Germany, they clearly benefited from the aura of the nuclear age, yet at the same time had to continuously reassert the eminently peaceful nature of their nuclear intentions.[326] This was to lead to interesting episodes such as the 'Göttingen Manifesto' against attempts by the German Federal Ministry of Defence to install American nuclear warheads on West German weapons systems.[327] The signatories were mainly from within Heisenberg's circle but also included 'southwestern' figures such as Hahn, Paneth,

324 Heisenberg himself did travel to several countries during the war, among them Hungary and other Axis Allies, neutral ones such as Switzerland, and also occupied ones such as Denmark and the Netherlands. These visits were arranged by German cultural institutions locally perceived as part of the Nazi propaganda apparatus, and while Heisenberg at the time thought he was being impartial and scientific when abroad, most scientists who had contact with him later expressed their dislike of Heisenberg's apologetic attitude to the German role in the war. On these travels, see Walker, Physics and Propaganda, 1992, 339–389. See also Rechenberg, The Early S-Matrix Theory, 1989, 551–578. After the war, the memory of these visits was one of the liabilities Heisenberg and his collaborators found most difficult to overcome. H. Walker, *The Quest for Nuclear Power*, 1989, 105–118.

325 Nevertheless, a minority of witnesses maintained a critical stance on the role of these 'good Germans.' See, for example, Lew Kowarski: interviews by Charles Weiner, 8 sessions, March 20, 1969–November 20, 1971. Transcripts, AIP, https://www.aip.org/history-programs/niels-bohr-library/oral-histories/4717-1. Last accessed 5/8/2019.

326 See, for example, Eckert, and Osietzki, *Wissenschaft für Macht und Markt*, 1989. Radkau, *Aufstieg*, 1983.

327 Elisabeth Kraus: *Von der Uranspaltung zur Göttinger Erklärung. Otto Hahn, Werner Heisenberg, Carl Friedrich Weizsäcker und die Verantwortung des Wissenschaftlers*. Würzburg: Königshausen & Neumann 2001. Robert Lorenz: *Protest der Physiker. Die »Göttinger Erklärung« von 1957*. Bielefeld: transcript Verlag 2011. Friedensinitiative Garchinger Naturwis-

and Maier-Leibnitz; but not Gentner, who disagreed with the public display of simplistic positions on a complex matter that ultimately was related to the orientation and funding of their own research.[328] One might add, however, that Gentner, unlike most of the signatories, did not need a public display of moral character to prove his worth to his international colleagues.[329]

Despite the Göttingen Manifesto, Heisenberg and his colleagues, who were themselves not particularly conservative, found their strongest and most enduring ally in the leading Bavarian party, the Christian Social Union (CSU), an ardent proponent of the coevolution of high modernization with traditional Catholic social values, and the closest possible geopolitical alignment with the United States.[330]

Heisenberg's move to Munich was part of Bavaria's broader success in the competition among the newly established federal states for dominance of the crucial federal institutions in what was now West Germany. Just as significant as this initial move was the long process of relocating the Max Planck Society's headquarters from Göttingen to Munich, justified by the appointment of Adolf Butenandt as Director of a new Max Planck Institute for Biochemistry (MPIB) in Munich-Martinsried, and shortly afterwards, as President of the Society. This was a double win for Bavaria, first, over Lower Saxony, where Göttingen now began its relative decline, from the center of the physical sciences and scientific organization to a provincial university city;[331] and, secondly, against the southwest of Germany, where Butenandt had been a professor in Tübingen.

---

senschaftler: *30 Jahre Göttinger Erklärung. Nachdenken über die Rolle des Wissenschaftlers in der Gesellschaft*. Marburg: Bund Demokratischer Wissenschaftlerinnnen und Wissenschaftler 1987. Gabriele Metzler: Kernphysik und Politik. Werner Heisenberg in der Wissenschafts- und Zeitgeschichte. Ein Forschungsbericht. *Historisches Jahrbuch* 115 (1995), 208–222. http://opac.ifz-muenchen.de/webOPACClient.ifzsis/start.do?Login=woifz&Query=10="BV011522186". Last accessed 10/30/2018.

328 Hoffmann, and Schmidt-Rohr, Wolfgang Gentner, 2006, 1–60, 38.

329 Gentner had in fact signed the *private* letter to Strauss which predated the public manifesto, which strengthens the argument that he was particularly opposed to making a public display of this disagreement, in contrast to the Göttingen 18. Alexandra Rese: *Wirkung politischer Stellungnahmen von Wissenschaftlern am Beispiel der Göttinger Erklärung zur atomaren Bewaffnung*. Vol. 835. Frankfurt am Main: Lang 1999, 192 (footnote 833).

330 See, for example, Jaromír Balcar: *Politik auf dem Land. Studien zur bayerischen Provinz 1945 bis 1972*. München: Oldenbourg 2004. Helmuth Trischler: Nationales Innovationssystem und regionale Innovationspolitik. Forschung in Bayern im westdeutschen Vergleich 1945 bis 1980. In: Thomas Schlemmer, and Hans Woller (eds.): *Bayern im Bund. Politik und Kultur im föderativen Staat 1949 bis 1973*. München: Oldenbourg 2004, 117–194.

331 Heisenberg's move to Munich was an early harbinger of the decline of the University of Göttingen in the late 20th century, relative to its counterparts in southern Germany. It did

ILLUSTRATION 84 From left: Otto Hahn, Carl Friedrich von Weizsäcker and Max von Laue in the 1950s, at the time of signing the Göttingen Manifesto with 15 other leading nuclear scientists

These moves also served to realign the regional allegiances of two competing factions in biochemistry, each now with its own geographical stronghold: Adolf Butenandt in Munich and Richard Kuhn in the southwest.[332]

---

preserve an edge in other sciences, maintaining several stellar Max Planck Institutes, such as Manfred Eigen's; and in time, it learned to turn this perception of its lesser standing to its own advantage, as founding institutes in the city came to be seen as a principal means to rescue it from the margins of science. But while Göttingen still hosts the university observatory and the Max Planck Institute for Solar System Research, it has never been able to recover the influence it had in astronomy and astrophysics up until the 1950s.

332 Aspects of this rivalry are being investigated as part of Jeffrey Johnson's current research project at GMPG on the relevant Max Planck Institutes' biochemical research networks in early postwar Germany, with an especial focus on Richard Kuhn and Adolf Butenandt.

Both scientists were leading candidates for the presidency of the Max Planck Society in 1960, and Butenandt's win was a clear defeat for the southwest in the competition for control of the central administration. This loss cemented the focal trend among the Max Planck Institutes in the southwest, namely to deepen their regional relationships and establish a rebellious counterpoint to what they saw as a growing Bavarian hegemony. Elsewhere in West Germany, too, from the 1960s onward, there was growing resentment—manifest in relations both with the central administration and the institutes controlled by Munich-based figures such as Heisenberg and Butenandt—of the major influence of Bavarian interests on the Max Planck Society. In the physical sciences, for researchers outside Bavaria, 'Munich' became shorthand for the expansionist interests of competing institutes that were presumed to be receiving preferential treatment from the administration and president. An increasing number of Scientific Members, mostly from outside Bavaria, were mobilized in response to this, throughout the 1960s; by Wolfgang Gentner, for instance, whose proposed strategy was to openly challenge the Munich hegemony. Gentner was expected to become president of the Max Planck Society in 1972, at the time a scenario of extreme concern to Heisenberg, who instead favored Reimar Lüst for this position.[333]

Finally, it should be emphasized that the political orientation of Gentner and his allies was more problematic than expected, particularly since preserving regional interests was the priority here. While the southwestern zone

---

333 In an interview, Reimar Lüst explicitly recalled the conversation with Heisenberg: "'Gentner cannot become… [D]on't do this to me, Mr. Lüst.' Well, that was the first time I realized all this, because I had a very close personal contact with Gentner, which, again, came through space research. And then Mr. Zähringer, [working] with Gentner, had an accident [1970], and through this we got closer […] he was so explicit […] but it's only now, since I read the book [Cathryn Carson], that some things have become clearer to me, how strong the animosities really were" [our translation]. Reimar Lüst: interview by Horst Kant and Jürgen Renn, Hamburg, May 18, 2010 (DA GMPG, ID 601068). In another interview, Lüst recalled that Heisenberg himself had taken him aside in Berlin, suggesting they take a walk together: "He explained that I was both young enough and old enough, and I had to be ready to run for president in November. He had heard that I had an offer from the industry, namely as a board member at Siemens. He said I could not do that to him; I had to stay with the Max Planck Society. In fact, I then rejected the offer from Siemens, without knowing whether I would be elected. My rival candidate, Wolfgang Gentner from Heidelberg, withdrew his candidature at the last moment. So, I was elected by the Senate in November." Reimar Lüst: interview by Hans von Storch and Klaus Hasselmann, December 2, 2000. Transcript. AIP, https://www.aip.org/history-programs/niels-bohr-library/oral-histories/33761. Last accessed 5/8/2019. See also Edingshaus, *Heinz Maier-Leibnitz*, 1986, 5–6.

ILLUSTRATION 85 Werner Heisenberg and Reimar Lüst in the 1960s

became the base for many German scientists considered to have behaved in an exemplary manner during the war, it also hosted high-profile problematic cases, among them, the two major figures in biochemistry within the Max Planck Society who both clearly had Nazi backgrounds: Adolf Butenandt at the University of Tübingen and Richard Kuhn, the other Director at Bothe's Institute for Medical Research in Heidelberg. While it is emphasized in participants' recollections that Kuhn and Bothe were enemies,[334] the two managed to tolerate each other for the greater good of securing a leading role for Heidelberg in the natural sciences, and frequently sat on committees together. The tension between them was partly behind the decision to create a new Institute for Nuclear Physics, as Gentner inherited the problems surrounding this complex relationship.[335]

---

334 Their relationship is treated extensively, for example, in Schmidt-Rohr, *Erinnerungen*, 1996, 26–28.

335 Scientists like Wolfgang Pauli were surprised that Gentner was appointed to return to Heidelberg in 1958 upon the death of Bothe, given Kuhn's presence there. See letter from Pauli to Jensen, 25.10.1957, in Wolfgang Pauli: *Wissenschaftlicher Briefwechsel mit Bohr, Einstein, Heisenberg u.a. / Scientific Correspondence with Bohr, Einstein, Heisenberg a.o. Teil/Part IV: 1957–1958*. Edited by Karl von Meyenn. Vol. 4. Berlin: Springer 2005, 580–581.

Interplay of Regional and Scientific Rivalries in Nuclear Research

Before the launch of Sputnik in October 1957, the relationship between Heisenberg's and Gentner's increasingly powerful factions manifested itself scientifically in a competition for leadership related to projects advancing the nuclear age. In Göttingen and Munich, these ambitions were first expressed in nuclear fission but were then frustrated by the selection of Karlsruhe as the site of West Germany's first home-built experimental nuclear reactor; in parallel to this, there was the brilliant trajectory of Biermann and his colleagues in plasma physics, which was articulated at the time as astrophysics and was ultimately conducted in the Max Planck Institute for Physics led by Heisenberg. There was an expectation that the scientists working in this field, as well as the new generation being trained in space plasmas, would later lead West Germany's thermonuclear fusion research program. Following the relocation of his institute to Munich in 1958, Heisenberg, allied with the Bavarian government, and what was then the Federal Ministry for Atomic Energy (Bundesministerium für Atomenergie, the former Ministry for Atomic Issues), likewise controlled by Bavarians, successfully initiated the gigantic Institute for Plasma Physics (IPP); this, an institute so large as not to be part of the Max Planck Society, was established as a private company, part of a first wave of institutes primarily funded by the 'Nuclear' Ministry.[336]

The Institute for Plasma Physics was one of the first examples of a general trend in Max Planck Institutes during the transition from the austerity of the postwar era toward a new regime in which West Germany successfully regained its status as an economic heavyweight. As will be seen in many examples throughout this study, with the new scale of resources available, institutes whose theoretical expertise had enabled them to maintain a foothold in their respective international scientific communities were able to mobilize this scientific capital in a move toward establishing much more expensive experimental facilities and programs in the same scientific fields.[337]

---

336 Boenke, *Entstehung und Entwicklung*, 1991, 1–3. Hohn, and Schimank, *Konflikte*, 1990. On the history of the Institute for Plasma Physics, see also Brigitte Röthlein, and Uwe Schumacher: *Max-Planck-Institut für Plasmaphysik. Garching bei München*. Edited by Generalverwaltung der Max-Planck-Gesellschaft. München: Generalverwaltung der Max-Planck Gesellschaft zur Förderung der Wissenschaften e.V. 1977. Breuer, and Schumacher, *Max-Planck-Institut für Plasmaphysik*, 1982. Arnulf Schlüter: Wozu Plasmaphysik? In: Generalverwaltung der Max-Planck-Gesellschaft (ed.): *Jahrbuch der Max-Planck-Gesellschaft zur Förderung der Wissenschaften e.V. 1970*. Göttingen 1970, 45–61.

337 Even adjusted for inflation, the budget of the entire Max Planck Society quadrupled between 1956 and 1966. However, the growth during this period was already feeling the effects of the Sputnik Shock, so will be discussed in more detail in Chapter 2.

Conversely, for scientists close to Gentner in the southwest of Germany, tapping into the 'nuclear age' came primarily through the close association with CERN, an institution that itself benefited from the optimistic aura of the 'nuclear,' strategically managing expectations around the scientific reality that researchers of particle physics were generally no longer interested in the type of work related to nuclear reactors or weapons. Politically, however, particle physics was presented as crucial for cultivating 'nuclear' expertise, and especially for maintaining this expertise in Europe; the urgency of this mission against 'brain drain' made CERN a precursor of European integration in competition with the United States.[338]

In southwestern Germany, cosmochemistry played a complementary role in opening access to scientific communities outside of the CERN core, particularly with the United Kingdom and the United States. Cosmochemistry in the 1950s and 1960s was clearly grounded in methods originating in nuclear and particle physics, both experimentally and theoretically. What the astrophysical connection allowed was research on a small scale that was still world class and based on expertise that could potentially be transferred to other initiatives in experimental physics. However, as in the case of Munich, the expectations before Sputnik were that space-related research would be a small-scale, temporary step or niche field likely to remain modest in contrast to research in nuclear and particle physics. This pre-Sputnik worldview underlay the founding of Gentner's Institute for Nuclear Physics in Heidelberg, where, up until the end of the century, astrophysical research was conducted primarily around questions that were relevant also from an experimental nuclear and particle physics perspective.

Similarly, the Institute for Chemistry in Mainz, later named the Otto Hahn Institute, maintained a foothold in methods and questions that could best be described as radiochemistry (based on Hahn and Meitner's work in the 1930s);[339] and it was not until the 1970s onward that it managed to detach itself from its original identity as a 'nuclear' institute, and switch its focus to

---

338 Armin Hermann et al.: *History of CERN. Launching the European Organization for Nuclear Research*. Vol. 1. Amsterdam: North-Holland 1987. Armin Hermann et al.: *History of CERN. Building and Running the Laboratory, 1954–1965*. Vol. 2. Amsterdam: North-Holland 1990. John Krige (ed.): *History of CERN*. Vol. 3. Amsterdam: North Holland 1996. See also Krige, *American Hegemony*, 2006.

339 On Lise Meitner's work at the Institute for Chemistry, see Ruth Lewin Sime: *From Exceptional Prominence to Prominent Exception. Lise Meitner at the Kaiser Wilhelm Institute for Chemistry*. Vol. 24. Berlin: Forschungsprogramm »Geschichte der Kaiser-Wilhelm-Gesellschaft im Nationalsozialismus« 2005.

questions more closely related to environmental research.[340] Tellingly, however, both the old 'nuclear' character and the more recent environmental facet of the Mainz institute were based on the same analytical methods, in radiochemistry and the mass spectrometry of small samples, and these gave the institute its strength in cosmochemistry throughout the second half of the 20th century.

Forced 'Clustering' via Large-Scale Nuclear Research

The first decade and a half of the Max Planck Society was characterized by the relative independence of figures such as Gentner and Heisenberg, each aligned with powerful political and regional interests. Nevertheless, there were occasions when their spheres of influence were forced to collaborate, as was the case when the Max Planck Society was not the dominant force. In this early stage, the participating scientists and research groups generally had to leave the Society altogether to conduct their research, but the emerging organizations still retained traces of the boundaries between the different factions.

The most prominent case of this was the development of the Institute for Plasma Physics. As was mentioned earlier, its founding was closely connected to Heisenberg's move to Munich and the failure to establish his experimental fission reactor there, which was instead built in Karlsruhe. For the formation of this federal facility, the Max Planck Institute for Physics gave up an entire research group headed by Karl Wirtz, who had been the central figure of Heisenberg's circles in the German efforts to build a nuclear reactor during the war.[341] Federal facilities such as Karlsruhe were meant to pool all the major

340 Carsten Reinhard: Max-Planck-Institut für Chemie—Mainz. In: Peter Gruss, and Reinhard Rürup (eds.): *Denkorte. Max-Planck-Gesellschaft und Kaiser-Wilhelm-Gesellschaft. Brüche und Kontinuitäten 1911–2011*. Dresden: Sandstein 2010, 256–265. This 'environmental turn' was part of a wider trend in Germany, where the Ministry of Research and Technology offered explicit incentives in the 1970s, to repurpose research institutes from the nuclear age to deal with environmental issues. See Hohn, and Schimank, *Konflikte*, 1990. That a growing political and social emphasis on environmental issues marked the rise of a new cluster in the Max Planck Society is treated in Gregor Lax: *From Atmospheric Chemistry to Earth System Science. Contributions to the Recent History of the Max Planck Institute for Chemistry* (*Otto Hahn Institute*), *1959–2000*. Diepholz: GNT-Verlag 2018. Gregor Lax: *Wissenschaft zwischen Planung, Aufgabenteilung und Kooperation. Zum Aufstieg der Erdsystemforschung in der MPG, 1968–2000*. Berlin: GMPG-Preprint 2020. Gregor Lax: Zum Aufbau der Atmosphärenwissenschaften in der BRD seit 1968. *NTM Zeitschrift für Geschichte der Wissenschaften, Technik und Medizin* 24/1 (2016), 81–107. http://nbn-resolving.de/urn:nbn:de:1111-201604011486 9. Last accessed 8/5/2016.

341 See, for example, Eckert, and Osietzki, *Wissenschaft für Macht und Markt*, 1989.

experts in the country, administered directly by what was then the 'Nuclear' Ministry.

When it came to building the Institute for Plasma Physics in the late 1950s, both the state of Bavaria and the Federal Ministry of Atomic Affairs headed by the Bavarians Franz Josef Strauss and, later, Friedrich Balke were at a high point of political influence. As one of the institutes conducting what was known as *Grossforschung*—the large-scale research initiated during this period as a national technological mission,comprising in this particular case the path toward controlled thermonuclear energy—the institute could at least implicitly lay claim to operating independently of the Max Planck Society, and use its focus on fundamental research to justify keeping outside interference at bay.[342] Still, that Max Planck figures wielded influence within the IPP was obvious: Werner Heisenberg signed personally as founder and, scientifically, the institute owed much of its theoretical basis to Ludwig Biermann and his collaborators. The great majority of the researchers at the Institute for Plasma Physics, including its first director Arnulf Schlüter, came from the Institute for Physics and Astrophysics.[343]

However, Bavaria and the Federal Ministry soon signaled that the new Institute for Plasma Physics was not going to remain exclusively the realm of Heisenberg's collaborators. Researchers with the relevant expertise from all around Germany were given the opportunity to participate in the new institute. This was particularly urgent, in part because the renowned expertise in plasma physics of the Max Planck Institute for Physics and Astrophysics was largely theoretical. The latter's researchers, hence, while still a significant majority within the new initiative, had also to be complemented by more experimentally oriented physicists. And so it was that Gentner's circles found their way into the Institute for Plasma Physics: Ewald Fünfer had trained with Bothe in Heidelberg in the 1940s and subsequently been sent to work with Heisenberg's reactor team during the war, acting as a bridge between the two groups.[344] Like many in Gentner's circles, he spent the first postwar years

---

342 On the organization of large-scale-research in Germany—in comparison to the USA, UK, and France, countries where new dedicated institutions had to be created (Atomic Energy Commission and NASA, USA; Atomic Energy Authority, UK; Commisariat à l'Energie Atomique, France) and no autonomously administered scientific institutions like the Max Planck Society and the Deutsche Forschungsgemeinschaft exist—as well as on the specific role of the Max Planck Society and its relationship with Big Science, see AMPG, II. Abt., Rep. 102, No. 437.

343 Boenke, *Entstehung und Entwicklung*, 1991.

344 Eckert, and Osietzki, *Wissenschaft für Macht und Markt*, 1989. Boenke, *Entstehung und Entwicklung*, 1991. On Fünfer's work during the war, see Walker, *The Quest for Nuclear Power*, 1989, 101–112.

at the military French-German Research Institute of Saint-Louis in France, working on ballistics problems, before completing his studies in Freiburg and then moving with Maier-Leibnitz to the Technical University of Munich in 1953. At the same time as Biermann's people in Göttingen were entering thermonuclear fusion via theoretical plasma astrophysics, Fünfer at the Technical University of Munich was already exploring small-scale experimental paths toward nuclear fusion.

Once plans for the Institute for Plasma Physics were underway, Maier-Leibnitz suggested that a team led by Fünfer should be involved.[345] This was strongly resisted by Heisenberg, who feared internal competition. But he was overruled by the central government and Bavarian authorities behind the new institute.[346] Nevertheless, in the early years, while the Institute for Plasma Physics still functioned within buildings of the Max Planck Society, Fünfer's group, nominally part of the same nuclear fusion project, was stationed on the premises of the Technical University near the 'Atomic Egg,' so straddling the internal border in Garching between the Technical University and the Max Planck Society, where the Institute for Plasma Physics was to be located. It was not until the IPP had been formally founded and moved to its permanent site that Fünfer's team would become an integral part of the institute.[347] Even before that, however, he took the wind out of Heisenberg's people's sails, just as Maier-Leibniz had done with the 'Atomic Egg,' by seeking to produce neutrons from thermonuclear fusion for the first time ever on German soil, in 1958, using what is known as the Pinch method.[348] Only in the course of the 1960s would

---

345 Eckert, and Osietzki, *Wissenschaft für Macht und Markt*, 1989, 120–125. Boenke, *Entstehung und Entwicklung*, 1991.

346 See, for example, Edingshaus, *Heinz Maier-Leibnitz*, 1986, 132 and 142.

347 The initial scientific leadership of the institute consisted of Heisenberg, Biermann, Schlüter, Gerhard von Gierke (all from the Max Planck Institute for Physics and Astrophysics), plus the aforementioned Fünfer from the Technical University, and another ally of Gentner's, Karl-Heinz Schmitter, as Director of the technical division coming from CERN. "Die Bestellung der Mitglieder der Wissenschaftlichen Leitung" München, 05.07.1960 IPP-ZA, Gesellschafter, Versammlungen und Beschlüsse, Protokolle: Gesellschafterbeschluss 1/1960 (ms., Original). Cited in Gerda Maria Lucha (ed.): *Dokumente zu Entstehung und Entwicklung des Max-Planck-Instituts für Plasmaphysik 1955–1971*. Garching: IPP 2005, 99. See also *Arnulf Schlüter und Rudolf Wienecke. Pioniere der Fusionsforschung*. IPP-Report, 16/22. Garching: Max-Planck-Institut für Plasmaphysik 2012.

348 Isabella Milch: Persönliches. Zum Tode von Ewald Fünfer. *Physik Journal* 51/10 (1995), 965–965. doi:10.1002/phbl.19950511014. Neutron detections from Pinch devices, which also contributed to the early prestige of British fusion research, soon became controversial, and the Pinch pathway was increasingly left aside in subsequent decades, as stellarators and especially tokamaks entered the race. In the early decades of fusion

ILLUSTRATION 86 Heinz Maier-Leibnitz with Ewald Fünfer in the 1960s

the theoretically trained physicists from Biermann's tradition catch up with and then surpass this experimental pathway by adopting the larger-scale stellarator and Q-machine reactors from the United States.[349] Over this decade, the two previously separate scientific traditions, one from the experimentally oriented southwest, the other, theoretically oriented, from Göttingen, merged to form a strong research identity at the Institute for Plasma Physics that continues to this day.

This was an early example of collaboration between these traditions being driven by external forces, and executed outside of the Max Planck Society itself. As will be described later, regardless of the relative success of the Institute for Plasma Physics in comparison with other plasma fusion projects of the period, the field itself entered a chronic decline once the grandiose expectations of fusion energy proved much more difficult to realize than anticipated.[350] By the late 1960s, the generation of large research institutes that

research, however, this was seen as a less complex pathway towards fusion. The collection of documents on the origin and development of the Max Planck Institute for Plasma Physics in the years 1955–1971 edited by Gerda Maria Lucha provides interesting insight of how this was seen at the time (https://www.ipp.mpg.de/59804/dokumentenband.pdf, p. 27. Last accessed 31/07/2021).

349 Robert Motley: *Q Machines*. Oxford: Elsevier Science 1975.

350 Weisel, Plasma Archipelago, 2017, 183–226.

ILLUSTRATION 87 Ludwig Biermann and Werner Heisenberg in 1967

included Karlsruhe and the Institute for Plasma Physics were in institutional crisis, and the institute was eventually reabsorbed in the Max Planck Society. But the coming together of research traditions at the Institute for Plasma Physics set a precedent that would soon be followed within the Max Planck Society itself, once outer space acquired a degree of political interest similar to that of the nuclear sciences prior to Sputnik. We can see that the Institute for Plasma Physics signaled how internal divisions within the Max Planck Society opened up opportunities for intervention by external political forces. In the future, it would prove preferable to coordinate internally and present a united front to external actors. As in the tradition of the Society since its foundation, this was viable particularly in cases that could be framed as 'fundamental research.'

CHAPTER 2

Space Age (1957–1980s): A Unique Opportunity for Expansion

This second chapter follows the enormous expansion of the space sciences around the world after the launch of Sputnik, as well as the uniquely constrained West German response; and it focuses on how the Max Planck Society maneuvered itself into a role of predominance in the space sciences, under these circumstances. Thanks to its strong scientific traditions and political backers, the Max Planck Society was singularly well placed to take advantage of the rising interest in the study and conquest of outer space: while guaranteeing a concerted emphasis on 'fundamental research' and international collaboration, it mobilized existing projects in plasma physics, cosmochemistry, and balloon-based cosmic rays, and joined in diverse space activities with the United States and various European countries. This entry into the space age paved the way to the Society's subsequent expansion into astronomy (the subject of the next chapter), and also allowed the scientific traditions of the early postwar era to diversify: dependency on 'nuclear' sociopolitical interests and funding was now succeeded by a focus on astrophysical subjects proper. As we will see in subsequent chapters, this reorientation ultimately became one of the vehicles propelling these longstanding traditions towards the most effervescent topics of 21st-century astrophysics.

1 'Sputnik Shocks'

Within only a few months of the launch of the Soviet satellite, the status of disciplines such as astronomy and astrophysics changed dramatically, as they now became integrated into the Cold War apparatus, just as experimental physics had been in 1945. Key players in this radical shift were those scientists around the world who had preexisting strengths and interests in the cosmic sciences, but had formulated their research in terms of 'nuclear' topics during the postwar years. Space exploration initiatives in the United States, Soviet Union, France, Britain, and other European countries would now become the model for the German MPI scientists described in the previous chapter, and, eventually, their collaboration counterparts, too. We describe this transition, from the predominantly 'nuclear' period up to 1957 to the nascent space age.

DOI:10.1163/9789004529137_004

All this unfolded still under Allied constraints on military technologies, which hindered the West Germans' construction of a fully national space launch capability.

A New Status for the Cosmic Sciences

The launch of the Soviet satellite Sputnik inaugurated the 'space age,' and it radically transformed the status of the cosmic sciences in the political and public arenas. During the first postwar decade, in order to access significant support, researchers rooted in varied traditions and interests had had to align themselves with nuclear-centered research environments. After October 4, 1957, in contrast, space acquired a sociopolitical import that extended support vastly beyond the nuclear worldview and even beyond expert scientific communities, to become a central sphere of competition between the two superpowers; this allowed the absorption of researchers coming from diverse fields and traditions that had previously remained outside of the generously funded nuclear research communities. And while in the first years of the space age the military-technical approaches and geopolitical strategies regarding outer space were inherited from the nuclear age, in the course of the 1960s the space age matured into a distinct logic based on the unique status of outer space that was agreed internationally. Meanwhile, astronomy and astrophysics developed significantly, thanks to the increase in support for all forms of science that resulted from the Western response to Sputnik, strengthened by spectacular astronomical discoveries throughout the 1960s, and the maturation of a much larger and diverse community of researchers.[1]

This section explores the tension between the contingent ideological impact of the 'Sputnik shock' itself, and the deeper, decade-long incorporation of the cosmic sciences into the Cold War system in the major western countries, periodically highlighting how these were reflected in the very unique West German scenario and the Max Planck Society. This provides the basis for exploration, in the following chapters, of specific internal scientific and institutional developments.

Sputnik provided an opportunity for Western scientific elites to augment their political power. Even though this first satellite was a rudimentary radio beacon of little scientific use, it served to catalyze an immense governmental expansion in space research and related areas through increased spending, state intervention, and centralization. This wave started in the United States and then spread to other Western countries.

1 David H. DeVorkin: The Space Age and Disciplinary Change in Astronomy. In: Steven J. Dick (ed.): *NASA's First 50 Years. Historical Perspectives*. Washington, D.C.: NASA 2010, 389–426.

A powerful, lasting legacy of this 'shock' was a rise in the status of scientists; this allowed them to take on powerful positions in areas related to space exploration which, until 1957, had been the preserve of the military.

These scientists' political interventions were particularly influential during the 1960s, and resulted in a boom in all forms of scientific research. But besides the spectacular effects of Sputnik on public opinion and science policy worldwide, the development of thermonuclear weapons carried on Intercontinental Ballistic Missiles (ICBMs) inaugurated a new stage of the Cold War; and the cosmic sciences were very particularly and profoundly affected, because of their close synergy with the technologies dominating this new stage of the global conflict. To begin with, there were the rockets themselves, which gave access to entirely new environments, and unhindered access to the cosmos beyond; but the synergy also encompassed techniques and instruments for tracking objects, which were astronomical in essence, as well as a particularly close overlap between detectors used at many wavelengths for military purposes, and their application in the examination of new astronomical bodies and phenomena. Cold War armament treaties even depended on the development and mass production of technologies instrumentally related to the cosmic sciences. As a consequence, the 'cosmic' elites in many countries circulated easily between scientific and military contexts and held influential advisory roles comparable to those held decades earlier by 'nuclear' physicists. In fact, as we detail later, many newcomers to astronomy in the 1960s had previously forged careers in fields connected to the 'nuclear' worldview. On the other hand, as clearly shown by the astronomer Martin Harwit, in the 25 years between 1954 and 1979, "most of the major cosmic phenomena were discovered by individuals prepared for careers other than astronomy," outsiders with an educational background and early experience with novel techniques for looking at the sky.[2]

Most important for the purpose of this book is the relationship between these global trends and their impact in the very particular environment of West Germany, a country with a mandatory subordinate status within the Western Alliance. Given how important military applications were in the development of space research in other large countries, the West German case provides a unique counterexample, and this explains to a large degree the unusual strength of the Max Planck Society: West German decision makers participated in the same discursive optimism regarding outer space as their

2 See Fig. 5 in Martin Harwit: Physicists and Astronomy—Will You Join the Dance? *Physics Today* 34/11 (1981), 172–187. doi:10.1063/1.2914355.

colleagues elsewhere, but, unlike them, were cut off from access to the enormous undercurrent of military applications which generally subsidized scientific developments. West German scientists had to find their viable research niches within these constraints, while also surviving in the politically problematic parallel resurgence of rocket development in Germany, a country teetering between initial but increasingly rogue attempts at national self-reliance within the Western Alliance, while also being steered by the Western Allies towards acceptance of its subordinate role as financial backer and industrial supplier within an integrated pan-European aerospace industrial landscape.

In this politically delicate environment, the rationale of 'fundamental research' crafted by the Max Planck Society in the first postwar decade resonated with the discourse of post-Sputnik scientific elites around the world.

Nuclear Annihilation and Outer Space

The objective challenge introduced by Sputnik was the threat of intercontinental ballistic missiles which, armed with nuclear warheads, could reach targets anywhere in the world. This menace eventually stabilized into the political equilibrium of the Cold War, based on mutual assured destruction. Reaching this balance, however, took over a decade, in a process that was closely intertwined with the first steps of the space age; and it had a profound impact, in particular on the cosmic sciences.[3]

Nuclear bombs carried on bomber airplanes were a threat that had been addressed in the first postwar decade by the development of supersonic jet aircraft, radar technologies, and anti-aircraft missiles, ICBMs, however, were conceived as unstoppable, and this became a defining feature of a new Cold War balance. During the first decade of the space age, the nuclear superpowers came to terms with mutual assured destruction through a series of political agreements, detailed below, which was necessary to make this standoff survivable. Events like the Cuban missile crisis of 1962 highlighted the risks of short-range nuclear weapons which reach their target very quickly. Consensus arose among the superpowers that the only viable balance in mutual assured destruction depended on incoming attacks being clearly identifiable well enough in advance to permit a response, as in the case of a ballistic missile's half-hour trip. Otherwise, first-use attackers would have the advantage of 'knocking out' their opponent before a significant retaliation could be launched.

3 Karsten Werth: A Surrogate for War—The U.S. Space Program in the 1960s. *Amerikastudien / American Studies* 49/4 (2004), 563–587.

This threat existed with nuclear weapons based on artificial satellites. Hence, through the early 1960s, the Americans and Soviets came to agree that nuclear weapons kept in orbit were best avoided. Based on these military considerations, the 'international,' 'peaceful,' and 'scientific' nature of outer space started to become established.

False alarms and mistakes could trigger doomsday, so a massive infrastructure to detect all rocket launches and follow their trajectories was crucial for the strategic nuclear standoff: surveillance satellites and rocket tracking technologies, often based on astronomical techniques, were involved in the new global balance of power from the outset. International treaties negotiated through the 1960s depended on nuclear deterrents that posed an overwhelming threat, but which were also mutually verifiable, under centralized political control, and able to be identified and assessed far enough in advance, if ever launched in an offensive. The new regime was crystallized in the Space Treaty of 1967, which defined outer space as international and 'peaceful,' but also left ample room for non-aggressive (largely, surveillance) technologies in orbit. This new regime was also reflected in the Non-Proliferation Treaty (NPT) signed in 1968, which explicitly restricted nuclear explosives (not just weapons) to the select 'club' of countries which already possessed them, and ruled out any exceptions for 'peaceful' nuclear explosions, including their use in outer space. Finally, the last item in this framework was the Anti-Ballistic Missile Treaty (ABM Treaty) signed in 1973, which prohibited the large-scale deployment of anti-missile technologies. In order to avoid further escalation of the missile race and disincentivize first strikes, the ABM upheld the unstoppable status of oncoming nuclear strikes, despite the technical feasibility of countermeasures.[4] All these treaties, and the years of negotiations preceding them, set the stage for the 'peaceful' and 'international' character of outer space that predominated from the late 1960s onwards.[5]

---

4 For Outer Space, Nuclear-Non-Proliferation and Anti-Ballistic Missile Treaties see the U.S. Department of State webpage https://2009-2017.state.gov/t/avc/trty/index.htm. Last accessed 2/3/2022. Anti-missile technologies themselves were not prohibited, but their deployment strictly limited. Still, this allowed for their continued development over the remainder of the Cold War, and threats to use them were a recurrent issue in the 1980s.

5 Even before signed treaties formalized these circumstances, outer space was undergoing denuclearization as a temporary effect of the negotiations toward the Test Ban Treaty, in which from 1957 to 1963 the Soviet Union and the United States sought to stabilize their nuclear duopoly, while addressing issues of public concern regarding the health effects of testing nuclear weapons in the atmosphere. During the brief 'thaw' that preceded the Cuban Missile Crisis there had even been talks of a complete ban on nuclear testing; but due to the tensions highlighted by the Cuban crisis, the final agreement of 1963 took into consideration

Verification of the terms and conditions laid down in these treaties depended on space-based technologies such as reconnaissance satellites, as well as methods to identify nuclear explosions themselves, such as radioactive trace 'sniffer' airplanes and orbiting gamma-ray detectors. Likewise, communications networks, global positioning systems, and other military infrastructures related to this threat-and-surveillance regime now flourished in outer space. Throughout the remainder of this book, we periodically encounter examples of how the development of these Cold War technologies contributed to instrumentation and infrastructures used also in scientific research.

The status of outer space that became established in the 1960s had an enormous impact on the way the space sciences, astronomy, and astrophysics developed over the next half century. Initially, still within the worldview of the nuclear age, leading scientific personalities appropriated space as one more arena in which to expand their preexisting interests, advocating for a future of nuclear rockets, routinized atomic explosions (both military and 'peaceful'),[6] and research oriented to phenomena epistemically linked to thermonuclear weapons such as plasmas. Space science in the early 1960s was an experimental 'nuclear' endeavor, an inquiry into the near-Earth environments fast becoming a sphere of operations for nuclear missiles and the first generation of civilian rockets.[7] Half a decade later, these environments were well characterized, while at the same time the 'international,' 'peaceful,' and non-nuclear status of space had been established. Scientific interest shifted

only those nuclear tests which could be easily detected beyond a country's borders, leaving room for continued underground testing. The test ban covered atmospheric and underwater tests, and also outer space. No atomic device has been exploded at high altitudes since 1962. See brief history and Treaty text at U.S. Department of State webpage https://2009-2017.state.gov/t/avc/trty/199116.htm. Last accessed 2/3/2022.

6 These were closely linked to the Project Plowshare initiative of the same era, which conducted 'peaceful' nuclear explosions between 1961 and 1973. In the new space age context, nuclear rocket propulsion was among the most publicized 'peaceful' use of nuclear explosions until the mid-1960s when space was denuclearized. Scott Kaufman: *Project Plowshare The Peaceful Use of Nuclear Explosives in Cold War America.* Ithaca, NY: Cornell University Press 2013.

7 Sometimes, scientific experiments were a low-hanging opportunity opened up by test rockets being filled with test materials other than the usual sand ballast. In the early 1960s, for example, Wernher von Braun exploded large quantities of water (86.000 Kg!) at high altitudes from his test rockets. These 'High Water' experiments, observed from the ground, helped characterize the plasma environment of the upper atmosphere. Andrew J Dunar, and Stephen P. Waring: *Power to Explore. A History of Marshall Space Flight Center, 1960–1990.* Washington, DC: National Aeronautics and Space Administration, NASA History Office, Office of Policy and Plans 1999, 228.

increasingly toward more distant exploratory pursuits, like unmanned interplanetary missions and space-based astronomy, while the political and public focus remained on manned spaceflight. Within this 'peaceful' framework, the tension between scientific research and human spaceflight settled into an uneasy compromise that still follows us to this day. But still, as we detail below, it was military interests and their industrial fulfillment which continued to drive most technological progress during this mature space age: technologies, infrastructures, and even knowledge developed in the Cold War context continuously spilled over into civilian and peaceful scientific enterprises.

Scientific Elites in the Post-Sputnik Cold War

Around the world, the transition to the space age was led by personalities with a 'nuclear' background. Thanks to their experience with the decision-making and funding structures of the early Cold War era, these scientists could guide specialists in the newly relevant fields toward the opportunities emerging after Sputnik, while continuing to hang on to their senior roles in decision-making committees for several decades more.

Such advisory roles were nothing new in activities derived from the military-industrial complex; but the scientific opportunities after October 1957 expanded far beyond research directly relevant to military applications. The 'Sputnik shock' led, for example, to the creation in the United States of a President's Science Advisory Committee (PSAC), via which senior scientists, largely from a background in radar and especially nuclear research within the Manhattan Project, had direct access to decision makers.[8] Political parties even began competing to foster and favor scientific-technical agendas: the 'missile gap' was a central narrative leading to the election of John F. Kennedy in 1960, and his government continued and expanded these scientific advisory roles, as did every new administration at least until the 1980s. And in parallel to

8 The creation of PSAC was decided in an October 15 meeting of the previously existing SAC, which included scientists like Isidor Rabi, Edwin Land, and James Killian. Deliberately left out were proponents of normalizing the use of nuclear weapons such as Edward Teller or Ernest Lawrence, in favor of those attuned to Eisenhower's conviction by 1957 that the purpose of nuclear weapons should be as deterrent. The PSAC included a majority of people who had been involved in either radar or nuclear weapons, but also included scientists coming from industry, academic administrators, and representatives of the major research organizations. It was a conspicuously elitist group and in its early years it was dominated by a 'Cambridge Mafia' which used the position to advocate for increasing the support of science in general, not just fields closely related to defense. See Zuoyue Wang: *In Sputnik's Shadow. The President's Science Advisory Committee and Cold War America*. New Brunswick, NJ: Rutgers University Press 2008, 74–85.

these public advisory roles came the creation of classified advisory committees dealing with issues of direct relevance to national security. Most relevant for the cosmic sciences was JASON,[9] a science advisory group of physicists which initially advised on matters directly related to the Cold War threats, playing a central role in technical advice on nuclear weapons and missiles during the establishment of the deterrent regime. The group was the intellectual arm of the Advanced Research Projects Agency (ARPA), newly created by the Pentagon in January 1958 (as successor to DARPA—Defense Advanced Research Projects Agency—founded in 1957). In view of its long-term advisory roles, JASON was crucial for the generational renewal of the committees still populated by Manhattan Project veterans. For younger generations, JASON was a mechanism of socialization into (and by) the scientific elites.[10] The key part of this socialization was the 'loss of innocence' that ensued from contributing to 'Strangelovean' enterprises with values so different from the ethos of fundamental science.[11] But crucially, these advisory groups insisted that their members should continue their scientific research careers as a main activity.

'Nuclear' advisors served on the committees that helped integrate new sciences into this Cold War framework, which was based on decades-long, nationwide planning that favored the centralization and rationalization of all endeavor. But the main difference between the pre-and post-Sputnik world was that scientists in such advisory positions managed to steer the conversation beyond research directly linked to the military-industrial complex, and advocate for much deeper state involvement in scientific research in general.

9 Ann K. Finkbeiner: *The Jasons. The Secret History of Science's Postwar Elite*. New York, NY: Viking Press 2006. See also, Ann Finkbeiner: JASON Past, Present, and Future. The World's Most Independent Defence Science Advisers. *Nature* 477 (2011), 397–399. doi:10.1038/477397a. While their main mission was to analyze nuclear missile exchange scenarios and propose technologies related to this challenge, the group later also gave tactical advice related to the Vietnam War; beyond the 1960s, the group extended its reach to new developments like molecular biology, and even economic and environmental matters. Contrary to the Cambridge-dominated PSAC, this parallel, classified group was initially dominated by Princeton-based scientists.

10 Edward Teller, Eugene Wigner and Hans Bethe are examples of the veteran generation who advised JASON. Younger members who crossed over to Astronomy and Astrophysics include Princeton-based Freeman Dyson and John Wheeler, as well as laser pioneer Charles Townes. Beyond an advisory group, their gatherings included their relatives over long summer retreats. The resulting dynamics was even described as a 'family,' metaphorically but also literally, as "the children grew up like cousins." See, Finkbeiner, *The Jasons*, 2006, 211.

11 Finkbeiner, *The Jasons*, 2006, xxviii.

As we show throughout this chapter, the discourse of planning and rational government was significantly boosted by Sputnik, and in scientific research this ideology allowed for intervention in fields that had never yet seen such state coordination. The expansion of this logic into new research specialties, as certainly occurred within the cosmic sciences, was led by such scientific advisors. To name but one foundational example: in the early 1960s, the first US Decadal Survey in Astronomy[12] was initiated by Manhattan Project veteran George Kistiakowsky, who had been both Eisenhower's Chair of the PSAC and, a few years earlier, co-artificer of the Single Integrated Operational Plan, which rationalized the plans for nuclear action in the age of ICBMs.[13]

The participation of 'nuclear' experts in the advent of the space age occurred at all levels. Particularly striking is how several crucial figures had actually begun their careers in astrophysics, then adopted a 'nuclear' identity and research programs after 1945; and now, these crypto-astrophysicists could instrumentalize Sputnik to return to their truly profound scientific interests. This pattern is evident in figures who appear repeatedly in this book, such as the Princeton-based Lyman Spitzer (Chapters 1 and

12 US decadal surveys, which still continue to this day, collect input from the ground-based astronomical community to coordinate research objectives and investments in research infrastructure. These decadal surveys in turn often drive astronomy plans in other countries around the world. National Academy of Sciences: *Ground-Based Astronomy. A Ten-Year Program. A Report Prepared by the Panel on Astronomical Facilities for the Committee on Science and Public Policy of the National Academy of Sciences*. Washington, D.C.: The National Academy Press 1964. See also Ground-Based Astronomy. A Ten-Year Program. A Report Prepared by the Panel on Astronomical Facilities for the Committee on Science and Public Policy of the National Academy of Sciences. *Science* 146/3652 (1964), 1641–1648. doi:10.1126/science.146.3652.1641.

13 George B. Kistiakowsky: *A Scientist at the White House. The Private Diary of President Eisenhower's Special Assistant for Science and Technology*. Cambridge, MA: Harvard University Press 1976. An external view of Kistiakowsky's role at PSAC is in: Roger L. Geiger: What Happened after Sputnik? Shaping University Research in the United States. *Minerva* 35/4 (1997), 349–367, 354. https://www.jstor.org/stable/41821079. Last accessed 5/24/2019. The SIOP, a still-classified plan of resources and action was the policy on which the mature Cold War standoff was based between 1961 and 2003.

5),[14] John Wheeler (Chapter 5),[15] and Freeman Dyson (Chapters 2, 3, and 5).[16]

Likewise significant, and occurring in parallel, was the mass migration to space-related fields of those researchers originally working in fundamental particle physics. Even though not necessarily involved in classified research, they had received generous funding throughout the 1950s, as a 'nuclear' enterprise. Sputnik coincided historically with the advent of the first generation of large particle accelerators at CERN, Dubna (Joint Institute for Nuclear Research, JINR), and Brookhaven National Laboratory (see Chapters 1 and 5), which replaced cosmic rays as a means to inquire into many fundamental physical questions. Particle physicists uninterested in, or unable to work

14 Spitzer had started his career as one of the first people in the United States with a Ph.D. in astrophysics, which led to a directorship in Princeton in 1946. During the next decade, however, he focused his interests on plasma astrophysics, which was relevant to both thermonuclear reactors and the hydrogen bomb, as we described in Chapter 1. Spitzer repeatedly failed to interest the astronomical community in an orbiting telescope, and his plans dating from the late 1940s could only be executed in the 1960s. Lyman Spitzer: interview by Joan Bromberg, March 15, 1978. Transcript, Niels Bohr Library & Archives, American Institute of Physics, College Park, MF USA, https://www.aip.org/history-programs/niels-bohr-library/oral-histories/4900. Last accessed 12/4/2020.

15 John Wheeler was a well-known theoretical physicist who in 1939 had developed with Niels Bohr a general theory of the mechanism of fission based on the liquid-drop model of atomic nuclei, later joining the Manhattan Project to work on the reactors that were needed to create plutonium for the atomic bombs. He was then invited to work with Edward Teller on the Matterhorn Project developing the H-bomb, and was a colleague of Lyman Spitzer at Princeton, and a leader of the secret theoretical group, while making substantial contributions to the theory of fundamental particles. At the same time, his 'hidden' interest was the theory of general relativity at a time when it was neglected by the mainstream (see other figures like Robert Oppenheimer). In the 1950s and '60s these interests finally came to the foreground, leading to his contributions to general relativity and relativistic astrophysics. John Archibald Wheeler: interview by Kenneth W. Ford, Session XI, March 4, 1994, https://www.aip.org/history-programs/niels-bohr-library/oral-histories/5908-9. Last accessed 12/4/2020. Session XII, 28 March 1994, https://www.aip.org/history-programs/niels-bohr-library/oral-histories/5908-12. Niels Bohr Library & Archives, American Institute of Physics, College Park, MD USA. More on Wheeler's involvement in the revival of general relativity will be outlined in the final chapter of this book.

16 Freeman Dyson started his scientific career in fundamental theoretical physics, while also participating in subjects as varied as nuclear reactor design and nuclear propulsion, which he continued after Sputnik within the Project Orion toward a nuclear-powered spaceship. His early interests, however, were in astronomy (he was even offered a position at the Greenwich Observatory in 1948, with prospects of becoming Astronomer Royal), and in the 1960s he made significant contributions to theoretical astrophysics, while his 'applied' contributions shifted to areas such as adaptive optics. Freeman J. Dyson: *Maker of Patterns. An Autobiography through Letters*. New York, NY: Liveright 2018.

with accelerators were driven toward astrophysical inquiries, as we explore in detail in Chapters 3 and 5. Large numbers of them mobilized their scientific capital toward research in the newly respectable cosmic fields, often in a combination of ground-based, airborne (balloon and aircraft), and space-based initiatives. Insofar they followed the lead of select American, European, and Soviet figures who played a defining role at the onset of the space age, such as James Van Allen,[17] Edoardo Amaldi,[18] Alexander Chudakov,[19] Patrick Blackett,[20] and Pierre Auger.[21] This generation of physicists had obtained their influential positions before the space age, in the course of remarkable careers in cosmic ray research at high altitudes, with balloons and rockets. In the West German case, this would have been the natural role for the recently deceased Erich Regener (see Chapter 1).[22]

Finally, by the late 1960s, the tide of researchers moving into space science, astronomy, and astrophysics swept up people from entirely separate fields: one telling example is Charles Townes, who started his career in radar, then became one of the pioneers of laser, while also engaging in senior classified advisory roles. In the late 1960s, Nobel Prize in hand, he completed his

17 David H. DeVorkin: *Science With A Vengeance. How the Military Created the US Space Sciences After World War II*. New York, NY: Springer-Verlag 1992. James A. Van Allen: What Is a Space Scientist? An Autobiographical Example. *Annual Review of Earth and Planetary Sciences* 18/1 (1997), 1–27. doi:10.1146/annurev.ea.18.050190.000245.

18 Michelangelo De Maria: *Europe in Space. Edoardo Amaldi and the Inception of ESRO*. ESA-HSR-5. Noordwijk, the Netherlands: ESA Publications Division 1993. Carlo Rubbia: *Edoardo Amaldi: Scientific Statesman*. Vol. 91–09. Geneva: CERN 1991. doi:10.5170/CERN-1991-009.

19 Sergei N. Vernov et al.: From Balloons to Space Stations. In: Yataro Sekido, and Harry Elliot (eds.): *Early History of Cosmic Ray Studies. Personal Reminiscences with Old Photographs*. Dordrecht: Springer 1985, 357–374. doi:10.1007/978-94-009-5434-2_34.

20 Bernard A. C. Lovell: Patrick Maynard Stuart Blackett, Baron Blackett, of Chelsea, 18 November 1897–13 July 1974. *Biographical Memoirs of Fellows of the Royal Society* 21 (1975), 1–115. doi:10.1098/rsbm.1975.0001. One of the leading mid-century figures in nuclear and particle physics research with the use of cosmic rays, and was also a very early proponent of radio astronomy in Britain. Initially a gradualist and critic of the panicked response to nuclear weapons and missiles, he later became a central artificer of the British technocratic state in the 1960s. Mary Jo Nye: *Blackett. Physics, War, and Politics in the Twentieth Century*. Cambridge, MA: Harvard University Press 2004.

21 Lars Persson: Pierre Auger—A Life in the Service of Science. *Acta Oncologica* 35/7 (1996), 785–787. doi:10.3109/02841869609104027.

22 As mentioned in Chapter 3, Regener had died in 1955, on the eve of the Sputnik launch, and while his disciples at the Max Planck Institute for Aeronomy (Ehmert, Pfotzer) tried to fill those positions, their scientific legitimacy was not comparable and the much more powerful scientists in Munich had more influence in shaping the West German response to the space age, as we see in the next section.

transition to observational astronomy, identifying there many opportunities in which his expertise and instrumental knowledge could open up new ways of exploring astronomical phenomena.[23]

In Section 2 we will see in detail how, in West Germany, it was scientists with similar profiles to—and often personal connections with—those leaders named above who led the Max Planck Society's expansion into the space age. In many respects, these German scientists were not unlike their European counterparts; the key difference was that the particular status of West Germany within the Western Alliance required them to deliberately tone down the links between space exploration, scientific research, and military applications.

Cold War Cosmic Sciences in the American Sphere

The majority of research activities in the cosmic sciences were not directly linked to military applications. Their incorporation in the Cold War system after Sputnik occurred as outwardly expanding circles determined by the degree of Cold War relevance: at the core were research activities favored by those influential scientists already part of the 'nuclear' complex, which could be conducted from rockets, satellites, and interplanetary probes. Primarily, these were cosmic ray and plasma-related phenomena, in the upper atmosphere and near-Earth space. At a further epistemic remove from military interests was space-based astronomy, which made the most of high-altitude rockets and satellites—and hence Cold War-related progress on both launchers and detectors—to gain access to wavelengths blocked by the atmosphere. Soon after, from the mid-1960s onwards, ground-based astronomy expanded greatly thanks to the synergy created by spectacular discoveries in radio astronomy and early space-based observations of cosmic, gamma, and X-rays, which gave rise to the need to cover the entire electromagnetic spectrum. Ground-based astronomy was much cheaper, durable, and upgradeable than space-based missions, and observations at some wavelengths and locations delivered the requisite quality within feasible budgets. While astronomical wavelengths each had distinct techniques and research traditions,

---

23 Townes, Charles Hard: interview by Suzanne B. Riess, 1991–92. Transcript, Selected oral histories from the UC Berkeley Oral History Center, Online Archive of California, http://ark.cdlib.org/ark:/13030/kt3199n627. Last accessed 12/4/2020. Townes, key person in the foundation of JASON, was already a Nobel Prize-winning physicist for his development of lasers. His entry point to astronomy were astronomical masers, and he soon branched out into infrared astronomy. In the 1970s he was the mentor of Reinhard Genzel, future Director of the Max Planck Institute for Extraterrestrial Physics (Chapter 4).

they increasingly needed one another. Furthermore, ground-based observatories kept their role as feeder pipeline to the astronomical profession, which remained a significant autonomous force and distinct tradition, only gradually merging with experimental physics from the 1960s onwards, to constitute our modern understanding of astrophysics. As we detail in this chapter and the next, until at least the 1960s, 'astrophysics' was clearly distinct from the discipline of astronomy, the latter having its own departments as well as control of the observatories. Astrophysicists, who began their careers in physics departments, advocated for much closer links between physical theory and astronomical observations, in the tradition of cosmic ray and particle physics, in contrast to the traditionally empiricist and instrument-focused approach of the astronomy professionals.

Theoretical astrophysics boomed in this era, transitioning from the nuclear- and plasma-centered interests of the 1940s and 1950s—detailed in previous chapters—toward the new approaches needed to explain the multitude of recent spectacular and novel astronomical observations. As the decade progressed, the abundance of freshly discovered phenomena to be explained, in combination with a declining emphasis on the directly nuclear-related aspects of astrophysics, propelled formerly esoteric subjects, such as relativistic astrophysics, into the mainstream.[24]

In most countries involved in the Cold War, however, the military connection remained largely in the background, thanks to the overlapping use of instruments in both military and civilian scientific activities. The forerunners of this instrumental connection were radio astronomers, whose specialty emerged directly from World War II radar development, as we explore in more detail in Chapter 3. The radio astronomy–radar connection was a dominant driver of instrumental innovation already before Sputnik, and thereafter continued to benefit most directly from the military-scientific link, as both contexts make use of similar antennas and detectors, often produced by the same contractors, while also sharing similar analytical and computational tools.[25] Sometimes even the same facilities were used for both astronomical and defense research purposes, especially in the early Cold War era when, for

24 See Chapter 5, Section 3 (gravitational waves).

25 Lovell, Bernard: The Effects of Defence Science on the Advance of Astronomy. *Journal for the History of Astronomy* 8 (1977), 151–173. https://journals.sagepub.com/doi/abs/10.1177/002182867700800301. Last accessed 5/24/2019.

example, many radio telescopes also featured transmitters.[26] From the 1960s onwards, ground-based radio astronomy continued to develop in synergy with technological developments coming from military contexts, including detectors and antennas for increasingly shorter wavelengths.[27]

Other fields of astronomy soon found their place in the military system; the first were airborne and space-based astronomy in wavelengths inaccessible from the ground. Many technical advances in infrared detection such as those used by MIDAS (Missile Defense Satellite System) later found their way into infrared astronomy, as did the KC-135 airplane-based infrared detectors, which later evolved into civilian uses such as the Kuiper Airborne Observatory.[28] In the mid-1960s, ARPA initiated the Vela program for the detection of nuclear explosions from outer space, an authorized 'technical means'[29] to enforce the Test Ban Treaty, the spectacular unexpected impact of which was the first-ever detection of the astrophysical phenomenon of gamma-ray bursts: short-lived bursts of gamma-ray light, the brightest and most energetic cosmic explosions known to occur in the Universe. The phenomenon was serendipitously discovered by the Vela defense satellites, originally intended to detect nuclear explosions from outer space.[30] The availability of large launchers made possible the first full-fledged space-based observatories, the High Energy Astronomy Observatories (HEAO), in wavelengths that are available only above the atmosphere. These gamma and X-ray observatories, conceived in the mid-1960s but launched throughout the 1970s, showcased the transition from an early 'nuclear'-focused exploration of outer space and the astronomical focus

---

26 See, for example, David Kaiser, and Benjamin Wilson: Calculating Times. Radar, Ballistic Missiles, and Einstein's Relativity. In: Naomi Oreskes, and John Krige (eds.): *Science and Technology in the Global Cold War*. Cambridge, MA: MIT Press 2014, 273–316. In Chapter 3 of this book we show how this was also the case of radar development in West Germany before the creation of a dedicated Max Planck Institute.

27 Chapter 3 will deal in more detail with the military aspects of radio astronomy.

28 S. D. Price: *History of Space-Based Infrared Astronomy and the Air Force Infrared Celestial Backgrounds Program*. AFRL-RV-HA-TR-2008-1039. Fort Belvoir, VA: Air Force Research Laboratory, Space Vehicles Directorate, Hanscom Air Force Base 2008, 365. doi:10.21236/ADA513643. Later flying observatories, like the German–American SOFIA, are direct descendants of these infrared detection systems.

29 See the chapter National Technical Means in Richard A. Scribner, Theodore J. Ralston, and William D. Metz: *The Verification Challenge. Problems and Promise of Strategic Nuclear Arms Control Verification*. Boston, MA: Birkhäuser 1985, 47–66.

30 Finkbeiner, *The Jasons*, 2006, 121–122. On the discovery of Gamma Ray Bursts, see J. T. Bonnell, and R. W. Klebesadel: A Brief History of the Discovery of Cosmic Gamma-Ray Bursts. *AIP Conference Proceedings*. Gamma-ray bursts. 3rd Huntsville symposium. Huntsville, Alabama (USA): AIP 1996, 977–980. doi:10.1063/1.51630.

a decade later; even former experts in nuclear propulsion were repurposed for high-energy astronomy.[31] This NASA initiative encouraged astronomers to "think big" as they "had the big rockets." At the time of their design, these would be the largest payloads in orbit. The experience with space-based observatories in shorter wavelengths, and the maturation of optical reconnaissance technologies, then led to the most famous of all space observatories, the Hubble Space Telescope, which had long been proposed by Lyman Spitzer (see Chapter 1), despite resistance from traditional optical astronomers. The built version was closely based on one of the serially produced Keyhole spy satellites.[32]

In the synergy between military applications and astronomy, expertise also sometimes circulated in the opposite direction, such as when the Hanbury Brown-Twiss interferometric technique was adopted for determining the size of reentering missile warheads in the early 1960s.[33]

Often, however, the interrelationship was more complex. In optical astronomy, one exemplary such development between the 1960s and 1980s was adaptive optics, which is used to counter the distortion caused by the atmosphere at visible and infrared wavelengths.[34] Since the 1990s, adaptive optics has made ground-based telescopes in the visible and near-infrared wavelengths comparable to space telescopes. In military contexts this technique was used for targeting and imaging objects accurately from the ground, as well as for viewing the ground clearly from satellites. One adaptive optics innovation in particular, laser guide stars, was developed during 'Star Wars' for the targeting of missiles and high-energy weapons. These innovations were then 'given back' to the astronomical community, where early attempts at the technique had originated in the 1960s, before becoming classified data.[35]

31 Dunar, and Waring, *Power to Explore*, 1999, 241–242.

32 Andrew J. Dunar, and Stephen P. Waring: The Hubble Space Telescope. *Power to Explore. A History of Marshall Space Flight Center, 1960–1990*. Washington, DC: National Aeronautics and Space Administration, NASA History Office, Office of Policy and Plans 1999, 473–525. Eric Chaisson: *The Hubble Wars. Astrophysics Meets Astropolitics in the Two-Billion-Dollar Struggle over the Hubble Space Telescope*. New York, NY: HarperCollins Publishers 1994.

33 Finkbeiner, *The Jasons*, 2006, 51–52.

34 For a good introduction, see Laird A. Thompson: Adaptive Optics in Astronomy. *Physics Today* 47/12 (2008), 24. doi:10.1063/1.881406.

35 In 1985, the French published in a scientific journal *Astronomy and Astrophysics*, for the first time in an astronomy context, so triggering the release of preexisting American developments in the field throughout the next decade. Charles Townes, himself one of the original developers of the laser, learned of guide stars via JASON and persuaded

The military applications in this case facilitated the fusion of distant instances of scientific expertise. Claire Max, working at the Lawrence Livermore National Laboratory, adapted sodium lasers originally designed for fusion research to experiments in developing guide stars, which were then implemented at the Lick Observatory, benefiting from Livermore being part of the University of California system.[36] The atmospheric layer used by these sodium guide stars was precisely that which had been studied a generation earlier, with the release of ionized alkaline metal clouds from sounding rockets.[37] As this example shows, in the American context, secret defense initiatives often gave rise to radical interdisciplinary crossovers useful for astronomy. Claire Max described the relationship between American astronomy and the military: "It's like a braid almost."[38]

Such military-scientific crossovers in the cosmic sciences were vastly more frequent in the United States and the Soviet Union, but there were also significant overlaps in the United Kingdom and especially in France, given its aspirations to military self-reliance. Many of the strengths of French astronomy happen to coincide with the techniques outlined above.[39] In the West

the military to declassify the technology for astronomers. Finkbeiner, *The Jasons*, 2006, 154–167.

36 On laser guide stars, see also C. Bruce Tarter: *The American Lab. An Insider's History of the Lawrence Livermore National Laboratory*. Baltimore, MD: Johns Hopkins University Press 2018, 265–267.

37 These include the experiments conducted by Jacques Blamont and Reimar Lüst, which were among the first space research activities at the Max Planck Institute for Extraterrestrial Physics. Jacques E. Blamont: Alkali Metal Cloud Experiments in the Upper Atmosphere. In: Johan A. M. Bleeker, Johannes Geiss, and Martin C. E. Huber (eds.): *The Century of Space Science*. Dordrecht: Kluwer Academic Publishers 2001, 189–202.

38 "It's like a braid almost [...] Academic and military scientists generally stay at arm's length, partly because of classification, partly because as pure and applied scientists their problems are often different, and partly because they're at different levels in the professional hierarchy. Max and Fugate both said the braiding continues, that the two formerly noncommunicating cultures have good relations, that they go to each other's conferences, that people who work on adaptive optics for the air force have moved over into the academic community. Fugate, whose military community was relatively small and secretive, said that before he gave that talk to the American Astronomical Society, he hadn't spent much time with astronomers: 'I've never run into a more closely knit, well-networked, everybody-knows-what-everybody's-doing kind of thing and everybody is willing to help everybody. It's a great group of people.'" Finkbeiner, *The Jasons*, 2006, 166–167.

39 For a more general treatment of such military applications, albeit for a popular audience, see Neil deGrasse Tyson, and Avis Lang: *Accessory to War. The Unspoken Alliance between Astrophysics and the Military*. New York, NY: W. W. Norton & Company 2018.

German case, after the mid-1960s, the relationship was necessarily more indirect, involving an additional degree of separation: working within European collaborations or through contact with foreign researchers, as in the case of the 'nuclear' fields in the first postwar decade, as we saw in Chapter 1. The lack of a comparable military demand for these technologies in Germany fostered the early internationalization of these fields; but still, on a smaller scale, contracting companies that built instrumentation did benefit from such scientific projects, later offering products based on them for commercial and military applications, as was the case, for example, with radio astronomical antennas and infrared detectors (see Chapters 3 and 4).

In general, however, the subaltern condition of West Germany demanded by the Allies made it particularly difficult for scientists to catch up with research in fields dependent on such instrumental developments. The easy solution was to collaborate with other countries, but this put them on an unequal footing, sometimes to the point of humiliation. The alternative was to carefully find instrumental niches that were feasible within West Germany, often thereby benefiting from its traditions in competitive instrument-building, in areas from antenna construction to optical manufacturing, as we see in subsequent chapters. But this signified that the results often were incremental improvements made possible by perfectionist manufacturing, which beyond the cultural stereotype, was often the only way forward when revolutionary new developments such as adaptive optics or interferometry were being supported in competing countries for their military potential.

Scientific Bandwagons and Educational Reform

The vast scale of expansion of the cosmic sciences after Sputnik notwithstanding, it was only one small part of the sweeping transformations ushered in by the Soviet satellite, which significantly changed attitudes to scientific research and the status of science in society in general.

The Eisenhower administration and space science pioneers like James Van Allen initially did not think much of Sputnik, seriously underestimating the impact that public opinion could have on the actual development of technologies and scientific research. In contrast, the Soviet announcement of Sputnik was interpreted beyond immediate government circles as a "technological Pearl Harbor," and the growing consensus across the entire political spectrum was to initiate a wide-ranging debate on the investment and reforms necessary to meet the Soviet challenge.[40] The first postwar years in Western Europe and the United States had seen attempts to revert to an idealized peacetime, with social structures and institutions resembling a prewar idyll. The Cold War apparatus was lavish, but spending on the whole had been limited to areas directly linked to the military challenge. This cutback was reflected in scientific research, too: prewar funding models based on private philanthropy persisted alongside education-centered, state-level funding in fields in which research could not credibly be framed as 'nuclear.'[41] These regressive developments in America even justified the mode in which scientific research was funded elsewhere in the non-Communist world. Most relevant for this book, during the first postwar decade, and even after restrictions were lifted in 1955, the constituent states of the Federal Republic of West Germany pushed to keep education and much of scientific research largely outside of the federal government's responsibility, the only exceptions being those areas of national priority which ended up under the direct purview of federal ministries, most notably the Ministry of 'Nuclear Affairs'.[42]

Up to October 1957, the Soviet Union was not regarded as a scientific or technological role model but, rather, as a menace in pursuit of territorial expansion and domestic infiltration. Soviet scientific and technological progress had been considered parasitical, originating largely among émigrés and in spying operations. This view changed radically after Sputnik; the Communist superpower was now recast as a trendsetter, a model for future living based on the 'scientific' organization of society. Technocratic admiration of the Soviet

40 Walter A. McDougall: *The Heavens and the Earth. A Political History of the Space Age*. 2nd ed. Baltimore, MD: Johns Hopkins University Press 1997, 141–156 (Chapter 6: "A New Era of History" and a Media Riot). Dunar, and Waring, *Power to Explore*, 1999, 24.

41 Roger L. Geiger, What Happened after Sputnik?, 1997, 349–367. The 'envy' that was created in those fields outside the nuclear complex in turn led to a differentiated ethos of non-nuclear disciplines which took an institutionally regressive turn. In an example directly relevant to this study, optical astronomers in the United States deliberately fell back on the model of private philanthropy that existed before the war. David H. DeVorkin: Who Speaks for Astronomy? How Astronomers Responded to Government Funding After World War II. *Historical Studies in the Physical and Biological Sciences* 31/1 (2000), 55–92. doi:10.2307/27757846. Quoted in Leandra A. Swanner: *Mountains of Controversy. Narrative and the Making of Contested Landscapes in Postwar American Astronomy*. Dissertation/PhD Thesis. Cambridge, MA: Harvard University 2013, 34.

42 Hans-Willy Hohn, and Uwe Schimank: *Konflikte und Gleichgewichte im Forschungssystem. Akteurkonstellationen und Entwicklungspfade in der staatlich finanzierten außeruniversitären Forschung*. Frankfurt am Main: Campus Verlag 1990.

Union dated back to the 1920s,[43] and the timing of Sputnik, shortly after the death of Stalin, facilitated a focus on positive traits that the West could imitate for the sake of its own survival. The years after 1957 saw the zenith of scientific approaches to government and scientific planning as espoused by Sovietologists and presidential advisors Max Millikan[44] and Walter Rostow, who mobilized their expertise and the opportunity afforded by Sputnik to advocate their planning-focused approach to government and economics. Beyond the United States, through foreign aid programs, the connection between the 'scientification' of society and material progress became the non-Marxist alternative for a teleological narrative of human development.[45]

The post-Sputnik interpretation of the first postwar decade in the West was that having privileged select realms such as nuclear science and military-

43 Thomas Parke Hughes: *American Genesis. A Century of Invention and Technological Enthusiasm, 1870–1970*. New York: Viking Press 1989, 295–352 Chapter 6: "Taylorismus + Fordismus = Amerikanismus."

44 Nils Gilman: *Mandarins of the Future. Modernization Theory in Cold War America*. Baltimore, MD: Johns Hopkins University Press 2003, 158–160. His most influential book mentions Sputnik in the introduction: Max F. Millikan, Universities National Bureau Committee for Economic Research, and Universities–National Bureau Committee for Economic Research: *National Economic Planning. A Conference of the Universities-National Bureau Committee for Economic Research*. New York City, NY: National Bureau of Economic Research 1967. Son of a physicist, Robert Millikan, Max had a particularly physicalist and planning-focused approach to economics. He was a close friend and ally of Lyman Spitzer. Lyman Spitzer: interview by Joan Bromberg, March 15, 1978. Transcript, Niels Bohr Library & Archives, American Institute of Physics, College Park, MF USA, https://www.aip.org/history-programs/niels-bohr-library/oral-histories/4900. Last accessed 12/4/2020.

45 Walt Whitmann Rostow: *The Stages of Economic Growth, a Non-Communist Manifesto*. Cambridge: Cambridge University Press 1960. Written right after Sputnik, the book advocates a roadmap for human progress that competed with the Marxist model, based largely on creating the precondition for a scientifically based society. See Kimber Charles Pearce: Narrative Reason and Cold War Economic Diplomacy in W. W. Rostow's "Stages of Economic Growth." *Rhetoric and Public Affairs* 2/3 (1999), 395–414. https://www.jstor.org/stable/41940179. Last accessed 5/29/2019. Rostow was foreign aid advisor to Kennedy and later national security advisor to Johnson's administration and he had significant impact on the mid-1960s space policy of the United States and its relations with other countries in this matter. See Audra J. Wolfe: *Competing with the Soviets. Science, Technology, and the State in Cold War America*. Baltimore, MD: John Hopkins University Press 2013. See also, Kevin V. Mulcahy: Walt Rostow as National Security Adviser, 1966–69. *Presidential Studies Quarterly* 25/2 (1995), 223–236. https://www.jstor.org/stable/27551419. Last accessed 5/29/2019.

relevant research was a failure in contrast to the model of generalized modernization and mobilization seen in the Soviet Union.[46]

One of the most significant social changes directly caused by Sputnik began in the United States with the National Defense Education Act of 1958, by which the American federal government expanded its influence to the previously decentralized realm of education, and augmented funding for all levels of education and research, even in areas far beyond the military-industrial complex.[47] These American initiatives were then quickly matched in other Western countries, which feared not only the Soviet vanguard, but also being left behind by the American response that followed on Sputnik. This response served to expand and democratize scientific careers throughout the industrialized world, which would in turn have an impact, a decade later, in fields such as astrophysics.[48]

Vannevar Bush's memorandum of 1945, which had led to the creation of the National Science Foundation (NSF) in the United States, already encompassed this generalized view of the role of education and scientific research for the military capability of the future.[49] But calls for generalized scientific mobiliza-

46 Roger L. Geiger, What Happened after Sputnik?, 1997, 349–367. The 1960 report "Scientific Progress, the Universities, and the Federal Government," chaired by Nobel Prize winner Glenn Seaborg, called for the involvement of the federal government in all fields of academic science. Seaborg soon after became the first scientist to be chairman of the Atomic Energy Commission, implementing expansionist research policies that contrasted with the approach of the conservative, pre-Sputnik, industry-oriented chairman Lewis Strauss. Daniel J. Kevles: *The Physicists: The History of a Scientific Community in Modern America*. Harvard University Press 1995, 390.

47 The NATO Science Committee was specifically established in 1957 to readdress the threat deriving from the growing quantity and quality of scientists and engineers in the Soviet Union, possibly creating an "educational imbalance" with Western science. John Krige: NATO and the Strengthening of Western Science in the Post-Sputnik Era. *Minerva* 38/1 (2000), 81–108. https://www.jstor.org/stable/41821156. Last accessed 12/7/2018.

48 For a great argument in the British case, see David Edgerton: *Warfare State. Britain, 1920–1970*. Cambridge: Cambridge University Press 2006, 229. "[declinist discourses] were central to the modernization project in British politics in the early 1960s. They did indeed result in new policy proposals and new policies. Among them were the extension of higher education, the reform of the higher civil service, the reform of the science policy machinery and the creation of the Ministry of Technology in 1964." Higher-level scientific education finally became more accessible to traditionally marginalized social groups. A British example is the radio astronomer Jocelyn Bell, co-discoverer of pulsars in 1968, who was able to pursue a scientific career because of post-Sputnik initiatives. See Jocelyn Bell Burnell: interview by David DeVorkin, 21 May 2000. https://www.aip.org/history-programs/niels-bohr-library/oral-histories/31792. Last accessed 12/1/2022.

49 United States. Office of Scientific Research and Development: *Science, the Endless Frontier*. Washington, DC: United States Government Printing Office 1945.

tion such as found in Bush's memorandum were interpreted very narrowly in the early nuclear age, and only fully flourished after October 1957. Bush called the Soviet satellite "one of the finest things that Russia ever did for us."[50]

In West Germany, these developments in the United States were appropriated by proponents of modernization of the educational system. West German education retained prewar features which reformers considered hierarchical, authoritarian, and elitist: for example, separating children at an early age and making it difficult for children from underprivileged backgrounds to access universities, all in a system where humanist scholarship was institutionally superior to the natural sciences. After Sputnik, reformers warned that this obsolescent educational system jeopardized the ability of West Germany to compete internationally, both economically and politically.[51]

In West Germany, the comparisons that Americans had made between their system and the Soviet one easily translated to a more immediate experience.[52] East Germany had already introduced radical educational reforms, including gender equality, measures to allay discrimination against people of poorer and less-educated backgrounds, and a heavy emphasis on technical and scientific education for all students. The objective of a scientifically educated general population was then even further encouraged after Sputnik.[53] Critics in the West called to mind the undemocratic aspects of East German educational efforts, such as active discrimination against 'bourgeois' families, and mixing political indoctrination and even military training into the school curriculum. But still, it was hard to ignore the very high quality of East German scientific and technical education: graduates of East German schools who emigrated to the West in the 1950s remained grateful for the scientific and technical quality of their education.[54]

50 McDougall, *The Heavens and the Earth*, 1997, 153.

51 Wolfgang Lambrecht: Deutsch-Deutsche Reformdebatten vor "Bologna". Die "Bildungskatastrophe" der 1960er Jahre. *Zeithistorische Forschung* 4/3 (2007), 472–477.

52 Georg Picht: *Die deutsche Bildungskatastrophe. Analyse und Dokumentation*. Olten: Walter Verlag 1964. The author described the West German separation system as an educational cul-de-sac or "Sackgassensystem der scharf voneinander getrennten Schularten."

53 The GDR leadership took on the discourse of Scientific and Technological Revolution (*Wissenschaftlich-technische Revolution*) which would demonstrate the superiority of socialism over capitalism. Lambrecht, Deutsch-Deutsche Reformdebatten vor "Bologna". Die "Bildungskatastrophe" der 1960er Jahre, 2007, 472–477, 474.

54 Two examples highly relevant to this book were the Max Planck scientists Joachim Trümper (Chapters 3 and 5), and Till Kirsten (Chapters 1, 2 and 5) who emigrated in the mid-1950s and in their interviews indicated that upon arrival in the West they ascertained that they were much further ahead in their scientific education than similarly

As with other issues described below, the fragmented federal structure of West Germany was blamed for the inability to keep up with the challenges of the modern world and competition with the East, and even after Sputnik, the reach of federal ministries into schools and university education remained limited in comparison even to the United States. The compromise that regulated the influence of federal resources in the early postwar era was the Königstein Agreement of 1949, but this had resulted from an emergency measure to support existing, struggling, pre-1945 research institutions and deliberately did not touch on educational matters. The need to move beyond the limitations of Königstein in West Germany came with the end of Allied restrictions in 1955 and the evident need for nuclear research institutes; and it led to the creation of a Ministry of Atomic Affairs, as mentioned in Chapter 1.[55] These 'nuclear' needs had sparked discussion of the national coordination of scientific research in general, and led to the creation of the Federal Science Council (*Wissenschaftsrat*) in September 1957, shortly before Sputnik, but only after long Bundestag deliberations in which the precarious institutional framework for research and education in scientific and technical fields had been exposed. Delegates to this new council were appointed on recommendation of the research organizations, including Max Planck Society, the DFG (*Deutsche Forschungsgemeinschaft*, German Research Foundation), and the WRK (West German Rector's Conference, the lobbying association of West German universities). These appointees were joined by others nominated by the federal ministries and the various states.[56] By the time the *Wissenschaftsrat* began meeting in 1958, the global wave of reforms sparked by Sputnik was well underway and its recommendations hence were reactions to the aforementioned global trends. One of the master moves by the Max Planck Society, in the post-Sputnik years, was to appoint Reimar Lüst to the Council in 1965. By then he was already the standard-bearer of the Society's forays into the space age and even served as Chair of the Council, from 1969 to 1972.

---

aged students. Till Kirsten: interview by Luisa Bonolis and Juan-Andres Leon, Heidelberg, October 24–25, 2017. DA GMPG, BC 601051. Joachim Trümper: interview by Luisa Bonolis and Juan-Andres Leon, Munich, August, 7–8, 2017, DA GMPG, BC 601036.

55 Thomas Stamm-Kuhlmann: Deutsche Forschung und internationale Integration 1945–1955. In: Rudolf Vierhaus, and Bernhard vom Brocke (eds.): *Forschung im Spannungsfeld von Politik und Gesellschaft. Geschichte und Struktur der Kaiser-Wilhelm-/Max-Planck-Gesellschaft*. Stuttgart: Deutsche Verlags-Anstalt 1990, 886–909. See also, Hohn, and Schimank, *Konflikte*, 1990.

56 Olaf Bartz: *Wissenschaftsrat und Hochschulplanung. Leitbildwandel und Planungsprozesse in der Bundesrepublik Deutschland zwischen 1957 und 1975*. Dissertation/ PhD Thesis. Köln: Universität zu Köln 2006, 41–42. http://kups.ub.uni-koeln.de/volltexte/2006/1879/. Last accessed 7/31/2015.

Throughout the crucial 1960s, the slow pace of reform on matters related to education further widened the gap between educational and non-educational research institutions. This proved favorable for organizations like the Max Planck Society,[57] which from 1957 quickly benefited from broader support from the existing federal ministries; and even more so, in the following decade, thanks to the national priorities set by the *Wissenschaftsrat*. However, the federal organization of West Germany, whose states insisted on retaining their limited influence on educational matters, steered much research of national relevance away from the universities throughout the first decade of the space age, more so than in other countries. Only in 1969, following constitutional reform, was responsibility for higher education passed from the Ministry of Scientific Research, founded in 1962, explicitly to the Federal Ministry of Education and Science (BMBW); but still, states continued to obstruct federal interventions, especially in directly educational matters. In consequence, the endeavor to unify teaching and research, led by the academic-turned-minister Hans Leussnik, lasted only three years: upon his retirement in 1972, a Ministry of Research and Technology (BMFT) was created, which functioned independently until the 1990s, dealing with research and development of more national relevance, such as nuclear affairs and aerospace, as well as taking charge of the so-called *Grossforschung* (large-scale research institutes) and a growing 'Blue List' of heterogeneous non-educational institutes.[58]

University research still benefited greatly from increased funding of the DFG after Sputnik, but this entity was located in the rival ministry, and support

57 The elites of the Max Planck Society even aimed to influence the educational reform movement from a scientific perspective by gathering allies for the creation of a Max Planck Institute for Education Research (*Bildungsforschung*, often translated as Human Development). Klaus Hüfner, and Jens Naumann: *Konjunkturen der Bildungspolitik in der Bundesrepublik Deutschland. Der Aufschwung (1960–1967)*. Vol. 1. Stuttgart: Klett 1977, 160. Heinz-Elmar Tenorth: *Geschichte der Erziehung. Einführung in die Grundzüge ihrer neuzeitlichen Entwicklung*. 5. Weinheim: Juventa 2010, 287.

58 The initial cohort of these was called the 'Königsteiner Institute' and reflected the initial pact of their co-financing at the state and federal level. Ministries kept adding new institutes, and only some, in nationally critical fields, were financed by the BMFT. See Hohn, and Schimank, *Konflikte*, 1990, 135–170 Kapitel 5: "Die Bund-Länder Institute." The status and financing of these 'Blue List' institutes remained contested until reunification, when they faced circumstances similar to those of the majority of non-educational research institutes in East Germany. This led to their unification under the name Leibniz Association, which is now a rising competitor to the Max Planck Society.

from the BMFT to universities, either directly or through the DFG, was a cumbersome process full of inter-institutional rivalries.[59]

Meanwhile, after 1969, Germany's Social Democrat (SPD) and Free Democrat (FDP) coalition made spectacular progress in expanding access to higher education. However, the expansion of tertiary research and the democratization of universities were treated as two separate issues, and were further complicated by the student protest movement that had exploded in the late 1960s. Influential scientists of the era, unsympathetic to student movements, used the opportunity to further widen the gap between universities and research-oriented organizations.[60]

The persistent asymmetry between education and scientific research in West Germany, coupled with institutional fragmentation up to the ministry level, further tilted the institutional advantages of the Max Planck Society, which was older than the research ministries themselves and remained independent of them, while at the same time benefiting from a significant lifeline from the Research Ministry (as the BMFT came to be known) in nationally relevant fields, which played an important role particularly in the cosmic sciences.

Civilian Space Programs and the 'Scientific' Framing of Space

A powerful, lasting legacy of the 'Sputnik shock' around the world was the rising status of scientists, which allowed them to take on powerful positions in areas related to space exploration, which until 1957 had been the preserve of the military. Pioneering space scientists had been proposing satellite launches at least since 1954, but these had fallen on the deaf ears of their military backers: satellites were already technically feasible, but the senior leadership considered them costly and of little benefit compared with the already viable suborbital flights.[61] The Eisenhower administration was also cautious

59 Hartmut Altenmüller: BMBW und BMFT. Fusionen und Teilungen. *Spektrum der Wissenschaft* 12 (1994), 127. https://www.spektrum.de/magazin/bmbw-und-bmft-fusionen-und-teilungen/821997. Last accessed 5/15/2019.

60 The laments about the detrimental role in research of the West German answers to the 1968 student movements are not restricted to Max Planck researchers. A representative perspective, comparing the German situation with that in France, can be found in: Karl Rawer: *Meine Kinder umkreisen die Erde. Der Bericht eines Satellitenforschers*. Freiburg im Breisgau: Herder 1986, 124–125. Rawer was based at the University of Freiburg (see Chapter 1).

61 The satellite project met relatively modest interest outside of scientific circles at the time. Incidentally, it was required that the satellite use Navy-developed rockets to prevent the potential embarrassment of the first satellite being launched by rockets coming from the Peenemünde veterans in Huntsville. See Dunar, and Waring, *Power to Explore*, 1999, 20.

about launching a satellite while the territorial status of outer space was still undefined. Only in the framework of the International Geophysical Year (IGY, 1957–58), an explicitly 'international' and 'scientific' endeavor, were American space scientists able to advance the launch of an artificial satellite under what was called Project Vanguard;[62] but even then, its funding and organization remained low priority and ultimately fell behind the Soviets. Still, this project demonstrated, even before Sputnik, that linking fundamental research and spaceflight could serve to legitimate them both. Moreover, the IGY satellite proposal was a chance to establish the idea that outer space was international, a precedent that would prove vital for the deployment of surveillance satellites.[63]

One of the key reforms following Sputnik was the creation in early 1958 of NASA, as a civilian federal agency in charge of coordinating the American space program.[64] One of its objectives was to centralize planning, so as to avoid any duplication of effort and, too, the rivalries that had arisen between several military rocket development programs. But another aim was to foster scientific research and international collaboration by constituting a separate civilian institutional framework for access to outer space, safely compartmentalized, away from classified activities. Both these objectives were based on the current wave of scientific management described above, but also clearly had public relations appeal: thanks to NASA, America's endeavors in outer space would take place in public view and remain accessible to external scientific researchers, in stark contrast to the Soviet program. NASA even insisted on the live broadcast of launches and related events.[65] Crucially, and in difference to earlier initiatives such as 'Atoms for Peace,' NASA was given considerable authority to instigate collaboration with scientific institutions abroad, as well as to focus on cooperation with friendly or neutral countries in possession of a significant scientific and technological base, so as to guarantee that all participants would gain from the exchange. NASA's collaborations were not a foreign

62 See Chapter 5, "The Satellite Decision," in McDougall, *The Heavens and the Earth*, 1997.

63 Krige, John: NASA's International Relations in Space. An Historical Overview. *NASA's First 50 Years. Historical Perspectives*. Washington, DC: NASA 2010, 109–150, 116–117. Jeroen van Dongen (ed.): *Cold War Science and the Transatlantic Circulation of Knowledge*. Vol. 1. Leiden: Brill 2015. McDougall, *The Heavens and the Earth*, 1997, 110, 121–124. Until 1963 the Soviets contested the legality of spy satellites, but these concerns were dropped as they became a crucial part of the nuclear test ban verification, and they had already caught up with the technology.

64 Robert R. MacGregor: Imagining an Aerospace Agency in the Atomic Age. *NASA's First 50 Years: Historical Perspectives*. Washington, DC: NASA 2010, 31–48.

65 Teasel Muir-Harmony: American Foreign Policy and the Space Race. *Oxford Research Encyclopedia of American History*, 2017. doi:10.1093/acrefore/9780199329175.013.274.

aid program, but a scientific collaboration framework that often dominated the pace of scientific developments in the fields that it touched, worldwide. The impact of this approach was colossal, in terms of the scale and quality of the ensuing collaborations.[66]

Furthermore, in 1960, one of NASA's great institutional accomplishments was its acquisition of the former Army Ballistic Missile Agency (ABMA) in Huntsville, Alabama—where Wernher von Braun and his team had led one of the most promising missile programs—which it repurposed as the Marshall Space Flight Center, in charge of the in-house development of America's civilian rockets, beyond the technologies developed up to that point in a military context. Although initially opposed to the move to a civilian setting, von Braun was made director of this first civilian rocket development center, so becoming NASA's public face. His center retained significant in-house capabilities through its first decade of operations, and contributed to NASA the organizational capacity to lead large projects, a legacy of its 1950s ('arsenal system') setting, thus assuring the agency a significant systemic advantage over other countries' space agencies, as well as the ability to deal from a position of authority with any military or commercial contractors.[67] On the other hand, the division of labor between these rocket developers and the scientists in charge of civilian scientific research was very clear, as the latter were externally based and developed their research payloads separately.[68] The Jet Propulsion Laboratory (JPL) in Pasadena, also previously under military command, and several civilian space-relevant institutions were likewise transferred to NASA.[69]

We described earlier how the denuclearized status of outer space was agreed on by the superpowers throughout the 1960s, as required for the new balance of mutual assured destruction. The creation of NASA as a civilian agency driven by scientific research preceded these treaties, and its early successes helped to legitimize the non-nuclear approach to outer space around the world. NASA's offers of scientific collaboration coincided in fact with the spirit of the resolutions of the United Nations General Assembly in the early 1960s, which called for the peaceful international exploration and use of outer

66 Krige, John, NASA's International Relations, 2010, 109–150, 115, 121–122 (Table), 132 (Table).

67 Dunar, and Waring, *Power to Explore*, 1999, 28–45. The 'arsenal system' went into decline in the 1970s as cost-cutting, consolidation, and the gradual retirement of the original German engineers set in. Afterwards NASA was more exposed to external expertise and pressure from industry.

68 Dunar, and Waring, *Power to Explore*, 1999, 227.

69 These had formerly been consolidated into the Army Ordinance Missile Command (AOMC): McDougall, *The Heavens and the Earth*, 1997.

space.[70] The United Nations General Assembly had been making calls for committees and deliberations on the peaceful use of outer space since the late 1950s, thus roughly following the path that had led to the International Atomic Energy Agency (IAEA) just a few years earlier. But in the case of outer space, while the UN efforts languished, much quicker progress was made through the scientist-led International Council of Scientific Unions, which instituted COSPAR, the Committee on Space Research.[71] The 'scientific' narrative leading cooperation in space became established in the early 1960s, which allowed national agencies and even individual research groups to cooperate on space matters, while minimizing bureaucratic and diplomatic intermediation. Similar situations developed with the two other main applications of outer space foreseen at the time, the World Meteorological Organization and the International Telecommunication Union.[72]

From the mid-1960s onwards, 'international collaboration on peaceful space exploration' became the dominant discourse.[73] But at the same time, the United States was able to maintain its leadership in many scientific fields covered by NASA, thanks to the vast underlying military-industrial complex, which shared technological and instrumental insights via experts and industrial contractors with parallel ongoing military activities.

NASA actually remained small in contrast with the military space programs in charge of missiles and spy satellites, and it was rarely the driver of developments in those areas; civilian and military agencies functioned in parallel, rather, while sharing a joint pool of contractors and experts and, occasionally, infrastructures.[74] NASA's different needs led to separate production cultures that sometimes complemented one another, but also often ran into conflict.

---

70 UNO Resolutions: 1085th Plenary Meeting, Sixteenth Session, 20 December 1961: International Cooperation in the peaceful uses of outer space (see documents at United Nations Digital Library https://digitallibrary.un.org/record/665195. Last accessed 1/25/2022); 1244th Plenary Meeting, Eighteenth Session, 17 October 1963: Question of general and complete disarmament [calling upon states to refrain from installing weapons of mass destruction in outer space] (see documents at https://digitallibrary.un.org/record/203960. Last accessed 1/25/2022).

71 Gerhard Haerendel et al. (eds.): *40 Years of Cospar*. Noordwijk: ESA Publications Division 1998.

72 See, for example, James Simsarian: Outer Space Co-Operation in the United Nations. *American Journal of International Law* 57 (1963), 854–867.

73 On the emergence of space science as a new field of scientific activity, see Homer Edward Newell: *Beyond the Atmosphere. Early Years of Space Science*. Vol. NASA SP-4211. Washington, DC: NASA 1980.

74 Infrastructure sharing was best avoided but often inevitable, even into the 1990s, as the Hubble Space Telescope illustrates. See Chaisson, *The Hubble Wars*, 1994.

Most generally, military developments were oriented toward reliability, durability, and mass production, which was also later the focus with commercial satellites. NASA's explorative and scientific focus demanded instead one-of-a-kind products, partly developed in-house, partly contracted out to industries for which they represented comparatively minor but cumbersome contracts.[75]

Finally, NASA as a civilian agency was expected to be the trailblazer for commercial applications in space, in the spirit of the 'Atoms for Peace' initiative, something that would later create tension with foreign collaborators, who perceived a conflict of interest in its limitation of activities to 'scientific purposes' while advancing domestic commercial goals. Their large-scale deployment and commercialization were meant to be led (often, in a public-private partnership) by corporations such as COMSAT,[76] for example, which provided the first network of communications satellites.

NASA inspired and, often, directly aided the creation of similar organizations in the major European countries.[77] But despite sharing a model, the resulting national institutional frameworks varied widely, due to their underlying industrial, political, and economic systems. Key here is that these agencies, even more than in the United States, highlighted their scientific research activities, and were often headed by scientists. France is perhaps the best example, where the national space research center (CNES, *Centre National d'Études Spatiales*) was first proposed in 1960 and came into existence in 1962. Thanks to the centralist tradition in France, its creation was quick, and its influence in spearheading technological developments in France and Europe was early and considerable. Its first director was the general and aviator Robert Aubinière, and it was led by scientific figures such as the cosmic ray pioneer Pierre Auger and balloon- and rocket-based researcher Jacques Blamont.[78] More than any other European country, France also benefited from its parallel ongoing military launcher developments, outlined below.

75 Dunar, and Waring, *Power to Explore*, 1999, 45.

76 David J. Whalen: *The Rise and Fall of COMSAT. Technology, Business, and Government in Satellite Communications*. London: Palgrave MacMillan 2014.

77 John Krige, Angelina Long Callahan, and Ashok Maharaj: *NASA in the World. Fifty Years of International Collaboration in Space*. New York: Palgrave MacMillan 2013, 23–50 Chapter 2: "NASA, Space Science, and Western Europe."

78 Pioneer of experiments with artificial plasma clouds injected in the ionosphere starting in 1959, Blamont had contributed to the development of the Veronique rocket, and in the 1950s started his academic career with atomic radiofrequency interaction topics; from 1957 onwards was also one of the directors of the Aeronomy Service of the CNRS. In the next chapter we will detail his close collaboration with Reimar Lüst and the Max Planck Institute for Extraterrestrial Physics. See, Blamont, Alkali Metal Cloud Experiments, 2001, 189–202.

A contrasting but similarly successful path to a civilian agency was pursued in Great Britain. Thanks to the UK's close civilian and military collaboration with the United States, NASA itself served in essence as a significant centralizing point for British scientific space activities, which were led locally by varied national agencies, depending on their uses. Scientific research was led by the Science Research Council (SRC), which was founded in 1965 as a consequence of Sputnik, but encompassed all fundamental research of national significance, including nuclear and particle physics, space research, astronomy, and the life sciences. The SRC funded and inter-networked research communities that remained dominated by the universities. Only very late, in 1985, was a dedicated British National Space Centre (BNSC) created.[79] The Britons' decentralized approach to their space program is still quite successful and provides a valuable parallel to the more anarchically decentralized West German case. Key to British success was that, while there was no dedicated central agency, the goals of the program were very clear and reflected the heavily scientific leadership of the SRC. Given the close military alliance with the US, British activities could focus on truly civilian and complementary aspects of spaceflight. Other agencies and industrial alliances fostered commercial interests, for example, pushing for a leadership role in communications satellites. An example of this successful decentralized coordination was the united front against spending resources on human spaceflight, and indeed there was no interest in sending a British astronaut throughout the entire 20th century. As we see below, this reflected the actual scientific consensus, even in the United States.

Sounding Rockets in Europe and West Germany

Despite their different structures, British and French civilian space programs during the first decades each maintained comparable national research capabilities based on small 'sounding' rockets that had been largely developed before Sputnik, initially for military purposes, and were later procured by their domestic industries. Over several decades, these small rockets, mostly incapable of putting objects into orbit, had the benefit of providing cheap, reliable access to outer space while being militarily unproblematic. Used creatively, they could lead to groundbreaking scientific experiments and observations. It was such small rockets which provided a significant basis for the

79 Only in 1994 did this multidisciplinary SRC split into smaller compartments, including a council on particle physics and astronomy, while space activities were transferred to the new Space Agency, which was later renamed UKSA.

early years of ESRO,[80] and in the French case they still brought significant expertise later used by Ariane and satellite programs.[81] While these nationally-based research rockets flourished, a proposed UN facility for sounding rockets led to the Thumba Equatorial Rocket Launching Station (TERLS) in India, to which NASA, CNES, SRC, and the Hydrometeorological Service of the USSR contributed.[82]

From the early 1960s onwards, small rockets supplied primarily by the United States, France, and the United Kingdom provided the suborbital launchers for research programs in countries with more modest capabilities, such as Italy, Netherlands, Sweden, Norway, or Switzerland.[83]

West Germany was a unique example, being in this 'user' category despite its size: by 1968, it had actually sponsored more sounding rocket launches than any other West European country, through a mix of French, American, and later, British vehicles.[84] When ESRO ended its sounding rocket program in the 1970s, German-sponsored launches, all with foreign rockets, dominated even further.[85] The development of domestic sounding rockets in Germany had been considered in the 1960s. Early in the decade, Berthold Seliger was developing and launching them successfully from Cuxhaven on the North Sea coast. But Seliger was soon mired in the Egypt scandal (detailed below), and the last rocket launch occurred in 1964.[86] After this embarrassing incident, West Germany outlawed the private production of missile-like devices, restricting them to large enterprises within collaborations with the state and other

80 John Krige, and Arturo Russo: *A History of the European Space Agency 1958–1987. The Story of ESRO and ELDO, 1958–1973*. Vol. 1. Noordwijk: European Space Agency 2000.

81 Matthew Godwin: *The Skylark Rocket. British Space Science and the European Space Research Organisation. 1957–1972*. Paris: Beauchesne 2007. Günther Seibert: *The History of Sounding Rockets and Their Contribution to European Space Research*. Noordwijk: ESA Publications Division 2006.

82 Simsarian, Outer Space Co-Operation, 1963, 854–867, 857.

83 For the history of space programs of such European Members States, see History Study Reports at https://www.esa.int/About_Us/ESA_Publications/ESA_historical_publications. Last accessed 02/05/2021.

84 Seibert, *History of Sounding Rockets*, 2006, 22 (Table).

85 Seibert, *History of Sounding Rockets*, 2006, 33. Seibert refers to the article: Gerhard Haerendel: Stand und Ergebnisse des deutschen Höhenforschungsprogramms. *Raumfahrtforschung* 1 (1976), 34. Haerendel claims in this article that sounding rockets constituted half of the entire German space science budget! In this regard, see also folder "ESRO Report Sounding Rocket Policy Study (Teile I-III), 1969" in Reimar Lüst's papers (AMPG, III. Abt., Rep. 145, No. 1248).

86 Lutz, Harald: Die vergessenen Raketenexperimente von Cuxhaven. *Sterne und Weltraum* 44/3 (2005), 40–45.

NATO countries. Simultaneously, the Federal Ministry for Scientific Research did not support the development of conventional (one-use and uncontrolled return) space launchers that were indistinguishable from missiles. There were attempts by industry to create a reusable, winged sounding rocket to work within these political constraints, but these proved impractical and West Germany ended up relying exclusively on foreign sounding rockets for the remainder of the century.[87]

Given the restrictions on both rocket development and launches within Germany, since 1965 the Bavarian DFL, (*Deutsche Forschungsanstalt für Luftfahrt*, one of the precursors of the DLR, German Aerospace Center), together with the Max Planck Society's Space Research Working Group (*Arbeitsgruppe für Raumfahrtforschung*; see later in this chapter), created the mobile rocket base MORABA (*Mobile Raketenbasis*) headquartered in Oberpfaffenhofen. From its inception, this group, which was later owned by the first attempt at a German aerospace agency, (the DFVLR, German Test and Research Institute for Aviation and Space Flight), provided mobile rocket launching infrastructure, such as mounting, ignition, communications, telemetry, and even the operation of the in-flight experiments. The rockets themselves were provided by foreign companies or research partners. MORABA could quickly deploy to airbases abroad (Norway, Sweden, Australia, USA, Canada, France, India, Brazil) providing West Germans with the closest thing possible to national launch capabilities.[88] Even after satellites became the dominant scientific platform, sounding rockets deployed by MORABA offered the possibility of a very quick reaction to interesting astrophysical phenomena, instead of the decades-long planning for a satellite mission.[89]

---

87 This referred to Project 621 by Dornier, with tests undergoing until its final cancellation in 1969. Regarding conventional sounding rockets, only in 2001 did the German DLR collaborate directly with Brazil for the development of the VSB-30 rocket, when the British Skylark was no longer produced. Alexandre Garcia et al.: VSB-30 Sounding Rocket. History of Flight Performance. *Journal of Aerospace Technology and Management* 3/3 (2011), 325–330. doi:10.5028/jatm.2011.03032211.

88 Seibert, *History of Sounding Rockets*, 2006, 23–24, 32. Alexander Schmidt, Andreas Stamminger, and Peter Turner: DLR's Mobile Rocket Base. 47 Years of Microgravity and Technical Experiments on Suborbital Flights. *65th International Astronautical Congress* (*IAC 2014*). Toronto 2014. https://elib.dlr.de/93678/. Last accessed 5/9/2019.

89 One key example was the quick deployment in Australia (within five months of first proposal) of an astronomical campaign led by the Max Planck Institute for Extraterrestrial Physics, following the 1987 Supernova explosion visible from the southern hemisphere: U. G. Briel et al.: Supernova 1987A in the Large Magellanic Cloud. *International Astronomical Union Circular* 4452 (1987), 1. https://ui.adsabs.harvard.edu/abs/1987IAUC.4452....1B/abstract. Last accessed 6/5/2019.

MORABA, in our view, epitomizes how space research in West Germany managed to establish scientific dominance while simultaneously maintaining a prudent distance from domestic rocket-building efforts. More than in other countries, its scientific research institutions played a major role, with the ability to choose between launching with entirely foreign collaborations, or with the DFVLR under conditions where the latter had scientifically subservient roles focused on the vehicles and support infrastructures. Something similar would develop with scientific satellite missions as well, where the satellites themselves and operational payloads were provided by the DFVLR and its industrial partners, while the scientific instrumentation was generally built by the participating research institutes themselves. Launches of German satellites were always provided by foreign national agencies or ESA.

As we detail later in this chapter, the process of creating something akin to a West German space agency extended over a decade, facing the hurdles of federal fragmentation, reluctant industries, and the diminishing possibility of a national launcher program. These uniquely West German constraints were reinforced by scientists who appropriated the discourse on fundamental scientific research and preferred using foreign rockets and infrastructure, leading to a late, fragmented, and hollowed-out West German space agency. When the (awkwardly named) DFVLR[90] finally started in 1969, it could not attract scientists of significant stature to be its directors, having to settle instead for local rocket experts.[91] Max Planck Institutes continued to benefit from direct support channels both to the federal ministries and the Max Planck Society itself, with the new DFVLR having little authority over them. Whether this agency had a supporting role or, as intended from its inception, a position of leadership, remained contested, especially by the Max Planck Institutes, which sought in subsequent decades to maintain their dominance.[92] The DFVLR had

90 The first name of the organization, *Deutsche Forschungs- und Versuchsanstalt für Luft- und Raumfahrt* (German Test and Research Institute for Aviation and Space Flight), reflected its fragmented background full of bureaucratic compromises, in detriment to a clean international brand such as NASA or CNES. The Max Planck Society often took advantage of these branding vacuums, in this case to be implicitly identified as 'German NASA.' Similar branding vacuums led many national academies in communist countries to identify the MPG as their counterpart.

91 The two candidates preferred for First Director were NASA-based Hermann Kurzweg, who preferred to stay in the United States, and Reimar Lüst (see next section) who preferred to stay at the Max Planck Society. Helmuth Trischler: *Luft-und Raumfahrtforschung in Deutschland 1900–1970. Politische Geschichte einer Wissenschaft*. Frankfurt am Main: Campus 1992, 497–498.

92 See, for example, Reimar Lüst, and Paul Nolte: *Der Wissenschaftsmacher. Reimar Lüst im Gespräch mit Paul Nolte*. München: C.H. Beck 2008, 61–62.

the mandate to be the operational coordinator in large-scale national missions such as scientific satellites, controlling the budgets and orchestrating the collaboration of industrial and scientific participants; but the scientific originators of the projects were based at research institutes which continued to wield significant authority and were better connected with international scientific networks and even foreign space agencies.

Large Rockets and the West German Nuclear Ambiguity in the Early Space Age

Since the return to sovereignty and alliance with NATO in 1955, the Adenauer administration aspired to become a significant voice within the Western Alliance. This led even to an intent to arm the West German military with tactical nuclear explosives.[93] Then, the advent of Intercontinental Ballistic Missiles transformed Europe's nuclear ambitions. In the years around 1957, the American–Soviet rivalry gravitated increasingly toward strategic bombing and mutually assured destruction. This meant that the threshold of use of nuclear weapons was rising toward an all-or-nothing standoff. Under these circumstances, fear arose in Western European countries that they might become sacrificial lambs in localized conflicts whenever the Americans chose not to escalate to a nuclear exchange. The invasion of Hungary by Warsaw Pact troops in 1956 recalled to mind the territorial ambitions of the Soviets; more important for the French and British, however, were the simultaneously occurring Suez crisis and loss of Vietnam, which demonstrated they could not always rely on American support.[94]

Both the British and French accelerated their moves toward nuclear independence in 1956. Britain had already exploded its own nuclear bomb in 1952, and as part of these efforts was already exploring missile delivery technolo-

93 In Chapter 1 we discussed the Göttingen Manifesto. Adenauer's mid-1950s ambitions were based on a pre-ICBM worldview, in which nuclear explosives might be used tactically, that is, in routine military operations mounted on artillery and missiles, or launched from airplanes. The Göttingen Manifesto of April 1957 mobilized against Adenauer's plans; and this activism was continued in subsequent decades by the circle of scientists around Werner Heisenberg's and Carl K. von Weizsäcker's circles. Their views were better aligned with the post-Sputnik regime embodied in the international treaties agreed over the first decade of the space age.

94 Wilfrid L. Kohl: The French Nuclear Deterrent. *Proceedings of the Academy of Political Science* 29/2 (1968), 80–94. doi:10.2307/1173251. Wilfrid L. Kohl: *French Nuclear Diplomacy*. Princeton, NJ: Princeton University Press 1971.

gies.[95] As early as 1954, the British had signed an agreement (Wilson-Sandys) for joint missile development with the Americans, thanks to which Britain began developing medium-range ballistic missiles very early (1955): called Blue Streak, they were to complement American work on long-range ICBMs. Still, the missile had to be able to carry a very heavy load, as it was intended to deliver a very powerful, preferably thermonuclear bomb, to compensate for its lack of precision (already in the mid-1950s they were testing fusion devices). This carrying capacity is what later made the development applicable for satellite launching.

Due to these pre-1957 efforts, of all Western countries, the British were least impressed by Sputnik on armament-related issues, as they already had an ongoing program that they considered proportionally adequate to their ambitions. In their case, what Sputnik provided was the opportunity to increase collaboration with the Americans on an even footing, thanks to their moment of perceived weakness, leading to their humorously called "declaration of interdependence":[96] in 1958 the two countries signed the Mutual Defense Agreement, through which they shared their nuclear deterrent. As a consequence, by 1960 the British had decided to interrupt their military Blue Streak program. This cancellation, however, came with the domestic problem of having to justify the resources already spent, and the solution was to try to spin off the missile as a civilian satellite delivery system. But Blue Streak could only work as a first stage of a larger satellite launcher, and no resources were available for a full-fledged national satellite or launcher program. The solution was to press for European integration on launcher development, leading to the birth of ELDO, which is detailed later in this chapter.

The situation in France was the opposite of Britain. The Americans had excluded the French from collaboration on nuclear and missile issues since before the end of the war, and the new British-American alliance specifically prevented the UK from trading such expertise with France, as had been the intention during their first attempts to enter the European Community.[97] The French, for their part, had been pursuing fully self-reliant nuclear capabilities

---

95 Charles N. Hill: *A Vertical Empire. The History of the UK Rocket and Space Programme, 1950–1971*. London: Imperial College Press 2001, 11–14; 69. At the beginning it was short-range missile program, so that the warhead could be carried most of the way by airplane, but then complete the last part of the journey as a missile.

96 Nigel J. Ashton: Harold Macmillan and the "Golden Days" of Anglo-American Relations Revisited, 1957–63. *Diplomatic History* 29/4 (2005), 691–723. https://www.jstor.org/stable/24915066. Last accessed 5/21/2019. David Edgerton: *The Rise and Fall of the British Nation. A Twentieth-Century History*. London: Allen Lane 2018, 735.

97 Ashton, Harold Macmillan, 2005, 691–723, 702.

since the 1940s. Sputnik and the British-American alliance further accelerated these intentions, and extended them to the development of delivery missiles. The French greatly accelerated their efforts after Sputnik, and made prodigious advances on a national military context until the 1970s. Their first nuclear weapons were to be delivered by Mirage airplanes, while medium-range delivery rockets were developed together with civilian ones through the 1960s:[98] between 1961 and 1965, the French military-led program introduced the series of 'precious stones' rockets (Agate, Topaz, Emerald, Sapphire, Ruby), leading both to satellite launchers and ICBMs. Medium-range rockets resulting from this program were operational by the end of the 1960s.[99] These were to be followed by truly long-range ICBMs in the 1970s, intended to reach anywhere in the world (the 'Tous Azimuts' program).[100] Once this domestic launcher technology was mature, the French national program applied this expertise to foster development of the European launcher Ariane.

The civilian outcome of the French national program was the orbital launcher Diamant, which launched the first French satellite, Astérix, in 1965. This small satellite made France the third country to reach orbit independently (Britain, Canada, and Italy had by then already sent their own satellites on American launchers).[101] A few years later, in 1970, an improved launcher under a binational agreement put into orbit the heavier Dial (*Diamant Allemand*), the second West German satellite (the first to be coordinated by the DFVLR),[102] which was also the first satellite launched from French Guyana.

98 An earlier generation of rockets, including 'Veronique,' had been developed in the pre-Sputnik era by teams that included many Peenemünde veterans. Deutsches Zentrum für Luft- und Raumfahrt e.V.: *50 Jahre DLR Lampoldshausen. 1959–2009*. Köln: Bernd Rölle 2009, 26–29.

99 Krige, and Russo, *European Space Agency I*, 2000, Vol. 1, 89. Jean About: *Les débuts de la recherche spatiale française. Au temps des fusées-sondes*. Paris: Institut français d'histoire de l'espace 2007.

100 Kohl, French Nuclear Deterrent, 1968, 80–94, 84.

101 McDougall, *The Heavens and the Earth*, 1997, 353. According to McDougall, through these launches "[…] the United States acquired its desired reputation as a fair and dependable provider of launch services for other nations, providing they restricted themselves to space science and released their data to all the world. In areas removed from strategic technology, the United States lived up to its principles of cooperation and openness in space."

102 Hervé Moulin: *La France dans l'Espace 1959–1979. Contribution à l'effort spatial européen*. Vol. 37. Nordwijk: ESA Publications Division 2006, 38–39. The scientific module was directed by Karl Rawer from Freiburg (see Chapter 1). In the next section we will see in more detail how French collaboration related to developments that led to the Max Planck Institute for Extraterrestrial Physics.

Shortly beforehand, in the wake of Algerian independence, the French had had to evacuate their traditional base at Hammaguir; but in any case, Guyana, being close to the equator, was a more advantageous site for geosynchronous satellite launches.

As the scale of research and development in Cold War armaments expanded, the French indirectly assisted the growth of their own military technologies by pursuing the strategic mobilization of other European countries in the framework of scientific and technological collaboration. Generally, this happened under ambiguous circumstances, such as when concerns circulated that EURATOM, created at the Messina Conference of 1955, while being apparently 'peaceful' was nonetheless understood to have military intentions, too.[103] Sometimes these intentions can be traced to explicit (although still secret) military collaboration agreements: very soon after the launch of Sputnik, German, French, and Italian representatives fast-tracked a joint project to develop nuclear weapons, the so-called 'nuclear flirtation.' The initiative circumvented West German restrictions, as controversial installations would be physically located in France. Crucial for this book, this 1958 collaboration plan already encompassed joint rocket development.[104]

EURATOM and subsequent secret agreements were typical examples of West German practices that raised suspicions in America and the Soviet Union during this period, regarding weapons and dual-use technologies: that they were tacitly facilitating their development within European collaborations (largely in France), while hiding these with their visible participation in the scientific and commercial uses of these technologies. West German nuclear and space policy between 1955 and 1969 has to be seen in the context of these continental ambitions and the reaction they provoked in the two superpowers.[105] The doubt only dissipated at the end of the decade with the signing of

---

103 The French, interested in the military-relevant aspects of atomic research, were its most enthusiastic supporters. Joachim Radkau: *Geschichte der Zukunft. Prognosen, Visionen, Irrungen in Deutschland von 1945 bis heute*. München: Carl Hanser Verlag 2017, 154 (cartoon). Joachim Radkau, and Lothar Hahn: *Aufstieg und Fall der deutschen Atomwirtschaft*. München: Oekom 2013, 113–116. EURATOM was jokingly referred as 'European Community for the Development of a French Atomic Bomb.'

104 Kohl, French Nuclear Deterrent, 1968, 80–94, 82. Kohl, *French Nuclear Diplomacy*, 1971, 54–60. Wolfgang Zank: Adenauers Griff nach der Atombombe. *Die Zeit* (7/26/1996). https://www.zeit.de/1996/31/Adenauers_Griff_nach_der_Atombombe/komplettansicht. Last accessed 6/3/2019.

105 On Adenauer's nuclear ambiguity and the American reaction, see William Burr, The Nuclear Nonproliferation Treaty and the German Nuclear Question, Parts I and II. National Security Archives, 2018: National Security Archive: Preoccu-

the Non-Proliferation Treaty (NPT) and the change of federal government in West Germany.

Accepting the NPT was part of a wider political, generational, and cultural shift in West Germany over the course of the decade, which in 1963 saw the retreat of Adenauer, who had opposed this treaty; followed in 1966 by a centrist Grand Coalition with the Social Democrats led by Kurt Georg Kiesinger, during which West Germany abstained regarding the treaty; and, finally, its ratification in 1969 under the new center-left coalition led by Willy Brandt.[106]

The era before this was characterized by intense collaboration with France on nuclear, rocket, and other defense-related issues, punctuated by calibrated retreats back to the United States' sphere.[107] This was a logic of European self-reliance within the Western Alliance against the immediate Soviet threat, and the political and technological infrastructure on which it was based was pursued most strongly by the Bavarian heavyweight Franz Josef Strauss (see Chapter 4), in his capacity as Federal Minister of Atomic Affairs (1955–56), then of Defense (1956–62), and later, of Finance (1966–69).

While the West Germans had already been excluded from direct involvement with French nuclear weapons development after the return of Charles

---

pations with West Germany's Nuclear Weapons Potential Shaped Kennedy-Era Diplomacy, 2/2/2018. https://nsarchive.gwu.edu/briefing-book/nuclear-vault/2018-02-02/german-nuclear-question-nonproliferation-treaty. Last accessed 6/10/2019.

106 The contrast with the previous generation's opinion is seen in how the two central political figures of the previous decade perceived the NPT as a humiliation: Franz Josef Strauss described the treaty as a "Versailles of cosmic dimensions" (Versailles von kosmischen Ausmaßen), and even the more cautious former chancellor Adenauer described this treaty as "a Morgenthau Plan squared" (Morgenthau Plan im Quadrat). For a detailed study of this inter-generational dynamics, see Tim Geiger: *Atlantiker gegen Gaullisten. Außenpolitischer Konflikt und innerparteilicher Machtkampf in der CDU/CSU 1958–1969*. München: Oldenbourg 2008.

107 Ibid. One significant arena of this balance related the type of nuclear reactors to be developed and their potential use for domestic nuclear weapons production. Americans favored the spread of reactors working with low-enriched uranium (about 4% U235), which needed enrichment technologies for their production that made Germany dependent on American supplies. The self-reliant alternative in the 1950s were reactors moderated with heavy water, which used natural uranium as fuel, but had the potential of producing plutonium. As would also happen with aerospace developments, American pressure combined genuine concerns regarding dual-use, with attempts to favor American industry by keeping Europeans dependent. The same debate, even without the military restriction aspect, was occurring also in France and EURATOM. See: Débats du Parlement européen sur la Communauté européenne de l'énergie atomique (18 October 1966). http://www.cvce.eu/obj/debats_du_parlement_europeen_sur_la_communaute_europeenne_de_l_energie_atomique_18_octobre_1966-fr-a1e7de68-53c9-434e-ae36-a62ecfc36a34.html. Last accessed 5/22/2019.

de Gaulle to the presidency in 1958, significant collaboration related to thermonuclear fusion and space-related technologies continued during the 1960s. De Gaulle's reluctance to share these technologies then revived pro-American inclinations in West Germany regarding both nuclear energy and outer space, establishing the longstanding practice of West Germany in these matters, attempting to find the best bargain among inter-Allied competition between American and pan-European (ultimately, French) interests.[108]

As part of this inter-Allied competition during the first decade of the space age, the American strategy regarding international collaboration in outer space included even offering to launch national satellites as symbolic gestures or as trading tokens in international politics.[109] That is how West Germany obtained the launchers for five of its first six national satellites.

It was emphasized, however, that these satellites would be for 'scientific' research, which again provided a distinct advantage for 'fundamental research' institutions like the Max Planck Institutes. The nascent aerospace industries still benefited, while being steered toward activities compatible with West Germany's subaltern status: the availability of foreign launchers incentivized the West German specialization in those aspects of spaceflight related to the militarily unproblematic upper stages and payloads. This also afforded the West German aerospace industry a distinctly scientific-to-commercial pathway, with which it expected to profit from the experience gained while building the scientific satellites, so as then to be able to sell upper stages and payloads with wider applications (commercial or even military) within legitimate pan-European aerospace industrial networks.

West German Roles within ESRO *and* ELDO

Shortly after Sputnik, in 1958, the International Council of Scientific Unions (ICSU) instituted a Committee on Space Research (COSPAR), which took advantage of the existing coordination work dating from the International Geophysical Year. The existence of this body incentivized nations around the world to appoint expert representatives and make initial moves toward their own national scientific programs. By 1959, NASA had announced through

108 Geiger, *Atlantiker gegen Gaullisten*, 2008.

109 Germany's first small satellites, Azur (study of Van Allen belts, solar particles, and aurorae) and Aeros 1 and 2 (state and behavior of the upper atmosphere and ionospheric F-region) were 'free' gifts, while the much larger Helios interplanetary probes (study of solar processes) were a return gesture in exchange for the US–German agreement to pay for the presence of American troops. Niklas Reinke: *The History of German Space Policy. Ideas, Influences, and Interdependence 1923–2002*. Paris: Beauchesne 2007, 112–113.

COSPAR the availability of its vehicles for bilateral scientific collaborations, which resulted in most of the first national European experiments and satellites described above. Simultaneously, however, there was talk about pan-European collaboration efforts, and it was established very early that this should be an entirely scientific and 'peaceful' organization: for example, it was decided early that NATO should not have a role in it. Instead, the organization was to be based on the successful model of CERN.[110] That same year, Edoardo Amaldi, who had been one of the founders of the Geneva-based laboratory, circulated a memorandum that is considered the foundational document of what would become the European Space Research Organisation (ESRO).[111] In this initial version, however, Amaldi's proposal for the space equivalent to CERN included the in-house development of the launcher technology needed to fulfill the organization's scientific objectives. This inclusion of launchers was justified by how, in Geneva, both particle detectors and the accelerators were built internally, while keeping industrial participation limited to a contractor role; as we saw earlier, this was also the NASA model.

One of the first proposals for a clear distinction between launchers and scientific research in space was advocated by someone influential in the mobilization of the Max Planck Society toward the space age, as we detail later in the chapter. Shortly after Amaldi's letter, Peter Meyer, a prominent cosmic ray physicist who had worked at the Max Planck Institute for Physics in Göttingen and was now in Chicago, sent a letter to the West German Atomic Minister (Alexander Hocker) in response to those plans. In this letter,[112] while praising the general idea of a CERN-like institution, Meyer emphasized the importance of keeping scientific research and launcher development entirely separate, as they [in his view] represented extremely different regimes. Launchers were a different matter than particle accelerators, as there was neither published literature nor a tradition of freely sharing the know-how for their construction. Rockets [he continued] would need to be developed from a very modest

110 Harrie Stewart Wilson Massey, and M. O. Robins: *History of British Space Science*. Cambridge: Cambridge University Press 1986, 109.

111 De Maria, *Europe in Space*, 1993. See also Michelangelo De Maria, Lucia Orlando, and Filippo Pigliacelli: *Italy in Space—1946–1988*. Noordwijk: ESA Publications Division 2003.

112 Peter Meyer letter to Alexander Hocker, July 10, 1959. Hocker was the German representative to CERN 1952–61, financial chairman of COPERS 1961–64, ESRO council member and chairman 1964–67 and its director general 1971–74. In: Helmuth Trischler: *Dokumente zur Geschichte der Luft- und Raumfahrtforschung in Deutschland: 1900–1970*. Köln: DLR, Deutsche Forschungsanstalt für Luft- und Raumfahrt 1993, 361–363. Peter Meyer will later play a role in the steps leading to the foundation of the Max Planck Institute for Extraterrestrial Physics (see Section 2).

starting point, which implied costly trial and error, and a long lag with respect to American and Soviet developments. Furthermore, launcher development entailed costs and infrastructural needs on a larger scale than the scientific experiments conducted with them, putting the scientists at a disadvantage while being at the mercy of the pace of rocket developments. Instead, for a scientifically dedicated research organization it would be easy, quick, and inexpensive to participate and even take a leading role in the emerging space sciences, where results were published, while taking advantage of an expected surplus of American rockets, and when the time came, of European launchers, which he considered European industries should be incentivized to initiate.

While most Europeans were in favor of a united launcher and research organization as proposed by Amaldi, a similar view to Meyer's was proposed by the British representatives, albeit for a different underlying reason, namely their intention to commercialize the Blue Streak rocket through the creation of a separate organization, the later ELDO.[113] Negotiations on this matter were ongoing simultaneously with those of ESRO, through 1960, even though the information circulating between both initiatives was not transparent. Ultimately, British insistence was the main reason that ESRO remained a purely scientific organization while ELDO focused on launcher development, in which British participation was dominant. The demarcation of 'scientific research' as something organizationally different from launcher 'development' was instrumentalized by the British, who argued ESRO would be a more independent and scientifically led organization, while ELDO necessarily involved political intervention and the need to include commercial applications in the main purpose of the organization.[114] But ultimately, 'fundamental' institutions like Max Planck Institutes were also beneficiaries of this rigorous distinction in the first decade of the space age, since they participated in ESRO under conditions similar to those of the other participating countries, while also benefiting from being junior partners within American-led collaborations. By the time the approach had changed toward the unitary agency that became ESA in the

113 Let's remember here that restrictions on the free publication of research evolved as the central argument used by the Max Planck Society to defend its scientific autonomy over the second half of the 20th century. These first emerged in the very context of its foundation in the late 1940s, and were vigorously reinforced in the mid-1950s around the possibilities made available by the expansion of nuclear power. Free publication of results also became one of the central boundaries regarding activities in areas with potential military dual-use, and later in discussions regarding the participation in research collaborations with private companies or more generally organizations with commercial intent. The space age was just one arena of these boundary-building efforts.

114 Massey, and Robins, *History*, 1986, 115–118.

early 1970s,[115] the status of fundamental research institutes had already been secured in the global ecosystem of space research collaborations.

The agreement to form ESRO was signed in December 1960, and preparatory committees and study groups started meeting and laying out plans a few months later. Over the next years there followed negotiations regarding the selection of headquarters, staff appointments, and, especially, the controversial problem of the scale at which activities were to be conducted at centralized sites and laboratories, as opposed to distributing them among the member countries' existing institutions.[116] West Germany obtained the space data analysis center in Darmstadt, which in 1967 evolved to become the European Satellite Operations Center. ESRO's first satellites launched in 1968 on American rockets. None ever flew on an ELDO launcher before the failure of that organization.

Negotiations for the creation of ELDO were simultaneously underway, and the final shape of this organization was agreed in 1961. By then, the highly compartmentalized nature of this launcher venture had been instituted, as it was designated that development of the large European rocket would be divided into rocket stages: the British contribution was the first stage, consisting of the already existing Blue Streak rocket; the French contribution was the second stage, a modified Veronique rocket called Coralie; and West Germany committed to the militarily unproblematic third stage, called Astris, which would carry the satellite payload to its final orbit operating in the vacuum. The participating West German institutions and industries were steered toward specializing in one particular niche of spaceflight which remains dominant to this day,

115 John Krige, Arturo Russo, and Lorenza Sebesta: *A History of the European Space Agency 1958–1987. The Story of ESA, 1973 to 1987*. Vol. 2. European Space Agency 2000.

116 Massey, and Robins, *History*, 1986, 131–133. Pierre Auger was elected to the largely ceremonial role of Director General. Albert W. Lines from the Royal Aircraft Establishment became the first Technical Director, and Reimar Lüst (see next section) was unanimously selected to the crucial role of Scientific Director. He, however, insisted the position be part-time, and continued to divide his time between the Max Planck Institute for Extraterrestrial Physics and ESRO. Upon the final establishment of the organization in 1965, Lüst gave his post over to Bert Bolin of Sweden, who would become a central player in the environmental and Earth system sciences. In July 1967, the ESRO Council elected Hermann Bondi as second Director General of ESRO, succeeding Auger. Bondi was an outstanding scientist and had already been chairman of the Space Committee in the British Ministry of Defence, "proving to be able to get the confidence of the government and also had learned to be realistic about their intentions [...] two qualities that were in those days a necessity for a DG of ESRO." Veerle J. Sterken: Sir Hermann Bondi: A Journey through His Life and the Early Endeavours of Europe into Space. *Acta Astronautica* 61/1 (2007), 514–525, 518. doi:10.1016/j.actaastro.2007.01.059.

focused on small, high-power engines that need to be tested in unique evacuated chambers that are still one of the specialties of the DLR testing site at Lampholdshausen, near Heilbronn.

The smaller participating countries also had their dedicated roles, such as satellite construction (Italy), radio guidance (Belgium), and telemetry (Netherlands). Australia was also a member, contributing its Woomera base. This excessive division of labor is now considered to have been the main cause for the technical failures of ELDO's first rockets, and for the subsequent dissolution of the organization itself, at the end of the 1960s. While the individual components worked well, the launches of the integrated rocket failed due to the lack of coordination between the different components.[117] The dream of a compartmentalized approach to European integration ended in 1972, soon followed by the creation of a new European Space Agency, ESA, the successor to ESRO, which assumed the tasks of building its own launcher under an entirely French leadership. By the early 1970s, it had become clear that for applications beyond scientific research, Europeans could not rely on launchers provided by the United States, because the superpower abused its dominant position to hinder European progress whenever this competed with its own interests. [118]

The Problem of German Rogue Rocketmen

Rocket development in West Germany followed a similar path to other dual-use cases treated in this book, such as nuclear energy (Chapter 1) and radar

117 This failure has become a frequently used example of the failures of compartmentalized approaches to European integration determined by politicians, but this interpretation itself has become uncritical mythology. Historians like John Krige have provided criticism which emphasizes rather the fact that the original intentions of ELDO were deeply rooted in British–French attempts at self-reliance with implied military aspects; by the mid-1960s, however, the constellation had changed, France had vetoed British entry to the Common Market, and this resulted in the said compartmentalization and active torpedoing of ELDO from all sides. Meanwhile, the technical capabilities were too uneven, as the British contribution worked flawlessly while the other stages were a continuous embarrassment. The later success of Ariane, rather than a success of European technical integration, also owed to French political stubbornness and the lucky contingency that the Space Shuttle, which diverted many American resources, was so disastrous. John Krige: The History of the European Launcher—An Overview. In: R. A. Harris (ed.): *The History of the European Space Agency. Proceedings of an International Symposium 11–13 November 1998, the Science Museum, London*. Noordwijk: ESA Publication Division 1999, 69–78.

118 Many European pioneers of Ariane even consider that the abusive decisions made by NASA in the post-Apollo era were the direct cause of the current independence of Europeans in space. Krige, History of the European Launcher, 1999, 69–78, 69–78.

(Chapter 3). In all these areas, German experts and their programs had advanced considerably during the war, and through the first postwar decade there was hope they might be able to jump back into the lead, once Allied restrictions were lifted. In the case of rocket development, Nazi Germany was a decade ahead of every other country, and had vast numbers of experts with knowledge and know-how, who had dispersed around the world after 1945. At the end of the war, obtaining these rocket experts was one of the central objectives of every Allied country. Many rocket experts remained abroad for the rest of their lives, and their former Nazi membership and wartime activities were usually not an obstacle.[119] A massive transfer of know-how and expertise occurred between 1945 and the early 1950s, and the first postwar generation of rockets in all the Allied countries had significant input from German teams and expertise. Some of these experts then started trickling back to West Germany during the 1950s, anticipating the end of Allied restrictions. The wartime capabilities had been completely dismantled and shipped abroad; but the technical expertise and ambition to participate again in rocket development began to be discussed again in the open, combining a fascination with both civilian and military aspects of space transport and exploration.[120] Key figure in this early institutionalization was Eugen Sänger, who in the mid-1950s founded a Society for Spaceflight (*Gesellschaft für Weltraumfahrt*) in Stuttgart, with the help of local industries, and—anticipating the space age, and already drawing on the fame of Wernher von Braun in the United States—published books for experts, politicians, and the general public.[121] Sänger was founding professor of the Research Institute for Jet Propulsion Physics (*Forschungsinstitut für Physik der Strahlantriebe*) in 1954. After Sputnik, Sänger mobilized the rising interest in spaceflight to accelerate his initially private initiative for a West German national rocket. Sänger himself was an expert in stratospheric airplanes, which were to crop up recurrently in the history of German aerospace. But for the immediate goal of rocket-based access to space, he founded the test site in Lampoldshausen and associated with the foremost domesti-

119 The delicate status of the German team in Huntsville, Alabama, is well described in: Dunar, and Waring, *Power to Explore*, 1999.

120 By that time, Wernher von Braun was already a notable public figure in the United States and in Germany, and his visions of spaceflight circulated in books and films. See, for example, Catherine L. Newell: The Strange Case of Dr. von Braun and Mr. Disney. Frontierland, Tomorrowland, and America's Final Frontier. *Journal of Religion and Popular Culture* 25/3, 416–429.

121 See, for example, Eugen Sänger: *Raumfahrt. Heute–Morgen–Übermorgen*. Düsseldorf: Econ-Verlag 1963.

cally available rocket scientist, Wolfgang Pilz, who had been one of the key contributors to the French Veronique rocket.[122]

In the few years after Sputnik, the Stuttgart-based intentions were for a fully national West German rocket which could put a satellite in orbit. But Sänger, Pilz, and their collaborators provided instead the founding cautionary tale of West German rocket development. Throughout their careers, these Stuttgart-based pioneers—whose engineering ethos stood in stark contrast to the Max Planck scientist, easily blurring the boundaries between scientific research, technological development, and military applications—maintained a swashbuckling, businesslike approach in their search for supporters.[123] The first ethical clash surfaced upon the creation of ELDO. The organization was strongly criticized by Pilz, who predicted its failure, while resenting how it threatened Germany's abilities do develop a fully national rocket. West German commitment to ELDO interrupted support for his rocket, after which his team continued its development in Egypt, at the invitation of Nasser. Their rocket test flights were announced in 1962, and also featured in Egyptian military parades that same year. In parallel, Berthold Seliger, a disciple of Sänger who had started a private company to develop and launch rockets from Cuxhaven, was found to have offered his domestically developed rockets to countries outside the NATO alliance. These episodes further tainted the potential for a national path to rocket development in West Germany. A public scandal exploded in Israel and West Germany in the very years when they were attempting to reestablish diplomatic relations. Sänger was sacked from his professorship in Stuttgart, but was hired at the Technical University in Berlin soon after, before dying prematurely in 1964. The Israelis bemoaned the inadequacy of West

---

122 Deutsches Zentrum für Luft- und Raumfahrt e.V., *50 Jahre DLR Lampoldshausen. 1959–2009*, 2009, 26–29. Sänger was already a prominent figure in the rocket-building community before his appointment in West Germany. See Eugen Sänger, and Heinz Gartmann: *Raketenantriebe. Ihre Entwicklung, Anwendung und Zukunft. Eine Einführung in das Wesen des Raketenantriebes, sowie Raketen- und Weltraumfluges*. Zürich: Schweizer Druck- und Verlagshaus 1952.

123 In many ways, this approach was reminiscent of early-20th-century flight pioneers. Their sometimes reckless, private-based approaches would gain legitimacy again only in the 21st century in the contemporary wave of American space privatization efforts. The controversial German entrepreneurial initiatives in Stuttgart are notably low-key in pre-2000 analyses of post-1945 German space policy, and the OTRAG episode (see next page) is not even mentioned: Niklas Reinke: *The History of German Space Policy. Ideas, Influences, and Interdependence 1923–2002*. Translated by Barry Smerin, and Barbara Wilson. Paris: Beauchesne 2007, 50. Trischler, *Luft-und Raumfahrtforschung in Deutschland 1900–1970*, 1992, 453.

German intervention, then started with targeted assassinations and the kidnapping of technicians. Pilz and his secretary were victims of a letter bomb, after which they retired to a low-profile life in Germany. Finally, Franz Josef Strauss himself agreed with Shimon Peres that rocket technicians should be offered employment in Germany, so as to prevent the continued proliferation of their expertise in the Arab world.[124]

But still, a near constant string of episodes traceable to this expertise occurred in the following decades. Lutz Kayser, a disciple of Sänger, was again subject to international condemnation when his private company OTRAG, which was based in Zaire in the 1970s, and in Libya in the 1980s, began to develop a cheap alternative to Ariane.[125] Throughout the Cold War, the Israelis, Soviets, and French were particularly concerned by these rogue rocketmen, and used public protests and diplomatic pressure to ensure that they were ostracized by the West German establishment.

Attempts to establish a West German national space agency unfolded thus against this backdrop of scandal, which made it more difficult than ever to leave behind the legacy of Peenemünde; and all the while, praise continued to be heaped on the 'well-behaved' West German scientist dedicated to fundamental research (but not to engineering) in a non-commercial, internationalist context. In space exploration in West Germany in the second half of the 20th century, a commitment to 'fundamental science' was not an abstract ideology, but a behavioral necessity periodically reinforced by politically charged episodes.

DFVLR and Aerospace Industries

In parallel to the return of such rogue space visionaries came the more respectable reconstruction of German airplane manufacturing after the end of Allied restrictions. This was based largely on reestablishing the most traditional prewar companies and growing the surviving flight research institutions to their full potential. The reconstruction of the aircraft sector (soon aerospace) followed an industry-centered approach steered initially by state gov-

124 Interview with Shimon Peres, in: Kersten Schüßler: Showdown am Nil. Der Mossad, Die Nazis, und die Raketen. Dokumentation. 45 Min. ARD 2018 (Film Documentary). This documentary is based on: Ronen Bergman: *Rise and Kill First. The Secret History of Israel's Targeted Assassinations*. New York, NY: Random House 2018, 63–84.

125 Deutsches Zentrum für Luft- und Raumfahrt e.V., *50 Jahre DLR Lampoldshausen. 1959–2009*, 2009, 60–62. Deutsches Zentrum für Luft- und Raumfahrt e.V., and Institut für Raumfahrtantriebe (eds.): *50 Jahre DLR Lampholdshausen*, 2009, 60–62. Oliver Schwehm: Fly Rocket Fly—Mit Macheten zu den Sternen. Spielfilm. 90 Min. Lunabeach TV und Media GmbH 2018 (Film Documentary).

ernments, and increasingly coordinated at the federal level. Key early champion was again Franz Josef Strauss, who envisioned a gradual pathway out of Allied restrictions: to have local industries find their feet first via contract work for Allied aeronautical companies, and then slowly work their way to a Europeanized industrial capability, in which West Germany would be deeply integrated. If successful, this approach would be economically competitive with the United States, guarantee European sovereignty in a crucial industrial sector, and reinforce a growing infrastructure for military aircraft production, which in turn would cross-subsidize civilian airplane development. The results of these ambitions in the aeronautical industry are now manifest in Airbus, the pan-European manufacturing consortium.[126]

After the launch of Sputnik, the West German federal government considerably augmented its involvement in space-related developments, seeking participation in already existing private and state-based organizations, and exerting pressure on the airplane industries to get involved. Before 1957, aerospace research, focusing on airplane development, had occurred in a decentralized way, condensed around different regional strongholds, each with their own industrial and political supporters. State interventionism after Sputnik was manifest in the pressures to centralize and coordinate their activities, while the two original expert communities, of rocket builders and aircraft manufacturers, remained clearly distinct. While initially, the entrepreneurial rocket scientists in Stuttgart took the lead, the Egypt scandals forced a radical change of approach, toward a state-supervised system based on large, established companies, and aligned with the international division of labor, by which Germany was to provide upper stages and payloads, not the rockets themselves.

At the height of the Egypt scandal, the private limited company Society for Space Research (*Gesellschaft für Weltraumforschung GmbH*) was created, with majority support from the federal state; it was a first attempt to centralize and coordinate aerospace efforts, namely by bringing all the surviving research establishments under one roof. Established companies, most prominently Bölkow in Bavaria, and Dornier in the southwest, were enthusiastic industrial partners in charge of the first and second German satellites respectively.

Owing to a patchwork of regional public-private partnerships and research associations in airplane-related research, a multipartite competition was

---

126 Horst Möller: *Franz Josef Strauß. Herrscher und Rebell.* München: Piper 2015. Strauss was a flight enthusiast since the early 1950s and later even became a licensed pilot.

already unfolding between the states of Bavaria, Baden-Württemberg, North-Rhine-Westphalia, and Lower Saxony, each of which sought to attract those institutions and experts still finding their way since the end of the war. This regional competition was similar to that described in Chapter 1 in the case of nuclear energy, the key difference here being the much more far-flung and preexisting industrial base, which helped reinforce the idea that aerospace development had fewer claims to being 'fundamental' research. In fact, as detailed below, the 'applied' status of aerospace research directly affected the Max Planck Society throughout the 1950s and 1960s, due to its contested stewardship of the Aerodynamic Test Station (AVA) and its close relationship with the Max Planck Institute for Fluid Dynamics.

The first decade of the space age in West Germany consisted in a slow, tortuous process of consolidation and integration of the different fragments of aerospace research institutes under a single roof, which was only accomplished definitely in 1969, with the establishment of the DFVLR.[127]

The dominant drivers of airplane and rocket developers were the longstanding tradition in engineering and the close backing both of industrial parties in the airplane industry and a federal ministry interested in pushing those industries toward spaceflight. Even disregarding the external pressures resulting from the rocket scandals mentioned earlier, there was already a stark cultural divide between fundamental research and aerospace development. This facilitated the transfer of the Aerodynamic Test Station of the Max Planck Society to the new DFVLR. And until the end of the Cold War, the stark division of labor, between the kind of 'applied' research conducted at the DFVLR, and that which was done at universities and Max Planck Institutes, was upheld. Only in the 1990s did these boundaries begin to dissolve.

During the Cold War decades, the Max Planck Institutes maintained a boundary with the (variously named) German space agencies based on the fundamental / applied divide; and the stronger MPIs were proud to maintain separate sources of funding. Tellingly, it was the weakest of all Max Planck Institutes, the Institute for Aeronomy, which maintained the closest links to the DFVLR; and after 1990, when the boundary between fundamental and applied dissolved, this institute lost its dominance in planetary exploration within a decade, because the newly expanded DLR started to compete in these activities.

127 The detailed process of this complex institutional merger is described in: Trischler, *Luft- und Raumfahrtforschung in Deutschland 1900–1970*, 1992.

1970s Onwards: Tension with NASA and the Maturation of European Space

The collaboration landscape with the Americans existed in a precarious balance, between scientific excellence and the instrumentalization of science for propaganda purposes. Scientists were aware of how they were being 'used,' but still found ways to make the most of the situation.[128]

In the first decade, this bargain afforded German researchers a fast route to outer space, mobilizing various forms of expertise, as described in the previous chapter. But dependence on the Americans proved problematic, especially as researchers tried to enter new fields. NASA would seek to prevent European researchers from doing research that competed with that of American teams, and simply keep them as junior partners. For example, Klaus Pinkau (a central figure throughout this book), realized in the late 1960s that Americans were steering him away from his ambitions in space astronomy, preferring that he and other Germans stick to "their known expertise" in increasingly outmoded space plasma research.[129] Until the success of Ariane rockets in the late 1970s, German researchers depended almost exclusively on American rockets to reach orbit; but their own experiments elicited support in West Germany for ESA and Ariane. Collaboration with the Soviet Union, a pathway opened up by France in the late 1970s, also served to break this monopoly.[130]

In any case, West German space science and astronomy in the 1960s and 1970s continued to be conducted to a significant degree with 'poor-man's space probes,' such as balloons, sub-orbital rockets, and analysis of substances from meteorites and space missions.

In dealing with the superpowers, the additional element of human spaceflight complicated matters. After all, the space race played out not just through Sputnik and subsequent probes and satellites, but also Yuri Gagarin's orbital flight of 1961, and the American challenge to put a man on the moon before

---

128 We will see in subsequent chapters that this often demanded keeping a low profile during the early proposal stages, when a project might be vulnerable to cancellations or even 'stealing' by competing American researchers; thus, superior capabilities and well-thought-out research programs were revealed only once already deployed, so as to take the senior partners by surprise. Cases throughout this book include barium cloud experiments, lunar sample analysis, and space telescopes like ROSAT.

129 Klaus Pinkau: interview by Helmuth Trischler, March 9, 2010. Transcript, Historical Archives of the European Union. Oral History of Europe in Space Collection, https://archives.eui.eu/en/oral_history/INT072. Last accessed 12/4/2020.

130 Erhard Keppler: *Der Weg zum Max Planck Institut für Aeronomie. Von Regener bis Axford—eine persönliche Rückschau*. Katlenburg-Lindau: Copernicus 2003, 35–36. Lüst, and Nolte, *Der Wissenschaftsmacher*, 2008, 205–207.

the end of the decade. Human spaceflight was the ideal proxy for symbolic competition between the superpowers.[131] On the surface was the calculated celebrity status of cosmonauts and astronauts and the general narrative of space exploration as a continuation of geographical discovery.[132] This tension between unmanned and manned spaceflight has existed since the onset of the space age and continues to this day, in various manifestations, in all countries. Over the course of the 1960s, scientists protested that the American space program disproportionately invested so much in the, in scientific terms, relatively uninteresting challenge of landing on the moon.[133]

Moreover, the presence of humans required more complex vehicles and communications infrastructures than those used by unmanned projects. And while the extra cost and complexity are obvious disadvantages, if they are intended as scientific platforms, these same features make them more attractive for the industries that build them, and they also better serve the purpose of cross-development of technologies with military applications. For example, after the end of Apollo, the Space Shuttle project required a large satellite network to guarantee permanent communication with the ground. This network now provides the communications infrastructure used for unmanned scientific activities, human spaceflight, and spy satellites, the Pentagon being the principal user and funder of the system.[134] This communications network puts NASA on a different level than the space programs of other countries, which

131 Wolfe, *Competing with the Soviets*, 2013.

132 The justification of human *versus* automated probes was always problematic. In the early years of rudimentary computers, it could be claimed that crews were indispensable for complex missions, and the long development times of space vehicles meant that a late 1960s decision, the manned Space Shuttle, dominated American spaceflight for the remainder of the century.

133 Kevles, *The Physicists*, 1995, 390. Quote by Meg Greenfield: "In Washington these days, the definition of a truly hip science adviser is one who knows that the moon money could be better spent on other scientific projects and who also knows that Congress won't appropriate it for any of them." See also W. D. Kay: *Defining NASA. The Historical Debate over the Agency's Mission*. Albany, NY: State University of New York Press 2005, 77. Only 3 out of 116 scientists were in favor of human spaceflight. Vannevar Bush himself was one of the most prominent critics.

134 NASA's Tracking and Data Relay Satellite System (TDRSS, https://www.globalsecurity.org/space/systems/tdrss.htm. Last accessed 3/20/2021), supports data transmission from spacecrafts at an extremely high rate. For its role in Hubble's operation and complications related to its classified activities, see also Chaisson, *The Hubble Wars*, 1994. Incidentally, the Tracking and Data Relay Satellite System (TDRSS) was also used as a radio telescope for the first even space-based Very Long Baseline: Interferometry (VLBI) observations in the 1980s. https://asd.gsfc.nasa.gov/blueshift/index.php/2016/07/25/thirty-years-of-space-vlbi/. Last accessed 12/4/2020.

rely on ground stations that can communicate with spacecraft only for brief segments of their orbit.[135]

The only comparative public engagement value of unmanned missions came through the race for 'firsts,' with unmanned probes reaching ever more distant parts of the solar system. Interplanetary probes provided spectacular scientific discoveries (such as determining the environments of the other planets in the solar system), while also pushing the limits of launch vehicles and communications systems. In contrast, orbital scientific missions, while being scientifically very productive, had a lower profile in the public imagination until the Hubble telescope of the 1990s. Even within unmanned spaceflight, a rivalry has developed between planetary scientists and space astronomers. These rivalries, which first emerged in the United States, were inherited by the corresponding scientific communities in ESA and European national institutions.[136]

Then, between 1969 and 1972, as lunar visits became routine in the public mind, the American civilian program shifted temporarily away from human spaceflight. Cost controls and higher expectations of tangible outcomes during the 1970s then yielded to the new Space Shuttle program mobilized by military and commercial interests, while also in search of a scientific rationale. The emphasis on human spaceflight created a case of 'the cart leading the horse': there was an ample supply of vehicles and astronauts, disproportionate to the relatively undeveloped scientific programs that they were supposed to

135 Incidentally, both ground-based and space-based communications antennas can be used for radio astronomy; conversely, radio telescopes can be used for communication with spacecraft, and some like the Parkes radio antenna in Australia did so routinely. This, however, created a point of tension between astronomers and space programs. For a German example, see Chapter 3 of this book. Early 1970s German space missions intended to use the Effelsberg Radio Telescope. The strong resistance by astronomers led to the creation of purpose-built antennas by the same company in Oberpfaffenhofen by the DLR. Keppler, *Max Planck Institut für Aeronomie*, 2003, 24–25.

136 See, for example, Malcolm Longair: interview by Robert W. Smith, June 14, 1984. Space Telescope History Project, National Air and Space Museum, Smithsonian Institution, https://sova.si.edu/record/NASM.1999.0035?s=0&n=10&t=C&q=oral+history+interview+with+Longair&i=0. Last accessed 12/4/2020. A German example of this rivalry, addressed in upcoming chapters, led to the increasing division of labor between the Max Planck Institutes for Aeronomy, Chemistry, and Nuclear Physics in planetary science on one hand, and the Institute for Extraterrestrial Physics, Radio Astronomy, and Astronomy on the other. As a result of this rivalry, for example, the Aeronomy Institute obtained significant funds outside of the MPG, from the DLR. Horst-Uwe: interview by Helmuth Trischler and Matthias Knopp, June 10, 2010. Transcript, Oral History of Europe in Space Collection, https://archives.eui.eu/en/oral_history/INT078. Last accessed 12/4/2020.

fulfill. This crystallized a tension which had begun with Apollo and persisted until the end of the century: Americans incentivized scientists from Allied countries, who often already had experience with efficient, unmanned rockets, to propose scientific projects for manned flights. Later in this chapter we will see examples of how this arrangement proved very fruitful in the case of the Apollo program. But especially since the early 1970s, coinciding with the failure of ELDO and more constrained NASA budgets,[137] European scientists were pressured instead to launch on the upcoming Space Shuttle, and to participate in programs which made sense only in the context of human spaceflight. One of the founding justifications for ESA in the 1970s, which again took up the challenge of developing a European launcher, was to counter these pressures.

But European countries had diverse understandings of sovereignty: the French set their priorities on independent access to space with the development of their Ariane program, which would satisfy their ambitions and military interests in rocket technology.[138] The British, whose accession to ESRO, ELDO and, later, ESA were also steps toward their eventual accession to the European Economic Community, focused their participation in the development of communications satellites.[139] West Germany got the short end of the stick: in the collaboration agreements made with ESRO for the post-Apollo era, their joint projects were channeled toward the Space Shuttle. The effect, given the power imbalance within the new ESA, was to leave West Germany and Italy with the least promising assignment, a scientific module for the Space

137 By the end of the 1960s, the postwar economic boom was coming to an end, and the Sputnik bonanza slowly adjusted accordingly. Americans landed on the moon and just a few years after, as public interest waned, NASA reduced its budget considerably. From the mid-1970s to this day, this reduced NASA budget has essentially remained unchanged, although adjusted for inflation, and it covers human spaceflight, scientific programs, and a growing presence in Earth sciences. Dunar, and Waring, *Power to Explore*, 1999, 135–178 Chapter 5: "Between a Rocket and a Hard Place: Transformation in Time of Austerity."

138 Moulin, *La France dans l'Espace*, 2006, Vol. 37.

139 Massey, and Robins, *History*, 1986. Douglas Millard: *An Overview of United Kingdom Space Activity 1957–1987*. Noordwijk: ESA Publications Division 2005. For space activities in other Member States see also Joost van Kasteren: *An Overview of Space Activities in the Netherlands*. Noordwijk: ESA Publications Division 2002. Bruno Philipp Besser: *Austria's History in Space*. Noordwijk: ESA Publications Division 2004. Jose M. Dorado, Manuel Bautista, and Pedro Sanz-Aránguez: Spain in Space. A Short History of Spanish Activity in the Space Sector. Edited by European Space Agency. *History Study Reports* 26 (2002). http://www.esa.int/esapub/hsr/HSR_26.pdf. Last accessed 5/2/2021. Further ESA History Study Reports about European space activities and other historical publications can be retrieved at https://www.esa.int/About_Us/ESA_Publications/ESA_historical_publications. Last accessed 2/5/2021.

Shuttle, called 'Spacelab.'[140] A significant portion of West Germany's aerospace budget, and its industries (led by ERNO of Bremen) became taken up with Spacelab, which also led to the creation of an astronaut center in Cologne, and to a West German being the first West European astronaut (but not cosmonaut).[141] A proportion of these Spacelab missions were even within the West German national program, not ESA, where they performed stunts such as taking over control during the part of the orbit that was in communication with Oberpfaffenhofen.[142] Spacelab also altered the scientific landscape by justifying large investments in microgravity research, including the ground-based DLR free-fall experiment facilities, and the copious TEXUS series of experiments based on sounding rockets. All these made Germany a 'leader' in a field that most countries and research organizations consider of modest scientific interest. Most of this research was funded through the DLR and ESA human spaceflight, with comparatively little Max Planck involvement.[143]

The program of 'scientific' modules for the manned program went to its next stage in the 1980s, with the Columbus laboratory, which was initially conceived as an independent ESA space station but was later merged with the International Space Station. All through this period, the scientific value of these costly efforts remained a point of conflict that divided scientists from

140 This module was seen in many circles as an American strategy to undermine Europeans' access to space by channeling their resources toward projects that were very costly and scientifically dubious. Spacelab was understood as a 'Consolation Prize' for the fact that Europeans, while forced to collaborate with NASA, would also not be allowed to participate in the development of the Space Shuttle itself. Dunar, and Waring, *Power to Explore*, 1999, 427–471 Chapter 11: "Spacelab: International Cooperation in Orbit." The European participation in Spacelab is interpreted as "second-fiddle position that the French particularly despised" in: McDougall, *The Heavens and the Earth*, 1997, 428.

141 Ulf Merbold flew on the first Shuttle mission that operated Skylab in November–December 1983. Merbold was actually employed by a Max Planck Institute (Metal Research, Stuttgart), but his entry into the astronaut program was a personal decision, having been selected from a very large pool of applicants. https://earth.esa.int/web/eoportal/satellite-missions/s/spacelab. Last accessed 3/4/2020. The Frenchman Jean-Loup Chretien had already spent a week on the Soviet space station Salyut 7 in June–July 1982, becoming the first Western European in space.

142 STS-61 (alternatively designated D-1) flew in 1985. STS-55 or D-2 only flew in 1993 because of the delays caused by the Challenger explosion. Astrid Becker: Zurück in die Zukunft. *Süddeutsche Zeitung* (5/5/2018). https://www.sueddeutsche.de/muenchen/starnberg/25-jahre-spacelab-d2-zurueck-in-die-zukunft-1.3967737. Last accessed 6/3/2019.

143 For a friendly account, see, for example, *Looking Up. Europe's Quiet Revolution in Microgravity Research.* New York City, NY: Scientific American Custom Publishing 2008. This issue includes contributions by Max Planck veterans Gerhard Haerendel and Gregor Morfill.

proponents of human spaceflight, further augmenting the distance between scientific researchers and the aerospace industry in Germany.[144]

Fortunately, because of the way in which ESA was set up, with a scientific program that is deliberately separate from its applications and human spaceflight aspects, competitive, highly reputable science remained compartmentalized within the core scientific program. While relatively small in budget compared to the other aspects of ESA, it did provide a stable platform for European scientists throughout the 'classical' space age, from the late 1960s to the end of the Cold War, which was less dependent on the political pressures and ebb and flow coming from NASA. Later in this chapter, we will see how throughout all these eras West German scientists could still take advantage of the opportunities offered. Key to their success was to choose collaborations that guaranteed a high scientific return, furthered their reputation in scientific circles, and even acted as springboards to other interesting projects not necessarily related to the space race. But, at times, and especially when ESA took on human spaceflight as a priority during the 1980s, debates regarding the funding of the scientific research conducted during such flights fragmented the West German scientific community, pitting longstanding colleagues against one another on the old question of whether the resources spent on research during those flights would not be better spent on unmanned planetary probes, astronomical satellites, or science done simply from the ground.[145]

Against all these tensions between national, European, and American pressures, an interesting development from the late 1970s onwards was the increase in collaboration with the Soviet Union. Such collaboration often comprised informal operations carried out without consultation with the Ministry of Foreign Affairs, generally entailing in-kind transfers, where no money changed hands. West German scientists provided scientific modules and questions, and the Soviets a platform on their unmanned spacecraft and their permanently manned space stations. Sometimes the transfer of export-restricted

144 Helmuth Trischler: *The "Triple Helix" of Space. German Space Activities in a European Perspective*. Noordwijk: ESA Publications Division 2002, 21–24.

145 Trischler, *The "Triple Helix" of Space*, 2002, 21–24. In the 1980s the German Physical Society issued a damning report against Human Spaceflight. Among its signatories were three leading Max Planck Researchers: Joachim Trümper (MPE), Klaus Pinkau (MPE–IPP), and Erhard Keppler (MPAe). This position has only somewhat dampened in the 1990s, as it was acknowledged that there were crucial *political* imperatives for human spaceflight in the context of post-Soviet collaborations. Trümmer einer Vision. *Der Spiegel* 32 (1992), 180–182. For the documents, see: Deutsche Physikalische Gesellschaft: Stellungnahmen zur Bemannten Raumfahrt. Homepage, 2018. https://www.dpg-physik.de/veroeffentlichungen/publikationen/stellungnahmen-der-dpg/bemannte-raumfahrt. Last accessed 6/6/2019.

technologies was an implicit part of the transactions. But overall, these collaborations with the Soviets became extremely fruitful, for example, establishing close personal contacts between researchers across the Iron Curtain. When the Soviet Union collapsed in the late 1980s, these personal scientific links were maintained, space-based collaborations continued, and some stellar scientists were even offered positions in the newly unified Germany. The general feeling was that these collaborations were a double-edged knife: on the one hand, they could be tense, owing both to political surveillance within a totalitarian state and, in particular, to working within areas shrouded in secrecy and related to national security; there were also difficulties arising from simple issues such as language barriers; yet on the other hand, the collaborations were very direct transactions without the bureaucracy and disadvantageous political status that Germans had experienced for decades in the European and American contexts. They could develop their instruments in peace and simply attach them to Russian missions.[146] We will see in later chapters how this Soviet and later Russian presence has been a steady and productive alternative channel for Max Planck researchers to keep strengthening their independence in relation to ESA, NASA, and internal German actors, too.

In their landmark history of the ESA, Arturo Russo and John Krige came to the synthetic judgment that space science was best described as "small science in a big context." In the case of West Germany and especially the Max Planck Society, the best way to describe their entry into the space age was that they figured out how to do small science in *other countries'* big contexts.

2 Reorientation of the Max Planck Society in the Early Space Age: Complementarity and Uncoordinated Competition

Thanks to preexisting expertise as well as the global connections forged during the first postwar decade, scientists at the Max Planck Institutes versed in all the traditions described in Chapter 1 were ideally placed to jump on the space

146 Horst-Uwe Keller: interview by Helmuth Trischler and Matthias Knopp, June 10, 2010. Transcript, Oral History of Europe in Space Collection, https://archives.eui.eu/en/oral_history/INT078. Last accessed 2/2/2020. Joachim Trümper: interview by Helmuth Trischler and Matthias Knopp, Munich, March 18, 2010. Transcript, Historical Archives of the European Union. Oral History of Europe in Space Collection, https://archives.eui.eu/en/oral_history/INT076. Last accessed 2/2/2020. Keppler, *Max Planck Institut für Aeronomie*, 2003.

age bandwagon. Each of these traditions used its particular expertise to position itself within international collaborations. Munich theoretical astrophysicists pivoted toward space-based plasma experiments. Southwestern cosmochemists, with their sample analysis expertise, participated in Apollo missions. The Max Planck Institute for Aeronomy in Lindau had the greatest stake in this new era, as its work had been the closest to what would later become space probes and satellites. Here, we foreshadow the problematic direct competition between Lindau and Munich, which will unfold more fully in subsequent chapters. The figure of Reimar Lüst emerges in Munich as someone originally from the plasma astrophysics tradition who then transitions to 'space,' collaborating on French and American rocket-based projects and serving as a delegate to the international bodies that created institutions such as ESRO and its successor ESA. Lüst then went on to become President of the Max Planck Society in 1973.

Max Planck Scientists and the Respectable Path to Outer Space

There is no doubt that the pivotal moment for astrophysics in West Germany was the launch of humankind's first ever orbiting object, the Soviet satellite Sputnik. The importance of the 'Sputnik shock' was perceived in all areas of scientific research, no matter how unrelated to the cosmic sciences. Scientists in the United States used Sputnik to argue for far greater investment in education and fundamental scientific research. Likewise, researchers in all other Western countries, inspired by the Sputnik challenge and already in scientific competition with the United States, quickly became immersed in the space race. This was the case in all scientific fields, but the ones that experienced the most significant boost were those that could be associated with outer space, or were instrumentally and methodologically linked with the technologies of intercontinental ballistic missiles and their logic of mutual assured destruction. We have seen how this applied especially to countries with the resources and ambitions to participate in the space race themselves, such as France and Great Britain, which, since the end of the war, had pursued their own nuclear ambitions as well as advanced rocket development projects with significant participation by German experts from the V-2 era. By the late 1950s, the size of the West German economy was commensurate with that of its two Western European allies. The Germans' prevailing question, starting with Sputnik and continuing throughout the rest of the 20th century, was to what degree the country might participate in the space race and associated scientific and technological developments, a real economic possibility, while cross-fertilization and cross-subsidization with military projects—which benefited these fields in all the other major Western countries—remained politically prohibited;

which amounted to insistently downplaying the elephant in the room that was the legacy of Peenemünde.

In the previous section we saw the troubled history of West Germany's recurrent attempts to build 'national' and even private rocket development programs, and how the geopolitical fragility of the Federal Republic seriously thwarted the success of such ambitions, even despite the availability of technical know-how and economic resources. The outcome, as in other technologies with significant military potential, was to integrate West Germany's rocket development capabilities into a European framework, and have them focus on the least sensitive aspects of such technologies, such as satellite payloads, upper stages, testing facilities, and the supply of specialized components for systems assembled elsewhere. In many ways, we will see how, in the framework of the country's integration into the space age, the approach of Germany's 'fundamental' scientists to scientific collaboration was a far better fit than that of its rocket experts.

For scientists at Max Planck Institutes (as elsewhere), Sputnik opened up a new, major area of public interest in science, one that quickly outpaced the prestige of the 'nuclear' which had propelled the Society forward in the first postwar decade. To be clear, 'nuclear' remained a crucial area of political support for science beyond 1957. Throughout the 1960s, it still retained most of its glamour, and in Germany it was in this decade that the nuclear aspirations dating from the 1950s could be fulfilled, in areas as wide as nuclear fission and fusion, as well as research that borrowed from this prestige, such as elementary particle physics, as well as plasma physics. But by the 1960s, outer space had gained easily as much prestige, and by the 1970s, rising environmental and political anxiety associated with anything 'nuclear' led to the wholesale reinvention of nuclear research, which was now to be conducted under the guise of other motivations, such as environmental issues, which also borrowed cultural themes, as well as actual scientific practices, from the space sciences.[147]

Also, the rising interest in outer space presented many 'nuclear' scientists with a solution to an existential problem, as plasma astrophysics in Munich

147 In Germany, there was considerable pressure for 'nuclear' institutes dating from the early postwar period to reconvert to environmental research in the 1970s. See: Hohn, and Schimank, *Konflikte*, 1990. This 'environmental turn' is being explored in the GMPG program by Gregor Lax. Gregor Lax: *From Atmospheric Chemistry to Earth System Science. Contributions to the Recent History of the Max Planck Institute for Chemistry* (*Otto Hahn Institute*), *1959–2000*. Diepholz: GNT-Verlag 2018. Gregor Lax: *Wissenschaft zwischen Planung, Aufgabenteilung und Kooperation. Zum Aufsteig der Erdsystemforschung in der Max-Planck-Gesellschaft, 1968–2000*. Berlin: GMPG-Preprint 2020.

clearly shows. Globally, a generation had made a career in space plasma astrophysics thanks to the promise of this expertise overlapping with thermonuclear questions that would, it was presumed, later provide the know-how for experimental fusion reactors and perhaps even an open door to nuclear weapons. However, already in the 1950s, the most scientifically inclined scientists in plasma astrophysics knew that research into fusion reactors would propel them away from the most interesting scientific questions in their own field of expertise. The prospect of an experimental plasma physics program, despite its economic and political importance (which many also doubted), seemed increasingly synonymous with a stagnation of their personal research interests. In a sense, the strategy of using astrophysics as the entry point to plasma physics backfired, in that many people then developed too deep an interest in astrophysics; and by the second half of the 1950s, it seemed as if an earlier 'nuclear' Faustian bargain was now pulling them into pedestrian laboratory plasma physics, where, instead of elegant general theories and fascinating phenomena, the work was mired in technology-specific details. This was certainly the case in the United States, and especially among the scientific peers of the German astrophysicists in Biermann's tradition.[148]

The Sputnik shock of 1957 suddenly opened up an entirely different level of justification for plasma astrophysics, and researchers in the Biermann tradition were exceptionally well placed to take advantage of it. During the first postwar decade (1946–56), the Max Planck Society had often justified much scientific work, especially in theoretical physics, as preliminary stages of what could be done once the country was back on its feet, the implicit expectation being what could be done in the nuclear sciences. Now, after Sputnik, all that was needed was to lobby for outer space as one of those fields well worth pursuing experimentally.

On May 26, 1961, Reimar Lüst (on whom more later) wrote to *Ministerialrat* Hans Karl Geeb, Undersecretary at the Ministry of the Interior. Speaking both as a member of a preliminary commission for space research and coordinating secretary of a scientific-technical working group, Lüst underlined some aspects which might be of interest for the Federal Republic of Germany. In particular, he emphasized how:

148 Gary J. Weisel: Properties and Phenomena. Basic Plasma Physics and Fusion Research in Postwar America. *Physics in Perspective* 10/4 (2008), 396–437. doi:10.1007/s00016-007-0371-1. Gary J. Weisel: The Plasma Archipelago. Plasma Physics in the 1960s. *Physics in Perspective* 19/3 (2017), 183–226. doi:10.1007/s00016-017-0205-8.

> With regard to the planned European Space Research Organisation (ESRO), it seems to me particularly important right now that the scientific work in this area in the Federal Republic is intensified and that, for example, development of instruments should be started immediately, which are to be deployed in sounding rockets and/or in satellites. This means that the necessary resources are already to be made available to a relevant extent for well-considered plans, so that such work can be carried out without hindrance and delay. Those countries which, immediately after the establishment of ESRO, are in a position to have ready-made scientific apparatus, will be in a particularly favorable position and will certainly most benefit from this organization in the later course.[149]

Already in early 1961, Lüst was thereby perfectly describing what would be one of the strengths of the future space activity of the Max Planck Institutes participating in extraterrestrial research.

Many of the scientific questions that could be asked came precisely from the realm of space plasmas. Some of these missions were fitted with basic detectors for particles and radiation coming from the Sun and extrasolar space.

During the first years of space exploration, roughly until the mid-1960s, the research that could be done was with rudimentary probes that often did not even reach orbit, analyzing the upper atmosphere and the most basic conditions of outer space. The first important scientific achievement of the space age was James Van Allen's discovery in spring, 1958, of an enormous population of energetic charged particles, mainly protons and electrons, trapped in the external magnetic field of the Earth, the so-called 'radiation belt.'[150] This phenomenon was discovered with very simple instruments, such as the Geiger counter, a typical tool of cosmic ray research, and confirmed previous theoretical investigations. It was immediately explored, artificially injecting particles from the detonation of small nuclear fission bombs at high altitudes.[151] Exploration of the Earth's magnetosphere (the region of space where the Earth's magnetic field dominates over the magnetic field of interplanetary space) was

149 AMPG, III. Abt., ZA1, No. 91.

150 James A. Van Van Allen, and Louis A. Frank: Radiation Around the Earth to a Radial Distance of 107,400 Km. 4659. *Nature* 183/4659 (1959), 430–434. doi:10.1038/183430a0.

151 For a survey of early experiments on the trapped particles see W.N. Hess: Energetic Particles in the Inner Van Allen Belt. *Space Science Reviews* 1/2 (1962), 278–312. doi:10.1007/BF00240580. and references therein.

later extended to space physics missions throughout the solar system, and magnetospheric physics has become a fundamental component of solar system astronomy and even a constitutive aspect of the physics of pulsars and other astrophysical systems. Very simple devices measuring the total electric charge of the particles arriving outside the magnetosphere were sent by the Soviets, thus suggesting that a flow was entering the instrument whenever it faced the Sun. More detailed observations, with a specific plasma probe designed by Bruno Rossi's group at MIT, were obtained by NASA's Explorer X in 1961, establishing the existence of a supersonic plasma in the space surrounding the Earth.[152] In 1962, Mariner 2 definitely detected a continuous flow of particles in interplanetary space.[153] These missions definitively confirmed Eugene Parker's theory of the solar wind (which had been inspired in part by Ludwig Biermann's research on comet tails, as we saw in Chapter 1),[154] and completely changed the view of outer space. The solar wind expansion carries with it the embedded magnetic field lines from the solar surface into interplanetary space, which is not void but filled and dominated by such plasma and magnetic fields, pervading the whole solar system.

The Soviet and American missions' spectacular confirmation of predictions made by Biermann almost a decade earlier[155] provided Munich researchers

---

152 H.S. Bridge et al.: Direct Observations of the Interplanetary Plasma. *Journal of the Physical Society of Japan* 17/Supplement A-II (1962), 553–559. The flight of Explorer X, which established the existence of a steady, albeit variable solar wind streaming past the Earth at supersonic speed, and the existence of a geomagnetic cavity, a region of space surrounding the Earth, which is shielded from the solar wind by the Earth's magnetic field, prepared the stage for the complete vindication of Parker's prediction of the supersonic expansion of the solar atmosphere.

153 C.W. Snyder, M. Neugebauer, and U.R. Rao: The Solar Wind Velocity and Its Correlation with Cosmic-Ray Variations and with Solar and Geomagnetic Activity. *Journal of Geophysical Research* 68/24 (1963), 6361–6370. doi:10.1029/JZ068i024p06361. C.W. Snyder, and M. Neugebauder: Interplanetary Solar-Wind Measurements by Mariner II BT. Proceedings of the Plasma Space Science Symposium. Held at the Catholic University of America Washington, D.C., June 11–14, 1963. Dordrecht: Springer Netherlands 1965, 67–90.

154 Eugene Parker: Dynamics of the Interplanetary Gas and Magnetic Fields. *Astrophysical Journal* 128 (1958), 664–676. doi:10.1086/146579. Eugene N. Parker: Coronal Expansion and Solar Corpuscular Radiation. In: C. C. Chang, and S. S. Huang (eds.): *Proceedings of the Plasma Space Science Symposium. Held at the Catholic University of America Washington, D.C., June 11–14, 1963*. New York, NY: Springer 1965, 99–114. Eugene N. Parker: A History of the Solar Wind Concept. In: Johan A. M. Bleeker, Johannes Geiss, and Martin C. E. Huber (eds.): *The Century of Space Science*. Dordrecht: Kluwer Academic Publishers 2001, 225–255.

155 Ludwig Biermann: Kometenschweife und solare Korpuskularstrahlung. *Zeitschrift für Astrophysik* 29 (1951), 274–286. https://ui.adsabs.harvard.edu/#abs/1951ZA.....29..274B.

and the Max Planck Society with the most significant scientific 'foot in the door' and hence an entry into the space age. Space plasma research was uniquely situated to take advantage of the interest in outer space, and lent a new air of legitimacy to plasma physics, which remained an interesting scientific field throughout the 1960s.[156] Suddenly, many phenomena which had been theorized or observed at a distance were available for direct experimental research, and new, unexpected phenomena like the Van Allen radiation belts were in the plasma physicists' realm of explanation.[157] Technologically and militarily, too, plasma physics was dearly needed during the space age, from the mundane calculation of the conditions faced by ballistic missiles reentering the atmosphere to the effects of nuclear explosions on the high atmosphere: during the early space age, between 1958 and 1964, American nuclear scientists from the Livermore weapons laboratory even exploded a series of nuclear devices in the high atmosphere to characterize the extreme plasma phenomena that resulted. As we saw in the last section, this was out-

---

Last accessed 10/30/2018. Ludwig Biermann: Physical Processes in Comet Tails and Their Relation to Solar Activity. In: P. Swings (ed.): *La Physique Des Comètes. Communications Présentées Au Quatrième Colloque International d'Astrophysique, Tenu à Liège Les 19, 20 et 21 Septembre 1952. Mémoires de La Société Royale Des Sciences de Liège. Quatrième Série.* Liège: Institut d'Astrophysique de l'Université de Liège 1953, 251–262. Ludwig Biermann: Solar Corpuscular Radiation and the Interplanetary Gas. *The Observatory* 77 (1957), 109–110. https://ui.adsabs.harvard.edu/abs/1957Obs....77..109B/abstract. Last accessed 8/14/2020. Ludwig Biermann: *The Plasma in Interplanetary Space.* Technical Report NASA-TN-D-1901. United States: NASA 1963. https://ntrs.nasa.gov/search.jsp?R=19630012233&hterms=Biermann&qs=N%3D0%26Ntk%3DAuthor-Name%26Ntt%3DBiermann%2C%2520L.%26Ntx%3Dmode%2520matchall. Last accessed 11/2/2017. Reimar Lüst: Interplanetary Plasma. *Space Science Reviews* 1/3 (1963), 522–552. doi:10.1007/BF00225270. David P. Stern: A Brief History of Magnetospheric Physics During the Space Age. *Reviews of Geophysics* 34/1 (1996), 1–31. doi:10.1029/95RG03508. See pp. 128–129 and Ch. 2 in John M. Logsdon (ed.): *Exploring the Unknown. Selected Documents in the History of the U.S. Civil Space Program. Space and Earth Science.* Vol. VI. Washington, D.C.: NASA 2004.

156 Ludwig Biermann: Relations between Plasma Physics and Astrophysics. *Reviews of Modern Physics* 32/4 (1960), 1008–1011. doi:10.1103/RevModPhys.32.1008. Lüst, Interplanetary Plasma, 1963, 522–552. Reimar Lüst: Extraterrestrische Forschung. *Mitteilungen der Astronomischen Gesellschaft* 16 (1963), 37–47. http://articles.adsabs.harvard.edu/pdf/1963MitAG..16...37L. Last accessed 8/28/2017. See also a description of early research at the Institute for Extraterrestrial Physics: Reimar Lüst: Extraterrestrische Physik. Über Arbeiten im Institut für Extraterrestrische Physik am Max-Planck-Institut für Physik und Astrophysik. *Die Naturwissenschaften* 52/19 (1965), 525–529. doi:10.1007/BF00645816.

157 Van Allen, Space Scientist, 1997, 1–27.

lawed by international treaty in the mid-1960s.[158] During the space race, when space missions became the focus of public attention and international law, there was a growing need for the 'scientific' justification of these launches, to maintain the rhetorical spin that space exploration was not only a nationalist, military-oriented pursuit. Scientific programs were needed to obtain results that demonstrated these scientific ideals. West German researchers, because of their country's self-proclaimed restriction to peaceful research, were uniquely suited to fill this niche.

Scientific Research Programs for the Space Age at Existing Max Planck Institutes

All of the institutes and scientific traditions mentioned earlier in this book benefited greatly from the new interest in outer space, but they differed in how they repositioned themselves in relation to this new social and political environment; the first to move were those who could most easily attach their preexisting research programs to outer space missions, thanks either to their theoretical insight, instrumental expertise, or ability to propose, design, and eventually build experiments to be conducted in outer space.

All the connected space activity underway at the time in America and the Soviet Union prepared the stage for a stronger dialogue between the Aeronomy Institute (Bartels and Dieminger), the Physics and Astrophysics Institute (Heisenberg and Biermann), and the Nuclear Physics Institute (Gentner).

As early as 1959, the Minister of Scientific Research, Siegfried Balke (see Chapter 1, Section 4), had arranged to meet Heisenberg, Biermann, and Reimar Lüst in Munich, to discuss plans for space projects; and over the following years, he actively pursued means to collaborate with the United States that would later lead to the first German space missions.[159]

However, the early expansionist ambitions—not only in space research—of the research ministry led by Balke prompted a backlash from the DFG, WRK,[160] and the Max Planck Society itself, which all feared an excessive loss of self-determination. It was this which led to the so-called 'Holy Alliance' of these three organizations. So, while generally the MPG was a very strong ally of the

158 See United Nations Treaties: Treaty Banning Nuclear Weapon Tests in the Atmosphere, in Outer Space and under Water. Submitters. United States of America United Kingdom of Great Britain and Northern Ireland Union of Soviet Socialist Republics. Moscow 05/08/1963. https://treaties.un.org/pages/showDetails.aspx?objid=08000002801313d9. Last accessed 12/7/2017.

159 Reinke, *The History of German Space Policy*, 2007, 56, 107.

160 The pressure group of West German universities mentioned earlier in the chapter.

new research ministry in select fields like nuclear and space research, it also defended its independence from the federal government throughout this era, even though it drew considerable funds from the research ministry, altering the early postwar formula of 50/50 contribution by the federal government and individual states.[161]

The Max Planck Society as a whole made early attempts to participate in activities related to this field through the creation of an extraterrestrial research group.[162] The research program presented by the Institute for Physics and Astrophysics in Munich was, on the one hand, a natural extension of previous interests in cosmic ray physics proper and in astrophysical plasmas, inspired by results obtained from the first successful satellites equipped with detectors for cosmic rays and ionizing radiation from the Sun.[163] During the pre-Sputnik era, cosmic ray research in various forms, particles and fields

---

161 Thomas Stamm: *Zwischen Staat und Selbstverwaltung. Die deutsche Forschung im Wiederaufbau 1945–1965*. Köln: Wissenschaft und Politik 1981, 225–243.

162 See draft of the letter to MPG President dated München and Lindau, December 30, 1959 announcing that the four Institutes were jointly building a working group for "extraterrestrial research." Following the Geophysical Year, the field entered a boom period in many countries, especially USA and USSR, and they were "urged by foreign colleagues that Germany should participate more actively" (AMPG, Abt. III, ZA1, No. 91). In a later document in the same folder (minutes of a meeting of the "Arbeitsgemeinschaft Extraterrestrische Forschung" held on 7 March 1961), in connection with research on meteorites it was mentioned that Gentner would represent the interests of Mainz in the group. The folder No. 91 of Biermann's papers (Arbeitsgruppe Extraterrestrial Physik 1959–1963) contains key material related to the early phase leading to the foundation of the Extraterrestrial Institute, as well as memoranda written by Reimar Lüst about Germany and space research, also including notes on the research activities. Preliminary steps before the official formation of the research group imply correspondence with Peter Meyer (contained in the same folder No. 91) and outlining all the plans related to the project. Since 1950 Meyer had been a member of the Institute for Physics, where he had worked on cosmic ray research and detection methods. In 1954, he had moved to Chicago where he began to collaborate with John Simpson, an experimental nuclear and cosmic ray physicist, on investigations of the variation of cosmic rays with solar activities. Simpson was also a leading scientist within the International Geophysical Year project and later became one of the members of the Space Science Board formed by the National Academy of Sciences in 1958 to support the start of NASA space activities. On January 10, 1963, Heisenberg and Biermann proposed that a Department for Extraterrestrial Physics be built, and launched with about 10 collaborators led by Reimar Lüst (see discussion in CPTS meeting minutes of 01.03.1963, 13–14.05.1963, AMPG, II. Abt., Rep. 62, No. 1740, 1741).

163 There was a collected volume of the same period resulting from a review meeting held in Paris in 1963 and devoted to recently acquired knowledge of the space environment, in which Biermann participated with a contribution on "New measurements of the interplanetary plasma and their interpretation": A. Ehmert (ed.): *The Space Environment. Le Milieu Spatial. Report on the Survey Meeting of Information on the Space Environment*

in space, and the relationship between Sun and Earth had been a common denominator in studies conducted in all those institutes promoting the formation of a research group on extraterrestrial topics.[164]

On the other hand, the recently discovered Van Allen Belts appeared to be a plasma made of protons and electrons mostly deriving from the Sun, trapped and held high around the Earth by the planet's magnetic field. Now that scientists were equipped to conduct experiments in space, as opposed to just making observations, an additional proposal was to use space probes to produce artificial plasma events which would allow exploration of the physical properties of the near-Earth space. Going to high altitudes, where the absorption of ultraviolet radiation by the Earth's atmosphere could be avoided, also allowed for the ultraviolet spectroscopy of comet tails related to Biermann's solar wind proposition.[165]

Paris, 27 September 1963. Wien: Springer 1964. See also Gotthard Gambke, Rudolf Kerscher, and Walter Kertz: *Denkschrift zur Lage der Weltraumforschung*. Wiesbaden: Franz Steiner Verlag 1961. Peter Fischer: *The Origins of the Federal Republic of Germany's Space Policy 1959–1965. European and National Dimensions*. Noordwijk: European Space Agency Publications 1994.

164 Georg Pfotzer: Kosmische Strahlung als Informationsquelle für Zustände und Vorgänge im Weltraum. In: Generalverwaltung der Max-Planck Gesellschaft zur Förderung der Wissenschaften e.V. (ed.): *Jahrbuch 1960 der Max-Planck-Gesellschaft zur Förderung der Wissenschaften e.V.* Göttingen 1960, 126–160. Biermann, Solar Corpuscular Radiation, 1957, 109–110. K. Goebel, P. Schmidlin, and J. Zähringer: Das Tritium-Helium- und das Kalium-Argon-Alter des Meteoriten "Ramsdorf." *Zeitschrift für Naturforschung A* 14/11 (1959), 996–998. doi:10.1515/zna-1959-1112. Hermann Wäffler: Die kosmische Strahlung. Ergebnisse und Probleme. In: Generalverwaltung der Max-Planck-Gesellschaft (ed.): *Jahrbuch 1959 der Max-Planck-Gesellschaft zur Förderung der Wissenschaften e.V.* Göttingen 1959, 202–222.

165 The possibility of such experiments on artificial comets were discussed in early 1960, as mentioned by Biermann in his closing comment in an article on the subject, submitted in July 1961. Ludwig Biermann et al.: Zur Untersuchung des interplanetarischen Mediums mit Hilfe künstlich eingebrachter Ionenwolken. *Zeitschrift für Astrophysik* 53 (1961), 226–236. http://articles.adsabs.harvard.edu/pdf/1961ZA.....53..226B. Last accessed 10/30/2018. On April 18, 1961, Biermann was already mentioning plans for the artificial comet tails experiment, as in a letter written to Rhea Lüst from the US (original in English): "Your husband brought the news from Stockholm that it has been proposed to carry out our project of artificial comet tails on a European basis at least as the connected observations from the ground are concerned. The European study group for space research urgently requires more details about our proposed experiments. For this reason, your husband and I would like you and Dr. Hermann Schmidt to look into this matter." Biermann also mentioned the recent results from the plasma probe built by Rossi's group at MIT (AMPG, Abt. III, ZA1, No. 91). The space probe Explorer X had performed the first in situ measurements of the solar wind at the boundary of the Earth's magnetosphere.

The post-Sputnik plans from the Aeronomy institute in Lindau were surprisingly similar, focusing on extraterrestrial plasmas and magnetic fields in space. In contrast, the Nuclear Physics Institute in Heidelberg focused instead on cosmochemical work related to cosmic rays and the noble gas content of samples from interplanetary space, the formation of elements in the universe, and interplanetary dust. In the postwar era, research fields such as meteorology, geomagnetism, and the effect of the Sun on the atmosphere were rapidly converging, while still immersed in the early 20th-century framework of 'cosmical physics.' The use of radio techniques for exploring the properties of the atmosphere and of the Earth found increasing application, in parallel with a growing expectation that radio exploration of the upper atmosphere was to be supplemented by measurements made in situ by means of automatic apparatus carried in rockets or balloons, a prewar German tradition. At the same time, the study of the astrophysical aspects of cosmic rays—which now included the knowledge of the constitution of the primary cosmic radiation—the growing awareness that electromagnetic phenomena were of great importance in cosmic physics, and the progress of individual sciences related to the Sun, Earth, and deep space, led to a new and even more comprehensive kind of 'cosmical physics,' during the 1950s, a productive framework eventually leading to the emergence of space science in the 1960s.[166] These activities were never restricted to the high atmosphere as they also studied, for example, cosmic rays and the interaction of the magnetic fields with particles coming from the Sun. A clear example of the disciplinary blur around the high atmosphere and near space was a meeting of the International Astronomical Union (IAU) organized in 1956 and including near-Earth topics in its realm of expertise, such as 'stellar magnetism,' 'solar and interplanetary magnetic fields,' and 'electromagnetic state in interplanetary space,'

Bridge et al., Direct Observations, 1962, 553–559. In a letter written to Biermann on August 25, 1960, Siegfried Balke—at the time German Federal Minister for Nuclear Energy—suggested that, for financing optimization, a combination with the plasma group should be also established (AMPG, III. Abt., Rep. 145, No. 1004). See also Lüst's report on Max Planck Society's involvement in international cooperation in space research prepared for the CPTS meeting of June 6, 1961, in Berlin, and typewritten page dated May 18, 1962, with preliminary research plans of the Department for Extraterrestrial Physics led by Lüst especially focusing on interplanetary medium, Earth-Sun and solar system, particles from the Sun, magnetic fields, artificial plasma clouds and comet tails in interplanetary space, and investigations with radio waves (AMPG, II. Abt., Rep. 66, No. 3047. Fol. 30 and 39–45).

166 Johan A. M. Bleeker, Johannes Geiss, and Martin C. E. Huber (eds.): *The Century of Space Science*. Dordrecht: Kluwer Academic Publishers 2001.

as well as discussions on 'magneto-hydrodynamics' and 'origin and structure of sunspots.' Much of the modern-day foundations of the 'plasma Universe'—the term Alfvén later coined to denote the cosmic space filled with high-energy particles, magnetic fields, and highly conducting plasmas[167]—can actually be traced to that 6th IAU Symposium, attended by the "Olympians" of the field, as a participant named them many years later.[168] And so 'cosmical' physics, a very old term used since the 19th century, was the basis for beginning to think in terms of what became 'space science': plasma phenomena discovered in the laboratory—which must be important also in the rest of the universe—could now be investigated with in situ measurements in accessible regions of the space surrounding the Earth and all other celestial bodies of our solar system.[169]

But in addition to these entry points, we will see in the next chapter how interest in space eventually led to one of the major expansions of the Max Planck Society, via the absorption of (ground-based and space-based) observational astronomy: a field that marked out a different path than that of the trends within the Society in the first postwar decade.

Following the CERN Model: ESRO and the Rise of Reimar Lüst

As we learned from Peter Meyer's letter earlier in this chapter, the first flurry of space-based research propositions in Germany came about in the wider context of a space strategy for Europe.[170] Space science in West Germany needed

---

167 Hannes Alfvén: The Plasma Universe. *Physics Today* 39/9 (1986), 22. doi:10.1063/1.881039. Hannes Alfvén: Plasma Universe. *Physica Scripta* T18 (1987), 20–28. doi:10.1088/0031-8949/1987/T18/002. The term "plasma Universe" was used in 1986 both by Alfvén and Anthony L. Peratt in titles of their contributions to a special issue on Space and Cosmic Plasma of the IEEE *Transactions on Plasma Science*, Vol. PS-14/6, published in December 1986. https://ieeexplore.ieee.org/xpl/tocresult.jsp?isnumber=4316609&punumber=27. Last accessed 6/5/2020.

168 The "Olympians and royalty of the astrophysical community" included Alfvén, Artsimovich, Biermann, Chandrasekhar, Cowling, Shklovsky, Schlüter, Spitzer, Burbidge, "and many other notables." Winston H. Bostick: Stockholm, August 1956, Revisited (Plasma Astrophysics). IEEE *Transactions on Plasma Science* 17/2 (1989), 69–75, 69. doi:10.1109/27.24610.

169 B. Lehnert (ed.): *Electromagnetic Phenomena in Cosmical Physics*. Cambridge, MA: Cambridge University Press 1958.

170 For materials on the organization of space research at a European level taking place in 1960 and on the preliminary organizational phase of space research within MPG, see correspondence between Gentner and Bartels and related documents in Gentner's papers, AMPG, III Abt., Rep. 68A, No. 158. See also Reimar Lüst: Weltraumforschung in der Bundesrepublik und Europa. *Weltraumforschung in der Bundesrepublik und Europa. Sonnenforschung*. VS Verlag für Sozialwissenschaften 1966, 7–29.

to be a multinational endeavor; the rocket technology and expertise needed for the launchers had been completely relocated to other countries after the war, and there was still strong international and domestic resistance to German self-sufficiency in space.[171] And as rocket-related expertise continued to be problematic in Germany, the more 'scientific' aspects of space exploration found favor. Germany benefited here from the path already prepared by CERN in the 1950s. By the end of the 1950s, it had become evident that this first case of pan-European collaboration was extremely successful. When the synchrotron producing protons of 28 GeV went into operation at CERN in fall 1959, it was far ahead of the accelerators with an energy of more than 1 GeV in service at the time in the United States (Berkeley and Brookhaven) and in the Soviet Union (Dubna). CERN's existence greatly stimulated the construction of accelerators in national institutions of the member states, but a large part of the work done at CERN laboratory was due to teams coming in from national institutes with their own equipment. Suddenly, CERN became a model that other scientific areas sought to reproduce, in areas as varied as observational astronomy,[172] space exploration, and later even molecular biology.[173]

Given its perceived importance at the time, space exploration was the first attempt to imitate CERN, with the creation in 1964, by ten Western European countries (plus Australia, where the first launch base was located), of the European Space Research Organisation (ESRO), whose statutory purpose was "to provide for, and to promote, collaboration among European States in space research and technology, exclusively for peaceful purposes."[174] The idea of a joint European space effort was a most natural step for scientists like Edoardo Amaldi and Pierre Auger (first Director General of ESRO), who had already been main actors in the process leading to the birth of CERN and who now turned their attention to scientific collaboration in post-Sputnik space.[175]

---

171 Helmuth Trischler: A Talkative Artefact: Germany and the Development of a European Launcher in the 1960s. In: Martin Collins, and Doug Millard (eds.): *Showcasing Space*. London: Science Museum 2005, 7–28.

172 Adriaan Blaauw: *ESO's Early History. The European Southern Observatory from Concept to Reality*. ESO 1991.

173 Georgina Ferry: *EMBO in Perspective. A Half-Century in the Life Sciences*. EMBO 2014.

174 Krige, and Russo, *European Space Agency I*, 2000, Vol. 1. p.198. The ten founding states were Belgium, Denmark, France, (Federal Republic of) Germany, Italy, Netherlands, Spain, Sweden, Switzerland, and the United Kingdom.

175 De Maria, *Europe in Space*, 1993. Krige, Russo, and Sebesta, *European Space Agency II*, 2000, Vol. 2. ESRO was merged with ELDO (European Launcher Development Organization) in 1975 to form the European Space Agency (ESA). Krige, and Russo, *European Space Agency I*, 2000, Vol. 1. The link with CERN was made explicit by holding the major confer-

The first and most important beneficiaries of the space age were in Munich, and are best personified by Reimar Lüst, who, as one the early postwar disciples of von Weizsäcker and Biermann, had long since become part of the space plasma research community, with periods in Chicago (1955–56) and Princeton (1956–57), on a Fulbright Scholarship, and later in New York (1959).[176] Even though the prestige in space plasmas accrued mostly to Biermann, it was the next generation, represented by Lüst, who were able to make the most of these new opportunities.

We have already followed Lüst's early career in plasma physics in Chapter 1. He had been involved with the European space science administration as a member of the *Commission Préparatoire Européenne de Recherches Spatiales* (COPERS). Now, in the course of the formation of ESRO, Lüst was appointed representative of West Germany and then, in turn, in this capacity, became a member of the subsequent national commissions seeking to create conditions under which Germans could benefit economically and scientifically from this new frontier.[177] Most importantly perhaps, in addition to

---

ence for its creation in Geneva, and by the participation of many CERN representatives in the creation and leadership of ESRO. For the launch of CERN and in particular on the role of Amaldi and Auger, see Chapter 1 in Armin Hermann et al.: *History of CERN. Launching the European Organization for Nuclear Research.* Vol. 1. Amsterdam: North-Holland 1987. In connection with this European trend to joint ventures, Klaus Pinkau emphasized "In Europe we think about cooperation, while the Americans think about who will be the best. America could afford that, because it had enough resources; if you let the best win, you destroy five runners-up. This is a very painful and uneconomical way to proceed [our translation]." Klaus Pinkau: interview by Helmuth Trischler, March 9, 2010. Transcript, Historical Archives of the European Union. Oral History of Europe in Space Collection, https://archives.eui.eu/en/oral_history/INT072. Last accessed 12/4/2020. See also Lorenza Sebesta: *US-European Cooperation in Space during the 1960s.* Noordwijk: ESA Publications Division 1994. Seibert, *History of Sounding Rockets*, 2006.

176 Reimar Lüst: Carl Friedrich Weizsäcker. Ein Doktorand erinnert sich. In: Klaus Hentschel, and Dieter Hoffmann (eds.): *Carl Friedrich von Weizsäcker. Physik, Philosophie, Friedensforschung.* Stuttgart: Wissenschaftliche Verlagsgesellschaft 2014, 263–270.Reimar Lüst: Ludwig Biermann. 13.3.1907-12.1.1986. *Berichte und Mitteilungen der Max-Planck-Gesellschaft.* München 1986, 78–81. See also the autobiographical volume Lüst, and Nolte, *Der Wissenschaftsmacher*, 2008.

177 Reimar Lüst: The European Space Research Organization. *Science* 149/3682 (1965), 394–397. doi:10.1126/science.149.3682.394. Ulf von Rauchhaupt: *To Venture Beyond the Atmosphere. Aspects of the Foundation of the Max Planck Institute for Extraterrestrial Physics.* Berlin: Max-Planck-Institut für Wissenschaftsgeschichte 2000. Ulf von Rauchhaupt: Coping with a New Age. The Max Planck Society and the Challenge of Space Science in the Early 1960s. In: Max-Planck-Gesellschaft zur Förderung der Wissenschaften (ed.): *Innovative Structures in Basic Research. Ringberg-Sympsium, 4–7 October 2000.* München 2002, 197–205.

tying national efforts to the pan-European organization, German researchers at Max Planck Institutes sought to diversify their partnerships by taking part in other national space programs, for example, with the United States, France, and Italy. In all these programs, they served as scientific specialists. In this context, Lüst quickly gained a leading role as 'ambassador' of the Max Planck Society in space affairs. Already in June 1961, he presented at a meeting of the MPG's Scientific Council a long report titled "Internationale Zusammenarbeit auf dem Gebiet der Weltraumforschung und die Beteiligung der Max-Planck-Gesellschaft" [International cooperation in space research and the participation of the Max Planck Society], in which he outlined the current international state of space research and the US plans for the future, in order to facilitate comparison with European plans, also in connection with the Max Planck Society. He also presented an overview of the projects for the European Space Research Organisation and reported on plans and activities within the Society.[178] Through his early involvement with ESRO, Lüst came into contact with scientists from other European countries and established scientific collaborations which later led to the first space experiments of his brand-new Institute for Extraterrestrial Physics.[179] Very early, from 1964, he participated also in official cultural exchanges with the Soviet Union, in particular with the Academy of Sciences, which led to the Russian plasma physicist Fedorovich Kolesnikov being invited to the Institute for Plasma Physics; and this internationalist profile would be a hallmark of his tenure as President of the Max Planck Society in the 1970s.[180] On the other hand, first as Scientific Director and Vice President

178 CPTS meeting minutes of June 6, 1961, AMPG, II. Abt., Rep. 62, No. 1737. In this regard, see also minutes of a meeting at the Ministry of Research on ESRO's research program in AMPG, III Abt., Rep. 145, Nachlass Reimar Lüst, No. 999–1002 (Bundesministerium für Wissenschaftliche Forschung), and N. 1004 (Deutsche Kommission für Weltraumforschung—Arbeitskreis "Astrophysik und Astronomie," 1963–1967).

179 Reimar Lüst: interview by Horst Kant and Jürgen Renn, Hamburg, May 18, 2010 (DA GMPG, ID 601068). Actually, at an ESRO meeting in London Lüst met Jacques Blamont, the French physicist who experimented with research rockets by evaporating sodium in the atmosphere in order to measure atmospheric winds, and Lüst was able to plan rocket research with him, as we will see later in this chapter. See material related to Lüst's extensive involvement with ESRO in his personal papers: AMPG, III. Abt., Rep. 145, No. 173, 260, 894, 971, 972, 980, 985, 986, 992, 1014, 1047–1049, 1054, 1055, 1060–1064, 1066–1068, 1090, 1248–1250, 1252.

180 In December 1962, a delegation of the Soviet Academy of Sciences had been invited by the Max Planck Society to visit West Germany for a couple of weeks. Präsidialbüro der Max-Planck-Gesellschaft (ed.): *Mitteilungen aus der Max-Planck-Gesellschaft zur Förderung der Wissenschaften. Heft 1-2/1963*. München 1963, 102–103. In 1963, Lüst visited Russian institutes and observatories as member of a Max Planck Delegation invited by

of ESRO, then, from 1984 to 1990, as third Director General of ESA, Lüst was constantly involved in space cooperation agreements (but he once said that cooperating with Member States at ESA was like "dancing with an octopus").[181]

At the same time, however, Lüst represented a national view of space research, stressing the importance of developing national capabilities independent of the nascent collaborations. International collaboration [he felt], would benefit most from a strong national base, as the French example clearly showed. This, at a time when German policymakers and scientists, including Werner Heisenberg, were vocal in their criticism of the patent imbalance between Germany's considerable economic contributions to European collaborations and German scientists and industries' lack of equivalent participation therein.[182] Defending this position gave Lüst powerful backing in West Ger-

the Russian Academy of Sciences. Ludwig Biermann, and Reimar Lüst: Jahresberichte astronomischer Institute 1963. München Max-Planck-Institut für Physik und Astrophysik. Institut für Astrophysik. *Mitteilungen der Astronomischen Gesellschaft Hamburg* 17 (1964), 180–186. http://adsabs.harvard.edu/abs/1964MitAG..17..180. Last accessed 4/13/2020. See also Reimar Lüst: Eindrücke von einer Reise in die Sowjetunion. Gegenbesuch einer Delegation der Max-Planck-Gesellschaft bei der Akademie der Wissenschaften der UdSSR. In: Generalverwaltung der Max-Planck-Gesellschaft (ed.): *Mitteilungen aus der Max-Planck-Gesellschaft. Heft 1–2*. Max-Planck-Gesellschaft 1964, 55–63. On Lüst's involvement in these journeys to USSR, see AMPG, Abt. III, Nachlass Reimar Lüst, Rep. 145, No. 890. Later, Lüst was one of the members representing MPG during the *Wissenschaftsrat*'s trip of Oct.25—Nov. 2, 1971 ("Bericht über eine Reise des Wissenschaftsrates in die Sowjetunion vom 25. Oktober bis 2. November 1971," Rep. 145, No. 959), which was related to an exchange in the field of university education, and dedicated to visits to relevant scientific centers, also with the USA, as a representative of the Bundesrepublik Deutschland (see also No. 960). After the trip of the delegation of the German Federal Minister of Science, Lüst felt that there was change in the atmosphere compared to his visit in 1963. A number of Russian institutes were very interested in collaborating with the Max Planck Society and very soon a delegation of German experts would travel to Russia to hold preparatory talks. About the problem of direct interaction with scientists, it was now possible to invite Russian scientists directly, without involving the official bodies. During his stay in Russia, Lüst established new fruitful relationships and invited Russian colleagues to the Max Planck Institute for Physics and Astrophysics thus creating the premise for future collaborations. About such trips see several folders in Lüst's papers in AMPG, III. Abt., Rep. 145, No. 1100–1106.

181 See Documents related to the history of space cooperation at the Historical Archives of the European Union (ESA.B.09-04.0201, https://archives.eui.eu/en/fonds/532919?item=ESA.B.09-04.02.01. Last accessed 5/23/2021). For obituaries of Lüst, see: European Space Agency: Professor Reimar Lüst (1923–2020). *European Space Agency*, 4/2/2020. https://www.esa.int/About_Us/ESA_history/Professor_Reimar_Luest_1923-2020. Last accessed 4/3/2020. See also, Roger Bonnet, and Gerhard Haerendel: Reimar Lüst (1923–2020). *Space Research Today* 208 (2020), 4–6. doi:10.1016/j.srt.2020.07.006.

182 Reimar Lüst: *Die Gegenwärtigen Probleme Der Weltraumforschung*. München: Oldenbourg; Düsseldorf: VDI-Verl 1964.

man political and industrial circles, and in some cases even allowed him to personally shape the nascent playing field, for example, by incentivizing industrial partners' first ventures in the aerospace business, which itself was strongly overseen by the federal ministry; its expert commissions were populated by circles close to Lüst.[183] And despite the tensions between Heisenberg and Gentner, Lüst, as the representative of a new, more pragmatic generation, did his best to establish a relationship with figures relevant to space exploration in the Max Planck Society, independently of his 'family' background. In 1967, he founded the Association for Extraterrestrial Physics (*Arbeitsgemeinschaft Extraterrestrische Physik*) to facilitate and promote in Germany scientific exchange in the emerging field of space research.[184]

Space Plasma Experiments and the New Institute for Extraterrestrial Physics

The most innovative step in the adaptation of the Max Planck Society to outer space came from deep within the expertise in space plasmas, with the design of experiments in the outer atmosphere that could connect with the theoretical insight of Biermann's tradition. These experiments could also be conducted with the most rudimentary rockets, which did not even need to enter into orbit, and were particularly well suited to collaboration with nascent national space programs throughout Europe.[185] It was proposed to release substances in the outer atmosphere, and then follow the path and

183 Rauck, Horst: interview by Helmut Trischler, June 19, 2010. Historical Archives of the European Union. Oral History of Europe in Space Collection, https://archives.eui.eu/en/oral_history/INT073. Last accessed 7/30/2019. Feustel-Büechl, Jörg: interview by Helmut Trischler on 09.04.2010. Historical Archives of the European Union. Oral History of Europe in Space Collection, https://archives.eui.eu/en/oral_history/INT065. Last accessed 7/30/2019.

184 Jörg Büchner (ed.): *Geschichte des Fachverbands Extraterrestrische Physik und der Arbeitsgemeinschaft Extraterrestrische Forschung*. Katlenburg-Lindau: Arbeitsgemeinschaft Extraterrestrische Forschung E.V. 2009, 1.

185 Moreover, as future MPE director Pinkau remarked, "When I came to Munich in 1965, Mr. Lüst had practically taken the lead in the German commitment to space research. But this, given his scientific interests, played out more in the magnetosphere and ionosphere physics sector. He had developed the method of ion clouds. Astrophysics in space was underdeveloped in Germany. Lüst himself was a plasma physicist, [Walter] Dieminger in Göttingen had conducted ionospheric physics, and therefore the German strengths were also in the universities, in meteorology, the upper atmosphere, or the ionosphere, and not so much in the field of astronomy. That fit for two reasons. First of all, these satellites were cheaper than astronomical satellites and easier to launch, and therefore better suited for a start; and secondly, it suited the Americans. As you know, we had two approaches in Germany. One was working with NASA, the other was

behavior of the ionized particles that resulted from their exposure to the conditions there.[186] Instrumental in this regard was the collaboration established between Lüst and the Frenchman Jacques Blamont, the pioneer of such cloud experiments,[187] during the aforementioned ESRO meeting:

> When he [Blamont] heard that I was planning barium cloud experiments, he said I should bring my experiment along for one of his rockets, he would manage to include it somehow. This led to my first experiments in high altitude research rockets.[188]

---

about European space exploration. Like many other nations, we've always had competition between the national program, which was essentially conducted with NASA, and participation in Europe" [our translation]. Klaus Pinkau: interview by Helmuth Trischler, March 9, 2010. Transcript, Historical Archives of the European Union. Oral History of Europe in Space Collection, https://archives.eui.eu/en/oral_history/INT072. Last accessed 12/4/2020, p. 8.

186 Ludwig Biermann et al.: Zur Untersuchung des interplanetaren Mediums mit Hilfe künstlich eingebrachter Ionenwolken. *Zeitschrift für Astrophysik* 53 (1961), 163–176. Gerhard Haerendel, and Reimar Lüst: Artificial Plasma Clouds in Space. *Scientific American* 219/5 (1968), 80–95. https://www.jstor.org/stable/24927565. Last accessed 5/16/2021. Jacques Blamont: The Beginning of Space Experiments in Munich. In: Gerhard Haerendel, and Bruce Battrick (eds.): *Topics in Plasma-, Astro- and Space Physics. A Volume Dedicated to Reimar Lüst on the Occasion of His 60th Birthday*. Garching: Max-Planck-Institut für Physik und Astrophysik 1983, 161–164. Reimar Lüst: Künstliche Wolken. Ein Mittel der Weltraumforschung. In: Generalverwaltung der Max-Planck-Gesellschaft (ed.): *Jahrbuch der Max-Planck-Gesellschaft zur Förderung der Wissenschaften e.V. 1968*. Göttingen 1968, 150–169. Gerhard Haerendel: Towards an Artificial Comet. In: Gerhard Haerendel, and Bruce Battrick (eds.): *Topics in Plasma-, Astro- and Space Physics. A Volume Dedicated to Reimar Lüst on the Occasion of His 60th Birthday*. Garching: Max-Planck-Institut für Physik und Astrophysik, Institut Extraterrestrische Physik 1983, 165–177. Reimar Lüst: Barium Cloud Experiments in the Upper Atmosphere. In: Johan A. M. Bleeker, Johannes Geiss, and Martin C. E. Huber (eds.): *The Century of Space Science*. Dordrecht: Kluwer Academic Publishers 2001, 179–187. Blamont, Alkali Metal Cloud Experiments, 2001, 189–202. Ulf von von Rauchhaupt: Colorful Clouds and Unruly Rockets: Early Research Programs at the Max Planck Institute for Extraterrestrial Physics. *Historical Studies of the Physical and Biological Sciences* 32/1 (2001), 115–124. doi:10.1525/hsps.2001.32.1.115. Rauchhaupt, Coping with a New Age, 2002, 197–205.

187 Blamont, Alkali Metal Cloud Experiments, 2001, 189–202.

188 Reimar Lüst: interview by Hans von Storch and Klaus Hasselmann. Niels Bohr Library & Archives, American Institute of Physics, College Park, MD USA, www.aip.org/history-programs/niels-bohr-library/oral-histories/33761. Last accessed 10/4/2020. Lüst, The European Space Research Organization, 1965, 394–397. Complementary investigations were those related to measurements of neutrons from the Sun—which could provide information on nuclear processes taking place in the solar photosphere—and to detection of neutrons generated in the Earth atmosphere from the interaction with primary

The first proposal for an ion cloud experiment in space with the aim of investigating the interplanetary medium was submitted by Lüst to ESRO as early as September 1962.[189] The first barium plasma clouds were observed in 1964 "on the evening night [*sic*] of the Sahara."[190] These observations would then be interpreted by specialists using the theoretical toolkit of plasma astrophysics. But only some 20 years later, in 1984–85, would Biermann's dream of an artificial comet be realized under the guidance of Gerhard Haerendel.[191]

ESRO's early space science program, which primarily addressed problems in plasma physics, consisted of small magnetospheric satellites launched between the late 1960s and early 1970s. Throughout the 1970s, solar system exploration with new satellites continued to be focused on magnetospheric research, and the first astronomy satellites were launched only between 1972 and 1978, making observations at ultraviolet, X-ray, and gamma-ray wavelengths. But during the 1970s, progress in laboratory studies of plasmas—also related to nuclear fusion research—as well as in the methods of transferring the results to cosmic conditions, together with the new empirical knowledge gained by in situ measurements in the magnetospheres and in the whole solar wind region, which at that time were the main objectives of the Voyager missions, drastically changed our understanding of the properties of cosmic plasmas. All this contributed to building the foundations of what Hannes Alfvén saw as a "paradigm transition," which he expressed in the term "plasma universe," to emphasize the fact that the plasma phenomena hitherto discovered in the laboratory and in the Earth's own space environment are fundamental

cosmic ray particles, and which are the sources of the high-energy protons of the internal Van Allen belt. Together with the barium clouds experiments, these measurements allowed the study of interactions between interplanetary plasma and magnetic fields in the near-Earth environment also in connection with the Sun–Earth relationship which was performed with the two Helios satellites, a joint venture of West Germany and NASA, launched in 1974 and 1976 for the purpose of studying the interplanetary medium from the vicinity of the Earth's orbit to 0.3 astronomical units, that is about 135.000.000 km. Both observed the dust and ion tails of at least three comets. See web page by Max-Planck-Institut für Sonnensystemforschung, http://www2.mps.mpg.de/de/projekte/helios/#e8#e8. Last accessed 2/3/2022.

189 "Ion cloud in the interplanetary space" (September 1962), Historical Archives of the European Union, S-16 COPERS-1191, https://archives.eui.eu/en/fonds/96512?item=COPERS-06.01-1191. Last accessed 6/20/2020. See also the related proposal "Ion cloud in the ionosphere," submitted in October 1964, ESRO-5634, https://archives.eui.eu/en/fonds/143134?item=ESRO.A-04.06-5634. Last accessed 2/3/2022.

190 Bonnet, and Haerendel, Reimar Lüst, 2020, 4–6, 5.

191 A. Valenzuela et al.: The AMPTE Artificial Comet Experiments. *Nature* 320 (1986), 700–703. doi:10.1038/320700a0.

also to the rest of the Universe, which consists almost entirely of matter in the plasma state.[192]

The proposal to create artificial comets, which would remain Ludwig Biermann's ultimate experimental aspiration for decades, was favorably received during the first discussions defining ESRO's scientific program.[193] In the context of these proposed experiments, and in his involvement in the preparation of European scientific space programs, Lüst became the key person around whom Heisenberg and Biermann managed to steer the Max Planck Society and the German federal government to support the transformation of the working group dedicated to "extra-terrestrial" research—initially meant to be a meeting point of different traditions in the Max Planck Society—into an Institute for Extraterrestrial Physics (MPE) within the Munich Institute for Physics and Astrophysics.[194]

---

192 Hannes Alfvén: Paradigm Transition in Cosmic Plasma Physics. *Geophysical Research Letters* 10/6 (1983), 487–488. doi:10.1029/GL010i006p00487.

193 Such discussions of the Scientific and Technical Working Group of the "Commission Préparatoire Européenne de Recherche Spatiale" were held in Stockholm in early April 1961 Krige, Russo, and Sebesta, *European Space Agency II*, 2000, Vol. 2, 42. In view of future cometary missions by means of space probes sent to larger distances from the Earth's orbit and from the ecliptic plane, which could not only study cometary physics but also use comets as probes for the solar wind and therefore contribute to a further understanding of the physics of interplanetary space, Biermann and his collaborator initiated in the early 1960s a systematic study of the observations of past comets. D. Antrack, Ludwig Biermann, and Reimar Lüst: Some Statistical Properties of Comets with Plasma Tails. *Annual Review of Astronomy and Astrophysics* 2/1 (1964), 327–340. doi:10.1146/annurev.aa.02.090164.001551. At the end of the 1960s, Biermann also estimated that a comet would have to be surrounded by an extensive cloud of hydrogen, sufficiently intense in the ultraviolet Lyman-line to be able to observe it from satellites outside the Earth's atmosphere. In fact, for the first time, a hydrogen cloud with an extension of more than 30 million kilometers, that is about fifty times the comet atmosphere visible from the ground, was observed in two bright comets in spring 1970 by a French and an American group. The Institute for Astrophysics was involved in the evaluation of these observations that made it possible to construct a much more accurate model of the constitution of a comet. Rhea Lüst, and Rudolf Kippenhahn: Max-Planck-Institut für Physik und Astrophysik. Institut für Astrophysik München. *Max-Planck-Gesellschaft. Berichte und Mitteilungen* 1/77 (1977), 1–64, 28.

194 On May 15, 1963, the Department for Extraterrestrial Physics was transformed into an Institute for Extraterrestrial Physics within the Institute for Physics and Astrophysics in Munich, with Reimar Lüst as its director (CPTS Minutes of 04.12.1963, II. Abt., Rep. 62, No. 2023). Interestingly, nearly in parallel with Biermann and Lust's initiative, in 1962 John Simpson and Peter Meyer (with whom Lüst had worked in Chicago during the 1950s) decided to establish a Laboratory for Astrophysics and Space Research within the Enrico Fermi Institute for Nuclear Studies at the University of Chicago, which was the

Capturing Space Research for the Max Planck Society and the 'Atomic' Ministry

The Institute for Extraterrestrial Physics was the first tangible result of the Max Planck Society's attempts to obtain a leading role in research related to outer space. As had been the case just a few years earlier with the Institute for Plasma Physics, when fusion research for 'peaceful purposes' took off, the MPE had its origins in a working group which was meant to coordinate activities in this research field, nationwide. But contrary to the case of the IPP, it was not clear at the beginning who were to be the main conversation partners. The IPP had clearly been the purview of the 'Atomic' Ministry, while, as was described earlier, it amalgamated, on the scientific side, a majority of researchers in Biermann's tradition with a select minority from other backgrounds, including Wolfgang Gentner's allies, so crystallizing factional rivalries into the early stages of the IPP (see Chapter 1). We can presume that, having learned from this experience in the extraterrestrial research realm, the most prominent figures in the Max Planck Society now consciously put aside their animosities, in order to emphasize the need for a strong role for the Society and a subservient one for the federal government. Space research, unlike experimental plasma physics, was to be conducted *within* Max Planck Institutes, led by scientists who maintained an emphasis on *fundamental* research. Max Planck heavyweights accordingly discussed which federal ministry should preferably fund space-based research by the Max Planck Society—the Ministry of the Interior, of Defense, or of what was then Atomic Energy—in the light of the scientists' past experience of these federal bodies and the degree of research freedom likely to be allowed them. Based on Wolfgang Gentner's experience of autonomy and freedom there, the 'Atomic' Ministry was first choice.[195] There was

first government-sponsored center of its kind in the US. The direct benefit of this to its future space missions induced NASA to fund a dedicated building, which was completed in 1964. Eugene N. Parker: John Alexander Simpson. November 3, 1916–August 31, 2000. *Biographical Memoirs. Volume 81*. Washington, DC: The National Academies Press 2002, 318–339. Eugene N. Parker: John Alexander Simpson. *Physics Today* 53/12 (2000), 83–84. doi:10.1063/1.4808481. In the exact same period, on February 15, 1965, the new building of the Institute for Extraterrestrial Physics was inaugurated. After welcome speeches by Heisenberg, Biermann, and Lüst, Pierre Auger, at the time Director of the European Space Research Organization, gave a long talk on radiations in space. Präsidialbüro der Max-Planck-Gesellschaft (ed.): *Mitteilungen aus der Max-Planck-Gesellschaft zur Förderung der Wissenschaften. Heft 4/1965*. München 1965, 181–205.

195 CPTS meeting minutes of 06.06.1961 AMPG, II. Abt., Rep. 62, No. 1737. The discussions in this meeting regarding which ministries to collaborate with showcase the longstanding priority at the MPG to defend its scientific autonomy, which was codified as being able to pursue research that led to publications in the open literature.

synergy too, at the federal level, as the 'Atomic' Ministry was already using the 'Sputnik shock' to catapult itself beyond the field of nuclear energy and into a wider nationwide role in every form of scientific and technological research deemed critical for the future of the country.[196]

But in any case, this space initiative, like the Institute for Plasma Physics a few years earlier, was dominated by Munich interests (both at the Institute for Physics and Astrophysics, and the federal ministry), and in fact the new institute was built in Garching, directly next to the Plasma Physics Institute. Both institutes ended up being built within only a few years of one another, and the Institute for Extraterrestrial Physics appropriated part of the scientific prestige and personnel that had been originally destined for experimental fusion research, starting with Reimar Lüst himself.

Southwestern Cosmochemistry in the Space Age

Gentner's Institute for Nuclear Physics in Heidelberg also benefited from the space age, albeit in a more indirect way. The first contrast is evident in how, instead of the model of founding new sub-institutes, the Institute for Nuclear Physics grew internally while maintaining its unity as an institute, remaining the largest single Max Planck Institute to be involved in astrophysics.[197] This allowed for greater fluidity between separate scientific fields, further facilitated by its eminently experimental tradition, whose stronghold was the large central technical workshops shared by all research units. Within the institute, the cosmochemistry tradition was in the minority in contrast to accelerator-based programs dedicated to the study of nuclear structure and related theoretical investigations. Informal accounts speak of a third of the institute being cosmochemistry. It was easy to connect cosmochemistry to the space age, in this case through acquired expertise in meteorites, the analysis of very small substance samples via mass spectrometry, and thus, participation in the global scientific community dealing with questions such as the formation of the solar system.[198] As was described earlier in Chapter 1, this tradition in cosmochem-

196 About the international perspective which was opening up for MPG, see Lüst's report at the Scientific Council meeting of June 6, 1961 on "Internationale Zusammenarbeit auf dem Gebiet der Weltraumforschung und Beteiligung der Max-Planck-Gesellschaft an dieser." CPTS meeting minutes of 06.06.1961, AMPG, II. Abt., Rep. 62, No. 1737.

197 See the Financial Appendix at the end of this book.

198 Heinrich Völk: interview by Luisa Bonolis and Juan-Andres Leon, Heidelberg, October 9–10, 2017. DA GMPG, BC 601037. Herbert Palme: Heinrich Wänke und die Erforschung des Mondes und der terrestrischen Planeten. In: Horst Kant, and Carsten Reinhard (eds.): *100 Jahre Kaiser-Wilhelm-/ Max-Planck-Institut für Chemie (Otto-Hahn-Institut). Facetten seiner Geschichte.* Berlin: Archiv der Max-Planck-Gesellschaft 2012, 203–239.

istry stood in close cooperation with the Max Planck Institute for Chemistry in Mainz, as well as other universities in the vicinity, like Heidelberg, Freiburg, and Bern.

Most interestingly, meteorite research, which had started out as empirical and descriptive in the early 1950s (determining the chemical composition of meteorites), had gradually acquired a much deeper astrophysical and cosmological layer of interpretation. The first significant step in this direction had started already with Paneth, during his exile in Durham, and continued in Mainz: he determined that the content of noble gases inside meteorites was related to how much they had been exposed to cosmic rays during their lifetime in outer space. While the helium measurements were done by Paneth in the mid-1940s, it was Carl A. Bauer who realized, in 1947, that the amount of helium found could only be produced by cosmic rays, and not just by radioactive decay.[199] In subsequent years Paneth expanded his method, now including mass spectrometry so as to ascertain the composition of the Helium-3 isotope, which was evidence of this cosmic ray bombardment; and he thus jump-started a more astrophysical branch of cosmochemistry.[200] The discovery of cosmogenic products in iron meteorites brought into the field nuclear physicists and cosmic ray physicists, who came to realize that meteorites contain not only a wealth of information concerning cosmic radiations and conditions in a recent past, but also information reaching back millions and even billions of years.[201] Gentner himself had moved in the early 1950s from nuclear-physics-oriented research to the problem of dating rocks through mass spectrometry, thereby developing new methods which he began using in 1955, in

199 David E. Fisher: *Much Ado About (Practically) Nothing. A History of the Noble Gases*. New York, NY: Oxford University Press 2010.

200 Friedrich A. Paneth, P. Reasbeck, and K.I. Mayne: Helium 3 Content and Age of Meteorites. *Geochimica et Cosmochimica Acta* 2/5 (1952), 300–303. doi:10.1016/0016-7037(52)90013-6. Friedrich A. Paneth, P. Reasbeck, and K.I. Mayne: Production by Cosmic Rays of Helium-3 in Meteorites. *Nature* 172/4370 (1953), 200–201. doi:10.1038/172200a0.

201 Heinrich Wänke, and Heinrich Hintenberger: Notizen. Helium und Neon als Reaktionsprodukte der Höhenstrahlung in Eisenmeteoriten. *Zeitschrift für Naturforschung A* 13/10 (1958), 895–897. doi:10.1515/zna-1958-1017. H. Wänke: Methoden und Probleme der kosmochemischen Forschung. In: Generalverwaltung der Max-Planck-Gesellschaft (ed.): *Jahrbuch der Max-Planck-Gesellschaft zur Förderung der Wissenschaften e.V. 1966*. Göttingen 1966, 168–197. H. Wänke: Meteoritenalter und verwandte Probleme der Kosmochemie. In: E. Heilbronner et al. (eds.): *Kosmochemie. Fortschritte der Chemischen Forschung*. Berlin: Springer 1966, 322–408. Friedrich Begemann: Noble Gases and Meteorites. *Meteoritics & Planetary Science* 31/2 (1996), 171–176. doi:10.1111/j.1945-5100.1996.tb02012.x. Friedrich Begemann: Edelgase in Meteoriten. In: Generalverwaltung der Max-Planck-Gesellschaft (ed.): *Jahrbuch der Max-Planck-Gesellschaft zur Förderung der Wissenschaften e.V. 1972*. Göttingen 1972, 59–82.

collaboration with Zähringer, to investigate the argon and helium content of iron meteorites originating from nuclear interactions of high-energy cosmic ray particles.[202]

In this way, as the space age began properly after Sputnik, it was immediately recognized that some of the questions that were to be researched in outer space could already be answered using meteorites or, as they came to be known, the 'poor man's space probe.'[203] Gentner's interests in archeochemistry and cosmochemistry, developed during the 1950s, launched investigation of solar system problems and of the formation of chemical elements in the universe. The old cosmic ray tradition initiated by Bothe was now moving toward a new basis in outer space itself.

For example, the composition of meteorites was the basis of a longstanding research tradition in theories of the origins and formation of the solar system, initially in Heidelberg and Mainz in the 1950s, but later also in Lindau (Aeronomy) and Garching (MPE), in the 1970s.[204] A close collaboration in cosmochemistry had been established with Brookhaven in the mid-1950s, when Gentner had sent his best disciple from Freiburg, Josef Zähringer, to be based there for several years before returning to Heidelberg as a Scientific Member. While in Brookhaven, Zähringer collaborated closely with Oliver Schaeffer, who would later become an External Scientific Member of the Nuclear Physics Institute.[205] Based on this early work with Brookhaven since the 1950s, German cosmochemists then became the Principal Investigators on

---

202 Wolfgang Gentner, and Josef Zähringer: Argon- und Heliumbestimmungen in Eisenmeteoriten. *Zeitschrift für Naturforschung A* 10/6 (1955), 498–499. doi:10.1515/zna-1955-0610. Wolfgang Gentner, and Josef Zähringer: Argon und Helium als Kernreaktionsprodukte in Meteoriten. *Geochimica et Cosmochimica Acta* 11/1–2 (1957), 60–71. See also Josef Zähringer: Isotope Chronology of Meteorites. *Annual Review of Astronomy and Astrophysics* 2 (1964), 121–148. doi:10.1146/annurev.aa.02.090164.001005. See also interviews with Begemann and Wänke in Ursula B. Marvin: Oral Histories in Meteoritics and Planetary Science. VIII. Friedrich Begemann. *Meteoritics & Planetary Science* 37/S12 (2002), B69–B77. doi:10.1111/j.1945-5100.2002.tb00905.x. Ursula B. Marvin: Oral Histories in Meteoritics and Planetary Science. IX. Heinrich Wänke. *Meteoritics & Planetary Science* 37 (2002), B79–B88. doi:10.1111/j.1945-5100.2002.tb00906.x.

203 Marvin, Oral Histories in Meteoritics, 2002, B69–B77, B71.

204 S. Pfalzner et al.: The Formation of the Solar System. *Physica Scripta* 90/6 (2015), 068001–068019. doi:10.1088/0031-8949/90/6/068001.

205 Oliver Schaeffer, and Josef Zähringer: Helium- und Argon-Erzeugung in Eisentargets durch energiereiche Protonen. *Zeitschrift für Naturforschung A* 13/4 (1958), 346–347. doi:10.1515/zna-1958-0413. Oliver Schaeffer, and Josef Zähringer: High-Sensitivity Mass Spectrometric Measurement of Stable Helium and Argon Isotopes Produced by High-Energy Protons in Iron. *Physical Review* 113/2 (1959), 674–678. doi:10.1103/PhysRev.113.674.

the scientific program to analyze mineral samples recovered from the Apollo manned missions to the moon.[206] Two decades of expertise in the traditions of Biermann's space plasma astrophysics and Gentner's cosmochemistry came together beautifully in 1969, when the lunar samples from Apollo 11, after their analysis with mass spectrometric methods, showed that the top surface layers of the moon were full of slowly deposited solar wind, which is not shielded by the weak lunar magnetic field and by the atmosphere, as in the case of our planet, and could thus provide information on the primary abundances of noble gases and their isotopes at the moment of the formation of these elements. The composition of cosmic rays only in part deriving from the Sun and penetrating the first thin strata of lunar rocks could be analyzed through the nuclear reactions they produced at the time of their impact.[207] Researchers from Mainz also participated independently in the analysis of American lunar samples.[208]

Most striking for physicists worldwide was how, in Heidelberg specifically, cosmochemistry could be linked to the deepest questions arising in particle physics, astrophysics, and cosmology. In addition to being a 'poor man's space probe,' extraterrestrial samples could be a poor man's particle physics experiment. In 1968, Heidelberg physicist Till Kirsten, then in Brookhaven, demonstrated the existence of a rare fundamental process, double-beta decay, which had been posited theoretically several decades earlier.[209] This was yet another fruitful result of the close relationship between Brookhaven and Heidelberg,

206 Josef Zähringer: Rätselhafte Mondproben. In: Generalverwaltung der Max-Planck-Gesellschaft (ed.): *Jahrbuch der Max-Planck-Gesellschaft zur Förderung der Wissenschaften e.V. 1970*. Göttingen 1970, 169–199.

207 Till Kirsten: interview by Luisa Bonolis and Juan-Andres Leon, Heidelberg, October 24–25, 2017. DA GMPG, BC 601051.

208 See, for example, Heinrich Wänke: 100 Gramm Mond nach Mainz. Interview mit Professor Heinrich Wänke vom Max-Planck-Institut für Chemie in Mainz. *Der Spiegel* 29 (1969), 105. http://www.spiegel.de/spiegel/print/d-45549231.html. Last accessed 3/1/2018.

209 Till A. Kirsten et al.: Experimental Evidence for the Double-Beta Decay of Te^{130}. *Physical Review Letters* 20/23 (1968), 1300–1303. doi:10.1103/PhysRevLett.20.1300. After a few years, Oliver A. Schaeffer, a radiochemist from Brookhaven National Laboratory, was named external scientific member of MPIK (CPTS meeting minutes of 22.04.1972, 20.06.1972, AMPG, II. Abt., Rep. 62, No. 1765, 1766). Schaeffer was an expert in radioactive spallation products resulting from the impact of high-energy particles on heavy nuclei, and Zähringer himself knew him since his postdoctoral stay in Brookhaven during the mid-1950s, which had been facilitated by Gentner's personal relationships from before the war. Zähringer on the other hand introduced there the technique of measuring stable isotopes of rare gases, so complementing Schaeffer's expertise. From the synergy of these two cultures, decay products could be counted not only for cosmochemical research but even for pure nuclear physics. These dating techniques used within Gentner's group have been

one that subsequently extended into neutrino research, a field that remained in obscurity elsewhere for one more decade, while the Brookhaven scientist Ray Davis conducted his now famous underground detection experiments that led to the so-called solar neutrino paradox. The observed discrepancy between the quantity of neutrinos received on Earth from the Sun as predicted by theoretical solar models and direct observations—the problem of missing neutrinos—became one of the most significant puzzles in astrophysics of the second half of the century.[210] As we will see later in this book (Chapter 5), the early involvement of Max Planck scientists from Heidelberg would lead directly to one of the first German-led international collaborations, created to solve this paradox.

Finally, and quite separately from the Munich initiatives, research institutes based in southwestern Germany were in charge of what would become Germany's second and third scientific satellites, the Aeros, which were dedicated to the study of the outer layers of the atmosphere, including the thermosphere and ionosphere. This project, which was delivered by American launchers, and marked the transition of Freiburg-based Karl Rawer (see Chapter 1) from ground-based ionosphere research to space missions, benefited from that close relationship of universities and other research institutes (Fraunhofer, Max Planck) so characteristic of the former French occupation zone, while also mobilizing industrial interests in Baden-Württemberg.[211] Peter Lämmerzahl and his team at the Max Planck Institute for Nuclear Physics developed the instrumentation with which the satellite would chemically analyze the upper atmosphere, setting the stage for a longstanding tradition in space-based mass spectrometry through which the Heidelberg and Mainz

---

described in detail in Till A. Kirsten: Gentner und die Kosmochemie. Hobby oder Symbiose? In: Dieter Hoffmann, and Ulrich Schmidt-Rohr (eds.): *Wolfgang Gentner. Festschrift zum 100. Geburtstag*. Berlin: Springer 2006, 177–208. See also Till A. Kirsten: Oliver Adam Schaeffer. 20.02.1919-11.11.1981. *Berichte und Mitteilungen Max-Planck-Gesellschaft. Jahresbericht 1981 und Jahresrechnung 1980, Nachrufe*. München 1982, 33–36. In Chapter 5, the Brookhaven–Heidelberg collaboration will be outlined in more detail as a premise leading to the GALLEX experiment for the detection of solar neutrinos.

210 See John N. Bahcall: Neutrinos from the Sun. *Scientific American* 221/1 (1969), 28–37. John N. Bahcall, and R. L. Sears: Solar Neutrinos. *Annual Review of Astronomy and Astrophysics* 10 (1972), 25–44. doi:10.1146/annurev.aa.10.090172.000325. John N. Bahcall, and Raymond Jr. Davis: Solar Neutrinos. A Scientific Puzzle. *Science* 191/4224 (1976), 264–267. doi:10.1126/science.191.4224.264.

211 See the technical Memorandum of October 1980 ("Ergebnisse des Aeros-Satellitenprogramms" summarizing data collected on the two Aeros missions edited by P. Lammerzahl, K. Rawer and N. Roemer: https://ntrs.nasa.gov/citations/19810012576. Last accessed 2/3/2022.

institutes participated in the coming decades in space missions such as the Pioneer probes to Venus and the Giotto deep space mission to Comet Halley, led by Ulf von Zahn,[212] and established an early foothold in atmospheric research by applying these techniques back on Earth; an expertise which would facilitate these institutes' early foothold in what is now called Earth system research, including crucial observations by Konrad Mauersberger confirming that depletion of the ozone layer was indeed occurring, because of the steady increase in the atmosphere of chemical compounds inducing dramatic ozone losses.[213]

In short, the early adaptation of the Max Planck Institutes of southwest Germany to the space age was achieved through them mobilizing their cosmochemistry tradition, in terms both of experimental expertise and a growing leadership in the formulation and interpretation of novel experimental questions with implications that combined subatomic particle physics and astrophysics. Then, over the next decades, these initial cosmochemical interests and areas of expertise secured footholds in various new fields, including fundamental particles and astroparticles (see Chapter 5), interplanetary missions, and environmental research.

The Max Planck Institute for Aeronomy in Search of a Space Age Identity

Finally, there is the story of how the Aeronomy Institute reacted to Sputnik. This institute had only been formally established a few years earlier, in the aftermath of Erich Regener's death and the consolidation of his group, now led by Julius Bartels, with Dieminger's ionosphere group in Lindau on the Harz mountains (Chapter 1). Based on the brilliant legacy of Erich Regener, this could have been the institute to benefit the most from the reorientation of social and political interests toward outer space.[214] In fact, even a few years before Sputnik, the institute had experienced an upsurge through its participation in the preparations for the International Geophysical Year, from which

212 See, for example, U. von Zahn et al.: The Upper Atmosphere of Venus during Morning Conditions. *Journal of Geophysical Research* 85/A13 (1980), 7829–7840. doi:10.1029/JA085iA13p07829. G. Schubert et al.: Structure and Circulation of the Venus Atmosphere. *Journal of Geophysical Research* 85/A13 (1980), 8007–8025. doi:10.1029/JA085iA13p08007.

213 Konrad Mauersberger: Measurement of Heavy Ozone in the Stratosphere. *Geophysical Research Letters* 8/8 (1981), 935–937. doi:10.1029/GL008i008p00935.

214 Walter Dieminger: Die Ionosphäre als Grenzschicht zwischen Erdatmosphäre und extraterrestrischem Raum. In: Generalverwaltung der Max-Planck-Gesellschaft der Max-Planck-Gesellschaft zur Förderung der Wissenschaft e.V. (ed.): *Jahrbuch 1954 der Max-Planck-Gesellschaft zur Förderung der Wissenschaften e.V.* Göttingen 1955, 42–80.

both Bartels' and Dieminger's sub-institutes profited, the latter even establishing a new observational station in the southern hemisphere, in the Tsumeb region of the former colony of German Southwest Africa (now Namibia), then under the control of South Africa.[215] This was also the German counterpart to the network of stations built by the French in their African colonies.[216]

Julius Bartels, mostly due to his prestige in magnetospheric research, gained earlier in Göttingen, (the cradle of German geophysics since the turn of the century), was initially as prominent in German and international commissions dealing with outer space issues as Reimar Lüst was to become.[217] However, at the same time, he was relatively slow to reorient his institute's research (precisely because of its involvement with the International Geophysical Year), and therefore failed to quickly take advantage of outer space. Furthermore, Bartels represented a research tradition in geophysics that dated from the prewar era, as was evident from his close collaboration with the British scientist Sidney Chapman, noted for his research in geophysics and one of the pioneers of solar-terrestrial physics.[218] This tradition increasingly antagonized the newer, plasma-physics framework of Alfvén and Biermann, who were much more closely aligned with the 'nuclear age.'[219]

---

215 Peter Czechowsky, and Rüdiger Rüster (eds.): *60 Jahre Forschung in Lindau. 1946–2006. Vom Fraunhofer-Institut zum Max-Planck-Institut für Sonnensystemforschung. Eine Sammlung von Erinnerungen*. Katlenburg-Lindau: Copernicus 2007. Julius Bartels, Walter Dieminger, and Alfred Ehmert: Max-Planck-Institut für Aeronomie in Lindau. In: Generalverwaltung der Max-Planck-Gesellschaft zur Förderung der Wissenschaften e.V. (ed.): *Jahrbuch 1961 der Max-Planck-Gesellschaft zur Förderung der Wissenschaften e.V. Teil II*. Göttingen 1962, 16–45.

216 Rawer, *Meine Kinder umkreisen die Erde*, 1986, 92–98.

217 See, for example, Julius Bartels: Weltraumforschung. Methoden und Ergebnisse. In: Generalverwaltung der Max-Planck Gesellschaft zur Förderung der Wissenschaften e.V. (ed.): *Jahrbuch 1962 der Max-Planck-Gesellschaft zur Förderung der Wissenschaften e.V.* Göttingen 1962, 19–50.

218 Chapman actually wrote Bartel's obituary when the latter died in 1964. Sydney Chapman: Julius Bartels. *Quarterly Journal of the Royal Astronomical Society* 6 (1965), 235–245. https://ui.adsabs.harvard.edu/abs/1965QJRAS...6..235. Last accessed 5/3/2020.

219 In 1939 Alfvén was the first to devise the technique that enables the complex spiral movement of a charged particle in a magnetic field to be calculated with relative ease. In considering such complex motion, Alfvén introduced the simplifying approximation of circular rotation about a 'guiding center' which was itself drifting along magnetic lines. He applied this principle to the study of magnetic storms and auroras, finding that particles in the Earth's magnetic field should move back and forth along the field lines, reflected from regions of increasing field strength. The concept of a magnetic mirror became important in work on controlled thermonuclear fusion requiring the confinement of hot plasmas whose contact would destroy the walls of any container.

As we will see next, a competition of interests soon emerged in the early 1960s between the Aeronomy Institute in Lindau and the Institute for Extraterrestrial Physics in Garching; a competition in which the stronger political, economic, and scientific forces were on the Bavarian side. For the Aeronomy Institute, the best asset in this competition would have been Bartels himself, but he unexpectedly died in 1964, triggering a decade of uncertainty at his institute.[220] His immediate but temporary successors were Regener's collaborators Pfotzer and Ehmert, both experts on matters related to cosmic rays and solar particles, and pivotal to drawing Biermann's attention to some of these issues in earlier years. And the person in Lindau who most prominently took charge of outer space questions was the young Erhard Keppler, who had trained in Weissenau with Regener in the postwar years.[221] Keppler, while never officially designated a Scientific Member of the Max Planck Society, was to become the counterweight to, and sometimes antagonist[222] of Reimar Lüst in matters of space science in the Max Planck Society during the 1960s and '70s. Once the West German federal government had established the funding mechanisms to support space research directly through the Ministry of Research and Technology, Keppler and his collaborators managed to effectively compete with the Extraterrestrial Institute in gaining the leading role in some of the space projects, including the first German attempt at a national satellite, called Azur, the main purpose of which was study of the cosmic ray particles trapped in the Van Allen Belts. This rather rudimentary satellite, to

---

These ideas were later useful also in interpreting such phenomena as the Van Allen radiation belt currents of electrons circulating in the Earth's magnetic field during magnetic storms. For an account of Alfvén's scientific achievements see R. S. Pease, and S. Lindqvist: Hannes Olof Gösta Alfvén. 30 May 1908–2 April 1995. *Biographical Memoirs of Fellows of the Royal Society* 44 (1998), 3–19. doi:10.1098/rsbm.1998.0001. Carl-Gunne Fälthammar, and Alexander J. Dessler: Hannes Alfvén. 30 May 1908–2 April 1995. *Proceedings of the American Philosophical Society* 150/4 (2006), 649–662. https://www.jstor.org/stable/i412827. Last accessed 3/27/2018. See also Luisa Bonolis: Hannes Olof Gösta Alfvén. *Lindau Nobel Mediatheque.* http://www.mediatheque.lindau-nobel.org/research-profile/laureate-alfveen#page=all. Last accessed 2/28/2018. The intellectual relationship between Biermann and Alfvén has also been described in Chapter 1.

220 S. K. Runcorn: Prof. J. Bartels. Obituary. *Nature* 203/4947 (1964), 814–815. doi:10.1038/203814a0. Karl-Heinz Glaßmeier, Manfred Siebert, and Emilio Herrero-Bervera: Bartels, Julius (1899–1964). Edited by David Gubbins. *Encyclopedia of Geomagnetism and Paleomagnetism.* Dordrecht: Springer 2007, 42–42. https://doi.org/10.1007/978-1-4020-4423-6_15. Last accessed 11/3/2018. A. Ehmert: In memoriam Julius Bartels. *Space Science Reviews* 3/1 (1964), 2–4. doi:10.1007/BF00226642.

221 The best source on Keppler is his autobiographical work Keppler, *Max Planck Institut für Aeronomie*, 2003.

222 See, for example, the Human Spaceflight debates described in Chapter 5.

be carried atop an American rocket, turned out to be a rather disappointing project, full of overcosts and delays: as in many early attempts at fully national space missions, an entire industrial sector had to be created for the completion of the project, and this was heavily subsidized but not sufficiently coordinated by the federal government. And throughout the completion of the satellite Azur, which involved not just the Institutes for Aeronomy and Extraterrestrial Physics, but also several universities, the rivalry of the two Max Planck Institutes became a salient issue. Lüst's institute benefited indirectly from NOT being in the major coordinating role in the making of this first satellite,[223] maintaining throughout the time of its construction, in parallel, a very strong research profile based on sounding rockets, while slowly acquiring the expertise to make its subsequent move toward space astronomy, which is the focus of Chapter 3.

In the mid-1960s, the Institute for Aeronomy attempted to create, in Munich-fashion, a third sub-institute called 'Space Physics,' to take direct advantage of the funds from the Research Ministry. But this move was not permitted by the Max Planck Society, which instead enforced a closer integration of all the different parts of the Aeronomy Institute into a single entity,[224] while at the same time it was a pending task for the coming years to find a director fit to run the Institute upon the imminent retirement of both Dieminger and Ehmert. The dominant logic in the search for a director was that this should be a person with a distinguished scientific career independent of the Institute, a requirement that disqualified Ehmert, Pfotzer, and especially its most promising figure, Erhard Keppler.

Then, in the late 1960s, conditions at the Aeronomy Institute further deteriorated, caused by the lack of leadership during the wave of reforms demanding

---

223 Reimar Lüst: interview by Helmut Trischler, September 8, 2010. Historical Archives of the European Union. Oral History of Europe in Space Collection, https://archives.eui.eu/en/oral_history/INT070. Last accessed 7/30/2019. A detailed account of the AZUR debacle is given in Johannes Weyer: *Akteurstrategien und strukturelle Eigendynamiken. Raumfahrt in Westdeutschland 1945–1965*. Göttingen: Schwartz 1993, 280–314. The causes of this failure were largely outside the responsibility of the project scientists, and instead the result of a lack of experience in space projects in Germany and the haphazard attempts at industrial coordination between the German federal government and the nascent industries in the space sector. Keppler himself admitted the difficulties with this first satellite project. Keppler and the Aeronomy Institute later proved their competence in space projects with many space probes starting with Helios. See Keppler, *Max Planck Institut für Aeronomie*, 2003. Chapters 3 and 4 are largely dedicated to Azur and Helios.

224 For discussions on Bartel's succession (appointment of Ehmert and Pfotzer as directors of the two sub-institutes) and the future of the Aeronomy Institute in Lindau see CPTS meeting minutes of 05.03.1965 and 22.06.1965, II. Abt., Rep. 62, No. 1745, 1746.

more employee participation (*Mitbestimmung*), which, according to senior Max Planck scientists, was laying waste to universities and scientific research institutes throughout the country.[225] The publication output of the institute practically ground to a halt in the early 1970s, and this increasingly came to the attention of the Max Planck Society and, especially, of the rival institutes. Keppler's team continued to yield very good work in the early 1970s, working toward the first German interplanetary probes, Helios I and Helios II, in which the Institutes for Nuclear Physics, Chemistry, Extraterrestrial Physics, and Astronomy participated, too, along with several universities and foreign partners. Through these, Germany became the third country in the world to send a probe beyond the Earth's orbit, albeit on an American rocket.[226] But still, until the major reform moment of 1973–75, which is discussed at the end of Chapter 3, the Institute for Aeronomy was the Max Planck Society's 'problem child' (*Sorgenkind*), menaced by the death of Bartels (in 1964), and of his

225 Institutional paralysis during the wave of activism in the late 1960s, toward greater democratic participation by staff and junior scientists, is a ubiquitous theme in many Max Planck Institutes and is brought up as a significant cause of disarray in several of them. A recurring theme in these narratives is how institutes with weak leadership or in transition were struck harder by these movements, and, for example, Heisenberg's own institute is described by Reimar Lüst as having been paralyzed by them, which situation was only resolved by measures taken by a newly appointed director, Léon Van Hove. At the Institute for Aeronomy, Keppler lamented that the acting directors let the matter get out of hand, and during the early 1970s, much time was lost on what he considered pointless political discussions, while critical indicators of productivity, such as scientific publications, plummeted. These indicators would later be mobilized by the Max Planck Society during enactment of the major reforms discussed later in this book. Reimar Lüst: Léon Van Hove and the Max-Planck-Institute for Physics. *Scientific Highlights in Memory of Léon Van Hove*. World Scientific 1993, 51–59. Keppler, *Max Planck Institut für Aeronomie*, 2003, 27. Future GMPG publications will deal with the issue of labor participation (*Mitbestimmung*) in richer detail and with a deeper historical perspective. See also, Juliane Scholz: *Partizipation und Mitbestimmung in der Forschung. Das Beispiel Max-Planck-Gesellschaft (1945–1980)*. Berlin: GMPG-Preprint 2019.

226 Herbert Porsche: HELIOS-Mission. Mission Objectives, Mission Verification, Selected Results. In: Burke, W.~R. (ed.): *The Solar System and Its Exploration*. ESA 1981, 43–50. The space probes Helios I and Helios II, an ambitious plan to dispatch space probes on a highly eccentric path round the Sun and penetrating into the orbit of Mercury, were constructed in duplicate by a consortium of firms headed by the West German aerospace manufacturer Messerschmitt-Bölkow-Blohm, and were sent into their orbits in the years 1974 and 1976 by American carrier rockets. Seven of the 11 scientific experiments carried on board were developed and controlled in German institutes and thus the Federal Republic of Germany thereby first established for itself the organizational and technological basis necessary for outstanding performance in space research. See documents related to the project in Lüst's papers (AMPG, III. Abt., Rep. 145, No. 1220, 1235).

temporary successor Ehmert (in 1971). In addition, there was the impending retirement of Pfotzer and Dieminger. But ultimately, the major problem in Lindau was the convergence of its research agendas with those of the much more powerful Institute for Extraterrestrial Physics, as described next.

Internal Rivalries

The key aspect of the early space age years at the Max Planck Society, roughly from 1957 to the late 1960s, was scientists' ability to quickly turn the opportunities of space research to their own advantage, by mobilizing any existing expertise that had direct or potential application in the new field. This was clearly the case with space plasmas in Munich, and also with cosmochemistry in Heidelberg and Mainz. The part of the Institute for Aeronomy that had originated in Regener's institute in Weissenau also jumped on the space bandwagon, with an expertise that was mainly instrumental (balloon-based instruments), and, too, what might be described as a 'territorial' claim to high-altitude research.[227] This last institute used such claims to leverage a role in the West German national space programs, but without any remotely comparable relationship to foreign agencies and organizations, unlike the plasma astrophysicists and the cosmochemists. Given this lack of a distinct scientific profile, they were set to enter into direct competition with the Institute for Extraterrestrial Physics in Garching—and were clearly at a disadvantage from the start.

The competition for the same scientific field and resources within institutes of the Max Planck Society started to become a problem in cosmic research in the mid-1960s, with the case of Aeronomy vs. Extraterrestrials. Before then, the founding scientific traditions were so distant that they naturally progressed toward very different kinds of projects and viewpoints, and connected with different global networks of expertise and validation. In the cases where several Max Planck Institutes worked on very similar topics, this was precisely a sign of their foundational closeness, constituting clearly defined 'families' of institutes and research fields, as in the case of Mainz and Heidelberg in the southwest (which collaborated closely in cosmochemistry), or of the growing array of sub-institutes created by 'cell division' out of Heisenberg's institute in Munich (working together in plasma-related fields). If anything, the rivalry

227 As we will see in the next section, these territorial claims continued to be mobilized by Max Planck Institutes, for example, with the division of responsibilities of institutes along wavelengths, or in the case of the Institute for Extraterrestrial Physics, the claim to be the site for space-based science in the Society. These territorial claims, however, especially at the boundaries, could also be challenged along disciplinary or instrumental lines.

that had emerged in the early decades between Heidelberg and Munich was caused precisely by their differing scientific traditions, leading to some degree of incommensurability in the appreciation of each other's scientific work. But this incommensurability had the advantage, in space research, of leading to widely different research programs.

Confrontation in Experimental Particle Physics, Collaboration in Space

There was, however, a precedent for how convergence of competing institutes in one scientific field could be disastrous, and it arose from the confrontations between Heidelberg and Munich not in cosmic research, but in nuclear and particle physics. As we saw in Chapter 1, in this field both institutes competed for the same federal resources, claiming the role of the foremost 'nuclear' research communities in West Germany. The disastrous competition included attempts by Heisenberg to block the creation of a new Institute for Nuclear Physics in Heidelberg in the 1950s, and also recurrent moves in the 1960s to block the influence and growth of CERN, or even to force it to be based in Munich.[228] Moreover, as a result of the growing personal hostility between Werner Heisenberg and the experimental physics community throughout the 1960s, even his own institute's particle research teams at DESY and at ESRO distanced themselves from him; and by the time Heisenberg retired at the end of the decade, they were only nominally part of the Munich institute.[229]

228 These episodes are well described in Cathryn Carson: Heisenberg als Wissenschaftsorganisator. In: Christian Kleint, Helmut Rechenberg, and Gerald Wiemers (eds.): *Werner Heisenberg, 1901–1976. Beiträge, Berichte, Briefe. Festschrift zu seinem 100. Geburtstag.* Leipzig: Verlag der Sächsischen Akademie der Wissenschaften zu Leipzig 2005, 214–222. And in Cathryn Carson: *Heisenberg in the Atomic Age. Science and the Public Sphere.* Cambridge: Cambridge University Press 2010. See also, Cathryn Carson: Beyond Reconstruction. CERN's Second-Generation Accelerator Program as an Indicator of Shifts in West-German Science. In: Helmuth Trischler, and Mark Walker (eds.): *Physics and Politics. Research and Research Support in Twentieth Century Germany in International Perspective.* Stuttgart: Steiner 2010, 107–130. And Helmuth Trischler, and Dieter Hoffmann: Wolfgang Gentner und die Großforschung im bundesdeutschen und europäischen Raum. In: Dieter Hoffmann, and Ulrich Schmidt-Rohr (eds.): *Wolfgang Gentner. Festschrift zum 100. Geburtstag.* Berlin: Springer 2006, 95–120.

229 Lüst, Léon Van Hove and MPG, 1993, 51–59. Heisenberg's stance on new developments in theoretical physics, most strikingly the move toward a Standard Model of elementary particles based on entities such as quarks, alienated him from newer generations of theoreticians, and reinforced his skepticism toward accelerator projects such as CERN. For some of these theoretical positions of his later years, see Werner Heisenberg: *Encounters with Einstein and Other Essays on People, Places, and Particles.* Princeton, NJ: Princeton

Gentner's faction in turn aimed for growing influence within the Max Planck Society, mobilizing the growing resentment against the hegemonic intentions of Bavaria and the institutes based there.[230] This factionalism led to what was for a while considered a major defeat for the Max Planck Society, namely the lost opportunity to control what would become the foremost German research center for particle physics, DESY. Discussion of the creation of a West German national particle accelerator took place in parallel to the creation of other large research institutes in the late 1950s, including the Institute for Plasma Physics. It was recognized at the time that these could not be an integral part of the Max Planck Society, as their scale was incomparably larger than that of any other Max Planck Institute, creating both internal and external problems regarding their weight in decision-making processes, as well as inviting external oversight and pressure on applied research that was then unwelcome in the Society.[231]

University Press 1989. Heisenberg's critical stance on quarks, and his general dissatisfaction with the current state of experimental particle physics is ubiquitous in all the essays in this publication.

230 One notable episode of this kind led to the establishment of the European Molecular Biology Laboratory in Heidelberg, essentially pulling the rug out from under Munich. This episode is proudly retold by the participants in: Ferry, *EMBO in Perspective*, 2014.

231 In May 1963 (CPTS meeting minutes of 13/14.05.1963, AMPG, II. Abt., Rep. 62, No. 1741), Gentner proposed to appoint as Scientific Member and Director Anselm Citron (1923–2015), who had been his student at Freiburg University immediately after the war, and had first joined the Cavendish Laboratory in Cambridge, taking part in research on accelerator physics. As part of the first staff physicists, Citron had contributed to the construction of CERN's first accelerator, the Synchro Cyclotron, later moving to work on the Proton Synchrotron, a machine with which it became possible to investigate artificially produced 'mesons,' leaving behind the field of nuclear physics and taking the next step forward: particle physics. Gentner's proposal was related to the intention of building within the Max Planck Society the new, more powerful proton accelerator being planned at the time and particularly discussed at ECFA (European Committee for Future Accelerators). The minutes of the meeting held on December 4, 1963 testify the ongoing discussion about having such big machines inside the MPG and even the possibility, being considered at the same time, of integrating within the Society DESY, the *Deutsches Elektronen-Synchrotron* research center in Hamburg. A similar discussion included the Institute for Plasma Physics. This project and Citron's appointment were blocked, and he instead later became Director of the Center for Nuclear Research at Karlsruhe, where he developed technologies which became instrumental for later accelerators. Echoes of this debate can be also found in the CPTS meeting minutes of 01.11.1963 and 09.06.1964, AMPG, II. Abt., Rep. 62, No. 1742, 1743. On these accelerator debates, see also Carson, Beyond Reconstruction, 2010, 107–130. Bernd-A. Rusinek: Europas 300-GeV-Maschine: Der grösste Teilchenbeschleuniger der Welt an einem westfälischen Standort? *Geschichte im Westen* 11/2 (1996), 135–153.

In the case of the IPP, where plasma physics was Munich's uncontested field of expertise, this resulted in a nominally independent institute which was nonetheless largely under the control of not just the Max Planck Society, but also the very specific family of institutes geographically adjacent to it and closely connected to Ludwig Biermann's theoretical astrophysical tradition.[232] Still, as was described in Chapter 1, the IPP epitomized a broad spectrum of national and regional interests, and even encompassed in a single location various scientific traditions that were to fuse over the following decades. Even the 'failure' of the IPP in its first decade, a part of the worldwide disappointment with the promises of nuclear fusion, could be mobilized to the benefit of the Max Planck Society, which in the early 1970s took direct control of the institute and reestablished there an eminently 'fundamental' research direction.

In the case of DESY, the opposite happened, and the organization drifted further and further beyond the reach of the Max Planck Society. Even in the early 1950s, there were already too many competing factions in Germany seeking to host the national accelerator, among which the Max Planck Institutes in Heidelberg and Munich could have been a dominant force, if only they had coordinated their efforts. Both Gentner and Heisenberg would have liked the DESY to acquire the dominance held ultimately by the IPP or, later, the national astronomical observatories; but instead, they ended up mired in several parallel fights for influence within the Max Planck Society as well as in particle physics directly, in the early 1960s, when the fate of DESY was being decided. This weakened the national position of the Society in this field, and also led Willibald Jentschke, the main driving force behind DESY, to take distance from both sides, while managing to keep DESY as another eminently 'fundamental' research center.[233]

232 The director (Arnulf Schlüter) and major scientific figures of the IPP came directly out of Biermann's plasma physics group, and the institute was on the same campus in Garching as would eventually host also the Institutes for Extraterrestrial Physics, Astrophysics, and Quantum Optics. The acceptance of this de facto control by the MPG led to its reabsorption into the Society in 1971. Susan Boenke: *Entstehung und Entwicklung des Max-Planck-Instituts für Plasmaphysik 1955–1971*. Frankfurt am Main: Campus Verlag 1991.

233 Erich Lohrmann, and Paul Söding: *Von schnellen Teilchen und hellem Licht. 50 Jahre Deutsches Elektronen-Synchrotron DESY*. Weinheim: Wiley-VCH Verlag 2009. Thomas Heinze, Olof Hallonsten, and Steffi Heinecke: From Periphery to Center. Synchrotron Radiation at DESY, Part I. 1962–1977. *Historical Studies in the Natural Sciences* 45/3 (2015), 447–492. doi:10.1525/hsns.2015.45.3.447. Thomas Heinze, Olof Hallonsten, and Steffi Heinecke: From Periphery to Center. Synchrotron Radiation at DESY, Part II 1977–1993. *Historical Studies in the Natural Sciences* 45/4 (2015), 513–548. doi:10.1525/hsns.2015.45.4.513. Victor F. Weisskopf, and Willibald Jentschke: *Die Zukunft der Elementarteilchenforschung. Das Deutsche Elektronen-Synchrotron (DESY) Eigenschaften und Forschungsmöglichkeiten*. Vol. 153. Wiesbaden: VS Verlag für Sozialwissenschaften 1965.

In fact, even before space research, particle accelerator research had managed to become one the responsibilities of what was then the 'Atomic' Ministry, taking advantage of the American model of the Nuclear Energy Commission and later Department of Energy; for in the early 1950s, particle physicists, with Gentner at the helm, had convincingly mobilized to make sure particle accelerators remained designated 'nuclear' research, as their funding as physicists would otherwise have depended on the much weaker and centralization-averse DFG.[234] This was part of the pattern evident throughout this book, by which, since the early 1950s, areas of truly national importance were funded by this ministry and the Max Planck Society, while smaller scale research—across the entire spectrum of the sciences (*Wissenschaften*), natural, social, and theological—was the purview of the DFG and the universities.

Then, as was mentioned earlier, in the decade after Sputnik, the Research Ministry, as it was now known, extended its domain to include outer space. In 1964 even, within the lines of battles with the DFG, a distinction based on altitude was established: the Ministry would now be in charge of research occurring above the D layer of the atmosphere, the lowest ionospheric region (at altitudes of about 70 to 90 km), that is, everything above what is considered meteorological phenomena.[235] At these altitudes ionized particles dominate, as well as all matters relevant to nuclear, elementary particles, and plasmas significant for the near-outer space research in which Max Planck Institutes were early world leaders. This practical demarcation by altitude legitimized the aforementioned direct axis between the Federal Ministry of Research and the Max Planck Institutes.

In this original division of responsibilities, ground-based astronomy was still part of the DFG's 'territory,' as it was conducted largely in the state-based universities and traditional state observatories. This would also soon radically change, as will be described in Chapter 3.

Remarkably, all these gains for the Max Planck Society occurred at a moment of particular internal disunity. Winning the race for outer space research was possible thanks to the relative complementarity of its fields of expertise, initially based on plasma physics (Munich), cosmochemistry (Heidelberg, Mainz), and high-altitude probes (Lindau).

234 Michael Eckert, and Maria Osietzki: *Wissenschaft für Macht und Markt. Kernforschung und Mikroelektronik in der Bundesrepublik Deutschland.* München: Beck 1989, 67.

235 Trischler, *Luft-und Raumfahrtforschung in Deutschland 1900–1970*, 1992, 442.

In contrast, in the case of accelerator-based particle physics, as we just saw, Max Planck Institutes ended up on a relatively equal footing with university institutes, and a step below DESY.[236]

For the coming decades, however, the Max Planck Society would face the challenge of coordinating the work of institutes with potentially competing interests. The negative experiences of the 1960s in other areas such as particle physics provided key insights for a coordinated division of labor in the cosmic sciences that would lead to their dominance in these fields.

Precedents such as this raised awareness of the need to better coordinate those areas in which several Max Planck Institutes overlapped in similar scientific fields. The major changes had to await the generational transition that followed Heisenberg's and Gentner's retirement and the presidency of Reimar Lüst. But first, we will look at the major expansion of the Max Planck Society, and the first 'cluster'-like coordination towards expansion, which came about through the absorption of observational astronomy in the 1960s.

236 For a description of the long path to the final juridical form of DESY, see Lohrmann, and Söding, *Teilchen*, 2009. Hohn, and Schimank, *Konflikte*, 1990. See also Eckert, and Osietzki, *Wissenschaft für Macht und Markt*, 1989, 63–73.

CHAPTER 3

Astronomical Revolution in the MPG (1960s–1980s): Completing the Wavelength Spectrum

Until the 1960s, observational astronomy was not considered a field of interest by the Max Planck Society, whose astrophysical pioneers were strongly oriented toward topics intersecting with the nuclear age. In West Germany, astronomy retained an aura of antiquatedness, and was based largely in observatories dating from previous centuries and still the purview of individual federal states. This changed radically after Sputnik, when astronomy underwent a revival around the world. Even before 1957, an astronomical revolution had been spearheaded by radio astronomy. This was the case also in Germany, where radar pioneers had built the first radio telescopes and forged an international reputation during the first postwar decade. The Max Planck Society, in its moment of most radical expansion, now absorbed these scientists and turned their projects into national infrastructures. This model was then repeated, with the absorption of the most promising observatory project in the traditional optically visible wavelengths, and, simultaneously, a major drive toward space-based astronomy in wavelengths inaccessible from the ground. In all these fields, the Max Planck Society grew by attracting external experts who, in addition to their flagship projects, continued to expand into adjacent wavelengths in subsequent decades, at their respective institutes. This absorption of astronomy led to a significant shift within the Max Planck Society itself, an institution where astrophysics had hitherto been dominated by theoretical plasma physicists in Munich, and experimental nuclear and particle physicists in Heidelberg. The growth of astronomy and its corresponding political influence led to a major reconfiguration of the disciplinary focus of several Max Planck Institutes in the 1970s, and this also signaled a transition from the space sciences of the early post-Sputnik era to the more differentiated astronomy, astrophysics, and planetary sciences of the coming decades.

1 Ground-Based Astronomy

The most significant transformation resulting from Sputnik in the Max Planck Society was the incorporation of observational astronomy as a research field. Up until 1957, there was a strong incipient research tradition in radio astronomy outside the Society. As with the other strong traditions, this one had a powerful political base, in North Rhine-Westphalia in the context of radar development, which reemerged as a dual-use technology after 1955. In the drive to expansion in astronomy, and taking advantage of regional rivalries, the Max Planck Society subsequently also absorbed the fledging project of what would become the Max Planck Institute for Astronomy, namely the construction of Germany's national optical telescopes, one in each hemisphere. After these two starters, the strategy and narrative of opening new wavelength windows became central to the Society's expansion, first internally, at the Institute for Radio Astronomy, and soon through additional directorships at many other institutes.

Post-Sputnik Absorption of Astronomy into the Max Planck Society

In 1957, no observational astronomy was conducted within the Max Planck Society.[1] Three decades later, this situation had been completely reversed, as it was felt that the Society meanwhile had a virtual monopoly on research in observational astronomy in West Germany, and, as critics indicated, also absolute control over the 'means of production' in the field.[2] Throughout these three decades, observational astronomers transitioned from being complete outsiders to the scientific traditions and organizational culture previously prevailing in the Max Planck Society to becoming central players, constituting a formidable 'core' of institutes, scientific members, decision makers, and allied economic and political forces. These observational astronomers came to challenge the political hegemony of the original scientific traditions and factions centered on Wolfgang Gentner in southwestern Germany and Werner Heisenberg in Bavaria.

As mentioned above, national dominance of fundamental research in Germany had been the ambition of the Max Planck Society in the postwar era in

1 There had been prewar attempts at a Kaiser Wilhelm Institute for Astronomy, significantly supported by its second President, the industrialist and amateur astronomer Carl Bosch. These, however, came at a moment of global financial crisis and by the time of World War II had not materialized. This prehistory had some symbolic influence on the siting of the Max Planck Institute for Astronomy in Heidelberg. See Dietrich Lemke: *Im Himmel über Heidelberg. 40 Jahre Max-Planck-Institut für Astronomie in Heidelberg (1969–2009)*. Edited by Archiv der Max-Planck-Ges. Vol. 21. Berlin 2011. More on Carl Bosch's influence in prewar German astronomy is described in: Juan-Andres Leon: *Citizens of the Chemical Complex. Industrial Expertise and Science Philanthropy in Imperial and Weimar Germany*. Dissertation. Cambridge, MA: Harvard University 2013.

2 Letter by Immo Appenzeller (AMPG, ZA 166, No. 57). Such criticism and its resolution are one of the main analytical axes of Chapter 4.

many fields, particularly in areas related to nuclear research, such as nuclear reactors and accelerators. But while this aspiration was thwarted in those areas, or at least only partially fulfilled, it was to be realized in full in a field that matured only later, in the post-Sputnik years. As we will see in Chapter 4, the Max Planck Society was so successful in appropriating the cosmic sciences in Germany that, by the end of the 1980s, attempts were being made by researchers and policy makers alike to devolve some of this concentration of scientific power to other organizations.[3]

As we saw in previous chapters, the Max Planck Society's initial move toward outer space was led by a generation for whom the cosmic sciences had emerged out of the nuclear age, methodologically and politically, mainly at institutes such as Astrophysics in Munich, Aeronomy in Lindau, and Nuclear Physics in Heidelberg. But in staking the Max Planck Society's claim to outer space, the leaders of these institutes then also fostered the absorption of observational astronomy, initially in radio and visible wavelengths. Over the next two decades, however, these new 'foreign bodies,' the observational astronomers, rose in prominence, and so were able to play the locals at their own game, on an equal footing; they accordingly gained more power and influence in the Society, and developed their own regional political and industrial support networks, as well as international partnerships, while also repeatedly emphasizing their independence from the old guard of plasma physicists and cosmochemists. Observational astronomy, for example, dominated research by the end of the century, even at the crown jewel of the cosmic sciences in the Max Planck Society, the Max Planck Institute for Extraterrestrial Physics, as the following chapters show. The book then culminates in the study of how even the two original 'nuclear' institutes in Munich and Heidelberg gradually mobilized their longstanding research traditions over a period of 50 years, to make world-class contributions to the three most innovative approaches to what is now called multi-messenger astronomy.

During the initial wave of expansion treated in this chapter, observational astronomers contributed a set of titans of their own to match Heisenberg and Gentner: for example, Otto Hachenberg and Peter Mezger at the Max Planck Institute for Radio Astronomy in Bonn, Hans Elsässer at the Max Planck Institute for (Optical) Astronomy in Heidelberg,[4] or Klaus Pinkau and Joachim Trümper at the Max Planck Institute for Extraterrestrial Physics. These actors

3 AMPG, ZA 166, No. 56, 57, 58, 59, 61.

4 Jakob Staude: Hans Elsässer. 29.3.1929–10.6.2003. In: Generalverwaltung der Max-Planck-Gesellschaft zur Förderung der Wissenschaften (ed.): *Max-Planck-Gesellschaft Jahrbuch 2004*. München: Max-Planck-Gesellschaft zur Förderung der Wissenschaften 2004, 111–112. Rolf

and their allies were behind many of most ambitious scientific projects ever undertaken in West Germany, and were the first ever able to hold their own in international competition, not by cleverly maneuvering their niche expertise in theory or instrumentation within larger collaborations (as had been the approach at the Institutes for Physics and Nuclear Physics), but by the sheer scale of their infrastructural projects and their ambition to become a world superpower in an entire scientific field based on their own observatories and instruments. Eventually, the prime objective was to build the most powerful telescopes in the world at every possible wavelength. The degree to which this and other aims were accomplished will provide a central storyline spanning this chapter and the next one.

Meanwhile, in this first section, we focus on showing how the introduction of astronomy and this ambition to dominate multi-wavelength projects in the 1960s was the first instance of strategic coordination among all the institutes of the Max Planck Society in the cosmic sciences, presenting a clear strategy for growth, as well as demarcating the natural domains of each institute. The expansion of institutes and new directors based on this observational wavelength logic successfully augmented the footprint of observational astronomers both in budget, number of researchers, and scientific members of the Society, leading to their growing influence in its decision-making bodies.[5] As will be described in the following chapters, it was only in later decades that contradictions inherent to this wavelength distribution logic came to the surface, as institutes, in expanding, increasingly stepped on each other's observational domains.

Finally, we will see how the growth of observational astronomy and its rationale of building national infrastructures benefited from the Cold War era mentality, in which astronomical gigantism was a race of its own among all the major countries, not just the United States and the Soviet Union, but

---

Schwartz: Otto Hachenberg. 1.7.1911-24.3.2001. In: Max-Planck-Gesellschaft zur Förderung der Wissenschaften (ed.): *Max-Planck-Gesellschaft Jahrbuch 2002*. Göttingen: Vandenhoeck & Ruprecht 2002, 863. Michael Kramer et al.: Peter Georg Mezger. 19.11.1928-09.07.2014. *Personalien 2014. Beileger zum Jahresbericht der Max-Planck-Gesellschaft*, 2015, 37–38.

5 In 1960, there were four institutes with a significant footprint in the cosmic sciences, with one sub-institute fully dedicated to this (Biermann's Institute for Astrophysics). By the mid-1990s, there were at least 11 institutes in the 'cluster,' of which six were fully dedicated to the cosmic sciences, and at least four of those primarily to observational astronomy. See the Financial Appendix at the end of this book for additional insights.

also the United Kingdom, France, and Japan.[6] In this race, observatory building often took precedence over the careful designation of scientific goals, so the outcomes varied from spectacular successes to mediocre disappointment, or even disastrous 'white elephants.' Furthermore, instrumental successes did not always translate into long-lasting scientific returns.

In terms of national dominance, one of the explanations for this effective monopolization is that Max Planck Society members had learned from the mistakes of the past decade, particularly when it came to internal rivalries in the face of nationally significant challenges such as nuclear energy. This lesson had been learned right at the time when 'Sputnik shock' impacted the cosmic sciences in the most direct manner possible, as we saw in previous chapters. Moreover, the shift to the space race occurred just as the West German *Wirtschaftswunder* (economic miracle) was getting into full swing, so that scientific research inspired by Sputnik benefited disproportionately from the new economic prosperity.[7] Most importantly, as shown earlier, because the expertise of scientists in Munich, Heidelberg, and, to a lesser degree, Lindau, had been accumulated during a period when astrophysics was the poor man's entry point to nuclear research, the Max Planck Society found itself with a considerable number of scientific experts in fields related to outer space; and it was their significant influence that would carry weight in decision-making pertaining to the national and international organizations now created to face the new challenges of outer space. As was described in the previous chapter, this privileged position was used to steer space-based research toward the Max Planck Institutes in Munich, Lindau, Heidelberg, and Mainz, as well as to guarantee their leading roles in international collaboration. In addition, the Max

6 For a wide review of the historical developments in modern astrophysics and cosmology, see Malcolm S. Longair: *The Cosmic Century. A History of Astrophysics and Cosmology*. Cambridge: Cambridge University Press 2006. For a retrospective of 20th-century scientific research in space with contributions by pioneers involved in the various disciplines, see Johan A. M. Bleeker, Johannes Geiss, and Martin C. E. Huber (eds.): *The Century of Space Science*. Dordrecht: Kluwer Academic Publishers 2001.

7 According to the Max Planck Society's budget plans, the total funds available to the Society grew around eight-fold between 1957 and 1975, inflation corrected. In the cosmic sciences, the rise was much more dramatic, as its weight within the MPG went from about 1 percent to 8 percent in the institutes fully dedicated to cosmic research, and about 12 percent to 24 percent of all the institutes with some presence in cosmic research (including also and especially nuclear research). For more details, see the Financial Appendix at the end of this book. For a more general view on the Society's dynamic growth during the years from 1955 to 1972 see Jaromír Balcar: *Wandel durch Wachstum in »dynamischen Zeiten«. Die Max-Planck-Gesellschaft 1955/57 bis 1972*. Berlin: GMPG-Preprint 2020.

Planck Society ended up moving into an activity hitherto unexplored: building astronomical observatories.

Before 1957, there was no interest in observational astronomy in the Max Planck Society, as illustrated in our first chapter by the case of Karl-Otto Kiepenheuer's solar astronomy institute in Freiburg, which remained an independent institution even beyond the postwar years.[8] In traditional optical astronomy, Germany had fallen significantly behind since the early 20th century. Research was still conducted by a decentralized constellation of state and university observatories with long histories, often extending back to the time of independent kingdoms.[9] In the postwar era, these observatories continued to be precariously funded by their corresponding federal states.[10] Optical astronomers had also been notably conservative in the early years of the 20th century, as is evident, for instance, in the widespread rejection of Einstein's theories by many optical astronomers, not only in Germany but also, for example, in the United States.[11] Even by purely observational standards, astronomy in Germany was considered to be in decline due to the unfavorable geographical location that combined frequently cloudy skies, too little elevation, and growing light and atmospheric pollution. So much was this the case that, since the 1920s, the best German observational astronomers had begun to migrate to other countries, such as, notably, Walter Baade, "arguably the most influential observational astronomer of the 20th century,"[12] who remained permanently at the Mount Wilson observatory, the world's largest astronomical facility in the first half of the century.[13]

---

8 See pp. 56–65 in Michael P. Seiler: *Kommandosache "Sonnengott". Geschichte der deutschen Sonnenforschung im Dritten Reich und unter alliierter Besatzung*. Frankfurt am Main: Verlag Harri Deutsch 2007.

9 In Lemke's *Im Himmel über Heidelberg* see, for example, Fig. 2.3–3 showing the light-collecting surface of telescopes with a diameter > 50 m in individual countries since 1945, showing how West Germany was well behind not only US and UK, but also France, Italy, and USSR. Lemke, *Himmel über Heidelberg*, 2011, Vol. 21. See also Dietrich Lemke, and Astronomische Gesellschaft (eds.): *Die Astronomische Gesellschaft 1863–2013. Bilder und Geschichten aus 150 Jahren*. Heidelberg: Astronomische Gesellschaft 2013, 3.

10 This dire situation is well illustrated in Hans-Heinrich Voigt et al.: *Denkschrift zur Lage der Astronomie*. Wiesbaden: Steiner 1962. It will be discussed in more detail later.

11 Leon, *Citizens of the Chemical Complex*, 2013. Chapter 3.

12 Norris S. Hetherington: Walter Baade: A Life in Astrophysics. *Physics Today* 55/11 (2002), 69–69. doi:10.1063/1.1535009.

13 After his education in Göttingen, Baade worked from 1919 to 1931 at the Hamburg Observatory at Bergedorf—at the time the largest telescope in Europe—and then moved to the USA. In 1933, he and the Swiss-born theoretical physicist Fritz Zwicky together proposed that cosmic rays could be produced in the supernovae explosion (a term they intro-

Theoretically informed observational astronomy had, however, remained a niche of excellence at a few sites in Germany, at the University of Göttingen, for instance, where Hans Kienle had trained a prominent generation of astronomers, including Heinrich Siedentopf and Otto Heckmann, who even studied relativistic questions.[14] Similarly, Albrecht Unsöld, who had obtained his PhD under Arnold Sommerfeld in Munich, was a rising star in Kiel from the early 1930s.[15] In the postwar era, these were the leading German astronomers, in Tübingen (Siedentopf), Hamburg (Heckmann), and Kiel (Unsöld), before the Max Planck Society had any interest in the field. Consequently, optical astronomy in Germany until the early 1960s was under the auspices of universities and federal states, with modest national funding channeled largely through the German Research Foundation (DFG).[16]

From these precarious bases, the first postwar generation of astronomers participated in the first attempt at European integration in this field, the

duced in 1931 to identify this new category of astronomical objects); and they advanced the view that such cosmic rays could represent the transition from ordinary stars into neutron stars, compact objects having a very small radius and consisting in their final stages of extremely closely packed neutrons. Walter Baade, and Fritz Zwicky: Cosmic Rays from Super-Novae. *Proceedings of the National Academy of Sciences* 20/5 (1934), 254–259. doi:10.1073/pnas.20.5.259. During the 1950s, Baade and Rudolph Minkowski identified the optical counterparts of several radio sources. Donald E. Osterbrock: *Walter Baade. A Life in Astrophysics*. Princeton, NJ: Princeton University Press 2001.

14 Otto Heckmann: Nachrufe. Hans Kienle. *Mitteilungen der Astronomischen Gesellschaft* 38 (1976), 9–10. https://ui.adsabs.harvard.edu/#abs/1976MitAG..38....9H. Last accessed 10/30/2018. Hans Elsässer: Nachrufe. H. Siedentopf. *Mitteilungen der Astronomischen Gesellschaft* 17 (1964), 33–41. https://ui.adsabs.harvard.edu/#abs/1964MitAG..17...33. Last accessed 10/30/2018. Hans-Heinrich Voigt: Nachruf auf Otto Heckmann. *Mitteilungen der Astronomischen Gesellschaft* 60 (1983), 9–12. https://ui.adsabs.harvard.edu/#abs/1983MitAG..60....9V. Last accessed 10/30/2018.

15 B. Baschek: Albrecht Unsöld (20. April 1905–23. September 1995). *Mitteilungen der Astronomischen Gesellschaft* 79 (1996), 11–15. https://ui.adsabs.harvard.edu/#abs/1996PhyBl..52..890B. Last accessed 10/30/2018. See also Albrecht Unsöld: Interview by Owen Gingerich, June 6, 1978. Transcript, AIP, https://www.aip.org/history-programs/niels-bohr-library/oral-histories/4924. Last accessed 1/4/2019. Richard Wielebinski: Albrecht Unsöld. His Role in the Interpretation of the Origin of Cosmic Radio Emission and in the Beginning of Radio Astronomy in Germany. *Journal of the Astronomical History and Heritage* 16/1 (2013), 66–80. http://adsabs.harvard.edu/abs/2013JAHH...16...67W. Last accessed 10/30/2018.

16 Voigt et al., *Denkschrift Astronomie*, 1962. This continued to be a problem in later decades for institutions outside the Max Planck Society: Heinrich J. Völk et al.: *Denkschrift Astronomie*. Weinheim: VCH 1987.

European Southern Observatory, which aimed for a joint observational site somewhere on the southern hemisphere with excellent climatic conditions.[17]

However, Siedentopf died unexpectedly in 1964, and the balance of power in Germany's contribution to the European collaboration shifted overwhelmingly northward, toward Heckmann in Hamburg, who was the first director of this European organization, and whose evolution we detail throughout this chapter and Chapter 4. There was growing pressure from rival German astronomers to create a national observatory. As we saw earlier in the book, the discussion of national versus European research had already taken place in the space sciences context, with ESRO,[18] and also with regard to the initiatives that led in 1959 to the creation in Hamburg of the research center German Electron-Synchroton (DESY, *Deutsches Elektronen-Synchrotron*), the national counterpart to CERN.[19] The role of observational astronomy in these debates can be considered to be largely derivative, benefiting from the decisions and institutional frameworks that had been established in other fields, to then argue for something similar in optical astronomy. This opened the door to what could be referred to as a German 'national' observatory.

In fact, through the next half a century, optical astronomers repeatedly made use of scientific arguments and debates which had previously occurred in other more forward-looking areas such as space exploration, particle physics, and even the other major ground-based observational field of radio astronomy. In Germany, discussion of a national optical observatory only became feasible in the 1960s after the foundations had been laid first by early research initiatives in outer space (possible only because of a national framework originating in the nuclear sciences) and, more directly, by the developments in radio astronomy, which at the time was considered to be more forward-looking, potentially revolutionary, and even more cost-effective.

---

17 Adriaan Blaauw: *ESO's Early History. The European Southern Observatory from Concept to Reality*. ESO 1991. See also ESO Historical Archives inventory: Adriaan Blaauw: *ESO Historical Archives (EHA). Inventory per December 1992*. Garching: European Southern Observatory 1992.

18 Reimar Lüst: *Aktuelle Probleme der Weltraumforschung. Festvortrag anläßlich der Jahresversammlung des Stifterverbandes in Wiesbaden am 11. Mai 1965*. Essen-Bredeney: Gemeinnützige Verwaltungsgesellschaft für Wissenschaftspflege 1965. Reimar Lüst: The European Space Research Organization. *Science* 149/3682 (1965), 394–397. doi:10.1126/science.149.3682.394.

19 See, for example, Claus Habfast: *Großforschung mit kleinen Teilchen. Das Deutsche Elektronen-Synchrotron DESY 1956–1970*. Berlin: Springer 1989. Erich Lohrmann, and Paul Söding: *Von schnellen Teilchen und hellem Licht. 50 Jahre Deutsches Elektronen-Synchrotron DESY*. Weinheim: Wiley-VCH Verlag 2009.

Radio Astronomy Enters the Max Planck Society

Before 1945, only the visible part of the electromagnetic spectrum was available for astronomical study, as it corresponds to the portion of light from outer space that manages to reach the ground through the atmosphere and can be seen with a telescope. Visible-light astronomy limited scientists to studying the Universe in a rather narrow wavelength interval, which is, fortunately, fundamental to observations via optical telescopes because stars, which all have very long lifetimes, emit a large proportion of their electromagnetic energy in the visible waveband.

Radio astronomy was the first of the new astronomies. Radio waves of extraterrestrial origin were discovered by Karl Jansky in the early 1930s but, for many years, this did not have an impact in the community of astronomers. It was not until the period from the late 1940s up to the end of the 1960s, when radio emission was discovered in a wide range of different astronomical objects, that radio astronomy become the true cutting edge of astronomical research.[20] This was a field that almost entirely developed as a result of World War II, when radar was developed. In fact, as we will see below, outside of Great Britain, the first postwar radio telescopes were repurposed German radars in the Netherlands and France.[21] Moreover, because of postwar restrictions on radar technology, radio astronomy could not be seriously practiced in West Germany during the first postwar decade.[22] Nevertheless, brilliant German astronomers such as Heinrich Siedentopf followed the field

---

20 Woodruff T. Sullivan III: *The Early Years of Radio Astronomy. Reflections Fifty Years after Jansky's Discovery*. Cambridge: Cambridge University Press 1984. Woodruff T. Sullivan III: The History of Radio Telescopes, 1945–1990. *Experimental Astronomy* 25/1 (2009), 107–124. doi:10.1007/s10686-009-9140-2. Wayne Orchiston (ed.): *The New Astronomy. Opening the Electromagnetic Window and Expanding Our View of Planet Earth. A Meeting to Honor Woody Sullivan on His 60th Birthday*. Dordrecht: Springer 2005. Gerrit L. Verschuur: *The Invisible Universe. The Story of Radio Astronomy*. 2nd ed. New York, NY: Springer 2007.

21 Astrid Elbers: *The Rise of Radio Astronomy in the Netherlands. The People and the Politics*. Cham: Springer 2016. B. R. Martin: Radio Astronomy Revisited. A Reassessment of the Role of Competition and Conflict in the Development of Radio Astronomy. *The Sociological Review* 26/1 (1978), 27–55. doi:10.1111/j.1467-954X.1978.tb00123.x.

22 See Control Council and Coordinating Committee of the Allied Control Authority: Enactments and Approved Papers of the Control Council and Coordinating Committee. Allied Control Authority, Germany (1945–1948). 9 Volumes. *Military Legal Resources. Federal Research Division. Library of Congress*, III. https://www.loc.gov/collections/military-legal-resources/?q=enactments. Last accessed 10/30/2018. On p. 108 of Vol. 3, the list of Prohibited Applied Scientific Research includes electromagnetic, infrared, and acoustic radiation having the purpose of detecting objects or obstacles or the determination of the position of vehicles, aircraft, ships, submarines, or missiles.

from the outset, so that whenever these restrictions were lifted, they could easily enter the field.[23] More importantly, a new generation of 'native' German radio astronomers began to be trained abroad, in the Netherlands, the United Kingdom, France, and, later, the United States. In fact, Americans were also very late in entering radio astronomy precisely because they were global leaders in gigantic optical telescopes.[24] In the USSR, too, radio astronomy grew out of wartime radar research programs, but, unlike in Europe, Australia, or the United States, Soviet radio astronomy remained largely within military-oriented and tightly controlled laboratories, with severe restrictions on publications in the open scientific literature.[25]

Like optical and infrared light, radio waves are able to pass through the atmosphere, and have the advantage that observations are not affected by clouds or rain. For these reasons, experts in radio astronomy during the early postwar period tended to be based in cloudy, low-lying European countries, making fast progress in areas unexplored by the Americans. It was not until the late 1950s onwards that Americans made a serious move into radio astronomy, rapidly building some of the world's largest instruments. This late move brought in its wake new institutional modalities adapted for the larger scale of this new generation of instruments; prior to this, American research was virtually impossible at a national level, and most observatories were owned by states, or privately owned. During the Cold War, it became increasingly urgent to fund scientific infrastructures on a national scale, and in astronomy, this resulted in a National Radio Astronomical Observatory (NRAO) in Green Bank, West Virginia[26] and, later, a National Optical Astronomy Observatory (NOAO) at the top of Kitt Peak, Arizona.[27]

West Germany, which was experiencing similar organizational problems because of scientific research having originally been the responsibility of each

---

23 Heinrich Siedentopf: *Methoden und Ergebnisse der Radioastronomie*. Vol. 6, 1954.

24 Woodruff Turner Sullivan III: *Cosmic Noise. A History of Early Radio Astronomy*. Cambridge: Cambridge University Press 2009. See also David Leverington: *Observatories and Telescopes of Modern Times. Ground-Based Optical and Radio Astronomy Facilities since 1945*. Cambridge: Cambridge University Press 2017.

25 Kenneth Kellermann (ed.): *A Brief History of Radio Astronomy in the USSR. A Collection of Scientific Essays*. Translated by Denise C. Gabuzda. Dordrecht: Springer 2012.

26 Kenneth I. Kellermann, Ellen N. Bouton, and Sierra S. Brandt: The Largest Feasible Steerable Telescope. In: Kenneth I. Kellermann, Ellen N. Bouton, and Sierra S. Brandt (eds.): *Open Skies: The National Radio Astronomy Observatory and Its Impact on US Radio Astronomy*. Cham: Springer International Publishing 2020, 461–531. doi:10.1007/978-3-030-32345-5_9.

27 Leverington, *Observatories and Telescopes*, 2017.

individual federal state, could look to the American national observatories as a model to follow in the 1960s. Several of the leading West German radio astronomers spent a part of their career at Green Bank, including Sebastian von Hoerner, who had been von Weizsäcker's student in Göttingen (Chapter 1), and, later, Peter Mezger, who had a doctorate in engineering from the Technical University Munich, obtained during his work with early German radars and radio telescopes, and had also trained in the Paris Observatory with captured German radars. As of the mid-1950s, German universities began to build their own radio telescopes, most notably in Kiel (Albrecht Unsöld), Tübingen (Heinrich Siedentopf), and Bonn (Friedrich Becker). In 1956, for example, still before Sputnik, Bonn had already constructed a magnificent 25 m-diameter radio telescope, the Stockert radio telescope or *Astropeiler*.[28] Simultaneously, in East Berlin, Otto Hachenberg was building the largest radio telescope antenna in the German-speaking world, a 36 m device completed in 1958.[29]

A Third Regional Pole in North Rhine-Westphalia

As we already saw in the case of Bavaria and of southwestern Germany as a whole, regional interests played a major role in shaping the early decades of the Max Planck Society. Next, we will describe how this unfolded in the northwestern state of North Rhine-Westphalia (NRW), which was at the time the most densely populated and industrialized part of West Germany. Due to historical peculiarities of Imperial Germany and the Weimar era, the Kaiser Wilhelm Society had been present there mainly through the heavily industrially based Institute for Coal Research and Institute for Iron Research.[30] This trend continued throughout the first postwar decade, when, although it was part of the British zone, the region did not add significant new institutes to the Max Planck Society. When, in the late 1950s, the final agreement was reached

28 Karl M. Menten: Leo Brandt. Pionier der Funkmesstechnik und Initiator der Radioastronomie in Deutschland. In: Bernhard Mittermaier, and Bernd-A. Rusinek (eds.): *Leo Brandt (1908–1971). Ingenieur–Wissenschaftsförderer–Visionär*. Jülich: Forschungszentrum Jülich, Zentralbibliothek 2009, 41–53. Richard Wielebinski: Fifty Years of the Stockert Radio Telescope and What Came Afterwards. *Astronomische Nachrichten* 328/5 (2007), 388–394. doi:10.1002/asna.200710766.

29 Richard Wielebinski: The New Era of Large Paraboloid Antennas. The Life of Prof. Dr. Otto Hachenberg. *Advances in Radio Science* 1 (2003), 321–324. doi:10.5194/ars-1-321-2003.

30 See, for example, Manfred Rasch: *Geschichte des Kaiser-Wilhelm-Instituts für Kohlenforschung 1913–1943*. Weinheim: VCH 1989. Sören Flachowsky: Von der Wagenburg der Autarkie zu transnationaler Zusammenarbeit. Der Verein Deutscher Eisenhüttenleute und das KWI/MPI für Eisenforschung 1917–2009. In: Helmut Maier (ed.): *150 Jahre Stahlinstitut VDEh 1860–2010*. Essen: Klartext 2010, 671–708.

between Federal Germany and its constituent states on their joint support of the new national institutes (both those of the Max Planck Society and a number of other major centers of research), the fact that North Rhine-Westphalia had been neglected was already a source of complaint, particularly in light of the immense resources it brought to these national bodies.[31] In the nascent West German federal system, Bavaria, one of the poorest states before 1945, had managed both to attract industries from the zone then under Soviet control and, through its conservative Christian Social Union party (CSU), to gain disproportionate influence at national level, also in scientific affairs, through the Federal Ministry for Nuclear Affairs. As described in earlier chapters, this led to a strategy to locate as many nationally funded research institutions as possible in Bavaria, which thus became a net recipient of federal funds.

At the other end of the spectrum was North Rhine-Westphalia, the largest net federal contributor, yet which in the early postwar decades saw all other federal states block its attempts to capture its fair share of national projects, the best illustration of which is that the first federally funded nuclear reactor ended up being built outside of NRW, in Karlsruhe. Even more so than Bavaria, North Rhine-Westphalia had striven for national dominance in new technological areas, but the new federal structure of the country was deliberately designed to counterbalance this. The solution in the case of such a powerful state was to develop large state-funded projects in parallel with the federal initiatives. The first notable example of this was a response to the aforementioned construction of the first German nuclear reactor. While Heisenberg's initial intention to build it in Bavaria was overwhelmed by non-Bavarian interests, in which the Rhinelander Konrad Adenauer played a direct role, the reactor was not constructed in North Rhine-Westphalia, but rather, as said,

31 The grievances concerning North Rhine-Westphalia being too big a net contributor to federally funded research located elsewhere is a major recurring topic in Hans-Willy Hohn, and Uwe Schimank: *Konflikte und Gleichgewichte im Forschungssystem. Akteurkonstellationen und Entwicklungspfade in der staatlich finanzierten außeruniversitären Forschung*. Frankfurt am Main: Campus Verlag 1990. On several occasions, North Rhine-Westphalia even threatened to stop funding federally led research. The specific example of the Effelsberg telescope was mentioned by Reimar Lüst, during the Roundtable "Astronomy and Astrophysics in the History of the Max Planck Society with a special focus on the Changes in the 'Cluster' of Astronomy and Astrophysics within the MPG," Max Planck Institute for the History of Science, October 21, 2016. *Research Program History of the Max Planck Society. Report 2014–2017*. Edited by Florian Schmaltz et al. 2014–2017. Berlin 2017, 108–109. See also Bernhard Mittermaier, and Bernd-A. Rusinek (eds.): *Leo Brandt (1908–1971): Ingenieur–Wissenschaftsförderer–Visionär*. Jülich: Forschungszentrum, Zentralbibliothek 2009.

near Karlsruhe, on the upper Rhine.[32] North Rhine-Westphalia's response was to build, almost at the same time, its own nuclear research reactors at the *Kernforschungsanlage Jülich*, a research center directly inspired by the main British atomic research center at Harwell.[33] For the next half century, Jülich continued to operate in parallel with Karlsruhe.

By the mid-1950s, even before Sputnik, the federal state of North Rhine-Westphalia began to place a particular emphasis on developing industries that facilitated its dominance in radio astronomy. Making radio telescopes in the first two postwar decades was primarily a construction challenge: large structures to create a static or, preferably, movable dish, coupled with the development of the electronic technology used for signal detectors. Both kinds of expertise were favored in what was then Germany's most industrialized state. Following a first generation of astronomical radio telescopes, there was also an explicit direct interest in the technological applications of dish antennas for communications as well as for the military. In early 1957, the *Forschungsinstitut für Hochfrequenzphysik* (Research Institute for High Frequency Physics) was created in North Rhine-Westphalia to develop these applications, largely supported by the company Telefunken, and sharing the same technology with the radio telescopes of the University of Bonn. In fact, when it was completed, the Stockert radio telescope operated to 50 percent as a radio telescope, and to 50 percent as an experimental radar for the applied research institute. The astronomer Wolfgang Priester was brought from Kiel to lead the research activities, and it was in this world of the Stockert antenna that Peter Mezger, one of the future Max Planck Institute directors, gained his expertise prior to his period abroad at Green Bank in the United States.[34]

Meanwhile, in East Berlin, in the late 1950s, Otto Hachenberg began to consider moving to the West, and a position was found for him in Bonn, where he

32 Joachim Radkau: *Aufstieg und Krise der deutschen Atomwirtschaft 1945–1975. Verdrängte Alternativen in der Kerntechnik und der Ursprung der nuklearen Kontroverse.* Reinbek: Rowohlt 1983. Michael Eckert, and Maria Osietzki: *Wissenschaft für Macht und Markt. Kernforschung und Mikroelektronik in der Bundesrepublik Deutschland.* München: Beck 1989.

33 John Cockcroft, first director of the Atomic Energy Research Establishment (AERE) at Harwell in the 1940s, was heavily involved in the establishment of Jülich under the initiative of Leo Brandt (more on him later). See Bernhard Mittermaier, and Bernd-A. Rusinek: *Leo Brandt (1908–1971). Ingenieur–Wissenschaftsförderer–Visionär. Wissenschaftliche Konferenz zum 100. Geburtstag des nordrhein-westfälischen Forschungspolitikers und Gründers des Forschungszentrums Jülich.* Jülich: Forschungszentrum Jülich 2009, 78.

34 Wielebinski, Fifty Years, 2007, 388–394.

arrived in 1961 shortly after the Berlin Wall was built.[35] In less than a decade, North Rhine-Westphalia had thus created the conditions for a third major focal point of political, industrial, and scientific interests converging in the cosmic sciences, thanks to a specific technical expertise, in this case, large antenna construction.

Leo Brandt and the Distinctive Engineering Tradition of Radio Astronomy in Bonn

It so happens that in North Rhine-Westphalia, the person who led the state's efforts to maintain supremacy in scientific and technological matters was one of the founders of radio astronomy, Leo Brandt. He, more than anyone, shaped the distinctive research tradition of the future Max Planck Institute for Radio Astronomy. As we will see, what distinguishes the people in Bonn is the pre-eminence of engineers, who played a very direct role in the design of their instruments, including many of the world's largest radio astronomical antennas. This engineering ethos was strongly reinforced by its close connection with powerful industrial and political interests in North Rhine-Westphalia, a political-scientific infrastructure shaped by Leo Brandt in the first two post-war decades.

Leo Brandt was born to a traditional family of social democrats, and came of age around the time of the rise of National Socialism. His political convictions constrained his work to private industry in the Nazi era, where he worked as a telecommunications engineer at Telefunken in the 1930s. Then, in the war, he was one of the key figures in the development of radar technology in Germany, including the famous Würzburg radars that would soon serve as the first generation of radio telescopes in the formerly occupied countries of Western Europe. Unlike many other key technical experts, Brandt did not move abroad after the war; instead, at a very early stage, he made a career for himself as defender of the interests of the new state of North Rhine-Westphalia, also becoming one of West Germany's most influential utopians of technocratic modernity.[36]

---

35 Conversations with Reimar Lüst during the Roundtable "Astronomy and Astrophysics in the History of the Max Planck Society with a special focus on the Changes in the 'Cluster' of Astronomy and Astrophysics within the MPG," Max Planck Institute for the History of Science, October 21, 2016. *Research Program History of the Max Planck Society*, 2017, 108–109.

36 Joachim Radkau: *Geschichte der Zukunft. Prognosen, Visionen, Irrungen in Deutschland von 1945 bis heute.* München: Carl Hanser Verlag 2017, 131–170. Leo Brandt: *Forschen und Gestalten. Reden und Aufsätze 1930–1962.* Köln: Westdeutscher Verlag 1962.

In the first postwar decade, under the protection of his friend, the left-wing CDU Prime Minister of North Rhine-Westphalia, Karl Arnold, Brandt maneuvered himself upward, from being in charge of the state's transportation network to the position of Secretary of State for Scientific Research, a role created especially for him and unique among the federal states. Throughout this decade, he denounced the Allied restrictions on research in fields such as atomic power and radar as serious impediments to Germany's path to modernity; and although his position actually required him to enforce these restrictions in NRW, he deliberately skirted them in order to facilitate research in many fields that would otherwise have been adversely affected.[37] While his persona was strongly connected to a technocratic brand of social democracy, Brandt remained a personal friend of Abraham Esau, for example, who had been in charge of both the nuclear and radar projects during the war, as plenipotentiary for nuclear physics within the Research Council of the Education Ministry. Brandt defended Esau during the postwar trials, in contrast to most scientific researchers who worked under him and who characterized him as a loyal participant in the Nazi apparatus.[38]

By the mid-1950s, Brandt had established an approach in North Rhine-Westphalia that consisted in attempting to capture as many prewar industries and research institutions moving from other parts of occupied Germany, and when this did not succeed, to create parallel entities in all the crucial scientific-technical fields. In addition to his work in radar/radio astronomy, which will be discussed in detail, he established a cyclotron and a computing center in Bonn, and relocated to North Rhine-Westphalia the *Deutsche Versuchsanstalt für Luftfahrt* (DVL), one of the precursors of the future federal aerospace agency (DLR). Most famously, Leo Brandt was the mastermind behind the nuclear research facility at Jülich, mentioned above, a direct response to the location of the national research reactor in Karlsruhe. By the early 1960s, the state of North Rhine-Westphalia counted over 40 different state-supported institutes

37 Mittermaier, and Rusinek, *Leo Brandt (1908–1971)*, 2009.

38 For a detailed, not entirely negative account of Esau's trajectory in the Nazi state, see Mark Walker: *German National Socialism and the Quest for Nuclear Power 1939–1949*. Cambridge: Cambridge University Press 1989. Esau and his deputy Kurt Diebner were arguably more effective for Germany's path toward nuclear fission than Heisenberg's team, but the latter outmaneuvered them in internal power struggles (see p. 130) to the point that Gerlach replaced Esau as chief administrator of both Heisenberg's and Diebner's reactor research projects. Esau was reassigned to radar research, where he began his close partnership with Brandt.

for scientific research in its *Arbeitsgemeinschaft für Forschung* (Association for Research) led by Brandt.[39]

Brandt's technocratic utopianism led him from radar to nuclear energy in the first postwar decade, but he was the reason that radio astronomy established itself primarily in Bonn over other competing sites in West Germany with stronger prewar traditions in astronomy, such as Kiel or Tübingen, which also both expressed an early interest in radio astronomy. Brandt did not go directly into radio astronomical research but instead supported the postwar career of Friedrich Becker as director of the university observatory in Bonn, the relocation of Wolfgang Priester from Kiel to Bonn, and most importantly, Otto Hachenberg's move from East Berlin to Bonn. This had been planned since the late 1950s, facilitated by their acquaintance during the war, when Hachenberg had shifted his career from astronomy to airplane radar development.[40] In the first postwar decade, it was an open secret that radio astronomy in North Rhine-Westphalia was a way to keep a foot in the door of radar technology. This is best expressed by the Stockert *Astropeiler* itself (an active radar built on a hill and mounted on a tall tower to have access to the horizon), which was deliberately constructed for a dual purpose: yet, while training a valuable first generation of German radio astronomers, it did not initially generate much scientific interest, functioning primarily as an experimental radar development site.[41]

Capturing Radio Astronomy for the Max Planck Society: Otto Hachenberg and Sebastian von Hoerner

How much a regional pole of support mattered in building scientific infrastructures quickly became apparent with the German project to construct

39 Mittermaier, and Rusinek, *Leo Brandt (1908–1971)*, 2009, 14.

40 Otto Hachenberg: interview by Woodruff T. Sullivan III, Bonn, February 22, 1973. NRAO Archives, http://www.nrao.edu/archives/Sullivan/sullivan_transcript_hachenberg_1973.shtml. Last accessed 1/4/2019. Peter Mezger: Interview by Woodruff T. Sullivan III, Bonn, November 22, 1973. NRAO Archives, http://www.nrao.edu/archives/Sullivan/sullivan_transcript_mezger_1973.shtml. Last accessed 1/4/2019.

41 Peter Mezger, who worked as an engineer on the *Astropeiler*, realized during his stay there that it was not very productive scientifically. His calibration work, however, was the early basis of his expertise in radio telescope construction, which was later perfected in Green Bank. Peter Mezger: interview by Woodruff T. Sullivan III, Bonn, November 22, 1973. For personal reminiscences on Green Bank, see also Jacob W.M. Baars: *International Radio Telescope Projects. A life among their designers, builders and users*. Rheinbach: Createspace Independent Publishing Platform 2013. Chapter 1 is entirely dedicated to the period in Green Bank.

the world's largest fully steerable radio telescope. By the early 1960s, a technological limit to the size of moveable radio telescopes had been reached because of the deformation that occurs in the parabolic dish antenna due to its own weight. Suddenly, a deeper level of theoretically informed design was needed. Sebastian von Hoerner, then in Green Bank, proposed a solution with his homologous antenna concept, in which the deformation occurring in the antenna is such that the perfect parabolic shape is maintained regardless of orientation.[42] Sebastian von Hoerner then became the candidate to replace Heinrich Siedentopf in Tübingen, who had unexpectedly passed away in 1964, and at the same time as he was invited to teach in Tübingen, he made a proposal to build the world's largest fully steerable radio telescope (160 m) based on his homologous design.[43]

In many ways, von Hoerner had a distinct profile shaped by his time in Göttingen as a student of von Weizsäcker's.[44] While always interested in astronomy and astrophysics, in Göttingen he developed an eminently theoretical expertise which combined a brilliant theoretical outlook with the use of the new calculating machines built by Billing's group, such as the G2 (see Chapter 1).[45] Von Hoerner then moved to his native Heidelberg for a short time, where he worked with Walter Fricke at the *Astronomisches Rechen-Institut* (ARI, Astronomical Calculation Institute) on the first N-body simulations of

---

42 Sebastian von Hoerner: Design of Large Steerable Antennas. *The Astronomical Journal* 72/1 (1967), 35–47. doi:10.1086/110198. On this subject, see also in Biermann's papers the folder of materials related to the foundation of the Institute for Radio Astronomy during the years 1964–67 (AMPG, III. Abt., ZA 1, No. 89) and similar material in Lüst's papers (AMPG, III. Abt., Rep. 145, No. 292, 862). The most detailed source on the career of von Hoerner are the interviews by Sullivan (Sebastian von Hoerner: Interviews by Woodruff T. Sullivan III, February 23, 1977 and August 20, 1979. NRAO Archives, http://www.nrao.edu/archives/Sullivan/sullivan_transcript_vonhoerner_1977.shtml. Last accessed 1/4/2019. About von Hoerner and the Max Planck Society plans to enter the field of radio astronomy, see also Kellermann, Bouton, and Brandt, Telescope, 2020, 461–531, 465–469.

43 Richard Wielebinski: Sebastian von Hoerner. *Mitteilungen der Astronomischen Gesellschaft* 86 (2003), 9–10. http://adsabs.harvard.edu/abs/2003MitAG..86....9. Last accessed 10/30/2018.

44 Sebastian von Hoerner: interviews by Woodruff T. Sullivan III, February 23, 1977, and August 20, 1979. NRAO Archives, http://www.nrao.edu/archives/Sullivan/sullivan_transcript_vonhoerner_1979.shtml. Last accessed 1/4/2019.

45 See, for example, Sebastian von Hoerner: Herstellung von Zufallszahlen auf Rechenautomaten. *Zeitschrift für angewandte Mathematik und Physik* 8/1 (1957), 26–52. doi:10.1007/BF01601153.

star clusters and obtained his *Habilitation* (post-doctoral teaching qualification) in 1959, with a thesis on the rate of star formation.[46]

In 1962, he was summoned to the National Radio Astronomical Observatory in Green Bank by its first director, the veteran optical astronomer Otto Struve. Von Hoerner was key to the success of Green Bank, which had very problematic early years characterized by poorly designed antenna constructions due to their excessively empirical engineering. Von Hoerner became an expert in the theoretical foundations of antenna structures, crucial for the scaling up of radio telescopes from a postwar diameter of tens of meters, to the entirely new terrain around the hundreds-of-meters mark. His approach was eminently theoretical, working out simple designs from first physical principles to build antennas that had never been attempted before. This was completely unlike the contemporary approaches of the generation of engineers who had constructed radio telescopes to date.[47] And, crucially, this absolute control of telescope design was also very new in astronomy: as we will see later, in optical astronomy, a separation between telescope makers and users persisted for much longer, so that the leading optical astronomers were rarely involved in key instrumental innovations themselves, with these often coming instead from industry or even amateurs. In radio astronomy, on the other hand, the leading radio astronomers were experts on either antenna construction or the electronics needed for their detectors.[48]

The appointment of von Hoerner in Tübingen and the construction of his 160 m radio telescope were likely to make Germany (and Tübingen) a world leader in astronomy, and so it was expected that the Volkswagen Foundation would donate resources for the latter. These plans, however, were in direct competition with those of North Rhine-Westphalia for radio astronomy in the state, led at the time by the recently immigrated Hachenberg. He quickly made a counterproposal to the Volkswagen Foundation for his own more traditional

46 Sebastian von Hoerner: Die numerische Integration des n-Körper-Problemes für Sternhaufen. I. *Zeitschrift für Astrophysik* 50 (1960), 184–214. http://adsabs.harvard.edu/abs/1960ZA.....50..184V. Last accessed 10/30/2018. Sebastian von Hoerner: Die numerische Integration des n-Körper-Problems für Sternhaufen. II. *Zeitschrift für Physik* 57 (1963), 47–82. http://adsabs.harvard.edu/abs/1963ZA.....57...47V. Last accessed 10/30/2018.

47 Jacob W. M. Baars: interview by Juan-Andres Leon, Bonn, February 5–7, 2018. DA GMPG, BC 601050.

48 Sebastian von Hoerner: interviews by Woodruff T. Sullivan III, February 23, 1977, and August 20, 1979. NRAO Archives, https://www.nrao.edu/archives/items/show/15272. Last accessed 1/4/2019.

and smaller 80 m telescope.[49] The ensuing clash between these projects highlighted the need to establish a national framework for building large astronomical facilities in West Germany, in light of the competition between different regions and scientific researchers. The two competing projects for the world's largest radio telescope forced West German science organizers at the recently created Federal Science Council (*Wissenschaftsrat*, see Chapter 2) to deliberate on the best framework for German national observatories, something that up to that point was nonexistent in astronomy. Given the very strong presence of Max Planck Society members and allies in this advisory body, and their closeness to both von Hoerner and Hachenberg, this seemed an ideal opportunity to enact the most radical expansion of the cosmic sciences within the Max Planck Society: turning the Max Planck Society into the vehicle for Germany's 'national' observatory projects, first in radio astronomy and, a few years later, in optical astronomy. This was a major departure from the outcome of discussions regarding the last wave of national facilities, namely the creation of DESY and the Institute for Plasma Physics as independent private organizations outside of the Max Planck Society. Nevertheless, within this 'national' framework, regional interests continued to dominate, and what was first viewed as a balanced solution to competition between von Hoerner and Hachenberg (and Baden-Württemberg versus North Rhine-Westphalia) took a definite turn in favor of the latter. Initially, the plans called for two large antennas to be created, one for Bonn and one for Tübingen, under a single Max Planck Institute distributed between the two sites.[50]

It soon became obvious, however, that the political pendulum would swing in the direction of Bonn under the direct influence of Leo Brandt, who put pressure on the Volkswagen Foundation to insist that all the infrastructure should be installed in North Rhine-Westphalia,[51] which in turn put strong

49 Wolfgang Priester: interview by Woodruff T. Sullivan III August 30, 1976. NRAO Archives, https://www.nrao.edu/archives/items/show/15130. Last accessed 1/25/2022; Peter G. Mezger: interview by Woodruff T. Sullivan, Bonn, November 22, 1973, Transcript. NRAO Archives, http://www.nrao.edu/archives/Sullivan/sullivan_transcript_mezger_1973.shtml. Last accessed 1/4/2019.

50 Wielebinski, Sebastian von Hoerner, 2003, 9–10.

51 Conversations with Reimar Lüst during the Roundtable "Astronomy and Astrophysics in the History of the Max Planck Society." Richard Wielebinski: The Effelsberg 100-m Radio Telescope. *Naturwissenschaften* 58/3 (1971), 109–116. doi:10.1007/BF00593099. During meetings of the CPT Section of the Scientific Council, discussions on the interrelated founding of the Institute for Radio Astronomy and of the Institute for Astronomy were intertwined (CPTS meeting minutes of 03.12.1964, 05.03.1965, 22.06.1965, 21.06.1966, AMPG, II. Abt., Rep. 62, No. 1744, 1745, 1746, 1747). See also Rolf Schwartz: *Chronik des Max-*

pressure on the Max Planck Society to base the institute in Bonn, as few Max Planck Institutes were based in NRW at the time, or even exceedingly few, relative to the financial contributions made by what was then West Germany's most industrialized and populous state.[52] The situation thus turned against von Hoerner, who was now increasingly expected to share a single radio telescope based on his principles but located in the rival federal state and controlled by Hachenberg. Instead of taking up his position in Tübingen in Baden-Württemberg, he remained in Green Bank for the rest of his scientific career.[53]

Over the next half decade, up until 1973, the Effelsberg radio telescope was built, becoming the first postwar case in which Germans controlled the world's most powerful scientific instrument in the cosmic sciences, and it was the showpiece of an industrial-scientific partnership that was expected to continue building the largest radio telescopes in the world for the next generation, while consolidating the dominance of North Rhine-Westphalia in the field. The actual telescope that was built was initially to be constructed by Krupp using a simplified, approximate version of von Hoerner's homology design. Ultimately, however, the telescope was built by a consortium: Krupp collaborated with the mechanical engineering company MAN (*Maschinenfabrik Augsburg-Nürnberg*), whose Gustavsburg branch in Wiesbaden was one of the most important builders of large structures in West Germany. When the Effelsberg radio telescope went into production, MAN had just completed construction of the Parkes Radio Telescope in Australia, one of the main inspirations for the Max Planck Institute for Radio Astronomy. Although MAN had not designed the Parkes, its expertise in the construction of large antennas was recruited for the consortium, leading to collaboration of the two industrial companies that built not only all of Germany's large radio telescopes but also

Planck-Instituts für Radioastronomie. Bonn: Siering 2010, 7. Max-Planck-Gesellschaft (ed.): *Max-Planck-Institut für Radioastronomie. Bonn*. München 1992.

52 Conversations with Reimar Lüst during the roundtable "Astronomy and Astrophysics in the History of the Max Planck Society."

53 The evolution of these debates can be seen in the committee reports "Gründung eines MPI für Radioastronomie" (1964-01-01 bis 1968-11-04) during the CPTS meetings of the period (meeting minutes of 03.12.1964, 05.03.1965, 22.06.1965, 21.06.1966, AMPG, II. Abt., Rep. 62, No. 1744, 1745, 1746, 1747). See also Richard Wielebinski, Norbert Junkes, and Berndt H. Grahl: The Effelsberg 100-m Radio Telescope. Construction and Forty Years of Radio Astronomy. *Journal of Astronomical History and Heritage* 14/1 (2011), 3–21. http://adsabs.harvard.edu/abs/2011JAHH...14....3W. Last accessed 10/30/2018.

the vast majority of its large communications antennas and military-related dishes.[54]

Like Effelsberg, all subsequent Max Planck Society radio telescopes were financed by private donations from the Volkswagen and Krupp Foundations, and constructed by the MAN/Krupp consortium, whose successor to this day ranks among the major antenna-building companies in the world, VERTEX Antennentechnik, owned by one of the world's largest global defense conglomerates.

The appointment of Hachenberg over von Hoerner also reinforced a distinct research tradition in Bonn built around an engineering ethos, as opposed to the more theoretical inclinations of the Göttingen tradition that von Hoerner was part of. Within the recently established framework of collegiate directorship for Max Planck institutes (see Chapters 1, 3, and 4), the Bonn institute was expected to have three equal directors, and those selected (Otto Hachenberg, Peter Mezger, and Richard Wielebinski) were both closer to the engineering tradition than von Hoerner would have been. The first addition, who would quickly overshadow Hachenberg himself to become one of the most powerful figures in the Max Planck Society, was Peter Mezger, who, as was mentioned above, started out as an engineer in Munich and was an apprentice in radio astronomy in France in the postwar years.

After intermittent employment with Siemens, Mezger was also recruited to work at Green Bank where, in the course of the 1960s, a whole contingent of Germans had been established, initially aided by the appointment of the emigré optical astronomer Otto Struve as its first director, the reason why von Hoerner had been offered a position there in the first place. In Green Bank, Mezger specialized in the detectors and antenna construction for increasingly shorter radio wavelengths: while the first generation of postwar radio telescopes worked in a range of wavelengths of tens of centimeters, Mezger was part of a new generation working on wavelengths under one centimeter, which require much more precise reflective surfaces, distinct detectors, and much higher, clearer geographical locations than the first generation of radio telescopes.[55]

54 Leverington, *Observatories and Telescopes*, 2017, 438–444. Peter G. Mezger: interview by Woodruff T. Sullivan, Bonn, November 22, 1973. Transcript. NRAO Archives, http://www.nrao.edu/archives/Sullivan/sullivan_transcript_mezger_1973.shtml. Last accessed 1/4/2019.

55 Peter G. Mezger: interview by Woodruff T. Sullivan, Bonn, November 22, 1973. Transcript. NRAO Archives, http://www.nrao.edu/archives/Sullivan/sullivan_transcript_mezger_1973.shtml. Last accessed 1/4/2019. Baars, *International Radio Telescope Projects*, 2013.

To complete the triumvirate in Bonn, there came the appointment of Richard Wielebinski, an engineer from an émigré Polish family settled in Australia. Wielebinski, the electronic detector engineer recruited from the Parkes telescope, was one of the highest profile 'foreign' directors hired by the Max Planck Society to date, and over the next ten years he became the one who would lead Effelsberg to decades of scientific productivity.[56]

It will be described later how Bonn also showcased conflicts brought about by the collegiate directorship of Max Planck Institutes, particularly when appointees were of the old guard generations who came of age before the war. Regardless of these conflicts, however, the directors in Bonn shared an engineering tradition focused on building many of the best, most innovative radio telescopes in the world, in clear contrast to the more theoretically embedded tradition from Göttingen/Munich and the experimental physics tradition of Germany's southwest.

Not only was the construction of Effelsberg a unique opportunity for the Max Planck Society to achieve world leadership in a scientific field, it also paved the way to an immense observational astronomy program that ended up accounting for most of the work at three of its largest Max Planck Institutes. Once the field of observational astronomy had gained recognition in the Society as a desirable scientific pursuit, and the organization had secured its status by running national projects, several other projects soon followed.

Optical Astronomy in Baden-Württemberg and at the European Southern Observatory

In optical astronomy, Hans Elsässer, a disciple of Siedentopf's who had been appointed head of the Königstuhl observatory in Heidelberg in 1962, negotiated the founding of a Max Planck Institute for Optical Astronomy, and the creation, following the Effelsberg model, of the first major German optical observatories located in favorable geographical locations in the northern and

---

56 Klaus Jäger: Schwarzschild-Medaille der Astronomischen Gesellschaft für Richard Wielebinski. *idw-Informationsdienst Wissenschaft*, 9/5/2017. https://idw-online.de/de/news680475. Last accessed 4/12/2018. For more on Wielebinski's early life and career, see Michael Globig: Zur Person. Richard Wielebinski. *Max Planck Forschung* 4 (2001), 98–103. For an overview of the main observational results see Wielebinski, Junkes, and Grahl, Effelsberg 100-m Radio Telescope, 2011, 3–21. See also Richard Wielebinski: Reminiscences of a Radio Astronomer. *Journal of Astronomical History and Heritage* 24/4 (2021), 1103–1122. https://ui.adsabs.harvard.edu/abs/2021JAHH...24.1103W. Last accessed 1/13/2022.

southern hemisphere.[57] Elsässer's meteoric rise, marked by his appointment in Heidelberg, was powered also by his authorship of the optical astronomy section of the memorandum on the future of astronomy, in which he had argued for the construction of large national telescopes.[58] This memorandum itself rode on the back of Sputnik, and was published a few years after a similar memorandum had established the research program for the space sciences.[59] In the early 1960s, Elsässer was also the representative of astronomers on the committees for space research initiated by the West German government.[60]

Unlike the privately funded radio telescopes, support in the case of optical astronomy came directly from the federal ministry; but the optical observatory project was executed by the same MAN/Krupp consortium behind the radio telescopes, with the Zeiss optical company in nearby Oberkochen in charge

57 For a general history of the Institute for Optical Astronomy, see Lemke, *Himmel über Heidelberg*, 2011, Vol. 21. Documents related to the founding of the Institute can be found in AMPG, II. Abt., Rep. 66, No. 365, 366, 375; Rep. 62, No. 447. Max Planck Society plasma physicists, including Biermann as well as Lüst, played a fundamental role in promoting the foundation of the Institute for Astronomy and the Institute for Radio Astronomy. Biermann had come in contact with radio astronomy already during the war, when he worked at Babelsberg Observatory. They invited Kurt Fränz, a pioneer of German radio astronomy, with whom he discussed early radio observations at a time when astronomers were far from being interested in the potential of radio waves for astronomy. Biermann, in particular, was interested in the propagation of radio waves through a plasma. Ludwig Biermann: interview by Woodruff T. Sullivan III, September 15, 1978. Transcript. NRAO Archives, https://www.nrao.edu/archives/items/show/896. Last accessed 2/4/2022. See also Biermann's memorandum on "Deutsche Südsternwarte" dated April 28, 1966, mentioning the importance of astronomy and research on quasars for the large-scale structure of the Universe and the questions on the structure of space and the nature of gravitation written when relativistic astrophysics was already becoming an established and quickly developing field, on the verge of exploding in connection with the upcoming breakthrough discovery of pulsars (AMPG, II. Abt., Rep. 66, No. 367, Fol. 283–287). This folder also contains general material related to the activities of the Institute for Astronomy and discussions on the establishment of ESO and the Calar Alto and Chile observatories.

58 Voigt et al., *Denkschrift Astronomie*, 1962. For early drafts of this *Denkschrift*, see AMPG, II. Abt., Rep. 66, No. 365, 375.

59 Gotthard Gambke, Rudolf Kerscher, and Walter Kertz: *Denkschrift zur Lage der Weltraumforschung*. Wiesbaden: Franz Steiner Verlag 1961.

60 For example, the scientific committee on research satellites in 1964 was already dominated by Max Planck interests, as it included Reimar Lüst from the Institute for Extraterrestrial Physics, and Bartels and Ehmert from the Institute for Aeronomy. In addition to Elsässer, who would soon become a Max Planck Institute director himself, the other two members were Martin Paetzold from Cologne and Fränz from Ulm. See Johannes Weyer: *Akteurstrategien und strukturelle Eigendynamiken. Raumfahrt in Westdeutschland 1945–1965*. Göttingen: Schwartz 1993. 297.

of the telescopes themselves.[61] As we saw earlier, Baden-Württemberg had been the site foreseen for the section of the Max Planck Institute for Radio Astronomy that would have been headed by von Hoerner in Tübingen, had Hachenberg and Bonn not come to dominate the project. Now Hans Elsässer was evening the score in this regional rivalry.

The foundational idea for the institute was to build two identical large telescopes of around 2.2 m in diameter, to be located at sites on the northern and southern hemisphere. Then, a third, gigantic telescope with a diameter of around 3.5 m (initially 4 m), to compete with the world's largest, would be housed in one of the observatories. The initial proposals hinted at a southern location for the giant telescope, where it would be competing directly with the ESO's. However, during the early planning phases, expert committees decided, against Elsässer's wishes, to locate it in the northern hemisphere, in order to reduce the construction and operational costs, to speed up its completion, and so it would serve as a complement to ESO. The chosen site was in Spain (still under the Franco dictatorship), in the southern province of Almería. The southern hemisphere observatory was to be located on the former colonial territory of German Southwest Africa, now Namibia, where observational conditions were better. The potentially problematic selection of this host country (then under South African rule) was not a consideration for Elsässer, who defended his choice based on what he felt were exclusively scientific criteria, in his view in contrast to ESO, which he thought had moved to Chile due to the interference of political considerations in scientific ones.[62] In the early 1970s, during the period of political instability between the Allende presidency and the beginnings of the military dictatorship, Elsässer appeared to have made the better choice. He was well aware of the political difficulties experienced by astronomers in Chile in the early 1970s that further justified the choice of an Africa location.

---

61 Zeiss Oberkochen played a major role in the rivalry between East and West Germany, as it had been established by members of Zeiss Jena invited to migrate to the American sector at the end of the war. The two companies competed in many fields throughout the Cold War, and Zeiss Jena even continued to provide optical telescopes and parts for West German observatories. The 1962 *Denkschrift* emphasizes the need for national telescopes in West Germany to also gain supremacy in a field where East Germans still had the upper hand. See Voigt et al., *Denkschrift Astronomie*, 1962. See also Armin Hermann: *Und trotzdem Brüder. Die deutsch-deutsche Geschichte der Firma Carl Zeiss*. München: Piper 2002.

62 This episode is recounted in Hans Elsässer: *Weltall im Wandel. Die neue Astronomie*. Reinbek bei Hamburg: Rowohlt 1989.

However, as will be seen repeatedly throughout this book, the Spanish observatory and its large telescopes are considered to be one of the largest *completed* failures of the Max Planck Society, while the second observatory planned for the southern hemisphere did not even come to fruition, and its telescope was eventually installed by ESO in Chile, as we describe later in this chapter and the next. What saved the scientific reputation of the Heidelberg institute, during the decades of observatory construction and subsequent disappointing scientific output, was the department or division of airborne and space-based infrared astronomy, which we discuss in more detail in Chapter 4. Elsässer himself had conducted his early scientific career in high-altitude astronomy on alpine stations, and this work favored the establishment in Heidelberg of a Department for Infrared Astronomy, which was directed by Dietrich Lemke and obtained funding of its own separately from the ministry's space science budget. At some point, it was attempted to turn it into an independent sub-institute under the name of 'Extraterrestrial Astronomy,' but such new sub-institutes were by then no longer standard Max Planck Society practice.[63] In any case, scientifically, this department remained relatively unaffected by the institute's observatory-building difficulties. As will be explained in Chapter 4, if anything, the main problem of this space research team based in Heidelberg was its competition with work carried out at the Max Planck Institute for Extraterrestrial Physics, and Heidelberg was quite successful here, as the Garching institute struggled for many years with its own attempts at a division of infrared astronomy.[64]

German Industrial Partnerships vs the ESO In-House Approach at CERN

With the astronomical institutes in Bonn and Heidelberg, a 'business model' in German national astronomical projects began to emerge, part of the reasoning behind which was to help national industries in technological fields develop products that could later be offered on a wider global scale, in this case, telescopes; and attempts were made in the late 1970s and 1980s to sell to third countries replicas of both radio and optical telescopes originally designed for the Max Planck Society. This even came to be encouraged by the Max Planck

63 The Max Planck Society's budgets, which are a good indicator of which units are considered independent sub-institutes, listed a Department for Extraterrestrial Astronomy during the 1970s. Ultimately, however, the Society was at the time moving toward unitary institutes with collegiate membership and this unit was reabsorbed into the unitary budget of the institute. AMPG: Haushaltspläne: II. Abt. Rep 69.

64 See Chapter 4.

Society in the 1970s, when institute mentors encouraged directors to commercialize their telescopes and observatories, placing a clause in the industrial contracts stating that the development costs of the instruments would be refunded if a second buyer was found. This eventually happened with the Iraqi National Observatory.[65]

One of the most enduring developments to come out of this partnership approach was the development of the glass-ceramic material Zerodur®, with which the telescope mirrors were made. Despite the disappointments of Calar Alto, and even the failure of Zeiss itself, which has now abandoned large telescope construction, the material, developed by the West German offshoot of Schott Mainz, continues to be used today for the world's largest optical and space-based telescopes, now led by international organizations. But the most profitable uses by far of this material are now non-astronomical, in microelectronic components requiring temperature stability.[66]

These attempts at a national optical observatory in Germany ran in parallel with the establishment of the European Southern Observatory which, as mentioned above, was modeled on CERN around the common need for an observation site in the southern hemisphere. On the German side, two significant astronomers participated in the early years of ESO: Heinrich Siedentopf in Tübingen and Otto Heckmann in Hamburg. Heckmann was in fact the first German to act as director of an international scientific organization, although its scale during his directorship was very small and his initial role involved little more than setting up meetings between representatives of the contributing countries.[67] Heckmann himself was a somewhat problematic choice of director because, while a brilliant astronomer who had, for example, defended general relativity against more ideological scientists during the Nazi era, he was also seen to have generally collaborated with this regime.[68] He carried this reputation into the postwar era, and it was only with the intervention

65 See Baars, *International Radio Telescope Projects*, 2013, 150–152. The observatory, still under construction was destroyed during the war with Iran.

66 Markus Voelter: *Once You Start Asking. Insights, Stories and Experiences from Ten Years of Reporting on Science and Engineering*, 2020, 278–281.

67 Blaauw, *ESO's Early History*, 1991.

68 Seiler, *Kommandosache*, 2007., 225. For example, when working as a visiting researcher at the Yerkes Observatory in Chicago—which under Otto Struve's direction in the 1930s and 1940s had become the main center of astrophysics in the US—Kiepenheuer was allowed to stay in the boarding house there together with researchers from other nations. Heckmann in contrast, due to his designation by Gerard Kuiper as having been loyal to the Third Reich, had to find private accommodations instead.

of American interests that all the European partners came to overcome their dislike of Heckmann and set up the organization.[69]

Nevertheless, in its first decade, it was a very loose institution deliberately kept small and decentralized by the constituent countries, which wanted to keep it largely as administrator of the astronomical sites where member countries would each install and control their own observatories.[70] For the first period, the major consideration was the selection of this site, which shifted from South Africa to Chile in consideration of both quality of the skies and the impending political isolation of the apartheid regime.[71] While these sites were being evaluated and comparisons made between South Africa and Chile (in which Hans Elsässer himself participated), Siedentopf, Elsässer's mentor, passed away. After this, Heckmann and Elsässer became adversaries, one in charge of a European project, the other of its national counterpart. For example, when consulted about a national observatory, or a Max Planck Institute for Astronomy, Heckmann deemed it unnecessary.[72]

Heckmann was instrumental in the choice of Chile which, as we will see later, turned out to be the most valuable asset for ESO. On the other hand, during the next phase of the organization, namely the venture to build a large telescope jointly with all member states, Heckmann failed as a result of attempting to concentrate all telescope manufacture using a local company in Hamburg. By the late 1960s, this led to Heckmann being ousted as director general of ESO, to be replaced by Adrian Blaauw from the Netherlands, who had been 'part-time' scientific director in the period 1968–69.[73]

Blaauw, who had a very close relationship to CERN, made the next crucial decision that would set ESO apart from its competitors. In an attempt to save the large telescope project, he created the Telescope Project Division on the CERN campus in Geneva, under an ESO-CERN cooperation agreement, and moved its manufacture directly to CERN, where teams also working on the

69 John Krige, presentation at the Workshop "Opening New Windows on the Cosmos: Astronomy and Astrophysics in the History of the Max Planck Society," Max Planck Institute for the History of Science, September 6–8, 2016. *Research Program History of the Max Planck Society*, 2017, 98–101.

70 Blaauw, *ESO's Early History*, 1991.

71 See, for example, Lodewijk Woltjer: *Europe's Quest for the Universe. ESO and the VLT, ESA and Other Projects*. Les Ulis: EDP Sciences 2006, 92–93.

72 Lemke, *Himmel über Heidelberg*, 2011, Vol. 21, 31. Heckmann's signed approval was needed for the institute and observatories, and Reimar Lüst had to do extraordinary diplomatic work to obtain it.

73 Blaauw, *ESO's Early History*, 1991, 8.

latest particle accelerator, the Super Proton Synchrotron (SPS), brought modern project management and manufacturing techniques to ESO's telescope, a relatively minor task for them.[74] The telescope was finished very quickly, on schedule and within budget, and, furthermore, it proved more innovative than what was offered by established industrial manufacturers.[75] Following the practice of CERN, instrument development was done in-house, and only when the specifications and instrumental design were ready were these contracted out to manufacturers. This was, for example, the opposite of the standard procedure of the Max Planck Institute for Astronomy, which contracted its three large telescopes out to Zeiss without much interest in a close co-development of the devices.[76]

The different approaches taken by ESO and the Max Planck Institute in Heidelberg resulted in the former's similarly sized large telescope being completed almost a decade earlier (1977 compared to 1986), while both initiatives had begun around the same time. By the time the large Calar Alto telescope started operations, ESO was finishing its much more innovative New Technology Telescope (NTT), which made the Max Planck telescope embarrassingly out of date in comparison.[77]

---

74 The ESO–CERN joint venture at the time was also instrumental in creating the first occasions where astroparticle physics—the new emerging discipline encompassing particle physics, cosmology, and astrophysics—could find a dedicated common space for discussion. Giancarlo Setti, and Léon van Hove (eds.): *Large-Scale Structure of the Universe, Cosmology and Fundamental Physics. First ESO-CERN Symposium, CERN, Geneva, 21–25 November 1983. Proceedings.* Garching: European Southern Observatory 1984. Giancarlo Setti, and Léon van Hove (eds.): *Cosmology, Astronomy and Fundamental Physics. Second ESO-CERN Symposium, ESO, Garching Bei München, 17–21 March 1986. Proceedings.* Garching: European Southern Observatory 1986. The first international school on astroparticle physics, organized in conjunction with the ESO-CERN symposia on cosmology and fundamental physics, was held at the 'Ettore Majorana Centre for Scientific Culture,' in Erice, Sicily, January 5–25, 1987. See also Christine Sutton: ESO and CERN: A Tale of Two Organizations. *CERN Courier* 52/8 (2012), 26–30. https://cds.cern.ch/record/1734856. Last accessed 5/4/2020.

75 Blaauw, *ESO's Early History*, 1991, 9.

76 Immo Appenzeller: interview by Juan-Andres Leon, August 2016. In fact, the technical expertise required to work with astronomical instruments in the early years of the MPIA/the Max Planck Institute for Astronomy continued to be provided by personnel from the neighboring state observatory (*Landessternwarte*).

77 Leverington, *Observatories and Telescopes*, 2017. The NTT had an altazimuth mount and was equipped with active optics to counteract the deformations of the system caused by gravity on very large telescopes. Both these innovations allowed the telescopes to be much lighter and, consequently, less expensive. Altazimuth mounts had been the standard in radio astronomy since the 1960s, and the concept of active optics is the optical

ESO's period in CERN, however, could not be permanent, as Germany insisted that it should remain the headquarters of the organization.[78] Because of Elsässer's rivalry with this organization,[79] the site considered for the headquarters was not Heidelberg, but rather Garching, where, as we will see in the next chapter, ESO worked much more closely with the Max Planck Institutes for Extraterrestrial Physics and for Astrophysics.[80]

Inter-institute Coordination and the Role of Institute 'Mentors'

The observatory institutes in Bonn and Heidelberg were fiercely independent within the Max Planck Society as well as from each other, as they both obtained external funding for their very expensive observatories, either through private donations (Bonn), or federal government funding (Heidelberg).[81] They also represented very different research traditions and distinct international partners. They even inherited the international rivalry between optical astronomers and radio astronomers. Despite all this, they shared a distinct philosophy of observational astronomy that gave precedence to general-purpose instrument construction over the close coupling of instruments and experiments with theoretical questions, in contrast to what had been the case in older traditions of the Max Planck Society in Heidelberg and Munich. After all, at the time, many discoveries in astronomy (particularly those related to radio astronomy) continued to be determined by the race toward new instruments coupled with sheer luck.[82]

---

equivalent—much more difficult to implement—of what Sebastian von Hoerner had suggested with his homologous design in radio telescopes. For these reasons, a former director of ESO informally referred to Calar Alto as the "last renaissance telescope."

78 Claus Madsen: *The Jewel on the Mountaintop. The European Southern Observatory through Fifty Years*. ESO 2012.

79 For more on this rivalry and the mediating role of Reimar Lüst, see the interview with Lüst conducted by Jakob Staude, published in the Annual Report (*Jahresbericht*) of the Max Planck Institute for Astronomy, 2009, pp. 121–23.

80 In fact, the buildings of ESO and the Max Planck Institute for Astrophysics in Garching (after this moved from Munich-Freimann) are neighbors, were built by the same company, and share the same architectural style. See Peter Gruss, Gunnar Klack, and Matthias Seidel (eds.): *Fehling+Gogel. Die Max-Planck-Gesellschaft als Bauherr der Architekten Hermann Fehling und Daniel Gogel*. Berlin: Jovis 2009.

81 See, for example, the Max Planck Society's budgets. The Effelsberg telescope was not even registered in the budgets, whereas the telescope funding of the Max Planck Institute for Astronomy (MPIA) was registered as large project investment from the Federal Ministry of Research and Technology. For more details, see the Financial Appendix at the end of this book.

82 Kenneth Kellermann, and B. Sheets: *Serendipitous Discoveries in Radio Astronomy. Proceedings of a Workshop Held at the National Radio Astronomy Observatory Green*

In addition to this epistemological proximity, both institutes shared one feature of the Max Planck Society which would become central in their first decades of activities: the figure of Günther Preiss as intermediary between the institutes and the president of the Society. One of the key characteristics during the presidency of Butenandt (see Chapter 1) was the Society's increasing monopoly on the relationships between its institutes and the 'outside world,' and these interactions were carefully managed at the highest level by the *Institutsbetreuer*, who acted as supervisors and liaison officers between the General Administration and the institutes.[83]

Throughout the most intense infrastructure-building period of the observatory institutes, Preiss was the link between directors and researchers at these institutes, with the interests, pressures, and influences coming from the presidency, the federal ministries involved, industrial partners, and financial supporters. Most salient among these were the interests from the 1960s to the 1980s in consolidating the industrial partnership between Krupp and MAN, which ultimately built most of their observatories, as well as the EISCAT (European Incoherent Scatter Scientific Association) installations for the Max Planck Institute for Aeronomy.[84] Preiss personally navigated the minefield of industrial contracts and international negotiations related to the first major presence of the Max Planck Society outside of Germany, with two observatories in Spain and one in Southwest Africa, as well as later in Arizona. The available primary sources and interviews indicate that in both Bonn and Heidelberg the institute's 'mentor' Preiss was seen as an ally in finding compromises between their point of view and these outside forces.[85] Similarly,

---

Bank, West Virginia on May 4, 5, 6, 1983. Green Bank, WV: National Radio Astronomy Observatory, Associated Universities 1983. http://library.nrao.edu/public/collection/02000000000280.pdf. Last accessed 3/21/2021.

83 Through their meticulous work and close involvement with their particular Max Planck Institutes, *Institutsbetreuer* (institute mentors), are the largest source of archival material for the AMPG. For every institute, there are meters and meters of their files (AMPG, II. Abt., Rep. 66).

84 Gerhard Haerendel: History of EISCAT. Part 4. On the German Contribution to the Early Years of EISCAT. *History of Geo- and Space Sciences* 7/2 (2016), 67–72. doi:10.5194/hgss-7-67-2016. For documents related to the EISCAT project and the draft agreement (1974–1975) of the EISCAT collaboration (France, Sweden, Norway, and Germany, MPG) see DA GMPG, BC 105528, 105537, 105538, 105539.

85 For examples of mediation by Preiss, see Baars, *International Radio Telescope Projects*, 2013, 33, 65, 79. Preiss was a lawyer, so had been wise enough to include in the contracts with Zeiss, MAN, and Krupp the provision that if, in the coming years, these companies should build telescopes based on those developed by the Max Planck Institutes, they would have to compensate the institutes for the original research and development costs

in the 1970s, Preiss would be crucial in setting up the GALLEX international collaboration at the Max Planck Institute for Nuclear Physics, which included a major donation by the Krupp Foundation, described in detail in Chapter 5.[86]

2 High-Energy Space-Based Astronomy

By the mid-1960s, there were initial attempts, internationally, to base astronomical observatories directly in outer space, a decades-old dream, as many wavelengths are blocked by the atmosphere even at mountain altitudes. This section follows the transition of the Max Planck Institute for Extraterrestrial Physics, from an early nuclear era focused on near-Earth space plasma experiments to the institution's increasing dedication to space astronomy. As with the ground-based astronomers, Max Planck leaders invited external pioneers in the field to become directors and participated in several revolutionary astronomical satellites in the gamma-ray and X-ray domains. Space-based astronomy was embedded in European collaboration as well as in competition with the United States. These satellite observatories then guaranteed further German—and hence Max Planck scientists'—participation in all the major missions in these fields, in Europe, the United States, and the Soviet Union. High-energy space-based astronomers differed significantly from their ground-based colleagues, having come from a tradition of experimental particle physics, and their appointment further shifted the center of gravity away from the plasma astrophysicists of previous decades.

Early Interest in Satellites at the Institute for Extraterrestrial Physics: Gamma-Ray Astronomy

From the late 1950s, Bruno Rossi and his collaborators at MIT—a group which had played a leading role in cosmic ray research since the early postwar years—were leaping into the dimension of space with visionary and challenging ideas about detecting cosmic gamma rays and X-rays from extrasolar sources, pioneering the birth of these new branches of astronomy. In 1961, the Explorer XI satellite carried into Earth's orbit the first gamma ray telescope

incurred. This provision led to the collaboration on building observatories in Iraq mentioned in this chapter.

86 Till Kirsten: interview by Luisa Bonolis and Juan-Andres Leon, Heidelberg, October 24–25, 2017. DA GMPG, BC 601051. Till Kirsten, personal collection of documents, DA GMPG, BC 600004, BC 600005.

built by MIT scientists William L. Kraushaar and George W. Clark.[87] In fall 1959, Rossi had also initiated a project for the detection of X-rays of extrasolar origin at American Science & Engineering (AS&E), a manufacturer of advanced X-ray equipment and related technologies, also specialized in detection of X-rays from bomb tests.[88] This project led to the unexpected discovery in June 1962 of Scorpius X-1, an object that emitted a thousand times more X-rays than the Sun, demonstrating the existence of a new class of stellar objects in which unknown physical processes were taking place.[89] Space research was opening new spectral regions as well as new regions of space to scientific investigation. The emergence of these new fields, requiring detection techniques drawn from experimental physics, was opening the domain of (high-energy) astrophysical research to cosmic ray physicists. From the early 1960s, increasingly sophisticated gamma ray space missions and satellites began to operate, and competing in this field became one of ESRO's main objectives.[90] As described in the previous chapter, this organization, the European Space Research Organization founded in 1964 by ten European nations and promoted by Edoardo Amaldi and Pierre Auger, was based on the successful model of CERN, among the main founding fathers of which Amaldi and Auger had numbered.[91]

In fall 1961, when a Department of Extraterrestrial Physics had just been established in the Max Planck Institute for Physics, Reimar Lüst was visit-

---

87 William L. Kraushaar, and George W. Clark: Search for Primary Cosmic Gamma Rays with the Satellite Explorer XI. *Physical Review Letters* 8/3 (1962), 106–109. doi:10.1103/PhysRevLett.8.106.

88 AS&E had been founded in 1958 by Martin Annis, Rossi's former student at MIT. Rossi was Chairman of the Board of Directors, which also included George W. Clark as a main scientific consultant of the Society. Rossi, who was not able to start such activity at MIT because his group was already too busy with the preparation of the solar probe and other commitments, notably the large cosmic ray shower array at Volcano Ranch, pushed toward the search for extra solar X-ray astronomy also because he hoped to redirect the company's activities toward more scientific and fundamental aims. Martin Annis: interview by Luisa Bonolis, Boston, MA, September 30, 2006.

89 Riccardo Giacconi et al.: Evidence for x Rays From Sources Outside the Solar System. *Physical Review Letters* 9/11 (1962), 439–443. doi:10.1103/PhysRevLett.9.439. They also found the existence of a diffuse X-ray background across the area of sky measured.

90 Volker Schönfelder: The History of Gamma-Ray Astronomy. *Astronomische Nachrichten* 323/6 (2002), 524–529. doi:10.1002/1521-3994(200212)323:6<524::AID-ASNA524>3.0.CO;2-Z.

91 Michelangelo De Maria: *Europe in Space. Edoardo Amaldi and the Inception of ESRO.* ESA-HSR-5. Noordwijk, the Netherlands: ESA Publications Division 1993. See also Lüst, The European Space Research Organization, 1965, 394–397.

ing professor at the Massachusetts Institute of Technology,[92] where he had an opportunity to follow, in person, the pioneering attempts taking place there, in the field of gamma and X-ray astronomy. At the time, Rossi's group was also preparing a further groundbreaking experiment devised to explore the conditions of near-Earth space plasmas and the Earth–Sun relation. The Earth-orbital satellite Explorer X launched in March 1961 was instrumented with two fluxgate magnetometers and the MIT plasma probe, which measured a steady flux of protons in the space around the Earth's magnetosphere and established the existence of a geomagnetic cavity, a region of space surrounding the Earth, which is shielded from the solar wind by the Earth's magnetic field.[93] These early measurements prepared the ground for the complete vindication of Eugene Parker's theory of the solar wind, having some of its main roots in Biermann's hypothesis based on the study of comet tails (see Chapters 1 and 2). On the basis of the experience gained in 1962 from a recent six-month stay at MIT and at the California Institute of Technology in Pasadena, Lüst suggested that no further time should be lost. Other countries, as well as other groups in the US, were entering the field. The Max Planck Society would be a particularly suitable framework for contributing to space activities with fundamental research in the broadest sense of the term.[94] The recent US results could be used as a source of information on plasmas and magnetic fields in the nearby interplanetary space for planning new space experiments, and this actually became a main activity at the Institute for Extraterrestrial Physics, which was actually established in May 1963, transforming Lüst's department into a sub-institute of the Max Planck Institute for Physics and Astrophysics.

The advent of the space age had been instrumental in accelerating the process of making astrophysics a respectable branch of physics. What Lüst had experienced in the US, in particular at MIT, was seeing new realms such as space science and gamma- and X-ray astronomy opened up by physicists—often migrating from cosmic ray physics—who were able to build their own

92 Ludwig Biermann: Jahresberichte deutscher astronomischer Institute für 1961 München, Max-Planck-Institut für Physik und Astrophysik, Institut für Astrophysik. *Mitteilungen der Astronomischen Gesellschaft* 15 (1962), 68–74, 69. http://adsabs.harvard.edu/abs/1962MitAG..15...68. Last accessed 10/30/2018. The following year, in 1962, Lüst visited Caltech in Pasadena.

93 Luisa Bonolis: From Cosmic Ray Physics to Cosmic Ray Astronomy. Bruno Rossi and the Opening of New Windows on the Universe. *Astroparticle Physics* 53 (2014), 67–85. doi:10.1016/j.astropartphys.2013.05.008.

94 See report on the situation of space research in Germany written by Lüst in 1962, after his second stay in the US (AMPG, II. Abt. Rep. 66, No. 3048, Fol. 21–26).

optical and electronic devices for research, setting the stage on which astronomy would eventually approach the scale of high-energy physics. With its deep roots in the Institute for Physics and Astrophysics in Munich, the Institute for Extraterrestrial Physics' strength lay both in theory and in the special skills and techniques typical of physicists.[95] In the period 1963–65, Lüst began to make plans to launch space science, but he also realized that space plasma physics "was too narrow a basis" for his institute;[96] and thus he decided to extend research activities to gamma and X-ray astronomy, too.[97] An aspect of the crucial role of Lüst's travels to the US in the early 1960s was emphasized by Klaus Pinkau:

> [Lüst] came to the conclusion that research in X-Ray astronomy in the US was so far advanced that he had no chance to catch up. In Gamma Ray astronomy on the other hand, and looking at the work of Kraushaar and Clark, Lüst considered that a new activity at MPE would have the chance of catching up with international standards and this is the reason why he thought that the new MPE could well do Gamma Ray Astronomy.[98]

By February 1964, Lüst had submitted to ESRO an experiment for the measurement of gamma rays, whose purpose was the determination of extrasolar

95 As emphasized by Martin Harwit, "Observational discovery comes on the heels of technological innovation, giving physics an increasingly dominant role in astronomy." Martin Harwit: Physicists and Astronomy—Will You Join the Dance? *Physics Today* 34/11 (1981), 172–187. doi:10.1063/1.2914355. See Figs. 4 and 5 showing how two-thirds of the major postwar discoveries in observational astronomy were made by physicists.

96 Joachim Trümper: Astronomy, Astrophysics and Cosmology in the Max Planck Society. In: André Heck (ed.): *Organizations and Strategies in Astronomy*. Dordrecht: Springer Netherlands 2004, 169–187, 75.

97 On preliminary work in view of the development of future devices for detection of X- and gamma rays from astrophysical sources, see the first annual report of the Institute for Extraterrestrial Physics (*Tätigkeitsbericht 1963–1965 des Instituts für extraterrestrische Physik am Max-Planck-Institut für Physik und Astrophysik. MPI-PAE Extraterr. 1 (22/66), Januar 1966*, 37–39), a copy of which can be found in BArch, B 196/7170. A scanned copy is available at http://www.mpe.mpg.de/303552/jb1963-1965.pdf. Last accessed 10/30/2018. The institute also announced a proposal on ultraviolet spectrophotometry for the detection of interstellar molecular hydrogen within an international collaboration planning the Large Astronomical Satellite (LAS), one of the first and main engagements of ESRO. See also R. Lüst, "Memorandum zur Weltraumforschung" (June 1964) where the possibility of extraterrestrial observations of the whole electromagnetic spectrum is, of course, mentioned among the future space activities (AMPG, III. Abt., ZA 1, No. 91).

98 Klaus Pinkau to Luisa Bonolis, October 15, 2016.

sources of high-energy gamma rays likely produced by "the interaction of primary charged cosmic rays with the interplanetary medium."[99] In 1965, the new field of gamma astronomy in Garching began in earnest with the appointment of Klaus Pinkau,[100] who had been an experimental cosmic ray physicist in Hamburg and Kiel with Erich Bagge, and in Bristol with Cecil Powell, as well as visiting scientist at Louisiana State University, with expertise in the detectors then used for elementary particle research with cosmic rays.[101] When CERN

99 See proposal "Extraterrestrial measurements of gamma-rays in the energy range above 50 MeV" (February 1964), with group leader Reimar Lüst (Historical Archives of the European Union, S-78, COPERS-1236, https://archives.eui.eu/en/fonds/96556?item=COPERS-06.01-1236. Last accessed 6/20/2020).

100 Larger-scale research on gamma ray astronomy began with the arrival of Klaus Pinkau at the institute on December 1, 1965. Ludwig Biermann, and Reimar Lüst: Jahresberichte astronomischer Institute für 1965, München Max-Planck-Institut für Physik und Astrophysik, Institute für Astrophysik und extraterrestrische Physik. *Mitteilungen der Astronomischen Gesellschaft* 20 (1966), 67–79. http://adsabs.harvard.edu/abs/1966MitAG..20...66. Last accessed 10/30/2018. For material related to research activity of the Institute for Extraterrestrial Physics in the 1960s see AMPG, III. Abt., Rep. 145, No. 230, 293, 771, 772, 773, 872, 874, 875, 887, 911, 992. Pinkau was also in charge of cosmic-ray research.

101 See Pinkau's publications up to 1966. Bagge had been visiting the United Kingdom in 1951 and had discussed the problem of cosmic rays with Powell and Patrick M. Blackett, the most influential UK scientists in the field, both Nobel laureates (Bagge to Biermann, March 30, 1951, AMPG, III. Abt., ZA 1, No. 1). At the end of the 1940s, heavy nuclei had been discovered to be a component of cosmic rays. Moreover, the recent discovery of the pion in cosmic rays by Powell's group with improved nuclear emulsions had solved a long-standing problem of cosmic ray research, contributing to the beginning of modern particle physics. Owing to his scientific interests, Bagge was fond of measurement methods and measurement techniques, and had developed in his institute a spark chamber with very fast rise time, important for short-term measurements, and now wanted to introduce the nuclear emulsions as part of his "collection of measurement techniques" (Klaus Pinkau: Interview by Helmuth Trischler, March 9, 2010. Transcript, Historical Archives of the European Union, Oral History of Europe in Space Collection, from now on HAEU, https://archives.eui.eu/en/oral_history/INT072. Last accessed 1/4/2019). In 1954 Pinkau had been given a fellowship to go abroad, and thus Bagge asked him to go to Bristol, work with Powell's group, and learn the nuclear emulsion technique. Pinkau remained in Bristol from 1955 to 1960 and worked there on his Master and PhD theses. At that time, "the hunt for new particles using the nuclear emulsion technique was nearing its end. What remained was the study of heavy and highly charged cosmic rays, and high energy interactions—the so-called 'jets.' Also, there were a number of 'soft cascades,' purely electromagnetic cascades that originated from high energy gamma rays that had entered the emulsion stack from outside as part of a large cosmic ray shower. No one had shown an interest to analyze them, and the entire material available was given to me as my field of work." Pinkau formulated a theory for the lateral distribution of cascade electrons which allowed the energy of primaries generating the shower to be determined.

and American facilities around the end of the 1950s shifted experimental particle physics overwhelmingly toward accelerators, Pinkau—like many of his colleagues in the field—saw the traditional use of cosmic rays as an efficient source of events for particle physics become obsolete, and so he found new applications for his instrumental expertise.

Interest was shifting to the already established study of the extensive cosmic ray showers generated in the atmosphere by high-energy primary particles, and this revived the study of their nature and origin, as messengers originating in extreme astronomical environments which could be studied directly by radio astronomers. In 1950, the two astrophysicists Hannes Alfvén and Nicolai Herlofson, along with Karl-Otto Kiepenheuer (see Chapter 1), had proposed a theory explaining the phenomenon of radio emissions as originating from ultrarelativistic electrons spiraling in weak interstellar magnetic fields and emitting synchrotron radiation (also known as *Magnetobremsstrahlung*, magnetic braking radiation).[102] Between the end of the 1940s and the early 1950s, only a few scientists, such as Alfvén, Biermann, Chandrasekhar, and later, Eugene Parker, realized the potential role of plasmas and magnetic fields in the Universe.[103] The connection established between cosmic rays (in particular their electron component) and cosmic radio emission of a synchrotron

---

This work became his PhD thesis (Klaus Pinkau to Luisa Bonolis, October 15, 2016). This allowed the Bristol group to use simple methods to determine the energies of the gamma rays originating in high-energy nuclear interactions arising from the neutral pion decay. Klaus Pinkau: Energy Determination of Electromagnetic Cascades in Nuclear Emulsions. *The Philosophical Magazine: A Journal of Theoretical Experimental and Applied Physics* 2/23 (1957), 1389–1392. doi:10.1080/14786435708243215.

102 Hannes Alfvén, and Nicolai Herlofson: Cosmic Radiation and Radio Stars. *Physical Review* 78/5 (1950), 616–616. doi:10.1103/PhysRev.78.616. Karl-Otto Kiepenheuer: Cosmic Rays as the Source of General Galactic Radio Emission. *Physical Review* 79/4 (1950), 738–739. doi:10.1103/PhysRev.79.738. The synchrotron mechanism at the time seemed mysterious and speculative to astronomers, but probably not to solar astrophysicists such as Kiepenheuer and Biermann, who were accustomed to think in terms of interactions between plasmas and magnetic fields in the Sun. Albrecht Unsöld, too, showed an early interest in radio astronomy observations and contributed to the interpretation of cosmic radio emissions. An overview of his work in the wider context of early discussions on the connection between radio emission and synchrotron radiation can be found in Wielebinski, Albrecht Unsöld, 2013, 66–80.

103 As mentioned in Chapter 1, Fermi had used Alfvén's theory of magnetohydrodynamic waves in plasmas to suggest a mechanism for the acceleration of cosmic rays by galactic magnetic fields embedded in plasma clouds. Enrico Fermi: On the Origin of the Cosmic Radiation. *Physical Review* 75/8 (1949), 1169–1174. doi:10.1103/PhysRev.75.1169. See also the later article Enrico Fermi: Galactic Magnetic Fields and the Origin of Cosmic Radiation. *The Astrophysical Journal* 119/1 (1954), 1–6. doi:10.1086/145789. Biermann had a copy of Fermi's 1949 article in mimeographed form before its publication (Correspondence

origin was shedding light on the possibility of acquiring information on cosmic rays far from the Earth, both inside our galaxy and beyond its limits. Thus, cosmic rays turned out to be a source of important astrophysical information, an essential ingredient of the Universe.

Many of these astrophysical processes were also expected to produce gamma rays, as pointed out by Philip Morrison in 1958.[104] At the end of the 1950s, when it had become increasingly clear that energy-releasing processes of a quite different type than the thermonuclear ones were of importance for the evolution of stars and galaxies, Morrison, who had studied at Berkeley under the supervision of Robert Oppenheimer, and had later worked at Los Alamos on the implosion problem for nuclear weapons, discussed the great potential of gamma ray astronomy. He pointed out that gamma radiation is more directly related to high-energy and nuclear processes than optical or radio emission, and yet does not share with high-energy charged particles the complete loss of information about the position of its source. The intensity of such fluxes later turned out to be far weaker than Morrison had predicted, yet his article was instrumental in raising interest in this new kind of astronomy—and in its connection with the origin of high-energy cosmic ray particles—and so gamma ray detection in outer space became a most promising field after Sputnik, as it was clear that the low intensity of cosmic gamma rays required space-based detectors.

Furthermore, there was a direct Cold War connection, as gamma ray detection in outer space used the same technology needed for detecting nuclear explosions, and much of the early work in cosmic gamma rays resulted from these bomb-detection satellites, as we mentioned earlier in Chapter 2.[105]

Biermann-Kiepenheuer, April–May 1949, AMPG, III. Abt., ZA 1, No. 2). See also already cited publications of the period 1948–53 by Biermann and Schlüter (Chapter 1), who focused earlier on the problem of interstellar magnetic fields and radio emission from the Sun. Interest of Biermann's group in developments involving astrophysical plasmas and galactic magnetic fields paved the way to the beginning of thermonuclear fusion research at the Institute for Physics and Astrophysics in Munich.

104 Philip Morrison: On Gamma-Ray Astronomy. *Il Nuovo Cimento* 7/6 (1958), 858–865. doi:10.1007/BF02745590. Klaus Pinkau: The Early Days of Gamma-Ray Astronomy. *Astronomy and Astrophysics Supplement* 120 (1996), 43–47. https://ui.adsabs.harvard.edu/#abs/1996A&AS..120C..43P. Last accessed 10/30/2018. Schönfelder, History, 2002, 524–529. F.W. Stecker: Gamma Ray Astrophysics. In: J.L. Osborne, and A.W. Wolfendale (eds.): *Origin of Cosmic Rays. Proceedings of the NATO Advanced Study Institute Held in Durham, England, August 26-September 6, 1974*. Dordrecht: Springer 1975, 267–334.

105 The first cosmic gamma ray bursts—extremely energetic explosions, in fact, the brightest electromagnetic events known to occur in the Universe—were actually identified in the

Pinkau was part of a distinct new scientific tradition in the Max Planck Society, which had grown around the hybrid field of balloon-based particle physics in the 1950s. These were the high-altitude experiments with photographic emulsions and, later, spark chambers, in which cosmic rays were not the main object of study but rather the source of high-energy subatomic events.[106] At the end of the 1950s, when the discovery of very high-energy cosmic particles detected by air shower arrays raised new questions about the astrophysical sources and acceleration mechanisms of the primary radiation, gamma ray astronomy was expected to be a leap forward in studies on the connections between cosmic particles and the emission of gamma rays. As the leading edge of particle physics shifted to CERN and other accelerator centers in the US, Pinkau moved to gamma ray astronomy—which represented for him the "astrophysical aspect of cosmic rays"—where a similar use of his spark chambers could be made.

In his capacity as technical director at ESRO in the period 1962–64, Lüst had met Jacques Labeyrie and Giuseppe Occhialini, with whom he discussed the proposal for a satellite experiment on gamma rays and cosmic particles in space.[107] In this regard, spark chambers were developed at MPE. Cosmic ray astronomy—and therefore gamma ray astronomy—was a field left open for research by cosmic ray scientists now that the path of high-energy interaction physics was closed by the advent of a new generation of more powerful

---

late 1960s/early 1970s from data recorded in the mid-1960s by the Vela satellites designed to detect gamma radiation pulses emitted by high-altitude nuclear detonations, a program initiated to verify the Limited Test Ban Treaty, banning nuclear weapon tests in the atmosphere, in outer space and under water. Ray W. Klebesadel, Ian B. Strong, and Roy A. Olson: Observations of Gamma-Ray Bursts of Cosmic Origin. *Astrophysical Journal* 182 (1973), L85–L88. doi:10.1086/181225. It became clear that the origin of such intense radiation is certainly related to catastrophic events such as supernovae collapsing to form neutron stars or even black holes occurring in distant galaxies. The gradual realization of the existence of violent events, both in stars and in galaxies, marked the entry into the realm of high-energy astrophysics, at energies beyond the reach of accelerators, and set the stage for further decisive events and developments that would inaugurate the era of relativistic astrophysics during the 1960s.

106 Morrison, On Gamma-Ray Astronomy, 1958, 858–865.

107 See the joint proposal "Multi-purpose detector for the study of electromagnetic and nuclear events" (December 1965), by the University of Milan, the Max Planck Institute for Extraterrestrial Physics and the Saclay Nuclear Research Centre (Historical Archives of the European Union, S-111, ESRO-5938, https://archives.eui.eu/en/fonds/142993?item=ESRO-5938. Last accessed 6/20/2020). Reimar Lüst, and Klaus Pinkau: Theoretical Aspects of Celestial Gamma-Rays. In: J.C. Emming (ed.): *Electromagnetic Radiation in Space. Proceedings of the Third ESRO Summer School in Space Physics, Held in Alpbach, Austria, from 19 July to 13 August, 1965*. Dordrecht: Springer 1967, 231–248.

accelerators. Studying cosmic rays and gamma radiation with balloons at high altitudes required a level of expertise in airborne instrumentation in Kiel that went far beyond what was available in Garching during the early years of Reimar Lüst's plasma cloud experiments.[108] In fact, upon his appointment in Garching, Pinkau brought with him his entire experimental team, which became a separate technical workshop from which true space-based astronomy would emerge.[109]

When Pinkau joined the Institute for Extraterrestrial Physics in 1965, it had become clear that satellites and spark chambers would be the tools to enable good high-energy gamma ray experiments to be conducted.[110] The early activities of Pinkau's group concentrated on setting up a competitive gamma ray astronomy research initiative, developing the instrumentation for it and establishing the international connections required. As recalled by Pinkau himself, in an interview in 2016, balloon work introduced to Kiel from Bristol, and later also to the Max Planck Institute for Extraterrestrial Physics,

> was not only important for Gamma and X ray astronomy as such, but also an important step in qualifying equipment for satellite experiments. We later stopped balloon launches in Germany and used the US possibilities.

He also emphasized that he had very little or no interaction with other teams at the Institutes for Physics, Astrophysics, and Extraterrestrial Physics, since by that time,

> their actual research work had developed in very different directions, using very different methods and applying to fields of science that were different and had little connection,

even if they were linked under the umbrella of the Max Planck Institute for Physics and Astrophysics; and he specified that:

108 See, for example, one of the last articles written by Pinkau in 1965 before moving to the Institute for Extraterrestrial Physics: Klaus Pinkau et al.: Balloon Experiment Using Spark Chambers and an Ionization Spectrometer. *Proceedings of the 9th International Cosmic Ray Conference. London, UK.* 1965, 821–823. http://adsabs.harvard.edu/abs/1965ICRC....2..821P. Last accessed 10/30/2018.

109 This traditional and experimental group is jokingly referred to as the 'Kiel Mafia' (Joachim Trümper: interview by Luisa Bonolis and Juan-Andres Leon, Munich, August 7–8, 2017. DA GMPG, BC 601036).

110 See also Pinkau, The Early Days, 1996, 43–47.

> At MPE, my group was strongly concentrated on gamma ray astronomy, and the international aspect of the large MIMOSA [Milano-Monaco-Saclay collaboration] and the Caravane Collaboration absorbed most of our free activities. These activities had little interaction with Plasma physics or any other topic of the Astrophysics Institute. Rather, we were interested in X-ray astronomy and finally helped to attract [Joachim] Trümper into the institute.[111]

And actually, once Trümper arrived as director at MPE in 1975, he had already devised plans for an ambitious X-ray astronomy project that would eventually lead to the ROSAT (ROentgen SATellite), one of the most successful X-ray astronomy missions of the past century.

Pinkau also recalled that they were very interested in infrared astronomy, because the process of inverse Compton scattering, by which very energetic electrons transfer some of their energy to photons, connects gamma rays, X-rays, photons of visible light, and infrared photons, and so these different astronomies have an internal connection:

> This interest finally led to attracting [Reinhard] Genzel into the institute. In this way, gamma ray astronomy within MPE influenced an even larger sector of its field of research.

And indeed, with the arrival of Genzel in the mid-1980s (described in more detail in Chapter 4), brand-new regions of the spectrum in ground- and space-based astronomy were opened at the Institute for Extraterrestrial Physics. Even earlier, when Trümper arrived at the institute, they stopped cosmic ray particle work completely in order to liberate manpower for his research activities. According to Pinkau, "Genzel later was supported for his infrared activities by manpower from the ion cloud group."[112] A further example of the mechanism

111 Klaus Pinkau to Luisa Bonolis, October 17, 2016. The Caravane Collaboration included a group of European research laboratories (Netherlands, Italy, Germany, France) that took responsibility for designing the large gamma-ray telescope for the satellite COS-B.

112 Klaus Pinkau to Luisa Bonolis, October 17, 2016. For an overview of developments and research activities during the period from 1965, when Pinkau arrived, up to 1975, see Pinkau's Report "Das Max-Planck-Institut für Extraterrestrische Physik, seine Planungsuberlegungen und Prioritäten im Jahre 1975," containing many graphs related to internal developments, from staff to publications to age statistics, including the number of projects over the years (AMPG, II. Abt., Rep. 26, No. 6). By the early 1970s, Pinkau had become so influential that Herbert Friedman, one of the most eminent US space scientists and member of the Space Science Board of the National Academy of Sciences,

of recycling internal expertise for launching a new research field dated back to 1961, when the whole group of experimental cosmic rays had been transferred from Heisenberg's Institute for Physics to the new Department for Extraterrestrial Research led by Reimar Lüst at the Institute for Astrophysics.[113]

In Garching, Pinkau became the most important scientific contributor to ESRO's first gamma ray observatory missions. Despite the relative weakness of West Germans in European collaborations at the time, at the Institute for Extraterrestrial Physics, Pinkau secured a primary scientific role for German researchers in space astronomy through expertise in detection techniques, which at the time consisted of spark chambers as used in particle physics research.[114] Later, Pinkau was the principal investigator of COS-B, the European Space Agency's first scientific satellite (rather than ESRO), launched in 1975 with a high-energy gamma telescope as payload, which provided the first map of the galactic gamma-ray emission.[115]

This was an early example of what became the periodic cross-fertilization of experimental techniques originating in particle physics or in the new fields of astronomy and astrophysics.[116]

---

invited him to be one of the 10–12 members of an international space science advisory group that should "operate on an international scale, bringing together scientists to focus on and survey cooperatively the problems, the opportunities, and the implications of space research, and to find ways to foster and promote wise and vigorous international scientific programs." At that time, it was becoming apparent that "the development and conduct of major space research programs will depend to a large extent on the pooling of national budgetary, scientific and technological resources. The scientific and technological gap which once existed between the launching and non-launching nations has been steadily closing to a point where the basis for significant expansion of cooperative programs is attractive and necessary to maintain viable space research programs." Herbert Friedmann to Klaus Pinkau, 12.13.1973, AMPG, III. Abt., Rep. 145, No. 230.

113 AMPG, III. Abt., ZA 1, No. 91. See also p. 30 in Max Planck Institute for Physics and Astrophysics, Institute for Extraterrestrial Physics, Report 1963–1965, MPI-PAE Extraterr. 22/66, January 1966, available at http://www.mpe.mpg.de/303552/jb1963-1965.pdf. Last accessed 4/22/2020.

114 As in the case of Gentner at CERN a decade earlier, instrumental expertise was the best way for German researchers to find senior roles in European collaborations.

115 K. Bennett et al.: Preliminary Results from the European Space Agency's COS-B Satellite for Gamma-Ray Astronomy. *NASA Conference Publication* 2 (1977), 27. https://ui.adsabs.harvard.edu/?#abs/1977NASCP...2...27B. Last accessed 11/20/2018. Wim Hermsen: COS-B Views on the Diffuse Galactic Gamma-Ray Emission and Some Point Sources. *Space Science Reviews* 49/1 (1989), 17–39. doi:10.1007/BF00173739.

116 Klaus Pinkau: Interview by Helmuth Trischler, March 9, 2010. Transcript, Historical Archives of the European Union, Oral History of Europe in Space Collection (from now on HAEU), https://archives.eui.eu/en/oral_history/INT072. Last accessed 1/4/2019.

Wavelength Completion in the 1970s and X-Ray Astronomy

By the early 1970s, observational astronomy was an activity conducted at three institutes, in Bonn, Heidelberg, and Garching. Most importantly for the future dominance of astronomy in the Max Planck Society, each of these institutes had been created around expertise with a particular wavelength: radio waves in Bonn, the visible range in Heidelberg, and high-energy radiation observable outside the atmosphere in Garching. This division of labor between the major wavelengths formed the basis of the expansion program in astronomy for the next generation, whose task would be to fill in the gaps between them. Thus, the Institute for Radio Astronomy in Bonn strove in its next projects for pioneering dishes and detectors for even smaller wavelengths, first with IRAM (Institut de Radioastronomie Millimétrique), in the millimeter range, in the 1970s (detailed in Chapter 4),[117] followed by the Heinrich Hertz telescope in the sub-millimeter range, in the 1980s.[118]

In Garching, at the other end of the spectrum, the race began in the most energetic gamma rays in the late 1960s, followed by its most successful project, the national X-ray satellite ROSAT built under the direction of Joachim Trümper from the late 1970s through the 1980s. X-ray astronomy was in fact another successful spin-off of cosmic ray research.[119]

---

117 Pierre Encrenaz et al.: Highlighting the History of French Radio Astronomy. 7. The Genesis of the Institute of Radioastronomy at Millimeter Wavelengths (IRAM). *Journal of Astronomical History and Heritage* 14/2 (2011), 83–92. http://www.narit.or.th/en/files/2011JAHHvol14/2011JAHH...14...83E.pdf. Last accessed 10/30/2018. See also minutes of the commission "Berufungsvorschläge für die Leitung des geplanten deutsch-französischen Millimeterwellen-Instituts of May 12, 1977 (AMPG, III. Abt., Rep. 68 A, No. 149).

118 Richard Wielebinski: The Development of Radio Astronomy from Metre to Sub-mm Wavelengths. *Acta Cosmologica* 23/2 (1997), 53–58. http://adsabs.harvard.edu/abs/1997AcC....23...53W. Last accessed 10/30/2018.

119 For a reconstruction of the development of X-ray astronomy in Germany, see Simone Jüngling: *Röntgenastronomie in Deutschland. Entstehungsgeschichte, Institutionalisierung und instrumentelle Entwicklungen*. Hamburg: Verlag Dr. Kovač 2007. Joachim Trümper: The History of X-Ray Astronomy in Germany. *Memorie Della Società Astronomica Italiana* 84 (2013), 493–500. http://adsabs.harvard.edu/abs/2013MmSAI..84..493T. Last accessed 5/11/2018. See also Riccardo Giacconi: History of X-Ray Telescopes and Astronomy. *Experimental Astronomy* 25/1–3 (2009), 143–156. doi:10.1007/s10686-009-9139-8. Riccardo Giacconi, and Harvey D Tananbaum: The High Energy X-Ray Universe. *Proceedings of the National Academy of Sciences of the United States of America* 107/16 (2010), 7202–7207. https://www.jstor.org/stable/pdf/25665331.pdf. Last accessed 10/30/2018. Riccardo Giacconi, a pioneer in X-ray astronomy in the US, who was awarded the Nobel Prize in Physics 2002, became a Scientific Member of the Institute for Astrophysics in June 1999 (minutes of 25/26 February 1999, 9 June 1999, 14/15 October 1999, AMPG, II. Abt. Rep. 62, No. 1747, 1748, 1749). Harvey Tananbaum, Ethan J. Schreier, and Wallace Tucker: Riccardo Giacconi

Together with Klaus Pinkau, Trümper had been the other main high-energy researcher within Erich Bagge's group between Hamburg and Kiel, becoming an expert builder of spark chambers and later developing an air shower experiment, which explored particle cascades generated by extremely energetic cosmic rays in the atmosphere, in the hope of determining the chemical composition of such primary particles and investigating high-energy processes within the showers.[120] Trümper was also interested in the origin of cosmic rays, but such experiments could provide no clues. The Crab Nebula pulsar (NP0532), actually recognized as such in 1968, opened a new perspective in this sense.[121] Since the Crab Nebula had for a long time been known to be a strong source of synchrotron radiation, covering the spectrum from the radio and optical range to X- and gamma rays, it appeared to be a possible source of high-energy particles. The hypothesis that losses in the ability of a pulsar to accelerate particles should result in a decrease of the X-ray signal led Trümper to apply in 1971 to the DFG (German Research Foundation, see Chapter 2) to obtain funds for an X-ray balloon experiment that would check theoretical predictions and observe other X-ray sources. In the meantime, Trümper had moved to Tübingen University, where he had been appointed to the Chair of Astronomy as successor to Heinrich Siedentopf:

> At the time [Trümper recalls] it was a monstrosity [*Ungeheuerlichkeit*] that a physicist—a nuclear physicist—had been appointed as Chair of Astronomy...[122]

In Tübingen he developed the balloon-borne High Energy X-Ray Experiment (HEXE).[123] The HEXE collaboration made 14 successful balloon flights from

(1931–2018). *Proceedings of the National Academy of Sciences* 116/26 (2019), 12587–12589. doi:10.1073/pnas.1902399116.

120 The role and impact of Bagge's cosmic ray group for the Max Planck Society will be further outlined in Chapter 5.

121 J. M. Comella et al.: Crab Nebula Pulsar NP 0532. *Nature* 221/5179 (1969), 453–454. doi:10.1038/221453a0.

122 Joachim Trümper: interview by Helmuth Trischler and Matthias Knopp, March 18, 2010. Transcript, HAEU, https://archives.eui.eu/en/oral_history/INT076. Last accessed 1/4/2019.

123 Trümper himself has described his path from cosmic ray physics to X-ray astronomy: "That fascinated me and I started working on pulsar models. Between 1967 and 1970, I gradually switched from nuclear physics to astrophysics […] In connection with my reflections on neutron stars (pulsars) I had made the plan to do X-ray astronomy. This became possible with the appointment to Tübingen. In 1971, we began to build up X-

1973 through 1987, discovering many new X-ray sources. Balloons were relatively inexpensive and still allowed some competitive results if compared with rockets in the case of neutron stars. When such balloon experiments were starting, Uhuru, the first satellite dedicated entirely to X-ray astronomy, launched in December 1970, provided the first comprehensive survey of the entire sky for X-ray sources.[124]

In 1969–70, Trümper had visited the Institute for Extraterrestrial Physics while still at Kiel University, and the foundations for more ambitious plans in X-ray astronomy were laid during this year.[125] It was quite clear for Trümper that, to pursue his goals—a big project like an X-ray satellite—he needed "the impact of a Max Planck Institute." The base in Tübingen—where he pursued balloon and rocket experiments, also working in connection with NASA and ESA satellites—was not sufficient for this.[126]

In 1975, when Trümper moved to Garching as director at the Institute for Extraterrestrial Physics,[127] he reunited with the formerly Kiel-based team already working there, now continuing the collaboration with the Tübingen group, organizing successful joint balloon expeditions, significantly enlarging and improving the instruments for HEXE. With their balloon program, their goal was to observe, with large-area counters, sources that had been discovered by the legendary Uhuru satellite in 1971–72. But Uhuru worked in the energy range of 2 to 6 keV, while their instruments could measure 20 to 200 keV, which

---

ray astronomy, initially with a balloon program that we were able to realize with the help of the German Research Foundation. A first highlight was the discovery of cyclotron resonance lines in the X-ray spectrum of the neutron star Hercules X-1, with which the magnetic field of a neutron star could be measured for the first time." Joachim Trümper: interview by Helmuth Trischler and Matthias Knopp, March 18, 2010. Transcript, HAEU, https://archives.eui.eu/en/oral_history/INT076. Last accessed 1/4/2019.

124 R. Giacconi et al.: An X-Ray Scan of the Galactic Plane from UHURU. *The Astrophysical Journal Letters* 165 (1971), L27–L35. doi:10.1086/180711. W. Forman et al.: The Fourth Uhuru Catalog of X-Ray Sources. *The Astrophysical Journal Supplement Series* 38 (1978), 357–412. doi:10.1086/190561.

125 Ludwig Biermann, and Reimar Lüst: Max-Planck-Institut für Physik und Astrophysik. Institut für Astrophysik und Institut für extraterrestrische Physik. *Mitteilungen der Astronomischen Gesellschaft* 29 (1971), 86–112, 86. http://adsabs.harvard.edu/abs/1971MitAG..29...86B. Last accessed 10/30/2018.

126 Joachim Trümper: interview by Helmuth Trischler and Matthias Knopp, March 18, 2010. Transcript, HAEU, https://archives.eui.eu/en/oral_history/INT076. Last accessed 1/4/2019.

127 Trümper was appointed Scientific Member and Director at the Max Planck Institute for Extraterrestrial Physics in 1974 (CPTS meeting minutes of 26/06/1963, 23/10/1963, 15/02/1964, AMPG, II. Abt., Rep. 62, No. 1969, 1970, 1971).

was "a very successful bread and butter program."[128] Their balloon experiments could thus extend the information to higher energies, leading in 1978 to new insight with a most important discovery, the first measurement of the magnetic field of a neutron star (Hercules X-1) using the cyclotron line emission.[129] These research activities, which were now relabeled 'high-energy astrophysics,' became more and more connected to what was being unveiled as the hot and energetic Universe, and laid the foundation for the further development of X-ray astronomy during the long preparation of the ROSAT satellite.[130]

---

128 Interview by Helmuth Trischler and Matthias Knopp, March 18, 2010. Transcript, HAEU, https://archives.eui.eu/en/oral_history/INT076. Last accessed 1/4/2019.

129 J. Trümper et al.: Evidence for Strong Cyclotron Line Emission in the Hard X-Ray Spectrum of Hercules X-1. *The Astrophysical Journal Letters* 219 (1978), L105–L110. doi:10.1086/182617. At that time, Trümper, who was already developing his project for an X-ray satellite (the future ROSAT), visited Moscow and had a chance to discuss the theory of such cyclotron line emission with Rashid Sunyaev, one of the most well-known theoretical physicists in astrophysics, a former student of Yakov Zeldovich: "But then, soon, at the beginning of the 1980s, the idea came up to make joint projects. The first wish on the Soviet side was to fly the X-ray telescope that we had already developed for ROSAT on the Salyut station. I immediately rejected this, because it was not feasible for political reasons at the time and would have seriously jeopardized our ROSAT plans. I suggested instead that our 32 cm telescope which was used in rocket experiments, should fly on the Salyut [the first space station to orbit the Earth]. But that was not possible for other reasons: the Russians had no star sensors. We had a development at MBB [Messerschmitt-Bölkow-Blohm, aerospace manufacturer], but star sensors were subject to the embargo. So this plan also had to be buried [...] In a third step, I then offered our balloon experiment, which we operated together with Tübingen, in modified form on the space station. The 'Mir-HEXE' was started in March 1987, on the Mir, the successor of Salyut space station. The collaboration was as easy as one could barely imagine after having worked with NASA and ESA." Joachim Trümper: Interview by Helmuth Trischler and Matthias Knopp, March 18, 2010. Transcript, HAEU, https://archives.eui.eu/en/oral_history/INT076. Last accessed 6/8/2019. The Mir- HEXE experiments, one of the four X-ray instruments operated on board the Kvant module docked to the Soviet space station Mir since April 1987. Their first target was the supernova 1987A. Rashid Alievich Sunyaev et al.: Detection of Hard X-Rays from Supernova 1987A. Preliminary Mir-Kvant Results. *Soviet Astronomy Letters* 13/6 (1987), 431. http://adsabs.harvard.edu/abs/1987SvAL...13..431S. Last accessed 12/13/2017. As we will see, in 1995, Sunyaev became Scientific Member and Director at the Max Planck Institute for Astrophysics.

130 The path to ROSAT has been widely described in Bernd Aschenbach, Hermann-Michael Hahn, and Joachim Trümper: *The Invisible Sky. Rosat and the Age of X-Ray Astronomy*. New York, NY: Springer 1998, 37–41. See also Joachim Trümper, Bernd Aschenbach, and Heinrich Brauninger: Development Of Imaging X-Ray Telescopes At Max-Planck-Institut Garching. *Proc. SPIE 0184, Space Optics Imaging X-Ray Optics Workshop*, (*9 August 1979*). Space Optics—Imaging X—Ray Optics Workshop. 1979, 12–19. doi:10.1117/12.957429. Joachim Trümper: Kosmische Röntgenquellen. In: Generalverwaltung der Max-

They also achieved good results with rocket experiments, however, as Trümper recalled:

> Rocket flights were much more expensive than balloon flights, that were financially within reach of DFG applications with an entity of a few 100,000 DM [Deutschmarks]. More importantly, missile observations lasted only about five minutes—far too short for neutron stars that I was particularly interested in. With rockets, it was practically only possible to make observations of the intense solar radiation, with funding going through ESRO and later the DFVLR.[131]

On the other hand, as also in the United Kingdom and the Netherlands, national programs were a necessary platform for a successful connection with ESA and NASA missions.

In the 1960s and 1970s a large mass of data of high scientific value was collected by stratospheric balloons and rocket experiments, despite the limitations on altitude (about 40 km), respectively on observation time (of only a few minutes). And while observations in the atmosphere and lower ionosphere could be made by relatively inexpensive rockets, in the case of gamma and X-ray astronomy, a good astronomical program required very expensive large satellites with high pointing accuracy and stability, which could observe X-ray sources over an extended period of time and make further important progress. In Tübingen, Trümper had already submitted a proposal to the DFG, for funding for a balloon experiment to study the spectra and time variability of the new X-ray sources discovered by the satellite Uhuru. While in Tübingen, they could do balloon experiments and rocket experiments and participate

---

Planck Gesellschaft zur Förderung der Wissenschaften e.V. (ed.): *Max-Planck-Gesellschaft Jahrbuch 1983*. Göttingen: Vandenhoeck & Ruprecht 1983, 81–91. Joachim Trümper: Bizarre Röntgenquellen im Kosmos. Erste Ergebnisse von Rosat. In: Generalverwaltung der Max-Planck Gesellschaft zur Förderung der Wissenschaften e.V. (ed.): *Max-Planck-Gesellschaft Jahrbuch 1991*. Göttingen: Vandenhoeck & Ruprecht 1991, 75–89. For a review of the X-ray missions up to the advent of the ROSAT era, see Hale Bradt, Takaya Ohashi, and Kenneth A. Pounds: X-Ray Astronomy Missions. *Annual Review of Astronomy and Astrophysics* 30 (1992), 391–427. doi:10.1146/annurev.aa.30.090192.002135. Herbert Gursky: Technology and the Emergence of X-Ray Astronomy. *Journal of Astronomical History and Heritage* 3/1 (2000), 1–12. http://www.narit.or.th/en/files/2000JAHHvol03/2000JAHH....3....1G.pdf. Last accessed 10/30/2018.

131 Joachim Trümper: interview by Helmuth Trischler and Matthias Knopp, March 18, 2010. Transcript, HAEU, https://archives.eui.eu/en/oral_history/INT076. Last accessed 5/8/2019.

in ESA and NASA satellites, but it was no coincidence that, in the year of his appointment as director at MPE, he submitted an application for an X-ray satellite to the Federal Ministry of Scientific Research (until 1962, the Ministry of Atomic Affairs), within the large-scale equipment program. Trümper had in fact accepted the directorship at MPE because it provided a much more powerful basis for carrying out satellite missions. In retrospect, Trümper remarked:

> EXOSAT, the first ESA X-ray satellite, was launched in 1983, XMM-Newton, the next ESA X-ray satellite was launched in 1999. In-between there are 16 years—half a scientific life! This is too diluted for an active scientific group or even for an institute. That is why we needed both the national program and also projects with NASA or the ESA or others. If we had stayed only with ESA, many things would not have succeeded, scientifically [our translation].[132]

And so, already in 1972, they started developing an X-ray telescope with the Zeiss company in Oberkochen.

Since the development of this satellite spanned more than two decades, its role in internationalization shifted over the years. The satellite initially fulfilled the ambitions for a national project or even 'infrastructure,' as far as this was possible for a West German space-based instrument: as we have described, its conception was the outcome of a longstanding research tradition spanning Kiel, Tübingen, and Garching. The satellite platform was provided by Dornier in Friedrichshafen, and communications with it were to be done from the German Aerospace Center at Oberpfaffenhofen, using the Weilheim antennas. As with all West German satellites, there was inevitably an international touch to the launch, in this case provided by NASA (initially as a Shuttle launch, but

132 In this regard, Trümper added that his scientific life was made to 65–70 per cent of national activities, first balloon and rocket experiments and, later, the big ROSAT project; about 30 percent of activities with ESA (EXOSAT and later XMM-Newton, the X-Ray Multi-Mirror Mission named after Isaac Newton); the same with NASA (Chandra and later Swift). Then there were bi-national collaborations with, for example, the Italian satellite BeppoSAX or the Soviet space station Mir. Joachim Trümper: interview by Helmuth Trischler and Matthias Knopp, Munich, March 18, 2010. Transcript, HAEU, https://archives.eui.eu/en/oral_history/INT076. Last accessed 12/4/2020.
See also Joachim Trümper: X-Ray Astronomy in Europe. In: T.D. Guyenne, and B. Battrick (eds.): *Twenty Years of the ESA Convention. Proceedings of an International Symposium, Held at Deutsches Museum, Munich, Germany, 4–6 September 1995*. Paris: European Space Agency 1995, 85–88. https://ui.adsabs.harvard.edu/abs/1995ESASP.387...85T. Last accessed 4/23/2019.

due to the Challenger explosion, a rocket instead). But by the 1980s, the expectations for internationalization had shifted, as we will see in further detail in Chapter 4. Several projects that were initially 'national' were, one could say, 'retrofitted' as international collaborations. The IRAM 30 m telescope is the best such example, as we will see in Chapter 4, and all three examples in Chapter 5 further illustrate the new era. In the case of ROSAT, a condition of its support by the West German government was that it attract enough international collaboration. Thus, the United States and the United Kingdom not only supplied third party funding, but also two instruments for the satellite launched in 1990: NASA's High Resolution Imager, built by the Smithsonian Astrophysical Observatory, and the Wide Field Camera, built by a British consortium led by the University of Leicester.

Since its launch in June 1990, "ROSAT has made history." It performed the first all-sky survey with an imaging telescope and radically changed our view of the Universe with high angular resolution and a sensitivity that was orders of magnitude better than previous X-ray surveys of the sky, a survey resulting in a catalogue that contained more than 150,000 individual sources, 25 times more than with all previous X-rays satellites together.[133]

Furthermore, ROSAT guaranteed the MPE's continued global leadership in X-ray telescopes to this day: in addition to contributing instrumentation to NASA and ESA telescopes (see below), the MPE has continued to strive for access to space independently of these organizations; firstly, with the

133 The ROSAT Bright Source Catalogue paper derived from the all-sky survey performed during the first half year (1990–91) of the ROSAT mission, cataloguing 18,811 sources, represented both the culmination of the ROSAT project's primary aim of surveying the whole sky at X-ray wavelengths with an unprecedented sensitivity, as well as a major step forward in our knowledge of the X-ray sky. W. Voges et al.: The ROSAT All-Sky Survey Bright Source Catalogue. *Astronomy and Astrophysics* 349/2 (1999), 389–405. http://cdsads.u-strasbg.fr/abs/1999A%26A...349..389V. Last accessed 10/21/2018. This paper was included in the special issue of *Astronomy & Astrophysics* celebrating the journal's first 40 years of publishing papers with a strong impact on the scientific community. Prominent members of the global astronomical community were asked to comment on the context in which these papers first appeared and the advances they had brought to their fields. M. G. Watson: ROSAT's View of the X-Ray Sky. Commentary on: Voges W., Aschenbach B., Boller Th., et al., 1999, A&A, 349, 389. *Astronomy & Astrophysics* 500/1 (2009), 581–582. doi:10.1051/0004-6361/200912206. ROSAT data have recently been reanalyzed with a new advanced detection algorithm in order to produce a new source catalog for the astrophysical community, also serving as a preparation for the forthcoming eROSITA all-sky survey: Th. Boller et al.: Second ROSAT All-Sky Survey (2RXS) Source Catalogue. *Astronomy & Astrophysics* 588/A103 (2016), 1–26. doi:10.1051/0004-6361/201525648. For a general overview of ROSAT's context, see Aschenbach, Hahn, and Trümper, *The Invisible Sky*, 1998, 165.

Abrixas satellite in the 1990s, which unfortunately was lost owing to failure of its German-built battery system (more on this in Chapter 4); and then the long-delayed but now extremely successful X-ray space telescope eROSITA (extended ROentgen Survey with an Imaging Telescope Array), one of the two instruments on board the joint German–Russian mission Spectrum-Roentgen-Gamma (SRG), successfully launched from Baikonur in July 2019—nearly 30 years after the ROSAT mission.[134] During its first all-sky survey, completed in June 2020, eROSITA detected over a million sources of X-rays, basically doubling in just six months the number of known sources discovered over the 60-year history of X-ray astronomy.[135]

Wavelength Completion and Coordination of the Different Institutes and Research Traditions

Finally, as already briefly mentioned, a parallel branch of research was initiated in airborne and space-based infrared astronomy in Heidelberg in the early 1970s—independently, in addition to the large optical telescopes—under the direction of Dietrich Lemke (further details in Chapter 4); and there was even talk of a separate sub-institute of 'Extraterrestrial Astronomy.'[136]

Each such project had a lasting impact on the form of the local expertise and specialized facilities for development and testing of these instrumental systems, and this in turn secured the institutes' participation in future projects, whenever research moved beyond the national framework to focus on international collaborations. One notable example is the X-ray testing facilities of the Institute for Extraterrestrial Physics. The first such facility, ZETA, was built to test the X-ray telescopes flown with rockets between 1979 and 1987, and improved over the years to meet the functional testing requirements for new projects. An X-ray beam line test facility named PANTER was subsequently built on the southwest outskirts of Munich, to test the mirrors for the final ROSAT satellite, and, later, a smaller facility called PUMA within the Max

134 X-Raying the Universe. 6. *Nature Astronomy* 4/6 (2020), 549–549. doi:10.1038/s41550-020-1137-9.

135 A. Merloni et al.: eROSITA Science Book. Mapping the Structure of the Energetic Universe. arXiv:1209.3114 [*Astro-Ph.HE*], 2012. http://arxiv.org/abs/1209.3114. Last accessed 11/21/2018. Andrea Merloni, Kirpal Nandra, and Peter Predehl: eROSITA's X-Ray Eyes on the Universe. *Nature Astronomy*, 2020, 1–3. doi:10.1038/s41550-020-1133-0.

136 The Max Planck Society's budget (see Financial Appendix), which is a good indicator of which units are considered independent sub-institutes, listed a Department for Extraterrestrial Astronomy in the early 1970s. Ultimately, however, the Society was at the time moving toward unitary institutes with collegiate membership and this unit was re-absorbed into the unitary budget of the institute.

Planck Institute for Extraterrestrial Physics.[137] These testing facilities, mostly used for the characterization of X-ray telescopes as well as for tests of detectors and other instruments, are accredited with the unparalleled precision attained by the telescopes of the MPE, also thanks to the radiation detectors developed at the *Halbleiterlabor*, the semiconductor laboratory of the Max Planck Society. PANTER, which meanwhile has over 40 years of experience in testing and calibrating X-ray optics, has gone on to play a crucial role in ground X-ray calibration in subsequent international projects such as EXOSAT (European X-Ray Observatory Satellite), BeppoSAX, XMM-Newton, the MPE instrument on Chandra, LETG (Low Energy Transmission Grating), the X-ray Telescope (XRT) on the Neil Gehrels Swift multi-wavelength space Observatory, eROSITA, etc.

The spread of observational astronomy over the entire electromagnetic spectrum (ranging from radio waves to high-energy photons with 10^{12} electron-volt energies) over the course of half a century, which made it possible to explore different aspects of the Universe, went hand in hand with a massive wavelength expansion logic, which enabled a growing number of Max Planck Institute directors to consider themselves observational astronomers. This formidable power base within the Society posed the first ever challenge to the dominance of the previous scientific traditions of plasma physicists and cosmochemists; so much so that by the mid-1970s, around the time of the appointment of Reimar Lüst as President of the Max Planck Society, and although he was representative of the generation of space plasma physicists, the power balance, expressed both through the number of scientific members and the scale of their projects, was shifting in favor of the astronomers, even at his own Institute for Extraterrestrial Physics.[138]

---

137 Built in 1980, under the direction of Heinrich Bräuninger, PANTER is a 123 m long vacuum tube of 1 m diameter, with an X-ray source system and a 12 m long test chamber of 3.5 m diameter. No more than 3–4 people are needed to operate this powerful tool, so direct and easy access for any kind of test was assured during all phases of ROSAT hardware development: "For the ROSAT mission we had years of 'PANTER time.' We have tested the mirrors and also all the instruments to death!" Trümper also recalled that he always insisted: "Do not tell me that it works in principle. We have to test that it works…." Joachim Trümper: interview by Luisa Bonolis and Juan-Andres Leon, Berlin, May 6–7, 2019. For a history of PANTER, see the album dedicated to Joachim Trümper: *Eine kleine grosse Welt*, July 24, 2001 (DA GMPG, BC 600003). We are particularly grateful to Joachim Trümper for allowing us an opportunity to consult such a special volume.

138 In the early 1960s, all scientific directors in the cosmic sciences researched in the traditions of cosmochemistry and plasma physics, and they were within 'nuclear' institutes, except for Biermann's Institute for Astrophysics. By the time of Lüst's presidency, in addition to the continued presence in plasma physics and cosmochemistry, there

This wavelength expansion that began in the mid-1960s is the first clear example of a well-articulated narrative advocating the coordination of scientific work among several Max Planck Institutes in order to guarantee their national dominance in a scientific field. It provided a clear scientific justification for continued growth (filling the wavelength gaps), and brought together a set of decision makers with a coherent common ground in the discipline of observational astronomy, who were superbly connected with a global network of research in the field.

We will see in the following chapters how this observational astronomy research program, based on building a national infrastructure, interacted with extant traditions in the cosmic sciences at the Max Planck Society. The growing importance of observational astronomy forced scientists in the Max Planck Society to reflect on the identity of the organization and the kind of research that best identified it. Astrophysics, and even the early Institute for Extraterrestrial Physics, had benefited from an ideology of putting theory first. This had made economic and political sense in the early postwar era, but continued to be fostered also through the 1960s, when theoreticians such as Schlüter and Lüst were appointed as directors of eminently experimental institutes. Even the experimentalist Gentner drew much of his legitimacy from his ability to link experimental initiatives with far-reaching theoretical questions in particle physics.

In contrast, the first generation of ground-based observational astronomers could best be described as large telescope builders, and in the case of radio astronomers, many had a strong engineering background. Space astronomers such as Pinkau and Trümper, meanwhile, came from an experimental particle physics tradition, so found themselves somewhere mid-way on this spectrum; yet as Max Planck Institute directors, their primary task was to build the best instruments in the world in their given wavelength. In this first generation of wavelength expansion, the emphasis was on large, general-purpose telescopes for sky-wide surveys.

Now, after a decade of expansion in astronomy, the gigantic observatory institutes of the Max Planck Society were staffed by directors and teams who

were already two entire Max Planck Institutes dedicated to observational astronomy (radio, millimeter, infrared, and optical wavelengths), plus the Institute for Extraterrestrial Physics, which had a footing in space-based gamma astronomy. Biermann's own Institute for Astrophysics was also expanding rapidly into relativistic astrophysics, as is described in detail in Chapter 5. For an overview of the Max Planck Society in the early 1970s, in coincidence with the turning point also marked by a change in the presidency, see M.R. Hoare: Max-Planck-Gesellschaft: A Model for "Small Science"? *Nature* 237/5352 (1972), 206–209. doi:10.1038/237206a0.

considered themselves instrument builders, and the cultural bias against them on the part of the older generation, particularly among the plasma physicists, would recurrently prove problematic.[139] This was compounded by differences in personal style: although belonging to different generations (Hachenberg had been trained in the 1930s, Elsässer and Mezger in the 1950s), the ground-based astronomers without a background in physics carried out their role of observatory builder and director in a leadership style not unlike that of a naval captain. This was a generation that considered a choleric temperament crucial to the success of its titanic projects.[140] Fortified by their regional and national sources of financial and political support, these scientists remained as independent as possible from the Max Planck Society, and especially from anything coming out of Munich. Some of them—Peter Mezger in Bonn, for one—were renowned for their antagonism towards Reimar Lüst during his presidency, and could afford to be, too, owing to the instrumental excellence of the Bonn institute and its pioneering telescopes.[141] Throughout the rest of the century, the observatory institutes in Heidelberg and Bonn, for example, rejected proposals to appoint directors with a theoretical background and programs, in contrast to the predominance of theoreticians in Munich/Garching and, increasingly, also at the Institute for Nuclear Physics in Heidelberg, and Aeronomy in Lindau.[142] Even the Institute for Plasma Physics and the Institute for Extraterrestrial Physics, which had started as eminently experimental endeavors, appointed some directors with a theoretical agenda.

The 'taming' of the astronomer directors in the Max Planck Society, in Bonn and Heidelberg, was a slow process, which began with Otto Hachenberg himself. Already in 1967–68, before his institute was inaugurated, the established personalities in the cosmic sciences, represented by Wolfgang Gentner, head

139 As we will see in the next section, one of the criticisms wielded against the Institute for Aeronomy was its lack of theoretical guidance, itself the result of a tradition dating back to Erich Regener in Weissenau, most of whose teams were apprentices with little or no contact with the broader scientific community, unlike the scale seen at other Max Planck Institutes. Aeronomy was, however, a weak institute in contrast to the fledging new astronomical initiatives.

140 Personal accounts of the choleric disposition of both Elsässer and Mezger are an integral part of the MPG mythology.

141 Jacob W. M. Baars: Interview by Juan-Andres Leon, Bonn, February 5–7, 2018. DA GMPG, BC 601050. Conversations with Reimar Lüst during the Roundtable "Astronomy and Astrophysics in the History of the Max Planck Society." See also Baars, *International Radio Telescope Projects*, 2013.

142 See, for example, Joachim Trümper: interview by Luisa Bonolis and Juan-Andres Leon, Munich, August 7–8, 2017. DA GMPG, BC 601036.

of the CPT section at the time, as well as by Reimar Lüst and Ludwig Biermann, were exchanging correspondence regarding the attitude of Hachenberg, who did not seem to accept the implications of being part of the Max Planck Society.[143] Hachenberg had successfully avoided a co-directorship with Sebastian von Hoerner and was now seeking to appoint a loyal disciple for the post. Given the contemporaneous reforms in the MPG, towards collegiate directorship, this was unacceptable, and the Society instead made sure to nominate two people with close links to NRAO and to von Hoerner: Peter Mezger, who ended up staying in Bonn for the rest of his career, and Peter Stumpff, a disciple of Biermann who did not accept the position, which was taken instead by Richard Wielebinski, one of the first non-German directors, whose trajectory was mentioned earlier in the chapter.

Hachenberg never quite accepted Mezger's presence at 'his' institute during the first decade of operations, and this thwarted Mezger's access to Effelsberg, who focused instead on building his own millimeter-wavelength telescopes. Another contrast with Hachenberg was Mezger's distant relationship with Bonn University (and others in the area). Gentner, Lüst, and Biermann had considered it inacceptable that Hachenberg retain a powerful directorial role at Bonn University in parallel to his Max Planck position, whereas in fact both the MPI and the university's Institute for Astronomy were put in the same building. Throughout the 1970s, Max Planck representatives pressed for a true collegiate directorship in Bonn, with a Board of Trustees (*Kuratorium*) and Scientific Advisory Board (*Fachbeirat*), as well as for clear boundaries between the Max Planck Institute and the university. Conveniently, the advent of these changes coincided with Hachenberg's retirement in 1977 from the MPI (although not yet from his work at the university). Hachenberg was initially replaced by someone from NRAO, Kenneth Kellermann (see Chapter 4), who, however, returned to his position in the United States after a few years.[144] Mezger, who survived this conflictive first decade, became the dominant figure in Bonn until his retirement in the late 1990s, and, too, a weighty presence in those MPG decision-making bodies dealing with matters of cosmic research.

At the Max Planck Institute for Astronomy in Heidelberg the situation was not significantly better. The initial plans there likewise called for collegiate

143 See, for example, Biermann Papers, AMPG, III. Abt., ZA1, Folder 18. Letters from Gentner to (MPG Secretary) Schneider (20.12.1967), Lüst and Biermann to Gentner (8.3.1968), and Gentner to Schneider (25.3.1968).

144 Kenneth I. Kellermann: interview by Woodruff Sullivan III. March 19, 1975. NRAO, https://www.nrao.edu/archives/items/show/14994. Last accessed 1/26/2022. In this interview, he hints at having had a tense relationship with Mezger dating from their time together at Green Bank.

directorship, and it was assumed that a second director would be needed to take charge of the instrumentation and scientific operations at the optical observatories foreseen. Promising personalities were offered this position—Peter Strittmatter, for example, whose career later intersected considerably with the MPG, as we will see in the next chapter—but they declined in view of the long delays. Eventually, a second director was found: Guido Münch had a strong track record at Yerkes, Mount Wilson, and Palomar, in the United States, and at the time of his appointment was relatively advanced in his prominent career.[145] Münch was a helpful balancing force to the choleric Elsässer, particularly in dealing with the Spanish counterparts during the construction and operation of Calar Alto; but he did not have the same authority at the Heidelberg institute and the Max Planck Society,[146] spending much of his time abroad instead. Elsässer in contrast held significant power in the MPG central organs, even becoming the first astronomer acting as head of the CPT section in the period 1976–79. The Heidelberg Institute for Astronomy had to wait until the 1990s for a truly collegiate directorship (a transition described at the end of Chapter 4).

It could be argued that the Max Planck Society had a bad hand even after the reforms of the 1960s, when dealing with founding directors, since they, having enabled the Society and West Germany to quickly rise to the challenges of the space age, often (rightfully) considered that these "owed them"; on subsequent generations, however, the MPG imposed terms and conditions that fostered a more collective approach to proceedings. In the case of the cluster of institutes conducting cosmic research, even after its post-Sputnik expansion for two additional decades, from the mid-1970s to the mid-1990s, collective decision-making at the Max Planck Society level depended on maintaining a fragile equilibrium between the modernized second-generation scientists representing nuclear physics in Heidelberg or the family of institutes in Munich, who in both cases identified as physicists, and the more idiosyncratic personalities of the astronomers and engineers in Heidelberg and Bonn.

---

145 Roland Gredel: Guido Münch (1921–2020). *Max-Planck-Institut für Astronomie*, 5/4/2020. http://www.mpia.de/aktuelles/mpia-news/2020-05-04-muench-en. Last accessed 8/16/2020. See also Guido Munch: interview by David DeVorkin, 7 July 1977, Transcript, AIP, https://www.aip.org/history-programs/niels-bohr-library/oral-histories/4789. Last accessed 3/21/2021.

146 Lemke, *Himmel über Heidelberg*, 2011, Vol. 21, 120–121.

3 Reconfiguration of the Astrophysical Sciences and Institutes

The next major coordination process that strengthened the monopoly of the cosmic sciences in the Max Planck Society was related to generational renewal and the shifting emphasis of scientific research. The initial 'space science' generation had focused on plasma physics problems, first theoretically and then experimentally. By the late 1960s, however, the future lay in space-based astronomy. Factional rivalry peaked around the election of the next Max Planck Society president in 1973, but when Reimar Lüst was elected, he worked toward reconciliation. This increased the circulation of scientists among the cosmic Max Planck Institutes as new directors were appointed, facilitating the division of scientific labor among them. Extraterrestrial Physics specialized further in space-based astronomy; space plasmas was concentrated in Lindau, and the institute there also moved into planetary exploration, together with the Mainz institute. Other plasma physicists became theoretical astrophysicists and inaugurated theoretical lines of research, for example, in Heidelberg. The enormous Institute for Plasma Physics was readmitted to the Max Planck Society and its infrastructure and institutional support mobilized for the benefit of the astrophysics institutes.

A Plasma Physicists' Diaspora

As was shown earlier using financial data, one characteristic of the cosmic sciences in the Max Planck Society is that they, unlike most other research fields, did not experience a period of 'stagnation' in the 1970s; instead, the growth sparked by the launch of Sputnik continued. Indeed, many of the largest projects in the field were completed during the 1970s and 1980s, and it was only in the final years of the Cold War that this growth significantly slowed, and then simply remained constant on a par with the Max Planck Society as a whole.[147]

However, there were significant changes in the way the cosmic sciences operated in the periods before and after the early 1970s. These coincided with several important factors but were also the culmination of a process that had been gathering momentum for the past decade. While the first decade after

147 From the mid-1980s, the budget of all the exclusively astrophysical institutes stabilized at around 8 percent of the MPG/Max Planck Society budget, whereby this figure rises to about 24 percent, if one adds the 'outer bounds,' i.e., all other institutes with some activity in astrophysics (MPP, MPIK, Aeronomy, but not IPP). The 'actual' figure in astrophysics is somewhere between the two. For more details, see the Financial Appendix.

Sputnik saw vast growth in the cosmic sciences, as these found their autonomy and justification beyond the nuclear sciences, initially they expanded in very different directions, as illustrated by the contrast between space plasma research in Munich, cosmochemistry in the southwest, and the observatory institutes in Bonn and Heidelberg. By the mid-1960s, however, there were increasing overlaps between the interests of various institutes, most strikingly between the Institute for Extraterrestrial Physics, in Garching, and the Institute for Aeronomy in Lindau, near Göttingen, regarding participation in space projects. By the early 1970s, there was a perceptible crisis in the Aeronomy Institute, where attempts to find a permanent replacement for Julius Bartels had failed for almost a decade. We return to the outcome of these at the end of the chapter.[148]

Generational Change

Simultaneously, the end of the 1960s saw one of the major turning points in the history of the Max Planck Society, owing to planning for the imminent retirement of Werner Heisenberg, which matter turned the spotlight on how best to organize the succession. Having gained influence and authority throughout the 1960s, Wolfgang Gentner now presided over the commission at the

148 A commission to appoint Bartels' successor was established in early June 1964 (CPTS meeting minutes of 09.06.1964, AMPG, II. Abt., Rep. 62, No. 1743). Discussions on the future of his Institute for Stratospheric Physics, one of the two branches of the Institute for Aeronomy, involved more in general the future of the institute itself. In parallel with the necessity of getting rid of well-worn research topics and most of its data-taking of a purely monitoring nature, one main weakness of the institute was related to its theoretical expertise. The appointments of Ian Axford as director, of Vytenis Vasyliunas as a member and of Jules A. Fejer as an external member provided the institute with a powerful theoretical potential which could enable it to fully exploit experiments such as SOUSY (SOUnding SYstem for atmospheric structure and dynamics), Ionospheric Heating experiments, EISCAT, the two Helios missions, beyond merely data collection. Documentation regarding the Institute for Aeronomy crisis is extensive and ubiquitous in the correspondence of scientific members of the Max Planck Society. A starting point is the CPTS meeting minutes reporting decisions of the committees in charge of the appointments of directors and scientific members, as well as discussions about the future of the institute: "Ernennung von Prof. v. Zahn zum WM und Direktor des MPI für Aeronomie," "Zukunft d. MPI f. Aeronomie/Ernennung Haerendel z. WM u. Direktor/Gründung MPI f. Meteorologie/Ernennung Vasyliunas z. WM" (CPTS meeting minutes of 08.02.1973, 26.06.1973, 15.02.1974, 18.06.1974, 25.10.1974, 09.05.1979, AMPG, II. Abt., Rep. 62, No. 1768, 1769, 1771, 1772, 1773, 1787); "Ernennung von Dr. H. Rosenbauer zum WM, Mitglied d. Kollegiums und Direktor am Institut des MPI für Aeronomie" (CPTS meeting minutes of 16.11.1976, 08.03.1977, AMPG, II. Abt., Rep. 62, No. 1779, 1780). A collection of documents on this problem can also be found in personal papers of leading figures such as Gentner (AMPG, III. Abt., Rep. 68 A, No. 153, No. 157).

Institute for Physics in Munich-Freimann tasked to find the next director.[149] This commission included international figures in particle physics, such as Victor Weisskopf, who was a professor at MIT and had served as Director General of CERN in the first half of the 1960s. On May 21, 1969, Weisskopf, who had been Heisenberg's post-doc student in 1931, wrote to Gentner that he felt "[…] it would be difficult for a theoretician to step into the shoes of Heisenberg," and added that an experimental physicist would be a more appropriate choice; Wolfgang Paul, he opined, "would be a perfect candidate."[150] In November, Heisenberg wrote to Weisskopf that the really important question for the future of the Institute for Physics would be how well research work there fit with other parts of the Institute for Physics and Astrophysics: "[…] *auf diese Einheit haben wir immer besonders Wert gelegt* (We have always attached a special importance to this unity)" [emphasis added]. Heisenberg also remarked,

> It is one of the basic principles of the Max Planck Society that its institutes should not simply participate in conventional research, but that scientific directions are promoted that are either too expensive, in order to be properly performed at universities, or to be tied to the person of a researcher who is to be given the opportunity to carry out his unconventional work vigorously. From these points of view one should also look for my successor.[151]

---

149 The committee was formed in 1968 in order to find "either a theoretical or experimental physicist" (CPTS meeting minutes of 04.11.1968, AMPG, II. Abt., Rep. 62, No. 1754), but after a whole year no decision had been taken (CPTS meeting minutes of 07.11.1969, AMPG, II. Abt., Rep. 62, No. 1757).

150 In the folder "Zukunft des Instituts für Physik im Max-Planck-Institut für Physik und Astrophysik" of Heisenberg's papers, see Victor Weisskopf to Gentner, 4.11.1968, AMPG, II. Abt., Rep. 62, No. 437. Since 1952, Wolfgang Paul was Professor at the University of Bonn and Director of the Physics Institute. In the late 1950s, he had built a 500 MeV electron synchrotron at the University of Bonn, the first strong-focusing machine operating in Europe, followed in 1965 by 2500 MeV synchrotron; in 1957, together with Willibald Jentschke and Wilhelm Walcher, had founded DESY in Hamburg and had always been in close contact with CERN, also as director of the nuclear physics division (1964–67), as member and later chairman of the Scientific Policy Committee and scientific delegate of Germany in the CERN–Council. He was awarded the Nobel Prize in Physics 1989 for the development of the ion trap technique. Ewald Paul: 50 Years of Experimental Particle Physics in Bonn. A Personal Recollection. *European Physical Journal H* 38/4 (2013), 471–506. doi:10.1140/epjh/e2013-30028-7. On February 11, 1970, von Weizsäcker wrote to Gentner stressing that the institute should conduct activities "that are not performed elsewhere and that will apparently become important in the future."

151 Heisenberg to Weisskopf, 13.11.1969, AMPG, II. Abt., Rep. 62, No. 437.

After one year, in June 1970, it was reported that the commission had been unable to find a successor and it had been decided: 1) To install Hans-Peter Dürr as provisional director; 2) To rename the commission "Future of the Institute for Physics"; 3) To include von Weizsäcker as member of the commission.[152]

Heisenberg retired as Managing Director on December 31, 1970, and Hans-Peter Dürr was installed as Provisional Director. Eventually, an interim period of leadership by the then Director of the Theory Division at CERN, Léon Van Hove, was arranged, in an attempt to repair the major breaches that had resulted in the past decade from the antagonistic relationship between Heisenberg and the particle physics community, including that in his own institute.[153] An excellent phenomenologist, Van Hove could well balance the theoretical and experimental traditions at the Institute for Physics, providing a perspective which had been sorely lacking, and that could truly answer questions arising from experimental physics. By that time, both theoretical and experimental groups at the institute were conducting a large proportion of their work in Geneva and thus a strong interaction with Van Hove himself

152 As of that date, Gentner became president of the new commission (CPTS meeting minutes of 10.06.1970, AMPG, II. Abt., Rep. 62, No. 1759). On the work of the commission for the future of the Institute for Physics and Astrophysics see also AMPG, II. Abt., Rep. 62, No. 375.

153 In July 1971, Van Hove accepted the appointment (Van Hove to Butenandt, July 15, 1971; Butenandt to Van Hove, June 30, 1971, AMPG, II. Abt., Rep. 62, No. 437, Fol. 22, 25), and in August he offered Haim Harari—then at the Department of Nuclear Physics of the Weizmann Institute of Science in Israel—a position as theoretical physicist at the Max Planck Institute: "The Institute's directorate is indeed very keen to have your advice also on the question of other appointments and some policy matters." But it was Harari's firm intention to remain in Israel and for this reason had not accepted other offers abroad, as he explained in his answer of early September: "I therefore cannot accept your kind invitation. Concerning our discussion of other possible candidates—my feeling is that the only way to attract top level people would be to organize a 'semester' or a one-year session at the Max Planck Institute and to simultaneously invite several excellent people together with a group of younger physicists to spend this time at that Institute. If a group including, say [Maurice] Jacob, [Albert] Schmid, [Giuliano] Preparata, [Holger] Nielsen, [Henry] Abarbanel, [Philip] Phillips, [Dieter] Schildnecht, [Louis] Michel, [Marco] Ademollo, [Jacques] Weyers, or any similar combination would agree to spend a semester or a year there it could give the place a tremendous push and *will put it again on the physics map of the world* [emphasis added]. Only such a shock treatment could help, as far as I can see from my distant observation point, and if the funds are available and the invitations to such a group of people can be sufficiently attractive, it may work." Van Hove to Harari, 23.08.1971 and Harari to Van Hove, 05.09.1971, CERN Archives, CERN-ARCH-SIS-095.

would result from all this. Van Hove proposed a *Kollegium* (Board of Directors, i.e., collegial directorship) at the institute, in which all scientific members should participate, and too, that he would lead it in the three-year reorganization phase, while continuing, in parallel, his scientific work at CERN.[154] He took up the directorship in Munich on October 1, 1971, approximately three years after the first official discussions about Heisenberg's successor.[155]

Reorganization of the institute further progressed in 1972, when Leo Stodolsky was called from Stanford, bringing on board a theoretician who would be able to focus in particular on a phenomenological approach. This kind of application of theoretical models to high-energy experimental physics would forge a beneficial bridge between experimental and theoretical groups at the institute.[156] Van Hove remained only until October 1, 1974, but stronger relationships with CERN had been established in this period, and both the theoretical and experimental groups had been reinforced. He continued to be connected with the institute as an External Scientific Member and became Director General at CERN in 1976.[157]

The early post-Heisenberg era also coincided with Biermann's retirement. During the meeting of the Scientific Council of June 26, 1973, Gentner stressed

154 CPTS meeting minutes of 23.06.1971, AMPG, II. Abt., Rep. 62, No. 1762. Van Hove had also proposed to invite Gerd Buschhorn as a new experimental physicist at the Institute, also in agreement with suggestions from the committee to reinforce the experimental group (Van Hove to Gentner, June 16, 1971, AMPG II. Abt., Rep. 62, No. 437, Fol. 212). The final decision on both these questions was eventually announced in October 1971, approximately three years after the first official discussions about Heisenberg's succession (CPTS meeting minutes of 22.10.1971, AMPG, II. Abt., Rep. 62, No. 1763).

155 CPTS meeting minutes of 22.10.1971, AMPG, II. Abt., Rep. 62, No. 1763. For the whole information on the work of the commission for the future of the Institute for Physics, see AMPG, II. Abt., Rep. 62, N. 437.

156 As of April 1, 1973, Stodolsky began to work in Munich (CPTS meeting minutes of 22.04.1972, 02.11.1972, 26.06.1973, AMPG, II. Abt., Rep. 62, No. 1765, 1767, 1769). In the early 1980s, Stodolsky and his group became closely involved in the emerging field of of astroparticle physics (personal communication with Luisa Bonolis, May 8–16, 2017). He was especially interested in developing new types of instruments to investigate such topics as dark matter, or to detect exotic particles like axions from the Sun, at a time when a few particle physicists paid some attention to solar nuclear reactions because of the rising problem of the missing solar neutrinos in the expected flux from the Sun, a problem identified in Ray Davis experiment in U.S. The detection of solar neutrinos and related puzzles will be widely discussed in the final chapter.

157 Norbert Schmitz: Léon van Hove. 10.2.1924-2.9.1990. *Berichte und Mitteilungen Max-Planck-Gesellschaft*, 1991, 99–102. See Reimar Lüst: Léon Van Hove and the Max-Planck-Institute for Physics. *Scientific Highlights in Memory of Léon Van Hove*. World Scientific 1993, 51–59.

that the problem of how to replace Biermann necessarily involved the larger question of the three sub-institutes' future development and collaboration. No decision was taken until the following February, when it was reported that the committee had decided to propose that the Senate appoint Rudolf Kippenhahn as Director of the Institute for Astrophysics, who had moved to the University of Göttingen after working for several years at the Institute for Astrophysics on the structure and evolution of stars (more on this in Chapter 4).[158] Hans Elsässer emphasized that he and fellow members of the commission had thoroughly examined the fundamental question of the institute's future: Should such an internationally recognized institute be continued? To dissolve it would fly in the face of policy clearly implemented in the previous years, when the Max Planck Society had founded two new Institutes, for Astronomy and for Radio Astronomy. A theoretical institute, stressed Elsässer, should complement the existing astronomical institutes. Alfred Seeger remarked that one might also consider developing theoretical groups in these two extant institutes, whereupon Klaus Pinkau explained the material and financial grounds for retaining a theoretical institute. The latter, the commission felt, should not relocate to Bonn or Heidelberg but remain in the Munich area.[159] In Chapter 4, we revisit how this line of argument periodically resurfaced during subsequent succession proceedings, regarding the status of the Institute for Astrophysics.

By the end of 1973, Adolf Butenandt's presidency was coming to an end, and it was expected that Gentner would become the next president of the Max Planck Society. However, once again, the unexpected happened: Gentner, of advanced age and in fact only five years younger than Heisenberg, had some health problems.[160] Reimar Lüst was appointed, therefore, while Gentner accepted the less demanding role of vice-president. Against all expectations, this did not reignite the strong rivalry between Heidelberg and Munich, but led rather, in Gentner's final active years, to a period of compromises intended to defuse this conflictive relationship as far as possible.[161] Many of

158 Kippenhahn became a Scientific Member at the Institute for Astrophysics in 1963 (CPTS meeting minutes of 13/14.06.1963, 01.11.1963, AMPG, II. Abt., Rep. 62, No. 1741, 1742). The following year he became professor at the University of Göttingen, but continued his intense collaboration with Biermann.

159 CPTS meeting minutes of 23.10.1973, 15.02.1974, AMPG, II. Abt., Rep. 62, No. 1770, 1771.

160 See letters from Gentner to Adolf Butenandt of October 10–11, 1971, announcing his withdrawal as presidential candidate for health reasons (AGMG, III Abt. Rep 68 A, No. 138).

161 By the mid-1970s, when high-energy accelerator physics became more clearly the aim of dedicated laboratories in Europe, the question of a division of labor between the

the moves toward greater coordination among the various Max Planck Institutes should be seen in the light of this reconciliation process orchestrated by Lüst and Gentner.[162]

While this rapprochement between the physics institutes in Munich and Heidelberg was underway, observational astronomers were separately gaining influence within the Max Planck Society. There was a display of independence on the part of the observational astronomers when the Effelsberg telescope came into service in the early 1970s, and the two space exploration institutes in Lindau and Garching sought to use it for communication with their spacecraft. Unexpectedly, the Bonn radio astronomers denied them access, emphasizing that the giant antenna was to be used exclusively for astronomical purposes.[163]

Simultaneously, the most expensive infrastructural project in astronomy in the history of the Max Planck Society was about to be developed in Spain and South Africa, in relative independence of the Max Planck Society's central administration, thanks to the availability of federal funding for the national optical observatories. In Garching itself, there was growing interest on Klaus Pinkau's part for expanding into X-ray astronomy, then seen as the next major frontier in space-based astronomy, and for which he sought to appoint his erstwhile colleague in Kiel, Joachim Trümper, as was mentioned in the previous chapter.

Institute for Physics and the Institute for Nuclear Physics in the realm of high- and low-energy physics was clearly outlined during a meeting of the 'Senatsausschuss für Forschungspolitik und Forschungsplanung' (Senate Committee on Research Policy and Research Planning) held on May 15, 1975 (AMPG, III. Abt., Rep. 68 A, No. 151). Otto Hahn's Institute for Chemistry, which had a long tradition of accelerator-based nuclear physics, was included in the discussion. Gentner explained how Heidelberg had mainly dealt with low-energy physics, also extending studies of the structure of nuclei to higher energies. Research work had always been performed in collaboration with the local university. Gentner also expressed the opinion that a reduction of activities in Heidelberg, particularly in view of the retirement of Hermann Wäffler (Director of the Nuclear Physics Department of the Max Planck Institute for Chemistry in Mainz since 1959), was not justified. His proposal was to maintain low- and medium-energy physics in Heidelberg and high-energy physics in Munich, as had always been done previously.

162 According to Heinrich Völk, when Lüst became president of the Max Planck Society, Gentner established with him "a very intimate relationship. Interestingly because the relationship of Gentner with Heisenberg was not so intimate, but somehow Lüst and Gentner managed to get along very well […]." Heinrich Völk: interview by Luisa Bonolis and Juan-Andres Leon, Heidelberg, October 9–10, 2017. DA GMPG, BC 601037.

163 This is described, for example, in Erhard Keppler: *Der Weg zum Max Planck Institut für Aeronomie. Von Regener bis Axford—eine persönliche Rückschau*. Katlenburg-Lindau: Copernicus 2003, 24–25. This impasse led to his initiative to build a separate antenna for communication with extraterrestrial probes in Weilheim, Bavaria, which is still in use to this day.

Meanwhile, at the Max Planck Institute for Nuclear Physics, another personal catastrophe occurred in 1970, when Gentner's closest collaborator in cosmochemistry, Joseph Zähringer, was killed in a road accident.[164] Attempts at finding a successor initially failed, extending these deliberations into Reimar Lüst's presidency. Discussions of the problem of replacing Zähringer were also tied up with the reorganization of the cosmochemistry department that he had led and, more generally, with the future of cosmochemistry in the Max Planck Society.[165]

At the same time, in Heidelberg, Gentner's advanced age and health problems and subsequent decision to not accept the presidency raised the major matter of his succession.[166]

---

164 See Gentner's obituary in Präsidialbüro der Max-Planck-Gesellschaft (ed.): *Mitteilungen aus der Max-Planck-Gesellschaft zur Förderung der Wissenschaften. Heft 6/1970*. München 1970, 346–348.

165 An initial attempt to replace Zähringer as leader of the Cosmochemistry Department was made in 1972 with Johannes Geiss (see Gentner to Butendandt, May 13, 1972, AMPG, III. Abt., Rep. 68A, W. Gentner Nachlass, No. 166/2-2). Geiss had been a pioneer in the field of isotope geochronology in the 1950s and during the 1960s had established a laboratory for extraterrestrial research at the Physics Institute of the University of Bern to study meteorites and samples of lunar soil also involved in measurements of the solar wind recorded by instruments of the Apollo mission on the moon. For this reason, he did not accept the offer (CPTS meeting minutes of 08.02.1973, AMPG, II. Abt., Rep. 62, No. 1768). A committee of experts including foreign members was formed who visited the Institute for Nuclear Physics (see also CPTES meeting minutes of 26.06.1973, AMPG, II. Abt., Rep. 62, No. 1769) as well as the Institute for Chemistry in Mainz and the Institute for Aeronomy, whose future was discussed at that same time. The final choice fell on Hugo Fechtig, but a further candidate was planned who should in particular tackle the theoretical aspects of cosmochemistry. On the subject of the future of cosmochemistry, Christian Junge remarked that the committee had recommended that both the groups in Heidelberg and Mainz should remain at the same level and that nothing in the organization of the two institutes should be changed. By early 1970s, the Cosmochemistry Department in Heidelberg included research groups working on cosmochronology and cosmic abundance of elements, interplanetary dust, mineralogy, geochemistry and chemistry of planetary material, physics and chemistry of the atmosphere. At the time, there were around 60 scientists working at the institute, 30 to 40 percent of them in cosmochemistry, and the rest in nuclear physics, solid state physics, and in the computer group (AMPG, III. Abt., Rep. 68A, No. 166/1-1, p. 3). During the first meeting of the ad hoc committee of the President of the Max Planck Society concerning the future of cosmochemistry in the Society held on March 12, 1973, it was general opinion that the cooperation between the Institute for Nuclear Physics and the Institute for Chemistry should be intensified. For a wide review on the research work carried on in the Heidelberg Cosmochemistry Department, see also AMPG, III. Abt., Rep. 68A, Folder No. 153.

166 No concrete decision had been taken about a successor to Gentner, and a committee to deal with this question was established. By 1974, around one-third of the research activi-

Finally, in Mainz, the decade-long search for a successor to Friedrich Paneth had ended in 1968 with the appointment of Christian Junge as Director of the newly founded Atmospheric Chemistry and Physical Isotopic Chemistry Department.[167] This choice not only averted the risk of the Mainz Institute being dissolved—which had been considered after potential successors

ties at the institute were dedicated to cosmochemistry and two-thirds to nuclear physics and associated fields. The *Kollegium* (collegial directorship) was in favor of appointing an experimental physicist, and the name of Bogdan Povh, who was already an External Scientific Member at the institute, was put forward (CPTS meeting minutes of 15.02.1974, 18.06.1974, AMPG, II. Abt., Rep. 62, No. 1771, 1772). Povh was a professor at the University of Heidelberg and his main research interests were nuclear and high-energy nuclear physics, which he pursued at CERN and at Lawrence Berkeley National Laboratory. He had also already been in contact with various groups at the Max Planck Institute for Nuclear Physics. A final decision proposing appointing Povh a Scientific Member and member of the *Kollegium* was taken in October of that year (CPTS meeting minutes of 25.10.1974, AMPG, II. Abt., Rep. 62, No. 1773). A broad outline of research work conducted at the institute at the time of Gentner's retirement can be found in Jürgen Kiko, and Ulrich Schmidt-Rohr: *Max-Planck-Institut für Kernphysik. Heidelberg*. Edited by Generalverwaltung der Max-Planck-Gesellschaft. München 1975. See also the Festschrift/ commemorative book published after his death in 1980: Generalverwaltung der Max-Planck-Gesellschaft (ed.): *Gedenkfeier Wolfgang Gentner*. München 1981.

167 The long process of finding a solution after Paneth's death involved discussions on the very future of cosmochemistry at the Institute for Chemistry. The first choice fell on Hans Suess, who had collaborated on the shell model of the atomic nucleus with future winner of the Nobel Prize in Chemistry Hans Jensen (CPTS meeting minutes of 02.06.1959, AMPG, II. Abt., Rep. 62, No. 1734). He had emigrated to the US in 1950, where, together with Nobel laureate Harold Urey, he had studied the abundance of elements in meteorites. He thus appeared to be the perfect candidate for this position, that Suess, however, was not ready to accept and so it appeared that it would not be possible to find an adequate successor to Paneth in the fields of research on meteorites and radiochemistry. In the following deliberations, the committee even decided to redirect research into organic or theoretical chemistry and move all meteorite research to Hintenberger's Department, but this met the disagreement of Karl Ziegler, as chemistry was already covered in other institutes (CPTS meeting minutes of 19.01.1961, AMPG, II. Abt., Rep. 62, No. 1736) and the successor was to focus on continuing the work of Paneth. Deliberations then focused on hiring the eminent physicist Rudolf Mössbauer, who plays an important role in several chapters of this book. It later transpired that there was a very low probability of Mössbauer accepting the position (CPTS meeting minutes of 01.11.1963, AMPG, II. Abt., Rep. 62, No. 1742). He did not in fact accept it and moved instead to the Technical University in Munich (CPTS meeting minutes of 09.06.1964, AMPG, II. Abt., Rep. 62, No. 1743). This unsolved problem of the directorship went on up to 1967–68 (CPTS meeting minutes of 07.04.1967, AMPG, II. Abt., Rep. 62, No. 1749), when, following Gentner's idea, a new research line was opened, calling upon Christian Junge who had the Chair of Meteorology at the University of Mainz (Dieter Hoffmann, and Ulrich Schmidt-Rohr (eds.): *Wolfgang Gentner. Festschrift zum 100. Geburtstag*. Berlin: Springer 2006, 50.). The new division was labeled "Chemistry of the Atmosphere" (CPTS meeting minutes of 23.02.1968, AMPG, II.

declined the offer to head scientific departments there—but also marked a significant twist of fate, as Junge identified primarily as a meteorologist. Under his directorship, the institute focused increasingly on what would come to be called the 'Earth system' sciences, and in later decades would even earn the institute the Nobel Prize for Chemistry 1995, which was awarded to Paul Crutzen, from 1980, Junge's successor.[168] Junge's appointment cast doubt on whether the southwestern institute would maintain its cosmochemistry tradition, yet Junge's scientific lineage and instrumental expertise fit this focus extremely well. During his time in the United States, he had worked closely with the research group led by Hans Suess (a former candidate for the Mainz directorship), which had pioneered the analysis of small radioactive samples in the atmosphere. This work in the 1950s was closely associated with the problem of nuclear explosions in the atmosphere, eventually leading to their pro-

---

Abt., Rep. 62, No. 1752). Over the years, decisions had also been taken to appoint Heinrich Hintenberger and Hermann Wäffler as independent Directors of the departments of mass spectroscopy and accelerator-based nuclear physics (CPTS meeting minutes of 02.06.1959, AMPG, II. Abt., Rep. 62, No. 1734) and Heinrich Wänke, who had worked with Paneth since the time they were both in UK, as the Scientific Member in charge of cosmochemistry (CPTS meeting minutes of 13/14.05.1963, AMPG, II. Abt., Rep. 62, No. 1741). On Paneth's group, see Ursula B. Marvin: Oral Histories in Meteoritics and Planetary Science. VIII. Friedrich Begemann. *Meteoritics & Planetary Science* 37/S12 (2002), B69–B77. doi:10.1111/j.1945-5100.2002.tb00905.x. H. Wänke: Meteoritenalter und verwandte Probleme der Kosmochemie. In: E. Heilbronner et al. (eds.): *Kosmochemie. Fortschritte der Chemischen Forschung*. Berlin: Springer 1966, 322–408.

168 Junge reorganized the institute into two departments: the first, led by Wänke, would conduct research on meteorites, the second, led by himself, would conduct research on chemistry of the atmosphere and isotopes (CPTS meeting minutes of 20.02.1969, AMPG, II. Abt., Rep. 62, No. 1758). Hintenberger's group continued separately until his retirement. Christian Junge: Die Entstehung der Erdatmosphäre und ihre Beeinflussung durch den Menschen. In: Generalverwaltung der Max-Planck Gesellschaft zur Förderung der Wissenschaften e.V. (ed.): *Max-Planck-Gesellschaft Jahrbuch 1975*. Göttingen: Vandenhoeck & Ruprecht 1975, 36–48. On Junge, see Paul J. Crutzen: Christian Junge. 2.7.1912-18.6.1996. *Jahresbericht der Max-Planck-Gesellschaft*, 1996, 196–199. Ruprecht Jaenicke: Christian Junge. The Pioneer of Aerosol Chemistry. In: Sem J. Gilmore et al. (eds.): *History & Reviews of Aerosol Science. Proceedings of the Second Symposium on the History of Aerosol Science, 13–14 October 2001, Portland, Oregon, USA*. Reston: American Association for Aerosol Research 2005, 37–47. Ruprecht Jaenicke: Die Erfindung der Luftchemie. Christian Junge. In: Horst Kant, and Carsten Reinhardt (eds.): *100 Jahre Kaiser-Wilhelm-, Max-Planck-Institut für Chemie (Otto-Hahn-Institut). Facetten seiner Geschichte*. Berlin: Archiv der Max-Planck-Gesellschaft 2012, 187–202. The long process to reform the MPIC is one of the focuses of Gregor Lax: *Von der Atmosphärenchemie zur Erforschung des Erdsystems. Beiträge zur jüngeren Geschichte des Max-Planck-Instituts für Chemie (Otto-Hahn-Institut), 1959–2000*. Berlin: GMPG-Preprint 2018.

hibition.[169] The radiochemical and mass spectrometry techniques used in this research were then increasingly applied to wider atmospheric environmental issues. Cosmochemistry, which used virtually the same methodologies, maintained its presence in Mainz until the end of the century. Links with the Heidelberg institute remained strong, not only in cosmochemistry but also in the atmospheric sciences, as the Max Planck Institute for Nuclear Physics likewise set up a department dedicated to isotope physics of the atmosphere.[170] This research tradition had existed in Heidelberg even before there was a dedicated research group under Frank Arnold (famous for his work on the depletion of the ozone layer); for example, Hugo Fechtig—trained, like Junge, under Suess in the United States—had been a researcher in Heidelberg and became director there in 1974. Around the same time, the Max Planck Institute for Nuclear Physics participated in the Aeros satellites built in Germany and launched by NASA in collaboration with Karl Rawer (Freiburg), and the DLR.[171] Hans Suess, mentor of many at this fruitful intersection between cosmochemistry and the new atmospheric and Earth system sciences, was a regular visitor to the institute in Mainz.[172]

Aeronomy at the Crossroads between Plasma Physics, and Planetary Science

The major problem, however, remained finding a permanent director for the Institute for Aeronomy, which was under threat owing to not only the long vain efforts to find a successor to Bartels, but also the impending retirement of Dieminger in 1975,[173] upon which his section of the institute was expected to close down—raising the specter of whether the entire institute would follow. The status and the future of the Aeronomy Institute were examined by

169 An account of the cross-fertilization between nuclear weapons and climate science can be found in Paul N. Edwards: Entangled Histories. Climate Science and Nuclear Weapons Research. *Bulletin of the Atomic Scientists* 68/4 (2012), 28–40. doi:10.1177/0096340212451574.

170 A detailed account of the transition toward the atmospheric sciences in Mainz as well as these MPG-wide movements toward an Earth system or environmental sciences cluster are discussed in Lax, *Von der Atmosphärenchemie zur Erforschung des Erdsystems*, 2018.

171 See Karl Rawer: *Meine Kinder umkreisen die Erde. Der Bericht eines Satellitenforschers.* Freiburg im Breisgau: Herder 1986, 58.

172 Herbert Palme: Cosmochemistry along the Rhine. *Geochemical Perspectives* 7/1 (2018), 1–116. doi:10.7185/geochempersp.7.1.

173 Walther Dieminger, Alfred Ehmert, and Georg Pfotzer: *Sonderheft Professor em. Dr. Walter Dieminger zum 70. Geburtstag am 7.7.1977. Ansprachen und Vorträge anlässlich seiner feierlichen Verabschiedung aus seinem Amt als Direktor des Max-Planck-Instituts für Aeronomie am 9. und 10.7.1975.* Edited by Julius Bartels. Berlin: Springer 1977.

a special committee that visited Lindau in summer 1973, as well as Mainz, Heidelberg, and Garching.[174] On page 1 of the final report it was remarked that:

> The Institute has recruited almost exclusively from its own students and is showing the signs of ingrowth/of lacking new blood, a phenomenon not unique in Lindau, but faced by similar institutes at many places [...] Ways must be devised to make possible a natural and regular turnover of scientific personnel of the Institute. Without such circulation and filtering process any laboratory that maintains a constant size is doomed to stagnation.

It was suggested (p. 2) that "[...] no attempt be made to transfer any existing groups from other MPIs to Lindau" as "the loss would be much larger than the potential gain for Lindau." A major recommendation was to give top priority "to outside appointments over internal promotions," in connection with the establishment of "a strong and active program for visitors and possibly a standing visiting committee to enhance and cultivate contacts with outside groups." The committee proposed strengthening ties with neighboring universities, following other institutes' example, wherever such contacts would be mutually beneficial. Moreover, universities should be encouraged "to use the unique facilities at Lindau in cooperative research projects." Strengthening the theoretical aspects of the work in all successful areas of the institute and coupling them closely with experimental endeavors was strongly recommended: "an attempt should be made to recruit staff members capable of creating new theories, rather than making detailed calculations on existing models." The final recommendations with respect to new directors or members of the institute were based on the "expectation that the future field of research will be aimed toward the exploration of solar-planetary relationships by various techniques." The committee also made it clear that it seemed impossible for the institute "to implement its proposed program without major reallocation of personnel and resources." Finally, "The choice of the new managing director is particularly crucial since it will be his [*sic*] task to restructure the Institute." These main points were followed by a very detailed analysis and critique of the work of the institute—and of its relationship to work carried out elsewhere—

174 Report by the Visiting Committee to the Max-Planck Institut für Aeronomie, Lindau, Germany, July 2, 1973, AMPG, III. Abt., Rep. 68A, No. 153. The folder also includes reports on the activities of the Cosmochemistry Department in Heidelberg and current and future activities at the Institute for Extraterrestrial Physics.

as well as some proposals for future programs. In particular, the committee stressed that (p. 3):

> The strong dichotomy between the two institutes of which the MPI Lindau is composed, and indeed between groups within each institute, is immediately obvious, a separation which is unfortunate and a hindrance to the scientific development. We note that other MPIs which we visited [Mainz, Heidelberg, and Garching], housing much more divergent groups within their organization, have managed to create a definite spirit of unity, working to their benefit.

After recalling the interests in common with Garching (energetic particle measurements from satellites, balloons, and rockets), and the difference in emphasis (astrophysical aspects prominent in Garching, and solar-terrestrial problems at Lindau), it appeared that (p. 9) "the level of the German space effort is large enough to accommodate these two groups and the resulting competition will be beneficial to both."

The report also stressed (p. 8) its belief that:

> the EISCAT facility will constitute a most important landmark in the development of atmospheric and magnetospheric physics and that the participation of the Federal Republic in the project will ensure for German scientists an interest in one of the major growth areas of upper atmosphere research.

By this time, due to the political liabilities of operating a research station on UN-embargoed territory, the institute's station in Tsumeb (see Chapter 1) had already been given to the South Africans.[175]

The crucial source of pressure in the Institute for Aeronomy, beyond the quality issues emphasized in the report, was that the expertise there too closely intersected with that of the Extraterrestrial Institute; moreover, the growing interest in both Garching and Mainz to create a new Max Planck Institute for Meteorology would in all likelihood further undermine Lindau's claim to a specific research subject in the atmosphere, the latter already attracting

175 See Reimar Lüst papers, travels in South Africa (AMPG, III. Abt., Rep. 145, No. 76). Even though this was de facto surrendered in 1970s, the official transfer is dated January 1, 1985, in Eckart Henning, and Marion Kazemi: *Chronik der Kaiser-Wilhelm-/Max-Planck-Gesellschaft zur Förderung der Wissenschaften 1911–2011. Daten und Quellen*. Berlin: Duncker & Humblot 2011.

more attention in Mainz. With growing interest by astronomers in the Max Planck Society to expand the Society's astronomy footprint by opening new wavelength windows, there was increasing pressure to redistribute research programs between Garching and Lindau so that the space plasma physicists would relocate to the latter. In the early 1970s, both Gerhard Haerendel and Heinrich Völk from the Max Planck Institute for Extraterrestrial Physics were invited to accept an appointment as Director of the Aeronomy institute, but both declined.[176] Eventually, the head of the search committee, plasma physicist Ian Axford, decided to take the job himself, after persistent encouragement from gamma ray astronomer Klaus Pinkau.[177]

Deliberations regarding the Institute for Aeronomy also addressed the rising interest in environmentally related research at the Max Planck Society. As mentioned above, atmospheric chemistry expert Christian Junge had been appointed director in Mainz in 1968, and the maturation of artificial satellites sparked much interest there in space-based atmospheric research. At

176 During the committee meeting at the Max Planck Institute for Aeronomy, Lindau/Harz of February 8, 1973, it was officially announced that Haerendel had communicated that he would not accept an appointment in Lindau (A. Weller to members of the committee "Max-Planck-Institut für Aeronomie, Linday/Harz," April 18, 1973. AMPG, III. Abt., Rep. 68A, No. 153, p. 2). The visiting committee previously mentioned had proposed Völk as a possible director at the Aeronomy Institute. Gerhard Haerendel had become a member of Lüst's research group for extraterrestrial physics in October 1961, immediately after having completed his PhD on the Van Allen belts, with Schlüter as supervisor. He himself claims to have been the first in Germany to write his doctoral thesis on a topic related to space. He in fact became the 'Haustheoretiker' of the group and had a leading role in the first barium and ion cloud experiments, successfully testing the technique in 1964, and took decisions on the preparation of the payloads for experiments on plasma clouds, which were performed with sounding rockets as part of a national project with ESRO, as a first step in preparation for subsequent participation in satellite experiments, the HEOS (Highly Eccentric Orbit Satellite) missions of 1969 and 1972. The ion cloud experiments continued to be his responsibility. A. Valenzuela et al.: The AMPTE Artificial Comet Experiments. *Nature* 320 (1986), 700–703. doi:10.1038/320700a0. Walter F. Huebner: Kometen, Sonnenwind und Wasserstoffkoma. In: Max-Planck Gesellschaft (ed.): *Ludwig Biermann. 1907–1986*. München 1988, 35–50. Haerendel became a Scientific Member in 1969 (CPTS meeting minutes of 26.06.1968, 20.02.1969, AMPG, II. Abt., Rep. 62, No. 1753, 1755) and then, in 1972, when Lüst was appointed President of the MPG/the Max Planck Society, Haerendel became Director of MPE/the Max Planck Institute for Extraterrestrial Physics; but in 1968, Lüst had reorganized the institute under a collegial directorship, although it was not divided into departments but organized in research groups (CPTS meeting minutes of 26.06.1968, 20.06.1972, AMPG, II. Abt., Rep. 62, No. 1753, 1766). See also Gerhard Haerendel: interview by Helmuth Trischler and Matthias Knopp, April 9, 2010. Transcript, HAEU, https://archives.eui.eu/en/oral_history/INT066. Last accessed 4/5/2020.

177 Klaus Pinkau: interview by Helmuth Trischler, March 9, 2010. Transcript, HAEU, https://archives.eui.eu/en/oral_history/INT072. Last accessed 5/8/2019.

one point, there were discussions about a possible reform of the Institute for Aeronomy along these lines. With the appointment of Axford, however, things took a different turn: a new Max Planck Institute for Meteorology was established, not least owing to the influence of scientists at the Institute for Chemistry in Mainz and the Institute for Extraterrestrial Physics in Garching.[178]

As with many foreigners appointed to Max Planck institutes in crisis,[179] Axford expected his move to Lindau to be a temporary solution but, in the end, stayed on as Director of the Institute for Aeronomy for 25 years. His appointment prompted the institute's further specialization in solar system research, with a strong emphasis on space plasmas related to the Sun–planet interactions.[180] Discussions also took place about the general development of the Institute for Aeronomy and coordination of its future activities with similar research taking place in Mainz, Heidelberg, and Garching. The outcome was a proposal that long-term projects in these institutes be harmonized as far as possible.

Several new directors were appointed, including Vytenis M. Vasilyunas from the United States,[181] and Helmut Rosenbauer, again from the Institute for

---

178 Lax, *Von der Atmosphärenchemie zur Erforschung des Erdsystems*, 2018.

179 In this book we have several examples of prestigious foreign directors who stay a few years, and whose appointment coincides with radical reforms. The list includes León Van Hove in Munich, Ken Kellermann in Bonn, and Steven Beckwith in Heidelberg. Other research clusters of the Max Planck Society have experienced a similar pattern. In the context of this book, Ian Axford in Lindau, and also Simon White in Garching (see next chapter), constitute exceptional cases who ended up staying permanently.

180 In a letter written on January 8, 1973 by Lüst to J. T. Jefferies (Institute for Astronomy at the University of Hawaii), about the choice of speakers for a joint discussion on the "The Solar Wind and its Interaction with the Interstellar Medium" at a meeting of the International Astronomical Union (IAU) commission No. 43 (Plasmas and Magnetohydrodynamics in Astrophysics, of which Lüst was the president), the former emphasized that Axford was "probably the most competent man" to work in this field. His latest review on this topic was defined by Davis Leverett of Caltech (Leverett to Lüst, November 28, 1972) as "the major paper on the subject that I have ever seen to date" (AMPG, III. Abt., Rep. 145, No. 957, Fol. 3 and 11). See also William Allan: Sir William Ian Axford. 2 January 1933–13 March 2010. *Biographical Memoirs of Fellows of the Royal Society* 59 (2013), 5–31. doi:10.1098/rsbm.2013.0007. Vytenis M. Vasyliunas: Sir Ian Axford FRS 1933–2010. *Astronomy & Geophysics* 51/3 (2010), 3.37–3.38. doi:10.1111/j.1468-4004.2010.51336_1.x. Vytenis M. Vasyliunas: Ian Axford. 07.01.1933–13.03.2010. Edited by Max-Planck Gesellschaft. *Jahresbericht Der Max-Planck-Gesellschaft* Beileger Personalien (2010), 15–17.

181 CPTS meeting minutes of 08.02.1973, AMPG, II. Abt., Rep. 62, No. 1768. A decision was also taken to offer the post of director to Ulf von Zahn, but Zahn did not accept the offer and so a new committee of experts was formed (CPTS meeting minutes of 26.06.1973,

Extraterrestrial Physics, who moved his entire experimental space plasma group to Lindau.[182] As for Gerhard Haerendel, Lüst's successor in Garching, despite having resisted relocation to Lindau, he still acted as a crucial overseer of the Institute for Aeronomy in matters of ionospheric research by becoming involved in the multinational EISCAT project, which conducted research on the lower, middle, and upper atmosphere and ionosphere using the incoherent scatter radar technique.[183] In contrast, Erhard Keppler, a disciple of Regener, Ehmert, and Pfotzer, and the most important space researcher at the Institute

---

AMPG, II. Abt., Rep. 62, No. 1769). In view of Ehmert's retirement, which would soon be followed by Dieminger's and Pfotzer's, the possibility of reorganizing the Lindau institute in a move toward research in meteorology was also discussed. It was suggested that the topic should be examined by Christian Junge himself in a "Memorandum zur Lage der Meteorologie in Deutschland" and especially within the Max Planck Society (CPTS meeting minutes of 15.02.1974, AMPG, II. Abt., Rep. 62, No. 1771). In the following meeting of the Scientific Council, Axford is mentioned as having accepted the position as director at the institute and beginning his activities in July (CPTS meeting minutes of 18.06.1974, AMPG, II. Abt., Rep. 62, No. 1772). At the same time, Vasyliunas, a well-known theoretician from Massachusetts Institute of Technology, was proposed as a Scientific Member, in response to the committee's proposal that a strong theoretical group should be created in order to establish a strong connection with the experimental groups (CPTS meeting minutes of 25.10.1974, 23.01.1975, AMPG, II. Abt., Rep. 62, No. 1773, 1774). Materials on the restructuring of the Institute for Aeronomy related to the years 1973–76 can be also found in AMPG, II. Abt., Rep. 66, No. 60 and No. 61.

182 See CPTS meeting minutes of 16.11.1976, 08.03.1977, AMPG, II. Abt., Rep. 62, No. 1779, 1780. Rosenbauer's arrival from the Max Planck Institute for Extraterrestrial Physics, with a strong experience in space missions (in particular, he had been principal investigator of the plasma experiment aboard the Helios solar probes) and related instrument building, is another example of the cultural influence of Biermann's tradition extending up to the recent Rosetta mission to the comet 67P/Churyumov-Gerasimenko in which Rosenbauer was responsible for the design and the scientific program of the lander Philae, which landed on the comet's nucleus in 2014, and for one of the most important instruments on board Philae. J.-P. Bibring et al.: The Rosetta Lander ("Philae") Investigations. *Space Science Reviews* 128/1 (2007), 205–220. doi:10.1007/s11214-006-9138-2. M.A. Barucci, and M. Fulchignoni: Major Achievements of the Rosetta Mission in Connection with the Origin of the Solar System. *The Astronomy and Astrophysics Review* 25/1 (2017), 1–52. doi:10.1007/s00159-017-0103-8.

183 In 1981, Rosenbauer petitioned to call the Norwegian Tor Hagfors, a radio astronomer and radar expert, who had pioneered studies of the interactions between electromagnetic waves and plasma, later becoming founding director of the multinational EISCAT facility (CPTS meeting minutes of 21.05.1981, 27.10.1981, AMPG, II. Abt., Rep. 62, No. 1793, 1794). The EISCAT project, which had its roots in the work with ionosondes at the Institute for Aeronomy and the work with barium plasma clouds at the Institute for Extraterrestrial Physics and which both institutes participated in, was instrumental also in relaunching the Institute for Aeronomy in the Axford era. Haerendel, History of EISCAT, Part 4, 2016, 67–72.

for Aeronomy, was never made a Scientific Member of the Max Planck Society but instead was given a new permanent position as 'Technical Director,' and continued to lead the institute's space missions.[184]

Since Haerendel remained in Garching, he was encouraged by Reimar Lüst to increasingly shift his focus from experimental space plasmas to astronomical, wavelength-based research, for which a small infrared team already existed. Haerendel explored this direction in the late 1970s, but the saturation of the wavelength-based division of scientific labor between the different Max Planck Institutes then started to show its limits; a well-developed infrared astronomy group already existed at the Institute for Astronomy in Heidelberg, and it protested this move.[185]

Moreover, plasma physicist migration away from Garching went far beyond Lindau. In the early 1970s, Gentner and Lüst convinced the theoretical plasma astrophysicist Heinrich Völk, as well as his collaborator Gregor Morfill, to move to Heidelberg as part of the Zähringer succession. Völk was appointed as one of the directors of the cosmochemistry section of the institute. There, he sought to redirect his work toward the theoretical interpretation of cosmochemical research on the evolution of the solar system, which was part of the joint research projects between Mainz and Heidelberg. But ultimately, he managed to maintain a foot in general astrophysics, which led to his involvement a decade later in the major push of his institute toward gamma ray astronomy, which will be discussed in the final chapter of this book.[186]

---

184 Keppler, *Max Planck Institut für Aeronomie*, 2003. In fact, he got on very well with Axford, but his relationship with people relocated from Munich such as Rosenbauer was always problematic, according to several episodes related to his interaction with Rosenbauer.

185 Gerhard Haerendel: interview by Helmuth Trischler and Matthias Knopp, April 9, 2010. Transcript, HAEU, https://archives.eui.eu/en/oral_history/INT066. Last accessed 5/8/2019. In 1984, Haerendel became the German national representative to the Committee on Space Research (COSPAR), and was elected COSPAR President replacing Ian Axford in 1994. See Haerendel's Preface in Gerhard Haerendel et al. (eds.): *40 Years of Cospar*. Noordwijk: ESA Publications Division 1998.

186 Even before the beginning of Lüst's presidency, in order to reinforce the theoretical side of cosmochemistry, the committee proposed to call Heinrich Völk from the Institute for Extraterrestrial Physics in Munich. At the same time, a similar proposal had been put forward by the Institute for Aeronomy. The final choice was left to Völk and he decided to go to Heidelberg (CPTS meeting minutes of 08.02.1973, 26.06.1973, 23.10.1973, 15.02.1974, AMPG, II. Abt., Rep. 62, No. 1768, 1769, 1770, 1771). Völk's opinion was that Gentner was a bit afraid "that his group would, so to say, narrow down too strongly, so that it would become a small appendix of this institute [for Nuclear Physics]. At the same time, there was a Cosmochemistry Department with many and very good chemists in Mainz, like Heinrich Wänke and so I think that Gentner wanted to get, so to say, access to space and

Astronomers in the Max Planck Society, who gained greatly in influence throughout the 1970s, appear to have exerted pressure in line with an implicit research hierarchy which considered deep space questions more interesting and fundamental than solar and planetary problems or plasma physics; these latter areas would have been seen as remnants of the early space age: redistribution of space plasma activities and near-space missions to Lindau and Mainz in the mid-1970s can be seen as a means for space astronomers and astrophysicists in Garching and Heidelberg (both nuclear physics and astronomy) to keep for themselves the research areas that they considered more interesting, while relegating to other institutes what they saw as the less glamorous solar system research, which bordered on geology rather than astrophysics. Lindau and Mainz scientists took up this challenge, however, and over the next two decades successfully turned their institutes into the major German base for interplanetary probes, and ultimately even found new international partners by collaborating with Soviet and, later, Russian space missions.[187]

The Institute for Plasma Physics (IPP): A Powerhouse for the Cosmic Sciences

The redistribution of expertise from plasma physics was not limited to the Institute for Extraterrestrial Physics, either. Throughout the 1970s, there were active attempts to make available to other Max Planck Institutes the expertise and technological developments that had originated at the Institute for Plasma Physics (IPP). As part of this move toward making the IPP more productive for the Max Planck Society, a company called Garching Innovation was founded for the purpose of commercializing the technological develop-

all of these things. And so, somehow, they decided to ask me to come here, and start something more in the direction of astrophysics […] I came here and I brought Gregor Morfill with me, who later went back to Garching and became finally also one of the directors there, at the Max Planck Institute for Extraterrestrial Physics, but he came with me originally. That was very nice, and so we did two things: one was just to generalize cosmic ray physics, because this had also to do with meteorite research […] then we started working on solar system formation questions; which was totally new for me and, but we had a good idea of how one could form a protoplanetary disc out of a collapsing molecular cloud, which forms a central star and that disc around it and, hopefully, planets out of it, and so forth…" Heinrich Völk: interview by Luisa Bonolis and Juan-Andres Leon, Heidelberg, October 9–10, 2017. DA GMPG, BC 601037.

187 See, for example, Keppler, *Max Planck Institut für Aeronomie*, 2003, 35–40. Peter Czechowsky, and Rüdiger Rüster (eds.): *60 Jahre Forschung in Lindau. 1946–2006. Vom Fraunhofer-Institut zum Max-Planck-Institut für Sonnensystemforschung. Eine Sammlung von Erinnerungen*. Katlenburg-Lindau: Copernicus 2007.

ments of the institute, and also of the Max Planck Society in general.[188] Also, the Institute for Radio Astronomy's new binational project IRAM (*Institut de Radioastronomie Millimétrique*, see next chapter), modeled on the *Institut Laue-Langevin* in Grenoble, obtained a team of detector experts from the IPP.[189]

At the same time, the Max Planck Society reasserted its authority and scientific worldview over the Institute for Plasma Physics. In 1970, after a disappointing decade as an independent entity,[190] the IPP was reabsorbed as an institute of the Max Planck Society, the implication being that the Society's scientific members could implement their vision of fundamental research at what had until then been primarily a reactor-building enterprise. At the end of this process, which included Arnulf Schlüter's retirement in 1981, the gamma ray astronomer Klaus Pinkau from the Institute for Extraterrestrial Physics was invited by Reimar Lüst to become Research Director of the IPP, which he remained until the end of the 1990s.[191]

This was a controversial move and caused disagreement with, for example, fellow director Joachim Trümper;[192] it signaled the end of the 1970s surge toward wavelength specialization, and a new tendency to appoint directors more on account of their general scientific authority and problem-solving skills, such as Klaus Pinkau was able to channel into his effective leadership of the gigantic IPP. But in leaving the Institute for Extraterrestrial Physics, Pinkau also jeopardized the balance there between the plasma physicists and the space astronomers.

Experimental plasma physics likewise had unforeseen positive consequences for deep space astronomy: in the mid-1970s, a major unexpected boost in gravitational wave research came from the IPP, after the Americans blocked its experiments in laser fusion research, deeming them too close to military applications.[193] The spin-off was a new Institute for Quantum Optics (MPQ)

188 Jaromír Balcar: *Instrumentenbau–Patentvermarktung–Ausgründungen. Die Geschichte der Garching Instrumente GmbH*. Berlin: GMPG-Preprint 2018.

189 Encrenaz et al., The Genesis of the IRAM, 2011, 83–92.

190 A disappointment stemming largely from the overblown expectations of thermonuclear fusion in the 1950s and 1960s. By the late 1960s, it was evident that the path to a thermonuclear power reactor, even if successful, would take many decades.

191 The appointment process can be followed in CPTS meeting minutes of 16.11.1976, 29.10.1980, 21.01.1981, 21.05.1981, AMPG, II. Abt., Rep. 62, No. 1779, 1791, 1792, 1793.

192 Joachim Trümper: interview by Luisa Bonolis and Juan-Andres Leon, Munich, August 7–8, 2017. DA GMPG, BC 601036.

193 Reimar Lüst: interview by Horst Kant and Jürgen Renn, Hamburg, May 18, 2010, DA GMPG, ID 601068.

in Garching, largely dedicated to laser technology, and as we explore in detail in the last chapter of this book, it quickly found one of its main scientific missions in gravitational wave detection by interferometric means. An entire experimental group dating back to the time of Heinz Billing at the Institute for Astrophysics was accordingly transferred to the MPQ.[194] This was the final step in turning what had been Biermann's multifaceted institute into a purely theoretical one. The work of this spin-off gravitational wave experimental group will be treated in more detail later, in Chapter 5.

194 See CPTS meeting minutes of 29.10.1980, AMPG, II. Abt., Rep. 62, No. 1791, reporting discussion on the continuation of the gravitational-wave experiment of Billing's Department at the Institute for Astrophysics.

CHAPTER 4

Internationalization (1970s Onwards): Infrastructural Disappointments and the New International Division of Labor

International collaboration was always key to Max Planck leadership; in the first postwar decade, astrophysicists and cosmochemists were frequent guests within much larger projects based in Allied countries. During the post-Sputnik boom, one of the objectives of the vast expansion was to be able to mobilize national strengths to obtain a stronger voice in international collaborations. The chapter takes up this process of internationalization as it matures, from the 1970s on, and becomes the main mode of research in the Max Planck Society, which it still is to this day. Unexpectedly, this was thanks not so much to German-owned large infrastructures but, rather, to the weight of long-standing scientific and technical traditions which brought to the global table the Germans' theoretical insights, innovative experimentation, and superior instrument-making capacities. German reunification and the end of the Cold War further accelerated the Max Planck Society's transition toward this 21st-century mode of scientific production. Reform in the 1990s coincided with the geopolitical shifts, as well as with the retirement of many of those leading Max Planck Institute directors who had led the wavelength expansion in the previous 30 years. Their successors de-emphasized the construction and ownership of observatories, focusing instead on scientific research within large collaborations, secure in the knowledge that their institutes' instrumental expertise would provide political leverage and a comparative advantage over their partners. Political pressures to relocate institutes to the former East Germany, or even to close them down, were successfully turned into opportunities for expansion, and ultimately, even the one most seriously under threat from these reforms, Biermann's original (theoretical) Institute for Astrophysics, found a reinvigorated mission within the cluster of Max Planck Institutes dedicated to cosmic research, as well as in the, by then, global powerhouse of Garching.

1 From National Infrastructures to International Collaborations

The giant telescopes and satellites of the 1960s were national projects, and several ended up becoming major disappointments, while by the 1970s, the parallel track of Europeanization began to bear fruit. Institutions such as the European Southern Observatory (ESO) and the Institute for Millimeter Astronomy (IRAM, founded by France and Germany) paved the way, and the Max Planck Society aimed to maximize its influence within such organizations. In parallel, from different starting points, all the observational institutes converged technologically on infrared astronomy, blurring wavelength as a demarcation between institutes and leading to intense inter-institute collaboration in the 1980s and '90s. As the large telescopes and astronomical satellites came to be built predominantly as international collaborations, and to operate as infrastructures, Max Planck Institutes reoriented much of their work, towards scientific publication on the one hand, and to instrument development on the other; and this afforded them privileged access within the new mode, namely the division of scientific labor. In this context, Max Planck Institutes innovated in many instrumental techniques, such as adaptive optics and interferometry, taking advantage of technical traditions and many decades of participation, in collaborations that initiated and benefited from these novel techniques.

National Infrastructural Disappointments

Around the mid-1980s, there was growing disappointment with the outcome of the wavelength expansion program that had driven both the expansion of the Max Planck Society in astronomy, and most specifically, the direct ownership of observatory infrastructure. Entering the global competition for large observatory projects at all wavelengths was one of the most ambitious tasks ever undertaken by the Max Planck Society to date. Many of these telescopes, such as the Effelsberg radio telescope, the IRAM 30 m-dish millimeter wavelength telescope (see later in this chapter), and the ROSAT space-based X-ray telescope, were some of the most brilliant and globally visible projects in the Max Planck Society.

As we will see throughout this chapter, the global organization of research was in constant flux from the 1970s onwards, and multi-partner international collaboration became the default. As described in Chapter 2, national space exploration programs had long since welcomed the financial support and political visibility that came with including international partners. But in the early decades, such 'invitations' were blatantly one-sided, with control remaining effectively in the hands of the senior partner (most prominently, NASA). In response, the first generation of West German national projects such as ROSAT sought to maximize national control. And despite the inclusion of international partners, which was not only politically and financially indispensable, but also the unavoidable basis for integration in the global community of space-based instrumentation and research, this aspiration to national control would persist.

Meanwhile, in ground-based astronomy, the first generations of telescopes in the Max Planck Society had been inexpensive enough to be built with West German resources, and even though they welcomed external users, there was likewise an obvious element of national control regarding the research produced with them. Even in cases like IRAM (detailed below), where the telescope was formally international, the constituent French and German observatories were self-contained and continued to act in line with national strengths and traditions for decades after their inauguration, easily contributing to their respective country's 'national' prestige in astronomical research.

As more projects came to fruition, however, it increasingly became obvious that such national control also meant financial and political responsibility for the embarrassing failures; and in West Germany, in particular, this altered the course of the Max Planck Society's ambition to control such expensive, long-term facilities.

A growing series of disappointments highlighted the high-risk and long-term commitment of creating an observatory infrastructure. The most visible of these, universally regarded as a traumatic experience, were the optical telescope projects of the Max Planck Institute for Astronomy. They became tangled up in an unfortunate political constellation that resulted in the cancellation of the southern hemisphere observatory, owing to its location in a territory deemed by the United Nations to be illegally occupied, namely the former German colony that is now Namibia. Meanwhile, its northern counterpart in Calar Alto, Spain, became mired in a long series of budget overruns and delays, such that by the time the instrument was completed in the mid-1980s, it was already obsolete. To make matters worse, the observatory turned out to be located on a site with mediocre climatic conditions.[1] For the Max Planck Society, the mere mention of Calar Alto is a telling reminder of its policy to no longer build observatories or large infrastructural projects.

Similarly, in the 1980s and '90s, the sub-millimeter telescope of the Max Planck Institute for Radio Astronomy, built in a 50/50 partnership with the

1 Tensions between the West German government and the Max Planck Society over the sites of the new observatories received attention from the international community; see, for example, Astronomy. Politics and Science. *Nature* 236/5346 (1972), 320–321. doi:10.1038/236320b0.

University of Arizona, was mired in a series of conflicts with Native Americans and environmentalists over the invasiveness of the observatory's mountain location.[2] By the time the telescope went into operation, it was further hampered by unexpected technical difficulties and the realization that (there, too!), the weather on the mountain was worse than expected. As a result of the delays, the observatory missed out on being the first to open up the new sub-millimeter wavelength windows.[3]

Both in Calar Alto and Mount Graham, as the observatories were delayed and it became apparent that they would fail to live up to expectations, conflicts between the Max Planck Institutes and their international partners intensified, which further poisoned their relationship well into the observatories' years of operation.

Even the successful observatories brought difficulties in their wake. The Effelsberg radio telescope, like Calar Alto, was meant to be a national facility open to all German researchers; but it faced persistent criticism for allocating different observation time slots to MPI and external users in Germany.[4] And in outer space, even incredibly successful satellite missions such as ROSAT revealed the limitations of the 'national' approach to astronomical infrastructure, given Germany's modest geopolitical footprint. Space missions would always be tied to the politics of launchers and communications networks, which remain outside of German 'national' control; and as ESA became increasingly self-reliant (thanks mostly to French ambitions with the development of the Ariane launchers), it seemed that a wholly sovereign astronomical program at the satellite scale would perhaps no longer be needed. As a final coup de grâce, in 1999, the last 'national' astronomical satellite of the MPE—the follow-up to ROSAT, called ABRIXAS (A BRoadband Imaging X-ray All-Sky Survey)—experienced a dramatic failure of the German-made battery system immediately after its launch, and the mission had to be abandoned.[5] In

2 Leandra A. Swanner: *Mountains of Controversy. Narrative and the Making of Contested Landscapes in Postwar American Astronomy.* Dissertation/ PhD Thesis. Cambridge, MA: Harvard University 2013.

3 Jacob W.M. Baars: *International Radio Telescope Projects. A life among their designers, builders and users.* Rheinbach: Createspace Independent Publishing Platform 2013.

4 For example, on radio astronomy, see Karl Menten: interview by Luisa Bonolis and Juan-Andres Leon, Bonn, February 5–8, 2018, DA GMPG, BC 601052, and, on optical astronomy, Dietrich Lemke: *Im Himmel über Heidelberg. 40 Jahre Max-Planck-Institut für Astronomie in Heidelberg (1969–2009).* Edited by Archiv der Max-Planck-Ges. Vol. 21. Berlin 2011. These issues were addressed in the 1987 *Denkschrift*, which will be discussed later: Heinrich J. Völk et al.: *Denkschrift Astronomie.* Weinheim: VCH 1987.

5 Alison Abbott: Battery Fault Ends X-Ray Satellite Mission. 6732. *Nature* 399/6732 (1999), 93–93. doi:10.1038/20029.

subsequent decades, fully national satellite projects became a rarity.[6] Owing to these disappointments, as well as to global transformations in scientific research practices, there was a significant shift in perspective: instead of owning the observational infrastructure, what the most successful Max Planck Institutes have since focused on, to guarantee a starring role in international collaborations, is controlling the *means of scientific production*,[7] for example, running their own workshops and instrumental testing facilities, as we detail later in this chapter.

From IRAM to ALMA: Success within Ground-Based International Collaborations

One of the most successful international collaborations in astronomy began at the Max Planck Institute for Radio Astronomy with Peter Mezger's project, a large (30 m-diameter) telescope for millimeter-wavelength astronomy. While initially conceived as the second large telescope of the Institute for Radio Astronomy in the early 1970s, it turned out to be the bargaining chip in a new form of international collaboration under the name *Institut de Radioastronomie Millimétrique* (IRAM).[8] Thanks to Mezger's early career contacts with the Paris observatory, he managed to combine his telescope project with concurrent French plans for an interferometric array in the millimeter wavelengths.[9] Development work for Mezger's telescope was carried out in parallel with negotiation of the IRAM treaty, so could be very quickly executed, once this was signed, saving face for the new organization, while the French contribution (although ultimately likewise very successful) was chronically delayed.

Mezger resisted French pressure to locate his telescope next to their planned interferometric array on the Plateau de Bure in the Alps near Grenoble, instead finding a more suitable, dry, high, and easily accessible location near a ski resort in southern Spain.[10] This location contrasted with the Bure site, which was difficult to reach and even experienced two fatal accidents

6 See, for example, Abbott, Battery Fault, 1999, 93–93. National 'research' projects usually have a large component of commercial and military applications, and the Max Planck Society does not participate in these.

7 Letter of Immo Appenzeller to Völk (AMPG, ZA 166, No. 57).

8 Extensive archival material related to IRAM (1973–93) can be found at AMPG, II. Abt., Rep. 85.

9 Karl Menten: interview by Luisa Bonolis and Juan-Andres Leon, Bonn, February 5–8, 2018, DA GMPG, BC 601052. Jacob W. M Baars: interview by Luisa Bonolis and Juan-Andres Leon, Bonn, February 5–7, 2018. DA GMPG, BC 601050.

10 Baars, *International Radio Telescope Projects*, 2013. Chapter 3 is entirely dedicated to IRAM.

in the 1990s, with the loss of dozens of lives.[11] As an international organization learning from the continuing mistakes of the Calar Alto Observatory, IRAM made an effort to be generously inclusive and encourage Spanish scientific participation in the project, increasingly extending the involvement of Spain's Instituto Geográfico Nacional (IGN), also in order to grow local expertise for the operation of the telescope. Completely built by the Max Planck Institute for Radio Astronomy, the telescope is a "fine piece of German engineering" constructed by the Krupp/ MAN partnership.[12] It came into operation on schedule in 1986 and is, to this day, by far the most productive single dish telescope working in the millimeter wavelength domain.[13]

However, as a founding member of the IRAM partnership, the Max Planck Society would also come to benefit enormously from the French contribution to the collaboration when it finally went into operation. Despite all the delays, the higher-risk French interferometric system paid off with its highly innovative approach, and was now easily available to Max Planck scientists. Together, the two IRAM sites beat all the competition in millimeter astronomy. Furthermore, it turned out that the control of infrastructure by international partners, each with their own decision-making bodies, was a much better way to deal with the conflicts of telescope time allocation and 'ownership,' such as came to plague both Effelsberg and Calar Alto. True, outside researchers frequently complained that Max Planck researchers and projects were unjustly given precedence in the national MPI observatories.[14] And even scientists

11 Sybille Anderl: Pforte zum Kosmos. *FAZ.NET* (11/3/2017). https://www.faz.net/-gwz-938nm. Last accessed 4/10/2018.

12 Baars, *International Radio Telescope Projects*, 2013, 45–56. Jacob W. M Baars: interview by Luisa Bonolis and Juan-Andres Leon, Bonn, February 5–7, 2018. DA GMPG, BC 601050. See also Pierre Encrenaz et al.: Highlighting the History of French Radio Astronomy. 7. The Genesis of the Institute of Radioastronomy at Millimeter Wavelengths (IRAM). *Journal of Astronomical History and Heritage* 14/2 (2011), 83–92. http://www.narit.or.th/en/files/2011JAHHvol14/2011JAHH...14...83E.pdf. Last accessed 10/30/2018. In this article, the relationship between IRAM and Spanish researchers is contrasted with the earlier 'neo-colonialist' approach of Calar Alto.

13 Karl Menten: interview by Luisa Bonolis and Juan-Andres Leon, Bonn, February 5–8, 2018, DA GMPG, BC 601052. The only other comparatively competitive (45 m-diameter) telescope was built by the National Astronomical Observatory of Japan in Nobeyama. The IRAM 30 m telescope's success was largely due to the constructive competition and collaboration within its European community. The impressive productivity of this telescope is quantified in Virginia Trimble, and Paul Zaich: Productivity and Impact of Radio Telescopes. *Publications of the Astronomical Society of the Pacific* 118/844 (2006), 933–938. doi:10.1086/505182.

14 This is one of the main topics diplomatically addressed by Völk et al., *Denkschrift*, 1987.

from other MPI departments had difficulty accessing Effelsberg, or influencing its instrumental development.[15] IRAM, on the other hand, epitomized the separation of observational infrastructure and scientific work, in similarity to high-energy physics in places such as CERN. The organization maintains an in-house technical team, while external researchers—including those from the member institutions Max Planck Society, the Centre National de la Recherche Scientifique (CNRS), and the Instituto Geográfico Nacional (IGN)—apply for observation time, and often use their own detection equipment. Over its several decades in operation, this model has proved able to adapt to innovations and technologies, to the point where the German–Spanish antenna remains to this day a very successful telescope, while the French site is undergoing a major upgrade under the name NOEMA (NOrthern Extended Millimeter Array).[16] Once finished, NOEMA will be the Northern Hemisphere counterpart to ALMA, with twelve antennas operating as an interferometer. In combination, the 30 m telescope in Spain and the NOEMA interferometer operate in four atmospheric wavebands, allowing simultaneous detection of thousands of spectral lines as well as continuum spectra from cold dust; and as described later in this chapter, all the millimeter-band telescopes of this family contributed significantly to the development and observations of the Event Horizon Telescope array.

In contrast to the earlier feeling that observatories were 'owned' by individual Max Planck Institutes, and despite the traditional wavelength-based distinction between scientific research projects, other MPIs, such as Astronomy (Heidelberg) and Extraterrestrial Physics (Garching), have made ample use of these facilities. Max Planck scientists from the interested institutes generally serve as delegates to the IRAM's governing bodies. Ultimately, this multi-institute relationship with IRAM is yet another clear example of what we call the 'clustered' nature of astrophysics in the Max Planck Society.

Furthermore, IRAM's predominance in millimeter astronomy was mobilized in the 1990s to help initiate a much larger multinational collaboration,

---

15 The lack of understanding between the departments of the Max Planck Institute for Radio Astronomy in its early days is seen by the authors as a missed opportunity. Technically, the strict time allocations of the Effelsberg telescope, coordinated with international collaborations in surveys and Very Long Baseline Interferometry (VLBI), made opening up access to other groups difficult. But this lack of cooperation had a fundamental background in the personal disagreements among the institute's first generation of directors, and particularly affected the electronic instrumentation development at the institute. This is further mentioned in: Jacob W.M. Baars: *International Radio Telescope Projects: A life among their designers, builders and users*. Rheinbach: Createspace Independent Publishing Platform 2013, 29–30.

16 Anderl, Pforte zum Kosmos, 11/3/2017.

the Atacama Large Millimeter/ Submillimeter Array (ALMA), an interferometric system of 64 antennas on a remote 5000 m-high plateau in northern Chile, developed to collect light reaching the Earth from some of the coldest and most distant objects in the Universe.[17] The dry climate and extreme elevation of the site provide the right conditions for detecting such faint signals. Two decades later, as the array went into operation, it featured many instrumental developments initiated at the Max Planck Institute for Radio Astronomy and IRAM, and half of its antennas (paradoxically, the American contribution to the project) are made by Vertex Antennentechnik, successor to the MAN/Krupp partnership that built most of the Max Planck observatories.[18] Moreover, the prolific scientific community of experts in millimeter and submillimeter astronomy fostered by IRAM has led to Europeans being very productive users of ALMA.[19]

Geopolitical Outsourcing: The European Southern Observatory in Garching

In fact, ALMA can be seen as a further step in the 21st-century trend to internationalization, building on the institutional frameworks both of IRAM, in radio astronomy, and of another organization, in optical astronomy, the European Southern Observatory (ESO, Chapter 3). From the 1970s onwards, the West German and Max Planck strategy was to move decisively towards fulfilling their astronomical ambitions largely through this kind of international collaboration. But for these to truly 'pay off,' the Germans needed to consolidate not just their scientific presence and economic gravitas, but also their political leverage over them.

In the early 1970s, a unique opportunity had presented itself when ESO, then headquartered at CERN in Geneva, was pressured to relocate to West

17 Karl Menten: interview by Luisa Bonolis and Juan-Andres Leon, Bonn, February 5–8, 2018, DA GMPG, BC 601052. Jacob W. M. Baars: interview by Luisa Bonolis and Juan-Andres Leon, Bonn, February 5–7, 2018. DA GMPG, BC 601050.

18 Karl M. Menten: Leo Brandt. Pionier der Funkmesstechnik und Initiator der Radioastronomie in Deutschland. In: Bernhard Mittermaier, and Bernd-A. Rusinek (eds.): *Leo Brandt (1908–1971). Ingenieur–Wissenschaftsförderer–Visionär*. Jülich: Forschungszentrum Jülich, Zentralbibliothek 2009, 41–53. 48. See also David Leverington: *Observatories and Telescopes of Modern Times. Ground-Based Optical and Radio Astronomy Facilities since 1945*. Cambridge: Cambridge University Press 2017.

19 Karl Menten: interview by Luisa Bonolis and Juan-Andres Leon, Bonn, February 5–8, 2018, DA GMPG, BC 601052.

Germany, its largest single financial contributor.[20] The Max Planck Society, in concert with the German Federal Government and Bavaria, generously offered a new seat in Garching, next to the Institute for Extraterrestrial Physics, and where the Institute for Astrophysics would also be relocated as part of the same operation.[21] This move concentrated most of the MPI astrophysical scientists in the Munich area in a single location, and literally next door to one of the major international research organizations in ground-based astronomy, whose director also became a frequent member of the Max Planck Society's governance bodies.

International collaboration within ESO solved the specifically German problem of having a weak footprint in foreign relations, and it allowed the Max Planck Society to circumvent one of its weaknesses, namely its lack of any explicit diplomatic authority.[22] Participating in ESO allowed what one might call *geopolitical outsourcing*. ESO's location in Chile had been the result of a tortuous learning process, over the second half of the 20th century, regarding the geographical siting of observatories. These increasingly have to be in high, dry locations. At the same time, however, the installation, maintenance, and operation of large infrastructure require easily accessible sites and a highly skilled local staff. All of this is often encumbered by the complex political, economic, and infrastructural challenges posed by certain locations or regimes.[23]

The Max Planck Society experienced its share of anxieties related to building observatories in foreign countries such as Spain and Namibia, as well as in Arizona in the United States. Even the future viability of American astronomy

---

20 Claus Madsen: *The Jewel on the Mountaintop. The European Southern Observatory through Fifty Years*. ESO 2012. See also Ch. IX in Adriaan Blaauw: *ESO's Early History. The European Southern Observatory from Concept to Reality*. ESO 1991.

21 This move was even against the wishes of its staff and director, who up until this point still worked at the original building of the Institute for Physics and Astrophysics in Munich-Freimann. Joachim Trümper: interview by Luisa Bonolis and Juan-Andres Leon, Munich, August, 7–8, 2017. DA GMPG, BC 601036.

22 The undefined diplomatic authority of the Max Planck Society has been often beneficial, as several other studies have shown: Thomas Steinhauser, Hanoch Gutfreund, and Jürgen Renn: *A Special Relationship. Turning Points in the History of German-Israeli Scientific Cooperation*. Berlin: GMPG-Preprint 2017. In the case of the building and operation of astronomical sites, however, this proved a considerable obstacle, as it invited continuous political scrutiny from the German Foreign Ministry.

23 See, for example, Baars, *International Radio Telescope Projects*, 2013. Leverington, *Observatories and Telescopes*, 2017. The entire book describes the scientific, technological, and political challenges of building and operating international observatories based on the experience of Jacob Baars with the Max Planck Institute for Radio Astronomy, IRAM, ALMA, and other international collaborations.

itself is increasingly challenged by growing local resistance to the building of astronomical observatories: while in the mid-20th century, astronomers were proud to raze mountaintops to build research campuses and access roads, environmental concerns now dictate a much more careful balancing of the costs and benefits of what is presented as an esoteric scientific endeavor. Furthermore, the remote location of observatory sites in areas with a dire colonial legacy often unfortunately leads to contested ownership issues, which can easily derail an observatory project.[24] Even if such challenges are eventually overcome, the years of delay may irreparably damage scientific projects and careers.

While the initial purpose of the European Southern Observatory was to provide a base for astronomical observations at a suitable geographical location and in ideal climatic conditions in the southern hemisphere, one of its key comparative advantages since the 1970s has been geopolitical. The organization, modeled on CERN, has faced the delicate task of negotiating the minefield of international relations, between the member countries in scientific collaboration and the host country, Chile: a long, conflictive relationship or 'marriage of convenience' yet one that has survived even a long dictatorship.[25] As early as the 1980s, the Max Planck Society grasped the advantages of outsourcing geopolitical matters; during the negotiations on making Garching its headquarters, it also decided that the southern telescope of the Max Planck Institute for Astronomy, initially planned for Namibia, should instead be hosted and operated by ESO in Chile.[26] Increasingly, as the large telescopes built by ESO came into operation, Max Planck scientists used them heavily, even more than their own Calar Alto.[27] Non-MPG German astronomers also preferred the ESO access conditions, which allowed them to sidestep the Max Planck Institutes' monopolistic tendences in astronomy, on which much more will be said later in this chapter.

By the end of the century, in the context of the planning for ALMA, the Max Planck Institute for Radio Astronomy even built a new, partly institute-owned observatory in collaboration with ESO in Chile. The Atacama Pathfinder Experiment (APEX), based on a prototype antenna constructed for the ALMA project,

24 Swanner, *Mountains of Controversy*, 2013. Swanner's dissertation follows in detail the controversies on Mauna Kea and Mount Graham, which have come to paralyze the ability to build new astronomical observatories and telescopes in the United States.

25 Madsen, *The Jewel on the Mountaintop*, 2012. Lodewijk Woltjer: *Europe's Quest for the Universe. ESO and the VLT, ESA and Other Projects*. Les Ulis: EDP Sciences 2006.

26 See, for example, Lemke, *Himmel über Heidelberg*, 2011, Vol. 21, 79–81. See also Leverington, *Observatories and Telescopes*, 2017.

27 Immo Appenzeller: interview by Juan-Andres Leon, Heidelberg, August, 2016.

was deliberately conceived by Karl Menten as a pilot 'experimental' observatory, justified primarily, not as an observational infrastructure, but as a site for the development and testing of instrumentation, techniques, and preliminary observations, which could then influence operations of the much larger-scale ALMA.[28] The APEX telescope was a replica of the Vertex design chosen by the American contribution to ALMA. This highlights the transition from the infrastructural approach of the Cold War era, which dictated the large early telescope projects of the Max Planck Society, towards an emphasis on scientific output and the development of instrumentation and techniques within large multinational collaborations. As we will see repeatedly, with the three case studies in the last chapter of this book, the Max Planck Society has since the 1990s increasingly focused its work and identity on 'pilot' projects of this type.[29]

During this same decision-making process in the late 1990s, which led to growing support for ALMA and the creation of APEX in Chile, the Max Planck Society ended its participation in Peter Mezger's Heinrich Hertz telescope in Arizona, and, more gradually, its investment in Calar Alto.[30] Nevertheless, the MPG's involvement in both observation sites dragged on a good while longer, owing to the very slow contractual procedure of returning Calar Alto to Spain,[31] which would be concluded only two decades later; and, in Arizona, the inertia of observatory-building activities was manifest in Heidelberg's MPIA's continued participation in another project on the Mount Graham site, the Large Binocular Telescope (LBT).[32]

---

28 See minutes of the meeting of the Executive Committee [*Verwaltungsrat*] 7.10.1998 (AMPG, II. Abt., Rep. 61, No. 185, Folder 6) presenting results of a discussion about astronomical research at the Max Planck Society, which took place at the Institute for Extraterrestrial Physics in Garching, on July 29, 1998 ("Astronomische Forschung in der MPG: MPI mit Interessen an bodengebundenen astronomischen Beobachtungen Koordinationsgespräch"). The following Scientific Members interested in coordinating ground-based astronomy were present: Appenzeller (MPIA), Genzel (MPE), Grewing (Tübingen), Hillebrandt (MPA), Hofmann (MPIK), Menten (MPIfR), Morfill (MPA), Schmitz (MPIfR), Trümper (MPE), Völk (MPIK), Weigelt (MPIfR), Wielebinski (MPIfR), and White (MPA).

29 Karl Menten: interview by Luisa Bonolis and Juan-Andres Leon, Bonn, February 5–8, 2018, DA GMPG, BC 601052.

30 Thomas Klein et al.: APEX beyond 2016. The Evolution of an Experiment into an Efficient and Productive Submillimeter Wavelength Observatory. *Society of Photo-Optical Instrumentation Engineers (SPIE) Conference Series* 10704 (2018), 107041V. doi:10.1117/12.2312687.

31 See, for example, Lucas Laursen: Cutbacks Kick off Kerfuffle over Spanish-German Observatory. *Nature News Blog*, 6/17/2013. http://blogs.nature.com/news/2013/06/calar-alto-cutback-kerfuffle.html. Last accessed 4/12/2018.

32 Increasing participation in the LBT was one of the pretexts used in 1998 to diminish involvement in Calar Alto, but this optical telescope was mired in the same site problems

Convergence on Infrared Astronomy

Meanwhile, specialization in astronomy at the Institute for Extraterrestrial Physics continued in the 1980s, and while some of this institute's most important successes were national satellites such as ROSAT, it had been a powerhouse of international collaboration since its early years, due to Germany's inability at the time to access space independently (Chapter 2), and subsequent reliance on European and American partners.

A further step leading the MPE towards observational astronomy was the decision of a commission of Max Planck Society astronomers and astrophysicists in 1984, to appoint Reinhard Genzel as Director of the Institute for Extraterrestrial Physics, at the age of 33, the youngest ever.[33] This was the latest among several moves at various Max Planck Institutes to tackle the wavelength gap in infrared astronomy, at a moment when space missions and detector technology in these wavelengths were maturing and opening up the potential for ground-based observations, too. At the time, the Infrared Astronomical Satellite (IRAS), the first cryogenically cooled telescope in space, launched in January 1983, inaugurated infrared space astronomy, conducting an all-sky survey at infrared wavelengths as a collaborative trial by United States (NASA), the Netherlands (Netherlands Agency for Aerospace Programmes, NIVR), and the United Kingdom (Science and Engineering Research Council, SERC). Over about 300 days, IRAS opened a new window onto the cold universe, collecting data consisting of gas and dust that is too cold to radiate visible light and can be associated with the early evolution of galaxies, stars in formation, or protoplanetary disks. Early in 1983, at the same time as IRAS was launched, the European Space Agency (ESA) decided, after years of careful studies, to select a space telescope for infrared light as its next scientific mission. Based on the latest innovations in infrared detector-technology, the Infrared Space Observatory (ISO) would provide detailed observations with greatly improved sensitivity and resolution, allowing a much more comprehensive perception of the 'infrared scenery.'

IRAS and infrared astronomy were also an instrumental gateway to other wavelengths which could finally answer the deepest cosmological questions. The Cosmic Background Explorer (COBE) satellite was being built in the early

---

as the Heinrich Hertz Telescope, as Karl Menten warned at the time, as the meeting's minutes show. After its delayed completion, there were still many technical difficulties with the LBT and its scientific productivity was relatively low compared to other telescopes in its class. See, for example, Alexandra Witze: Teething Troubles at Huge Telescope. *Nature News* 499/7457 (2013), 133. doi:10.1038/499133a.

33 CPTS meeting minutes of 18.09.1984, 12.6.1985, AMPG, II. Abt., Rep. 62, No. 1803, 1805.

1980s, and its technology was heavily based on the IRAS experience. Its objective was to measure the diffuse infrared and Cosmic Microwave Background radiation from the early universe (CMB), and after its delayed launch in 1988, it actually provided key evidence, namely that CMB has a near-perfect blackbody spectrum, which supported the big bang theory as cosmological model for explaining the origin, properties, and evolution of the observable Universe and initiated the era of cosmology conducted as precision science. This was acknowledged by the award of the Nobel Prize 2006 to John Mather and George Smoot, Principal Investigators of the two COBE instruments mapping the spectrum of, and variations in the CMB.[34]

Considering the rapid improvements in infrared detector technology, the successful IRAS mission, and ESA's selection of ISO in 1983—which was immediately followed by a call for scientific experiment and mission proposals—it was decided that the small infrared group at the MPE, then led part-time by Gerhard Haerendel (Chapter 2) should have its own dedicated director. The range of future research with infrared and adjacent submillimeter astronomy had the potential to open up new perspectives and lead to wider involvement in international projects.[35]

The young Reinhard Genzel, chosen to promote the still relatively new field of infrared astronomy, represented that third generation of Max Planck directors who circulated more easily among the different institutes and scientific traditions of the Society: his own long trajectory had begun at birth, quasi, for he was the son of one of the most prominent MPI directors, Ludwig Genzel, of the Institute for Solid State Research in Stuttgart, an expert in infrared spectroscopy. The younger Genzel had studied in Bonn, conducting his research initially with Peter Mezger at the Institute for Radio Astronomy, and then gaining his doctorate at the same institute with Dennis Downes, the local leader of the interstellar maser group. He then pursued his postdoctoral work at the Harvard-Smithsonian Centre for Astrophysics, later being offered a Miller Fellowship to join the group of Charles Townes at the University of

34 COBE carried three instruments: a Diffuse Infrared Background Experiment (DIRBE), to search for the cosmic infrared background radiation, a Differential Microwave Radiometer (DMR), to map the cosmic radiation, and a Far Infrared Absolute Spectrophotometer (FIRAS), to compare the spectrum of the CMB with a precise black body radiation. J. C. Mather et al.: Measurement of the Cosmic Microwave Background Spectrum by the COBE FIRAS Instrument. *The Astrophysical Journal* 420 (1994), 439–444. doi:10.1086/173574.

35 CPTS meeting minutes of 27.06.1984, 18.9.1984, 12.06.1985, AMPG, II. Abt., Rep. 62, No. 1802, 1803, 1805. During these meetings such discussions were also related to reports of the commission that was to decide whether to propose appointing Genzel as director at MPE, in order to promote the field of infrared and submillimeter astronomy.

California, Berkeley, to where he relocated in 1980. There he worked first on far-infrared spectroscopy and then carried out research in infrared and submillimeter astrophysics, in collaboration with Charles H. Townes, one of the trio awarded the Nobel Prize 1964, for its invention of the maser and laser.[36] Townes, who had had a passion for astronomy since his university days, had developed the maser "partly with astronomy in mind" and later "played with astronomy a number of times," until he became professor at Berkeley and decided to develop an astrophysical program there; he was especially intrigued by radio waves coming from the center of our galaxy, the Milky Way,[37] which was actually the first radio source identified after Jansky's discovery of diffuse galactic radio emission in the early 1930s. By the end of the 1950s, it could be considered definitely established that Sagittarius A, the discrete radio source near the border of the constellation Sagittarius, was coinciding with the galactic center.[38] In 1974, a very intense radio source was discovered within the Sagittarius A radio complex,[39] and it was later given the name Sagittarius A* (Sgr A*). However, no optical, infrared, or X-ray counterpart to Sgr A* could be identified, and its nature remained a mystery, but it was later established that it was extremely bright, compact, and less than the size of our solar system.

In 1969, soon after the discovery of neutron stars, the British astrophysicist Donald Lynden-Bell had proposed that quasars are powered by accretion disks around black holes as massive as 100 million suns or more, and conjectured that many if not most ordinary galaxies host a massive black hole at their center. He discussed in detail the physical properties of the accretion disks that would encircle them, definitively shifting attention to the environment surrounding such still unknown objects.[40] The Milky Way was to be no exception,

---

36 See Reinhard Genzel: Autobiography of Reinhard Genzel. *The Shaw Prize*, 9/9/2008. https://www.shawprize.org/prizes-and-laureates/astronomy/2008/autobiography-of-reinhard-genzel. Last accessed 8/15/2020.

37 Charles Hard Townes, A Life in Physics: Bell Telephone Laboratories and World War II; Columbia University and the Laser; MIT and Government Service; California and Research in Astrophysics. Interview by Suzanne B. Riess, 1991–92. Transcript, UC Berkeley Oral History Center, Online Archive of California, https://oac.cdlib.org/ark:/13030/kt3199n627/?brand=oac4. Last accessed 6/8/2019.

38 W. M. Goss, and R. X. McGee: The Discovery of the Radio Source Sagittarius A (Sgr A). In: Roland Gredel (ed.): *The Galactic Center, Astronomical Society of the Pacific Conference Series*. The Galactic Center. Astronomical Society of the Pacific 4th international meeting, March 10–15, 1996. La Serena 1996, 369. http://adsabs.harvard.edu/abs/1996ASPC..102..369G. Last accessed 5/8/2020.

39 Bruce Balick, and Robert L. Brown: Intense Sub-Arcsecond Structure in the Galactic Center. *The Astrophysical Journal* 194 (1974), 265–270. doi:10.1086/153242.

40 Donald Lynden-Bell: Galactic Nuclei as Collapsed Old Quasars. *Nature* 223/5207 (1969), 690–694. doi:10.1038/223690a0.

and in a very influential paper, published in 1971 with Martin Rees, Lynden-Bell argued for the existence of a supermassive black hole in the center of our own galaxy, and proposed key observations to explore the nature of such a compact object.[41] Until the 1970s, the black hole was still a novel concept, studied by specialists in Einstein's general theory of relativity, and there was certainly no serious evidence that it really existed.

Meanwhile, in Germany, radio emission from the center of the Milky Way had long been a specialty of the Max Planck Institute for Radio Astronomy, and the Effelsberg telescope had in fact been positioned to have access to this part of the sky. This tradition was then also followed by Peter Mezger with the newer, millimeter wavelength telescopes.[42] But Genzel's more specific interest in the question of whether the Milky Way's center harbors a supermassive black hole went back to his Berkeley period. In fall 1967, after arriving in Berkeley, Townes had built a maser amplifier and microwave spectrometer in order to use a radio telescope to search for molecules in the center of the Galaxy, and pioneered infrared astronomy and precision infrared spectroscopy.[43]

Since no light can escape a black hole, it must be detected indirectly. But the Galactic Center is hidden behind dense gas and dust clouds and only becomes visible in the infrared light. In the early 1980, Townes and his group had already found that the interstellar gas in the vicinity of the radio source Sagittarius A* at the Galactic Center was moving very fast, as observed from Earth, suggesting that there was a non-stellar central mass concentration of a few million solar masses, most likely in the form of a massive black hole.[44] The development of

---

41 Donald Lynden-Bell, and M. J. Rees: On Quasars, Dust and the Galactic Centre. *Monthly Notices of the Royal Astronomical Society* 152/4 (1971), 461–475. doi:10.1093/mnras/152.4.461.

42 Peter G. Mezger: The Center of the Galaxy. In: A. Reiz (ed.): *Research Programmes for the New Large Telescopes, Proceedings of the ESO/SRC/CERN Conference Held 27–31 May, 1974 in Geneva, Switzerland.* Research Programmes for the New Large Telescopes. 1974, 79–107. http://adsabs.harvard.edu/abs/1974rpnl.conf...79M. Last accessed 5/8/2020. See also Mezger's later review article including an introductory outline on studies of the Galactic Center up to the early 1990s. Peter G. Mezger, Wolfgang J. Duschl, and Robert Zylka: The Galactic Center: A Laboratory for AGN? *The Astronomy and Astrophysics Review* 7/4 (1996), 289–388. doi:10.1007/s001590050007.

43 At that time, there was a widespread belief among astronomers that such molecules could not survive in space, but Townes's group discovered three-atoms combinations and others discovered even more complex molecules, providing evidence for the existence of chemical reactions taking place in stars.

44 Reinhard Genzel et al.: OI and OIII in SGR A. Neutral and Ionized Gas at the Galactic Center. *AIP Conference Proceedings* 83 (1982), 72–76. doi:10.1063/1.33504. Reinhard Genzel et al.: Far-Infrared Spectroscopy of the Galactic Center—Neutral and Ionized Gas in the

innovative detectors allowed infrared/submillimeter spectroscopic and radio interferometry measurements of the gas clouds' dynamics in the inner core of the Galaxy, which made a "convincing case" that the mass distribution at the Galactic Center was "more concentrated than a spherical isothermal stellar cluster."[45] Still, the possibility that ionized gases might be affected by non-gravitational forces made this evidence not compelling. From 1981, Genzel had become professor at Berkeley, specializing in infrared and submillimeter astronomy and studies of the interstellar medium,[46] and after his appointment in 1985 as Director of the MPE, Genzel's continued his collaboration with Townes, and the latter was made an External Scientific Member of the institute in 1987. The problem of determining the mass distribution in the center of the Galaxy became the starting point of Genzel's long-term study on the motion of stars near the center, which decades later provided persuasive evidence for the presence of a supermassive black hole, as will be described in more detail later.[47] A decade after his appointment in Garching, Genzel was also given a part-time position at Berkeley, becoming part of the new modality of internationalization of Max Planck directors common from the 1990s on.[48]

In the 1980s, the astronomical institutes of the Society, which had neatly divided their research along wavelength windows, ended up all converging on infrared astronomy, a field that had remained relatively underdeveloped in Germany, due to the military applications of infrared detectors, which made

---

Central 10 Parsecs of the Galaxy. *The Astrophysical Journal* 276 (1984), 551–559. doi:10.1086/161644.

45 Using the newly designed spectrometers, the group examined a region of hot, rarefied ionized gas extending out from the galactic center and a disk of cool neutral gas and dust stretching for tens of light years. The measured velocities of gas and dust showed that they are held in orbit around the center of the Galaxy by the gravitational pull of a mass about four million times greater than the mass of the Sun. The group concluded from the motions of interstellar gas in the vicinity of the compact central radio source Sagittarius A* (SgrA*) at the center of the Milky Way that there was a non-stellar central mass concentration of about four million solar masses, most likely in the form of a massive black hole. M. K. Crawford et al.: Mass Distribution in the Galactic Centre. *Nature* 315 (1985), 467. doi:10.1038/315467a0.

46 Genzel, Autobiography of Reinhard Genzel, 9/9/2008.

47 For a summary of the investigations carried out by the group on the central nucleus of the Galaxy up to 1987, see the review article Reinhard Genzel, and C. H. Townes: Physical Conditions, Dynamics, and Mass Distribution in the Center of the Galaxy. *Annual Review of Astronomy and Astrophysics* 25 (1987), 377–423. doi:10.1146/annurev.aa.25.090187.002113.

48 Reinhard Genzel, conversation with the authors, June 2018.

them unavailable outside of the United States.[49] Initially, this was a new major problem of overlap in different research fields, and there was conflict in the 1970s and early 1980s between the Max Planck Institute for Astronomy in Heidelberg, which had an existing section dedicated to airborne and space-based infrared astronomy, and the Institute for Extraterrestrial Physics, which unsuccessfully attempted to maintain a monopoly on space-based research.[50] Over the 1980s, the situation ended up inverting the original aims of the institutes, with the Institute for Astronomy now coordinating space-based infrared research, and the Institute for Extraterrestrial Physics focusing, under Genzel, more on ground-based observations, with the prospect of becoming one of the major users of the new generation of optical/infrared telescopes built by ESO in Chile. Meanwhile, approaching from the other end of the wavelength continuum, the Institute for Radio Astronomy likewise entered infrared astronomy through the development of detectors, while Peter Mezger was also a very early proponent of what would become NASA's aircraft-based Stratospheric Observatory for Infrared Astronomy (SOFIA), a Boeing 747SP aircraft extensively modified to carry a 2.7 m infrared telescope, operating at an altitude of about 12 km with observation periods of a few hours.[51] This was eventually built and operated as a collaboration between NASA and the DLR, with decisive input from Max Planck Institute scientists from Garching, Heidelberg, and Bonn.[52]

49 Reinhard Genzel, conversation with the authors, May 2017. Ever since the first infrared detectors were developed in World War II (initially in Germany), the technology has been driven by infrared radiation emitted by hot objects; it has applications such as heat-seeking missiles, night vision, or the detection and identification of missiles and other flying objects. For a historical overview, see A. Rogalski: History of Infrared Detectors. *Opto-Electronics Review* 20/3 (2012), 279–308. doi:10.2478/s11772-012-0037-7. In the 1970s and '80s, Germany participated in the development of an infrared space telescope GIRL (German Infrared Laboratory), which was to be flown on the space shuttle. The program was canceled in 1985, as the costs of flying on the space shuttle rose and exceeded the budget estimate. But still, the participating company MBB delivered a similar device for the American SDI or 'Star Wars' initiative. See Lemke, *Himmel über Heidelberg*, 2011, Vol. 21, 97.

50 Gerhard Haerendel: interview by Helmuth Trischler and Matthias Knopp, April 9, 2010. Transcript, HAEU, https://archives.eui.eu/en/oral_history/INT066. Last accessed 1/13/2019.

51 Karl Menten: interview by Luisa Bonolis and Juan-Andres Leon, Bonn, February 5–8, 2018, DA GMPG, BC 601052.

52 Unfortunately, SOFIA was chronically delayed and only started operating in the 2010s. Markus Völter: Alfred Krabbe, Thomas Keilig, Christian Fischer, Dörte Mehlert, and Zaheer Ali. 189—SOFIA Part 1, Basics. Podcast. 2015. Omega Tau. http://omegataupodcast.net/189-sofia-part-1-basics/. Last accessed 10/31/2018. Markus Voelter: *Once You Start Ask-*

In the late 1980s came the ultimate convergence on infrared space astronomy of all the observational astronomical institutes, as well as the Institute for Nuclear Physics in Heidelberg, through ESA's Infrared Space Observatory (ISO), the largest and most complex satellite ever built in Europe up to that time. This satellite reflected the maturation of the multinational approach to space astronomy advanced by ESA, at a time when the submillimeter and far-infrared wavelength band was still a largely unexplored part of the electromagnetic spectrum.[53] The selection of ISO's scientific payload was completed in 1985 and consisted of four instruments: a camera (ISOCAM), an imaging photo-polarimeter (ISOPHOT), a long wavelength spectrometer (LWS), and a short wavelength spectrometer (SWS), each of which was built by an international consortium of scientific institutes and industry headed by a Principal Investigator. Two out of these four were West German contributions, one from the Institute for Extraterrestrial Physics,[54] and another a collaboration between the Max Planck Institutes for Radio Astronomy, Astronomy, and Nuclear Physics led by Dietrich Lemke at the Institute for Astronomy.[55]

---

ing. Insights, Stories and Experiences from Ten Years of Reporting on Science and Engineering, 2020, 19–52. For further details on the rivalries between Heidelberg and Garching in infrared astronomy related to SOFIA, see Gerhard Haerendel: interview by Helmuth Trischler and Matthias Knopp, April 9, 2010. Transcript, HAEU, https://archives.eui.eu/en/oral_history/INT066. Last accessed 12/4/2020.

53 While design and development of ISO were underway, and COBE was beginning its successful mission, in order to move forward on the space program, as emphasized by Genzel, it was clear that the "challenging goals of technological developments and the necessity of keeping the program realistic financially" required "a coherent international effort." Thomas G. Phillips, and Reinhard Genzel: An International Program for Submillimeter and Far-Infrared Astronomy from Space. In: B. H. Kaldeich (ed.): *From Ground-Based to Space-Borne Sub-Mm Astronomy. Proceedings of the 29th Liège International Astrophysical Colloquium Held at the Institut D'Astrophysique, Liège, Belgium, 3–5 July 1990*. Paris: European Space Agency 1990, 407–414, 407.

54 The Max Planck Institute for Extraterrestrial Physics built the Short Wavelength Spectrometer (SWS), which provided the first detection of water-molecule absorption lines in an expanding shell of a star. C.M. Wright et al.: ISO–SWS Observations of Pure Rotational H2O Absorption Lines Toward Orion–IRc2. *Astronomy and Astrophysics* 358 (2000), 689–700. http://adsabs.harvard.edu/abs/2000A&A...358..689W. Last accessed 10/31/2018.

55 As recalled by Lemke, the main role of Max Planck scientists in ISO, and later, in the Herschel Space Observatory and in the James Webb Space Telescope, which is intended as one of the successors to the Hubble Telescope, had deep roots in his long stay as a young researcher in the leading research group in infrared astronomy at the University of Arizona in the United States, where he had an opportunity to work at the forefront of research. Without this experience, he stressed, later participation in such scientific missions would have been unthinkable; an example, of course, that can be extended to

Eventually launched in 1995, ISO fulfilled the need for sensitive infrared observatories, allowing for the detailed spatial and spectroscopic study of specific targets, following its 1980s predecessor IRAS, mentioned earlier. ISO doubled the number of catalogued astronomical sources by detecting about 500,000 infrared sources, in particular revealing for the first time the central core of our Galaxy.[56]

Question-Oriented Integration of Theory and Observation in Astronomy

Around the time of completion of the large observatory projects of the wavelength expansion era, many of the research programs at the Max Planck Institutes could be described as extremely productive in the quantity of observations and data at every wavelength, but were not necessarily tied to an inquisitive, theoretical, question-raising program.

This began to change in the 1980s, with the line of work which had been hinted at early by Peter Mezger in Bonn, and followed up as a lifetime pursuit by his disciple Reinhard Genzel, first at Berkeley and then in Garching: devising astronomical observation programs to confirm the existence of, and characterize, the supermassive black hole at the center of our galaxy.[57] The answer to such a fundamental question is related to the understanding of 'active galactic nuclei' (AGNs), compact regions at the center of a galaxy—the most powerful and steady sources of luminosity in the Universe—emitting energy in the form of radio, optical, X-ray, or gamma radiation or high-speed

a great number of young Max Planck researchers in other institutes. Lemke also emphasized how these scientifically and productive research stays in America were possible for Max Planck Society scientists, in contrast to colleagues at other institutions in West Germany. Dietrich Lemke, personal communication with Jürgen Renn, September 20, 2016.

56 Catherine Cesarsky, and Alberto Salama (eds.): *ISO Science Legacy. A Compact Review of ISO Major Achievements*. Dordrecht: Springer 2005. Frank J. Low, G. H. Rieke, and R. D. Gehrz: The Beginning of Modern Infrared Astronomy. *Annual Review of Astronomy and Astrophysics* 45 (2007), 43–75. doi:10.1146/annurev.astro.44.051905.092505. Dietrich Lemke: The Short History of Infrared Space Telescopes. *Astronomische Nachrichten* 330/6 (2009), 562–567. doi:10.1002/asna.200911217. G. H. Rieke: History of Infrared Telescopes and Astronomy. *Experimental Astronomy* 25/1–3 (2009), 125–141. doi:10.1007/s10686-009-9148-7. Dietrich Lemke, and M. Kessler: The Infrared Space Observatory ISO. In: G. Klare (ed.): *Reviews in Modern Astronomy*. Berlin: Springer 1989, 53–71. See also Joachim Trümper: interview by Luisa Bonolis and Juan-Andres Leon, Munich, August, 7–8, 2017. DA GMPG, BC 601036.

57 On the massive black hole residing in the center of the Milky Way, see the review article Tal Alexander: Stellar Processes near the Massive Black Hole in the Galactic Center. *Physics Reports* 419/2–3 (2005), 65–142. doi:10.1016/j.physrep.2005.08.002.

particle jets. Some of them are classified as quasars, and understanding their evolution in the early Universe is crucial because, as observations have shown, many—probably most—galaxies harbor supermassive black holes in their nuclei and might have been quasars in their early history.

Towards the end of the 1980s, Genzel and Townes, in discussing "the case for and against a massive black hole," had concluded that

> The current evidence for a massive ($\approx 10^6$ solar masses) black hole from the observed radiation phenomena and the gas and stellar dynamics is substantial but not fully convincing.[58]

For a direct proof of the 'black hole' paradigm, it was necessary to determine the characteristic mass concentration and to show the existence of an 'event horizon,' the boundary defining the region of space around the black hole, beyond which nothing, not even light, can escape.[59] When Genzel moved from Berkeley to Munich, it was clear that, to make progress, measurements had to be done at a smaller scale and, especially, that it was necessary to use stars, instead of gas, to really probe the gravitational potential on the scale of the event horizon, inferring it from spatially resolved measurements of their motions in close orbit around the candidate black hole. This meant higher resolution imagery in the infrared, to overcome the blurring of the Earth's atmosphere and get sharper images. The quest for the massive black hole in the Galactic Center became one of the central research themes promoted by Genzel at MPE, together with studies of active galactic nuclei and star formation in galaxies at high redshift, with very sensitive infrared instruments developed at the institute for increasingly precise observations. It provided a uniquely accessible laboratory for studying in detail the connections and interactions between a massive black hole and the stellar system in which it grows; moreover, stars moving very rapidly near the center could be used to probe the dark mass and to test how gravity works near a supermassive black hole. By the early 1990s, observational evidence for a dark central mass concentration at the core of our Galaxy had been steadily growing for over two decades. At that time, Genzel's group began to conduct a program to study the properties of the central nuclear stellar cluster, which they carried out via near-infrared high spatial resolution measurements, using the MPE speckle camera SHARP

58 Genzel, and Townes, Physical Conditions, 1987, 377–423, 419.

59 Later in this chapter, in the section dealing with VLBI, we follow an entirely different approach to ascertaining the presence of a black hole.

(System for High Angular Resolution Pictures), at the ESO New Technology Telescope (NTT) in La Silla, Chile.[60] Significant progress in the following years strengthened the evidence for the presence of a central mass of the order 10^6 solar masses, but it was not yet considered compelling by many researchers in the field. A "first conclusive evidence" for a massive black hole in the center of the Galaxy could be presented in 1998, based on the study of the velocity field of stars and gas orbiting the 'black-hole candidate' at unprecedented resolution, to determine the form of the gravitational potential, "probably the most unambiguous method" for carrying out such a proof.[61] In early 2005, the new instrument SINFONI went into operation at the ESO facilities on Paranal, having as one of its prime targets the Galactic Center region.[62] Together with its follow-up, GRAVITY,[63] these instruments afforded further advances in resolu-

60 A central scientific goal of the SHARP experiment was the imaging of the central stellar cluster in order to measure the proper motions of stars in its vicinity and answer the key question of whether (or not) the Galactic Center contains a massive black hole of about 10^6 solar masses. First results of the first five years of work with SHARP were published in 1996. A. Eckart, and Reinhard Genzel: Observations of the Galactic Center with SHARP: First Stellar Proper Motions. In: Roland Gredel (ed.): *The Galactic Center, Astronomical Society of the Pacific Conference Series.* 4th international meeting jointly organized by the European Southern Observatory (ESO) and Cerro Tololo Inter-American Observatory (CTIO), held March 10–15, 1996 in La Serena, Chile, San Francisco. Astronomical Society of the Pacific. 1996, 196–202. http://adsabs.harvard.edu/abs/1996ASPC..102..196E. Last accessed 5/9/2020.

61 Andreas Eckart, and Reinhard Genzel: First Conclusive Evidence for a Massive Black Hole in the Center of the Milky Way. In: Friedrich W. Hehl, Claus Kiefer, and Ralph J.K. Metzler (eds.): *Black Holes: Theory and Observation.* Berlin: Springer-Verlag 1998, 60–68. doi:10.1007/978-3-540-49535-2_3.

62 SINFONI consisted of two major combined components: the near-infrared integral field spectrograph SPIFFI in conjunction with the adaptive optics system MACAO, a system correcting the distortions of the light beams from the telescope induced by the atmospheric turbulence, before they are directed towards the common focus at the VLT interferometer (VLTI). F. Eisenhauer et al.: SINFONI in the Galactic Center: Young Stars and Infrared Flares in the Central Light-Month. *The Astrophysical Journal* 628/1 (2005), 246–259. doi:10.1086/430667.

63 S. Gillessen et al.: GRAVITY: A Four Telescope Beam Combiner Instrument for the VLTI. *Optical and Infrared Interferometry II.* SPIE Astronomical Telescopes + Instrumentation. San Diego, CA: International Society for Optics and Photonics 2010, 77340Y. doi:10.1117/12.856689. At the perihelion-passage the star S2, within the cluster orbiting close to the black hole, moves at a velocity of about 7650 km/s (that is about 0,026 the velocity of light, *c*) and thus the gravitational redshift and the orbital precession can be realistically detected in the spectra and through precise measurements of the positions of the star near pericenter, which would take place in 2018. For this reason, in 2005 they proposed to ESO to build the novel instrument GRAVITY, combining the light of all four 8 m VLT

tion, leading to an extremely strong case for the existence of the supermassive black hole based on measurements of stellar orbits.[64]

By 2010, analysis of the orbits of more than two dozen stars within the very dense cluster orbiting Sagittarius A*, provided detailed information on the distribution, kinematics and physical properties of this nuclear star cluster and of the hot, warm, and cold interstellar gas interspersed in it. They were able to show that "the empirical evidence for the existence of a central massive black hole of about 4×10^6 solar masses is compelling."[65] Instrumental in this dedicated long-term effort was the use of European Southern Observatory telescopes and the co-development of instruments, which made it possible to enter an era of observational black hole physics. The highly elliptical, 16-year-period orbit of the star S2 around the massive black hole candidate Sagittarius A* is a sensitive probe of the gravitational field in the galactic center. For 25 years the group has monitored the radial velocity and motion on the sky of S2, mainly with the SINFONI and NACO adaptive optics instruments on the ESO Very Large Telescope, and since 2016, with the instrument GRAVITY.[66] The conclusion has been that "[t]he S2 data are inconsistent with pure Newtonian dynamics," demonstrating that the gravitational potential is dominated by a compact object of about 10^6 solar masses.[67] Recent progress in testing the paradigm that a supermassive black hole resides at the center of our galaxy added further evidence, allowing observation of the accretion disk: clumps of gas swirl around at about 30 percent of the speed of light on a circular orbit, just outside the black hole event horizon, the innermost stable orbit

---

telescopes, each assisted by adaptive optics, which would afford the required precision to detect the general relativity effects and probe physics close to the event horizon of the black hole.

64 S. Gillessen et al.: Monitoring Stellar Orbits around the Massive Black Hole in the Galactic Center. *The Astrophysical Journal* 692/2 (2009), 1075–1109. doi:10.1088/0004-637X/692/2/1075. Reinhard Genzel, F. Eisenhauer, and S. Gillessen: The Galactic Center Massive Black Hole and Nuclear Star Cluster. *Reviews of Modern Physics* 82/4 (2010), 3121–3195. doi:10.1103/RevModPhys.82.3121.

65 Genzel, Eisenhauer, and Gillessen, The Galactic Center Massive Black Hole, 2010, 3121–3195, 3181.

66 Gravity Collaboration et al.: First Light for GRAVITY. Phase Referencing Optical Interferometry for the Very Large Telescope Interferometer. *Astronomy and Astrophysics* 602 (2017), A94. doi:10.1051/0004-6361/201730838.

67 R. Abuter et al.: Detection of the Gravitational Redshift in the Orbit of the Star S2 near the Galactic Centre Massive Black Hole. *Astronomy and Astrophysics* 615 (2018), L15. doi:10.1051/0004-6361/201833718.

close to the point of no return.[68] After having monitored the star's radial velocity and motion over nearly 30 years, the GRAVITY collaboration reported the first detection of the general relativity Schwarzschild precession in S2's orbit around the compact radio source Sagittarius A* at the center of our galaxy. As predicted by Einstein's theory, the orbit is shaped like a rosette and not like a simple, stationary ellipse, as predicted by Newton's theory of gravity, an effect first seen in the orbit of Mercury around the Sun, and explained by Einstein's theory of gravity about a hundred years ago. This observational breakthrough is thus further strengthening the evidence that Sagittarius A* must be a supermassive black hole of 4 million times the mass of the Sun.[69]

The year 2019 also saw the publication of the first ever image of the event horizon—or 'shadow'—of a black hole, an even larger one, with about a billion solar masses, sitting at the center of the galaxy M87. The picture was captured by the Event Horizon Telescope detailed later in this chapter, a network of eight ground-based radio telescopes around the globe, which together operate as a single instrument, creating a virtual, Earth-sized telescope.[70]

Almost exactly one hundred years after the final formulation of Einstein's general theory of relativity, its most elusive predictions have been confirmed: the existence of gravitational waves and of black holes, the most extreme objects in the Universe. The Nobel Prize in Physics 2020 was awarded thus: one half to Roger Penrose for the theoretical proof that such exotic objects must exist; and one half split equally between Reinhard Genzel and Andrea Ghez for their convincing observational evidence, collated over decades, that the compact object at the center of our galaxy is indeed a supermassive black hole.

---

68 A. Amorim et al.: Detection of Orbital Motions near the Last Stable Circular Orbit of the Massive Black Hole SgrA*. *Astronomy & Astrophysics* 618/L10 (2018), 1–15. doi:10.1051/0004-6361/201834294. See also A. Amorim et al.: Test of the Einstein Equivalence Principle near the Galactic Center Supermassive Black Hole. *Physical Review Letters* 122/10 (2019), 101102. doi:10.1103/PhysRevLett.122.101102.

69 R. Abuter et al.: Detection of the Schwarzschild Precession in the Orbit of the Star S2 near the Galactic Centre Massive Black Hole. *Astronomy & Astrophysics* 636 (2020), L5. doi:10.1051/0004-6361/202037813.

70 The individual telescopes involved are: ALMA, APEX, the IRAM 30 m telescope, the IRAM NOEMA Observatory, the James Clerk Maxwell Telescope, the Large Millimeter Telescope, the Submillimeter Array, the Submillimeter Telescope, the South Pole Telescope, the Kitt Peak Telescope, and the Greenland Telescope. The Max Planck Institute for Radio Astronomy and the *Institut de Radioastronomie Millimétrique* (IRAM), are part of the Event Horizon Telescope consortium. The Event Horizon Telescope Collaboration: First M87 Event Horizon Telescope Results. I. The Shadow of the Supermassive Black Hole. *The Astrophysical Journal Letters* 875/1 (2019), L1–L17. doi:10.3847/2041-8213/ab0ec7.

Interestingly, (as analyzed earlier in the chapter), most work in this field would be conducted using not MPI-owned telescopes but external observations and those of the European Southern Observatory in Chile. This was one further example of how, from the 1980s onwards, perhaps the most spectacular scientific achievements would come not from owning and operating observational infrastructure, but rather from question-oriented, theory-inspired scientific programs supported by the development and refinement of the novel instrumentation and techniques needed to address such difficult questions more and more efficiently. As we will see in the following chapter, this transition to theoretically grounded, question-based research has afforded the Society its most spectacular achievements in astrophysics to date. This period of transition to theoretically oriented observation even opened up the opportunity for some plasma physicists to return to the Institute for Extraterrestrial Physics, as was the case of Gregor Morfill, who, after being brought to Heidelberg, together with Heinrich Völk, was appointed as Director in Garching in 1983, to fulfill a similar role, providing theoretical interpretation of the increasing flood of information from the institutes' observational missions.[71]

Instrumentation in the Production Chain of Globalized Astronomical Research

As we have just seen in this chapter, the major forces behind the globalization of scientific research were well underway even before the end of the Cold War; such moves were modelled on the success of CERN since the 1950s, and gradually applied to further scientific areas. At the national scale, too, the Brookhaven collegiate model, on which CERN had been based (Chapter 1), had long inspired local counterparts such as the West German DESY, which provided the large, long-term facilities for particle acceleration, but left the actual research to separate, smaller-scale teams led by the member universities and research institutes.

It was no accident that this model of research spread considerably from the 1980s onwards, in parallel with the liberalization and globalization of the

71 The rapidly increasing flood of data from satellite experiments in all areas (space plasma physics, X-ray, and gamma ray astronomy) led to the need to create a theory group working in close contact with the observers. In 1983, Morfill was called back to the Institute for Extraterrestrial Physics for his theoretical competence and his interest both in the near Earth and in deep space, in particular, for his ability to interact both with theoreticians and experimenters, which would be a strong stimulus for the research at the institute (CPTS meeting minutes of 07.10.1983, 27.06.1984, 18.09.1984, AMPG, II. Abt., Rep. 62, No. 1800, 1802, 1803).

economy. One of the pillars of the post-1990s globalized economy has been the deliberate dislocation of infrastructure, construction, and operational costs from their economic usage, in direct opposition to the previous model of vertical and national integration, which was applied, at least to some degree, also to astronomical observatories in the Cold War era, just as it was to services such as railways, electrical grids, roads, and the Internet itself; and this enabled activities requiring very different time scales and magnitude of investment to be separated, such that some were now subject to competition, while an even, stable playing field in the common interest of all participants in the 'game' was (at least in theory) maintained.[72]

In European astronomy, in particular, this new understanding of scientific infrastructures went hand-in-hand with its internationalization. Initially, in the 1950s, ESO was meant only to administer a geographical site, upon which the different countries built their own observatories. Subsequently, during the 1960s and 1970s, ESO engulfed observatory construction as well as the design and operation of the telescopes themselves. Finally, from the 1990s onwards, the final step came with the adoption of practices coming from space-based research at the time of the construction of its new facility in Paranal. Under the directorship of Riccardo Giacconi,[73] it instituted a model of remote operation and 'service mode,' under which (in principle) the scientists could submit their research instructions and have them carried out entirely by ESO telescopes and staff in Chile. Its expanded mission, now with multiple observatories and a modularized approach to astronomical observation, came with a new name: The European Organization for Astronomical Research in the Southern Hemisphere.[74]

---

72 Explicit characterizations of 'research infrastructures' are rare, but see the discourse within which the terminology appears in policy documents of the past 30 years. For an example, see European Strategy Forum on Research Infrastructures: *Roadmap 2018. Strategy Report on Research Infrastructures*. Milan: ESFRI 2018. http://roadmap2018.esfri.eu/media/1066/esfri-roadmap-2018.pdf. Last accessed 8/15/2020. An academic analysis of this trend, beyond the brief episodes treated in this chapter, can be found in: Katharina C. Cramer, and Olof Hallonsten (eds.): *Big Science and Research Infrastructures in Europe*. Cheltenham: Edward Elgar Publishing 2020. For further examples and a novel interpretation, see David Baneke: Let's Not Talk About Science: The Normalization of Big Science and the Moral Economy of Modern Astronomy. *Science, Technology, & Human Values* 45/1 (2020), 164–194. doi:10.1177/0162243919846600.

73 Harvey Tananbaum, Ethan J. Schreier, and Wallace Tucker: Riccardo Giacconi (1931–2018). *Proceedings of the National Academy of Sciences* 116/26 (2019), 12587–12589. doi:10.1073/pnas.1902399116. A more specific treatment of Giacconi's work in X-ray astronomy can be found in Chapter 3.

74 Madsen, *The Jewel on the Mountaintop*, 2012, 317–325.

This was the same model as had been set up years earlier for the Hubble Space Telescope, for example, where Riccardo Giacconi had been director before moving to ESO. But in the 1980s, the European Southern Observatory had likewise begun experimenting with remote observation, due to the high costs of travel as well as political instability in Chile, and it established a direct satellite link between La Silla and Garching.[75] Giacconi would later implement an even stricter scheme at ALMA, his final directorship of an international organization. Increasingly, the same companies that provide infrastructural management for space activities also operate observatories like Paranal.[76]

In practice, however, the model borrowed from space-based research was never completely adopted. One of the unavoidable features of space-based observatories is that all the devices on board are installed before the launch, and remain the same for the duration of a mission, which at most wavelengths should ideally be at least a decade, but now often lasts three decades or more. Add to this the long development time for space missions, and the instruments on board very quickly fall an entire generation behind the state-of-the-art science on the ground. The risks and costs of an upgrade, either manned or unmanned, are comparable to those for launching a new satellite, and still at least an order of magnitude above the total cost of a full-fledged observatory on the ground.[77]

Besides their much lower overall cost, ground-based observatories benefit greatly from the diverse lifespans of their various components, which allow for separate development and funding models for each. The observatory sites have a centuries-long timescale and face largely geopolitical cultural and environmental challenges, as described earlier in the chapter. The observatories and telescopes built on them typically remain state of the art for around 30 years, and face challenges largely in the realms of large-scale engineering and large telescope optics and design. Finally, and crucially, the instrumentation

---

75 Madsen, *The Jewel on the Mountaintop*, 2012, 116–117.

76 OHB: *Faszination Raumfahrt. Geschäftsbericht 2018.* Bremen 2016. https://www.ohb-system.de/files/images/mediathek/downloads/OHB_GB_2016_dt_s.pdf. Last accessed 10/15/2020.

77 The most recent example of the problems of space telescope missions is the James Webb Space Telescope (JWST); an infrared facility that was originally scheduled to launch in 2014 and was successfully launched on 25 December 2021. Many of its instrumental components are now more than a decade behind the state of the art, and its rising costs have damaged many other astronomy projects. See Lee Billings: Space Science: The Telescope That Ate Astronomy. *Nature* 467/7319 (2010), 1028–1030. doi:10.1038/4671028a. Alexandra Witze: Delays Mount for NASA's $8-Billion Hubble Successor. 7712. *Nature* 559/7712 (2018), 16–17. doi:10.1038/d41586-018-05567-2.

that processes the light captured by the telescopes may have a much shorter life cycle, often less than a decade, and its cost is of an order of magnitude such as to be easily covered by research funding organizations and small inter-institutional collaborations. The challenges here tend to be much more of the electronic variety, requiring one-of-a-kind solutions to specific research inquiries.

So, while funding for long-term aspects of a project depends on political programs and diplomacy, the instrumental scale can benefit directly from institutions' own budgets and especially from the increasingly common competitive grants. And while throughout the 20th century these were almost always nationally based, the European Research Council began making a dent in that funding landscape in the early 2000s, and even expects applicants to represent a variety of nations and institutions (see below and in Chapter 5).[78]

This extends likewise to instruments that improve a telescope's operation. In fact, the boundary between instruments for scientific analysis and those for telescope improvement are in practice very blurred, as they both operate on the same 'raw product,' the light captured by the telescope.

Moreover, skillful instrumental design can often allow older telescopes and observatories to be retrofitted, bringing them back up to date by inserting state-of-the-art instrumentation. Instruments also sometimes travel from telescope to telescope, either periodically, or as part of a life cycle, and drift in line with the specialization of instruments and observatories over time.

Crucially, these instruments at the forefront of engineering and scientific design are not instantly standardized or put into operation, and in practice remain their developers' intellectual property for years, either contractually or as embodied tacit knowledge. This affords the producers a dual advantage over the 'users' or 'clients,' who were the kind of researcher idealized by the large infrastructural operators.

Privileged access to the telescopes (which actively seek to attract the best instrumentation) is unavoidable for installation testing and calibration. At the same time, these early testing or 'engineering' runs often give a head start to the most spectacular scientific discoveries. Moreover, the rights to the intellectual property ensuing from this instrumentation may remain exclusive for years, depending on the patent terms and conditions, and control lies with its developers/producers, even if it is indispensable to projects involv-

78 Veera Mitzner: *European Union Research Policy. Contested Origins.* London: Palgrave Macmillan 2020. doi:10.1007/978-3-030-41395-8.

ing other interested users around the world who wish to participate.[79] There is increasing tension, today, with the international infrastructure organization (ESO, ALMA) pressing for open access and the infrastructuralization of instruments, whereas the reality is that the scientific cutting-edge can most often be found at the small and pilot scales, often in highly customized, quasi artisanal devices. Since the 1980s, individual Max Planck institutes have become further convinced that instrumentation development is one of their strong points.

European Adaptive Optics (AO)

'Adaptive optics' technology, (on the global development of which, see Chapter 2), perhaps represents the defining moment in the Max Planck Society's participation in instrumental research and development of this sort.

As early as 1974, the world-renowned physicist Freeman Dyson was invited to spend a sabbatical year at the Max Planck Institute for Physics and Astrophysics in Munich, during which he published his perspective on the nascent field of adaptive optics. Moreover, he was invited to ESO (then still in Geneva, but already determined to move to Garching), where he presented papers on the subject. A few years later, however, in the early context of Star Wars, the United States classified this hitherto open scientific development, so putting an end to it in most academic settings.[80]

Throughout the 1980s, many aspects of this technology were nonetheless further developed by astronomers tied to the defense establishment, in places like the United States and France. As we saw in Chapter 2, laser guide stars were one such development.

The astronomical potential of adaptive optics is to greatly improve the quality of images in wavelengths that manage to reach the ground. This is a realm of 'visible' optical astronomy, but crucially, also of some wavelengths of infrared astronomy. In addition to the importance of infrared wavelengths for military applications, the wavelength is the obvious application for adaptive

---

79 For examples of the preferences given to instrument contributions to large telescope projects, see the examples of the Giant Magellan Telescope. GMT Scientific Advisory Committee: *GMT Scientific Advisory Committee. Operations Concept White Paper*, 2012. http://www.gmto.org/Resources/GMT_SAC_Operations_White_Paper.pdf. Last accessed 10/8/2010. and Gran Telescopio Canarias: P. L. Hammersley, and J. M. Rodríguez Espinosa: *Guaranteed Time for PI Instruments on the GTC*. 1, 2005. http://www.gtc.iac.es/instruments/media/GTpolicy.pdf. Last accessed 10/8/2020.

80 Freeman J. Dyson: *Selected Papers of Freeman Dyson. With Commentary*. Cambridge, MA: American Mathematical Society 1996, 41–44. F. J. Dyson: Photon Noise and Atmospheric Noise in Active Optical Systems. *Journal of the Optical Society of America* 65 (1975), 551–558. http://adsabs.harvard.edu/abs/1975JOSA...65..551D. Last accessed 8/13/2020.

optics, because the technology is more effective at the longer wavelengths than at the visible ones. The appointment of Reinhard Genzel as Director of the Institute for Extraterrestrial Physics must also be seen in this context: namely his link to Berkeley, his mentor Charles Townes, and the nearby research ecosystem that was developing adaptive optics, while making advances in ground-based infrared astronomy.[81] The first European telescope used by Genzel was the 1987 ESO New Technology Telescope (NTT) in La Silla; at the time, the next-generation telescope, the one that allowed Europeans to overtake American astronomy, was going to be the VLT in the new location of Paranal. While under planning since the 1980s, the construction of the new observatory coincided with the declassification of adaptive optics in the early 1990s. The result was a crash program by ESO to develop this technique for its existing and upcoming telescopes, retrofitted to their already fixed design.[82] Besides the development of the actuator system that changes the shape of a mirror surface inside an instrument on the optical path (not the telescope itself), the plans also included a guide star like the one developed by the Americans, which in turn required purpose-built lasers.[83]

ESO's solution was heavily dependent on the Max Planck Society. The first generation of the custom lasers was developed by the MPE modifying a commercial dye laser model, and the adaptive optics technologies were then tested and perfected at, of all places, the Calar Alto observatory in Spain. Since successful completion of the guide stars pilot phase, subsequent generations have been industrially produced in the Munich area, and installed in Hawaii's Keck

81 Genzel, Autobiography of Reinhard Genzel, 9/9/2008.

82 The Coming of Age of Adaptive Optics. *ESO Press Release* (10/23/1995). https://www.eso.org/public/news/eso9527/. Last accessed 8/13/2020. The size of telescope mirrors, as in the case of VLT at ESO or the Keck Telescope in Hawaii, has continuously increased to enhance optical resolution, which, however, is thwarted by turbulences in the Earth's atmosphere causing distortion of the wavefront emitted by astronomical objects. Space-based telescopes like Hubble or the James Webb Space Telescope are a straightforward solution to this problem, but their size and scope is limited by the weight of the mirrors, in addition to the high cost of launching and operating them compared to ground-based facilities. For ground-based telescopes, instead, blurring of an image can be corrected by applying the adaptive optics technology, which improves the performance of optical systems by compensating the distortion in a wavefront with sophisticated deformable mirrors controlled by computers that correct in real time the blur caused by local atmosphere.

83 A reference star (*guide* star) that is very close to the object under investigation is used to measure and correct the blurring caused by the local atmosphere. Whenever suitable stars are not available, artificial star images can be created by shining a powerful laser beam into the Earth's upper atmosphere. Such laser guide stars make a much larger fraction of entire sky accessible to adaptive optics imaging.

telescopes as well as in European projects. And like many other developments in astronomy, this technology has promising commercial applications, such as optical communications and tracking with satellites and space probes, usages that are of obvious military interest, too.[84] The development of adaptive optics and laser guide stars proved a novel use of the struggling Calar Alto observatory in the 1990s, so aiding the transition from its primarily observational purpose, dominated by the Max Planck Institute for Astronomy in Heidelberg, to a new lease of life as a testing facility for new technologies; and this benefited other institutes of the Max Planck Society as well as the German scientific and commercial aerospace sector.[85]

In addition to this point of access to ESO's VLT, which went on to become the source of Reinhard Genzel's successes categorizing the galactic center, the Max Planck Society also participated, in parallel, in instrumentation that uses the adaptive optics developments for the LBT on Mount Graham, Arizona, this time in a collaboration between the Institutes for Astronomy (Heidelberg) and Radio Astronomy (Bonn). Thanks to the delayed start of the LBT (due to political struggles detailed earlier in the chapter), its Italian-built adaptive optics is integrated more deeply into the telescope's design, acting directly on its secondary mirrors.[86] Finally, the next-generation telescope from ESO, the ELT (Extremely Large Telescope)—which, when it opens in the mid-2020s, will be by far the world's largest, at 39 m in diameter—benefits from all these latest developments in adaptive optics, without which the ELT would be useless: the design integrated adaptive optics from the start, and features eight guide stars (provided by the same Munich-area company), the number required given that the atmosphere 'seen' by the telescope varies along its vast diameter.

---

84 Richard I. Davies et al.: ALFA: First Operational Experience of the MPE/MPIA Laser Guide Star System for Adaptive Optics. *Adaptive Optical System Technologies*. Astronomical Telescopes and Instrumentation. Kona, HI: International Society for Optics and Photonics 1998, 116–124. doi:10.1117/12.321747. A Quirrenbach, and W Hackenberg: The ALFA Dye Laser System. In: N. Hubin, and H. Friedmann (eds.): *Laser Technology for Laser Guide Star Adaptive Optics Astronomy*. ESO 1997, 126–131.). Toptica Photonics: Laser Guide Star. Improving the Resolution of Telescopes with Artificial Stars. https://www.toptica.com/applications/astronomy-geology/laser-guide-star/. Last accessed 8/13/2020. Davies et al., ALFA, 1998, 116–124.

85 This 'pilot facility' approach is the model that also justified many observatory-like projects in the 1990s and early 2000s, including GEO600, MAGIC, H.E.S.S., and APEX (see also Chapter 5).

86 Ralph Hofferbert et al.: LINC-NIRVANA for the Large Binocular Telescope: Setting up the World's Largest near Infrared Binoculars for Astronomy. *Optical Engineering* 52/8 (2013), 081602. doi:10.1117/1.OE.52.8.081602. See also: Lemke, *Himmel über Heidelberg*, 2011, Vol. 21, 215–219.

Instrumentation Research

The 1987 *Denkschrift* (Memorandum on the Future of Astronomy in West Germany, see next section), advocated for a next-generation national telescope to replace Calar Alto, this time outside of the Max Planck Society. Reunification quickly put an end to such ambitions and rather catalyzed German integration into international collaborations. Perhaps most indicative of this new approach was the development of the most pivotal among the first generation of instruments to be installed at ESO's VLT: the FORS cameras and spectrographs (Focal Reducing Imager and Spectrograph). These were developed under the new modality of *Verbundforschung* (science and industry research partnership: see later in this chapter), led by the Heidelberg State Observatory (in which the MPIA had originated in the 1960s). This instrumentation provided an early point of access for German astronomers to what, since first commissioned, has come to be considered the world's best optical telescope of the early 21st century. This was accompanied in the first decade of operation by several other VLT instruments led by the MPIA: one was the telescope's mid-infrared interferometer called MIDI (MID-Infrared-Interferometric instrument), for which efforts were transferred from initial plans for Calar Alto to the new, much more powerful and better situated telescopes in Chile. MIDI was built in collaboration with two other German institutions (Kiepenheuer Institute in Freiburg, Thuringian State Observatory) as well as with French and Dutch partners. The other instrument, the CONICA near-infrared camera (COudé Near-Infrared Camera), started in collaboration with the Institute for Extraterrestrial Physics, and later expanded into a German–French partnership that resulted in the interferometric instrument NACO (NAOS+CONICA), the first to take advantage of the (French-provided) adaptive optics at the VLT.[87]

By 2020, German astronomers at the Max Planck Society, universities, observatories, and other institutions had provided two to three generations of instruments, being directly involved in more than half of the instruments

87 Bundesministerium für Bildung und Forschung (BMBF) (ed.): FORS—Das Arbeitspferd am Very Large Telescope. *Stark im Verbund: Naturwissenschaftliche Grundlagenforschung an Großgeräten.* Bonn 2009, 49–50. Ch. Leinert et al.: MIDI—the 10 Mm Instrument on the VLTI. *Astrophysics and Space Science* 286/1 (2003), 73–83. doi:10.1023/A:1026158127732. N. Ageorges, L.E. Tacconi-Garman, and C. Lidman: One Year of NACO Operations. In: Wolfgang Brandner, and Markus E. Kasper (eds.): *Science with Adaptive Optics. Proceedings of the ESO Workshop Held at Garching, Germany, 16–19 September 2003.* Berlin: Springer 2005, 53–61. On all these developments at the VLT, see also: Lemke, *Himmel über Heidelberg*, 2011, Vol. 21, 210–215.

required for the world's most productive optical telescope.[88] To this must also be added ESO's in-house ('facility') instruments, which involve a sizable portion of German-made parts stemming from the country's membership in the organization and benefiting particularly from the presence of ESO's headquarters in Garching.

The quick turnover in scientific instrumentation is necessary owing both to the now faster pace of scientific discovery and the intellectual property implications of working in global infrastructures. Access to pioneer-stage instruments, so-called PI Instruments, guarantees more extensive control over the observational process, data post-production, and publication. In the 21st century, this privileged access is a differentiating factor in a field where smaller institutions and researchers can apply for telescope and instrument time as 'users,' in line with the egalitarian philosophy that stands nominally behind these new generations of telescopes. The mandate to allow external access to these telescopes and instruments—and to the data originating from them, even if collated by other teams—is increasingly a source of tension between the heavyweights (including Max Planck Institutes), and a much larger number of weaker research institutions, in a growing number of countries around the world. Battles over data priority and ownership are also significant in contemporary projects such as ALMA and CTA (see Chapter 5), which monopolize access to particular wavelengths. In practice, PI Instruments continue to constitute the spearhead of research, with external access to 'facility' instruments and older PI instruments (after an initial period of exclusivity) being lower in the pecking order. [89]

This instrumental specialization within international infrastructures has likewise occurred in space missions, where this form of participation was

88 The list of VLT instruments at ESO (https://www.eso.org/public/teles-instr/paranal-observatory/vlt/vlt-instr/. Last accessed 2/4/2022) contains the following items with German involvement: 4LGSF (Laser Guide Star Facility, Industry: TOPTICA Photonics); CRIRES+ (Cryogenic high-resolution InfraRed Echelle Spectrograph+, Thüringen State Observatory, University of Göttingen); ERIS (Enhanced Resolution Imager and Spectrograph, MPE); FORS 1+2 (Focal Reducer and low dispersion Spectrograph, Heidelberg State Observatory, Munich University Observatory, Göttingen University Observatory); KMOS (K-band Multi Object Spectrograph, Munich University Observatory, MPE); MUSE (Multi-Unit Spectroscopic Explorer, Göttingen Astrophysics Institute, Potsdam Astrophysics Institute); SPHERE (Spectro-Polarimetric High-contrast Exo-Planet Research, MPIA); GRAVITY (MPE, MPIA, University of Cologne); MATISSE (Multi-AperTure mid-Infrared SpectroScopic Experiment, MPIA, MPIfR); AMBER (Astronomical Multi-BEam Recombiner, MPIfR); NACO (MPIA, MPE); MIDI (MPIA, Kiepenheuer Institute, Thüringen State Observatory); SINFONI (MPE).

89 On the privileges given to PI instruments in large telescopes, see footnote 72, above.

pioneered several decades ago: in the X-ray domain, the next-generation satellites XMM-Newton and Chandra mainly featured German instrumentation,[90] whereas eROSITA, the ROSAT successor finally launched in 2019, features a fully German-made telescope yet is mounted on a Russian platform administered by (Rashid Sunyaev's) Institute for Cosmic Research;[91] and ESA's INTEGRAL satellite in the gamma ray domain, and the American space telescopes EGRET and Fermi[92] are likewise multinational constructions.

As mentioned earlier in the chapter, all observational astronomy MPIs in the infrared domain had a heavy instrumental presence in space missions from the late 1970s on, as well as in the German–American airplane-based observatory SOFIA, which, even though it was delayed for decades, can still participate in the latest discoveries because, unlike space missions, it can be equipped with newly developed instruments.[93]

Below the infrared, in the millimeter and submillimeter wavelength domains, the Max Planck continued its successful presence in IRAM, while increasingly shifting weight from Peter Mezger's 30 m single dish in Spain towards the potential of the interferometric array on the Plateau de Bure, which itself was the European platform for the expansion into ALMA. In all these, an ongoing point of access was the traditional link between Bonn and the large antenna manufacturers (see Chapter 3). But the MPI presence in IRAM and APEX is also further justified by the Society's capacity to develop,

---

90 The Max Planck Institute for Extraterrestrial Physics had a main contribution on the XMM-Newton mission. The Institute contributed to the telescope development and test, developed the EPIC-pn CCD detector (one of the instruments on board of XMM) and operates the XMM-Newton Survey Science Centre, selected by ESA to ensure that the scientific community can access data accumulated by the mission. MPE also contributed the Low Energy Transmission Grating (LETG) on the Chandra X-ray Observatory, in collaboration with the Space Research Organisation Netherlands in Utrecht.

91 See the public website https://www.mpe.mpg.de/eROSITA and the science portal dedicated to eROSITA https://erosita.mpe.mpg.de. Last accessed 1/26/2022.

92 See the webpage dedicated to INTEGRAL (German Space Agency): https://www.dlr.de/rd/desktopdefault.aspx/tabid-2448/3635_read-5473/. The German contribution to FERMI (the Gamma-ray Burst Monitor) is described at https://www.mpe.mpg.de/617954/Fermi-GBM. Last accessed 1/26/2022.

93 One of the advantages of instrument development is the ability to stay up to date scientifically, even if the platforms that use them have been delayed. This compartmentalizes the political damage of infrastructural delays, away from the instrumental developers and users of the telescopes. SOFIA's instrument GREAT, for example, detected in 2016 the oldest possible molecules in the Universe. Rolf Güsten et al.: Astrophysical Detection of the Helium Hydride Ion HeH^{+}. *Nature* 568/7752 (2019), 357–359. doi:10.1038/s41586-019-1090-x. Rolf Güsten et al.: Astrophysical Detection of the Helium Hydride Ion HeH +. 7752. *Nature* 568/7752 (2019), 357–359. doi:10.1038/s41586-019-1090-x.

there, instruments and techniques that are subsequently deployed on a wider scale, as we will see below.[94]

Moreover, at these millimeter wavelengths, as well as in the longer wavelength domain of traditional radio astronomy, astronomical observations can be recorded as electronic signals. This recording and subsequent combination and analysis of astronomical observations foster a unique modality of global collaboration and instrumental development that we detail next.

VLBI: Globally-Distributed Astronomical Interferometry

A unique form of international collaboration in astronomy occurred through development of the particular technology of interferometry, in which the wave signals from several telescopes are combined to gain information, with resolution that is, for many purposes, equivalent to an instrument of the diameter spanning the distance between two telescopes.[95] The natural starting point for this technique was in long wavelength radio astronomy, where the ingenious procedure became an alternative to the race for ever larger single-dish telescopes and, for more than a decade of its development, was the subject of the Nobel Prize in 1971, awarded to Martin Ryle from Cambridge, UK. Initially, interferometric observations were conducted with sets of telescopes whose signals were combined via cable. Soon after, longer distances or 'baselines' were made possible by radio communication with the telescopes. With 1960s technology, at long wavelengths, it also became possible to record the signals separately and combine them at a later time, allowing for many possible baselines around the world for a given wavelength, as long as the observations were recorded at exactly the same time. The technology was first explored in the Soviet Union in the early 1960s, with the Evpatorija antenna built in Crimea for communication with Soviet space probes. During his visit there, it was suggested to Bernard Lovell, artificer of the Jodrell Bank antenna (also used at the

---

94 IRAM and APEX benefit from their service as access platforms to gain a firm footing in ALMA, currently the world's largest astronomical infrastructure. Due to its enormity, ALMA is heavily standardized, with very strict serialization demands for its instruments and mandates for transparency in its data handling. For smaller organizations like a Max Planck Institute, the regimented setting of ALMA needs to be complemented by more flexible and private facilities. A similar logic will hold in similar-scale projects at other wavelengths such as CTA, where the previous generation facilities will continue to play a role as pilot facilities (see Chapter 5).

95 For an overview of the history of interferometry in radio astronomy, see: K. I. Kellermann, and J. M. Moran: The Development of High-Resolution Imaging in Radio Astronomy. *Annual Review of Astronomy and Astrophysics* 39/1 (2001), 457–509. doi:10.1146/annurev.astro.39.1.457.

time for deep space communications), that his and the Soviets' antennas be used for interferometry experiments.[96] The idea came to nothing until later, from 1969 on, when Crimea and the MIT Haystack Observatory began collaborating.[97] Thus, the technology was marked from the start by a spirit of international collaboration, even across the Iron Curtain, as well as by technical overlap with contemporary developments in space exploration and geolocation technologies. In parallel, yet independently of this 1960s Soviet–British initiative, similar experiments were underway among various radio observatories in the US, including NRAO's Green Bank, MIT's Haystack, and the Naval Observatory in Annapolis. Canadians, who initially had worked independently with telescopes spanning their country, soon joined in the American collaboration, while British scientists continued separately to make significant contributions to the field.

Besides the means of recording signals, a key instrument in this technology was high-precision timekeeping, as the signals from different sites had to be time-stamped for simultaneity. For this purpose, there were atomic clocks based on hydrogen masers, developed since the 1950s as an offshoot of wartime radar technology. A third key instrument was the correlator, a computational process by which the signals from the two sites could be combined and analyzed. In the second half of the 1960s, the NRAO standardized the tape recorders, atomic clocks, and correlators needed for VLBI (Very Long Baseline Interferometry), and distributed them among observatories spanning ever longer distances.[98] This type of interferometry initially relied on retrofits—existing telescopes were equipped with or adapted to these newer instruments—while the latest radio telescope observatories were planned with them in mind from the outset.

VLBI functions in its particular time scale of slow, incremental technological development strongly anchored in local expertise and traditions, but uniquely regimented by strict standardization that ultimately allows its inevitably global coordination. Moreover, VLBI, operationally speaking, necessarily lags behind the corresponding developments in directly linked interferometry, which faces similar technological challenges, but not the added dif-

96 Ibid. see also L. Matveyenko: Early VLBI in the USSR. *Astronomische Nachrichten* 328/5 (2007), 411–419. doi:10.1002/asna.200710763.

97 J. M. Moran: Thirty Years of VLBI: Early Days, Successes, and Future. *International Astronomical Union Colloquium* 164 (1998), 1–10. doi:10.1017/S0252921100044353.

98 The observations with Crimea used this American-made equipment, which faced hurdles at both ends due to the restrictions on exporting high technology to the Soviet Union. Kellermann, and Moran, The Development, 2001, 457–509, 481.

ficulty of recording signals simultaneously at independently operated observatories far apart in the world. VLBI efforts have consequently lagged behind those in direct interferometry at the same wavelength for about a decade.

The technology also faces the persistent problem of having to strictly schedule simultaneous observation times among multiple telescopes, which interferes with their other uses. In the United States, this led to the creation in the 1980s of a dedicated group of telescopes, the Very Long Baseline Array (VLBA);[99] but this was a one-time luxury in the lower wavelengths and could not be repeated for every new wavelength window opening up in the second half of the century. So, throughout its existence, VLBI has generally been an add-on to new telescopes that have their own independent research programs; these benefit from access to new wavelengths to which to apply the technology; but depend on the generosity of the host observatories to fulfill the strict scheduling and instrumental adaptations necessary for the duration of the observations.

The long-term technological challenge for directly linked astronomical interferometry is to incorporate shorter wavelengths. This requires, on the one hand, higher resolution and accuracy when recording the wavelengths, and on the other, exponentially more computational power for the correlators used to convert the raw signals into useful astronomical information. Directly linked interferometry matured in the radio wavelengths in the 1970s at places such as Westerbork in the Netherlands (14 telescopes) and Socorro (VLA, Very Large Array, 27 telescopes); and only since the late 1980s did interferometry succeed in the millimeter domain, thanks to the French contribution to IRAM, the Plateau de Bure (6 telescopes, now expanded to 12); to its American counterparts in California (23 telescopes) and Hawaii (10 telescopes); to the Japanese Nobeyama (6 telescopes); and to the Australian ATCA (Australia Telescope Compact Array, 5 telescopes). ALMA (Atacama Large Millimeter Array, 66 telescopes), completed in the 2010s, is the latest accomplishment in this quest for shorter wavelengths and bigger arrays.[100]

---

99 Kenneth I. Kellermann, Ellen N. Bouton, and Sierra S. Brandt: VLBI and the Very Long Baseline Array. In: Kenneth I. Kellermann, Ellen N. Bouton, and Sierra S. Brandt (eds.): *Open Skies. The National Radio Astronomy Observatory and Its Impact on US Radio Astronomy*. Cham: Springer 2020, 391–459. doi:10.1007/978-3-030-32345-5_8.

100 John Carpenter: Introduction to Radio Facilities: Millimeter and Submillimeter Interferometers. Academia Sinica, 2016. https://events.asiaa.sinica.edu.tw/school/20160815/talk/jcarpenter0817.pdf. Last accessed 10/8/2020. For a historical picture, see: Wm. J. Welch: Millimeter and Submillimeter Interferometry. In: Graeme D. Watt, and Adrian S. Webster (eds.): *Submillimetre Astronomy. Proceedings of the Kona Symposium on Millimetre*

In the optical infrared domain, interferometric observatories matured only from the late 1990s onwards with the Keck telescope, ESO's Paranal, and, (a decade late), the LBT (Large Binocular Telescope). At these shorter, near-visible wavelengths, the technology depends not on electronic signatures, but solely on the direct optical combination of the incoming light from several telescopes.

In Germany, there was a path dependency regarding VLBI that ensued from the initial focus on giant optical telescopes and single-dish radio telescopes. This meant that during the decades of national infrastructure construction, no attempt was made to compete in the domain of directly linked interferometric telescopes until quite late. The first significantly German interferometric telescope (25 percent) was the LBT, which went into operation only in the second decade of the 21st century. Rather, German astronomers seeking a presence in interferometric observatories relied from the outset on international collaborations, as when Max Planck researchers, for example, benefited from the French-built IRAM interferometric array.

At the same time, however, there was a particular synergy between large, single-dish observatories and the VLBI, which made the Max Planck Institute for Radio Astronomy a central player in the development and use of this technology from the 1970s onwards.

In West Germany, at the time when the Effelsberg radio telescope was under planning and early construction, VLBI usage became a key interest. Plans dating from 1968 attest that collaboration between Effelsberg, Jodrell Bank (UK), and Onsala (Sweden) was a distinct possibility. Following the 'business model' mentioned in Chapter 3, the Krupp/ MAN consortium even offered to build replicas of Effelsberg in other locations, and at one point Otto Hachenberg's counterpart, Olof Rydbeck, seriously considered building such a clone in Sweden. Onsala was then already participating in the first intercontinental experiments in VLBI, with the aforementioned research groups in the United States.[101]

By the time of its inauguration, Effelsberg was a founding member of the VLBI consortium together with Jodrell Bank, Onsala, and Bologna, and, later, Westerbork, while associate members included Nançay (France), Torun (Poland), and the Space Research Institute of Moscow (IKI), with its Crimea

---

and Submilimetre Astronomy, Held at Kona, Hawaii, October 3–6, 1988. Dordrecht: Springer 1990, 81–86.

101 Olof E. H Rydbeck: *Femtio år som rymdforskare och ingenjörsutbildare. Från skånska horisonter till fjärran galaxer* (*Vol 2: 1951–1989*). Vol. 2. Göteborg: Chalmers tekniska högsk 1991, 743–753.

antenna.[102] VLBI was notably international and benefited both from the period of Cold War 'detente' and long-time collaboration in this field to forge a bridge between astronomer communities on both sides of the Iron Curtain; and, as we will see in Chapter 5, this had interesting implications for other fields, such as ground-based, gamma ray astronomy. By the 1980s, there were three complementary consortia (American, West European, Eastern Bloc), with about 50 locations around the world.[103] By 2004, at least four interconnected networks (in America, Europe, and Australia) provided open access to astronomers, and included dishes in associated areas, such as South Africa and China.[104] Ultimately, the technology even extended to a radio telescope based in outer space, the Russian RadioAstron, on the Spektr-R astronomical satellite launched in 2011.[105]

Early VLBI benefited from usage of geodesy technology, that is, the accurate measuring of the Earth. Here, the positions of far-away radio sources (such as quasars) are the known quantity used to infer the exact locations of the observing radio telescopes. The geodetic use of VLBI was operationalized in the late 1960s and continues, to this day, to be a significant non-astronomical use of radio astronomical observatories, although dedicated geodetical VLBI observatories were built too, in the following decades. In fact, the correlator of the Max Planck Institute for Radio Astronomy has, since its inception, been partly dedicated to the calculation of geodetic observations.[106]

Traditionally, one of the three directors of the Max Planck Institute for Radio Astronomy has carried the responsibility (and diplomatic burden) of

---

102 Richard Wielebinski: Coordination of VLBI Observations. In: C. Jaschek, and C. Sterken (eds.): *Coordination of Observational Projects in Astronomy. Proceedings of an International Conference Held in Strasbourg, November 23–26 1987*. Cambridge: Cambridge University Press 1988, 91–96, 92. E. Preuss: The Beginnings of VLBI at the 100-m Radio Telescope. In: E. Ros et al. (eds.): *Proceedings of the 6th European VLBI Network Symposium*. Bonn 2002.

103 Wielebinski, Coordination, 1988, 91–96.

104 M. A. Garrett: Ground Based VLBI Facilities—the European and Global VLBI Network. In: Franco Mantovani, and Andrzej Kus (eds.): *The Role of VLBI in Astrophysics, Astrometry and Geodesy*. Dordrecht: Springer 2005, 403–413.

105 Y. Y. Kovalev et al.: The RadioAstron Space VLBI Project. *2014 XXXIth URSI General Assembly and Scientific Symposium (URSI GASS)*. 2014 XXXIth URSI General Assembly and Scientific Symposium (URSI GASS). Beijing 2014. doi:10.1109/URSIGASS.2014.6929994.

106 The current correlator is operated jointly by the MPIfR and the *Bundesamt für Kartographie und Geodäsie* (BKG) in cooperation with the *Institut für Geodäsie und Geoinformation der Universität Bonn* (IGG). Simone Bernhart et al.: The Bonn Astro/Geo Mark IV Correlator. *Annual Report*, 2008, 201–204. Axel Nothnagel, Wolfgang Schlüter, and Hermann Seeger: Die geodätische VLBI in Deutschland. *Zeitschrift für Geodäsie* 129/4 (2004), 219–226.

the VLBI: firstly, Otto Hachenberg, then Kenneth Kellermann, a VLBI pioneer. After his departure this was done non-exclusively by other directors until Anton Zensus, director from 1997, fully recommitted his department to VLBI. This continuous VLBI membership not only coordinates projects with existing telescopes, but facilitates access to new projects in recently opened wavelength domains as the technology matures. Between 1977 and 1993, the MPIfR hosted the European network's correlator supercomputers, a function now served at the Joint Institute for VLBI in Europe (JIVE), based in Dwingeloo, Netherlands. This was in turn transformed in 2014 into what is known as a 'European Research Infrastructure Consortium' (ERIC) of the European Union.[107]

Moreover, the foothold in VLBI served as a springboard into other aspects of interferometry, even at wavelengths traditionally outside the realm of Bonn: in 1989, the Institute for Radio Astronomy appointed Gert Weigelt to specialize in infrared and visible interferometry, an expertise that led to participation in the interferometry instrumentation for both the VLT (MATISSE) and the LBT (LINC-NIRVANA).[108]

Given the pioneering participation in VLBI, a significant long-term reason for the MPIfR's ambitions to build radio telescopes at each new possible wavelength was rooted in the scientific potential deriving from them eventually linking up (although perhaps decades later) with the VLBI network. This came about eventually with the IRAM telescope built in Spain, then its French counterparts near Grenoble, and later, the submillimeter Heinrich Hertz telescope in Arizona, and the APEX in Chile. The unfinished millimeter antenna for the Iraq National Observatory bombed in the 1980s would also have been a likely participant.[109]

This was also the case with longer wavelengths: the association with VLBI facilitated the Max Planck Society's participation in LOFAR and the forthcoming Square Kilometre Array (SKA).[110]

An interesting development that highlights this half-century-long specialty of VLBI in Bonn culminated in the first observation, in 2017, of the 'shadow'

107 J. Anton Zensus, and Eduardo Ros: European VLBI Network: Present and Future. *Proceedings of Science*. 12th European VLBI Network Symposium and Users Meeting—7–10 October 2014. Cagliari 5, 001. doi:10.22323/1.230.0001.

108 Thomas Becker, and Philip Rosin: *Geschichte der Universität Bonn. Die Natur- und Lebenswissenschaften*. Vol. 4. Göttingen: Vandenhoeck & Ruprecht 2018, 318.

109 Jacob W. M Baars: interview by Luisa Bonolis and Juan-Andres Leon, Bonn, February 5–7, 2018. DA GMPG, BC 601050.

110 M. A. Garrett et al.: LOFAR, E-LOFAR and Low-Frequency VLBI. arXiv:0902.2534 [*Astro-Ph*]. European VLBI Network IX Symposium. Bologna 2008. http://arxiv.org/abs/0902.2534. Last accessed 8/15/2020.

of a black hole itself, which was made public in April 2019.[111] Millimetre-VLBI started to mature at the MPIfR with the appointment of Anton Zensus as director in 1997. He grew the team of scientists around Ivan Pauliny-Toth, Arno Witzel, David Graham, Richard Porcas, Thomas Krichbaum, and others into a full department. He brought major VLBI activities to Bonn including Space VLBI and the Global Millimeter VLBI Array (GMVA), which paved the way toward observations on event horizon scales with the EHT. Among the many postdocs he attracted to his department was Heino Falcke.[112,113,114]

The advances in millimeter-wavelength interferometry which made this proposal possible had benefited significantly from contributions of the Max Planck Institute for Radio Astronomy, starting with its participation in pioneering intercontinental 7 mm observations in the mid-1980s, pushing the limits of the Effelsberg telescope. In subsequent years, these efforts advanced towards shorter wavelengths, thanks to IRAM and the longstanding participation of interferometry expert Thomas P. Krichbaum.[115] The first successful, long-distance VLBI experiment at 1.3 mm wavelength was conducted with the IRAM dishes in Spain and France in 1994–95. These observations were already pointing to the possible black hole at the center of the Milky Way.[116] In 1998, Falcke who had become aware of the potential based on these early promises and was a frequent guest in Arizona, organized a conference there, under the name 'The Central Parsec of the Milky Way Galaxy.'

Soon after, the first transatlantic interferometric observations at millimeter wavelengths were conducted in 1999, using IRAM's 30 m dish in Spain and

111 For a detailed history of this episode centered on the American contributions see Seth Fletcher: *Einstein's Shadow. A Black Hole, a Band of Astronomers, and the Quest to See the Unseeable*. New York, NY: Ecco 2018. Further details were clarified in an interview with Heino Falcke by Luisa Bonolis, Roberto Lalli, and Juan-Andres Leon, Berlin, August 22, 2019.

112 An alternative autobiographical account of the events described in this section can be found in Heino Falcke: *Light in the Darkness. Black Holes, The Universe and Us*. San Francisco, CA: HarperOne 2021.

113 Heino Falcke, Fulvio Melia, and Eric Agol: Viewing the Shadow of the Black Hole at the Galactic Center. *The Astrophysical Journal Letters* 528 (2000), L13–L16. doi:10.1086/312423.

114 M. Miyoshi, S. Kameno, and H. Falcke: A Proposal for Constructing a New VLBI Array, Horizon Telescope 289 (2003), 33–36.

115 Thomas P. Krichbaum: Millimeter-VLBI with a Large Millimeter-Array: Future Possibilities. In: Peter A. Shaver (ed.): *Science with Large Millimetre Arrays*. Berlin, Heidelberg: Springer Berlin Heidelberg 1996, 95–102. doi:10.1007/978-3-540-69999-6_12.

116 Thomas P. Krichbaum et al.: VLBI Observations of the Galactic Center Source Sgr A* at 86 GHz and 215 GHz. *Astronomy and Astrophysics* 335, L106–L110.

several sites in the United States, including the Heinrich Hertz submillimeter telescope in Arizona, at a time when the MPIfR had already decided to give this away.[117] It was in this context that Falcke, with Fulvio Melia and Eric Algol, refined the 1979 predictions of Jean-Pierre Luminet of the shadow around a supermassive black hole with a more realistic emission model.

The American side of the transatlantic interferometric efforts was led by Sheperd Doeleman at MIT's Haystack Observatory (an early pioneer of VLBI), who closely collaborated with Zensus' department to make progress towards millimeter interferometry. Continuing with this work, observations made in 2007 obtained features consistent with the presence of an observable black hole.[118] These results, then called an 'event-horizon-scale structure,' already satisfied scientific experts in the field, which was balanced with caution by other experts against over-interpretation. But an essentially single-axis interference pattern between Arizona, California, and Hawaii is far removed from what the wider community and the public are willing to accept as an *image* of a black hole. While the MPIfR continued investing in mm-VLBI developments and operations of the GMVA, Doeleman accordingly took on the difficult task of organizing the global VLTI campaigns that would improve this *image*. After successful publication of his impressive results of 2007, Doeleman moved beyond the stream of grant applications that had sustained his team's work to date, many of which were for technical improvements rather than scientific observations, and instead submitted a proposal to the Decadal Survey Committee, the organization that sets the 'Ten Year Plan' priorities in astronomy in the United States, defending the feasibility of imaging a black hole within the decade.[119]

Among the listed co-authors were many of the best-known names in the astronomical community, and directors of many of the world's leading research centers (see author list in footnote). Doeleman did not invite Heino Falcke, however, on the pretext that he was a theoretician, so that "it was not

117 Sheperd S. Doeleman, and Thomas P. Krichbaum: Status of VLBI Observations at 1 MM Wavelength and Future Prospects. In: A. Greve, and T. P. Krichbaum (eds.): *Proceedings of the Second Millimeter VLBI Science Workshop, Held at IRAM in Granada (Spain) on May 27–29, 1999*. St. Martin d'Heres: IRAM 1999, 73.

118 Sheperd S. Doeleman et al.: Event-Horizon-Scale Structure in the Supermassive Black Hole Candidate at the Galactic Centre. 7209. *Nature* 455/7209 (2008), 78–80. doi:10.1038/nature07245.

119 Sheperd S. Doeleman et al.: *Imaging an Event Horizon: Submm-VLBI of a Super Massive Black Hole. A Science White Paper to the Decadal Review Committee*, 8. https://arxiv.org/abs/0906.3899. Last accessed 3/23/2021. While in this document the center of the Milky Way is the main object of study, the potential of the center of M87 is already contemplated. It was this more distant, but much larger black hole at the center of M87 that was first successfully imaged as a result of the 2017 campaign.

clear at the time what Heino's role would be in the project."[120] Doeleman and Falcke's relationship had deteriorated over the decade: the MIT group saw itself as the technical pioneer taking all the risks and working hard, while Falcke, in its view, had a more diversified career because he often sprang around between topics. In any case, throughout all the necessary technical developments, the Bonn institute had maintained a key presence in the project through the institute's VLBI expert Thomas P. Krichbaum (one of the co-authors of the Decadal proposal), and the tutelage that came with the directorship in Bonn and VLBI leadership role and continuous investments of Anton Zensus.[121] Moreover, many of the millimeter-wavelength telescopes that would perform the observations had been the 'babies' of Peter Mezger and his successor at the MPIfR, Karl Menten: IRAM, the Heinrich Hertz Telescope (by then given away), APEX, and, indirectly, ALMA.

An environment of intellectual exchange and collaboration had been continually sustained, furthermore, between Arizona and Bonn in the 1990s, when the HHT (Heinrich Hertz Submillimeter Telescope) was still partly owned by the Max Planck Society. Arizona had had a powerful influence on researchers like Falcke.[122]

The tense nature of international collaborations in astronomy is well exemplified by how the American-led project turned into a global partnership. After the tortuous maturation of the technology at the MPIfR and Haystack Observatory throughout the first decade of the 21st century, a global collaboration was set in motion explicitly for observing the shadow of a black hole, once an outcome could be guaranteed.[123] And while the VLBI technical experts at the MPIfR in Bonn were part of the team since the beginning, Falcke (meanwhile at Radboud University in the Netherlands) together with

120 Robert Gast: Bild des Schwarzen Lochs: Das Monster zeigt seine Zähne. *Spektrum der Wissenschaft* (*online*). https://www.spektrum.de/news/das-bild-des-schwarzen-lochs/1638154. Last accessed 8/19/2019.

121 Anton Zensus: Radio Astronomy / VLBI. Research. *Max Planck Institute for Radio Astronomy*. https://www.mpifr-bonn.mpg.de/research/vlbi. Last accessed 3/23/2021.

122 Heino Falcke: interview by Juan-Andres Leon, Roberto Lalli, and Luisa Bonolis, Berlin, August 23, 2019.

123 The decadal proposal concluded thus: "We emphasize that the path forward is clear, and recent successful observations have removed much of the risk that would normally be associated with such an ambitious project. Details of the technical efforts required to assemble this 'Event Horizon Telescope' will be described elsewhere, but no insurmountable challenges are foreseen." Doeleman et al., *Imaging an Event Horizon*.

theoretician Luciano Rezzolla at Frankfurt University, and Michael Kramer, director at the MPIfR in Bonn, obtained a grant from the European Research Council[124] for their 'Black Hole Cam' project.[125] This enabled them to join the project.

Owing to the global nature of VLBI, it was necessarily an initiative that would join forces, rather than work in parallel, with the Haystack project. Very soon after, during Haystack's negotiations with individual observatories on installation of the atomic clocks and data collection equipment, ALMA exerted pressure, saying it would participate in the trials only if the European project were included.[126] This led to a larger consortium being negotiated in 2014, the so-called Event Horizon Telescope Collaboration (EHTC), in which the processing load of the data was shared between Haystack and the Max Planck Institute for Radio Astronomy. In this new, higher-level organization of 13 organizations, Zensus chaired the EHT Board, Doeleman acted as project director, and Falcke as Chair of the Scientific Council overseeing the scientific objectives of the consortium.[127] The first successful observational campaign was conducted in April 2017, with participation from the following observa-

124 We will see in Chapter 5 how, at the beginning of the 21st century, recently established pan-European research funding organizations deliberately prioritized projects that needed support to be competitive, and then collaborated with American projects on a more equal basis than would otherwise have been possible.

125 Falcke, Heino: interview by Juan-Andres Leon, Roberto Lalli, and Luisa Bonolis, Berlin, August 23, 2019.

126 Access to ALMA for a VLBI observational run was particularly challenging, regardless of any political hurdles. While the multi-telescope array had originally been conceived also with VLBI in mind, and space for the equipment existed, the necessary technical adaptations for a so-called Phased Array had been cancelled for budget purposes. Markus Voelter: *Once You Start Asking: Insights, Stories and Experiences from Ten Years of Reporting on Science and Engineering*, 2020, 339. All these investments and the adaptations to Haystack's equipment would be necessary for the Event Horizon Telescope. They would be feasible only if both the American and European members of the ALMA partnership agreed and requested it of the observatory.

127 The complex bureaucratic structure of this collaboration had been negotiated at the 2014 meeting of the EHT at the Perimeter Institute for Theoretical Physics in Waterloo, Canada. Fletcher, *Einstein's Shadow*, 2018.
The 13 stakeholder institutions were: Academia Sinica Institute of Astronomy and Astrophysics (Taiwan), University of Arizona, University of Chicago, East Asian Observatory, Goethe-Universität Frankfurt, IRAM, Large Millimeter Telescope (Mexico), Max-Planck-Institut für Radioastronomie, MIT Haystack Observatory, National Astronomical Observatory of Japan, Perimeter Institute for Theoretical Physics (Canada), Radboud University and the Smithsonian Astrophysical Observatory.

tories: ALMA, APEX, IRAM, Submillimeter Telescope (MPIfR's former Heinrich Hertz Telescope), Large Millimeter Telescope (Mexico, built by MAN), James Clerk Maxwell Telescope (UK-built in Hawaii, now owned by the East Asian Observatory), Submillimeter Array (USA-Hawaii), and South Pole Telescope (USA-South Pole). Several more observatories have joined the trials since then.[128]

After two difficult years of processing, the spectacular image showing the shadow of the supermassive black hole in the center of M87 made it into the global press in 2019.

This latest episode, the culmination of a 30-year process of technical development and theoretical insight, highlighted the tensions underlying this new kind of global research conglomerate (see also Chapter 5), for there were separate 'unveiling' ceremonies, at the NSF in Washington and at the European Commission in Brussels. The Event Horizon Telescope, like other global-scale projects of the 21st century, demonstrates how research is inevitably becoming global yet simultaneously retaining a heterogenous internal structure, resulting from the tension between researchers, funding agencies, and the owners of research infrastructures; all of these, and their shifting alliances, exert novel forms of pressure that shape the outcomes of scientific projects as well as the distribution of credit among the national institutions and individual researchers.

2 Historical Change and Resilience in Times of Hardship at the End of the Century

The end of the Cold War was a turning point which shifted the relative position of power of the cosmic sciences globally. But in Germany, in particular, this was further magnified by the challenges brought about by German reunification. Even before the fall of the Berlin Wall, the German scientific community had recommended a reshuffle to revert the excessively dominant position of the Max Planck Society. The rapid and unexpected reunification of the country then tested these plans to the limit, intensifying regional demands and financial pressures on the Society. Yet, despite the succession crises at several institutes, closures were averted and instead there was an expansion eastward. Amid these financial and regional pressures, however, projects such as a planned gravitational wave interferometer had to be scaled down.

128 For a current list of participating observatories, see the EHTC website: Event Horizon Telescope. https://eventhorizontelescope.org/home. Last accessed 3/23/2021.

Denkschriften (*Memoranda*)

Looking back at the second half of the 20th century in German astrophysics, as in this present study, we see that there are very clear historical milestones. Between 1945 and 1957, scientists in the Max Planck Society addressed astrophysical questions mostly because these overlapped with nuclear questions and expertise. The launch of Sputnik changed this situation radically, and within a few years, the Society expanded to become the dominant force in astronomy, astrophysics, and space-based research in Germany. This upswing was so radical that even after the 1970s, when the German economy experienced considerable slowdown and many of the MPG's research fields faced stagnating financial support, the cosmic sciences continued to grow both in terms of their budget and, in particular, proportionally, compared to scientific research as a whole. In fact, the largest projects in this field were concluded through the 1970s and 1980s, shortly before the end of the Cold War.[129]

Bookending this period of great expansion between 1957 and the late 1980s were two landmark publications commissioned by the German astrophysical community: the memorandum on astronomy (*Denkschrift Astronomie*) of 1962, respectively of 1987.[130] These two documents mark the major changes in cosmic science eras in Germany, and played a key role in determining how these disciplines variously clustered (or not) at the height of the Cold War and beyond. They are also indicative of how, a generation later, the astronomical community had to adapt in order to survive, and are testimony to the rise of a new regime of scientific production dominated by international collaboration, and in which the fundamental physical sciences saw their prestige diminished, in contrast to other disciplines such as the life sciences, digital revolution, and materials sciences, which are much more closely linked to commercialization.

Within these two *Denkschriften* lie many keys to the peculiar success of the Max Planck Society in these fields. The 1962 Astronomy Memorandum

---

129 According to the Max Planck Society's *Haushaltspläne* (Budget Plans), the year with the highest proportional expenses in the cosmic research cluster was 1977, due to the costs of Calar Alto: either 14 percent or 27 percent of the total expenses of the MPG, depending on whether one includes in the count institutes that only partially conduct cosmic research, such as the 'nuclear' ones in Munich and Heidelberg. After the end of the large infrastructural projects, the proportion of the budget allocated to the purely astrophysical institutes of the MPG stabilized at around either 8 percent or 20 percent, depending on this same definition. For more details, see the Financial Appendix at the end of this book.

130 Hans-Heinrich Voigt et al.: *Denkschrift zur Lage der Astronomie*. Wiesbaden: Steiner 1962. Völk et al., *Denkschrift*, 1987.

had followed an earlier memorandum on space research,[131] and was clearly piggybacking on the success of Sputnik to push forward the most ambitious astronomical projects, calling for coordination on a national scale, and pointing out that West Germany's potential strengths could place it at the forefront of the construction of leading observatories and telescopes across the entire spectrum of wavelengths; this, in an era when building national capabilities, across the board, was a vital means for Germany to regain a competitive edge among the Western industrial nations, both in scientific research and new hi-tech fields. The Max Planck Society, through the participation of Biermann, Lüst, Bartels, Gentner, and others of its scientists with an interest in space research, made sure that it would be entrusted with control of the national infrastructures required to accomplish these goals.

In the early 1960s, it was mainly the 'nuclear age' physicists who opened the door to astronomy in the Society—despite their persistent disunity with other fields of research, and their regional rivalries—by absorbing the radio telescope projects in North Rhine-Westphalia and the optical telescope projects of Hans Elsässer, then at the Heidelberg observatory. Throughout the subsequent decades, as these astronomical observational projects matured and were joined by space-based astronomical observatories, a new autonomous circle of observational astronomers gained in influence to the point that, by the mid-1980s, it was the theoretical astrophysicists, the legatees of the space plasma tradition which had started it all, whose influence was dwindling. Parallels in the United States show how this growing influence of observational infrastructure focusing on multipurpose instruments was as deeply rooted in a Cold War logic as plasma science, since general purpose instruments are also those with the easiest dual-use potential.[132] Back in the 1970s, the inversion of scientific prestige and influence led plasma physicists to reinvent themselves as either general purpose theoretical astrophysicists, or in the case of space experimentalists, to move into planetary science, areas which at the time were largely led by questions from the plasma astrophysics that had first been studied in the first decade of the 'space age,' in the Earth's high atmosphere.[133]

---

131 Gotthard Gambke, Rudolf Kerscher, and Walter Kertz: *Denkschrift zur Lage der Weltraumforschung*. Wiesbaden: Franz Steiner Verlag 1961.

132 David Kaiser, and Benjamin Wilson: Calculating Times. Radar, Ballistic Missiles, and Einstein's Relativity. In: Naomi Oreskes, and John Krige (eds.): *Science and Technology in the Global Cold War*. Cambridge, MA: MIT Press 2014, 273–316.

133 The *Denkschrift Planetenforschung* of 1977 was the first time that this field featured independently of astrophysics or space science, but all the scientific questions were oriented towards space plasma questions applied to the near space, atmosphere, and magnetos-

The events of the late 1980s, however, pulled the brakes on this trend. Globally, the physical sciences, including astrophysics, began to see a serious decline, resulting from the end of the Cold War and the attrition of accelerator-based particle physics. In Germany, this crisis was compounded by the disappointing outcomes at some of the recent observatory projects, such as Calar Alto, in the visible wavelengths, and Mount Graham, in the submillimeter range. The German astronomical community outside of the Max Planck Society (in universities and state observatories) had become frustrated by the uneven playing field on which they were expected to train the younger generations of astronomers and astrophysicists, without having access to the resources and 'means of production' for significant research in the field.[134]

There was vocal criticism in the late 1980s—expressed in the *Denkschrift*, the Memorandum on German Astronomy of 1987—of how observational astronomy, now with a full range of observational capabilities in all wavelengths, was neglecting the scientific potential of interpreting the data produced, in an era when astronomy could no longer depend on the low-hanging fruit of sky surveys or expect to discover unexpected sources in new wavelengths. Increasingly, good astronomy was becoming more similar to fields such as experimental particle physics, in which the interaction of theory and experiment is an established tradition and instruments are designed from the outset for the purpose of answering sophisticated and highly specific questions. This was the case, for example, in the neutrino detection experiments that we follow in detail in Chapter 5. Even in observational astronomy, research activities were moving increasingly in this direction. With the appointment of Reinhard Genzel in 1984, for example, the Max Planck Society was already pursuing a question-oriented program to characterize the center of the Milky Way.

The memorandum had initially been called for by observational astronomers hoping to push forward the next generation of large infrastructures, and to redress the imbalance in access to these afforded the Max Planck Society and other German institutions respectively. The plans originally neglected theoretical astrophysics and new, non-electromagnetic areas of research, such as cosmic rays, neutrinos, or gravitational waves. Ultimately, however, given the changing circumstances and internal disagreements, several theoretical astrophysicists—including Rudolf Kippenhahn from the Insti-

---

phere environments surrounding other worlds in the solar system. Karl-Wolfgang Michel et al.: *Denkschrift Planetenforschung*. Edited by Deutsche Forschungsgemeinschaft. Boppard: Harald Boldt Verlag 1977.

134 AMPG, III. Abt. ZA 166 No. 57. Letter from Immo Appenzeller to Völk.

tute for Astrophysics in Garching, Peter Biermann (son of Ludwig Biermann), then at the Institute for Radio Astronomy in Bonn, Gregor Morfill in Garching, and Heinrich Völk in Heidelberg—became arbiters owing to their 'impartial' status as theoreticians.[135] Inspired by their global experience of a far closer interaction of theory and experiment, both in astrophysics and in experimental particle physics, they urged that the mission of the Max Planck Society and German university institutes be radically reinterpreted, so as to enable these actors to work together on projects that united theory, experiment, and instrumentation, while yet reducing their focus on large national observational infrastructure; for in the present era, [they argued], multinational observational infrastructures, the European Southern Observatory, for instance, had proven their ability to provide German researchers with the raw materials for scientific research, and perhaps in a more impartial way than even the Max Planck observatories.[136]

The theoretical astrophysicists also ensured that the new astronomy program would include the latest research fields, such as neutrinos, gravitational waves, and ground-based gamma ray astronomy,[137] which the majority of observational astronomers, following their colleagues in the United States, fiercely resisted.[138] In 1986–87, when the *Denkschrift* was drafted, it was "not yet the time for astroparticle physics."[139] Still, in being "so to say, fair to fields which were not the conventional fields of astronomy," the authors were helping raise awareness of the fact that early-Universe cosmology, high-energy processes in astrophysics, neutrinos, dark matter, dark energy, gravitational waves, and other novel topics were about to propel astronomy and astrophysics into the 21st century.

Presidential Commission for the Future of Astronomy and Astrophysics

German reunification in 1990 followed quickly on this *Denkschrift*, and brought with it a major new challenge for the Max Planck Society. We have already described how this was not a good moment for cosmic research in

---

135 Heinrich Völk: interview by Luisa Bonolis and Juan-Andres Leon, Heidelberg, October 9–10, 2017. DA GMPG, BC 601037.

136 To this day, however, the German representation in the governance bodies of ESO is still dominated by Max Planck representatives.

137 These three research fields constitute the case studies treated in Chapter 5.

138 For an example of this resistance in gravitational waves, see H. Völk's papers: Völk to Martin Harwit, 12.10.1987, AMPG, III. Abt., ZA 166, No. 59.

139 Heinrich Völk: interview by Luisa Bonolis and Juan-Andres Leon, Heidelberg, October 9–10, 2017. DA GMPG, BC 601037.

the Society, due to the ongoing difficulties with the observatories in Calar Alto and Mount Graham. Furthermore, as the *Denkschrift* indicated, the West German astronomical community was seeking to limit the Society's blatant monopoly on these fields, even before reunification came to loom on the horizon. And there were two other reasons that the early 1990s were a period of particular weakness: in the late 1980s, a corruption scandal emerged in the administration of the, until that moment, still monolithic Max Planck Institute for Physics and Astrophysics, of which the Institute for Extraterrestrial Physics was also part. The complex organization of this behemoth had created supervisory vacuums, thanks to which the administrator of the Institute for Astrophysics had managed to embezzle funds for six years, before being caught.[140] To make matters worse, this pretty much coincided with the retirement of Rudolf Kippenhahn, sole Director of the Institute for Astrophysics. The Max Planck Society was spurred to action by this affair, and determined to finally disentangle the products of the 'cell division' from which had ensued three fully independent Max Planck Institutes in the Munich area.[141] But while the Physics Institute in Munich-Freimann (then renamed Heisenberg Institute), and the Institute for Extraterrestrial Physics in Garching were each assured a very secure and independent future, this separation put the Institute for Astrophysics (MPA) at immediate risk. Further below, we describe the lengthy process of finding a successor to Kippenhahn and of ensuring that an independent MPA would survive into the 21st century. What is crucial at this point is that the process took place exactly in parallel with German reunification, and that what began as a discussion on the future of one institute eventually led to a complete reassessment of the role of cosmic research in the Max Planck Society. In 1991, a Presidential Commission (*Präsidentenkommission*), convened to advise the President Hans Zacher, was asked to make recommendations on the support of astronomy and astrophysics in the Max Planck Society, and to assess the medium to long-term development opportunities for institutes in this research section. The 'external constraints' were the limited opportunities for funding, both in terms of specific institutes and from a cross-institutional (we would say, cluster-wide) point of view, which

---

140 Dietrich Lemke, and Astronomische Gesellschaft (eds.): *Die Astronomische Gesellschaft 1863–2013. Bilder und Geschichten aus 150 Jahren*. Heidelberg: Astronomische Gesellschaft 2013, 82.

141 A proposal to abolish the subdivision of the institute into three sub-institutes was discussed on February 15, 1990, during a meeting of the institute's Board (AMPG, II. Abt., Rep. 62, No. 377, Fol. 10). The Senate approved the proposal for such division on March 8, 1991 (AMPG, II. Abt., Rep. 60, No. 127).

required that research tasks regarded as particularly important and promising, in view of their international standing and potential development, be clustered and given top priority. Taking into consideration the rising scale and costs of research instruments in astronomy and astrophysics, there was tacit agreement, when setting up this commission, that new topics could be taken up only if other, less promising directions, were abandoned; and that anything said by the commission would be interpreted in this light. The commission was once again chaired by Heinrich Völk, who had recently edited the *Denkschrift*, and now had to navigate an objective assessment under the proverbial dangling sword of Damocles.[142]

While German reunification was the main challenge at the time, the commission made sure to protect the interests of theoretical astrophysicists, including the newly independent Max Planck Institute for Astrophysics,[143] while also promoting the foundation of a new Max Planck Institute for Gravitational Physics (Albert Einstein Institute), in Potsdam.[144] At the same time,

---

142 For material related to the work of the Presidential Commission, see AMPG, II. Abt., Rep. 62, No. 17 and GVMPG, BC 218421, BC 218422, BC 108504. In addition, many of the members of the commission also contributed to Völk's edited volume, Heinrich J. Völk (ed.): *Facetten der Astronomie*. Leipzig: Barth 1993.

143 The three institutes had in fact always been autonomous from a scientific and organizational point of view, each having its own Scientific Advisory Board. On the other hand, Gerd Buschhorn, Director of the Institute for Physics, expressed the opinion that this division was moving in the opposite direction to trends at the time in big scientific centers such as CERN or Fermilab, where, conversely, high-energy physics and astrophysics sectors were merging. Instead of organizational division, [he argued], closer scientific links should be established (CPTS meeting minutes of 05.06.1991, 23.10.1991, 07.02.1992, 03.06.1992, 16.10.1992, AMPG, II. Abt., Rep. 62, No. 1823, 1824, 1825, 1826, 1827).

144 The founding of the Albert Einstein Institute with Jürgen Ehlers and Bernard F. Schutz as founding directors will be discussed in more detail in Chapter 5. It was intertwined with the problem of the future of the Max Planck Institute for Astrophysics and with the closure both of the Zentralinstitut für Astrophysik (ZIAP) of the former German Academy of Sciences of the German Democratic Republic (GDR), and the Hans-Jürgen Treder's Einstein-Laboratory for Theoretical Physics in Potsdam, as recommended in 1991 by the German Council of Science (*Wissenschaftsrat*) (GVMPG, BC 108504, Fol. 320–328). Material related to the Presidential Commission's involvement in the dissolution of ZIAP can also be found in BC 218422. A report on Astronomy in East Germany (Im Auftrag des Wissenschaftlichen Rates des ZIAP im Zusammenwirken mit allen astronomischen Einrichtungen auf dem Gebiet der ehemaligen DDR), edited by Hans-Erich Frölich (Babelsberg) and Siegfried Marx (Tautenburg), also contained a review of research activities at ZIAP (GVMPG, BC 218422, Fol. 353–384). See also Joachim Trümper: interview by Luisa Bonolis and Juan-Andres Leon, Munich, August, 7–8, 2017. DA GMPG, BC 601036. Aspects of the genesis of the Einstein Institute are discussed in Hubert Goenner: Some Remarks on the Early History of the Albert Einstein Institute. arXiv:1612.01338 [physics.hist-ph] 2016. doi:10.48550/arXiv.1612.01338.

the commission put an end once and for all to the Society's ambitions to operate large scientific infrastructures such as astronomical observatories. In the future, Max Planck Institutes were to be medium-sized centers of excellence pursuing question-oriented research programs, largely within international collaborations. Here, both universities and the Max Planck Institutes would participate on an equal footing, within the framework of the *Verbundforschung* (science and industry research partnership) called for both in the *Denkschrift* itself and by the Council of (West) German Observatories (*Rat Westdeutscher Sternwarten*; after 1990, *Rat Deutscher Sternwarten*), then led by Gregor Morfill, Völk's close collaborator and Director of the Institute for Extraterrestrial Physics.[145]

This represented a major shift for the cosmic sciences, which has since then initiated no major infrastructural project within the MPG. Even in the three scientific fields in which the Max Planck Society became a world leader, (neutrino detection, gravitational waves, and ground-based gamma ray astronomy, as detailed in Chapter 5), its ventures were mostly referred to as 'experiments,' not as 'observatories.' And the expectation was that these 'experiments,' once their missions were fulfilled, would be dismantled or given away to their international partners.

Financially speaking, the relative size of the core of astronomy and astrophysics in the Max Planck Society remained constant in the 1990s, 'locking in' their expenditures as a proportion of the total MPG budget. But this apparent stability actually represented a radical internal transformation, inasmuch as space-based programs continued to grow, theoretical astrophysics remained constant, and the observatory infrastructure declined in importance.[146]

Meanwhile, the crossover from particle physics to astrophysics continued in the 'nuclear' institutes in Heidelberg and Munich. During the 1990s, both venerable traditions of cosmochemistry and space plasma astrophysics came to an end in Heidelberg, Mainz, and Garching. Only their indirect influence remains, in the form of an instrumental legacy now in the service of other scientific activities, such as interplanetary probes and atmospheric environmental research in the case of cosmochemistry, and of cosmic rays and gamma

---

145 The BMBF project funding was established in 1989 as one of the funding schemes proposed by the *Denkschrift*, the Memorandum on German Astronomy, of 1987. German Astronomical Society: BMBF Project Funding Review Board of the BMBF (Bundesministerium Für Bildung und Forschung) Project Funding Scheme Astrophysics and Astroparticle Physics. *German Astronomical Society*. http://www.astronomische-gesellschaft.org/en/rds/bodies/bmbf?set_language=en. Last accessed 4/10/2018.

146 For details of the financial evolution of individual institutes and the cluster, see the Financial Appendix at the end of this book.

ray astronomy in the case of plasma physics. In fact, what is now known as planetary science, that is, research conducted with space probes into the other planets and smaller bodies of the solar system, is still led scientifically by questions and methodologies related to plasma physics and cosmochemistry. The conceptual frameworks for these were developed in the early decades of spaceflight in places like Garching, Lindau, Heidelberg, and Mainz.[147] This extends to experimental nuclear fusion in which, after half a century, the Max Planck Society's gamble of committing long term to Princeton's stellarator design continues to pay off for the Institute for Plasma Physics.

German Reunification

In Germany, the end of the Cold War and the reunification radically altered the scientific research landscape, catalyzing a change in the organization of research that had been under discussion earlier in the 1980s, notably the need to more clearly separate scientific research activities from their large-scale infrastructures.

As was described in detail above, the major changes in the cosmic sciences cluster of the Max Planck Society had been set in motion in the mid-1980s and expressed in the 1987 *Denkschrift*. The proposals set out in this memorandum were considered by the MPG's Presidential Commission, chaired by Heinrich Völk, which reached its conclusion on their implementation in 1993. Yet the commission had been convened now to respond, not only to the ideas put forward several years previously, but also to the growing political pressure on the Max Planck Society to help incorporate East German scientific research into the unified German research system.[148] All research fields covered by the Max Planck Society were under immense pressure to take over institutes from the 'new states' of unified Germany or to establish new institutes there; and there were even threats to close facilities in the West and 'banish' their staff eastwards.[149]

The Society's position at the very moment of reunification was to flex its muscles and defend to the max its privilege, excellence mandate, and scientific independence as the most powerful and prestigious body on the West German research landscape; (which, hardly surprisingly, was criticized as arrogant by many in the East and the West). Rather than absorb institutes and researchers from the GDR, the Society was determined to expand its footprint into the new

147 Michel et al., *Denkschrift Planetenforschung*, 1977.

148 See GVMPG, BC 108504, Fol. 164–172.

149 Mitchell G. Ash: *Die Max-Planck-Gesellschaft im Kontext der Vereinigung 1989–1995*. Berlin: GMPG-Preprint 2020.

federal states on its own terms, by founding entirely new institutes. The pressure to absorb GDR entities was to be resisted. The principle would be to find reasonable solutions for existing research sites, not within the MPG, but by exerting political influence. Aid for contemporary GDR researchers was preferably to take the form of working groups under the tutelage of individual Max Planck Institutes for a five-year period. Such groups would benefit promising young researchers and be based at East German universities, in an attempt to re-instill in them a research profile that under the GDR regime had largely been lost to the Academy of Sciences. The latter organization would be dissolved and the fate of its many institutes individually assessed by committees set up by the now, pan-German *Wissenschaftsrat* (Scientific Council).[150]

Several clusters of the Max Planck Society quickly mobilized to use this external pressure as a pretext for significant expansions that had long been on the horizon. This was the case, for example, of research institutes in the material sciences, in environmental research, and in the social sciences and humanities. What happened in all of these fields, despite the plans nominally calling for the takeover of existing East German institutes, was that West German scientific communities relocated to new settings in the East, at best incorporating some staff from East German institutes as a temporary measure, while awaiting generational replacement. Ultimately, only one Max Planck Institute became the direct successor to an existing GDR facility (in Halle), and even then, its senior staff came mainly from West Germany. The end effect was a major expansion of existing research clusters.

The cosmic sciences did not immediately benefit from this expansion, as German reunification coincided with their major identity crisis, mentioned above. In addition to this challenge, many of their institutes in the early 1990s faced the impending retirement of the longtime director, not only Kippenhahn in Garching, but also Elsässer in Heidelberg, and Mezger in Bonn.[151] Institutes

---

150 See, for example, Hans F. Zacher: Die Max-Planck-Gesellschaft im Prozeß der deutschen Einigung. In: Max-Planck-Gesellschaft zur Förderung der Wissenschaften e.V. (ed.): *Max-Planck-Gesellschaft Jahrbuch 1991*. Göttingen: Vandenhoeck & Ruprecht 1991, 11–23.

151 A scheme of the retirement of directors of various institutes from 1991 to 2017 was enclosed in President Zacher's invitation letter of April 29, 1991, to candidate members of the committee, along with a list of questions to be discussed by the commission and the aforementioned report on Astronomy in GDR (GVMPG, BC 108504, Fol. 282–319). The first meeting took place in Heidelberg on July 12, 1991. Both Max Planck and foreign scientists were members of the committee (see list in GVMPG, BC 218422, Fol. 258–259). Some active Scientific Members working in the field of Astronomy/Astrophysics were requested by the Presidential Commission to discuss the list of topics submitted by President Zacher (Fol. 12–57).

facing their most serious existential crisis ever were about to be left with no one at the helm. Moreover, the German government was pressuring the MPG to take over East German institutes, such as the Institute for Cosmos Research in Adlershof, and the aforementioned ZIAP, the Central Institute for Astrophysics in Potsdam. As in other fields, the Max Planck Society maneuvered, here too, to safeguard its reputation and elite status as a hub of 'fundamental research,' arguing against its incorporation of these facilities, which instead ended up elsewhere, integrated, in the case of ZIAP, into what came to be known as the 'Blue List,' or even, in the case of Adlershof, into the German Aerospace Center (DLR).[152]

Unlike the Max Planck Society, DESY, Germany's largest accelerator center—the most important for particle physics—did absorb its major East German counterpart, the Institute for Particle Physics in Zeuthen (now DESY Zeuthen).[153] Since then, the Zeuthen site has expanded considerably to become an international center for astroparticle physics, focusing on gamma-ray and neutrino astronomy, as well as on theoretical astroparticle physics. DESY is Europe's biggest partner in the neutrino telescope IceCube, located in the Antarctic.

Instead of taking over existing facilities, the cosmic sciences expanded eastward primarily by establishing a new Max Planck Institute in Golm, near Potsdam, so safeguarding the city's traditional scientific profile while remaining entirely separate from what was to become the Leibniz Institute for Astrophysics.[154] The new MPI for Gravitational Physics (or Albert Einstein Institute) was initially an expansion of Jürgen Ehler's research group at the Institute for

152 Goenner, Some Remarks, 2016. Hubert Goenner, and F.W. Hehl: Zur Gründung des Albert-Einstein-Instituts für Gravitationsphysik. *Physikalische Blätter* 47/10 (1991), 936–936. doi:10.1002/phbl.19910471015. For Potsdam and Adlershof, see Joachim Trümper: interview by Luisa Bonolis and Juan-Andres Leon, Munich, August, 7–8, 2017. DA GMPG, BC 601036. See also Keller, Horst-Uwe: Interview by Helmuth Trischler and Matthias Knopp, June 10, 2010. Transcript, Historical Archives of the European Union. Oral History of Europe in Space Collection, https://archives.eui.eu/en/oral_history/INT078. Last accessed 12/4/2020. Keller explains how the DLR, successor to the East German Institute for Cosmic Research (*Institut für Kosmosforschung*) ended up competing directly with the planetary science research conducted at the Max Planck Institute for Aeronomy.

153 Ulrike Behrens et al.: Zeitreise. Vom Institut X zum DESY—eine deutsche Geschichte. Edited by Deutsches Elektronen-Synchrotron (DESY). DESY 2012. https://pr.desy.de/sites/sites_desygroups/sites_extern/site_pr/content/e104098/e104108/Broschuere_DESY_Zeuthen_Zeitreise_Web_ger.pdf. Last accessed 8/15/2020.

154 Goenner, and Hehl, Albert-Einstein-Instituts für Gravitationsphysik, 1991, 936–936. Goenner, Some Remarks, 2016. Joachim Trümper: interview by Luisa Bonolis and Juan-Andres Leon, Munich, August, 7–8, 2017. DA GMPG, BC 601036.

Astrophysics in Munich into a full-fledged institute. Then, as we will see in the next chapter, once gravitational wave research gained momentum in the 1990s, its mission broadened to also include the theoretical work necessary for the multinational experimental endeavor in gravitational waves, which finally led to the merger (completed in 2002) with what had originally been the Hannover branch of the Max Planck Institute of Quantum Optics. Since this branch institute originated in Heinz Billing's gravitational waves research group, this was in a sense a 'family reunification': both were successors to a scientific tradition originating with Werner Heisenberg and Ludwig Biermann, whose migration had begun in Berlin, passed through Göttingen, and ended in Munich. Now a younger generation had taken the reverse route, at the end of the century, moving northward to Lower Saxony, and eastward into the Berlin area.

Another offshoot of this scientific tradition was secured with the establishment of a new branch of the Institute for Plasma Physics (IPP) in Greifswald, in the far northeast of the former German Democratic Republic, previously the site of an East German nuclear power plant. Given the status of the IPP as a so-called *Grossforschung* research facility funded to over 90 percent by the German federal government, resistance to 'banishment' in this case was futile; and the new IPP Greifswald is now one of the most important employment and technology drivers of the state of Mecklenburg-Western Pomerania.[155] In fact, this move eastwards, although initially protested, ended up becoming one of the major scientific flagships of the Max Planck Society, post-reunification: the technological research conducted in Greifswald is now held to rank among the most promising experiments on the long path to a functional thermonuclear reactor. The latest of these reactors, the famed Wendelstein 7-X, a supercomputer-designed stellarator which went into operation in 2016, can be traced directly back to the scientific tradition of Biermann, Schlüter, Lüst, and Billing, who excelled in the theoretical and computational understanding of plasma physics. When they visited Lyman Spitzer in Princeton in the early 1950s, they began work on the stellarator design with their 'Wendelstein' series, named in honor of Spitzer's secret Project Matterhorn (see Chapter 1). The IPP stubbornly pursued this line of research while the rest of the world was lured in the 1970s toward the Soviet tokamak design (as the IPP was too, in fact; and it continues to pursue it in Garching). While extremely

155 Angela Merkel: Rede von Bundeskanzlerin Dr. Angela Merkel beim Besuch des Max-Planck-Instituts für Plasmaphysik (IPP) am 3. Februar 2016 in Greifswald. *Bulletin der Bundesregierung 14–1*, 2/5/2016. https://www.bundesregierung.de/Content/DE/Bulletin/2016/02/14-1-bk-besuch-max-planck-institut.html. Last accessed 4/12/2018.

complex in shape, compared to tokamaks, the stellarator has grown in the 21st century into a promising alternative approach to thermonuclear fusion. For decades, this design had seemed justifiable only as a sort of 'backup' on the spectrum of worldwide experimental programs, one with which to avoid the risk of committing the venture to a single-track reactor development with tokamaks. This, all at a stage where there is still no energy-producing fusion reactor on the horizon; rather, they are still figuring out all the details concerning how a burning plasma behaves, and developing the corresponding technologies based on this knowledge. And in the early 21st century, the stellarator approach has gained visibility in contrast to the delays and problems of the upcoming largest tokamak ITER (International Thermonuclear Experimental Reactor), which is expected to go into operation in 2025. With Wendelstein 7-X, currently the world's largest stellarator,[156] the Max Planck Society is the undisputed world leader on this alternative path to gaining a better perspective on the still unknown challenges of reaching the 'holy grail' of thermonuclear fusion, the *break-even* point, that is, the critical temperature and density where fusion reactions produce more power than is used to create them.[157]

Finally, in cosmic research, much aid during the first years after reunification took the form of working groups established in East German universities for a five-year period under the tutelage of individual Max Planck Institutes: in Potsdam, a group on nonlinear dynamics run by the MPE; in Jena, two groups, on dust clouds and star formation (MPIfR), and gravitational theory (MPA). The MPE additionally set up a temporary external site in Berlin to advise the plasma research activities of the Institute for Cosmic Research in Adlershof; and the Max Planck Institute for Astronomy participated in the reorganization and modernization of the Tautenburg Observatory (with the largest telescope in unified Germany), as well as the older Sonneberg Observatory, which had both been separated from the GDR's ZIAP.[158]

A look at the financial evolution of the cosmic sciences cluster in the Max Planck Society shows that, in contrast to many other research fields, the move eastward was financially 'neutral' and, from the mid-1980s to the beginning of the 21st century, the amount of money allocated to the cosmic sciences as a proportion of the entire MPG budget remained remarkably stable.[159] It was

156 G. Grieger et al.: Modular Stellarator Reactors and Plans for Wendelstein 7-X. *Fusion Technology* 21/3P2B (1992), 1767–1778. doi:10.13182/FST92-A29977.

157 Daniel Clery: The Bizarre Reactor That Might Save Nuclear Fusion. *Science*, 10/21/2015. doi:10.1126/science.aad4746.

158 Lemke, *Himmel über Heidelberg*, 2011, Vol. 21. Lemke, *Himmel über Heidelberg*, 2011, Vol. 21. Lemke, and Astronomische Gesellschaft, *Die Astronomische Gesellschaft*, 2013.

159 See the Financial Appendix at the end of this book.

only because the Max Planck Society experienced renewed growth globally, after reunification, that growth was sustained in the cosmic sciences. As this period saw eastwards expansion and the foundation of new facilities, the size of the MPG in West Germany remained largely unchanged. There was even considerable pressure to reduce its footprint in the West: at different points in the 1990s, there were deliberations about closing down several institutes after their respective director's retirement (more details at the end of the chapter), such as the Institute for Astrophysics in Garching,[160] and even the Institute for Astronomy in Heidelberg.[161] In the case of all these institutes, however, their temporary weakness was counterbalanced by the overall strength of the cluster in general, and the institutes in question were protected by scientists from more established 'astro' institutes, such as Extraterrestrial Physics and Radio Astronomy, as well as by astrophysicist allies in the 'nuclear' institutes in Munich (MPP), Garching (IPP), and Heidelberg (MPIK).

However, the Institute for Aeronomy in Lindau was a particular case. As we saw in the previous chapter, this institute had already been seriously threatened with closure in the 1970s, but was then saved by the appointment of Ian Axford. With the impending retirement of Axford and the other directors appointed around the same time, the institute's fate was once again up for debate, and further complicated by the then particularly difficult economic demands on the Max Planck Society and the state of Lower Saxony. President Hubert Markl at one point even announced the closure of the institute. This was, however, averted thanks to an international outcry in the scientific community, and to the state of Lower Saxony's wish to maintain a strong scientific presence in the Göttingen area. Yet over the next decade, the institute was scaled down and its name changed to Solar System Research, and eventually it was moved from its historic Cold War-era location in Lindau and integrated into the Göttingen University campus.[162] Finally, in the framework of strategic moves to strengthen the scientific and technological landscape of the former GDR, the Göttingen instrument-building activities for solar system research

160 Alison Abbott: German Astronomers Fight Rumoured Closing of Institute. *Nature* 358 (1992), 267–267. doi:10.1038/358267a0. Alison Abbott: Max Planck Institute Gets New Life. *Nature* 360 (1992), 6. doi:10.1038/360006b0. See also G. Jerke: Welche Zukunft hat das MPI für Astrophysik? *Physikalische Blätter* 48/7–8 (1992), 514, 605. doi:10.1002/phbl.19920480704.

161 Presentation by Dietrich Lemke, GMPG Astrophysics Workshop, September 2016. See also Lemke, *Himmel über Heidelberg*, 2011, Vol. 21.

162 Toni Feder: Max Planck Institute for Aeronomy Averts Closure. *Physics Today* 50/5 (1997), 51–52. doi:10.1063/1.881823.

probes were discontinued and consolidated instead at the DLR institutes in Berlin-Adlershof.[163]

In the end, no cosmic research institute was closed down. Instead, the new directorships in these institutes, (also including the succession of Peter Mezger in Bonn), were further aligned with the new globalized era, with a research agenda oriented to international collaboration. The pressure of German reunification was ultimately mobilized by the more powerful figures in the cosmic sciences cluster, to take the agenda in the direction set by the 1987 *Denkschrift*. The final outcome of this process, dating from 1998, included the decisions to ultimately give away the Calar Alto and Mount Graham telescopes.[164]

The bitter disappointment surrounding these two failed observatories also hampered the aspirations voiced by the Garching group since the mid-1980s, to compete directly with the Americans by building a full-sized gravitational wave observatory, with a 3 km tunnel facility under the Harz mountains.[165] As described in Chapter 5, the Max Planck Society did still play a key role in the most significant astrophysical development of the 21st century by focusing on building a smaller, pilot-scale facility near Hannover (GEO600). But a Nobel Prize was likely lost because the Society did not host the large-scale detectors.[166]

---

163 Horst-Uwe Keller: interview by Helmuth Trischler and Matthias Knopp, June 10, 2010. Transcript, Oral History of Europe in Space Collection, https://archives.eui.eu/en/oral_history/INT078. Last accessed 1/26/2022.

164 Not all large infrastructures were abandoned. As part of the strategy to divest from Calar Alto—a long process, expected to end in 2018—emphasis was placed instead on increased participation in the Large Binocular Telescope (LBT), in collaborative research with other German institutions still seeking a substitute for the LBT, which had been sacrificed to German reunification. See, for example, Lemke, *Himmel über Heidelberg*, 2011, Vol. 21. This was despite the LBT's location on Mount Graham, subject to most of the same difficulties that had already doomed Peter Mezger's brilliantly designed telescope. Karl Menten, the incoming director who decided to abandon this telescope, instead built the pilot 'experiment' APEX as a replacement, together with ESO in Chile (Karl Menten: interview by Luisa Bonolis and Juan-Andres Leon, Bonn, February 5–8, 2018, DA GMPG, BC 601052). Back in 1998, Menten was the main dissenter to speak out against an enlarged presence in the LBT project which, twenty years later and after many delays, is again considered to be a rather disappointing infrastructure. See minutes of the meeting of the Executive Committee [*Verwaltungsrat*] 07.10.1998 (AMPG, II. Abt., Rep. 61, No. 185, Folder 6). For the recent problems with the LBT, see Witze, Teething Troubles, 2013, 133.

165 Gerd Leuchs et al.: *Proposal for the Construction of a Large Laser Interferometer for the Measurement of Gravitational Waves*. MPQ 131. Garching: Max-Planck-Institut für Quantenoptik 1987.

166 In addition to the next chapter, see also our contribution: Luisa Bonolis, and Juan-Andres Leon: Gravitational-Wave Research as an Emerging Field in the Max Planck Society:

Although focusing on a pilot facility while renouncing a very costly, uncertain infrastructural project was understandable given the historical context, German scientists now regret that, in the 1990s, under the financial constraints of reunification and in light of the astronomical observatory failures, the gravitational wave 'baby' ended up being thrown out with the bathwater.

3 Into the 21st Century: A New Role for the Max Planck Institute for Astrophysics

This section centers on the most difficult rescue of the decade, namely that of the Institute for Astrophysics (MPA) dating back to Ludwig Biermann's arrival in Göttingen in the late 1940s, and by then the most veteran among the entire cosmic research institute cluster. While already facing doubts about the contemporary significance of 'theoretical astrophysics,' the institute now had to confront the rising predominance of observational astronomy in all wavelength domains within the ensemble of institutes conducting larger-scale cosmic research in the Society. German reunification increased pressure to relocate or close down Max Planck Institutes, while a local institutional crisis and independent institutes in the Garching area made the small theoretical institute particularly vulnerable. The resulting institutional debates reached beyond this particular institute, however, to question how cosmic research was to be conducted in the Max Planck Society overall, and specifically to ask what the function of a theoretical institute within this constellation should be in the 21st century. The solutions devised to save the institute further strengthened both the Society's 'clustering' approach to cosmic research and its international connections, and helped propel appointments and reforms at other institutes in crisis, leading them into the new century.

Continuity vs Adaptation

Given all the developments outlined in this chapter, the most notable divide by the end of the 1980s was perhaps not between the various observational astronomy institutes, but rather between them and the astrophysical institutes with a more theoretical angle. Since Heinrich Völk's arrival at the Institute for Nuclear Physics in Heidelberg in the mid-1970s, the cosmochemistry section

The Long Roots of GEO600 and of the Albert Einstein Institute. In: Alexander S. Blum, Roberto Lalli, and Jürgen Renn (eds.): *The Renaissance of General Relativity in Context.* Basel: Birkhäuser 2020, 285–361. doi:10.1007/978-3-030-50754-1_9.

had expanded to accommodate a growing emphasis on theoretical interpretation, as indeed had been the intention behind his appointment there. Then there was the original Institute for Astrophysics founded by Biermann and from which the Institute for Plasma Physics and the Institute for Extraterrestrial Physics had emerged in the 1960s. This comparably smaller institute was directed by Biermann until the mid-1970s, and subsequently by Rudolf Kippenhahn, as mentioned in Chapter 3. After studying mathematics at the University of Erlangen-Nuremberg, Kippenhahn had become an assistant at the Remeis Astronomical Institute in 1951, and came into contact with Biermann in 1956, whereupon he was invited to work at the Institute for Astrophysics while Lüst was in the United States. Kippenhahn began work there on the structure and evolution of stars, using the G2 machine for star model simulations and also devising, with Stefan Temesváry, a computer program called the 'star model construction kit' (*Sternmodellbaukasten*), which was a predecessor of modern stellar evolution programs.[167] But the real turning point came in 1961–62, when Kippenhahn was able to spend 12 months in the USA, first with Martin Schwarzschild at Princeton, and then at the California Institute of Technology, where he discovered the new and much more powerful industrially produced computers and the methods developed by Louis Henyey for automatic solution of the equations of stellar evolution.[168] Back at

---

167 At the time, Fred Hoyle and Martin Schwarzschild, (who had emigrated from Germany in 1934), had for the first time used computers in Princeton to do massive simulation work on the evolution of stars, confirming that when a star exhausts its hydrogen content, the helium core gradually grows, eventually transforming the star into a red giant. F. Hoyle, and M. Schwarzschild: On the Evolution of Type II Stars. *The Astrophysical Journal Supplement* 2 (1955), 1–40. doi:10.1086/190015. As a follow-up to this work, Kippenhahn had the task, together with Temesváry, of repeating Hoyle and Schwarzschild's calculations with improved methods. In a contribution written on the occasion of Biermann's 75th birthday, Kippenhahn explained that "The two had calculated sets of solutions for the different areas of the star and then put together a complete solution by hand. At that time, we wanted to treat the star on the computing machine G2 developed at the institute rather as a whole." However, their project was too ambitious, given the limited memory capacity of computers at the time. The typescript, which can be found in Lüst's papers (AMPG, III. Abt., Rep. 145, No. 1221), was to be published in the volume Christoph Schneider (ed.): *Forschung in der Bundesrepublik Deutschland. Beispiele, Kritik, Vorschläge*. Weinheim: Chemie 1983. On this work, see also Trefftz, Eleonore: Obituary—Temesvary, S. *Quarterly Journal of the Royal Astronomical Society* 27/1 (1986), 129–130. http://adsabs.harvard.edu/abs/1986QJRAS..27..129T. Last accessed 1/13/2020.

168 The codes developed by Henyey were tested on computers used for research on thermonuclear weapons (Martin Schwarzschild: interview by William Aspray, November 18, 1986, Transcript, Charles Babbage Institute, https://conservancy.umn.edu/handle/11299/107629. Last accessed 12/4/2020). Louis G. Henyey: The Evolution of Stars Near the Main

the MPA, he started studying the structure and evolution of stars, together with Emmi Hofmeister and Alfred Weigert, and applied an upgrade of the Henyey methods he had learned in the USA to the computing machines developed by Billing. After becoming professor at the University of Göttingen, Kippenhahn continued working with Biermann and, between 1964 and 1968, wrote with his collaborators a series of articles on the structure and evolution of very compact stars.[169] All this contributed to establishing the tradition of hydrody-

---

Sequence. *Publications of the Astronomical Society of the Pacific* 68/405 (1956), 503–504. doi:10.1086/126986. Louis G. Henyey et al.: A Method for Automatic Computation of Stellar Evolution. *Astrophysical Journal* 129 (1959), 628–636. doi:10.1086/146661. Peter H. Bodenheimer: Louis George Henyey. February 3, 1910-February 18, 1970. In: National Academy of Sciences (ed.): *Biographical Memoirs*. Washington, DC: The National Academies Press 1995, 169–189. See also letters exchanged between Biermann and Louis G. Henyey during 1959 and 1960 about Stefan Temesváry's sojourn at the Berkeley Astronomical Department, where Henyey was leading a research group in the field of stellar evolution (AMPG, III. Abt., ZA 1, No. 41). See Rudolf Kippenhahn: interview by Owen Gingerich, June 18, 1978. Transcript, AIP, https://www.aip.org/history-programs/niels-bohr-library/oral-histories/5091. Last accessed 11/4/2020.

169 Emmi Hofmeister, Rudolf Kippenhahn, and Alfred Weigert: Sternentwicklung I. Ein Programm zur Lösung der zeitabhängigen Aufbaugleichungen. *Zeitschrift für Astrophysik* 59 (1964), 215–241. https://ui.adsabs.harvard.edu/#abs/1964ZA.....59..215H. Last accessed 10/31/2018. Emmi Hofmeister, Rudolf Kippenhahn, and Alfred Weigert: Sternentwicklung II. Die Wasserstoff-brennende Phase eines Sterns von 7.0 Sonnenmassen. *Zeitschrift fur Astrophysik* 59 (1964), 242–260. https://ui.adsabs.harvard.edu/#abs/1964ZA.....59..242H. Last accessed 9/19/2018. Emmi Hofmeister, Rudolf Kippenhahn, and Alfred Weigert: Sternentwicklung III. Die Helium-brennende Phase und die Cepheidenstadien eines Sterns von 7.0 Sonnenmassen. Mit 6 Textabbildungen. *Zeitschrift fur Astrophysik* 60 (1964), 57–75. https://ui.adsabs.harvard.edu/#abs/1964ZA.....60...57H. Last accessed 9/19/2018. Rudolf Kippenhahn, H. C. Thomas, and A. Weigert: Sternentwicklung IV. Zentrales Wasserstoff- und Heliumbrennen bei einem Stern von 5 Sonnenmassen. Mit 4 Textabbildungen. *Zeitschrift fur Astrophysik* 61 (1965), 241. https://ui.adsabs.harvard.edu/#abs/1965ZA.....61..241K. Last accessed 9/19/2018. Rudolf Kippenhahn, H. C. Thomas, and A. Weigert: Sternentwicklung V. Der Kohlenstoff-Flash bei einem Stern von 5 Sonnenmassen. *Zeitschrift fur Astrophysik* 64 (1966), 373–394. https://ui.adsabs.harvard.edu/#abs/1966ZA.....64..373K. Last accessed 9/20/2018. A. Weigert: Sternentwicklung VI. Entwicklung mit Neutrinoverlusten und thermische Pulse der Helium-Schalenquelle bei einem Stern von 5 Sonnenmassen. *Zeitschrift fur Astrophysik* 64 (1966), 395–425. https://ui.adsabs.harvard.edu/#abs/1966ZA.....64..395W. Last accessed 9/20/2018. Emmi Hofmeister: Sternentwicklung VII. Zur Entwicklung von Sternen mit 5 und 9 Sonnenmassen. *Zeitschrift fur Astrophysik* 65 (1967), 164–184. https://ui.adsabs.harvard.edu/#abs/1967ZA.....65..164H. Last accessed 9/20/2018. H.-C. Thomas: Sternentwicklung VIII. Der Helium-Flash bei einem Stern von 1. 3 Sonnenmassen. *Zeitschrift für Astrophysik* 67 (1967), 420–455. https://ui.adsabs.harvard.edu/#abs/1967ZA.....67..420T. Last accessed 10/31/2018. K. von Sengbusch: Sternentwicklung IX. Die erste hydrostatische Kontrakionsphase für einen Stern von 1 solar mass. *Zeitschrift für Astrophysik* 69 (1968), 79–111. https://ui.adsabs.harvard.edu/#abs/1968ZA.....69...79V. Last accessed 9/20/2018.

namic numerical modeling, which later led to a long-term program of research into the nature of accretion disks and accretion flows of plasma, dust, and particles—and especially of the strong and directed outflows of gas known as jets—around astrophysical objects such as stars or black holes. It also set the stage for the growing role of relativistic astrophysics at the Institute for Astrophysics and paved the way for the appointment in 1971 of the eminent relativist Jürgen Ehlers, so laying the foundations for a research group on general relativity and gravitation, for the Albert Einstein Institute, built much later in Potsdam/Golm, and for the formation there of one of the world's outstanding general relativity groups (further details in Chapter 5).

After Wolfgang Hillebrandt arrived at the Institute for Astrophysics in 1978, to lead a new group researching nuclear astrophysics, Kippenhahn's pioneering computer codes were used to confront models with recent supernova observations.[170] In particular, the group used such codes at the time of the explosion of Supernova 1987A, a unique opportunity to confront models of core-collapse and Type II supernova observations with neutrino emission from a forming neutron star or further formation of a black hole.[171] Supernova research then became a leading field at the Institute for Astrophysics, carried out also through numerical simulation and data modeling, which were playing an increasingly crucial role in astrophysics, as tools spanning the two traditional poles of theory and observation. But rapid advances in the computer industry made Billing's series of homemade machines obsolete from the

170 On the use of computers in astronomy, see Rudolf Kippenhahn: Als die Computer die Astronomie eroberten. In: Siegfried Röser (ed.): *Cosmic Matter*. Weinheim: Wiley-VCH 2008, 1–14.

171 See, for example, Wolfgang Hillebrandt: Supernova 1987A. Core Collapse Models and Neutrinos. *ESO Workshop on the SN 1987A, Garching, Federal Republic of Germany, July 6–8, 1987, Proceedings (A88–35301 14–90)*. Garching (Germany): Federal Republic of Germany, European Southern Observatory 1987, 301–312. http://adsabs.harvard.edu/abs/1987ESOC...26..301H. Last accessed 10/31/2018. When working on models for thermonuclear combustion which could be used directly to interpret observational data, Hillebrandt set out, together with Ewald Müller, to develop models for the core-collapse supernovae, studying the influence of rotation and magnetic fields, and calculating the gravitational wave signal and the production of heavy elements in the explosion. Ewald Müller: Computational Problems in Supernova Simulations. *Computer Physics Communications* 44/3 (1987), 271–277. doi:10.1016/0010-4655(87)90082-8. Supernova modeling was further advanced by Thomas Janka's studies on neutrino losses from a forming neutron star and how their emission is ultimately responsible for the explosion of objects like the Supernova 1987A. H.-T. Janka, and W. Hillebrandt: Neutrino Emission from Type II Supernovae. An Analysis of the Spectra. *Astronomy and Astrophysics* 224/1–2 (1989), 49–56. http://adsabs.harvard.edu/abs/1989A%26A...224...49J. Last accessed 10/31/2018.

1960s onwards,[172] and Hillebrandt's group focused in later years on gravitational wave experiments instead, as will be detailed in the last chapter of this book.

The appointment of Jürgen Ehlers to lead a theoretical research group on relativistic astrophysics and cosmology led overall to a more in-depth focus on theory at the Institute for Astrophysics.[173] Up to that time, during the 30 years under Biermann, the main objective had been to carry out fundamental research in theoretical astrophysics, although several individual researchers or groups had been involved also in observational programs. This trend continued after Biermann's retirement in 1975, under Kippenhahn's leadership. The strength of the institute—as emphasized by Kippenhahn himself, immediately before his retirement in 1991—lay "in the development of new methods to investigate physical processes in astronomical objects."[174] This concerned hydrodynamics and plasma physics, in particular, yet also the supplementary fields of general relativity, nuclear physics, radiative transfer, and computer simulation. It was largely to these activities that the institute owed its international standing. The research projects comprised work on fundamental physical processes as well as on mathematical methods to obtain solutions. Consequently, it was easier

> to understand the various activities of groups in the context of their favorite methods rather than with respect to the astronomical objects they [were] interested in, because the latter may change quite often.

Typical examples, in this regard, were the transfer of methodological expertise between topics that may have differed superficially, such as research on the magnetohydrodynamics of sunspots on the properties of accretion disks around compact astrophysical objects, for example, neutron stars or black holes. In turn, others might have changed their scientific interests from accretion disks to the formation of large-scale structures in the early Universe. Their

---

172 Heinz Billing: Schnelle Rechenmaschinenspeicher und ihre Geschwindigkeits- und Kapazitätsgrenzen. In: Generalverwaltung der Max-Planck Gesellschaft zur Förderung der Wissenschaften e.V. (ed.): *Jahrbuch 1962 der Max-Planck-Gesellschaft zur Förderung der Wissenschaften e.V.* Göttingen 1962, 52–79.

173 Ehlers is discussed in more detail in Chapter 5. The petition to bring him to the institute appears in a letter from Heisenberg and Biermann to President Adolf Butenandt, 31.10.1969, AMPG, III. Abt., Rep. 93, No. 1667.

174 Rudolf Kippenhahn, Memorandum for the future development of the Institute (meetings minutes of the Presidential Commission "Institute for Astrophysics, and astronomy research," 1988–93, AMPG, II. Abt., Rep. 62, No. 17, Fol. 297–305).

expertise in the underlying theoretical framework made such changes of topic possible and easy. This typical style of operations was reflected also in the structure of the institute: research groups formed and dissolved in response to specific scientific questions, and there was flexibility at all levels.[175]

By the mid 1980s, however, given the growing predominance of observational astronomy in the Max Planck Society, doubts arose as to whether an entire institute dedicated to theoretical astrophysics should continue to exist after the retirement of its sole director, Kippenhahn, which was foreseen for the early 1990s.[176] The same question had emerged already in connection with Biermann's retirement, before Kippenhahn was appointed as his successor (Chapter 3). Now, however, the problem of finding new scientific leadership was strongly entangled with more general issues regarding the future of astronomy and astrophysics in the Max Planck Society overall. The main question that arose in the case of this single dedicated institute, was whether theoretical work should be carried out there alone, or within each of the observational institutes. It also called back into question whether the distinction between universities and Max Planck Institutes still held: Why could this theoretical work not be done at a university instead? All this took place in parallel with German reunification, at a time when the Max Planck Society was deeply involved in founding new institutes in East Germany, an operation requiring extra funds which could be raised only by saving money at the extant institutes.

175 Kippenhahn's Memorandum outlined in detail the mode of operation of the institute, the scientific and organizational perspectives, the scientific outlook, administrative questions, and future leadership (AMPG, II. Abt., Rep. 62, No. 17, Fol. 297–305). By 1974, research fields at MPA included the gravitational wave experiment, numerical computing, quantum chemistry, stellar structure and evolution, interplanetary matter, comets, theoretical high-energy astrophysics, general relativity and relativistic astrophysics, plasmas and turbulence theory. Some of these had been traditionally researched since the early years, while others had been started later. New fields such as nucleosynthesis, galaxies and extragalactic objects, and cosmology had been added in 1989 (GVMPG, BC 108504, Fol. 186).

176 Preliminary discussions on the future of the Institute for Astrophysics following Kippenhahn's retirement can be found in CPTS meeting minutes of 19.10.1988, 01.02.1989, 07.06.1989, AMPG, II. Abt., Rep. 62, No. 1815, 1816, 1817. A committee was set up in February 1990 (CPTS meeting minutes of 08.02.1990, AMPG, II. Abt., Rep. 62, No. 1819), which in June decided to propose four possible candidates in the field of theoretical astrophysics (CPTS meeting minutes of 20.06.1990, AMPG, II. Abt., Rep. 62, No. 1820). See also discussions on forming an Appointments Committee to find a new director, in AMPG, II. Abt., Rep. 1, No. 430.

The Threat of Closure and Dispersion

The critical situation at MPA made it a natural target for such savings. Closure and the dispersion of its staff to the other astronomically oriented Institutes became a menacing option to solve the pressing financial problems.

The initial discussions regarding the 'post-Kippenhahn' future of the Institute for Astrophysics took up some of the arguments used in the mid-1970s, when he had succeeded Biermann. Back then, decision makers primarily from the observational institutes had defended it against closure, preferring to keep a separate institute dedicated to theory rather than integrate theoretical subdivisions into each astronomical institute.[177] This was not necessarily altruistic, since the astronomical institutes, particularly in Bonn and Heidelberg, had been resisting calls to appoint theoreticians within their institutes for several decades already, which the closure of the institute in Garching could have facilitated. In contrast, by the 1980s, theoreticians originally from Biermann and Lüst's tradition of plasma physics had already settled at primarily experimental institutes, Heinrich Völk at Nuclear Physics in Heidelberg, for example, Ian Axford at Aeronomy, and Gregor Morfill at Extraterrestrial Physics. Ludwig Biermann's son Peter had a working group within the department of Peter Mezger at the Max Planck Institute for Radio Astronomy; but this was far from being a directorship.[178]

The development of astrophysics in the coming decade, and perhaps even longer, was clearly going to be dominated by the findings of large observatories (optical/infrared terrestrial telescopes, mm- and sub-mm telescopes, space telescopes in the optical, X-ray, and infrared wavelength range), and not by the development of theoretical methods detached from observations; a theoretical institute should accordingly cooperate with institutes dedicated to observation and base its activities on such data.[179] Observational astronomy had a clear preponderance in the large experimental facilities operated by the Max Planck Society, so theory required greater promotion. According to Lüst,

> [a] Max Planck Institute for Astrophysics has a right to exist only if it plays a leading role, with outstanding achievements in the fields of theoretical astrophysics *also* for observational astronomy [our emphasis].[180]

---

177 See discussions following Biermann's retirement described in the previous chapter.

178 During the earliest discussions regarding the retirement of Kippenhahn, Peter Biermann was briefly considered as a successor in Garching (CPTS meeting minutes of 19.10.1988, 01.02.1989, 07.06.1989, AMPG, II. Abt., Rep. 62, No. 1815, 1816, 1817).

179 Heinrich Völk to H.-A. Weidermüller, March 4, 1990, AMPG, II. Abt., Rep. 62, No. 17, Fol. 280.

180 R. Lüst to H.-A. Weidermüller, February 13, 1990, AMPG, II. Abt., Rep. 62, No. 17, Fol. 287.

All of this indicated significant scientific reorientation of the Institute for Astrophysics, and was closely related to the search for a suitable leading scientific personality as Kippenhahn's successor. On the other hand, as Kippenhahn stressed

> an independent theoretical institute has to be able and willing to respond fast and flexibly to new discoveries. In particular in rapidly developing fields like astronomy and astrophysics, such an institute has to gather and combine expertise in a variety of seemingly disconnected areas: This goal can only be reached if, as in the past, Scientific Members as well as staff members are appointed according to their quality and the changing scientific priorities and not necessarily in order to maintain particular areas of research.[181]

At the time, of course, all this had to be complemented by new theoretical tools in view of the new generation of outstanding telescopes, such as ROSAT, ESO-VLT, the Compton Gamma Ray Observatory, as well as other mm- and submillimeter telescopes. In 1989, the institute's recognition of this fact was reflected in the decision to appoint to its directorate Rolf-Peter Kudritzki, a leading expert in stellar spectroscopy; but this did not fully settle the problem of Kippenhahn's succession.[182] The new challenge would be to close the existing gap between 'pure theory' and observation by fostering close interaction with observational astronomy in all the wavelengths domains. In the late 1980s, extragalactic astronomy, including cosmology, had increasingly become an active field of research at the institute, and the appointment of a new scientific member with the corresponding research interests appeared necessary to establish a firm basis for this type of extragalactic work. The proximity of ESO, the excellent computing resources shared with the Institute for Plasma Physics, and the other research institutes (particularly the Institute for Extraterrestrial Physics) on the same Garching site held additional appeal for guest scientists, who were convinced that the MPA program was one of the most valuable on offer.

---

181 AMPG, II. Abt., Rep. 62, No. 17, Fol. 302.

182 Rolf-Peter Kudritzki had been an External Scientific Member since 1983 (CPTS meeting minutes of 01.02.1983, AMPG, II. Abt., Rep. 62, No. 1969). The proposal to appoint Kudritzki as Scientific Member and Director of the Institute was approved in October 1989 (CPTS meeting minutes of 09.10.1989, AMPG, II. Abt., Rep. 62, No. 1818; see also related material in documents produced by the commission "Developments at the Institute for Physics and Astrophysics" in the period 1984–90, AMPG, II. Abt., Rep. 62, No. 376).

The initial idea of amalgamating the Institute for Astrophysics with the Institute for Extraterrestrial Physics, which were after all in adjacent buildings, would be "unfortunate," according to Subrahmanyan Chandrasekhar—at the time, one of the most revered astrophysicists in the world—as it would reverse

> the independent organization of experimental and theoretical institutes, in the physical sciences, an innovation first introduced in Germany after the First World War and widely adopted subsequently in other countries of Western Europe and, most particularly, in the United States.

Chandrasekhar had followed the work of several members of the institute for many years, and observed in particular that it would be

> a great pity, if not a positive discouragement, for the efforts of these brilliant scientists to be diluted with the 'big science' of space ventures.

He liked to believe that there could still be "a place in modern science where individual scientists can pursue their interests free of the noise and bustle always associated with space efforts."[183] In any case, given the lack of theory at ESO and the program-oriented approach at the Extraterrestrial Institute, the Astrophysics Institute alone could provide the necessary broad-based theory for the Munich astronomical community. All this, coupled with the strength in particle physics at the Heisenberg Institute in Munich-Freimann, represented a unique scientific program that could attract visitors who were broadening their horizon to include the emerging field of particle astrophysics.[184]

Avoiding a Merger with the MPE

Saving the institute from dissolution, however, next raised the question of why it should be independent of the MPE next door. Wolfgang Hillebrandt, who

183 Chandrasekhar to Heinz Staab, President of the Max Planck Society, April 25, 1990, AMPG, II. Abt., Rep. 62, No. 17, Fol. 141.

184 It was actually this double reason that led David N. Schramm to visit Munich in 1987–88 (see letter from David Schramm to Heinz Staab, April 11, 1990, AMPG, III. Abt., Rep. 62, No. 17, Fol. 146). Schramm, who was one of the world's leading astrophysicists and an authority on the Big Bang model of the formation of the Universe, was killed in 1997, at the age of only 52, when the twin-engine plane he was piloting crashed outside of Denver. He is considered a pioneer and a leading figure in the merging of particle physics, nuclear physics, and astrophysics in the study of the early Universe. Edward W. Kolb, and Michael S. Turner: David N. Schramm (1945–97). 6666. *Nature* 391/6666 (1998), 444. doi:10.1038/35044.

had taken over provisional management of the institute after Kippenhahn's retirement, and was at the time also Chairman of the German Astronomical Society,[185] was alarmed by the fact that "the whole world" seemed to think it 'a fait accompli' that the institute would "lose its independence" and be "united" with the Institute for Extraterrestrial Physics, without its members having been consulted on the matter, or even informed. Hillebrandt emphasized the "good health" of the institute, while also recalling that its Scientific Advisory Board had clearly deemed it currently "one of the world's preeminent astrophysical institutes" and home to "one of the world's outstanding theoretical groups" with "the potential, in the near term, to claim to be the world's leading Institute for Theoretical Astrophysics." The merger idea was therefore to be rejected:

> [...] if we ask, from a purely scientific point of view, what is the advantage of an independent theoretical Institute for Astrophysical Research over a theoretical department of an Institute for Extraterrestrial Physics, the answer is simple: it is the higher quality of independent research.

Hillebrandt was convinced that the more general management issues of the 'Garching institutes' could not be solved by a merger, because the real problem lay in "extraterrestrial physics," and in the difficulty of keeping pace with the enormously expanding research projects; what was urgently required was "reinforcement in the administrative sector, independently of whether or not a theory department should be included." He clearly stated that if such arguments were to play a role in the discussion,

> one should also make clear that for the sake of a partial reorganization of the administration of a large institute, a well-functioning small institute would be destroyed.[186]

Concerns about the future of the Institute for Astrophysics were expressed in spring 1990 by other senior and influential members of the world scientific

185 Hillebrandt had been elected Chairman of the Astronomical Society at the end of September 1990. See his contribution on the restructuring of research institutions and universities in the field of astronomy and astrophysics in the new federal states during German reunification. Wolfgang Hillebrandt: Kraft schöpfen aus dem Wandel: Die deutsche Astronomie in der Zeit der Wiedervereingung. In: Dietrich Lemke (ed.): *Die Astronomische Gesellschaft 1863–2013*. Heidelberg: Astronomische Gesellschaft 2013, 127–141.

186 Hillebrandt to Staab, April 27, 1990, AMPG, II. Abt., Rep. 62, No. 17, Fol. 135–138.

community, such as William A. Fowler, Eugene N. Parker, and Martin Rees, who not only reiterated general opinion about the status of the institute within the world scientific community, but also expressed, each from their own standpoint, views quite similar to those already well explained by Hillebrandt.[187]

The Max Planck Society thus had a responsibility to the global theoretical astrophysics community, namely to appoint a new directorship commensurate with the acclaimed status of a world-leading institute for theoretical astrophysics. On one hand, the fact that several key members of the institute, including the director, were approaching retirement raised concerns about continuance of the 'Biermann–Kippenhahn model'; but on the other, such circumstances were an opportunity to bring in new blood and fresh perspectives at the senior level. The "tremendous potential of the Institute" had been emphasized in the international *Fachbeirat* (Scientific Advisory Council) report on the institute's research activities: "There is probably no other

187 AMPG, II. Abt. Rep. 62, No. 17, Fol. 122–129. The influential astrophysicist Martin Rees, who had also been a member of the evaluation commission at the Institute, commented that, as stated in previous reports of said commission, "the Institute had a first-rate international reputation, and had maintained its strength in all the areas initiated by Biermann, also evolving in new directions. The physical location near ESO ("already one of the world's major observatories [and which] with the completion of the VLT may well become the very best in the world") and near the Institute for Extraterrestrial Physics, guaranteed its staff "optimum access to first-rate data, as well as easy contact with experts in observational techniques and interpretation." He also added that it was important to bear in mind that "a theoretical institute differs in its style of work from most experimental ones, and especially from, for instance, the neighboring Institute of Space Research. Theorists tend not to work in large teams, nor do they generally pursue very long-term projects. The most active theorists tend to be opportunistic in what I think is a good sense of the word. They exploit new computational techniques as these become available, and are responsive to observational data and new discoveries. One needs a good mix of senior and junior people in a compact and interactive environment" (Martin Rees to Heinz Staab, May 8, 1990, Fol. 131–132). The prominent solar astrophysicist Eugene Parker drew on examples from his long experience of advisory boards in the US, and explicitly warned against not paying proper attention "to the problems created by the reorganization, wherein a small group was placed under the aegis of a larger": when a smaller organization is absorbed into a larger one, the bigger group "tends to tell the smaller group what to do, and, when there is a financial pinch, tends to pinch the smaller group hardest." The risk, according to Parker, could be that the best scientists in the small group, who have their own ideas as to what is most productive, become seriously frustrated and so "start looking for jobs elsewhere." It was Parker's opinion that the "basic laws of administrative bureaucracy" seem to require close supervision, but "the effective pursuit of science does not survive close supervision and control" (Eugene Parker to Heinz Staab, May 7, 1990, Fol. 133–134).

place in the world which can provide so many research positions for theoretical astronomy, with such excellent working conditions."[188] The stakes were high, moreover, for the entire Garching 'cluster' had the potential to become a European center of astronomy and astrophysics rivalling even renowned and traditionally strong institutions, such as Cambridge, UK. However, it was also evident that attaining such preeminence would crucially depend on how the directorship vacancy were filled.

The Appointment Committee led by Hans-Arwed Weidenmüller (a particle phenomenologist from the MPI for Nuclear Physics), was hard at work discussing the future of the MPA and especially the problem of finding a successor to Kippenhahn.[189] In October 1990, it had suggested appointing the British astrophysicist Simon White, whom it hoped would strengthen activities in extragalactic physics—hitherto not one of the MPA's strong points—and so gain a more competitive edge, internationally.[190] White's research had focused on gravitational interactions, especially as related to galaxy clustering in the early Universe and to galaxy formation and the evolution of large-scale structures—an area of expertise within the new, cutting-edge fields of astrophysics—and he ranked among the world's leaders also in the computational aspects of this problem and would thus be able to exploit the outstanding computational resources available in Munich.[191] White's broad

188 See p. 7 of the "Seventh Report of the *Fachbeirat* of the Max Planck Institut für Astrophysik," held in Garching on May 9–11, 1990, AMPG, II. Abt., Rep. 62, No. 17, Fol. 11.

189 Initially, a so-called *Stammkommission* (core committee), led by H. Weidenmüller, had been formed specifically to discuss the future of the Institute for Astrophysics after Kippenhahn's retirement. Members of the *Stammkommission* were involved in discussions related to the question of theoretical astrophysics in the Society, as the future of the Institute for Astrophysics of course played a special role in such considerations (H. Völk to members of the *Stammkommission*, July 31, 1991, GVMPG, BC 218422, Fol. 241). See also description of the Institute's activities and organization, in addition to remarks on the presidential list of questions in Friedrich Meyer's report (GVMPG, BC 108504, Fol. 189–192).

190 The decision to nominate Simon White, and to appoint a new director at the Institute for Astrophysics, had been taken in October 1990 by the committee supervising the institute's future development (CPTS meeting minutes, 02.10.1990, AMPG, II. Abt., Rep. 66, No. 1821).

191 See, for example, Simon D. M. White: The Early ISM and Galaxy Formation. In: Harley A. Thronson Jr., and J. Michael Shull (eds.): *The Interstellar Medium in Galaxies*. Dordrecht: Springer 1990, 371–386. Simon D. M. White, and Luiz A. N. da Costa: Real and Imaginary Clusters. In: David W. Latham (ed.): *Large-Scale Structures and Peculiar Motions in the Universe*. 1991, 285–298. https://ui.adsabs.harvard.edu/#abs/1991ASPC...15..285W. Last accessed 9/19/2018. Arif Babul, and Simon D. M. White: Quasar-Modulated Galaxy Clustering in a Cold Dark Matter Universe. *Monthly Notices of the Royal Astronomical Society* 253/1 (1991), 31P–34P. doi:10.1093/mnras/253.1.31P. Simon D. M. White, R. Pallavicini, and

interests and his ability and experience in observational programs would positively influence collaboration with the Institute for Extraterrestrial Physics and with the strong program in extragalactic research at the European Southern Observatory.[192] The British-born White was officially nominated but the whole procedure was unfortunately interrupted in spring 1991, when he moved from the University of Arizona to the University of Cambridge (UK), and quickly became deeply involved there in collaborative research with other astronomical institutions in Europe.[193]

As we will see below, only in spring 1994 was White able to accept the appointment. And so, from 1990 to '94, the Max Planck Society continued to grapple with the problem of finding a new director for the Institute for

M. R. Kundu: Radio Flares and Magnetic Fields on Weak-Line T Tauri Stars. *Astronomy and Astrophysics* 259 (1992), 149–154. https://ui.adsabs.harvard.edu/#abs/1992A&A...259.149W/abstract. Last accessed 9/19/2018. Simon D. M. White, and Dennis Zaritsky: Models for Galaxy Halos in an Open Universe. *The Astrophysical Journal* 394/1 (1992), 1–6. doi:10.1086/171552. Guinevere Kauffmann, Simon D.M. White, and Bruno Guiderdoni: The Formation and Evolution of Galaxies Within Merging Dark Matter Haloes. *Monthly Notices of the Royal Astronomical Society* 264/1 (1993), 201–218. doi:10.1093/mnras/264.1.201. Since 1994, White has co-led the Virgo Consortium, dedicated to simulating the growth of cosmic structure on supercomputers and, since 1997, the institute has hosted the German data center for ESA's Planck satellite mission. See Simon D. M. White, and Volker Springel: Fitting the Universe on a Supercomputer. *Computing in Science & Engineering* 1/2 (1999), 36–45. http://ieeexplore.ieee.org/document/753045/. Last accessed 10/31/2018.

192 CPTS meeting minutes of 02.10.1990, AMPG, II. Abt., Rep. 62, No. 1821.

193 White had been nominated in 1991, but he was not able to begin his activity until September 1994 (CPTS meeting minutes of 07.02.1991, 05.06.1991, 08.06.1994, AMPG, II. Abt., Rep. 62, No. 1822, 1823, 1832). As White explained to Hillebrandt, "The attractiveness of the MPA and of the opportunities associated with a position as Scientific Member are so great that it was very difficult for me to abandon them until it became clear that I don't really have any other choice at the present time" (Simon White to Wolfgang Hillebrandt, March 7, 1991, AMPG, II. Abt., Rep. 62, No. 17, Fol. 149). The following year, White became inaugural Director of a European Association for Research in Astronomy established by combining resources of the Cambridge's Institute of Astronomy, the Leiden Observatory, and the Paris Institute of Astrophysics, and which was to engage in a broad research program covering studies of the Sun, stars, and interstellar gas, quasars, and galaxies, and the development of structure in the Universe. All this further postponed the possibility of accepting the directorship in Munich. See also, White to President Hans Zacher, March 8, 1991, AMPG, II. Abt., Rep. 62, No. 17, Fol. 603. By fall 1992, the problem of finding suitable candidates had not yet been solved, Simon White having explained that he could not accept the appointment at that time (see related documents in AMPG, II. Abt., Rep. 62, No. 1195). In October, a final attempt to find a successor to Kippenhahn was launched; the *Stammkommission* (core committee) now became a *Suchkommission* (search committee), and included new members.

Astrophysics; but, in spite of all its efforts, none of the potential or invited candidates was found suitable, and frustration and doubt about the possibility of ever settling the matter grew. In the meantime, from June 1, 1991, Hillebrandt took over as Acting Director.

In 1991, the Max Planck Society was already supporting the state of Brandenburg in its plan to found a large 'Blue List' institute as successor to the ZIAP (*Zentralinstitut für Astrophysik*) of the former GDR's Academy of Sciences; initially, the support consisted in establishing up to three working groups on plasma astrophysics there.[194] This was to counteract the *Wissenschaftsrat*'s recommendation that the ZIAP be shut down. A merger of the Max Planck Institute for Astrophysics with a Blue List institute and its subsequent relocation to Potsdam were further ideas under discussion.

In 1992, with rumors already spreading like wildfire in the international scientific community,[195] letters of solidarity were sent expressing the opinion that the closure of this "unique and ideal institute [...] would be a great tragedy."[196] The MPA was considered "a preeminent, world-class institution, comparable to leading theoretically oriented institutions such as the University of Cambridge and Princeton University," and it also played an important role in the training and education of theoreticians in Germany and throughout the world via its postdoctoral and visiting appointment programs:

> Closure of MPA, a productive institute with vigorous programs at the forefront of astrophysical research, will seriously impede the future development of theoretical astrophysics in Germany.[197]

---

194 Gerhard Haerendel, "Vorschlag zur Einrichtung einer Projektgruppe für Plasmastrophysik in Potsdam," September 13, 1991 (GVMPG, BC 218422, Fol. 58–63.

195 Abbott, German Astronomers, 1992, 267–267. Abbott, Max Planck Institute, 1992, 6. See also Jerke, Welche Zukunft, 1992, 514, 605.

196 Yoji Osaki, University of Tokyo, to Hans-Arwed Weidermüller (Chair of the Appointment Committee for the Kippenhahn succession), July 27, 1992, AMPG, II. Abt., Rep. 62, No. 17, Fol. 504. Others were "astonished and could not understand at all why that institute, which has been so productive scientifically and has been one of the leading institutes in the field of astrophysics, should suddenly face a fate of being closed down in the midst of its vigorous activity [...] It would be a great loss [to] the astrophysical research of the world if the MPI for Astrophysics should happen to be closed down" (Yutaka Uchida, University of Tokyo, to H. Weidenmüller, September 9, 1992, AMPG, II. Abt., Rep. 62, No. 17, Fol. 506).

197 Ronald E. Taam, Northwestern Universiy, to H. Weidermüller, July 28, 1992, AMPG, II. Abt., Rep. 62, No. 17, Fol. 508.

Joachim Trümper of the MPE was of the same opinion; he would very much regret the closure of the Institute for Astrophysics, in view of the excellent work that had been and would be done there. He, too, was convinced that theoretical astrophysics in Germany would be severely weakened at a time when astronomical observations in many fields had reached a top level internationally, and were providing new impulses also for theory.[198] Last but not least, thanks to the Max Planck Institute for Astrophysics, the Institute for Extraterrestrial Physics, and the European Southern Observatory, Garching had developed over the previous years into one of the most important centers of astrophysical research in the world. Among those research institutions, the Institute for Astrophysics was the only one dedicated exclusively to the investigation of astrophysical processes. Not only was the institute seen as an ideal complement to the two more instrumental-oriented neighboring institutions, but according to the Council of German Observatories:

> [w]ithout the presence of the MPA, it would hardly have been possible to recruit ESO as the most important European astrophysical research organization for Garching. Breaking the MPA out of the existing configuration in Garching would certainly significantly reduce the attractiveness of Garching for national and international institutions in this research field, and in the long term, could jeopardize Garching as an international center in this field.[199]

Theoretical Astrophysics within the Global Reforms of Cosmic Research in the Max Planck Society

In the meantime, discussions about the future of the Institute for Astrophysics and, more generally, the mid- and long-term perspectives for astronomy and astrophysics research in the Max Planck Society, were facilitated by the Presidential Commission chaired by Heinrich Völk (see above, in this chapter). It had been initiated in 1991 by President Hans Zacher at the time of reunification, in order to answer a list of specific questions and to make recommendations on the further development of astronomy and astrophysics.[200] Due

198 Jerke, Welche Zukunft, 1992, 514, 605.

199 AMPG, II. Abt., Rep. 62, No. 17, Fol. 537.

200 Internal members of the Society on the commission were Völk, Genzel, and Trümper; other permanent members were the Vice President Herbert Walther and Weidenmüller. See related reports of the commission in the CPTS meeting minutes of 05.06.1991, 23.10.1991, 07.02.1992, 03.06.1992, 16.10.1992, 03.02.1993, 16.06.1993, 19.10.1993, 03.02.1994, AMPG, II. Abt., Rep. 62, No. 1823, 1824, 1825, 1826, 1827, 1828, 1829, 1830, 1831. Material

to the unusually high number of retirements foreseeable in that decade, the commission was also asked to submit recommendations on the development of the Max Planck Institutes related to that research area. As the commission underlined, astronomy was in a period of rapid development, and in terms of instruments, the great potential of new detection technologies in infrared/submillimeter astronomy and high-energy astrophysics was not yet exhausted, nor that of interferometric methods and the elimination of air turbulence (adaptive optics) in ground-based optical astronomy. Such advances in these and other fields were revolutionizing observational astronomy, and would doubtless continue to do so in the next decade, thereby contributing to fundamental shifts in certain scientific views of the Universe. These appeared to be the greatest development opportunities for the MPIs. But it was also recognized that there were promising opportunities for development also in

---

on the activity of the commission is in Rep. 62, No. 17. The commission was asked by President Hans Zacher to discuss and comment on a series of questions such as "1. In which areas of astronomy and astrophysics are there particularly promising development opportunities for basic research, which is typically to be conducted at Max Planck institutes? Where should priorities be set? 2. Where are overlaps and where can gaps be identified between the research directions of the astronomically and astrophysically oriented Max Planck Institutes? Should overlaps be eliminated by concentrating areas of work [in] one institute or are they justified from a scientific point of view? Are there gaps that should and could be closed? Are there any imbalances in the current promotion of the various areas of work which should be corrected? 3. Should the Max Planck Society provide additional funding for research in certain spectral ranges? Which other work, if any, could be restricted or abandoned in order to make this possible? 4. Should the Max Planck Society participate in national and international projects for the development of next-generation large telescopes (VLT)? How, on what projects and to what extent, if any, should this be done? 5. What is the assessment of the importance of international cooperation for the further development of astronomical-astrophysical research in the Max Planck Society? What are the areas of activity which, in the context of such cooperation, offer specific opportunities to the institutes of the Society and on which they should therefore concentrate their efforts? 6. What are the consequences for the Max Planck Society in view of the situation of astronomical and astrophysical research in the new federal states? 7. Is the cooperation of the astronomical-astrophysical Max Planck Institutes with the universities in need of improvement and what measures could be taken to improve it? 8. Are structural changes and shifts of emphasis in the tasks of the Max Planck Institutes deemed necessary? To what extent should the institutes continue to perform service tasks? Could the funding instrument of the junior research groups, which until now has only been established in the institutes of the Biological-Medical Section, also be of use to the astronomical-astrophysical Max Planck Institutes? 9. Does an increased integration of the theory into the observational institutes seem necessary, and what conclusions could be drawn from this for the future task of the Max Planck Institute for Astrophysics?" (AMPG, II. Abt., Rep. 62, No. 1823).

the fast-growing fields of very-high-energy gamma, neutrino, and gravitational wave astronomy. Moreover, it also clearly emerged at the end of the 1980s, that combining elementary particle physics and cosmology was likely to bring new impulses for modern astrophysics. Such synergy did indeed develop fully in the 1990s, forging common scientific goals and new connections between theory, observation, and experiments, and providing novel insight into the macroscopic and microscopic components of the Universe. However, all this had as 'boundary conditions' the financial constraints the Society was facing at that time. Therefore, despite the simultaneous eastward expansion of the Max Planck Society, this fast-developing scientific field of fundamental research had to be seen within the confines of a broader framework, namely the German reunification process then underway.

In October 1992, when the commission presented its report to President Zacher, it was clear that the rapid development of cosmic research prevented the expansion program being determined long term with any certainty; such stormy developments would be challenging even under normal circumstances. And the financial situation of the Max Planck Society had become more pressing since the commission was convened in spring 1991. This aggravated its task, but it was nonetheless important to not lose sight of the longer-term perspective. On the contrary, a potential crisis could be handled [the report stated], only if requisite decisions were taken in the short term within the framework of a future-oriented overall concept. In this respect, the commission assumed a normal development without substantial global growth. This meant in any case that the institutes would be able to participate as before in the scientific use of devices that already existed or were under construction. The strong interdependence of institutional and project funding in astronomy would be an essential and necessary feature of this research area, also in the future. In the postwar period, such funds were made available primarily by the federal Research Ministry and private foundations, and the Max Planck Society functioned as a provider of research facilities for the scientists at universities and state institutes. If the Society were not to be limited to the development of additional instruments for large-scale equipment outside of its institutions, especially abroad, it would continue to require project funding.[201]

Overall, the Presidential Commission saw in astronomy and astrophysics "a field of basic research of special rank," with a

201 CPTS meeting minutes of 16.06.1993, II. Abt., Rep. 62, No. 1829. See section. 4.3 of the agenda, and especially a two-page summary of the consultation results prepared by the commission.

> strong, foreseeable growth potential, *to which the Max Planck Institutes contribute significantly at present and should also contribute in the future* within a substantially constant outside framework [our emphasis],

and it advised the President "to pursue this commitment." The Presidential Commission also expressed its views on the question of "integrating theory into observational institutes" but considered "the strong concentration of theory in the MPI for Astrophysics (MPA) as unsatisfactory." The question of obtaining an outstanding theoretician with pronounced leadership qualities as director of the institute was regarded "as a prerequisite for the existence of a (theoretically oriented) MPA." In any case, if a suitable leader could not be found, the commission would consider it right to take ongoing retirements at the institute "*as an opportunity to establish theoretical working groups under the leadership of newly appointed directors at the observing institutes*" [our emphasis].[202] However, members of the MPA's Scientific Advisory Board "unanimously opposed" such a merger, on a number of grounds. In particular, it was remarked that,

> [i]t would be more difficult to attract outstanding theoreticians to a large institute, in which the direction of research is dictated by the needs of a big experimental group.

Moreover, it was remarked that at times of financial stringency, or during the emergencies that might arise in very large experimental programs, there would be "a strong temptation" to extract research positions from any group working on a long-term perspective.[203]

In the meantime, in order to gain a broad overview of the current problems and long-term perspectives in the field of theoretical astrophysics—and to have a basis for the forthcoming decisions on the future commitments of the MPG in this research sector—the Scientific Advisory Board invited 36 international leading experts in the field of astrophysics/astronomy (none of them over the age of 50) to participate in its forthcoming conference on 'Future Trends in Theoretical Astrophysics,' to be hosted by ESO in Garching, in early

202 See the *Fachbeirat* (Scientific Advisory Council) report of May 1990 (AMPG, II. Abt., Rep. 62, No. 17, Fol. 5–15, Fol. 11). Some members of the Presidential Commission considered the proposed merger of the Institute of Astrophysics with the Astrophysical Institute in Potsdam an unviable compromise (AMPG, II. Abt., Rep. 62, No. 17, Fol. 404). See general discussions on the ZIAP question in October 1991 (Fol. 431–433).

203 AMPG, II. Abt., Rep. 62, No. 17, Fol. 5–15, Fol. 11.

September 1993, and which it hoped would bring forth new ideas and possibly also new recruits for scientific projects at the Institute for Astrophysics.[204]

Following the failure to appoint White, the Appointment Committee had looked elsewhere, and in October 1992, had recommended Rashid Sunyaev, one of the world's leading astrophysicists and a member of the Russian Academy of Sciences, who had long been in close collaboration with the Garching scientists, most prominently with Joachim Trümper on X-ray astronomy projects, as described in Chapter 3.[205] The 'Future Trends' conference in Garching helped turn the spotlight on Sunyaev again, and the decision to appoint him as Scientific Member and Director at the Institute for Astrophysics was taken sometime between fall 1993 and early 1994. Crucial for the MPA was that besides his activities in space-based observational astronomy, Sunyaev had grown up within the great Russian tradition of relativistic astrophysics founded in the early 1960s by Yakov Borisovich Zeldovich; and not only was he a pivotal figure in Russian space science but also was associated with a number of fundamental findings in theoretical astrophysics and physical cosmology.[206] This was the kind of theoretician the Max Planck Society needed, one who could help diversify research activities and build bridges

204 CPTS meeting minutes of 16.06.1993, AMPG, II. Abt., Rep. 62, No. 1829. On the organization of said conference, see also Rep. 62, No. 12, Fol. 334–337. About the role of the conference in this context and the whole process leading to these calls, see also Joachim Trümper: interview by Luisa Bonolis and Juan-Andres Leon, Munich, August, 7–8, 2017. DA GMPG, BC 601036.

205 CPTS meeting minutes of 16.06.1993, 19.10.1993, 03.02.1994, 19.10.1994, AMPG, II. Abt., Rep. 62, No. 1829, 1830, 1831, 1833. Contacts with Rashid Sunyaev and Boris Zeldovich went back to 1975, when Joachim Trümper began to visit Moscow very frequently, in connection with the establishment of collaborations with Russian scientists, especially with Sunyaev himself at the Institute for Space Research, leading to the aforementioned Mir-HEXE experiment. Trümper himself had promoted Sunyaev's appointment as director at the Max Planck Institute for Astrophysics to solve the leadership problem (AMPG, II. Abt., Rep. 62, No. 1195. Fol. 14). Simon White, too, believed that it was important to do everything possible to attract Sunyaev as Scientific Member, as he suggested during the procedure for his own appointment (White to Zacher, May 20, 1994, AMPG, II. Abt., Rep. 62, No. 12, Fol. 129). He continued to fully support this nomination after accepting the position at the Institute for Astrophysics in March 1994 (White to Zacher, May 10, 1994, No. 12, Fol. 169).

206 Measurements of the so-called Sunyaev–Zeldovich effect, according to which clusters of galaxies produce distortions in the cosmic microwave background (CMB), provide fundamental information both in an astrophysical and a cosmological context. Rashid Alievich Sunyaev, and Yakov Borisovich Zel'dovich: The Spectrum of Primordial Radiation, Its Distortions and Their Significance. *Comments on Astrophysics and Space Physics* 29 (1970), 66–73. http://articles.adsabs.harvard.edu/cgi-bin/get_file?pdfs/ComAp/0002/1970ComAp...2...66S.pdf. Last accessed 10/31/2018.). In the early 1970s, after becoming

between the theoretical work traditionally pursued at the Institute for Astrophysics and the observational work of other institutes.

Sunyaev had played a decisive role in the foundation and advancement of high-energy astrophysics and X-ray astronomy in the USSR (and later, Russia), also leading the team which built and operated the X-ray observatory aboard the Kvant-1 module of the Mir space station, for which the Institute for Extraterrestrial Physics provided the instruments.[207] Its main achievement and most exciting result was the first-ever detection of X-ray emission from a supernova, the Supernova 1987A in the Large Magellanic Cloud.[208] Like other newly appointed Max Planck directors of this period, Sunyaev did not abandon his previous commitments; after the move in November 1994, he continued to pursue his space-based observational agenda with the Institute for Space Science in Moscow, while specializing in theoretical activities in Garching. When the eROSITA, the MPE's contribution to the Russian Spektrum-RG space telescope, was finally launched in 2019 (see Chapter 3), Sunyaev was in the unique position of having been involved with this German project not officially, as a Max Planck Director, but rather as the Principal Investigator on the Russian side.[209]

---

interested in X-ray sources, Sunyaev also developed with the Russian astrophysicist Nikolai Shakura a model describing the accretion of matter onto a black hole, in what became a seminal paper on the structure and observable properties of our principal window onto black hole growth: Nicolai Ivanovich Shakura, and Rashid Alievich Sunyaev: Black Holes in Binary Systems. Observational Appearance. *Astronomy and Astrophysics* 24 (1973), 337–355. http://adsabs.harvard.edu/abs/1973A%26A....24..337S. Last accessed 10/31/2018.

207 In the period 1974–82, Sunyaev was head of the Theoretical Astrophysics Laboratory in the new Department of Theoretical Astrophysics at the Space Research Institute of the USSR Academy of Sciences, which he had organized with Zeldovich. In 1982, Sunyaev created there the new Department of High Energy Astrophysics, which became responsible for the international orbital X-ray observatory operating for six years on board the Mir-Kvant space station, and consisting of Soviet, British, Dutch, and German X-ray instruments. Sunyaev's Department was also responsible for operation and analysis of scientific data obtained by Granat, an international X-ray satellite launched in 1989.

208 Rashid Alievich Sunyaev et al.: Detection of Hard X-Rays from Supernova 1987A. Preliminary Mir-Kvant Results. *Soviet Astronomy Letters* 13/6 (1987), 431. http://adsabs.harvard.edu/abs/1987SvAL...13..431S. Last accessed 12/13/2017. R. A. Sunyaev et al.: Hard X-Rays from Supernova 1987A. *Advances in Space Research* 10/2 (1990), 47–53. doi:10.1016/0273-1177(90)90117-I.

209 Sunyaev was one of the main promoters of the Spectrum-RG project from the very beginning, when it started in 1987 as an international collaboration. Max Planck Institute for Astrophysics: Successful Launch of the Spectrum-RG Mission, 7/13/2019. https://www.mpa-garching.mpg.de/720407/news20190713. Last accessed 3/23/2021.

Luckily, in early 1994, Simon White had finally communicated that he was ready to accept the appointment he had been offered in 1990. He was officially appointed Scientific Member and Director at the Institute for Astrophysics in March 1994 and took up his position in September that year. With White's and Sunyaev's arrival, the research program dramatically broadened to include topics in physical cosmology and theoretical high-energy astrophysics. As recommended by the Appointment Committee,[210] White installed a new management system—with collegiate leadership by a Board of Directors (*Kollegium*, including White, Sunyaev, and Hillebrandt), and a Managing Directorship rotating every three years—and reorganized the institute, introducing radical internationalization.[211]

Regional Strongholds and Internal Migration of Researchers at the End of the Century

In the end, the new Max Planck Institute for Astrophysics was able to stay in Garching by very strongly justifying its regional ties. The branch of computationally heavy theoretical astrophysics pursued by both Biermann and Kippenhahn and historically decisive throughout the institute's entire existence was firmly rooted in the calculational resources of the Garching area, which by the 1990s were manifest in the shared supercomputing facilities that existed there, thanks to the presence of the Institute for Plasma Physics. White's appointment made sense only in combination with the supercomputing facilities in Garching. Likewise, Sunyaev's hybrid profile as both one of the world's most influential theoreticians and the experienced director of one of the world's leading experimental and observational cosmic research institutions in Moscow would be fruitful only in the proximity of both the Institute for Extraterrestrial Physics and the European Southern Observatory. This was not the case with the theoretical gravitational physics pursued by Ehlers. As we detail in the next chapter, the experimental trials on gravitational waves had been transferred to Hannover since the 1980s. This made gravitational physics a more 'mobile' activity than others in Garching, and the Max Planck Society's major contribution to saving Potsdam's historical strength in

210 During the meeting of the *Stammkommission* of October 1, 1990, Reimar Lüst had recommended a collegiate leadership, a proposal supported by the whole committee (AMPG, II. Abt., Rep. 62, No. 1195, Fol. 217–218).

211 See White's Welcome Note: *MPA 50 Years. 50th Anniversary Brochure of the Max Planck Institute for Astrophysics*. Edited by Mona Clerico. Garching: Max-Planck-Institut für Astrophysik 2008. https://www.mpa-garching.mpg.de/69639/MPA_englisch.pdf. Last accessed 10/31/2018.

astrophysics after the dissolution of the GDR Academy of Sciences ended up being the creation there of the Max Planck Institute for Gravitational Physics (Albert Einstein Institute). Jürgen Ehlers personally took the initiative, proposing the creation of this institute in Potsdam both to the President of the MPG and the state of Brandenburg, drawing for justification in particular on Einstein's historical connections with the city and several precedents, such as the GDR-era Einstein Laboratory.[212] The rapid success of this maneuver somewhat protected the future of the MPA in Garching from the immediate pressures of German reunification, while simultaneously securing yet another regional cosmic stronghold.[213] Collaboration with the preexisting community continued through very close institutional involvement with the reestablished Potsdam Astrophysics Institute (IAP), where Günther Hasinger, a disciple of Joachim Trümper, was director from 1994 to 2001. Thanks to his directorship, cosmic researchers from the former GDR and their next generation of disciples were able to play a significant role in the scientific bonanza that came with ROSAT, whose launch also coincided with German reunification. Hasinger would then become Trümper's successor in Garching in 2001, before going on, from 2008 to 2011, to succeed Klaus Pinkau's successor Alexander Brandshaw as Research Director at the Institute for Plasma Physics, which by then had two operational branches, in Garching and Greifswald.[214]

Finally, the reshuffle in expertise between Garching, Hannover, Potsdam, and Greifswald had additional implications at another of the traditional regional strongholds of cosmic research. At the Max Planck Institute for Astronomy in Heidelberg, the 1990s was a time of deep crisis.[215] To the disappointment of Calar Alto was added the retirement of its founder Hans Elsässer, and then, in 1997, the speedy departure of his de facto successor Stephen Beckwith, who went on to direct the Space Telescope Science Institute in Baltimore, the scientific headquarters of the Hubble Space Telescope. Beckwith's arrival in Heidelberg had coincided with German reunification in 1990. He was originally meant to replace its other director, Guido Münch, with Elsässer still at the helm. However, the terms on which Beckwith was willing to accept an

---

212 Hillebrandt, Kraft schöpfen, 2013, 127–141. Hubert Goenner: General Relativity and the Growth of a Sub-Discipline "Gravitation" in Germany. *The European Physical Journal H* 42/3 (2017), 395–430. doi:10.1140/epjh/e2017-70057-4. Goenner, Some Remarks, 2016.

213 CPTS meeting minutes of 05.06.1991, AMPG, II. Abt., Rep. 62, No. 1823. See also discussion of 07.02.1992, No. 1825.

214 As of 2020, Hasinger is ESA's Director of Science.

215 All these developments in Heidelberg are described in further detail in Lemke, *Himmel über Heidelberg*, 2011, Vol. 21, 127–143.

appointment made clear that he would be the one to spearhead modernization of the institute and lead it into the 21st century. Elsässer's health problems, even before his official retirement in 1997, made Beckwith de facto Managing Director through the 1990s reform period. The succession process for Elsässer himself began in those years, but by 1997, when Beckwith left, all those who had been offered the position had declined and the selection was still underway.

As in many other Max Planck Institutes undergoing post-founding-generation upheaval, Beckwith instituted a more horizontal leadership style with collegiate directorship (which, however, materialized in practice only after his successors were appointed), as well as an emphasis on integration into the global research community and a renewed focus on publishing the institute's scientific findings; not least, he endeavored to make the most of Calar Alto by modernizing its instrumentation, opening it up to external participants, and, crucially, highlighting its value as an instrumental research site in fields such as adaptive optics (see earlier in the chapter). Thanks to these initiatives, there was a marked increase in collaboration also with other Max Planck Institutes, especially Extraterrestrial Physics in Garching, and Radio Astronomy in Bonn. Astronomical observations, in contrast, were increasingly outsourced to external telescopes. And shortly before leaving, Beckwith was also the main proponent of participation in the Large Binocular Telescope (LBT) on Mount Graham, Arizona, even despite the persistent problems that the MPIfR had had on this site with the Heinrich Hertz telescope. Participation in the LBT opened up many opportunities for international exchange and instrumental developments, and also secured a continuing role for the Heidelberg-based institute within the German astronomical community. All this, despite the significant delays in production of the multinational (25 percent German) telescope. But within a global strategy of participation in cosmic research, the LBT can be seen to have played a significant role in diversifying access to astronomical observatories, securing independent, parallel projects to those conducted via ESO. This collaboration also maintained a foothold in a decades-long and sometimes difficult collaborative relationship with Arizona, meanwhile for a good half century. In fact, MPI's counterpart in the Arizona collaboration, Peter Strittmatter, had been invited to become a second director in Heidelberg in 1974, but declined at the time, and his career eventually led him to Arizona, and made him the main protagonist of the ambitious projects (and debacles) on Mount Graham.

Upon Beckwith's quick departure for the United States, the Max Planck Institute for Astronomy was provisionally directed by Immo Appenzeller from the nearby state observatory in which the MPIA had originated in the first place, and with which opportunities for collaboration had now become more egalitarian, thanks to the 1990s reforms. As mentioned earlier, it was the Heidelberg State Observatory, not the Max Planck Institute, which coordinated the construction of Paranal's most significant first-generation instrument, FORS. In a process culminating in 2005, the Heidelberg State Observatory was made an integral part of the University of Heidelberg's Center for Astronomy, which further unified the area's multiple, formerly dispersed sites of astronomical research by integrating the *Astronomisches Recheninstitut*. In parallel to the VLT, this observatory also led the development, together with the Max Planck Institutes for Astronomy and Extraterrestrial Physics, of the near-infrared LUCIFER camera spectroscope—officially renamed LUCI in 2012—for the LBT.[216]

By the end of the century, the directors appointed in Heidelberg, Hans-Walter Rix and Thomas Henning, together had a track record that reflected all of these changes in the 1990s, indeed, that had been facilitated by the many different Max Planck Institutes' collective coordination of their activities throughout this difficult decade. Rix had studied under Simon White at the University of Arizona in the early 1990s, and made a career in galactic evolution (including entities like distant quasars and supermassive black holes), combining theoretical insights with observations from the most recent telescopes and instrumentation, and including novel methodologies such as gravitational lensing. Shortly after White's arrival in Garching, Rix (then a postdoc at Princeton) moved to the neighboring Max Planck Institute for Extraterrestrial Physics, from where he was appointed as professor in Arizona. Shortly after, in 1998, he was selected (from among 12 candidates) and appointed as successor to Beckwith.[217]

Thomas Henning, for his part, was the Max Planck Society's most prominent 'catch' from among the collaboration networks established during the process of German reunification.[218] The Max Planck Institute for Radio Astronomy had taken under its tutelage a working group in Jena, on dust clouds and star formation. It was headed by Henning, who had initially trained

216 The instrument was eventually renamed LUCI in 2012, more appropriate since the mountaintop is shared with the Vatican. https://www2.mpia-hd.mpg.de/homes/LBTBWEB/LBT/luci.html. Last accessed 2/4/2022. For the early plans, see H. Mandel et al.: LUCIFER-LBT NIR Spectroscopic Utility with Camera and Integral Field Unit for Extragalactic Research. *Optical and Infrared Spectroscopy of Circumstellar Matter, ASP Conference Series, Vol. 188*. 1999, 321.

217 For further insights on the career trajectory of Rix, see Heinz Horeis: Hans-Walter Rix: Der Jung-Star. *Bild Der Wissenschaft* 11 (2004), 46.

218 Lemke, *Himmel über Heidelberg*, 2011, Vol. 21, 133–136.

in plasma physics at the East German University of Greifswald, working on topics of star formation, one of the traditional areas of expertise in the MPG, with a long lineage, from Weizsäcker and Lüst to Völk; and during the late 1980s and '90s, Henning pivoted this theoretical expertise increasingly towards observational programs. From his position in the Max Planck working group at the University of Jena (1991–96), Henning established an internationally recognized position in this field, based mainly in the infrared realm and invigorated in the 1990s by astronomical observations of planetary systems around other stars. The Jena programs interpreted data from the new generation of telescopes stemming from global collaborations such as ISO (see earlier in the chapter), and used this entry point to participate at the vanguard of the decade's key instrumental developments. Henning's group participated in the development of adaptive optics and interferometry in optical wavelengths in Calar Alto, LBT, and VLT. From Jena, Henning had also participated in the scientific planning for the LBT on Mount Graham, as well as SOFIA, where the MPIfR and MPE also participated (see also earlier in the chapter). Then in 2001, Henning became the first person to have conducted his academic career within former East Germany and become a Max Planck director in the cosmic research cluster. It should not be forgotten that a tradition of close collaboration and expertise in optical astronomy has linked Heidelberg and Jena (plus the Cold War offshoots in Oberkochen and Mainz) since the late 19th century.

CHAPTER 5

Global Leadership in Emerging Fields: Toward Astro-Particle Physics, Relativistic Astrophysics, and Multi-Messenger Astronomy

This last chapter breaks with the bird's-eye view to engage in depth with three episodes that best highlight the intense interrelationship of longstanding scientific traditions in the Max Planck Society and global leadership in scientific fields. All three case studies have in common their roots in traditions that date from before Sputnik and that benefited from unique features of the Max Planck system, such as interdisciplinarity, embeddedness in international collaboration, and strong theoretical, experimental, and instrumental expertise. These all facilitated the rise of astro-particle physics and multi-messenger astronomy in Europe, in contrast to the difficulties experienced by their American counterparts, and fostered Max Planck scientists' early participation in the entirely new field of gravitational wave astronomy. But the growing scale of scientific infrastructures and shifts in conditions at the end of the Cold War also heralded the constraints that Max Planck scientists would face in the 21st century, given that their scientific and technological achievements are meanwhile interwoven with vast multinational research organizations, where successes are not easily accredited.

1 Three Case Studies

This last chapter of the book will highlight three case studies—solar neutrinos, gravitational waves, and ground-based gamma astronomy—that directly exemplify how the longest-standing scientific traditions in the Max Planck Society, which have existed since the postwar era and whose roots go back even to the 1920s and '30s, were especially effective in determining how the Society would successfully enter brand new research fields such as gravitational wave astronomy and astro-particle physics, a cross-disciplinary field at the intersection of particle physics, astronomy and cosmology that emerged in the 1970s and '80s aiming to answer fundamental questions related to the history of the Universe.

Zooming in

Up to this point, our study has followed prominent threads in the history of the cosmic sciences in the Max Planck Society, constructing an explanatory framework centered around the interplay of various scientific traditions with the leading sociopolitical forces of the time, such as the atomic age and the space age. We have documented how, within this context, independent Max Planck Institutes learned to coordinate with one another their respective activities. In tracing the evolution of such scientific traditions, this survey has been organized as a loosely chronological and political narrative. We have focused on the 'big picture,' without attempting to provide an exhaustive account of all activities in the field; so far, we have also deliberately not examined any particular scientific development in close detail, only alluding to the most relevant cases.

This last chapter of the book, however, will engage in depth with the intense interrelationship of these longest-standing scientific traditions in the Max Planck Society, highlighting along the way three episodes that were key to its early presence—and its achievement of global leadership—in nascent fields of scientific interest in astrophysics and high-energy physics, and that even contributed to entirely new fields such as gravitational wave astronomy and astro-particle physics. This cross-disciplinary, multidimensional field, linking cosmology and particle physics, emerged in the 1970s and '80s and was widely acclaimed by theoreticians studying the foundations of cosmology, nuclear/particle physics, and gravity, as well as by astrophysicists, astronomers, space physicists, and experimental particle and nuclear physicists.[1]

The choice of these case studies—solar neutrinos, gravitational waves, and ground-based gamma astronomy—is by no means based on a hierarchical evaluation of research in the Max Planck Society, but rather, most directly exemplifies how the longer-lasting traditions which have existed since the postwar era—and whose roots go back even to the 1920s and '30s—were especially effective in determining how the Society would successfully enter new scientific fields.

---

1 From here on, we use the more common terms 'astroparticle physics' or 'particle astrophysics.' For a preliminary survey of the field, see Vanessa Cirkel-Bartelt: History of Astroparticle Physics and Its Components. *Living Reviews in Relativity* 11 (2008), 2–58. doi:10.12942/lrr-2008-2. Brigitte Falkenburg, and Wolfgang Rhode: *From Ultra Rays to Astroparticles. A Historical Introduction to Astroparticle Physics*. Dordrecht: Springer 2012.

The Deep Roots of Astroparticle Physics and Gravitational-Wave Research in the Max Planck Society

As mentioned above, the three selected case studies are rooted in research traditions dating from before Sputnik and that benefited from the unique features of the Max Planck system, such as interdisciplinarity, embeddedness in international collaboration, and strong theoretical, experimental, and instrumental expertise. In all of them, people and practices dating back to either Göttingen/Munich or Heidelberg served as springboards for, (in its most positive sense), opportunistic expansion into the most novel scientific questions of the day, with an impact lasting well into the 21st century, and even facilitating the rise of astroparticle physics and multi-messenger astronomy in Europe, in contrast to the difficulties experienced by the disciplines' American counterparts. Within an increasingly globalized scientific landscape, it also happened that institutes belonging to different traditions entered the same field, alternately competing against and collaborating with one another, as in the case of neutrino physics and ground-based gamma ray astronomy. All the cases in this chapter follow a distinct pattern: instead of attempting to compete on an equal footing with their counterparts in the United States or even in France or the United Kingdom, Max Planck researchers increasingly identified the features that made their system unique, and considered how, in the increasingly global landscape, it could best take advantage of scientific competition and collaboration.[2] These features were as follows:

- An existing foothold in crucial international collaboration projects, which had first been gained through niche expertise in theory, small-scale experimentation, or the development of instruments.
- The presence in several of the strongholds of the Max Planck Society, such as Munich and Heidelberg, of technical expertise and the capacity to develop radically new, medium-scale experimental systems, without the numerous hurdles that this type of research needs to overcome in other countries. These strongholds control flexible, highly competent, in-house technical workshops and maintain industrial contacts relevant to their specific requirements. These favorable conditions afforded further expansion as well as the retention, sometimes even attraction to Max Planck Institutes, of the global experts in the field.
- An opportunistic trigger: the ability to identify, thanks to preexisting immersion in international collaborations, when leadership at the next

2 The issue of international scientific collaborations was the main focus of the Senate meeting of November 1987 (Minutes of the 117th Senate meeting of 11.19.1987, in Munich, AMPG, II. Abt., Rep. 60, No. 117).

scale of an experimental program could be achieved by Max Planck scientists, either in parallel with competitors or by taking over an international scientific research program. In all three cases mentioned in this chapter, the truly groundbreaking work had been done by others, sometimes over several decades, exceeding all the expectations and the opposition of their own scientific communities. Max Planck researchers accompanied such struggles, but from the wings, making a decisive move only once the pioneering work had been seen by those with deep expert knowledge of the field to show promising results, but had not yet been accepted by the mainstream scientific community which, in other countries, would determine whether the next scale of research should be funded. While their counterparts struggled to obtain the funding for scaling up their pioneering work, a good amount could be done in-house in Max Planck Institutes, and their early dominance in this scaling-up stage then allowed the rapid formation of collaborations with European partners in which Germans had the dominant instrumental role. In such pioneering enterprises, the relationship with the Americans was friendly but ambivalent: they all supported one another as dark horses against skeptics in their respective scientific communities; but at the same time, all sides played up the competition with other projects to highlight the urgency of their plans, and in Europe particularly, highlighted the opportunity for overtaking the Americans, a major justification in the eyes of their financial backers.

- The cases treated in this chapter all feature decades-long searches for new, previously undetected signals, for which incremental improvements of several orders of magnitude in sensitivity and cost were expected. Measurements during the intermediate stages often only provided 'upper limits,' that is, null results, or non-detection at some statistical significance.[3] Even more striking, false positives which could otherwise have derailed whole research fields, here rather provided the initial spark for revolutionary developments to come, by attracting a welcome flow of attention and resources. The long-term, independent support of the Max Planck Society and its ability to rely on in-house technical expertise and local networks of academic and industrial partners, combined with the theoretical expertise to confidently guide the search for something previously undetected

3 On the problem of negative results, see P. Astone, and G. Pizzella: Upper Limits in the Case That Zero Events Are Observed: An Intuitive Solution to the Background Dependence Puzzle. hep-ex/0002028. *1st Workshop on Confidence Limits, CERN, Geneva, Switzerland, 17–18 Jan 2000*. Geneva: CERN 2000, 199–205. doi:10.5170/CERN-2000-005.199.

allowed these projects to remain on the global leading edge for a vast range of experimental scales, up to financial limits of around 100 million USD.

Once the scale of research enterprises approached the billion dollar mark, at which global consolidation into just a few large-scale projects occurs, the outcome of Max Planck projects was quite heterogeneous, as the cases here illustrate: in some they obtained global dominance (ground-based gamma astronomy), in others they only narrowly averted defeat (gravitational-wave search), and on average, (also in neutrino research), they ended up on the front line of the enormous, flexible, horizontal collaborations that dominate much of 21st-century science.

Owing to how the Max Planck Society was formed in the postwar era, the scientific traditions involved in this discussion were led by physicists of both the theoretical (in Göttingen and, later, in Munich), and the experimental (in Heidelberg) varieties. This, in contrast to observational astronomy, which entered the Max Planck Society during the 1960s and followed an explicit logic of 'catching up' with the other major countries via well-trodden paths of incremental developments, such as telescope size, detector sensitivity, and optimized production techniques—what one might call 'conservative innovation.' As we have seen in previous chapters, such efforts were led by people and research traditions which were brought into the Max Planck Society only after Sputnik.[4]

This last chapter instead illustrates how conceptually 'difficult' fields, dealing with concepts and technologies that were far from the mainstay of most astronomers, astrophysicists, and particle physicists, entered the mainstream only after a long process, including decades of embryonic conceptualization and experimentation, eventually leading to experimental results which convinced a skeptical mainstream not only of the validity and scientific promise of the projects, but also of their financial and institutional feasibility. But this external skepticism also benefited these marginal new fields, lending them

4 The early generation of directors in ground-based observational astronomy came from different work cultures, namely optical astronomy (Heidelberg) and engineering (Bonn); these were soon supplemented by directors in space-based astronomy, many of whom started their careers in nuclear/particle physics and plasma research. These were closer to the earlier Max Planck traditions. Both ground-based and space-based astronomy were, however, much more directly determined by geopolitical considerations of the space race than the examples treated in this chapter. In ground-based astronomy, the choice of location, often abroad, combined with national competition to build the world's largest instruments. In space-based astronomy, West Germany's peculiar position in international politics, which precluded independent access to outer space, overdetermined the maneuvering options of its scientists. See Chapters 3 and 4.

the particular dynamism of solidarity among competitors in different countries, aware that, while they were all racing to be the first and best in entirely new branches of research, they were all in it together, up against resistance from the more powerful, mainstream communities of accelerator physicists, observational astronomers, and space-based researchers.

The developments that follow—which initially, from the 1960s to the '80s, were perceived as esoteric scientific quests—have during the time of writing this study advanced to become the core of the new 'multi-messenger' approach, which makes use of different cosmic messengers of the fundamental forces in nature—the electromagnetic, gravitational, weak and strong nuclear forces—to explore and understand the most violent phenomena in the Universe. The term multi-messenger astrophysics for this new field—a natural extension of traditional multi-wavelength astronomy—expresses the intimate connection between high-energy cosmic rays, astrophysical neutrinos, photons in multiple wavelength bands, and gravitational waves: the possibility of coincident observations of signals from diverse carriers can reveal inherently complementary information that is otherwise hidden, so finding the answers to some of the most important problems in astrophysics, and leading to the discovery of new phenomena by merging data from the world's different detection sites.[5] And more than half a century after their humble beginnings, and in the face of much skepticism from astronomers themselves, these traditions managed to establish that their instruments and sites are now also called 'telescopes' and 'observatories.' Multicomponent-based programs and experiments covering all four messengers in a broad energy range and using different techniques and detectors have currently become a main task for fundamental science, involving large international scientific collaborations working with big instruments in space and on the ground, that produce vast amounts of observational data to be analyzed and interpreted.

2 The Solar Neutrino Puzzle: Heidelberg between Cosmochemistry and Astroparticle Physics

The first newly emerging field benefited directly from the research tradition of cosmochemistry in southwestern Germany, introduced in Chapter 1. Through experimental techniques of mass spectroscopy and small sample radiochemistry, scientists from Freiburg and, later, the Max Planck Institute for Nuclear

5 Péter Mészáros et al.: Multi-Messenger Astrophysics. *Nature Reviews Physics* 1/10 (2019), 585–599. doi:10.1038/s42254-019-0101-z.

Physics in Heidelberg were able to collaborate with Brookhaven National Laboratory, where they met Ray Davis, father of solar neutrino detection and the 'solar neutrino deficit' paradox. Researchers from Heidelberg, led by Till Kirsten, improved the instrumentation and, in the 1970s, were even able to overtake the Americans by setting up the GALLEX collaboration, the next-generation experiment in which Germans, Italians, the French, and Israelis worked together with indirect support from the USA and the Soviet Union. Two decades after its conception, in the early 1990s, came the experimental results from GALLEX, which were part of the 'Decade of the Neutrino' that culminated in Nobel Prizes for the founders of the field. This leadership guaranteed a subsequent foothold in neutrino research even as it evolved away from cosmochemistry toward the electronic detection methods which have now become a central aspect of neutrino-based multi-messenger astronomy.

American Origins

The first newly emerging field benefited directly from the research tradition of cosmochemistry in southwestern Germany, introduced in Chapter 1. An intrinsically interdisciplinary research tradition with internationally recognized strengths in instrumental techniques, such as radiochemistry and mass spectrometry, was coupled with the ability to combine specific experimental approaches with the deepest theoretical questions of the time, something that had been the trademark of this Heidelberg tradition since the early collaboration initiated by Walther Bothe and Wolfgang Gentner in the 1930s.

Through experimental techniques of mass spectroscopy and small sample radiochemistry, scientists from Freiburg and, later, the Max Planck Institute for Nuclear Physics in Heidelberg were able to collaborate with Brookhaven National Laboratory, where Ray Davis, since the mid 1960s, had pioneered solar neutrinos detection using radiochemical techniques.[6] Researchers from Heidelberg, led by Till Kirsten, improved the instrumentation and, in the 1970s, were even able to overtake the Americans by setting up the GALLEX collaboration, the next-generation experiment in which Germans, Italians, the French, and Israelis worked together with indirect support from the US and the Soviet Union. The cosmochemical path to neutrino research, which had initially been paved in collaboration with scientists in the United States, turned into one of the earliest examples of German researchers leading an international collaboration project, which at one point in the 1980s was the only spearhead in an

6 Kirsten, Till: interview by Luisa Bonolis and Juan-Andres Leon, Heidelberg, October 24–25, 2017. DA GMPG, BC 601051.

otherwise neglected form of research falling right into the chasm between the interests of particle physics and astrophysics. Thanks to this groundbreaking work, in subsequent decades, the collaboration became part of a rapidly growing mosaic of interlocking research enterprises that led to a series of Nobel Prizes from the 1990s on. This cosmochemical tradition gradually gave way to the now prevalent forms of neutrino research based on electronic detection methods derived from experimental particle physics, which were adopted not only in Heidelberg, but also, in another manifestation of the most ancestral rivalry within the Max Planck Society, by researchers in Munich (see Chapter 1).

The neutrino plays a vital role in nuclear physics, particle physics, and astrophysics, and so intensive experimental efforts have always been conducted to elucidate its properties. For about 25 years, the neutrino was only a theoretical entity. Postulated by Wolfgang Pauli in 1930 to explain the apparent failure of conservation laws in nuclear beta decay,[7] it was incorporated by Enrico Fermi in his successful theory of this process and taken to be convincing evidence of its existence.[8] Neutrinos can travel several light years without interacting with matter because of the weakness of their interactions with other particles. For this reason, a neutrino was viewed as an 'undetectable particle' for many years and only in 1956 was the detection of a free (anti)neutrino finally announced by Frederick Reines and Clyde L. Cowan, who had used the newly developed technology of organic liquid scintillators, positioning a large tank of water near a nuclear reactor, a very intense source of antineutrinos.[9] With this experiment—heralding the beginning of the era of neutrino detection—the status of the neutrino changed drastically: it was no longer a hypothesis, a mere theoretical construct, but a very solid fact.

By that time, thanks to many years of technological and scientific developments, it had become possible to think about detecting reactions triggered by neutrinos. The Sun, too, is a good source of neutrinos, which are produced by nuclear reactions taking place in its core. Detailed elaboration of the proton-proton chain (*pp* chain) converting hydrogen to helium as a main mechanism

---

7 Wolfgang Pauli: On the Earlier and More Recent History of the Neutrino. In: K. Winter (ed.): *Neutrino Physics*. Cambridge: Cambridge University Press 1991, 1–25.

8 Enrico Fermi: Versuch einer Theorie der Beta-Strahlen. *Zeitschrift für Physik* 88 (1934), 161–171. doi:10.1007/BF01351864.

9 The reactor was expected to produce fluxes to the order of 10^{12}–10^{13} antineutrinos per second per cm^2. After months of data collection, they accumulated data on about only three antineutrinos per hour in their detector. Clyde L. Cowan et al.: Detection of the Free Neutrino. A Confirmation. *Science* 124/3212 (1956), 103–104. doi:10.1126/science.124.3212.103.

for the energy production in stars like the Sun, showed in 1958 that a small fraction of these neutrinos, those from the decay of beryllium-7 and boron-8, should be energetic enough to be detectable through their interaction with the nucleus chlorine-37 to form argon-37,[10] the radiochemical method of neutrino detection suggested by Bruno Pontecorvo between 1945 and 1946,[11] and by Luis Alvarez in 1949.[12] In the late 1950s and early '60s, it was pointed out that detection of solar neutrinos could be a direct way of testing theoretical solar models.[13]

In the early 1960s, Ray Davis Jr., a radiochemist from Brookhaven, took up the challenge and devised an experiment in the one-mile-deep Homestake Mine in South Dakota to detect the flux of these energetic neutrinos coming from the Sun. This was done using a huge tank of carbon tetrachloride (CCl4, used in the past as dry cleaning fluid) as target material and locating it deep underground—to minimize cosmic ray interactions producing reactions that could mimic neutrino capture—and then painstakingly extracting and counting with a small Geiger counter the tiny amounts of argon-37 atoms that were produced by the very rare interaction of neutrinos from the Sun with chlorine-37 atoms. The enormous volume—which would be characteristic of all future neutrino detectors—could overcome the problem of the very small probability for a neutrino to interact with chlorine nuclei of the radiochemical detector.[14] The basic reason for doing solar-neutrino experiments was to test *quantitatively* the theories of nuclear energy generation in stars and of stellar

---

10 Alastair G.W. Cameron: Nuclear Astrophysics. *Annual Review of Nuclear Science* 8/1 (1958), 299–326. doi:10.1146/annurev.ns.08.120158.001503. William A. Fowler: Completion of the Proton-Proton Reaction Chain and the Possibility of Energetic Neutrino Emission by Hot Stars. *The Astrophysical Journal* 127 (1958), 551–556. doi:10.1086/146487.

11 Luisa Bonolis: Bruno Pontecorvo. From Slow Neutrons to Oscillating Neutrinos. *American Journal of Physics* 73/6 (2005), 487–499. doi:10.1119/1.1852540. Frank Close: *Half-Life. The Divided Live of Bruno Pontecorvo, Physicist or Spy*. New York, NY: Basic Books 2015.

12 Luis W. Alvarez, A Proposed Experimental Test of the Neutrino Theory, Report UCRL-328, University of California Radiation Laboratory, April 18, 1949.

13 John N. Bahcall: Solar Neutrinos. Theoretical. *Physical Review Letters* 12/11 (1964), 300–302. doi:10.1103/PhysRevLett.12.300. Bahcall was the first to fully develop a solar model that included all the physical parameters needed to calculate the solar neutrino flux at the Earth. He continued to work over the years on refining his calculations and made a key contribution to understanding of the significance of the discrepancy between the result of the chlorine experiment and Standard Solar Model predictions.

14 By 1954, Davis had already used a large tank of carbon tetrachloride in the basement of one of the Savannah River reactors, at that time the most intense antineutrino source in the world, to try to detect reactor antineutrinos; and he established that they *did not interact* with chlorine nuclei. Raymond Jr. Davis: Attempt to Detect the Antineutrinos from a Nuclear Reactor by the Cl^{37}(Anti ν, e^-)A^{37} Reaction. *Physical Review* 97/3 (1955),

evolution. The photons that are the subject of conventional astronomy come from the outermost layers of a star, whereas neutrinos, because of their large mean free path, can reach us directly from the deep interior of a star, where the nuclear reactions responsible for energy generation and stellar evolution occur. Davis himself originally set up his experiment in the context of solar physics, seeking to determine whether the measured neutrino flux coincided with the theoretical predictions made by solar and nuclear astrophysicists. However, the initial results, published in 1968, came as a surprise: the neutrino capture rate in the detector showed that the upper bound on the solar neutrino flux was two to three times smaller than expected on the basis of the Standard Solar Model.[15] The deficit in the solar-neutrino flux resulting from the first large-scale experiment designed to detect neutrinos from the interior of the Sun was thus a blow to the then acknowledged theory of solar-type stellar physics and became known as the *solar neutrino problem*.

Over the following 20 years, Davis's chlorine-argon experiment, which could be considered as the beginning of neutrino astrophysics, remained the only experiment providing data on solar neutrinos. For many years, these experiments faced resistance from the mainstream of accelerator-based particle physicists, who expected it would be too difficult to obtain reliable data due to the experimental design, which was also perceived as unglamorous chemistry carried out in abandoned mines, in contrast to their well-established work with nuclear reactors and particle beams in accelerators.[16]

766–769. doi:10.1103/PhysRev.97.766. Davis's 'failed' experiment, as well as the successful detection of the ghost particle by Cowan and Reines, provided the first experimental evidence that an *anti-neutrino*—the particle which, according to the theory, is emitted in nuclear processes taking place in a nuclear reactor—is actually different from a *neutrino*, the particle which can instead trigger the expected reaction with chlorine nuclei, as studied by Davis, and which is released during nuclear fusion reactions in the Sun's core.

15 See Davis and Bahcall's back-to-back articles: Raymond Jr. Davis, Don S. Harmer, and Kenneth C Hoffman: Search for Neutrinos from the Sun. *Physical Review Letters* 20/21 (1968), 1205–1209. doi:10.1103/PhysRevLett.20.1205. John N. Bahcall, Neta A. Bahcall, and Giora Shaviv: Present Status of the Theoretical Predictions for the 37Cl Solar-Neutrino Experiment. *Physical Review Letters* 20/21 (1968), 1209–1212. doi:10.1103/PhysRevLett.20.1209. Such results were confirmed up until 1994, during the period when Davis's experiment continued to run. Raymond Davis: A Review of the Homestake Solar Neutrino Experiment. *Progress in Particle and Nuclear Physics* 32 (1994), 13–32. doi:10.1016/0146-6410(94)90004-3.

16 Bahcall, John N., and Raymond Jr. Davis: The Evolution of Neutrino Astronomy. *Publications of the Astronomical Society of the Pacific* 112/770 (2000), 429–433. doi:10.1086/316545. For a mid-1970s account of the reaction of the scientific community to the Homestake experiment, see also Trevor Pinch: *Confronting Nature: The Sociology of Solar-Neutrino Detection*. Dordrecht: Springer Science & Business Media 2013.

An Unexpected Paradox between Astrophysics and Particle Physics

To address the difficult theoretical implications of his research results—which were steadily improved, as the unwanted background effects and uncertainties from nuclear decay-counting statistics of the few argon atoms produced in the tank were constantly lowered—Davis found his greatest ally in John Bahcall, a solar astrophysicist from Princeton, whose systematic work over the years proved that the low flux found in the solar neutrino experiments of Davis and others could not be explained by errors in the Standard Solar Model.[17] Over the many years that Davis's experiment was conducted, during which time results trickled down only slowly, the measured flux continued to be only around one-third of the theoretical prediction, launching almost three decades of scientific debates that covered a wide range of epistemological questions, from the accuracy of the experimental system to the validity of the nuclear fusion theories of the Sun, and ultimately, whether the problem may have originated in the nature of neutrinos themselves.

In 1967, before the first results were published by Davis,[18] Bruno Pontecorvo had anticipated the solar neutrino problem, pointing out that neutrinos could oscillate between different states, and thus solar neutrinos (electron neutrinos) might transform into muon neutrinos during their journey from the Sun to the Earth. This phenomenon would lead to an observed deficit of neutrinos in chlorine-based experiments which could only detect neutrinos of a specific lepton flavor, electron neutrinos.[19] The phenomenon of neutrino oscillation

17 John N. Bahcall: Two Solar Neutrino Problems. *Nuclear Physics B—Proceedings Supplements* 43/1–3 (1995), 41–46. doi:10.1016/0920-5632(95)00447-H.

18 Davis, Harmer, and Hoffman, Neutrinos from the Sun, 1968, 1205–1209. Bahcall, Bahcall, and Shaviv, Present Status, 1968, 1209–1212.

19 Bruno Pontecorvo: Neutrino Experiments and the Problem of Conservation of Leptonic Charge. *Pis'ma v Zhurnal Èksperimental'noi i Teoreticheskoi Fiziki* 53 (1967), 984–988. http://inspirehep.net/record/51319?ln=en. Last accessed 11/10/2017. In 1962, it was experimentally established that a second type of neutrino (the muon neutrino) exists, which is paired with the muon in the same way the already known (electron) neutrino is paired with the electron: the neutrino involved in nuclear beta-decay and the one in muon-decay are thus two different particles. G. Danby et al.: Observation of High-Energy Neutrino Reactions and the Existence of Two Kinds of Neutrinos. *Physical Review Letters* 9/1 (1962), 36–44. doi:10.1103/PhysRevLett.9.36. Reactions induced by electron neutrinos (and electron anti-neutrinos) produce electrons (or positrons), while events triggered by muon neutrinos (or muon anti-neutrinos) produce muons (or anti-muons), all members of the lepton family, which do not undergo strong nuclear interactions. The first evidence of the existence of the tau particle, introducing the third generation of the lepton family, with which the tau neutrino is associated, a third neutrino flavor, was announced in 1975. Martin L. Perl et al.: Evidence for Anomalous Lepton Production in e^+-e^- Annihilation. *Physical Review Letters* 35/22 (1975), 1489–1492. doi:10.1103/PhysRevLett.35.1489.

from a type (flavor) to another—a pure quantum mechanical phenomenon—could take place only if neutrinos had a mass, while neutrinos are assumed to be massless in the Standard Model, the Standard Theory of elementary particle physics describing not only the microscopic forces, but also the nature of the basic constituents of matter. In this case there turn out to be three 'families' of elementary particles called quarks and leptons, neutrinos belonging to the latter in three 'flavors' depending on their relationship to the three heavy leptons: electron, muon, and tau. If neutrinos had mass, they would alternate among these.[20] The neutrinos that have jumped from one flavor to another would then escape detection by Davis's radiochemical detector, as this was geared only toward detecting one kind of neutrino, electron neutrinos produced in the solar core.[21] At the end of the 1960s, the first controversial—and disappointing—results of Davis' solar neutrino experiment set the stage for the next decade, during which the necessity of clearing up the question about lepton charge conservation and the number of neutrino types (neutrino oscillations) became increasingly a pressing problem, in particular for the future of solar neutrino astronomy. Most people believed that a possible solution for the discrepancy could be an astrophysical problem, that is, the consequence of the Standard Solar Model providing an inadequate description of the internal workings of the Sun. But perhaps our understanding of the neutrino itself was at fault, as suggested by Pontecorvo: it may have been a problem in the

20 Pontecorvo, who had devised the chlorine-argon radiochemical method for detecting neutrinos in 1945–46, put forward the idea of neutrino flavor oscillations in different forms from the 1950s onward. Neutrinos, like the charged particles electron, muon, and tau (and their antiparticles) to which each neutrino type is associated (three types of neutrinos: electron-type/muon-type/tau-type), are leptons, that is, elementary particles that do not undergo strong interactions. During the transition from one type to another, lepton numbers (i.e., the number of leptons minus the number of corresponding antileptons in each reaction) would not be conserved, contrarily to what is implied by the Standard Theory of elementary particle physics. The fact that the lepton number is not a conserved quantum number in the phenomenon of neutrino transition from one flavor to another leads to neutrino mixing, and consequently, oscillations of neutrinos, and thus to physics beyond the Standard Theory, as was suggested by Gribov and Pontecorvo one year after the first chlorine results were published by Davis. Vladimir N. Gribov, and Bruno Pontecorvo: Neutrino Astronomy and Lepton Charge. *Physics Letters B* 8/7 (1969), 493–496. doi:10.1016/0370-2693(69)90525-5. For Pontecorvo's contributions to neutrino physics, see Samoil M. Bilenky: Bruno Pontecorvo and the Neutrino. *Physics-Uspekhi* 57/5, 489–496. doi:10.3367/UFNe.0184.201405g.0531.

21 In 1957, Pontecorvo put forward for the first time the idea of neutrino oscillation suggesting that *antineutrinos* from the reactor might transform into *neutrinos* and be able to trigger Davis's detector. But this fruitful idea was only due to rumors reaching Pontecorvo that Davis had actually observed such events, which eventually were not confirmed.

Standard Theory of Particle Physics, according to which neutrinos are massless particles, as had been assumed by Pauli and Fermi. But massless neutrinos cannot oscillate.

The issue of neutrino mass and the solar neutrino deficit thus established itself as a question at the intersection of astrophysics and particle physics, becoming one of the most controversial proposals of fundamental physics at the time.[22] The physical and cosmological implications of this possibility could be enormous, as massive neutrinos were initially proposed as the most natural particle candidate for dark matter,[23] invisible unknown matter distinct from ordinary matter such as protons and neutrons, making up about 27 percent of the Universe, and whose presence can only be inferred from gravitational effects on visible matter. And so, not only could they play a role in the development of structure in the Universe (galaxies, clusters, etc.), but they were one of the first recognized illustration of the close relationship that exists between cosmology and elementary particle physics: cosmology could be used to put constraints on the properties of neutrinos and particle theory could have important consequences for cosmology.[24]

In the early hot and dense Universe, interactions between elementary particles were essential, determining the structure of the Universe we see today. A new class of theories proposed during the 1970s, called the Grand Unified Theories (GUTs), suggested that all interactions (strong, electromagnetic, and weak, but not gravity) are unified at a large energy scale and neutrino masses would be inversely proportional to this scale.[25] As neutrino masses and mixing could represent a probe into the physics at GUT energy scales, the questions of neutrino mass and of solar neutrino flux grew in importance, both in high energy physics and cosmology.

---

22 For an overview of the problem in the late 1960s and early 1970s, see John N. Bahcall, and R. L. Sears: Solar Neutrinos. *Annual Review of Astronomy and Astrophysics* 10 (1972), 25–44. doi:10.1146/annurev.aa.10.090172.000325. For an historical overview, see Donald H. Perkins: The Remarkable History of the Discovery of Neutrino Oscillations. *The European Physical Journal H* 39/5 (2014), 505–515. doi:10.1140/epjh/e2014-50037-4.

23 Semen Solomonovich Gershtein, and Yakov Borisovich Zeldovich: Rest Mass of Muonic Neutrino and Cosmology. *Soviet Journal of Experimental and Theoretical Physics Letters* 4 (1966), 120–122. http://adsabs.harvard.edu/abs/1966JETPL...4..120G. Last accessed 8/8/2020. For a discussion about neutrinos as candidates for dark matter on a cosmological scale, see also Virginia Trimble: Dark Matter in the Universe. Where, What, and Why? *Contemporary Physics* 29/4 (1988), 373–392. doi:10.1080/00107518808213765.

24 Michael S. Turner: Neutrinos and Cosmology. *AIP Conference Proceedings* 72/1 (1981), 335–355. doi:10.1063/1.33010.

25 Andrzej J. Buras et al.: Aspects of the Grand Unification of Strong, Weak and Electromagnetic Interactions. *Nuclear Physics B* 135/1 (1978), 66–92. doi:10.1016/0550-3213(78)90214-6.

Neutrinos had always provided "the major consummated connection between particle physics and astrophysics and cosmology."[26] Solar neutrino detection, as a challenge both for astrophysics (test of the Standard Model of the Sun and of the stars) and for particle physics (the observed deficiency of solar neutrinos might be due to the oscillation phenomenon, a test for physics beyond the Standard Theory of elementary particles), became one of the major contributors to the emergence of particle-astrophysics. The identity of the new field materialized in conferences held from the early 1980s on, where a widely diversified physics community had the chance to discuss and explore the conceptual links between theoretical and experimental particle physics, nuclear astrophysics, and fundamental topics, such as the early Universe, its large-scale structure, dark matter and dark energy, and cosmic background radiation.[27] The Big Bang, and the very early Universe with its high temperatures and particle densities, became a "hot laboratory for the nuclear and particle physicist," in the words of Zeldovich, one of the founding fathers of particle cosmology.[28]

Heidelberg's Privileged Position in Experimental Cosmochemistry

But to solve the solar neutrino paradox, another series of solar neutrino-counting experiments on a different scale was needed in order to detect the most abundant but low-energy flux of neutrinos from the dominating *pp* chain of thermonuclear reactions occurring inside the Sun, converting hydrogen into helium starting from the fusion of two protons, which is responsible for about

---

26 David N. Schramm: Neutrinos and Cosmology. *Nuclear Physics B—Proceedings Supplements* 38/1–3 (1995), 349–362. doi:10.1016/0920-5632(94)00769-R.

27 In bringing together experts in the fields of nuclear and particle physics, astrophysics, and cosmology, the international conference *Weak and Electromagnetic Interactions in Nuclei*, organized in October 1986 by the Max Planck Institute for Nuclear Physics, in conjunction with the 600th anniversary of the University of Heidelberg, testifies with its wide program to the early and deep involvement of the Institute in the novel trend connecting the laws of microphysics, astrophysics, and cosmology. Hans Volker Klapdor (ed.): *Weak and Electromagnetic Interactions in Nuclei. Proceedings of the International Symposium, Heidelberg, July 1–5, 1986*. Berlin Heidelberg: Springer 1986. In this regard, see also a slightly later volume exploring the close connections between micro (nuclear and particle physics) and macro physics (astrophysics and cosmology) induced by the weak interactions, and paying special attention to neutrinos. Klaus Grotz, and Hans Volker Klapdor: *The Weak Interaction in Nuclear, Particle and Astrophysics*. Bristol: Adam Hilger 1990.

28 Yakov Borisovich Zeldovich: The Universe as a Hot Laboratory for the Nuclear and Particle Physicist. *Comments on Astrophysics and Space Physics* 2 (1970), 12–17. http://adsabs.harvard.edu/abs/1970CoASP...2...12Z. Last accessed 9/19/2020.

99 percent of the energy production.[29] Only the *pp* neutrino species could be predicted accurately. This is almost solar model independent and, consequently, more significant for testing the hypothesis that fusion of hydrogen powers the Sun. It could therefore serve as a 'known source' in the long-baseline neutrino oscillation experiment, the Sun–Earth distance of about 1.5×10^8 km. A significant deficit would provide a further experimental proof, and explain electron neutrino disappearance in terms of oscillations to a different flavor, i.e., the neutrinos would not in fact disappear but just change into a different type, so escaping detection.

However, *pp* neutrinos were not accessible in Davis's experiment, because their energy is below the threshold of the neutrino reaction converting nuclei of chlorine-37 into radioactive argon-37. But detection of the *pp* neutrinos could still be done using the radiochemical methods, with a tank full of fluid, deep underground, but containing the much rarer substance of gallium. This experiment would detect solar *pp* neutrinos by employing a reaction in which the impinging neutrino would transform a nucleus of gallium-71 into a nucleus of germanium-71 plus an antineutrino. The lower threshold of this neutrino capture reaction would allow detection of *pp* neutrinos.[30]

It was known that a realistic experiment would require tenths of tons of this aluminum-affine metal with the required radiochemical purity, but gallium was very expensive and thus the construction of specific gallium plants to extract it at an industrial level would require an investment to the order of 100 million dollars. Moreover, even before anyone could dare to ask for funding, a series of open questions would have to be answered, some of which related to the development of a suitable low-level counting procedure for germanium-71.[31] The questions arising included whether it was possible to

---

29 In a proton-proton cycle, four hydrogen nuclei (protons) are fused, combining to form one helium nucleus. A small percentage of the original mass is lost in the process, mainly by conversion into heat energy, but some energy escapes in the form of neutrinos.

30 The capture of neutrinos by gallium-71 to produce germanium-71, an isotope with an 11-day half-life (the time needed for half the neutrons to decay) had a threshold of 233 keV, which was ideal for observing neutrinos from the *pp* reaction. This gallium method meant it was possible to detect the neutrinos from *all* solar thermonuclear reactions, including the initial proton fusion chain, which represents more than 98 percent of the neutrinos produced in the Sun, as preliminarily discussed in John N. Bahcall, and Raymond Jr. Davis: Solar Neutrinos. A Scientific Puzzle. *Science* 191/4224 (1976), 264–267. doi:10.1126/science.191.4224.264.

31 When very small activities of radionuclides are to be measured by direct observation of the radioactive decay, a certain amount of effort is required to choose and adapt detector systems, in order to attain high counting sensitivity and keep instrument background as low as possible.

establish a committed international network of top scientists with the relevant expertise and support from their agencies, as well as whether there was a suitable underground laboratory available.[32]

It is at this stage of developments that Heidelberg's scientists entered the business of solar neutrinos, but there were deep roots that made this possible in the first place.

Since the 1950s, Wolfgang Gentner had been collaborating most closely with American researchers at Brookhaven National Laboratory (BNL), where his disciple, the cosmochemist Joseph Zähringer, had been the first in a line of visitors to the Chemistry Department, among them, later, also Till Kirsten.[33] German cosmochemists were valued in Brookhaven for their expertise in mass spectrometric detection of extremely small quantities of stable rare gas isotopes. In Heidelberg, geochemical investigations by mass spectrometry had been used in particular to determine the half-life for the double-beta decay of tellurium-130 in connection with studies on the xenon and krypton isotopic composition of meteorites.[34] From 1966, during his postdoc stay at BNL, Kirsten collaborated with Oliver Schaeffer, working on the double-beta decay

32 Till A. Kirsten: Radiochemical Solar Neutrino Experiments and Implications. *Physica Scripta* 2000/T85 (2000), 71–81, 52. doi:10.1238/Physica.Topical.085a00071.

33 See, for example, articles published by Zähringer and Kirsten with Oliver Schaeffer while they were at Brookhaven: Oliver Schaeffer, and Josef Zähringer: Helium- und Argon-Erzeugung in Eisentargets durch energiereiche Protonen. *Zeitschrift für Naturforschung A* 13/4 (1958), 346–347. doi:10.1515/zna-1958-0413. Oliver Schaeffer, and Josef Zähringer: High-Sensitivity Mass Spectrometric Measurement of Stable Helium and Argon Isotopes Produced by High-Energy Protons in Iron. *Physical Review* 113/2 (1959), 674–678. doi:10.1103/PhysRev.113.674. Till A. Kirsten et al.: Experimental Evidence for the Double-Beta Decay of Te^{130}. *Physical Review Letters* 20/23 (1968), 1300–1303. doi:10.1103/PhysRevLett.20.1300.

34 Till A. Kirsten, W. Gentner, and O. Müller: Isotopenanalyse der Edelgase in einem Tellurerz von Boliden (Schweden). *Zeitschrift für Naturforschung A* 22/11 (1967), 1783–1792. doi:10.1515/zna-1967-1116. Ordinary beta decay occurs when a neutron in an unstable nucleus spontaneously turns into a proton emitting an electron (known as a beta particle) and an antineutrino. The proton is retained in the nucleus; the electron and the antineutrino are ejected. Since the atom gains a unit of positive charge, it moves up one slot in the periodic table. The half-life of this process is just a few thousandths of a second, while for a neutron outside the nucleus the half-life is about 10 minutes. In the double-beta decay, the process occurs when two neutrons *simultaneously* convert into protons, emitting two electrons and two antineutrinos. The half-life of this second decay process varies from one isotope to the next, but is much longer than the standard beta decay, being at least 10^{18} years. A third form of beta decay, theoretically hypothesized in the 1930s, should take place if the two emitted antineutrinos can annihilate each other—provided they are their own antiparticles—so that neither will escape outside the nucleus. This is the so-called *neutrinoless* form, a very rare process that can only

problem, a phenomenon which largely involves questions related to the nature of the neutrino.[35] In Brookhaven, Kirsten was impressed by Davis's work and

take place if neutrinos, in contrast to all the other known fermions, are Majorana particles, that is, they coincide with their own antiparticles. In this case, the two neutrinos emitted in the double-beta decay process can annihilate, leaving behind only the two electrons as results of the decay process. Its existence is still to be established (only limits on its half-life have been reported over the years, in different, increasingly refined and challenging experiments). For this reason, investigations of these processes can provide valuable information on the properties of the neutrino. Postwar experiments on the search for double-beta decay were resumed in the late 1940s, using Geiger, proportional, and scintillation counters, but a much higher sensitivity could be reached at that time with geochemical experiments consisting in the separation of nuclei resulting from double-beta decay processes from nuclei of ancient minerals of known geological age (up to several billion years), and their subsequent isotopic composition analysis through mass spectroscopy. It is still an open question in neutrino physics, whether the neutrino is a Dirac particle or a Majorana particle. In the former case, the neutrino and its antineutrino are different particles, in the latter case the neutrino is its own antiparticle like the neutral pion, the π^0. The Max Planck Institutes for Physics and Nuclear Physics are both currently involved in the GERDA (Germanium Detector Array), an underground international experiment at the Gran Sasso laboratories, using germanium detectors to investigate the neutrinoless double-beta decay process, that can occur *only if neutrinos and their antiparticles are identical and have a mass.*

35 The double-beta decay transition from selenium-82 into krypton-82 was studied for the first time by Gentner, Kirsten, and Schaeffer (the authors thanked Davis for "valuable discussions"). Till A. Kirsten, W. Gentner, and O. A. Schaeffer: Massenspektrometrischer Nachweis von ββ-Zerfallsprodukten. *Zeitschrift für Physik* 202/1 (1967), 273–292. doi:10.1007/BF01331214. The following year, the existence of the process was proved through the decay of tellurium-130 into xenon-130 with a half-life of about 10^{21} years: Kirsten et al., Double-Beta Decay, 1968, 1300–1303. In these articles, neutrinoless double-beta decay was also discussed, but it is impossible with geological methods to distinguish between neutrinoless and two-neutrino double-beta decay. Later experiments, conducted in Heidelberg by Klapdor-Kleingrothaus, who specifically searched for neutrinoless double-beta decay, used different methods. Then, in 1986, Mike Moe, working with Fred Reines at the University of California, Irvine, finally succeeded with the first *real-time direct* observation of two-neutrino double-beta decay in the laboratory, using selenium-82 as a target and deriving a half-life value which roughly matched the one obtained from the Max Planck group's analyses. In August 1987, 20 years after the Heidelberg/Brookhaven results, and during which skepticism had circulated within the physicists' community, they could claim a definitive laboratory observation of two-neutrino double-beta decay. S. R. Elliot, A. A. Hahn, and M. K. Moe: Search for Double Beta Decay in ^{82}Se. In: T. Kitagaki, and H. Yuta (eds.): *Proceedings of the 12th International Conference on Neutrino Physics and Astrophysics, Sendai, Japan, June 3–8, 1986.* Singapore: World Scientific 1986, 93–100. S. R. Elliott, A. A. Hahn, and M. K. Moe: Direct Evidence for Two-Neutrino Double-Beta Decay in ^{82}Se. *Physical Review Letters* 59/18 (1987), 2020–2023. doi:10.1103/PhysRevLett.59.2020. These results were also presented at an International Workshop on Neutrino Physics held in Heidelberg in October 1987: S. R. Elliott, A. A. Hahn, and M. K. Moe: A Direct

followed with great interest his efforts to improve his low-level proportional counters to detect the very small activities from radioactive decay of nuclides that could also be applied to solar neutrino experiments. The same capabilities in low-level counting and determination of minute quantities of stable rare gas isotopes by mass spectrometry were applied over the following years in a collaboration between BNL and the Max Planck Institute for Nuclear Physics in Heidelberg, in NASA's Apollo Lunar Sample Analysis Program (see Chapter 2), which meant the two institutes remained in contact. During this time, they discussed the possibility of using gallium, which would allow detection of the major neutrino flux coming from the Sun.

For a long time, this remained a dream because of the high cost of industrial production but, all the same, they kept an eye on the problem.

Gallium Experiment Proposals and American Failure

The gallium solar neutrino experiment became a practical method for observing the proton-proton (*pp*) reaction neutrinos when industry began extracting ton quantities of gallium as a by-product of the manufacture of aluminum. Producing gallium in quantity was motivated by the need for gallium to produce various electronic devices. In 1974, research on chemical procedures for extracting germanium from gallium began at Brookhaven National Laboratory in collaboration with the University of Pennsylvania. A similar effort was started in the Soviet Union, where a gallium experiment had been proposed by Vadim Kuzmin at the Lebedev Physics Institute in Moscow already in the mid-1960s, as a means of observing the low energy neutrinos from the basic proton-proton fusion reaction, whose flux is accurately calculated from solar models, and is essentially independent of many factors that influence the calculations of the boron-8 neutrino flux which represented the main source for Davis' chlorine-argon experiment.[36]

Laboratory Measurement of Two-Neutrino Double Beta Decay in ^{82}Se. In: Hans Volker Klapdor, and Bogdan Povh (eds.): *Neutrino Physics. Proceedings of an International Workshop Held in Heidelberg, October 20–22, 1987*. Berlin Heidelberg: Springer 1988, 213–219. doi:10.1007/978-3-642-73679-7_20.

36 Vadim A. Kuzmin, and Georgii T. Zatsepin: On the Neutrino Spectroscopy of the Sun. *Proceedings of the 8th International Cosmic Ray Conference, December 2–14, 1963, Jaipur, India.* Bombay, India: Commercial Printing Press 1965, 1023–1024. https://ui.adsabs.harvard.edu/#abs/1965ICRC....2.1023K. Last accessed 1/6/2019. Kuzmin was the first to suggest, in 1965, the reaction scheme related to the transformation of gallium-71 into germanium-71, whose low threshold of 233 keV would allow the detection of *pp* neutrinos, by far the most abundant solar neutrinos, but with a very low energy <420 keV. Vadim A. Kuzmin: Detection of Solar Neutrinos by Means of the $Ga^{71}(\nu,e^{-})Ge^{71}$ Reaction. *Journal of Exper-*

Moreover, in the meantime, Davis had found a significant neutrino deficit in his Homestake experiment and, around the mid-1970s, Davis and Bahcall had begun to put forward the idea that another experiment was required "to settle the issue of whether our astronomy or our physics is at fault."[37] A measurement of the solar proton-proton neutrino flux was regarded as a critical test of our knowledge of neutrino physics and the fusion processes in the interior of the Sun. At that time, this possibility was being taken into consideration in Heidelberg, as Till Kirsten wrote to Davis:

> we have some more or less speculative ideas about a gallium or bromine solar neutrino experiment and very much need your judgment and advice in order to facilitate a decision whether we should get serious [...].[38]

The first official reference to these "speculative ideas" can be found in the Annual Report of the Max Planck Society for the year 1977, where investigations using cosmochemical methods were mentioned in connection with the possibility of identifying interactions of solar neutrinos with Earth nuclei.[39] In early January 1978, a Solar Neutrino Workshop devoted to the status and future of solar neutrino research took place in Brookhaven, with an emphasis on possible new experiments.[40] Till Kirsten participated, and a collaboration to develop a gallium radiochemical detector was set up between Heidelberg and Brookhaven, which included the University of Pennsylvania and the Insti-

---

imental and Theoretical Physics 22/5 (1966), 1051–1052. http://www.jetp.ac.ru/cgi-bin/e/index/e/22/5/p1051?a=list. Last accessed 4/30/2019.

37 John N. Bahcall et al.: Proposed Solar-Neutrino Experiment Using ^{71}Ga. *Physical Review Letters* 40/20 (1978), 1351–1354. doi:10.1103/PhysRevLett.40.1351.

38 See timeline of "Early events" related to the preliminary phase in Till Kirsten's papers (Till Kirsten, private collection, DA GMPG, BC 600004). We are very grateful to Till Kirsten for giving us an opportunity to consult such relevant documents related to the early phase of the Heidelberg solar neutrino project. From conversations with Kirsten we derived great insight into the development of the GALLEX experiment (Till Kirsten: interview by Luisa Bonolis and Juan-Andres Leon, Heidelberg, October 24–25, 2017. DA GMPG, BC 601051).

39 Max-Planck-Gesellschaft zur Förderung der Wissenschaften (ed.): *Max-Planck-Gesellschaft Jahrbuch 1977*. Göttingen: Vandenhoeck & Ruprecht 1977, 494.

40 G. Friedlander: Report on an Informal Conference on the Status and Future of Solar-Neutrino Research. *Comments on Astrophysics* 8 (1978), 47–54. http://adsabs.harvard.edu/abs/1978ComAp...8...47F. Last accessed 1/5/2019.

tute of Advanced Study in Princeton.[41] The discrepancy between the results of the chlorine-37 solar neutrino experiment and the predictions made using the standard model of the solar interior increasingly suggested that either some basic aspects of the standard theory of stellar evolution were wrong, or that neutrinos produced in the interior of the Sun did not reach the Earth, at least not in the form or quantity in which they are emitted. The idea was to demonstrate the feasibility of a gallium experiment that could distinguish between these two broad classes of explanation. The possibility that neutrinos could oscillate from one flavor to another, or even decay before they reached the Earth, was also considered.[42] In 1979, the premises were laid for a joint MPI/ BNL gallium experiment in the Homestake mine, where Davis was conducting his own hunt for solar neutrinos.[43] As an initial step, because of the high costs, a 'pilot experiment' with 1.5 tons of gallium was planned, which it was announced would be underway by the end of that year.[44] It aimed to demonstrate that all steps in the planned experiment, from the extraction of the germanium-71 atoms produced by interaction of solar neutrinos with gallium to counting them, would be feasible at the level necessary for a full-scale experiment.[45]

---

41 A draft of the proposal was signed by Till Kirsten on February 21, 1978: "Zum Forschungsprogramm SOLARE NEUTRINOS" am MPI für Kernphysik, Abteilung Kosmochemie (Isotopengruppe)" (AMPG, III. Abt., Rep. 68 A, No. 166/1.1). The proposal stressed how the expertise of the Cosmochemistry Department in low-level counting techniques, mass-spectrometry, and neutron reactions analysis would be an excellent basis for participating in such a project, even if this meant that the lunar sample investigations would be restricted. The end of the document mentions Oliver Schaeffer's sabbatical year, beginning in late 1978 and to be spent in Heidelberg, in order to organize the collaboration. The participation of Wolfgang Hampel, as head of the Low-Level Laboratory, would be fundamental for such an experiment looking for rare events, where the background identification plays a key role, requiring the development of detectors for extremely low count rates. Hampel eventually became a leading figure in the solar neutrino experiment. Wolfgang Hampel: Particle Physics. Can the Gallium Detector Solve the Solar Neutrino Problem? *Nature* 308/5957 (1984), 312. doi:10.1038/308312a0.

42 Bahcall et al., Solar-Neutrino Experiment, 1978, 1351–1354.

43 See minutes of the first meeting of the solar neutrino collaboration, BNL, February 1–2, 1979 (Till Kirsten, private collection, DA GMPG, BC 600004). Pilot counting experiments, foreseen in both Brookhaven and Heidelberg, were discussed.

44 Within the pilot experiment, counting techniques were investigated in the course of 1980. The group was formed by Kirsten, W. Hampel, G. Heusser, M. Hübner, J. Kiko, O. A. Schaeffer, and R. Schlotz.

45 Neutrino capture in gallium-71 leads to germanium-71, which then decays back to gallium-71 by electron capture with a half-life of 11.4 days. Fifty tons of gallium as gallium trichloride solution would be needed for one neutrino capture per day (only neutri-

Collaboration between Heidelberg and Brookhaven continued during the following two years, and the pilot experiment operating at Brookhaven National Laboratories was completed in 1983, having demonstrated the feasibility of a full-scale gallium detector.[46] But there was also the problem of finding a suitable underground laboratory, because in fact there was not sufficient space at the Homestake site used by Davis. The search in North America, including Canada, had not worked out, and nor had options in Germany (Asse salt mine near Salzgitter).

In any case, the US Department of Energy (DOE) decided not to fund the project, the evident potential of the pilot experiment notwithstanding. Funding for the full-scale project was denied in the United States overall, owing

---

nos deriving from *pp* or *pep* reactions). A photo of the proportional counter specially developed to observe decay of the few germanium-71 atoms produced by reactions triggered in the gallium tank by neutrinos was reproduced in the 1981 Report. Max-Planck-Gesellschaft zur Förderung der Wissenschaften (ed.): *Max-Planck-Gesellschaft Jahrbuch 1981*. Göttingen: Vandenhoeck & Ruprecht 1981, 536. The international team collaborating on the pilot experiment had the following members: R. Davis, B. T. Cleveland, G. Friedlander, S. Katcoff, J. K. Rowley, and J. Weneser (Brookhaven National Laboratory); T. Kirsten, W. Hampel, G. Heusser, M. Hübner, J. Kiko, O. A. Schaeffer, and R. Schlotz (Max Planck Institute for Nuclear Physics, Heidelberg); I. Dostrovsky and Y. Eyal (Weizmann Institute of Science, Rehovot); J. N. Bahcall (Institute for Advanced Study, Princeton); K. Lande, W. Frati, R. I. Steinberg (University of Pennsylvania, Philadelphia). Assuming there was no problem with funding, they were supposed to obtain the necessary amount of 50 tons of gallium in stages of 10 or 15 tons per year over the next three or four years. The 1.5 ton gallium (2 million USD) were eventually financed by the Max Planck Society, indicating that the Society had full confidence in the relevance of such an enterprise. In the final section of his contribution to the conference proceedings, Hampel discussed the prospects for the detection of neutrinos emitted from collapsing stars: Wolfgang Hampel: Low-Energy Neutrinos in Astrophysics. In: Ettore Fiorini (ed.): *Neutrino Physics and Astrophysics*. Boston MA: Springer 1982, 61–79. The pilot experiment was concluded in 1983 and results were presented at the conference on Solar and Neutrino Astronomy held in August 1984 in Lead, US. Wolfgang Hampel: The Gallium Solar Neutrino Detector. In: M.L. Cherry, K. Lande, and W.A. Fowler (eds.). Solar Neutrinos and Neutrino Astronomy. 23–25 August 1984, Lead, SD, USA. American Institute of Physics 1985, 162–174. doi:10.1063/1.35172.

46 In this phase, Israel Dostrovsky and his colleagues from the Weizman Institute of Science in Rehovot, representing Israel at GALLEX, made a major contribution (we thank Till Kirsten for this remark). A report on the status of the gallium solar neutrino experiment conducted by the collaboration team, updated to early 1983, can be found in the paper presented at the conference *Science Underground*, held in Los Alamos in 1982. W. Hampel: The Gallium Solar Neutrino Detector. *AIP Conference Proceedings*. American Institute of Physics 1983, 88–96. doi:10.1063/1.33928. Until 1985, the solar neutrino experiment was not mentioned in the Annual Reports of the Institute for Nuclear Physics, except for one sentence in the general short description of the research fields conducted at the institute: "Aufbau eines Experiments zur Messung solarer Neutrinos."

to what was later described by Bahcall as a typical problem of interdisciplinary research at the time: the astrophysicists recommended that it be funded as a particle physics experiment, and the particle physicists expected it to be funded from the astrophysics budget, and so "DOE could not get the nuclear physics and the particle physics sections to agree on who had the financial responsibility for the experiment."[47] Kirsten later commented that, despite John Bahcall's influential help, the funding effort failed

> most probably because the whole conception of radiochemical neutrino experiments had the image of being exotic, at best. More often, it triggered late party amusement at conference banquets.[48]

In the meantime, in striking contrast to what was happening in the US, the collaborative effort in the Soviet–American Gallium Experiment (SAGE) was going ahead under the leadership of Vladimir Gavrin, Georgii Zatsepin (from the Institute for Nuclear Research of the Russian Academy of Sciences) and Thomas J. Bowles (Los Alamos National Laboratory). SAGE then went into operation in 1986 at the Baksan underground facility for neutrino physics in the Northern Caucasus.[49]

Europe Goes It Alone: GALLEX Outcomes and the Beginning of the Neutrino Decade

The failure of the American–German attempt at a joint solar neutrino project led to a new collaboration formed in Europe and aiming for a full-scale experiment with 30 tons of gallium in the underground Gran Sasso National Laboratory in Italy (LNGS, Laboratori Nazionali del Gran Sasso), whose creation had been strongly backed by Antonino Zichichi at the end of the 1970s, and which was then being built under the Gran Sasso massif, not far from Rome.[50] This unique and timely infrastructure—built at the dawning of

47 Bahcall, John N., and Davis, Evolution of Neutrino Astronomy, 2000, 429–433, 431.

48 Kirsten, Radiochemical, 2000, 71–81, 53.

49 Vladimir N. Gavrin: The Russian-American Gallium Experiment SAGE. *Physics-Uspekhi* 54/9 (2011), 941–948. doi:10.3367/UFNe.0181.201109g.0975.

50 Antonino Zichichi was President of the National Institute for Nuclear Physics from 1977 to 1982, and in the late 1970s, when a tunnel under the Gran Sasso mountain was under construction, as part of the highway connecting Rome to the Adriatic Sea, he saw this as a unique opportunity for the excavation of the large halls of an underground laboratory, which would also have an excellent connection to the road network. Antonino Zichichi: The Gran Sasso Project. *AIP Conference Proceedings* 96/1 (1983), 52–64. doi:10.1063/1.33925. Alessandro Bettini: *The Gran Sasso Laboratory, 1979–1999. A Vision Becomes Reality.*

astroparticle physics—has since enabled Italy to take a leading role in this field. From 1984, the leading theoretical physicist Nicola Cabibbo, President of the National Institute for Nuclear Physics (INFN), strongly supported solar neutrino research as a major topic for the nascent laboratory, as recalled by Kirsten:

> I explained my issue to Nicola Cabibbo. From there on I had an ally. I now had the valuable asset that in the difficult negotiations for funding and collaboration formation I could argue: Yes, we do have a place to go in Europe: Gran Sasso.[51]

The initial problem in forming the GALLEX collaboration was in fact "to activate the key factors needed to conduct the experiment," solving what Kirsten called "a circular problem":

> If you need an annual world production of gallium, to get industry interested you must convince them that you are not crazy and that you can pay for it. For this, competent and influential players have to convince their home institutions to support the activity and to lobby for money acquisition at governments and foundations. But how can you ask for that unless you know where to go with your experiment—deep underground being an indispensable prerequisite?

Essential to solving the puzzle were the support of European funding institutions and foundations and, too, the

> enthusiastic support of the pioneers that got the LNGS underground laboratory going: Nicola Cabibbo, Luciano Maiani and Enrico Bellotti,

---

Assergi: INFN 1999. Lucia Votano: Origin and Status of the Gran Sasso INFN Laboratory. *Modern Physics Letters A* 29/36 (2014), 1430040. doi:10.1142/S0217732314300407.

51 Till Kirsten, personal communication with the authors (August 29, 2019). From Ettore Fiorini, Kirsten learned about Zichichi's project to build the Gran Sasso underground laboratory, for which excavations had started in 1982. Kirsten had known Fiorini a long time, both being pioneers of double-beta decay experiments (even on the same isotope, tellurium-130, with different techniques). In 1984, Fiorini mediated the first connection between Kirsten and the National Institute for Nuclear Physics (INFN), in particular with Nicola Cabibbo, universally known for his seminal theoretical work on the weak interaction. Cabibbo was President of INFN until 1993, his successor being the theoretical physicist Luciano Maiani, who was President until the end of 1998, when he became CERN Director-General.

> the first director of LNGS, member and—at the same time—great supporter of GALLEX in the critical initial phase, when LNGS still was in *statu nascendi*.[52]

Gallium funding was eventually secured through joint funding by the German Federal Ministry for Research and Technology, the Alfred Krupp von Bohlen und Halbach Foundation, and the Max Planck Society. The Italian National Institute for Nuclear Physics financed the underground Gran Sasso National Laboratory which would host the experiment, and the French made available their high-flux reactors, with which an artificial neutrino source was obtained in 1993 in order to conduct detection tests.[53]

Major goals of the gallium experiment, GALLEX, would be:

> to provide the first experimental proof that the Sun produces its energy by nuclear fusion, to limit or identify neutrino mass differences through eventual electronic-neutrino disappearance between Sun and detector via neutrino oscillations, to identify the cause of the 'solar neutrino puzzle' posed by the chlorine solar neutrino experiment.[54]

The Sun was becoming a powerful laboratory for exploring physics beyond the Standard Theory of elementary particles, not only to investigate the nature of the neutrino, but also to search for other weakly interacting particles like

52 Till A. Kirsten: GALLEX/GNO. Context and Recollections. In: Mikko Meyer, and Kai Zuber (eds.): *Solar Neutrinos. Proceedings of the 5th International Solar Neutrino Conference, Dresden, Germany, 11–14 June 2018*. Singapore: World Scientific 2019, 47–68. doi:10.1142/11384.

53 The agreement with Brookhaven National Laboratory implied that they would take care of providing the enriched isotope chromium-50 to be used at Grenoble and Saclay to produce chrome-51 and test the neutrino capture process, but their funding request to the US Department of the Interior—supported by Rudolf Mössbauer—was not accepted. Unfortunately, the lack of enriched chromium source would be "a real tragedy," which might have caused the French to withdraw from the agreement, as Ettore Fiorini wrote in a letter to Nicola Cabibbo, proposing that the whole question of funding and retrieving the source (from Oak Ridge Laboratories) could be handled by the Italian collaboration, with financial support from INFN (Fiorini to Cabibbo, President of INFN, September 25, 1989, Till Kirsten, private collection, DA GMPG, BC 600004).

54 Wolfgang Hampel et al.: Results of Ultra-Low Level ^{71}Ge Counting for Application in the Neutrino Experiment at the Gran Sasso Underground Physics Laboratory. In: F.C. Jones, J. Adams, and G.M. Mason (eds.): *Proceedings from the 19th International Cosmic Ray Conference, La Jolla, USA, August 11–23, 1985*. NASA. Goddard Space Flight Center 1985, 422–425. http://adsabs.harvard.edu/abs/1985ICRC....5..422H. Last accessed 10/31/2018.

the so-called 'solar cosmions,' which the Sun was supposed to have accreted from the dark matter of the Galaxy. In this case, such captured particles could alter the Sun's thermal profile and thus change the predicted neutrino flux, so solving the solar neutrino problem and the missing mass problem of the Galaxy.[55]

Against this backdrop, while new fundamental questions were crying out for answers, favoring the blossoming of non-accelerator physics, the first GALLEX meetings were held in Milan, in February 1985, and in Heidelberg, in October of that year.[56] The Heidelberg cosmochemists, led by Till Kirsten, had a unique constellation of factors in their favor. This ambitious project would shift the emphasis of the research program of the isotope group in Heidelberg toward a considerably broader perspective. Also involved in the project was their close ally Rudolf Mössbauer from the Technical University of Munich, who had studied under Bothe in the 1950s and now had his own Nobel Prize (1961).[57] With this combined expertise, including Kirsten and Mössbauer, the

---

55 D. N. Spergel, and W. H. Press: Effect of Hypothetical, Weakly Interacting, Massive Particles on Energy Transport in the Solar Interior. *The Astrophysical Journal* 294 (1985), 663–673. doi:10.1086/163336. W. H. Press, and D. N. Spergel: Capture by the Sun of a Galactic Population of Weakly Interacting, Massive Particles. *The Astrophysical Journal* 296 (1985), 679–684. doi:10.1086/163485. G. B. Gelmini, L. J. Hall, and M. J. Lin: What Is the Cosmion? *Nuclear Physics B* 281/3–4 (1987), 726–735. doi:10.1016/0550-3213(87)90424-X. As already outlined in Chapter 3, at that time theoreticians at the Max Planck Institute for Physics in Munich—notably Leo Stodolsky and his group—were investigating the problem of dark matter and also developing instruments which could be used to detect axions from the Sun, hypothetical particles considered excellent dark matter candidates. Georg Raffelt, and Leo Stodolsky: New Particles from Nuclear Reactions in the Sun. *Physics Letters B* 119/4 (1982), 323–327. doi:10.1016/0370-2693(82)90680-3. Andrzej K. Drukier, and Leo Stodolsky: Principles and Applications of a Neutral-Current Detector for Neutrino Physics and Astronomy. *Physical Review D* 30/11 (1984), 2295–2309. doi:10.1103/PhysRevD.30.2295.

56 A detailed schedule of the topics discussed at the first and second meetings can be found in the document "AGENDA—FIRST GALLEX-MEETING, February 19–21, 1985" and in "2nd GALLEX-meeting, October 10–11, 1985, in Heidelberg, MPI Kernphysik." See list of all the meetings from 1985 to 1997. A proposal to the Bundesministerium für Forschung und Technologie/German Federal Ministry for Research and Technology (see a copy of the proposal "Messung der Sonnenneutrinos mit einem Gallium-Detektor") had been presented in December 1984 (Till Kirsten, private collection, DA GMPG, BC 600004).

57 Mössbauer's discovery of the effect later named after him, which inaugurated a new tool for precision spectroscopy, had been the last success of Bothe's Institute for Physics at the MPI for Medical Research, which he led until his death in 1957. Rudolf L. Mössbauer: Kernresonanzabsorption von Gammastrahlung in Ir191. *Die Naturwissenschaften* 45/22 (1958), 538–539. doi:10.1007/BF00632050. Chapter 1 details Mössbauer's connections with Heidelberg, which remained close also after his appointment to the Technical University of Munich.

Max Planck people in Heidelberg took over the leadership of the radiochemical gallium experiment that came to be known as GALLEX. It was the first case of an international collaboration of this type led by Germans, which was named 'The European GALLEX collaboration,' a designation to be used for all public presentations representing the work of the team.[58] A key element was the support of the Italian National Institute for Nuclear Physics, namely provision of infrastructure in the fledging new Gran Sasso Laboratory shielded by about 1,400 meters of dolomite rock, the first large underground facility to be exclusively devoted to fundamental research.[59]

The laboratory was completed in early 1987, while construction and preparation of the GALLEX experiment went on from 1986 to May 1991, when the first solar neutrino recording started. It was among the first experiments to be conducted at Gran Sasso Laboratory.[60] The project team also included scientists from other European countries and there was significant involvement on the part of Israel, further strengthening Heidelberg's traditional link to the Weizmann Institute of Science mentioned earlier in this study. In 1986, agreement was reached also on the participation in the GALLEX project of a group from Brookhaven National Laboratory.[61] As emphasized later by Kirsten during the inauguration ceremony,

58 For a snapshot of the GALLEX Collaboration at the beginning of the 1990s, see P. Anselmann et al.: Solar Neutrinos Observed by GALLEX at Gran Sasso. *Physics Letters B* 285/4 (1992), 376–389. doi:10.1016/0370-2693(92)91521-A. P. Anselmann et al.: Implications of the GALLEX Determination of the Solar Neutrino Flux. *Physics Letters B* 285/4 (1992), 390–397. doi:10.1016/0370-2693(92)91522-B.

59 The then President of INFN, Nicola Cabibbo, was instrumental in the approval process, supporting solar neutrinos as a major research area in the Gran Sasso National Laboratory. An official letter from Cabibbo to Kirsten was sent on July 30, 1985, confirming that the GALLEX experiment had been approved by the Gran Sasso Scientific Committee, given the great importance that INFN attached to the success of GALLEX (Till Kirsten, private collection, DA GMPG, BC 600004).

60 Another experiment being installed there, and operational since 1989, was the MACRO (Monopole, Astrophysics, and Cosmic Ray Observatory) large-area detector, designed to search for super heavy magnetic monopoles (cosmic relics from the early Universe predicted by Grand Unified Theories), high-energy gamma and neutrino cosmic sources, and, more in general "rare exotic phenomena in the cosmic radiation." The MACRO collaboration included researchers from 10 Italian universities and laboratories, 6 US universities and one Moroccan university. The MACRO Collaboration: MACRO, a Large-Area Detector at the Gran Sasso Laboratory. *Il Nuovo Cimento C* 9/2 (1986), 281–292, 282. doi:10.1007/BF02514848. A further experiment, the neutrino observatory LVD (Large Volume Detector), mainly designed to study low energy neutrinos from gravitational stellar collapse, operated at the Gran Sasso Laboratory from June 1992.

61 See "Memorandum of Understanding," signed by Gerhart Friedlander (BNL) and Till Kirsten on May 24 1986 (Till Kirsten, private collection, DA GMPG, BC 600004).

> GALLEX is a fine example for a smoothly functioning fruitful international collaboration, in particular for the potential of European nations if they combine their skills and resources without national egoism. Only by joining forces was it possible to get GALLEX going.[62]

The new radiochemical solar neutrino experiment, GALLEX, was specifically developed with the main objective of achieving a clear distinction between the astrophysical and the particle physics solution to the solar neutrino problem.[63] Its scientific purposes were rooted in using refined radiochemical techniques for detection of neutrinos from the Sun, but its scope quickly widened and evolved within the emergent field of particle astrophysics, whose endeavor to detect cosmic neutrinos had long since been a major focus.[64]

Construction and test operations in the underground Gran Sasso National Laboratory lasted from 1986 to 1991, when GALLEX began recording the first solar neutrino flux, taking data over the course of six years.[65]

---

62 Document "INFN-LNGS. GALLEX Solar Neutrino Experiment. Inauguration Ceremony. November 30, 1990" (Till Kirsten, private collection, DA GMPG, BC 600004).

63 Till A. Kirsten: Das GALLEX-Sonnenneutrino-Experiment. *Mitteilungen der Astronomischen Gesellschaft* 68 (1987), 59–70. https://ui.adsabs.harvard.edu/#abs/1987MitAG..68...59K. Last accessed 4/30/2018. Theoretical related issues also connected to the nature of neutrinos were also investigated at the Institute for Nuclear Physics: K. Grotz, H. V. Klapdor, and J. Metzinger: Microscopic Calculation of Neutrino Capture Rates in ^{69,71}Ga and the Detection of Solar and Galactic Neutrinos. *Physical Review C* 33/4 (1986), 1263–1269. doi:10.1103/PhysRevC.33.1263. Hans Volker Klapdor: From Nuclear Physics to Fundamental Questions of Particle Physics, Cosmology and Reactor Physics. *Iadernaia Energiia* 25 (1987), 74–137. https://ui.adsabs.harvard.edu/#abs/1987IadEn..25...74K. Last accessed 10/14/2018.

64 Christian Spiering: Towards High-Energy Neutrino Astronomy. A Historical Review. *The European Physical Journal H* 37/3 (2012), 515–565. doi:10.1140/epjh/e2012-30014-2.

65 The inauguration ceremony took place on November 30, 1990. In his presentation, in the name of the whole collaboration, Kirsten emphasized the many meanings that GALLEX—which was expected to be a really important scientific adventure—had for him. He mentioned three of them: "[...] it is not a formal body but a vivid association of scientists, engineers, and technicians, individuals which are driven by their curiosity to get a deeper insight into the fundamental laws of nature [...] To achieve this goal, they are forced to give up some of their individualism for the common goal, and they do so voluntarily in respect for each other." As second goal, Kirsten mentioned the importance for Europe of such a collaboration and then outlined that "GALLEX is a challenge and by no means a 'simple' experiment. It is not a technocratic exercise where you plan, construct, push the button, and get a programmed output. Instead, it extends into new frontiers of experimental techniques, like separating single atoms of a reactive chemical element out of a reservoir of 10^{30} atoms, a ratio like a grain of salt dissolved in all oceans of the Earth; or, to detect the decay of an individual single atom at very low energy and not to

In 1992, the GALLEX experiment could claim to have observed for the first time the primary *pp* neutrinos, constituting nearly the entirety of the solar neutrino flux.[66] As recalled by Kirsten, this result, announced at the *Neutrino 92* conference held in Granada, Spain, in June 1992,

> implied a definite contribution from *pp*-neutrinos and thus, their discovery. This converted 'what nobody doubted to know about how stars produce their energy' *into an experimental fact*.[67]

Indeed, such discovery represented a significant test of the hypothesis that hydrogen fusion powers the Sun. In the foreword to the conference proceedings, coeditor Angel Morales judged that "the first GALLEX results will mark a cornerstone in neutrino history."[68] GALLEX continued to operate until January 1997. At the end of data-taking, the result was only 60 percent of what had been hoped, significantly (6 sigma) below the Standard Solar Model prediction and, hence, the disappearance of bulk (sub-MeV) neutrinos was established at > 99 percent confidence level.[69]

However, in early 1994, the solar neutrino situation, with results from the Homestake experiment, the Kamiokande experiment, GALLEX, and SAGE, was still ambiguous, as emphasized by David N. Schramm: the differences between observations and the Standard Solar Model might "still be due to either astrophysical inputs or new neutrino physics."[70] But from around 1994 onward, both GALLEX and SAGE, the gallium experiment carried out as of end of 1989 by the SAGE collaboration in the underground laboratory in the

---

be swamped by backgrounds ever present at this level." Till Kirsten, private collection, DA GMPG, BC 600004.

66 Anselmann et al., Solar Neutrinos, 1992, 376–389. Anselmann et al., GALLEX Determination, 1992, 390–397. The most important results from GALLEX were outlined in the 1993 Annual Report: Generalverwaltung der Max-Planck Gesellschaft (ed.): *Max-Planck-Gesellschaft Jahrbuch 1993*. Göttingen: Vandenhoeck & Ruprecht 1993, 437–444. See also material related to the GALLEX project in AMPG, II. Abt. Rep. 66, No. 1990, 1991, 1992, 1993, 1994.

67 Kirsten, Radiochemical, 2000, 71–81, 55.

68 Kirsten, Radiochemical, 2000, 71–81, 55.

69 Till A. Kirsten: Solar Neutrino Experiments: Results and Implications. *Reviews of Modern Physics* 71/4 (1999), 1213–1232. doi:10.1103/RevModPhys.71.1213. Wolfgang Hampel et al.: Final Results of the ^{51}Cr Neutrino Source Experiments in GALLEX. *Physics Letters B* 420/1–2 (1998), 114–126. doi:10.1016/S0370-2693(97)01562-1.

70 David N. Schramm, and Xiangdong Shi: Solar Neutrinos: Solar Physics and Neutrino Physics. *Nuclear Physics B—Proceedings Supplements* 35 (1994), 321–333. doi:10.1016/0920-5632(94)90271-2.

North Caucasus Mountains—which employed very different chemical procedures[71]—began to give very similar solar neutrino results, in striking disagreement with Standard Solar Model predictions, so providing additional evidence for electron neutrino disappearance.[72] Neutrino flavor changes remained as the only viable possible consistent explanation for this evidence.[73] With these results, GALLEX and SAGE significantly contributed to making the 1990s the 'decade of the neutrino,'[74] inaugurated in 1987 by the first ever detection of a burst of neutrinos from the explosion of a supernova in the Large Magellanic Cloud, the first supernova since 1604 visible to the naked eye—an extraordinary event marking the birth of neutrino astronomy.[75]

---

71 V. V. Kuzminov: The Baksan Neutrino Observatory. *The European Physical Journal Plus* 127/9 (2012), 113. doi:10.1140/epjp/i2012-12113-0.

72 Early outcome from SAGE showed a strong discrepancy with the average value found by GALLEX. A. I. Abazov et al.: First Results from the Soviet-American Gallium Experiment. *Nuclear Physics B Proceedings Supplements* 19 (1991), 84–93. doi:10.1016/0920-5632(91)90191-G. A. I. Abazov et al.: Search for Neutrinos from the Sun Using the Reaction $^{71}Ga(\nu_e,e^-)^{71}Ge$. *Physical Review Letters* 67/24 (1991), 3332–3335. doi:10.1103/PhysRevLett.67.3332. A chronology of results from both experiments in 1994 showed a rising trend in SAGE results; see fig. 3 in P. Anselmann et al.: GALLEX Results from the First 30 Solar Neutrino Runs. *Physics Letters B* 327/3 (1994), 377–385. doi:10.1016/0370-2693(94)90744-7. See also articles reporting measurements up to 1993: J. N. Abdurashitov et al.: Results from SAGE (The Russian-American Gallium Solar Neutrino Experiment). *Physics Letters B* 328/1 (1994), 234–248. doi:10.1016/0370-2693(94)90454-5. J. N. Abdurashitov et al.: Results from SAGE. *Nuclear Physics B—Proceedings Supplements* 48/1 (1996), 299–303. doi:10.1016/0920-5632(96)00264-2. D. Vignaud: The GALLEX Solar Neutrino Experiment. *Nuclear Physics B—Proceedings Supplements* 60/3 (1998), 20–29. doi:10.1016/S0920-5632(97)00498-2.

73 There was actually another explanation: neutrino decay. But nobody really believed in it, especially since observation of neutrinos from Supernova 1987A spoke against it. We are grateful to Christian Spiering for this comment. At that time, discussions on what appeared to be "the last hope for an astrophysical solution to the solar neutrino problem" led to the conclusion that the *standard neutrino* solution to the solar neutrino problem was strongly disfavored (or excluded) in favor of a *non-standard neutrino*, i.e., one with properties beyond the Standard Theory; and thus, "the last hope turned out to be a no-hope case." Vadim L. Berezinsky, G. Fiorentini, and M. Lissia: LAST HOPE for an Astrophysical Solution to the Solar Neutrino Problem. *Physics Letters B* 365/1 (1996), 185–192, 191. doi:10.1016/0370-2693(95)01241-9.

74 Till A. Kirsten: Results from Solar-Neutrino Experiments. *Il Nuovo Cimento C* 19/6 (1996), 821–833. doi:10.1007/BF02508123. GALLEX Collaboration: Results of the Whole GALLEX Experiment. *Nuclear Physics B. Proceedings Supplements* 70/1–3 (1999), 284–291. doi:10.1016/S0920-5632(98)00438-1. J. N. Abdurashitov et al.: Solar Neutrino Flux Measurements by the Soviet-American Gallium Experiment (SAGE) for Half the 22-Year Solar Cycle. *Journal of Experimental and Theoretical Physics* 95/2 (2002), 181–193. doi:10.1134/1.1506424.

75 S.E Woosley, and M.M Phillips: Supernova 1987A! *Science* 240/4853 (1988), 750–759. doi:10.1126/science.240.4853.750. The need to reduce cosmic ray muon-induced background

Complementary Competitors: Real-Time Detectors Jump in

The radiochemical chlorine and gallium neutrino detectors were the first generation of large-scale solar neutrino experiments. Along with the Japanese Kamiokande, a water Cherenkov detector installed beneath one kilometer of earth and rock in the Kamioka Mine, they were in the early 1990s the only operational experiments, contributing to advances in our understanding of neutrino properties and in identifying the solar neutrino problem, as well as providing key elements along the path to its solution. Kamiokande had initially been devised by Masatoshi Koshiba for the search for the proton decay predicted by Grand Unified Theories, and it was modified in order to detect solar neutrinos. Kamiokande II went into operation in 1986, and on February 23, 1987, it happened to be sensitive enough to detect neutrinos produced by the Supernova 1987A, the most spectacular experimental outcome from this cataclysm.[76] In fact, as we see later in this chapter, this supernova considerably boosted all the nascent fields treated in this chapter; but in all cases other than this neutrino detection—which marked the birth of extra-solar system neutrino astronomy—it was a 'missed opportunity,' as the other experiments were not yet ready to take advantage of the rare phenomenon.

After 1990, the emphasis in solar neutrino research shifted from solar physics to the particle physics realm, as the most likely cause of the missing electron neutrinos was *new neutrino physics*. In the meantime, after nine years of successful solar neutrino recording, Kamiokande was upscaled and replaced by the larger Super-Kamiokande imaging Cherenkov detector, under the guidance of Yoji Totsuka and Yoichiro Suzuki. In 1998, it showed evidence for the oscillations of atmospheric neutrinos, produced as decay products in hadronic showers resulting from collisions of cosmic rays with nuclei in the upper atmosphere.[77] The deficit in the observed ratio of the flux of muon to electron flavor atmospheric neutrinos, which was inconsistent with expectations based on calculations of the atmospheric neutrino flux, could be

---

effects in very large underground detectors primarily built to detect proton decay—with a lifetime less than 10^{32} years, as predicted by early Grand Unified Theories—made them good detectors of neutrinos from the Supernova 1987A: Kamiokande in Japan recorded 11 events, the Irvine-Michigan-Brookhaven (IMB) in Ohio 8 events, and Baksan in the Caucasus 5 events.

76 K. Hirata et al.: Observation of a Neutrino Burst from the Supernova SN1987A. *Physical Review Letters* 58/14 (1987), 1490–1493. doi:10.1103/PhysRevLett.58.1490.

77 Takaaki Kajita, and Superkamiokande and Kamiokande collaborations: Atmospheric Neutrino Results from Super-Kamiokande and Kamiokande—Evidence for ν_μ Oscillations. *Nuclear Physics B—Proceedings Supplements* 77/1 (1999), 123–132. doi:10.1016/S0920-5632(99)00407-7. Y. Fukuda et al.: Evidence for Oscillation of Atmospheric Neutrinos. *Physical Review Letters* 81/8 (1998), 1562–1567. doi:10.1103/PhysRevLett.81.1562.

explained by neutrino oscillations between muon neutrinos and tau neutrinos, providing indication for a small but non-zero mass for neutrinos.[78]

But the turning point was *real-time* neutrino detectors, for these see the neutrino interactions *event by event* by transmitting data to real-time monitoring and processing systems which analyze them, so yielding several neutrino parameters, in particular the oscillation pattern. In this way, without recourse to the Standard Solar Model, they are able to provide definitive proof of neutrino flavor oscillation.

At the turn of the millennium, the Sudbury Neutrino Observatory (SNO) *real-time* Cherenkov detector for boron-8 neutrinos was being completed at the Creighton Nickel Mine in Sudbury, Canada. The advent of the SNO experiment marked a breakthrough in solar neutrino physics. All the previous attempts had been *electron-neutrino disappearance experiments*. Oscillations produce neutrinos of different flavors and thus *neutrino appearance* experiments should be able to observe neutrinos of flavor different from the electron neutrinos produced by the Sun. The SNO detector was able to provide *a direct proof* that the neutrinos from the Sun were not disappearing on their way to Earth. Instead, a part of the solar electron neutrinos had transformed into a different flavor, and they were captured with a different identity when arriving at the Sudbury Neutrino Observatory. These measurements revealed the existence of a large flux of muon and/or tau neutrinos in the flux coming from the Sun, and since all neutrinos generated deep inside the Sun are created with the electron flavor, the results clearly demonstrated that neutrinos can oscillate from one type to another, if oscillations have sufficient time to develop. The 'missing solar neutrinos' of previous experiments (that were sensitive only to the electron flavor) were not really missing at all, but only unobservable, being present as *muon* or *tau neutrinos*, which are not detected by the chlorine and gallium radiochemical experiments.[79]

---

78 Electron neutrinos and muon neutrinos are produced mainly by the decay chains of charged pions produced in interactions between cosmic rays and atmospheric nuclei. The observed flux ratio of muon neutrinos (+ muon antineutrinos) and electron neutrinos (+ electron antineutrinos) showed a deficit of muon-neutrino events. In 2001, very precise information was provided by results based on more than 18000 solar neutrino events, increasing the number of previously reported events by an order of magnitude. Super-Kamiokande Collaboration et al.: Solar 8B and Hep Neutrino Measurements from 1258 Days of Super-Kamiokande Data. *Physical Review Letters* 86/25 (2001), 5651–5655. doi:10.1103/PhysRevLett.86.5651.

79 SNO Collaboration: Direct Evidence for Neutrino Flavor Transformation from Neutral-Current Interactions in the Sudbury Neutrino Observatory. *Physical Review Letters* 89/1 (2002), 011301. doi:10.1103/PhysRevLett.89.011301.

The combined results of the Super-Kamiokande and SNO measurements showed that new physics is required to describe the propagation of solar neutrinos, and that the Standard Solar Model prediction can be verified to high accuracy—provided that the electron neutrino mixes significantly with the muon neutrino and the tau neutrino. Building on the contribution made by all the previous experiments—now integrated with key results from SNO—finally made it possible to determine the corresponding oscillation parameters.

Both the Super-Kamiokande and SNO *real-time* experiments provided convincing smoking-gun evidence for the process of neutrino oscillation that many physicists had long regarded as an "intellectual luxury."[80] Such a scenario definitely implied a non-zero mass for the neutrino, clearly showing that the Standard Theory explaining the framework of elementary particles cannot be the complete theory of the fundamental constituents of the Universe.

The year 2002 became the *annus mirabilis* of solar neutrino physics. Convincing data from SNO had validated the existence of neutrino oscillations and Ray Davis and Masatoshi Koshiba were awarded the Nobel Prize "for pioneering contributions to astrophysics, in particular for the detection of cosmic neutrinos."[81] By that time, the first-generation experiments like Kamiokande, GALLEX, SAGE, (and since 2007, Borexino), had widely established the existence of the phenomenon of neutrino oscillations, providing a strong motivation for another Nobel Prize for neutrinos to be awarded in 2015 to Takaaki Kajita (Super-Kamiokande Collaboration, Japan) and Arthur B. McDonald (SNO collaboration, Canada) *"for the discovery of neutrino oscillations, which shows that neutrinos have mass."*[82] This discovery, more than 40 years after the prediction of the phenomenon by Bruno Pontecorvo—who had unfortunately passed away in 1993—has profound implications for our understanding of the Universe.[83] The study of solar neutrinos had made a fundamental contribution both to astroparticle and to elementary particle physics, providing both

80 Kirsten, Solar Neutrino Experiments, 1999, 1213–1232. Kirsten, Radiochemical, 2000, 71–81. Till A. Kirsten: Solar Neutrino Spectroscopy. *Nuclear Physics B—Proceedings Supplements* 87/1–3 (2000), 152–161. doi:10.1016/S0920-5632(00)00655-1.

81 The Nobel Prize in Physics 2002. Nobelprize.org. Nobel Media AB 2014.: The Nobel Prize in Physics 2002. https://www.nobelprize.org/nobel_prizes/physics/laureates/2002/. Last accessed 3/31/2018. Davis and Koshiba both shared the prize with Riccardo Giacconi, who had pioneered X-ray astronomy in the early 1960s together with Bruno Rossi. The latter, who had been his mentor, had unfortunately passed away in 1993.

82 The Nobel Prize in Physics 2015, NobelPrize.org. https://www.nobelprize.org/prizes/physics/2015/summary/. Last accessed 6/8/2019.

83 As clearly explained by Frank Close, Pontecorvo, who devoted much of his later career to studying the neutrino, for which he was given the sobriquet "Mr. Neutrino," would have deserved at least three Nobel Prizes related to neutrinos. Close, *Half-Life*, 2015, 7.

a test of solar models and relevant indications of the fundamental interactions among particles. In the process, this race, together with the detection of the burst of neutrinos from the Supernova 1987A, created the entirely new field of neutrino astrophysics and paved the way to completely new scenarios and to physics beyond the Standard Theory of particle physics.[84] From then on, the neutrino field focused on precisely determining the fundamental properties of this unique particle.

The Beginning of Astroparticle Physics in Heidelberg

In the meantime, toward the end of the 1990s, the Institute for Nuclear Physics in Heidelberg was completely reorganizing its research activity. The field of nuclear physics was focused now on two main topics: many-body dynamics of atoms and molecules and the synergy between particle physics and astrophysics, including the area of non-accelerator physics, which was now one of the components of the Institute's program on nuclear and particle physics. The Heidelberg–Moscow experiment on neutrino-less double-beta decay was fully operational at Gran Sasso Laboratory from 1996, dominating the scene in the 1990s and contributing to the changing scenario in Heidelberg at the turn of the century.[85] This program definitely marked that shift from 'cosmophysics' to high-energy astrophysics underway at the Heidelberg Institute for Nuclear Physics, also owing to participation in the HEGRA experiment for high-energy gamma astronomy and the planning of its successor project, the new imaging Cherenkov telescope system H.E.S.S. (see later section on TeV gamma astronomy). Together with the solar neutrino measurements, these fields had become a main focus of astrophysical research at the institute. Moreover, a new phase of solar neutrino exploration was opening with the advent of

84 K. Lande: Experimental Neutrino Astrophysics. *Annual Review of Nuclear and Particle Science* 29 (1979), 395–410. doi:10.1146/annurev.ns.29.120179.002143. Spiering, Towards High-Energy Neutrino Astronomy, 2012, 515–565.

85 The nuclear double-beta decay research was expanding its broad potential—already explored with the Heidelberg–Moscow experiment—into the search for new physics beyond the Standard Theory of particle physics. Two new experimental proposals could increase the sensitivity for double-beta decay and dark matter search: the underground setup GENIUS (GErmanium in liquid NItrogen Underground Setup) and the HDMS (Heidelberg Dark Matter Search) experiment. In addition, the technology of producing and using enriched high purity germanium detectors produced for the Heidelberg–Moscow experiment found application in high-resolution gamma-ray astrophysics using balloons or satellites. See, for example, S. I. Svertilov et al.: Hard X-Ray and Gamma-Ray Spectrometer of High Resolution and Sensitivity on Board the International Space Station (ISS). *Advances in Space Research* 25/3–4 (2000), 901–904. doi:10.1016/S0273-1177(99)00864-9.

Borexino (the Italian diminutive of the preliminary project BOREX (BORon solar neutrino EXperiment), a new challenging next-generation experiment conceived in the late 1980s, in which the Max Planck Institute for Nuclear Physics would play a major role.[86] Unlike GALLEX, Borexino could provide a *real-time view* of the core of the Sun through direct detection of the neutrino interactions with the target nuclei of the liquid scintillator target, enabling sub-MeV solar neutrino spectroscopy for the first time.[87] Built and operated by a large international collaboration again located at the Gran Sasso National Laboratory, Borexino laid the foundations for the analysis of the still unexplored sub-MeV energy region, isolating for the first time the monoenergetic beryllium-7 neutrinos.[88] This would be essential to test the stability and consistency of the standard explanation of the oscillation mechanism, either confirming or disproving the presence of discrepancies between theory and experiments.[89] At the turn of the new millennium, the focus of solar neutrino research was shifting from the relatively high-energy boron-8 neutrinos (maximum energy 14.1 MeV) to the low-energy beryllium-7 neutrinos, and to those produced by other nuclear reactions in the Sun, having energies less than or of the order of 1 MeV. Together with Super-Kamiokande and the Sudbury Neutrino Observatory, the planned Borexino experiment would be at the cutting edge of *solar and particle physics*.

One of the *unique* features of the Borexino detector was the very low radioactive background.[90] Very low background levels were in fact required

86 This new challenging experiment, a large-volume liquid scintillator detector, viewed by about 2000 photomultipliers, characterized by a very low background level due to an unprecedented radio-purity of the detector material (liquid scintillator), would make it possible to study *the entire spectrum of solar neutrinos* from very low energies.

87 The experiment's goal was the direct measurement of the flux of beryllium-7 solar neutrinos of all flavors via neutrino-electron scattering in an ultra-pure scintillation liquid. See also a description of the experiment in Max-Planck-Gesellschaft zur Förderung der Wissenschaften (ed.): *Max-Planck-Gesellschaft Jahrbuch 2001*. Göttingen: Vandenhoeck & Ruprecht 2001, 515–520.

88 Member countries (including several different institutions) were Germany, Russia, France, Italy, the US, the UK, and Poland.

89 G. Alimonti et al.: Science and Technology of Borexino. A Real-Time Detector for Low Energy Solar Neutrinos. *Astroparticle Physics* 16/3 (2002), 205–234. doi:10.1016/S0927-6505(01)00110-4.

90 Gianpaolo Bellini, and J. A. Villar: The Borexino Experiment and the Results of the Counting Test Facility. Edited by A. Morales, and J. Morales. *Nuclear Physics B—Proceedings Supplements* 48/1 (1996), 363–369. doi:10.1016/0920-5632(96)00273-3. G. Alimonti et al.: A Large-Scale Low-Background Liquid Scintillation Detector. The Counting Test Facility at Gran Sasso. *Nuclear Instruments and Methods in Physics Research Section A.*

to detect beryllium-7 neutrinos, but what was even more challenging was detecting the signal of neutrinos from the rare proton-electron-proton (*pep*) reaction and of neutrinos from the even less common carbon, nitrogen, oxygen (CNO) fusion cycle, a major energy source of large and late stage stars.[91] Running continuously since 2007, the Borexino experiment, using about 300 ton of ultra-pure organic liquid scintillator, has measured beryllium-7 neutrinos, *pep* neutrinos,[92] boron-8 neutrinos, and *pp* neutrinos, confirming the theoretical prediction for all neutrinos formed in the multistage nuclear fusion process, providing information about neutrino oscillations and providing the most complete real-time insight into the core of our Sun so far.[93] The study of solar neutrinos has been completed by the first detection ever, with a high statistical significance, of neutrinos from the CNO cycle, the Sun's second fusion process. With this achievement, all the theoretical predictions on how energy is generated within the Sun have been experimentally verified, closing an era commenced in the 1930s, with the first theories on the nuclear fusion mechanisms that are active in stars. The CNO process, though sub-dominant in the Sun, plays a key role in understanding the fundamental properties of all stars larger than our Sun in the Universe, where the majority of energy is generated in the CNO cycle. Their size, temperature, brightness, and lifetime are determined by the concentration levels of carbon, nitrogen, and oxygen, acting as

---

Accelerators, Spectrometers, Detectors and Associated Equipment 406/3 (1998), 411–426. doi:10.1016/S0168-9002(98)00018-7.

91 Its expected interaction rates were a few counts per day in a 100-ton target, while the main background of cosmogenic and radiogenic origin is one order of magnitude more intense. Detection of neutrinos from the CNO cycle has important implications in astrophysics, as it provides direct evidence for the nuclear process that is believed to fuel massive stars, with more than 1.5 solar masses.

92 Gianpaolo Bellini et al.: First Evidence of pep Solar Neutrinos by Direct Detection in Borexino. *Physical Review Letters* 108/5 (2012), 051302. doi:10.1103/PhysRevLett.108.051302. See also Borexino's contribution in the special *Nuclear Physics B* issue (vol. 908, July 2016) celebrating the Nobel Prize in Physics 2015, which included, among others, "members from large experimental collaborations with major results in the last ten years." Gianpaolo Bellini: The Impact of Borexino on the Solar and Neutrino Physics. *Nuclear Physics B* 908 (2016), 178–198. doi:10.1016/j.nuclphysb.2016.04.011.

93 G. Ranucci et al.: Overview and Accomplishments of the Borexino Experiment. *Journal of Physics: Conference Series* 675/1 (2016), 012036. doi:10.1088/1742-6596/675/1/012036. For the simultaneous precision spectroscopic measurements of solar neutrinos from all the reactions belonging to the *pp* nuclear fusion chain and their implications for both solar and neutrino physics, see The Borexino Collaboration: Comprehensive Measurement of pp-Chain Solar Neutrinos. *Nature* 562/7728 (2018), 505–510. doi:10.1038/s41586-018-0624-y.

catalysts and intermediate products in the cycle in which a total of four hydrogen nuclei ultimately combine to form a helium nucleus.[94]

3 The Quest for Gravitational Waves

This second emerging field was the result of the research tradition in theoretical astrophysics in Göttingen and Munich. The 1960s saw an explosion of interest in the new field of relativistic astrophysics, boosted by the unveiling of the violent Universe by radio astronomy and by spectacular astrophysical discoveries. Then came the decisive push through the pioneering experiments of Joseph Weber, who claimed to have detected gravitational waves (1969). Munich scientists quickly entered the field with a three-branched approach: experimental detection, statistical analysis of the results, and a solid theoretical footing in general relativity, owing to the appointment of the renowned relativist Jürgen Ehlers. This initial strength then allowed them to shift toward the new method of laser interferometry, taking advantage of expertise at the nearby Max Planck Institute for Plasma Physics. In the 1970s and 1980s, this effort was led by an itinerant group of experts circulating through institutes in the Munich area, facilitating the transition from resonant bars towards laser interferometry and its innovation at increasingly large scales, eventually finding a dedicated site in Hanover in the early 1990s. Resistance from the worldwide astronomical community and financial constraints resulting from German reunification then compelled the project to 'Europeanize' and, ultimately, to downsize its proposed experiment to pilot scale. The German approach never developed into a fully-scaled detector, putting the emphasis instead on perfecting experimental systems and building excellence in technology and instrumental innovation. In parallel, Ehlers founded an institute for gravitational physics in Potsdam, and soon both branches were unified as the Albert Einstein Institute of the Max Planck Society, one of the central contributors to the detection of gravitational waves in 2015.

The 'Renaissance' of General Relativity, Quasars, and Relativistic Astrophysics

This second emerging field was the result of the research tradition in theoretical astrophysics in Göttingen and Munich. Since the early 1960s, Ludwig

94 M. Agostini et al.: Experimental Evidence of Neutrinos Produced in the CNO Fusion Cycle in the Sun. 7835. *Nature* 587/7835 (2020), 577–582. doi:10.1038/s41586-020-2934-0.

Biermann's Institute for Astrophysics actively participated in the explosion of interest in the new field of relativistic astrophysics, which had been triggered by the discovery of quasars and boosted by the growing interplay of astronomy and astrophysics with general relativity, which was fast becoming "one of the most active and exciting branches of physics," based on the premises laid down in the post-World War II period by the process dubbed "Renaissance of general relativity."[95]

Technological progress during World War II had opened new horizons in the study of astronomy and the advent of radio astronomy had revealed that much in the Universe is of an explosive nature and that violent events occur within galactic nuclei. Astrophysicists had tried to understand the source of the tremendous energy stored in cosmic rays and the magnetic fields of some powerful radio galaxies.[96] The realization that the energy released within strong radio sources can exceed an energy equivalent of millions of solar masses led William Fowler and Fred Hoyle to explore the possibility that

---

95 Clifford M. Will: The Renaissance of General Relativity. In: P.C.W. Davies (ed.): *The New Physics*. London: Cambridge University Press 1989, 7–33, 7. Recent scholarship has shown that the revival of the field started already in the 1950s, mainly due to two factors: the discovery of the untapped potential of general relativity—as theorized by Einstein—as a tool of theoretical physics, and the emergence of a real community of relativists and cosmologists. Alexander Blum, Roberto Lalli, and Jürgen Renn: The Reinvention of General Relativity. A Historiographical Framework for Assessing One Hundred Years of Curved Space-Time. Isis 106/3 (2015), 598–620. doi:10.1086/683425. Alexander Blum, Roberto Lalli, and Jürgen Renn: The Renaissance of General Relativity. How and Why It Happened. Annalen Der Physik 528/5 (2016), 344–349. doi:10.1002/andp.201600105. Alexander Blum et al.: Editorial Introduction to the Special Issue "The Renaissance of Einstein's Theory of Gravitation". *The European Physical Journal H* 42/2 (2017), 45–105. doi:10.1140/epjh/e2017-80023-3. Roberto Lalli: *Building the General Relativity and Gravitation Community During the Cold War*. Cham: Springer International Publishing 2017. doi:10.1007/978-3-319-54654-4. See also how the status of general relativity transformed from the mid-1920s to 1970, through an analysis of the dynamics of the collaboration networks of scientists working in the field. Roberto Lalli, Riaz Howey, and Dirk Wintergrün: The Dynamics of Collaboration Networks and the History of General Relativity, 1925–1970. *Scientometrics* 122 (2020), 1129–1170. doi:10.1007/s11192-019-03327-1. Issues related to Section 3 are discussed in greater detail in our chapter, Luisa Bonolis, and Juan-Andres Leon: Gravitational-Wave Research as an Emerging Field in the Max Planck Society: The Long Roots of GEO600 and of the Albert Einstein Institute. In: Alexander S. Blum, Roberto Lalli, and Jürgen Renn (eds.): *The Renaissance of General Relativity in Context*. Basel: Birkhäuser 2020, 285–361. doi:10.1007/978-3-030-50754-1_9.

96 Geoffrey R. Burbidge: Estimates of the Total Energy in Particles and Magnetic Field in the Non-Thermal Radio Sources. *Astrophysical Journal* 129 (1959), 849–852. doi:10.1086/146680.

> at the centers of the galaxies there are *star-like objects* with masses ranging from about 10^5 up to about 10^8 solar masses for abnormal galaxies [our emphasis].

Fowler and Hoyle's opinion was that

> only through the contraction of a mass of 10^7–10^8 solar masses to the relativity limit can the energies of the strongest sources be obtained.[97]

This article appeared in August 1962, but in the meantime, Hoyle and Fowler took a further step. In February 1963, they argued that nuclear energy could not be the key to the problem, being unable to maintain sufficient internal pressure to provide support against gravity for such massive astrophysical objects, and observed that *gravitational energy*, instead, could be of decisive importance for bodies in that range of masses. The energies demanded by the strong sources were "so enormous as to make it clear that the relativity limit must be involved." As this limit was approached "general relativity must be used" [our emphasis].[98]

> The conclusion was now clear; that at a certain stage of its contraction (at about the size of the whole solar system) a very massive object would *implode* catastrophically, in about 100 seconds.[99]

Soon after, in the following March, Fowler and Hoyle's suggestions appeared to materialize in the "star-like" objects—celestial bodies with a very large red-shift and corresponding unprecedented large radio and optical luminosities—whose identification was announced in four consecutive articles in *Nature*.[100]

---

97 Fred Hoyle, and William A. Fowler: On the Nature of Strong Radio Sources. *Monthly Notices of the Royal Astronomical Society* 125/2 (1962), 169–176, 170. doi:10.1093/mnras/125.2.169.

98 William A. Fowler, and Fred Hoyle: Nature of Strong Radio Sources. *Nature* 197/4867 (1963), 533–535, 535. doi:10.1038/197533a0.

99 Fred Hoyle: The Nature of Cosmic Radio Sources. *New Scientist* 17/332 (1963), 681–683, 682.

100 C. Hazard, M.B. Mackey, and A.J. Shimmins: Investigation of the Radio Source 3C 273 By The Method of Lunar Occultations. *Nature* 197/4872 (1963), 1037–1039. doi:10.1038/1971037a0. M. Schmidt: 3C 273. A Star-Like Object with Large Red-Shift. *Nature* 197/4872 (1963), 1040–1040. doi:10.1038/1971040a0. J.B. Oke: Absolute Energy Distribution in the Optical Spectrum of 3C 273. *Nature* 197/4872 (1963), 1040–1041. doi:10.1038/1971040b0. J.L. Greenstein, and T.A. Matthews: Red-Shift of the Unusual Radio Source. 3C 48. *Nature* 197/4872 (1963), 1041–1042. doi:10.1038/1971041a0.

The most pressing problem in astrophysics at the time became how to explain the mechanism whereby such 'superstars,' the most bizarre and puzzling objects ever observed through a telescope to date, and which proved to be the most powerful energy sources in the sky, managed to radiate away the energy equivalent of five hundred thousand suns in short order. In recognition of their small size, they were called quasi-stellar radio sources, soon renamed *quasars*.[101]

The connection with Fowler and Hoyle's proposed mechanism involving gravitational collapse—a purely relativistic phenomenon at the time not yet fully understood—turned a spotlight on the bonds between general relativity, astronomy, and astrophysics. In December 1963, the international Symposium on Gravitational Collapse and other Topics in Relativistic Astrophysics took place in Dallas, organized by three relativists: Ivor Robinson, Alfred Schild, and Engelbert Schücking. Bringing together optical and radio astronomers, theoretical astrophysicists, and general relativists, it marked the start of a new era bridging the gap between the still exotic world of general relativity and the realm of astrophysics. Moreover, it officially opened the discussion on topics ranging from neutron stars to the possibility of gravitational collapse or a singularity in space-time, setting the stage for a dialogue between different scientific communities. This conference, the first of a long series of Texas Symposia, officially launched the brand-new field of relativistic astrophysics, merging two seemingly distant fields, so far removed that the organizers had to invent a new label for it.[102] This event took place at a time when the complex process developing since the aftermath of World War II, which had set in motion the 'renaissance' of Einstein's theory after a long period of stagnation, was being completed. After remaining cut off from mainstream physics for a generation, this formerly dispersed field was attracting an increasing number of practitioners, becoming the basis for the standard theory of gravitation and cosmology. New connections were now on the verge of being established with astrophysics and physical cosmology, through which general relativity would enter its 'astrophysical turn,' becoming an established branch of physics.[103]

---

101 Hong-Yee Chiu: Gravitational Collapse. *Physics Today* 17/5 (1964), 21. doi:10.1063/1.3051610.

102 Ivor Robinson, Alfred Schild, and Engelbert Schücking (eds.): *Quasi-Stellar Sources and Gravitational Collapse, Proceedings of the 1st Texas Symposium on Relativistic Astrophysics*. Chicago, IL: Chicago University Press 1965.

103 Alexander Blum, Roberto Lalli, and Jürgen Renn: Gravitational Waves and the Long Relativity Revolution. *Nature Astronomy* 2 (2018), 534–543. doi:10.1038/s41550-018-0472-6. See also, Alexander S. Blum, Roberto Lalli, and Jürgen Renn (eds.): *The Renaissance of General Relativity in Context*. Vol. 16. Basel: Birkhäuser 2020. doi:10.1007/978-3-030-50754-1.

During the 1960s, the detection of the cosmic microwave background radiation by Arno Penzias and Robert Woodrow Wilson[104]—together with the interpretation, by Robert Dicke and his associates, of such radiation as a relic of the Big Bang—and, too, the discovery of pulsars that were immediately identified as neutron stars, became part of a wider scenario connecting the rise of the 'golden age' of general relativity[105] to the birth of relativistic astrophysics. In providing the first definite proof of the existence of these highly compact stars—previously only theoretical entities—this discovery radically widened the perspective, firmly establishing the belief that strong gravitational fields may be of key importance for quasars, for violent events in the nuclei of galaxies, for supernova explosions and remnants, for the death by collapse of very massive stars and, in general, for the very compact astrophysical objects that were beginning to populate the Universe in the 1960s. Toward the end of the decade, black holes—exotic objects hitherto having only a purely theoretical status—became serious, albeit much debated, astrophysical hypotheses.[106] The discovery of pulsars did settle the existence of neutron stars as endpoints of the stellar evolution of massive stars, and had the effect that

> rather less was heard about the inherent absurdity of the more radical end-state, especially after Wheeler had dignified it with a name: 'black hole.'[107]

104 Arno A. Penzias, and Robert W. Wilson: A Measurement of Excess Antenna Temperature at 4080 Mc/s. *Astrophysical Journal* 142 (1965), 419–421. doi:10.1086/148307.

105 See chapter with this name in Kip S. Thorne: *Black Holes & Time Warps. Einstein's Outrageous Legacy*. New York, NY: Norton 1994, 258–299. Thorne actually identified the 'golden age' of general relativity with the explosion of interest in black hole research.

106 The term 'black hole' began to circulate and was officially launched by John Wheeler in 1968: John Archibald Wheeler: Our Universe. The Known and the Unknown. *American Scientist* 56/1 (1968), 1–20. However, it is not clear who used it first, although it appears that it circulated as early as September 1963, during the first Texas Conference, as reported in the January 24, 1964, issue of *Life* magazine by Al Rosenfeld, *Life*'s science editor, who had heard the term mentioned at the symposium. The story is told in Tom Siegfried: 50 Years Later, It's Hard to Say Who Named Black Holes. *Science News* (2013). https://www.sciencenews.org/blog/context/50-years-later-it's-hard-say-who-named-black-holes. Last accessed 7/19/2018.

107 Werner Israel: Black Hole 2000: The Astrophysical Era. *Publications of the Astronomical Society of the Pacific* 112/771 (2000), 583–585, 583. doi:10.1086/316557.

Setting the Stage for a Gravitational Wave Experiment at Biermann's Institute for Astrophysics

In Germany, Ludwig Biermann's sub-institute at the Max Planck Institute for Physics and Astrophysics was uniquely well situated to make contributions to this revolution in relativistic astrophysics. The longstanding commitment at the Max Planck Institute for Astrophysics to study of the structure and evolution of stars, also conducted with computer simulations, developed into research on very dense stars such as white dwarfs or neutron stars.[108] In addition to expanding from already dominant fields, entirely new perspectives and research pathways opening up in the astrophysical sciences were mirrored by research activities conducted at Biermann's Institute for Astrophysics.[109] From 1964, the young researcher Peter Kafka began to work on topics related to general relativity and cosmological questions in Munich, also related to the existence of quasars, the most exciting topic of the time.[110] He investigated the problem of gravitational collapse in general relativity, but in particular he explored the space-time distribution of the quasars and radio galaxies as deduced from observational evidence.[111] From radio astronomical observations it appeared that there were relatively more quasars at larger distances, and so that must mean they were more common in the early life of the Universe. This could be explained as an effect of its evolution: if their redshifts

108 See, for example, a letter by Kippenhahn to Biermann, referring to white dwarfs, collapsing stars, binary systems, and mentioning the problem that for the study of such non-linear dynamical effects one needed a powerful computing machine and that they had further perfected their own program on the evolution of stars, being at the forefront compared with other groups (Kippenhahn to Biermann, July 10, 1968, AMPG, III. Abt., ZA 1, No. 18).

109 Ludwig Biermann, and Reimar Lüst: Jahresberichte astronomischer Institute für 1964, München Max-Planck-Institut für Physik und Astrophysik, Institute für Astrophysik und extraterrestrische Physik. *Mitteilungen der Astronomischen Gesellschaft* 18 (1965), 57–66, 61. https://ui.adsabs.harvard.edu/abs/1965MitAG..18...57. Last accessed 10/30/2018.

110 Peter Kafka later recalled that at the time he did his 'Diplom' in Physics, Arnulf Schlüter, who had become Director of the Institute for Plasma Physics, had developed an interest in general relativity and asked him to work on this topic for his dissertation. Quasars were discovered at that time and so it became quite clear that general relativity would play a growing role in astrophysics and in cosmology; and as an expert on such topics, Kafka got a position at the Max Planck Institute for Astrophysics. Peter Kafka: interview by Michael Langer, March 27, 1999. Live-Gespräch-Sendung "Zwischentöne," Deutschlandfunk, http://www.gegen-den-untergang.de/zwischentoene1999.html. Last accessed 11/4/2018.

111 Ludwig Biermann, and Reimar Lüst: Jahresberichte astronomischer Institute für 1965, München Max-Planck-Institut für Physik und Astrophysik, Institute für Astrophysik und extraterrestrische Physik. *Mitteilungen der Astronomischen Gesellschaft* 20 (1966), 67–79, 71. http://adsabs.harvard.edu/abs/1966MitAG..20...66. Last accessed 10/30/2018.

were of cosmological origin, quasars—whose very nature was still a subject of debate—must have existed only very far away in time and space, contradicting the perfect cosmological principle, which was at the core of steady-state cosmology.[112] The counting of radio quasars as recognized sources at cosmological distances might thus help to confirm the existence of the Big Bang model, ruling out the steady state model of the universe, according to which the expanding universe would maintain a constant average density, with matter being continuously created to form new stars and galaxies, implying the notion of a Universe on average homogeneous and isotropic in space, and constant in time. However, there was disagreement about the meaning of relations, between the observed numbers, redshifts, and brightness of quasars, and in the abstract of his article in *Nature*, Kafka remarked that "no decision can be made, from a statistical count of quasars, between *steady state* and other cosmological models."[113]

In the same 1960s scenario, in which the Max Planck Institute for Astronomy in Heidelberg and Max Planck Institute for Radio Astronomy in Bonn

112 The static Universe proposed by Fred Hoyle and, independently, by Hermann Bondi and Thomas Gold, rejected the idea of an initial singularity, maintaining that a steady-state Universe could be compatible with the drifting apart of galaxies if new matter (approximately one hydrogen atom per cubic kilometer per year) were continuously generated in the intergalactic space. Since the mid-1950s, complete, new catalogues of radio sources had shown that the number of intense sources increased with distance, while, according to the steady-state theory, they were expected to be uniformly distributed throughout the Universe. Apparently the most distant objects of the Universe, quasars, had an impact in cosmology. If the high redshift of observed quasars was of cosmological origin, it meant that they were at distances such that the Universe was much younger than it is now, when the radio waves were emitted. This implied that galaxies produced more radio waves in the past, and thus began to call attention to the conflict between the Big Bang as a theory of cosmic expansion from a hot early Universe and the steady-state cosmology, according to which the observable Universe is basically the same on the large scale at any given time, a view called the "Perfect Cosmological Principle." An intense controversy developed between proponents of different theories of the Universe, as discussed in Helge Kragh: *Cosmology and Controversy. The Historical Development of Two Theories of the Universe*. Princeton, New Jersey: Princeton University Press 1996.

113 Peter Kafka: How to Count Quasars. *Nature* 213/5074 (1967), 346–350. doi:10.1038/213346a0. In his article, Kafka also mentioned having used a method programmed on a computer and announced that details would be provided in a forthcoming internal report of the Institute for Astrophysics (Quasars and Cosmology. Institutsbericht MPI-PAE/Astro 2/67). For a discussion on such debate, see, for example, Andrey Georgievich Doroshkevich, Malcolm S. Longair, and Yakov Borisovich Zeldovich: The Evolution of Radio Sources at Large Redshifts. *Monthly Notices of the Royal Astronomical Society* 147/2 (1970), 139–148. doi:10.1093/mnras/147.2.139.

were both finally founded in the 1960s, and while gamma ray and X-ray astronomy activities were in progress at the Institute for Extraterrestrial Physics in the Munich suburb of Garching, new conditions for the interaction between nuclear physics, astrophysics, cosmology, and optical and new astronomies were being created, widening the scope and context of what was being relabeled as the field of 'cosmic physics.'

Pulsars, Black Holes, and Other Possible Sources of Gravitational Waves

According to Einstein's theory of general relativity, accelerated masses produce gravitational waves, which propagate at the speed of light through the Universe. The existence and physical properties of gravitational radiation became central to various research agendas as one of the important open questions addressed by the general relativity and gravitation community emerging from the mid-1950s onward, when "the availability of appropriate notions of what a gravitational wave is allowed physicists to put forward heuristic arguments for their existence and detectability."[114] Towards the end of the 1950s, gravitational radiation became a key focus of theoretical studies in general relativity, reinforcing Joseph Weber's interest at the University of Maryland. Encouraged by John Wheeler, one of the leading figures in the renaissance of general relativity in the US, Weber accepted the challenge and pioneered the quest for the experimental detection of gravitational waves from astronomical sources. For several years, however, his experiments remained an isolated example. Weber had also mentioned as possible sources "events which might be associated with supernovae, neutron stars or closely spaced binaries."[115] His work in turn inspired interest in such astrophysical objects as possible sources of gravitational waves. At that time, Freeman Dyson had pointed out that the usual formula giving the gravitational-wave energy flux from a binary star, leads—in the extreme relativistic case of a close binary collapsing system formed from a pair of neutron stars—to the prediction of a huge output of radiation. The powerful burst of gravitational waves should be detectable by Weber's existing equipment.[116] This remark gave an extra stimulus to the pioneering experimental work of Weber, also prompting the

---

114 Blum, Lalli, and Renn, Gravitational Waves, 2018, 534–543, 534.

115 Joseph Weber: Remarks on Gravitational Experiments. *Il Nuovo Cimento* 29/4 (1963), 930–934, 934. doi:10.1007/BF02827954.

116 Freeman J. Dyson: Gravitational Machines. In: Alastair G.W. Cameron (ed.): *Interstellar Communication*. New York, NY: Benjamin Press 1963, 115–120, 119.

physics and astrophysics communities to consider gravitational radiation—whose physical reality was becoming evident—as a phenomenon of great potential importance in the physical world. More generally, gravitational radiation was being considered during the 1960s also as a possible mechanism for both the dissipation and transfer of energy in the domain of relativistic astrophysics,[117] and spinning compact objects, too, were candidate sources of gravitational waves. Pulsars were thus quickly recognized as promising sources of detectable gravitational waves, attracting wider attention to Weber's ongoing efforts.[118]

In mid-June 1969, Weber published his famous article claiming to have observed coincidences on gravitational radiation detectors based on resonating metal bars separated by a distance of about 1,000 km at Argonne National Laboratory near Chicago and at the University of Maryland: "There is good evidence that gravitational radiation has been discovered."[119] The announcement immediately caused a sensation in the physics community. Soon, gravitational waves—as well as hard X-rays and gamma rays—would be envisaged by John Wheeler and Remo Ruffini as one of the most promising ways to detect black holes.[120] In 1970, Franck Zerilli analyzed the problem of the pulse of gravitational radiation given off when a star falls into a black hole and Stephen Hawking's prescient article of 1971 even discussed gravitational radiation resulting from the collision of two black holes.[121]

Between Theory and Experiment in Munich: The Appointment of Jürgen Ehlers and Billing's Resonant Bar

The Max Planck Institute for Astrophysics quickly reacted to the new exciting perspective opened by Weber's claims. His article was published in the June 16

117 Vladimir B. Braginskii: Gravitational Radiation and the Prospect of Its Experimental Discovery. *Soviet Physics Uspekhi* 8/4 (1966), 513–521. http://stacks.iop.org/0038-5670/8/i=4/a=R02. Last accessed 7/24/2018.

118 Weber himself estimated the expected fluxes of gravitational radiation from such objects. Joseph Weber: Gravitational Radiation from the Pulsars. *Physical Review Letters* 21/6 (1968), 395–396. doi:10.1103/PhysRevLett.21.395.

119 Joseph Weber: Evidence for Discovery of Gravitational Radiation. *Physical Review Letters* 22/24 (1969), 1320–1324, 1324. doi:10.1103/PhysRevLett.22.1320.

120 Remo Ruffini, and John Archibald Wheeler: Introducing the Black Hole. *Physics Today* 24/1 (1971), 30–41. doi:10.1063/1.3022513.

121 Frank J. Zerilli: Gravitational Field of a Particle Falling in a Schwarzschild Geometry Analyzed in Tensor Harmonics. *Physical Review D* 2/10 (1970), 2141–2160. doi:10.1103/PhysRevD.2.2141. S. W. Hawking: Gravitational Radiation from Colliding Black Holes. *Physical Review Letters* 26/21 (1971), 1344–1346. doi:10.1103/PhysRevLett.26.1344.

issue of *Physical Review Letters* and by July there was a telephone conversation between Weber and Biermann, who was at the time in the United States, where he was a regular visitor every year. During the call, Biermann expressed his strong interest in Weber's experiments, which he had most probably discussed with his collaborators immediately before leaving Munich.[122] The characteristics of Weber's gravitational wave antennae were immediately studied by Hermann Ulrich Schmidt, while Kafka explored in detail the possible consequences of the gravitational waves "supposedly discovered by Weber."[123]

By early summer of 1969, both Biermann and Heisenberg were working toward intensifying research on gravitation theory and relativistic astrophysics.[124] They shared the common aim to invite the renowned relativist Jürgen Ehlers to spend a long period of time at their Max Planck Institute. Ehlers had studied general relativity with Pascual Jordan, one of the pioneers of quantum physics, who had formed a research group in this field in Hamburg back in the early 1950s, which was one of the seeds fertilizing the renaissance of general relativity. After obtaining his PhD and habilitation (German postdoctoral lecturing qualification) with Jordan, Ehlers had moved to Syracuse University in 1961, where he had worked several years with Alfred Schild's group.[125] He had now a professorship at the University of Texas, Austin, and

122 Biermann to Weber, March 19, 1970, AMPG, III. Abt., ZA 1, No. 48. A telephone call was made between Aspen, Colorado, where Weber spent part of his time (as acknowledged in his articles), and Boulder, Colorado, where Biermann had spent the months of July and August in 1969, giving lectures. Ludwig Biermann, and Reimar Lüst: Max-Planck-Institut für Physik und Astrophysik. Institut für Astrophysik und Institut für extraterrestrische Physik. *Mitteilungen der Astronomischen Gesellschaft* 28 (1970), 79–105, 79. http://adsabs.harvard.edu/abs/1970MitAG..28...79B. Last accessed 10/30/2018.

123 These research activities, together with Biermann's studies on some characteristics of the density of pulsars were announced in the new section of the Annual Report entitled "Relativistische Astrophysik, Quasare und Pulsare." Biermann, and Lüst, Report 1969, 1970, 79–105, 86–87. See also, Peter Kafka: Discussion of Possible Sources of Gravitational Radiation. *Mitteilungen Der Astronomischen Gesellschaft* 27 (1969), 134–138, 138. https://ui.adsabs.harvard.edu/#abs/1969MitAG..27..134K. Last accessed 11/3/2017.

124 See minutes related to the meeting of June 9, 1969, of the search commission for Heisenberg's successor in AMPG, II. Abt., Rep. 62, No. 437, Fol. 273.

125 Jordan had even favored Ehlers as his own successor in Hamburg (see related correspondence between Jordan and Heisenberg during winter 1967–68 in Heisenberg's papers, AMPG, III. Abt. Rep. 93, No. 1745). Ehlers had emerged as a candidate successor to Heisenberg in spring 1969, when the search committee had not yet decided whether a theoretical or an experimental physicist should lead the Institute for Physics after Heisenberg's retirement. During discussions about the possibility of appointing a theoretician, in particular an expert in general relativity, it was also mentioned that Einstein's theory had somewhat receded into the background at universities in Germany, something that Jor-

held visiting professorships in Germany.[126] At the time, Ehlers had just published a broad survey of the state of cosmology in relation to the impact of the recent discoveries of quasars, pulsars, the cosmic microwave background, and the beginning of experimental gravity physics.[127] It became Biermann and Heisenberg's ambition to have him back in Germany.

During this effervescent wave of new astrophysical phenomena in the late 1960s, things moved quickly. Biermann proposed that Ehlers should move to the Max Planck Institute for Astrophysics[128] and, at the end of October, Heisenberg and Biermann sent a joint letter to Adolf Butenandt, then President of the Max Planck Society, in which they emphasized how during the last year general relativity and the gravitation question had become relevant at the Institutes for Astrophysics and for Physics, especially in relation to gravitational waves and neutron stars. For this reason, the Munich institutes would strongly benefit from the presence of a renowned relativist like Jürgen Ehlers.[129] Ehlers became a Scientific Member of the Institute for Astrophysics on June 1, 1971.

---

dan had pointed out on several occasions. In May–June 1969, Gentner (who presided over the commission tasked to find Heisenberg's successor) and Jordan exchanged correspondence on this question, and Ehlers's name was definitely the most favored, according to the opinion of several relativists (Gentner to Jordan, May 13, 1969, and Jordan to Gentner, May 19 and June 2, 1969, AMPG, II. Abt., Rep. 62, No. 437, Fol. 42–59).

126 Bruce Allen et al.: Jürgen Ehlers. 29.12.1929-20.05.2008. *Jahresbericht der Max-Planck-Gesellschaft. Annual Report 2008*. Max-Planck-Gesellschaft zur Förderung der Wissenschaften e.V. 2009, 24–26.

127 Jürgen Ehlers: Probleme und Ergebnisse der modernen Kosmologie. *Mitteilungen der Astronomischen Gesellschaft* 27 (1969), 73–86. http://adsabs.harvard.edu/abs/1969MitAG.27...73E. Last accessed 4/25/2018.

128 Minutes of the 15th meeting of the Board of Trustees (Kuratorium) of the Max Planck Institute for Physics and Astrophysics, 17.03.1970, AMPG, II. Abt., Rep. 66, No. 3069.

129 L. Biermann and W. Heisenberg to Adolf Butenandt, October 31, 1969, AMPG, III Abt., Rep. 93, No. 1667. On November 7, a commission to appoint Ehlers as a Scientific Member of the Institute for Astrophysics was formed. The same commission was also involved in Heisenberg's succession (CPTS meeting minutes of 07.11.1969, AMPG, II. Abt., Rep. 62, No. 1757). Documents clearly show how both Heisenberg and Biermann's scientific interests would benefit from having Ehlers at the institute in order to build a bridge between unified field theory and gravitation theory, also in connection with new related interests in astrophysics and the idea of creating a working group on gravitational wave experiments. This would thus also create a better relationship between theory and experiment (minutes of the 15th meeting of the Board of Trustees (Kuratorium) of the Max Planck Institute for Physics and Astrophysics, 17.03.1970, AMPG, II. Abt., Rep. 66, No. 3069). On February 9, 1971, during the meeting of the CPT Section of the Scientific Council, it was communicated that Ehlers had accepted, and that he would take up his position on June 1, 1971 (AMPG, II. Abt., Rep. 62, No. 1761). On March 3, the Senate confirmed the appointment, remarking that Ehlers' visit to Munich had shown that his presence would be of

In the meantime, in late November 1970, the possibility of starting a gravitational wave experiment was being seriously considered by Biermann's group.[130] With this incursion, the Institute for Astrophysics would also move into experimental astrophysics based on a strong theoretical standpoint, a process that we have shown is characteristic of the Munich family of institutes. In parallel with intense theoretical work on general relativity and relativistic astrophysics, planning continued for the gravitational wave experimental activity at the Max Planck Institute for Astrophysics, also involving Heinz Billing, who was still leading the computing group but now successfully returned to physics. Wheels were put in motion and work began in earnest in 1971, when the gravitational wave experiment had its own specific section in the Annual Report.[131] The aim was "to confirm or disprove the existence of gravitational pulses suggested by Weber as an explanation of his results."[132] With the arrival of Ehlers in June 1971, the new Department for Gravitation Theory and Relativistic Astrophysics was established.

International collaboration was an inherent aspect of this experimental enterprise: for a coincidence experiment, similar to the one conducted by

---

the greatest importance for both the Institute for Physics (Hans-Peter Dürr's theoretical group) and the Institute for Astrophysics, as well as for the Institute for Extraterrestrial Physics.

130 See Biermann to Gerhard Börner, November 26, 1970, AMPG, III. Abt., ZA 1, No. 20. Hermann Ulrich Schmidt, who was spending some time at the National Solar Observatory at Sacramento Peak in New Mexico, wrote to Biermann about a discussion he was having with Ehlers, Weber, and Remo Ruffini about beginning a gravitational wave experiment in Munich. See also Biermann's answer (Schmidt to Biermann, November 26 1970, and Biermann to Schmidt, December 8, 1970, AMPG, III. Abt., ZA 1, No. 21).

131 They specified that the decision to repeat Weber's gravitational-wave experiment had been taken because of both its great astrophysical significance and the still pending difficulties in evaluating Weber's findings. The prerequisites for this were particularly favorable at the Institute, as the necessary engineering and electronic experiences were available at the Numerical Calculators Division, while the local astrophysicists would be able to handle the theory and the statistical problems, and the addition of Ehlers would guarantee the close connection with the general theory of relativity. It was further emphasized how Weber's detector could be improved. Ludwig Biermann, and Reimar Lüst: Max-Planck-Institut für Physik und Astrophysik. Institut für Astrophysik und Institut für extraterrestrische Physik. *Mitteilungen der Astronomischen Gesellschaft* 31 (1972), 323–350, 326. https://ui.adsabs.harvard.edu/#abs/1972MitAG..31..323B. Last accessed 4/29/2018.

132 Heinz Billing et al.: Results of the Munich-Frascati Gravitational-Wave Experiment. *Nuovo Cimento, Lettere* 12/4 (1975), 111–116, 111. doi:10.1007/BF02790471. As the Munich setup was planned to be as close as possible to Weber's experiment, both Billing and his new assistant, Walter Winkler, visited Weber at the University of Maryland for two weeks in January, in order to become familiar with his antenna and obtain all the information that would be useful for their future work.

Weber, they needed a second antenna, far from Munich. They were lucky, because, independently from them, a German colleague, the electronics engineer Karl Maischberger, and the physicist Donato Bramanti had also begun to work on a Weber-type gravitational wave antenna at the European Space Research Institute (ESRIN) in Frascati, near Rome, with which the institute had already interacted in recent years.[133] While intending to be as close as possible to the original experiment, they still made several improvements, which made their detector—together with the similar one built in Frascati—"the most sensitive room-temperature bar experiment at that time."[134] The Munich resonant bar—a long aluminum cylinder reproducing Weber's setup, and that would ring at a certain frequency in response to a gravitational wave—began operating in October 1972.[135] The aim was to test whether the pulses of gravitational radiation reported by Weber were detectable in coincidence between Munich and Frascati. The first negative results, in conflict with Weber's, were presented in June 1973, in Paris, at the International Colloquium on Gravitational Waves and Radiation.[136] By that point, the Munich–Frascati pair was a respected participant among the growing number of locations around the world where gravitational waves were being hunted experimentally.

Triggered by Weber's announcement that he had observed coincident pulses between resonance gravitational wave detectors located at the University of Maryland and at Argonne National Laboratory, other groups had also started experiments to analyze and test Weber's results: in the United Kingdom,[137] in the United States at IBM Research Center in Yorktown Heights (New

133 Donato Bramanti, and Karl Maischberger: Construction and Operation of a Weber-Type Gravitational-Wave Detector and of a Divided-Bar Prototype. *Nuovo Cimento, Lettere* 4/17 (1972), 1007–1013. doi:10.1007/BF02757124.

134 Walter Winkler: History of Physics (19). Fundamental Research for the Development of Gravitational Wave Detectors in Germany. *SPG Mitteilungen* 54 (2018), 14–18, 15. https://www.sps.ch/fileadmin/doc/Mitteilungen/Mitteilungen.54.pdf. Last accessed 5/30/2018.

135 Both Munich and Frascati built detectors as close to Weber's as possible, including a close match with his resonant frequency of 1660 Hz. Donato Bramanti, Karl Maischberger, and Donald Parkinson: Optimization and Data Analysis of the Frascati Gravitational-Wave Detector. *Lettere al Nuovo Cimento* 7/14 (1973), 665–670. doi:10.1007/BF02728048.

136 Peter Kafka: On the Evaluation of the Munich-Frascati Weber-Type Experiment. In: Y. Choquet-Bruhat (ed.): *Ondes et Radiations Gravitationelles*. International Conference, Paris, France, June 18–22, 1973. Paris: Centre National de la Recherche Scientifique 1974, 181–200.

137 The group in Glasgow (James Hough, Jon R. Pugh, Roger Bland), was at that time led by Ronald W. P. Drever, who later became a member of the team initially running the LIGO project, after working for some time in parallel on Glasgow and US projects. On early work in Glasgow and interaction with German scientists, see Ronald W. P. Drever: inter-

York), and Bell Laboratories in Holmdel (New Jersey),[138] in Japan,[139] and in the Soviet Union, where discussions on gravitational wave detection began already around 1960 and led to experimental efforts repeating the search for coincident signals on separated Weber-type antennae.[140] Results obtained were negative.[141]

The Munich–Frascati experiment reported results of the first 150 days of coincident data in 1975,[142] and in March 1976, after 580 days of total useful observation time, the detectors were dismantled and the experiment stopped.[143]

Jumping on the Laser Interferometer Bandwagon

In the early 1970s, the Max Planck Institute for Astrophysics considerably expanded its research activities and the growing number of international visitors corresponded to a similar flux of internal members visiting scientific

---

view by Shirley K. Cohen, June 1997. Transcript, Caltech Archives, http://resolver.caltech.edu/CaltechOH:OH_Drever_R. Last accessed 4/19/2021. Drever mentioned their friendly competition: "And we felt a bit envious, because they seemed to have more people, more money, more of everything. And everything they built was so beautifully built, and ours was kind of thrown together […] they were very friendly. We got on very well with them […] I wouldn't say we were jealous, but we envied them."

138 James L. Levine, and Richard L. Garwin: Absence of Gravity-Wave Signals in a Bar at 1695 Hz. *Physical Review Letters* 31/3 (1973), 173–176. doi:10.1103/PhysRevLett.31.173. J. A. Tyson: Null Search for Bursts of Gravitational Radiation. *Physical Review Letters* 31/5 (1973), 326–329. doi:10.1103/PhysRevLett.31.326. D. H. Douglass et al.: Two-Detector-Coincidence Search for Bursts of Gravitational Radiation. *Physical Review Letters* 35/8 (1975), 480–483. doi:10.1103/PhysRevLett.35.480.

139 Hiromasa Hirakawa, and Kazumichi Narihara: Search for Gravitational Radiation at 145 Hz. *Physical Review Letters* 35/6 (1975), 330–334. doi:10.1103/PhysRevLett.35.330.

140 Vladimir B. Braginskii, Ya B. Zel'Dovich, and V. N. Rudenko: Reception of Gravitational Radiation of Extraterrestrial Origin. *Soviet Journal of Experimental and Theoretical Physics Letters* 10 (1969), 280–283. https://ui.adsabs.harvard.edu/#abs/1969JETPL..10..280B. Last accessed 10/25/2018.

141 Vladimir B. Braginskii et al.: Search for Gravitational Radiation of Extraterrestrial Origin. *Journal of Experimental and Theoretical Physics Letters* 16/3 (1972), 108–112. http://www.jetpletters.ac.ru/ps/1759/article_26757.shtml. Last accessed 6/12/2018. In December 1974, also the Glasgow group reported a negative result. James Hough et al.: Search for Continuous Gravitational Radiation. *Nature* 254/5500 (1975), 498–501, 501. doi:10.1038/254498a0.

142 Billing et al., Results, 1975, 111–116. The Frascati-Munich group claimed to have "set the lowest limits so far obtained for the rate of incoming short gravitational pulses stronger than a few times 10^5 erg/cm^2 Hz at frequencies around 1660 Hz." The same frequency band had been used by Weber.

143 Peter Kafka, and Lise Schnupp: Final Result of the Munich-Frascati Gravitational Radiation Experiment. *Astronomy and Astrophysics* 70 (1978), 97–103, 97. http://adsabs.harvard.edu/abs/1978A&A....70...97K. Last accessed 11/18/2017.

centers abroad, as evident from the Annual Reports at the time. This happened in parallel with the explosive developments in astrophysics and cosmology, strongly supported by the rapidly evolving field of the new astronomies, whose birth had been fueled by the advent of the space age. These new technological windows also promised to facilitate studies on astrophysical processes that only seemed possible within the framework of general relativity. For instance, black holes and the search for their observational evidence, theories of quasars, neutron stars, compact X-ray sources, and high-energy phenomena in galactic nuclei, the physics of high-density and nuclear matter, and the distribution of quasars in the universe were becoming popular subjects addressed at conferences. The possibility of emission of gravitational wave pulses, as well, was proving the crucial role which relativistic gravity could play in these frontier astrophysical phenomena. Just a decade before, it was difficult to find an application of relativistic gravity to astrophysics outside of cosmology. On theoretical grounds, the role which general relativity must play in resolving such issues as the end point of stellar evolution and the importance of gravitational collapse as a source of energy had been anticipated since the late 1930s by pioneers such as Robert Oppenheimer in the US and Lev Landau in USSR,[144] and in the 1950s by Wheeler,[145] Hoyle, and others.[146] Their ideas were now being proven relevant to observation. The objects whose properties these theories predicted were white dwarf and neutron stars, but there was every reason to believe that the other objects predicted by relativistic gravity should also exist, notably black holes.

The first wave of gravitational wave experiments—as well as the discovery of pulsars and the longstanding aim to detect gravitational radiation pulses produced in catastrophic collapse of stars resulting in supernovas or black holes—had prompted other researchers to propose alternative detectors claiming sensitivities rather better than those of Weber's original experiments, but within about an order of magnitude of them.[147]

---

144 Luisa Bonolis: Stellar Structure and Compact Objects before 1940. Towards Relativistic Astrophysics. *The European Physical Journal H* 42/2 (2017), 311–393. doi:10.1140/epjh/e2017-80014-4.

145 John Archibald Wheeler et al.: Some Implications of General Relativity for the Structure and Evolution of the Universe. *Proceedings, 11ème Conseil de Physique de l'Institut International de Physique Solvay. La Structure et l'évolution de l'univers. Rapports et Discussions. Brussels, Belgium, June 9–13, 1958*. Bruxelles: R. Stoops 1958, 96–148.

146 Hoyle, The Nature of Cosmic Radio Sources, 1963, 681–683.

147 A first obvious approach involved using larger bars of aluminum—or of new types of material—and cooling them down to very low temperature (2 K or less, near the absolute zero) to reduce thermal noise, measuring their oscillations by totally new types

Impinging gravitational waves cause free test bodies to exhibit displacements which are proportional to their distance. It is this extremely small change in separation which has to be experimentally detected against a background of perturbing influences such as thermal and seismic vibrations. While these groups were concentrating on the problem of reduction of the background effects, an alternative approach to improving sensitivity could be obtained by increasing the displacement caused by the wave. As the variation of the distance between the test masses induced by the passage of a gravitational wave is proportional to the distance between the masses, one must increase the separation between the test masses as much as possible. It is this extremely small change in separation which has to be experimentally detected against a background of perturbing influences such as thermal and seismic vibrations. The basic idea behind this new approach was to continually compare the lengths of the arms of optical interferometers by bouncing laser beams between pairs of mirrors at the ends of each arm, and then making the two beams converge on a point and overlap. In the absence of gravitational waves, the beams' electromagnetic oscillations cancel one another out. If there is a space-time disturbance caused by gravitational waves, the arms change length, and the laser beams no longer cancel one another out: light is detected.[148] Any relative distance changes in two optical paths at right angles

---

of mechanical/electrical transducers. Developments in this direction had been proceeding for several years at Stanford University, at Louisiana State University, and at Sapienza University in Rome. Another very challenging proposal from a technical point of view came from Braginsky's group at Moscow University. Instead of bars, they were experimenting the possibility of building relatively small gravitational wave detectors using single sapphire crystals weighing only a few kilograms, which were supposed to be very efficient in discriminating between thermal noise and gravitational wave pulses.

148 In a Michelson laser interferometer, a laser beam will be split into two identical beams by a partially reflecting mirror, with one beam reflected at 90 degrees from the first, but preserving the original frequency. Each beam travels down an arm of the interferometer and both are reflected back and merged into a single beam before arriving at the photodetector. As long as the arms do not change length while the beams are traveling, light waves will stay perfectly aligned, canceling out in the recombined beam (totally destructive interference). Gravitational waves cause space to stretch in one direction and squeeze in a perpendicular direction simultaneously. For this reason, one arm of an interferometer will lengthen, while the other one shrinks, and constructive interference pattern will be observed in the photodetector. If one arm gets longer than the other, one laser beam will take longer to return, creating a phase difference between the two beams which will affect the interference pattern, showing that something happened to change the distance traveled by one or both laser beams. The interference pattern can be used to measure precisely how much change in length occurred and to extract information. The longer the arms of an interferometer, the smaller the measurements they can make. But this is

to one another would detect displacements, due to a gravitational wave propagating in a direction normal to the plane of the system. The advantage of this method is that the mirrors, acting as test masses, can be placed kilometers apart, so that a gravitational wave induces larger relative motions. The changes in optical path might be further increased by reflecting each beam back and forward many times between each pair of masses to enhance displacement sensitivity. But a most important feature of interferometer antennae—which are potentially more sensitive than resonant-bar detectors—is that they are *inherently broadband*, being also sensitive over a much wider range of frequencies than had been practicable with bar detectors, and can detect and measure the wave forms of all classes of sources. However, laser systems of course had also the disadvantage of being technologically more complex and, in particular, *more expensive* than bars. The pioneer of this technique was Robert L. Forward, a disciple of Weber who was the first to build a small size interferometer in the late 1960s at Hughes Aircraft Company Research Laboratories in Malibu, and to put into operation the first prototype detector in 1971,[149] improving it until 1978. He demonstrated that this idea could work in practice but did not obtain funds to move to a more sophisticated instrument.

Simultaneously, Rainer Weiss from MIT, especially intrigued by pulsars,[150] had since the end of the 1960s been actively exploring the idea of laser interferometry as a better chance of detecting gravitational waves, starting a very detailed theoretical analysis of the ultimate sensitivity and of the noise sources of an interferometer. After the failure of a first attempt in 1972, Weiss sent another funding application to the National Science Foundation (NSF) in August 1974, proposing the construction of a prototype interferometer with arms nine meters in length.

---

an incredibly tiny effect, as gravitational waves, for example, can just change the length of a 4 km arm interferometer by 1/1000th the width of a proton, that is, 10^{-18} m. The trick is thus to create a longer light path which amplifies the gravitational-wave input to detectable amplitude: as long as the wave is passing, laser light in each arm bounces back and forth between the two mirrors hundreds of times before being recombined after such multiple passes. Nevertheless, detection of such small effect also implies that filtering out all possible sources of noise is one of the most challenging tasks for this investigative technique.

149 G. E. Moss, L. R. Miller, and R. L. Forward: Photon-Noise-Limited Laser Transducer for Gravitational Antenna. *Applied Optics* 10/11 (1971), 2495–2498. doi:10.1364/AO.10.002495. Forward had gained his PhD in physics from the University of Maryland in 1965, collaborating in the building and operation of Weber's bar antenna.

150 Harry Collins: *Gravity's Shadow. The Search for Gravitational Waves*. Chicago, IL: University of Chicago Press 2004, 274.

This is the point at which the concentration of expertise in gravitational topics, theoretical and experimental, became decisive for Munich: because of Kafka's deep involvement in the analysis and evaluation of Weber's experiment, he was asked to be one of the reviewers of the project. As a theoretician, he felt like an outsider in the experimental side of the field (—"I didn't understand much about the experimental possibilities [...] I had to talk to the experimentalists anyhow"), and decided to circulate the proposal among the experimental groups in Munich.[151] It was unavoidable that they discussed all these things in detail as they were actually planning to upgrade their experiment, investigating the possibility of designing an antenna that was to be kept at very low temperatures—near absolute zero—to reduce thermal noise, in parallel with other technical improvements to achieve better sensitivity. However, the Munich/Garching group was so enthusiastic about Weiss's plans that they immediately determined it would be possible to replicate the interferometric experiment using in-house resources, even if this meant starting from scratch again and exploring a whole new technology. In the meantime, Weiss did not get the money from the National Science Foundation, and so the original American project was delayed, while the Munich group quickly moved forward with the new project. At the same time, the Americans themselves would use the Germans' success (and the fact that the project proposal had been inspired by Weiss's leaked proposal) to receive funding in the end, and over the next decades, to an ever-increasing extent, they eventually took back control over the largest effort in gravitational wave detection experiments.[152] Walter Winkler, who worked at the Munich project from the very beginning, recalled: "Rai Weiss stated in this respect: LIGO would not have been funded without the results from the Munich/Garching group."[153]

In the meantime, the discovery of the first pulsar in a close binary system, in 1974, had opened up new possibilities for the study of relativistic gravity.[154] The decrease of the orbital period (obtainable from the observed time variation of the pulsar period), while the two stars gradually spiral closer to one another

---

151 Collins, *Gravity's Shadow*, 2004, 276–277. See also, Peter Kafka: interview by Harry Collins, available at http://sites.cardiff.ac.uk/harrycollins/webquote/. Last accessed 6/10/2018. According to Collins's interview with Robert L. Forward (also available at the same URL), Maischberger at ESRIN, in Italy, was involved as a reviewer, too, and he immediately thought of carrying out the interferometric experiment himself.

152 Peter Kafka: interview by Harry Collins, available at http://sites.cardiff.ac.uk/harrycollins/webquote/. Last accessed 5/5/2019.

153 Walter Winkler, personal communication with the authors, March 23, 2019.

154 Russel A. Hulse, and Joseph H. Taylor: Discovery of a Pulsar in a Binary System. *Astrophysical Journal* 195 (1975), L51–L53. doi:10.1086/181708.

as gravitational waves carry energy away—a consequence predicted by Einstein's theory—would thus constitute a test for the existence of gravitational radiation.

An Itinerant Gravitational-Wave Group

In March 1976, while observations with the Weber-type resonant antennae ended, a 3 m interferometer was being built by the Munich group.[155] In December 1978, the Ninth Texas Symposium on Relativistic Astrophysics, which had become the principal international meeting where relativists and astrophysicists met and discussed recent research, was held in Munich.[156] For the first time in the history of these series of meetings, a Texas Symposium was held not just outside Texas but also outside the continental United States. The Texas Symposium held in Munich was also the occasion of the first public announcement of the experimental evidence for the reality of gravitational radiation damping in the binary pulsar discovered by Hulse and Taylor, which was published shortly afterwards.[157]

Discussions on continuing research on the gravitational wave experiment with laser interferometry were ongoing, also in view of Heinz Billing's retirement in 1982.[158] Ludwig Biermann officially retired in March 1975, but continued to be active at the institute. In promoting the gravitational wave exper-

155 The group comprised Billing, Kafka, Maischberger, Schnupp, and Winkler. Their Weber-type coincidence experiment had been run between July 1973 and February 1976. Kafka, and Schnupp, Final Result of the Munich-Frascati Gravitational Radiation Experiment, 1978, 97–103, 103.

156 Jürgen Ehlers, J. J. Perry, and M. Walker (eds.): *9th Texas Symposium on Relativistic Astrophysics. 14–19 December 1978, Munich*. New York, NY: New York Academy of Sciences 1980.

157 Joseph H. Taylor, L. A. Fowler, and P. M. McCulloch: Measurements of General Relativistic Effects in the Binary Pulsar PSR1913 + 16. *Nature* 277/5696 (1979), 437–440. doi:10.1038/277437a0.

158 The fate of Billing's group, still named "Numerische Rechenmaschinen," came under discussion from March 1977 (CPTS meeting minutes of 08.03.1977, AMPG, II. Abt., Rep. 62, No. 1780). The committee specified that the preliminary gravitational wave experiment was of fundamental importance and should be continued (CPTS meeting minutes of 22.06.1977, 01.02.1978, AMPG, II. Abt., Rep. 62, No. 1781, 1783). During the meeting of the CPT section on October 29, 1980, it was reported that "Mr. Walther agreed to take the group into the Institute of Quantum Optics" (which would be founded soon after), CPTS meeting minutes of 29.10.1980, AMPG, II. Abt., Rep. 62, No. 1791. Reimar Lüst, then President of the Max-Planck Society, "was obviously ready to support the research after Billing's retirement in 1982." Walter Winkler, personal communication with the authors, April 4, 2019.

iment, he had added a last fruitful item to his rich and enduring legacy.[159] The heroic era of gravitational wave experiments at the Institute for Astrophysics was coming to an end and at the same time the development of laser interferometers was changing globally the scale of gravitational wave experiments. A prototype 3 m gravitational wave antenna was in the preliminary phase of testing, in view of the more ambitious project for a 30 m antenna.[160] By October 1980, a decision had been taken to transfer the gravitational wave experiment group to the Max Planck Institute of Quantum Optics, which was founded on January 1, 1981.[161]

---

159 Biermann became Emeritus on March 31, 1975. In the Annual Report, signed by his successor Kippenhahn, his instrumental role during almost 30 years at the Institute for Astrophysics in opening and promoting new research fields—ultimately leading to the foundation of new institutes—was emphasized, and the dynamic effect of his establishment of the Department for Gravitation Theory and Relativistic Astrophysics and promotion of the Munich gravitational experiment was recalled. Rudolf Kippenhahn, and Klaus Pinkau: Max-Planck-Institut für Physik und Astrophysik. Institut für Astrophysik und Institut für extraterrestrische Physik. I. Institut für Astrophysik; II. Institut für extraterrestrische Physik. Reports 1975. *Mitteilungen der Astronomischen Gesellschaft* 39 (1976), 112–134, 112. http://adsabs.harvard.edu/abs/1976MitAG..39..112K. Last accessed 4/29/2018.

160 Albrecht Rüdiger et al.: The Garching 30-Meter Prototype and Plans for a Large Gravitational Wave Detector. In: Melville P. Ulmer (ed.): *13th Texas Symposium on Relativistic Astrophysics. Chicago, December 14–19, 1986*. Singapore: World Scientific 1987, 20–22. https://ui.adsabs.harvard.edu/1987txra.symp...20R/abstract. Last accessed 10/29/2019.

161 In fact, the roots of the Max Planck Institute of Quantum Optics dated back to the establishment on January 1, 1976, of a Laser Research Group at the Max Planck Institute for Plasma Physics (IPP), a result of an agreement between the German Federal Ministry for Research and Technology, as it was called at the time, and the Max Planck Society. The aim of the group was to work on the development of high-power lasers and their application to plasma physics, chemistry, spectroscopy, and other fields. This issue was discussed at the meetings of the Max Planck Society's 'Senatsausschuss für Forschungspolitik und Forschungsplanung' (Senate committee on research policy and research planning) in 1975 (see copies of the minutes in AMPG, III. Abt., Rep. 68 A, No. 151). The committee discussing the future of this group and its transformation into the Institute of Quantum Optics (with Herbert Walther, Karl-Ludwig Kompa, and Siegbert Witkowski as Directors of the three departments: Laser Physics, Laser Chemistry, and Laser Plasmas), was formed on June 14, 1978, and during the CPT Section meeting of May 5, 1979, the final formal decision was unanimously taken (CPTS meeting minutes of 14.06.1978, 30.01.1979, 09.05.1979, AMPG, II. Abt., Rep. 62, No. 1784, 1786, 1787). In 1981, the research group was given separate status as the Institute of Quantum Optics and the Research Group on Gravitational Waves became involved with the development of laser interferometers. The group at IPP initially had 46 members and quickly grew to 105, so that the space made available by IPP, including additional barracks, soon became too small. In 1986, when the institute moved to a dedicated new building, there were 184 staff members. See preface

In May 1982, when the gravitational wave group became officially part of the Max Planck Institute of Quantum Optics in Garching, construction of a new prototype interferometer, which would have a 30 m path, had already started; construction was completed in mid-1983, but improvements continued to be made over the years.[162] Weiss himself very well expressed the valuable efforts made by the group:

> So, the Max Planck group actually did most of the very early interesting development. They came up with a lot of what I would call the practical ideas to make this thing better and better.[163]

In the meantime, an Italian–French collaboration was being established in view of a project for an interferometric antenna.[164] It was led by Adalberto Giazotto, working at the University of Pisa from 1982, and by Alain Brillet working on laser interferometry at CSNSM (*Centre de Sciences Nucléaires et de Sciences de la Matière*) in Orsay.[165]

The group at Glasgow University, too, had moved towards the development of techniques for the detection of gravitational radiation using optical interferometry since 1975. As in Garching, the strategy had been based on developing the monitoring instrumentation on prototype detectors of small arm length, in

---

and Section 3.2.10, entitled "Messung von Gravitationswellen—eine Revolution in der Astronomie?" in Max-Planck-Gesellschaft: Max-Planck-Institut für Quantenoptik Garching b. München. *Max-Planck-Gesellschaft. Berichte und Mitteilungen* 6/86 (1986), 1–129.

162 A description of the laser interferometric project related to that stage of activities can be found in Heinz Billing et al.: The Munich Gravitational Wave Detector Using Laser Interferometry. In: Pierre Meystre, and M.O. Scully (eds.): *Quantum Optics, Experimental Gravity, and Measurement Theory*. Conference "Quantum Optics and Experimental General Relativity", held August 16–29, 1981 in Bad Windsheim, Federal Republic of Germany. Boston, MA: Springer 1983, 525–566. doi:10.1007/978-1-4613-3712-6_23. See also, Albrecht Rüdiger et al.: Gravitational Wave Detection by Laser Interferometry. In: I. Ursu, and A.M. Prokhorov (eds.): *Lasers and Applications*. Bucharest: CIPPress 1983, 155–179.).

163 Reiner Weiss: interview by Shirley K. Cohen, May 10, 2000. Transcript, Caltech Archives, https://resolver.caltech.edu/CaltechOH:OH_Weiss_R. Last accessed 1/19/2019.

164 Carlo Bradaschia et al.: The VIRGO Project: A Wide Band Antenna for Gravitational Wave Detection. *Nuclear Instruments and Methods in Physics Research Section A: Accelerators, Spectrometers, Detectors and Associated Equipment* 289/3 (1990), 518–525. doi:10.1016/0168-9002(90)91525-G.

165 Adele La Rana, and Leopoldo Milano: The Early History of Gravitational Wave Detection in Italy. From the First Resonant Bars to the Beginning of the Virgo Collaboration. In: Salvatore Esposito (ed.): *Società Italiana Degli Storici Della Fisica e Dell'Astronomia. Atti Del XXXVI Convegno Annuale / Proceedings of the 36th Annual Conference. Napoli 2016*. Pavia: Pavia University Press 2017, 185–196. doi:10.23739/9788869520709/c17.

the hope that the sensitivity to gravity waves could be improved fairly rapidly by scaling up the length of the arms, without making major changes to the instrumentation by which the length difference was monitored. In Glasgow they had built and were further developing a system with an arm 10 m in length,[166] and were considering the possibility of building a larger detector with an arm approximately 1 km in length.[167]

Meanwhile, in Garching, after encouraging progress with the 30 m prototype, the group was stepping up efforts in order to prepare for a big leap in size: a full-sized laser interferometer with arms 3 km in length.[168] Both the British and German groups had gained considerable experience in the design and operation of prototype versions of interferometric detectors since the early 1970s, but in 1988 it became clear that the British proposal for a 1 km antenna would not be financed by the Science and Engineering Research Council (SERC).[169] In Germany, preliminary investigations for this ambitious

166 N. A. Robertson et al.: Passive and Active Seismic Isolation for Gravitational Radiation Detectors and Other Instruments. *Journal of Physics E: Scientific Instruments* 15/10 (1982), 1101–1105. doi:10.1088/0022-3735/15/10/032. Ronald William P. Drever et al.: Gravitational Wave Detectors Using Laser Interferometers and Optical Cavities. Ideas, Principles and Prospects. In: Pierre Meystre, and Marlan O. Scully (eds.): *Quantum Optics, Experimental Gravity, and Measurement Theory*. Conference "Quantum Optics and Experimental General Relativity", held August 16–29, 1981 in Bad Windsheim, Federal Republic of Germany. Boston, MA: Springer US 1983, 503–514. doi:10.1007/978-1-4613-3712-6_21. In 1979, Ronald Drever took up a part-time appointment to Caltech, and full-time later, in 1983, leaving James Hough as the Glasgow leader. At Caltech Drever started a project which was eventually funded.

167 James Hough et al.: The Development of Long Baseline Gravitational Radiation Detectors at Glasgow University. *Gravitation, Geometry and Relativistic Physics. Proceedings of the "Journées Relativistes" Held at Aussois, France, May 2–5, 1984*. Berlin: Springer 1984, 204–212. doi:10.1007/BFb0012592. James Hough et al.: *A British Long Baseline Gravitational Wave Observatory*. Design study report GWD/RAL/86-001. Glasgow: Rutherford Appleton Laboratory, Glasgow University 1986.

168 The concept of the large antenna was described in Karl Maischberger et al.: *Vorschlag Zum Bau Eines Grossen Laser-Interferometers Zur Messung von Gravitationswellen*. MPQ 96. Max-Planck-Institut für Quantenoptik 1985. Plans for the large detector were described also in Rüdiger et al., The Garching 30-Meter Prototype and Plans for a Large Gravitational Wave Detector, 1987, 20–22. Karl Maischberger et al.: Status of the Garching 30-Meter Prototype for a Large Gravitational Wave Detector. In: Peter F. Michelson, Hu En-ke, and Guido Pizzella (eds.): *International Symposium on Experimental Gravitational Physics. 3–8 August 1987. Guangzhou, China*. Singapore: World Scientific 1988, 316–321. https://ui.adsabs.harvard.edu/1988egp..conf..316M/abstract. Last accessed 10/29/2019.

169 Serious financial problems in the UK led to fierce competition with projects put forward by the astronomy/astrophysics community, also because at the end of the 1980s, many conventional astronomers were still very suspicious and did not consider gravitational

project, led by Gerd Leuchs at MPI of Quantum Optics, were financed by the German Federal Ministry for Research and Technology (BMFT) for the period 1987–89.[170] In a context of funding difficulties and increased emphasis on international collaborations, the Garching project for a 3 km interferometric gravitational wave detector resurfaced in 1989 as a joint German–British proposal,[171] strongly encouraged by the two funding agencies BMFT and SERC.[172] As stated in the preface to the proposal, it was expected that

> all the long baseline detectors to be built [the LIGO project and the Italian–French Virgo project] will operate *as part of a coordinated world-wide network* [our emphasis].

At that time, the prospects for the realization of a big interferometer looked excellent.[173] From 1990, the gravitational wave project at the Max Planck Institute of Quantum Optics in Garching was led by Karsten Danzmann, who had come back from Stanford University, where he had moved in 1982 after gaining his PhD at the Technical University in Hanover.[174]

In 1991, the German–British project, now named GEO, was presented as an interferometer with arms each 3 km in length, to be built near Hanover, in

---

waves as something worth funding. This also happened in the US, where astronomers and astrophysicists felt that the LIGO project was competing for their resources. On this question see Collins, *Gravity's Shadow*, 2004, 500–504.

170 See AMPG, II. Dpt., Rep. 66, No. 3122, 3853, 3868.

171 James Hough et al.: *Proposal for a Joint German-British Interferometric Gravitational Wave Detector.* MPQ 147. Garching: Max-Planck-Institut für Quantenoptik 1989. http://eprints.gla.ac.uk/114852/7/114852.pdf. Last accessed 4/18/2018.

172 In Appendix A of the proposal, the two groups presented the results of a 100-hour period of coincident observation using the two prototypes at Garching and Glasgow (the 30 m and the 10 m arm length), which had been solicited by BMFT and SERC to show that such detectors could be operated in the production fashion by the two teams working together. It was the first time that two detectors had been run continuously in a data-taking mode, demonstrating the potential for long-term operation of laser interferometric detectors.

173 Funds for the preliminary investigations of the German project were provided by BMFT and searches for a suitable site went on during the second half of the 1980s. See documents on the financing of a grant for "Voruntersuchungen für den Bau eines großen Laserinterferometers zur Messung von Gravitationswellen" starting in November 1987 and ending in December 1990, and for a second tranche covering the period from January 1, 1990, to December 31, 1992 (AMPG, II Abt., Rep. 66, No. 3853, 3868 and Rep. 68, No. 65).

174 Gerd Leuchs, who had led the Garching group from 1985 to 1989, moved to work in industry and later became Director of the Max Planck Institute for the Science of Light.

the German state of Lower Saxony.[175] In summer 1992, a 3 km GEO interferometer was still part of a list of the detectors at that scale being planned in the world: the French–Italian 3 km Virgo (comprising nine groups from both countries) to be built near Pisa; the American 4 km LIGO (Laser Interferometer Gravitational-Wave Observatory) project (approved in fall 1991) with scientists at MIT and Caltech; and a more recent Australian collaboration proposing AIGO, a 3 km detector near Perth (not yet approved at the time). They were meant not to be in competition with one another, on the contrary, "a worldwide network of four detectors," each "crucially dependent on the others," would be required "to fully unravel the information contained in the signals with respect to the source direction, time structure, and polarization."[176]

However, in spite of ongoing contacts between the European groups during the second half of the 1980s, the Italian–French collaboration and the British–German venture had not merged into a real pan-European joint effort, a European network of gravitational-wave telescopes that might have followed and matched the successful example of effective cooperation in the CERN enterprise.[177]

---

175 Karsten Danzmann et al.: The GEO—Project a Long-Baseline Laser Interferometer for the Detection of Gravitational Waves. In: J. Ehlers, and G. Schäfer (eds.): *Relativistic Gravity Research With Emphasis on Experiments and Observations. Proceedings of the 81 WE-Heraeus-Seminar Held at the Physikzentrum, Bad Honnef, Germany 2–6 September 1991*. Berlin: Springer 1992, 184–209. doi:10.1007/3-540-56180-3_9.

176 Karsten Danzmann: Laser Interferometric Gravitational Wave Detectors. In: R.J. Gleiser, C.N. Kozameh, and O.M. Moreschi (eds.): *Proceedings. 13th International Conference on General Relativity and Gravitation. Cordoba, Argentina, June 28-July 4, 1992*. Bristol: Institute of Physics 1993, 3–19, 19. https://inspirehep.net/record/348400. Last accessed 1/24/2019.

177 According to the Italians, "In spite of a few European meetings, and a good collaboration with German and British colleagues through 2 European grants, we were actually pushed in the direction of a bi-national project, rather than a joint two-detector European project, by the fact that the German team at Garching and the British team (mainly at Glasgow) were pushing their own national projects, and feared that the settlement of a European collaboration would delay their acceptation." Carlo Bradaschia: VIRGO 20th Anniversary. *H—The Gravitational Voice. Special Anniversary Edition* (2009), 2–12, 6. http://www.ego-gw.it/public/hletter/doc/h_Special_Edition.pdf. Last accessed 1/29/2019. Both the tension existing between national ambitions and efforts towards international collaboration, and the problem of the lack of a really coordinated gravitational-wave community at a European level, played a negative role in this phase. On the question of why European leading groups in the field of gravitational-wave interferometry did not join forces to build a European observatory with at least two detectors at kilometer scale see Adele La Rana: The Origins of Virgo and the Emergence of the International Gravitational Wave Community. In: Alexander S. Blum, Roberto Lalli, and Jürgen Renn (eds.):

Retreat from Full-Scale Experimentation: Rescuing Excellence through Technology and Instrumental Innovation

By 1992, the existence of gravitational waves had been demonstrated by the motion of the double neutron star system PSR 1913+16, in which one of the stars is a pulsar emitting electromagnetic pulses at radio frequencies at precise, regular intervals, as it rotates. Arrival-time measurements of the radio signals running since 1974 showed an orbital-motion decay consistent with the gravity-wave emission according to general relativity, with an accuracy better than 0.5 percent.[178] This timely result would lend further momentum to ongoing plans to build large-scale, ground-based laser interferometers.

However, following the fall of the Berlin Wall in November 1989, after nearly three decades in existence, the Reunification Treaty was signed by the two German states in August 1990, and the German Federal Ministry for Education and Research (BMFT) took a stance that was justified both by the critical situation ensuing from German reunification and the challenge of assuming responsibility for the process of restructuring East German science:[179] the ambitious dream of a 3 km interferometer was definitely not to be considered a priority with respect to other planned physics projects in which BMFT had programmed huge investments since the mid-1980s.[180]

The Renaissance of General Relativity in Context. Cham: Springer International Publishing 2020, 363–406. doi:10.1007/978-3-030-50754-1_10. See also La Rana, and Milano, The Early History of Gravitational Wave Detection in Italy, 2017, 185–196. For an analysis of the main causes of the failure to establish a network of three long-based antennas in Europe, see Adele La Rana: EUROGRAV 1986–1989: The First Attempts for a European Interferometric Gravitational Wave Observatory. *European Physical Journal H* 47/1 (2022), 1–23. doi:10.1140/epjh/s13129-022-00036-x.

178 Joseph H. Taylor et al.: Experimental Constraints on Strong-Field Relativistic Gravity. *Nature* 355/6356 (1992), 132–136. doi:10.1038/355132a0. The following year, the Nobel Prize in Physics 1993 was awarded to Russell A. Hulse and Joseph H. Taylor for the discovery of the binary pulsar PSR 1913+16. Subsequent observations and interpretations of the evolution of the orbit, had opened up "new possibilities for the study of gravitation." Nobel Media AB 2020, https://www.nobelprize.org/prizes/physics/1993/summary/. Last accessed 1/2/2020.

179 Bernhard A. Sabel: Science Reunification in Germany: A Crash Program. *Science* 260/5115 (1993), 1753–1758. www.jstor.org/stable/2881355. Last accessed 11/17/2019.

180 In mid-1990, BMFT formed a multidisciplinary commission led by the theoretical physicist Siegfried Großmann, which was supposed to make recommendations on fundamental research in Germany. A 124-page report was officially released in April 1992. Notwithstanding the *special sympathy* with which the commission regarded the large-scale experiment of a gravitational-wave detector "because of its novel scientific objectives," also acknowledging its "*special charm*" due to its innovative approach to gravitation, "the smallness, possibly the still-undetectability, of the effect," was also highlighted

German reunification had changed the circumstances in a truly dramatic fashion. With plans underway for similar large-scale antennas both in Europe and the US, the German and British teams who had pioneered research in the field since the early 1970s, and had long since collaborated owing to their respective commitment to building prototype interferometers, were deeply disappointed and struggled to find an alternative strategy, such as that pursued by Karsten Danzmann, who led the gravitational wave project at the Max Planck Institute of Quantum Optics in Garching. A reduction in arm length would cut down the detector cost considerably, making the plan to build a much smaller facility a realistic aim for the British–German teams. Max Planck scientists thus joined forces with British researchers to build the smaller GEO600 experiment, a gravitational-wave antenna with arms 600 m in length.[181] GEO600 itself, construction of which began in September 1995, was not explicitly funded as a detector, but obtained money from different sources, even from the BMFT, for each innovative technology development.[182]

---

in the report. A copy of the pages in the commission's report related to the project of building a detector for gravitational wave astronomy (pp. 76–78) can be found in the Archives of the Max Planck Society in Munich (Akten der Registratur und des Archivs der Max-Planck-Gesellschaft, MPI für Quantenoptik, Gravitationswellenexperiment III, 1991–1997, Aktenzeichen 18140907, Barcode 233163, Fol. 373–380). BESSY II, the upgraded new electron storage ring producing synchrotron radiation for materials research purposes, to be built in Berlin, was instead considered a "high priority" initiative, and already in July of that year the project got the "green light." By contrast, fundamental particle physics did not receive favorable treatment, but a financial horizon which "should not enlarge, nor back off in the next few years." As for gravitational waves, the commission had not recommended "immediate implementation, but swift prosecution with intensive scientific discussion" E. Dreisigacker: Grundlagenforschung—quo vadis? Großmann-Kommission legt Empfehlungen vor. *Physikalische Blätter* 48/5 (1992), 372–374, 374. doi:10.1002/phbl.19920480514.

181 Karsten Danzmann et al.: *GEO 600. Proposal for a 600 m Laser-Interferometric Gravitational Wave Antenna.* MPQ-190. Garching: Max-Planck-Institut für Quantenoptik 1994. Karsten Danzmann et al.: GEO 600—A 600m Laser Interferometric Gravitational Wave Antenna. In: Eugenio Coccia, Guido Pizzella, and Francesco Ronga (eds.). First Edoardo Amaldi Conference on Gravitational Wave Experiments. World Scientific 1995, 100–111. doi:10.1142/9789814533652. Harald Lück, and the Geo600 Team: The GEO600 Project. *Classical and Quantum Gravity* 14/6 (1997), 1471–1476. doi:10.1088/0264-9381/14/6/012.

182 In the period 1993–2000, this strategy was supported by Hermann Schunck, then Director of the BMFT and responsible for fundamental research, especially for Physics, who was able to channel "leftover" money from other projects that had not been able to spend it, in order to finance specific GEO600 needs justifiable as "standalone projects" such as vibration isolation, data acquisition, novel optics, laser stabilization, and novel vacuum system design (Hermann Schunck: Written interview by Adele La Rana, May 14, 2019, and personal communication with the authors, November 20, 2019).

In 1993, Danzmann became professor at the University of Hanover as well as Director of the Institute for Atomic and Molecular physics, and from 1994 he continued, in parallel, to lead the project as leader of the Hanover branch of the Max Planck Institute of Quantum Optics. In March 1991, the Max Planck Institute for Physics and Astrophysics had been split up into three independent institutes: the MPI for Physics in Munich (Werner-Heisenberg-Institut), the MPI for Astrophysics, and the MPI for Extraterrestrial Physics, the last two in Garching.[183] In 1995, a Max Planck Institute for Gravitational Physics—named after Albert Einstein, the physicist who developed the theory of general relativity (Albert-Einstein-Institut, AEI)—was founded, with Directors Jürgen Ehlers and Bernard F. Schutz, the latter also remaining part-time in Cardiff.[184]

183 Minutes of the 127th Senate meeting in Frankfurt am Main, 08.03.1991, AMPG, II. Abt., Rep 60, No. 127.SP, pp. 23–24.

184 The foundation of the Max Planck Institute for Gravitational Physics has its roots in the period of German unification and in the role and commitments of the Society in the new Federal States, which were discussed in a long report by the MPG President Hans F. Zacher during the CPT Section meeting of October 2, 1990 (CPTS meeting minutes of 10.02.1990 AMPG, II. Abt., Rep. 62, No. 1821). Following the recommendation of the German Council of Science and Humanities (*Wissenschaftsrat*) to close the Einstein-Laboratorium of the Academy of Sciences in Potsdam, the Max Planck Society was involved to advice about the establishment of a thematically new Albert Einstein Institute for Gravitational Physics. A memorandum for the reorganization of the Einstein-Laboratorium into an International Einstein Center had already been formulated by Hubert Goenner and Friedrich Hehl in February 1991 and announced in Hubert Goenner, and F.W. Hehl: Zur Gründung des Albert-Einstein-Instituts für Gravitationsphysik. *Physikalische Blätter* 47/10 (1991), 936–936. doi:10.1002/phbl.19910471015. For a detailed reconstruction see Hubert Goenner: Some Remarks on the Early History of the Albert Einstein Institute. arXiv:1612.01338 [physics.hist-ph] 2016. doi:10.48550/arXiv.1612.01338. However, in July 1991, Hans Zacher, president of the Max Planck Society, formed a working group led by Ehlers in order to examine the scientific concept, equipment and funding of such institute and asked Ehlers to prepare a memorandum (see a copy of this document within the CPTS meeting minutes of 16.10.1992, AMPG, II. Abt., Rep. 62, No. 1827). Between July and September 1992, a decision was taken to support Ehlers' proposal to found a Max Planck Institute for Gravitational Physics and to form a special committee to examine all relevant conceptual ideas and scientific aspects related to the project (Zacher to Gerhard Wegner, president of the CPT Section, 18 September 1992, AMPG, II. Abt., Rep. 62, No. 87, Fol. 498). The final version of the memorandum, approved by members of the founding committee, who recommended that Ehlers apply for the founding of the Institute for Gravitational Physics, was sent by Ehlers to President Zacher on 10 June 1992 (AMPG, II. Abt., Rep. 62, No. 205, Fol. 161–167). Material related to the work of the committee for the foundation of an Institute for Gravitational Physics can be found in AMPG, II. Abt., Rep. 62, No. 87, 205, 206, 207, and also in Rep. 57, No. 482, 483, 1228. Works of the committee were reported during several meetings of the CPT Section (CPTS meeting minutes of 07.02.1992, 03.06.1992, 16.10.1992, 03.02.1993, 16.06.1993,

Immediately after its foundation, Hermann Nicolai was appointed as third Director of AEI.[185] Initially located in a temporary seat in Potsdam, where it began operations in April 1995, the institute moved to its new building in Potsdam-Golm in 1999. The creation of a new astrophysics-oriented Max Planck Institute in the new *Bundesländer*, following German reunification, resulted from a further 'cell division' in the Munich area,[186] and increased the dominance of the Max Planck Society in the astronomical-astrophysical research fields. At the time, the strong pressure to move research institutes eastwards was particularly felt in Munich, which in addition to the AEI, was forced to continue the expansion of its flagship Institute for Plasma Physics to Greifswald on the East German Baltic Sea coast. The epicenter of gravitational-wave research also moved away from the Munich area, albeit for altogether different reasons already in place before reunification: while Bavaria at the time concentrated the specific expertise in gravitational waves, the laser technologies involved in their detection were pioneered in Lower Saxony, among a collection of institutes in Hanover and nearby Braunschweig. At the moment of generational change, regional actors led by Danzmann's mentor Herbert

---

19.10.1993, 03.02.1994, 08.06.1994, 09.02.1995, AMPG, II. Abt., Rep. 62, No. 1825, 1826, 1827, 1828, 1829, 1830, 1831, 1832, 1834). Details about Ehlers' role and other aspects related to the phase preceding the actual decision to found the new institute—also connected to the above-mentioned difficult period of the Institute for Astrophysics following Kippenhahn's anticipated retirement—can be found in Wolfgang Hillebrandt: Kraft schöpfen aus dem Wandel: Die deutsche Astronomie in der Zeit der Wiedervereingung. In: Dietrich Lemke (ed.): *Die Astronomische Gesellschaft 1863–2013*. Heidelberg: Astronomische Gesellschaft 2013, 127–141. For a related discussion on the German unification phase, see E. Dreisigacker: Außeruniversitäre Forschung in den östlichen Bundesländern vor Neuanfang. *Physikalische Blätter* 47/8 (1991), 763–767. doi:10.1002/phbl.19910470808. For the evolution of research on general relativity in Germany and its eventual institutionalization in the form of a dedicated research institute, see Hubert Goenner: General Relativity and the Growth of a Sub-Discipline "Gravitation" in Germany. *The European Physical Journal H* 42/3 (2017), 395–430. doi:10.1140/epjh/e2017-70057-4.

185 CPTS meeting minutes of 09.02.1995, 21.06.1995, 19/20.10.1995, 8/9.02.1996, AMPG, II. Abt., Rep. 62, No. 1834, 1835, 1836, 1837. Ehlers, Schutz, and Nicolai led research activities respectively in general relativity, relativistic astrophysics, and quantum gravity/unified theories. Numerical relativity and computer simulations, also related to collapsing relativistic binaries and their associated gravitational waves, were an active part of the research activity since the very beginning of AEI, see, for example, Bernard S. Schutz: Max-Planck-Institut Für Gravitationsphysik (Albert-Einstein-Institut). *Mitteilungen Der Astronomischen Gesellschaft* 82 (1999), 649–656. https://ui.adsabs.harvard.edu/abs/1999MitAG..82..649S/abstract. Last accessed 12/1/2019.

186 Joachim Trümper: Astronomy, Astrophysics and Cosmology in the Max Planck Society. In: André Heck (ed.): *Organizations and Strategies in Astronomy*. Dordrecht: Springer Netherlands 2004, 169–187.

Welling and backed by funders such as the Volkswagen Foundation exerted their influence to win the location of the interferometer (then with arms 3 km in length).[187] Even though this full-scale project did not come about, the expertise continued to be focused in the area, and during the period of transition, when the founding generation of researchers in Bavaria reached retirement age, the new positions to replace them were created in Lower Saxony.

In 1995, in parallel with the founding of the Albert Einstein Institute, the construction of the 600 m-long gravitational wave detector GEO600 started in Ruthe, a site 20 km south of Hanover. Soon afterwards, this activity became one of the main research focuses of the new Institute, following the decision to transform the preexisting research center at the Max Planck Institute of Quantum Optics, based in Hanover and led by Karsten Danzmann, into a branch of the Albert Einstein Institute. The founding of a 'center of excellence' for gravitational wave research thus brought both experimental and theoretical activities under the same roof.[188]

In 2001, Danzmann was promoted to Director of the Laser Interferometry and Gravitational Wave Astronomy Division, the first of the two divisions planned when the Quantum Optics branch in Hanover became officially part of the Albert Einstein Institute, which has since maintained sites in both Potsdam and Hanover.[189]

---

187 Interview with Karsten Danzmann, March 29, 2018, Deutsche Physikalische Gesellschaft e. V., Stern-Gerlach-Medaille 2018, available at https://www.youtube.com/watch?v=tNTB74bFGuc. Last accessed 2/23/2020. Danzmann is considered part of the so-called 'Welling Laser Family' (*Laserfamilie Welling*), and just a few years before, Welling had consolidated the region's footprint in this field with the establishment of the Laser Zentrum Hannover e.V. Gerd Liftin, and Jürgen Mlynek: Zum 80. Geburtstag von Herbert Welling. *Physik Journal* 8/10 (2009), 45. https://www.pro-physik.de/restricted-files/98106. Last accessed 4/14/2020. Liftin, and Mlynek, Herbert Willing, 2009, 45. For the latest account of his career, see "Grosses Verdienstkreuz für Professor Herbert Welling," Presseinformation des Niedersächsischen Ministeriums für Wissenschaft, 31.8.2019, available at https://www.mwk.niedersachsen.de/startseite/aktuelles/presseinformationen/grosses-verdienstkreuz-fur-professor-dr-herbert-welling-180202.html. Last accessed 2/23/2020.

188 In June 2000, the founding of a center for gravitational-wave research was discussed during a meeting of the CPT Section. As stressed by Bernard Schutz, the whole operation would assure participation of the Max Planck Society, with a cutting-edge role, in the outstanding projects EURO and the laser-interferometric detectors LIGO and LISA (the latter, the Laser Interferometer Space Antenna mission, a giant interferometer to be placed in space). A committee was formed to examine the whole plan (CPTS meeting minutes of 07.06.2000, 19/20.10.2000, 15/16.02.2001, AMPG, II. Abt., Rep. 62, No. 1851, 1852, 1853).

189 CPTS meeting minutes of 15/16.02.2001, 20.06.2001, 18/19.10.2001, AMPG, II. Abt., Rep. 62, No. 1853, 1854, 1855.

While the German 3 km interferometer project had to be put aside in favor of the smaller GEO600, the American proposal for the Laser Interferometer Gravitational-Wave Observatory (LIGO), consisting in two widely separated longbased installations (4 km arms) within the United States, was funded, as was the Italian Virgo.[190] The Virgo project for a 3 km interferometer was approved between 1992 and 1994 by the *Centre National de la Recherche Scientifique* (CNRS) and the *Istituto Nazionale di Fisica Nucleare* (INFN), eventually leading to the construction of the Virgo interferometer at Cascina, near Pisa, beginning in the late 1990s.

In 1997, the British–German collaboration finally entered into partnership with LIGO, becoming part of the worldwide network of gravitational wave detectors and contributing to the next generation of US detectors with new advanced technologies.[191] A collaboration linking the three LIGO detectors in the US with its partner GEO600 in Germany and the Virgo detector in Italy was established in early 2007. Many of the technologies developed at GEO600 thus became instrumental in enabling the unprecedented sensitivity of LIGO and Virgo.

On September 14, 2015, at 09:50:45 UTC, 100 years after Einstein formulated the field equations of general relativity, the two detectors of the Laser Interferometer Gravitational-Wave Observatory simultaneously observed a transient gravitational wave signal matching the waveform predicted by general relativity for the inspiral and merger of a pair of black holes of about 30 solar masses each. The signal caused the mirrors at the ends of each interferometer's 4 km

190 The LIGO construction proposal was approved by the National Science Board in 1990, and in 1992 the LIGO cooperative agreement for the management of LIGO was signed by NSF and Caltech, while construction at the chosen sites Hanford and Livingston began between 1994 and 1995.

191 B. P. Abbott et al.: Detector Description and Performance for the First Coincidence Observations between LIGO and GEO. *Nuclear Instruments and Methods in Physics Research Section A. Accelerators, Spectrometers, Detectors and Associated Equipment* 517/1 (2004), 154–179. doi:10.1016/j.nima.2003.11.124. Katherine L. Dooley et al.: GEO 600 and the GEO-HF Upgrade Program: Successes and Challenges. *Classical and Quantum Gravity* 33/7 (2016), 075009. doi:10.1088/0264-9381/33/7/075009. Since 2001, when the Hanover branch of the Max Planck Institute of Quantum Optics merged with the Albert Einstein Institute, GEO600 has been operated by AEI within the international collaboration with the Leibniz University of Hannover (which had been actively involved in the program through Karsten Danzmann), the University of Glasgow, and the University of Wales in Cardiff, and is now part of the worldwide network of gravitational wave detectors, including LIGO in the US, Virgo in Italy, and KAGRA (Kamioka Gravitational Wave Detector) in Japan, which was completed in October 2019.

arms to oscillate with an amplitude of about 10^{-18} m, roughly a factor of a thousand smaller than the classical proton radius. It was the first direct detection of gravitational waves after decades of experimental efforts and the first ever observation of a binary black hole merger,[192] a crowning achievement in the long process of the renaissance of general relativity.[193]

Many of the instrumental innovations that eventually led to the first 2015 detection of gravitational waves using the LIGO detectors had been pioneered by the Max Planck Institute researchers.[194] They also

192 LIGO Scientific Collaboration & Virgo Collaboration et al.: Observation of Gravitational Waves from a Binary Black Hole Merger. *Physical Review Letters* 116/6 (2016), 061102. doi:10.1103/PhysRevLett.116.061102.

193 Blum, Lalli, and Renn, Gravitational Waves, 2018, 534–543.

194 As an example, we cite here a series of articles related to the first realization by the German group of innovative detector technologies which made Advanced LIGO and Virgo so sensitive. Jun Mizuno et al.: Resonant Sideband Extraction: A New Configuration for Interferometric Gravitational Wave Detectors. *Physics Letters A* 175/5 (1993), 273–276. doi:10.1016/0375-9601(93)90620-F. Gerhard Heinzel et al.: An Experimental Demonstration of Resonant Sideband Extraction for Laser-Interferometric Gravitational Wave Detectors. *Physics Letters A* 217/6 (1996), 305–314. doi:10.1016/0375-9601(96)00361-1. D. Schnier et al.: Power Recycling in the Garching 30 m Prototype Interferometer for Gravitational-Wave Detection. *Physics Letters A* 225/4 (1997), 210–216. doi:10.1016/S0375-9601(96)00893-6. Gerhard Heinzel et al.: Experimental Demonstration of a Suspended Dual Recycling Interferometer for Gravitational Wave Detection. *Physical Review Letters* 81/25 (1998), 5493–5496. doi:10.1103/PhysRevLett.81.5493. Gerhard Heinzel et al.: Automatic Beam Alignment in the Garching 30-m Prototype of a Laser-Interferometric Gravitational Wave Detector. *Optics Communications* 160/4 (1999), 321–334. doi:10.1016/S0030-4018(98)00654-3. Harald Lück et al.: Correction of Wavefront Distortions by Means of Thermally Adaptive Optics. *Optics Communications* 175/4 (2000), 275–287. doi:10.1016/S0030-4018(00)00468-5. Andreas Freise et al.: Demonstration of Detuned Dual Recycling at the Garching 30 m Laser Interferometer. *Physics Letters A* 277/3 (2000), 135–142. doi:10.1016/S0375-9601(00)00704-0. Gerhard Heinzel et al.: Dual Recycling for GEO 600. *Classical and Quantum Gravity* 19/7 (2002), 1547–1553. doi:10.1088/0264-9381/19/7/343. Harald Lück et al.: Thermal Correction of the Radii of Curvature of Mirrors for GEO 600. *Classical and Quantum Gravity* 21/5 (2004), S985–S989. doi:10.1088/0264-9381/21/5/090. Hartmut Grote et al.: Automatic Beam Alignment for the Mode-Cleaner Cavities of GEO 600. *Applied Optics* 43/9 (2004), 1938–1945. doi:10.1364/AO.43.001938. Kasem Mossavi et al.: A Photon Pressure Calibrator for the GEO 600 Gravitational Wave Detector. *Physics Letters A* 353/1 (2006), 1–3. doi:10.1016/j.physleta.2005.12.053. Frank Seifert et al.: Laser Power Stabilization for Second-Generation Gravitational Wave Detectors. *Optics Letters* 31/13 (2006), 2000–2002. doi:10.1364/OL.31.002000. Stefan Hild et al.: DC-Readout of a Signal-Recycled Gravitational Wave Detector. *Classical and Quantum Gravity* 26/5 (2009), 055012. doi:10.1088/0264-9381/26/5/055012. Hartmut Grote et al.: First Long-Term Application of Squeezed States of Light in a Gravitational-Wave Observatory. *Physical Review Letters* 110/18 (2013), 181101. doi:10.1103/PhysRevLett.110.181101. H. Wit-

played a key role in the computational tasks related to the detection efforts.[195]

Events such as the first one detected by the LIGO collaboration, which was given the name GW150914, are invisible for traditional astronomical instruments, as any signal other than gravitational waves is emitted near the merging black holes. But then, on August 17, 2017, four decades after Hulse and Taylor discovered the first neutron star binary, the Advanced LIGO and Advanced Virgo observatories made their first direct detection of a swell of gravitational waves from the coalescence of a neutron star binary system, which was followed after 1.7 seconds by a burst of gamma rays detected by the orbiting Fermi Gamma-Ray Space Telescope and INTEGRAL observatory.[196] The detection of this new gravitational-wave signal (GW170817) offered a novel opportunity to directly probe the properties of matter in the extreme conditions found in the interior of these stars, while the unprecedented joint gravitational and electromagnetic observation of this astronomical cataclysm marked the beginning of a new era in multi-messenger astrophysics.[197]

---

tel et al.: Thermal Correction of Astigmatism in the Gravitational Wave Observatory GEO\hspace0.167em600. *Classical and Quantum Gravity* 31/6 (2014), 065008. doi:10.1088/0264-9381/31/6/065008.

195 The development of highly accurate analytical and numerical models of gravitational-wave sources—in particular of gravitational waves that neutron stars or black holes generate in the final process of orbiting and colliding with each other—have allowed extraction of astrophysical and cosmological information from the observed waveforms. These waveform models are then implemented and employed in the continuing search for binary coalescences. To significantly increase the probability of identifying gravitational waves in LIGO and Virgo data, the search for burst-like events in turn requires detailed knowledge of the expected signals from different sources and such search tools are sensitive because of systematic development in the algorithm and methods which can be used as templates to filter the data. Numerical relativity simulations with supercomputers not only play an important role in predicting gravitational waveforms that are used for gravitational wave detection, but allow, in general, exploration of general relativistic phenomena and other high-energy phenomena, such as gamma-ray bursts and stellar core collapse, or mass ejection with related nucleosynthesis processes.

196 B. P. Abbott et al.: Gravitational Waves and Gamma-Rays from a Binary Neutron Star Merger. GW170817 and GRB 170817A. *The Astrophysical Journal Letters* 848/2 (2017), L13. doi:10.3847/2041-8213/aa920c.

197 B. P. Abbott et al.: Multi-Messenger Observations of a Binary Neutron Star Merger. *The Astrophysical Journal Letters* 848/2 (2017), L12. doi:10.3847/2041-8213/aa91c9. This paper was coauthored by almost 4000 scientists from more than 900 international institutions, using 70 ground- and space-based observatories.

4 From Cosmic Rays to Ground-Based Gamma-Ray Astronomy

This final emerging field is the complex result of the evolution, throughout the entire 20th century, of the question about the origin and nature of cosmic rays. Until the 1960s, cosmic ray particles were one of the key research areas in experimental nuclear and particle physics, part of all the research traditions mentioned in Chapters 1–3. From the late 1950s onward, however, ground-based cosmic-ray research declined, as most of its stellar researchers moved toward accelerators or jumped on the Sputnik bandwagon to become space scientists. In the following three decades, cosmic rays were studied at less prestigious institutions, such as in Kiel, which nonetheless obtained results in the early 1980s that attracted worldwide attention. A new generation of accelerator-based particle physicists from both the Max Planck Institute for Physics and the Max Planck Institute for Nuclear Physics then began collaborating with Kiel, which was crucially also joined by a community of Armenians from the Yerevan Physics Institute, who had pioneered the innovative technique of stereoscopic Cherenkov Atmospheric Imaging, to detect the light and image the cascades of subatomic particles generated by cosmic gamma rays. This technique turned out to be its most promising feature, finalizing this tradition's leap towards ground-based gamma-ray astronomy. Armenian success with Cherenkov telescopes, increasingly supported by Max Planck scientists, sparked competition between two Max Planck Institutes, in Munich and Heidelberg, to become world leaders in what promised to become an entirely new form of ground-based astronomy, thereby absorbing the Armenian scientists. The Max Planck Institutes then built the most successful telescopes of the subsequent generation, MAGIC and H.E.S.S., while competing both with each other and with other global players. Thanks to their complementary double presence in the field, the two Max Planck Institutes won the race towards ground-based, gamma-ray telescopes, leading to the global Cherenkov Telescope Array (CTA) collaboration with over 100 telescopes, which the Americans then entered as junior partners.

Cosmic Rays as an Entity between Particle Physics and Astrophysics

Finally, the third example of an emerging field in this volume is the entirely new field of Very-High-Energy (VHE) gamma-ray astronomy with ground-based Imaging Atmospheric Cherenkov Telescopes (IACT), a most sensitive technique for the observation of the most energetic form of gamma rays. After the detection of solar and supernova neutrinos, the very-high-energy gamma ray photons recorded by ground-based Cherenkov telescopes were the second new window opened by astroparticle physics.

Like the programs that led to neutrino research and the search for gravitational wave signals, treated earlier in this chapter, ground-based gamma-ray astronomy was a development rooted not in traditional astronomy, but rather in experimental physics, namely in both the Munich and Heidelberg research traditions. In fact, to this day, the field's technological and cultural practices remain deeply tied to the particle physics community. Furthermore, we will see how this new field benefited from several other non-astronomical traditions such as classical cosmic ray research and plasma physics. Like in the two previous cases, ground-based gamma-ray astronomy had a long latency period of several decades of slow improvements, which only started to bear fruit in the 1980s in the United States and the Soviet Union. But this case is more complex than the other two studied, and much more articulated, involving cosmic-ray physics, high-energy physics at accelerators, high-energy astrophysics and astronomy. In particular, it also includes a gradual shift in the very object of research, initially focused on cosmic-ray particles, only later transitioning towards gamma rays, having the smallest wavelengths and the most energy of any wave in the electromagnetic spectrum. These photons, which are thought to be correlated with the acceleration sites of cosmic rays, are born in the most extreme environments of the known Universe: supernova explosions, active galactic nuclei, and gamma-ray bursts.

Technologically, this shift in the object of research also implied a shift from the exclusive use of arrays of detectors for ground-level recording of extensive cascades of the ionized particles and electromagnetic radiation produced by cosmic rays interacting with nuclei in the upper atmosphere (Extensive Air Showers), towards the development and deployment of techniques that use the atmosphere itself as a detector. In this case, the by-products of the interaction of cosmic rays and gamma rays with atmospheric particles can be traced via the Cherenkov light that they emit in the air and which is bright enough to be picked up by large focusing mirrors on the ground. With this information, the nature, energy, and direction of the primaries can be calculated, including their time of arrival, in order to determine the variability in the source emission.

The field of gamma-ray astronomy is closely linked to the physics of cosmic rays. The communities that contributed to the developments outlined in this last part of the chapter originated in cosmic ray research, a tradition that used particles coming from outer space to inquire into fundamental physical processes in their interactions with matter. After a slow start in the 1930s, these experimental studies came to be of central importance for the branch of fundamental physics which deals with the ultimate structure of matter. New elementary particles, such as the positron, the muon, the pion, as well as the kaons and also certain hyperons, were discovered in cosmic rays. From the early 1930s to the mid-1950s, the study of elementary particles and the interactions of these particles at high energies was the basic part of cosmic ray physics. Experiments were often conducted at high altitudes—in airplanes and balloons, or on mountaintops—so as to capture the extensive showers of particles generated by 'primary' cosmic rays in the atmosphere. As the energy of the incoming primary increases, secondary particles can also reach sea level. Until the mid-1950s, cosmic ray particles were one of the key research areas in experimental physics, part of all the research traditions mentioned in Chapter 1. However, these cosmic-ray-based communities were dealt a blow by the advent of accelerators of higher energy, which were a more dependable source, providing controlled, intense, high-energy particle beams, and hence repeatable, statistically solid experiments.[198] From the late 1950s onward, ground-based cosmic ray research declined, as most of its stellar researchers moved toward particle accelerators or jumped on the Sputnik bandwagon to became space scientists, or even began pioneering gamma and X-ray astronomy. In parallel, the center of gravity of this traditional aspect of cosmic ray physics moved into the high-energy range (energies above 10^{10} eV), still out of the reach of contemporary accelerators, to be investigated by large arrays of detectors, so as to increase the chance of catching those rare events originated by the most energetic particles in the Universe.

At the same time, the astrophysical perspective on cosmic rays became reinforced. While the 'little science' program conducted by the cosmic ray physicists began to be overshadowed by the advent of accelerator-based research, investigations of the primary cosmic rays during the 1940s had increasingly raised questions connected with the rays' origin and potential importance for astrophysics, which had been traditionally part of the field since its very beginning. In this period, it was definitely established that the bulk of the high-energy primary cosmic rays are mainly protons and, to a lesser extent, helium and heavier atomic nuclei, solving a problem which had preoccupied scientists since Hess's balloon flight of 1912 showed that some type of 'high-

198 In 1953, the seminal international cosmic-ray conference was held at Bagnères-de-Bigorre, at the foot of the Pyrenees dividing France from Spain, where the famous French high-altitude Pic du Midi Observatory was located. During this meeting, the word 'hyperon' was announced for the first time, and the 'invasion' of accelerators was explicitly mentioned. James W. Cronin: The 1953 Cosmic Ray Conference at Bagnères de Bigorre. The Birth of Sub Atomic Physics. *The European Physical Journal H* 36/2 (2011), 183–201, 197. doi:10.1140/epjh/e2011-20014-4.

energy radiation' from outer space is constantly bombarding the Earth.[199] Starting from the 1950s, the transition to cosmic ray astrophysics diverted scientists' attention from the interest in cosmic rays shown by high-energy physics, and such astrophysics studies became a great source of information about the energy spectrum and detailed chemical composition of the primary radiation. Investigations of the nuclear aspects of cosmic rays continued to evolve also because of the growing postwar role of nuclear astrophysics, whose realm spans explanations of the huge energy output from stars and other cosmic objects, while also providing a coherent picture of the abundance of the nuclides in the Universe and their evolution through time and space. Big-Bang nucleosynthesis—the formation of nuclei of helium, lithium, and deuterium from neutrons and protons in the very early Universe—would become a way to understand both particle physics and cosmology. At the same time, the discovery of the existence of primary particles characterized by energies much higher than could be previously imagined revived interest in the sources of cosmic rays and the mechanisms able to accelerate them at such super-high energies, on the whole one of the key questions of 20th century physics. Thus, the very existence of the primary cosmic radiation presented itself as an astrophysical and cosmological problem of great interest and importance, while the field underwent an evolution which gradually changed its old character, and its practitioners began to look at cosmic rays more and more as astrophysicists.[200] Since their discovery, these particles had been also considered an important probe with which to explore the space beyond the Earth's atmosphere, that is, the interplanetary space; this latter perspective being the complex result

199 Marcel Schein, William P. Jesse, and E. O. Wollan: The Nature of the Primary Cosmic Radiation and the Origin of the Mesotron. *Physical Review* 59/7 (1941), 615–615. doi:10.1103/PhysRev.59.615. Phyllis Freier et al.: Evidence for Heavy Nuclei in the Primary Cosmic Radiation. *Physical Review* 74/2 (1948), 213–217. doi:10.1103/PhysRev.74.213. The latter observations were conducted with a cloud chamber and stacks of nuclear emulsions flown on a high-altitude balloon.

200 The evolution of speculations and theories in cosmic ray origin up to the 1960s can be followed in a volume containing 76 of the most outstanding original papers in the field and including translations of some of the important Russian papers. Stephen Rosen (ed.): *Selected Papers on Cosmic Ray Origin Theories*. New York: Dover 1969. On the transition to cosmic-ray astrophysics in the 1950s, see Vitaly L. Ginzburg: On the Birth and Development of Cosmic Ray Astrophysics. In: Yataro Sekido, and Harry Elliot (eds.): *Early History of Cosmic Ray Studies. Personal Reminiscences with Old Photographs*. Dordrecht: Springer Netherlands 1985, 411–426. doi:10.1007/978-94-009-5434-2_37. Vitalii L. Ginzburg: Cosmic Ray Astrophysics. History and General Review. *Physics-Uspekhi* 39/2 (1996), 155–168. doi:10.1070/PU1996v039n02ABEH000132.

of the evolution, throughout the entire 20th century, of questions about the origin and nature of cosmic rays.

New discoveries coming from astronomy at the beginning of the 1950s also considerably assisted the birth of cosmic-ray astrophysics. After the development of radio astronomical methods, it became possible to obtain information about cosmic ray activity far away from the Earth, through observation of radio synchrotron emission produced by ultra-relativistic electrons accelerated within magnetic fields at a distance from the Earth, which also occurs in the optical and X-ray part of the spectrum. Up to that time, all hypotheses concerning the origin of cosmic rays were almost purely speculative and there was no real hope of investigating cosmic rays beyond the limits of the solar system.[201] With the development of radio astronomy, and also of cosmic electrodynamics, which focuses on astrophysical and space plasma phenomena in which electromagnetic interactions play an essential role, it became possible to determine the character of the energy spectrum of relativistic electrons in different regions of our Galaxy and far beyond it. The question of cosmic ray origin became a truly astrophysical problem and transcended the realm of mainly hypothetical constructions. Moreover, radio astronomy demonstrated that cosmic rays are a universal phenomenon, as they are present in the interstellar space of the galaxy and of other galaxies, in quasars, in supernova remnants, and collectively they constitute an important energetic and dynamic factor.

Then, with the launch of Sputnik, outer space appeared to be a better platform than balloons, airplanes, and short-duration rocket flights for the study of the 'primary' cosmic rays before they interact with the atmosphere. Solar cosmic rays, their generation, Earth-bound motion, and effect on processes taking place in the near-Earth space, could now be studied, too, along with the giant, doughnut-shaped swaths of magnetically trapped, highly energetic charged particles that surround the Earth—the Van Allen radiation belts—the first of which was discovered in January 1958 by Explorer 1, the first satellite launched

201 At a time when no electrons had yet been detected in the primary cosmic radiation, suggestions about radio waves being emitted by ultra-relativistic electrons moving in interstellar magnetic fields (see Chapters 1 and 3) was taken up and developed by Vitaly Ginzburg and Iosif Shklovskii, who showed that radio-astronomy could give quantitative information about cosmic rays in distant regions of the Universe, vastly improving the chance of understanding their origin. These deductions are outlined in a paper summarizing the results of Ginzburg's earlier work. Vitaly L. Ginzburg: The Nature of Cosmic Radio Emission and the Origin of Cosmic Rays. *Il Nuovo Cimento* 3/1 (1956), 38–48. doi:10.1007/BF02745509. On USSR theories on cosmic synchrotron radiation, see Ginzburg, Birth and Development, 1985, 411–426.

by the United States.[202] Artificial satellites and space probes, together with the general progress in solar geophysics and physics, and the rapid development of radio astronomy and astrophysics, led to the appearance of a large number of investigations of all these interconnected questions, including the composition, energy spectrum, and spatial distribution of the primary cosmic rays on the Earth, the effect of the interplanetary medium and interplanetary magnetic fields, and other phenomena affecting variations in cosmic rays on the Earth and within the solar system.

Such studies of the chemical and isotopic composition and, in particular, of the 'energy spectrum' of the primary cosmic rays—strongly linked to their origin and mechanisms of acceleration—became intertwined with the appearance of observational gamma-ray astronomy, the branch of high energy astrophysics that studies the cosmos in gamma-ray photons, the most energetic form of electromagnetic radiation. Charged particles are continuously bent by magnetic fields embedded in interstellar and intergalactic plasmas, so that their direction is almost completely uncorrelated with that of their sources. This is why the quest for such cosmic accelerators should mostly rely on photons that, being electrically neutral, do not undergo deflections in galactic and intergalactic magnetic fields, and might provide valuable and otherwise unattainable information on the hidden sources. As seen in Extensive Air Showers in the atmosphere, decay processes of charged cosmic rays—that occur as protons, electrons, heavier nuclei, and their antiparticles—can ultimately produce high-energy photons, and thus possible sources of cosmic rays (like supernovae and their remnants, interactions of energetic electrons with cosmic magnetic fields, or places where cosmic rays can be confined, such as the Galaxy) were expected to be visible in gamma-ray astronomy.[203] And so,

---

202 During the 1950s, Van Allen had launched rockets and 'rockoons'—rockets carried aloft by balloons—to carry on cosmic ray experiments above the atmosphere, and from 1956, had proposed the use of satellites for cosmic-ray investigations. The launch of Explorer 1 and its scientific payload was the culmination of his work for the 1957–58 International Geophysical Year. Van Allen's simple cosmic-ray experiment consisted of a Geiger-Müller counter and a tape recorder. Follow-up experiments on further Explorer missions established that there were two belts of radiation circling the Earth. See the first article summarizing results already presented on official occasions or in scientific reports: James A. Van Van Allen, and Louis A. Frank: Radiation Around the Earth to a Radial Distance of 107,400 Km. 4659. *Nature* 183/4659 (1959), 430–434. doi:10.1038/183430a0.

203 High-energy electrons and positrons interacting with magnetic fields or low-energy photons produce gamma rays by so-called *leptonic* processes, while *hadronic* processes can produce gamma rays through the decay of neutral mesons originated by high-energy protons and nuclei interacting with matter through nuclear interactions.

while high-energy electrons and positrons can be indirectly detected by their synchrotron radiation, the really energetic photons produced by cosmic-ray interactions with matter and radiations fields now appeared to be the clue to gaining more direct insight into the acceleration sites and the targets where such high-energy interactions take place, thus broadening the potential to study cosmic rays in the Universe, and opening a window onto investigations of phenomena in extreme astrophysical environments and processes. Cosmic-ray astrophysics could therefore be based on the study of primary cosmic rays near the Earth, as well as on the new fields of radio astronomy and gamma-astronomy.

However, the Earth's atmosphere is entirely opaque not only to many bands of the electromagnetic spectrum, but also to all cosmic ultraviolet, gamma, and X-rays; it is transparent only to electromagnetic radiation in the radio and optical regimes. A direct detection of such emissions is impossible without going beyond the atmosphere. Beginning in the early 1960s, gamma radiation generated in outer space in interactions between cosmic rays and interstellar matter was observed first by sending telescopes aloft with balloons that took them to within a few grams of residual atmosphere, and then with satellites experiments, which make accessible gamma-rays below a certain energy (about 20 GeV), that are quickly dying out at the top of the atmosphere. On the other hand, at significantly higher energies, the photon flux falls off very rapidly and satellite instruments lose sensitivity due to their limited collection area. The most energetic gamma rays (at GeV–TeV energies), messengers of the relativistic Universe which provide us with much information—on the conditions prevailing in remote regions, such as magnetic and electric fields or matter and radiation densities; on the acceleration mechanisms of charged particles and their distribution; and in particular, on the sources and mechanisms of their production—cannot be observed with balloon—or satellite-borne detectors. Fortunately, observations of GeV–TeV gamma rays can be carried out by ground-based experiments using the atmosphere itself as part of a giant gas detector: incoming gamma rays interact with the atmosphere, producing an electromagnetic shower, a broad distribution of secondaries that can be recorded by ground-based detectors spread over tens of thousands of square meters.[204] The problem is that only a tiny fraction of extensive air showers are initiated by gamma rays. Charged cosmic rays outnumber ultrahigh-energy

204 Interactions between GeV–TeV photons result in secondary relativistic electron-positron pairs which lose energy, emitting bremsstrahlung radiation producing new high-energy photons. Further pairs are produced in the cascading process, which in turn create new gamma rays and so on, until the energy of the impinging photon is redistributed as the

gamma rays by many orders of magnitudes and so it is instrumentally and epistemically challenging to discriminate the effects of the interaction of gamma rays with the atmosphere from those of energetic particles of matter which constitute a significant background and limit the sensitivity of such measurements. In any case, the techniques developed to detect high-energy cosmic rays and gamma rays from the ground were developed in line with a single tradition, with the epistemic focus shifting between them throughout the century. As a consequence of this shift, ground-based researchers using detector arrays specialized in aspects that it would be impossible to pursue anywhere else, namely those higher-energy particles that could not be produced using the existing accelerators, as well as collection areas wider than what could ever be set up in outer space. And this new focus increasingly directed attention toward very-high-energy gamma-ray astronomy, and consequently, to its full integration into high-energy cosmic ray astrophysics.

Cosmic Ray Research Traditions

Research on cosmic rays had arrived at the prewar Kaiser Wilhelm Society in the 'golden age,' the 1930s, courtesy of three of the world's most respected physicists: Erich Regener, Walter Bothe, and Werner Heisenberg, the founders of the 'fundamental research' traditions research traditions described in detail in Chapter 1. Regener and Bothe had a common trajectory insofar as each was admitted to the KWG under the auspices of its then President, Max Planck, who thus provided an alternative arrangement to their difficult circumstances in universities during the early Nazi era. Within the KWG, Regener investigated cosmic radiation in ionization chambers, as well as Geiger-Müller counters in the atmosphere with balloons and under water; and at Friedrichshafen, on the shores of Lake Constance, he developed a research station for studies of the stratosphere, from where he could also continue his altitude-based cosmic-ray research and, during the war, also contribute to pioneering military rocket instrumentation for measuring the ultraviolet radiation from the Sun. Walther Bothe was appointed Director of the Institute for Nuclear Physics at the Kaiser Wilhelm Society for Medical Research, where he continued his detector-based tradition of nuclear research, which innovated many of the instruments he himself used for the detection of cosmic rays; and he was among the first to become a member of the Uranium Club at the start of the war, thanks to his expertise in nuclear physics and accelerators.

shower develops in the atmosphere, until only low-energy electrons and positrons are produced, which lose energy only through ionization processes.

Werner Heisenberg entered the KWG directly, in the context of the wartime Uranium Club, but he had been making cosmic rays one of his central areas of research since the early 1930s in Leipzig.[205] It was in Leipzig that he launched the career of Erich Bagge, a key player in the chain of developments outlined here. Bagge's relationship with cosmic rays dated back to his doctoral studies with Heisenberg in the late 1930s.[206] A theoretician, Heisenberg continued to closely follow developments even during the war, when the German nuclear project was absorbing all his energies, and cultivated cosmic ray studies, which provided the clue to understanding nuclear processes taking place at energies greater than those provided by radioactive sources or early accelerators. Throughout the war, in parallel to the nuclear efforts, Heisenberg and his team, now at the Kaiser Wilhelm Institute for Physics in Berlin-Dahlem, dedicated much of their open scientific activities to cosmic rays.[207] After the war, cosmic ray research continued in Heisenberg's new Institute in Göttingen,[208] where an active experimental group worked at mountain laboratories or with balloons, before moving into accelerator physics in the second half of the 1950s. After returning from internment at Farm Hall in January 1946, together with the other members of the original Göttingen staff (seven of the ten detained at Farm Hall), Bagge became Heisenberg's assistant, and in 1948, was appointed to the Chair of Physics at Hamburg University; but, as we will see later, he continued to cultivate cosmic ray studies, which in postwar Germany were also

---

205 Helmut Rechenberg: *Werner Heisenberg—Die Sprache der Atome. Leben und Wirken—Eine wissenschaftliche Biographie. Die "Fröhliche Wissenschaft" (Jugend bis Nobelpreis)*. Berlin: Springer-Verlag 2009, 12.4. Since the very early 1930s, Heisenberg had an exchange of correspondence on cosmic ray physics with Bruno Rossi, as discussed in Luisa Bonolis: International Scientific Cooperation During the 1930s. Bruno Rossi and the Development of the Status of Cosmic Rays into a Branch of Physics. *Annals of Science* 71/3 (2014), 355–409. doi:10.1080/00033790.2013.827074.

206 See Bagge's articles between 1939 and 1941 on nuclear processes in cosmic rays, closely related to his Habilitation dissertation on nuclear processes and heavy particles in cosmic rays (Chapter 1).

207 For research activities on cosmic rays during the war period, see Werner Heisenberg (ed.): *Vorträge über Kosmische Strahlung*. Berlin: Springer-Verlag 1943.

208 Werner Heisenberg: Kosmische Strahlungen und Atomphysik. *Jahrbuch 1951 der Max-Planck-Gesellschaft zur Förderung der Wissenschaften e.V.* Göttingen: Max-Planck-Gesellschaft zur Förderung der Wissenschaften 1951, 229–263. Werner Heisenberg (ed.): *Kosmische Strahlung. Vorträge gehalten im Max-Planck-Institut für Physik Göttingen*. 2nd ed. Berlin: Springer 1953. See Chapter 1 for Heisenberg's publications in the field between the 1940s and early 1950s.

the sole means to study nuclear and subnuclear processes.[209] In the foundational years of the Max Planck Society, cosmic rays actually played an even greater role as one of the few remaining experimental activities permissible in the context of 'nuclear' research. Regener relaunched the high-altitude balloon research program at his research facility, which had recently become associated with the new Max Planck Society and in 1952 became known as the Institute for the Physics of the Stratosphere. As administrator of the Kaiser Wilhelm Institute site in Hechingen, he had also provided a base for his disciple Erwin Schopper, whose balloon-based cosmic ray research especially aimed to observe high-energy nuclear disintegrations.[210] Heisenberg's institute, while still in Göttingen, maintained mountain stations for cosmic ray research on the Zugspitze (Germany's highest peak), and the Wendelstein, both in Bavaria. And Walther Bothe's disciple Wolfgang Gentner, based in Freiburg in the first postwar decade, created a cosmic ray observatory atop the Schauinsland, in the Black Forest, which was later inherited by his new Max Planck Institute for Nuclear Physics in Heidelberg.

During the 1950s, both Heisenberg and Gentners' teams maintained experimental research groups on cosmic rays, but they were also aware that the future of physics lay in particle accelerators and also, after Sputnik, in outer space. They were both involved in starting CERN research activities in Geneva, and as Director of the newly formed Max Planck Institute for Nuclear Physics in Heidelberg (successor to Bothe's Institute), Gentner promoted the construction of a Tandem accelerator dedicated to advanced nuclear research. So, at these two Max Planck Institutes, as in many other physics institutes around the world, the 1950s was a decade of 'retraining,' away from cosmic rays and toward particle accelerators, while still maintaining a provisional foothold in their old cosmic research stations.[211] Meanwhile, in 1954, Erich Regener died and his successor Georg Pfotzer continued the high-altitude cosmic ray tradition. The younger Schopper, however, left for the University of Frankfurt,

209 Bagge was also asked to prepare a chapter on the origin and nature of cosmic rays for the well-known series *Ergebnisse der Exakten Naturwissenschaften*. Erich Bagge: Ursprung und Eigenschaften der kosmischen Strahlung. In: S. Flügge, and F. Trendelenburg (eds.): *Ergebnisse der Exakten Naturwissenschaften*. Berlin: Springer-Verlag 1949, 202–262. doi:10.1007/978-3-662-25834-7_6.

210 See, for example, Erwin Schopper: Early History of Shock Waves in Heavy Ion Collisions (The Frankfurt Group). In: Walter Greiner, and Horst Stöcker (eds.): *The Nuclear Equation of State: Part A: Discovery of Nuclear Shock Waves and the EOS*. Boston, MA: Springer US 1989, 427–446. doi:10.1007/978-1-4613-0583-5_33.

211 As mentioned in Chapter 1, Klaus Gottstein was one of the protagonists of this evolution at Heisenberg's Institute. Klaus Gottstein: interview by Luisa Bonolis and Juan-Andres Leon, Munich, September 7, 2017. DA GMPG, BC 601006.

where he became Director of the Institute for Nuclear Physics. There he pursued nuclear and accelerator physics and later contributed to space-based cosmic-ray research.

As was described in Chapter 2, Sputnik provided an enormous new path of expansion for these three research traditions, so that, in both Heidelberg and Munich, the two most powerful Max Planck Institutes increasingly focused their research on either experimental particle physics with accelerators on the ground, or space-based research, initially in fields such as space plasmas and cosmochemistry, while in Lindau, the airborne cosmic ray tradition progressed over time more in the direction of planetary science and solar system research. All these space traditions had close epistemic, instrumental, and methodological links to earlier cosmic ray research, but from the 1960s onwards, they deliberately left behind their 'cosmic' legacy as part of the general reorientation to the more topical questions relating to the scientific use of outer space for astronomy, astrophysics, and fundamental particle physics.

In Heidelberg, the experimental line of cosmic ray research proper closed down in 1964,[212] while in Munich, the experimental cosmic ray group at the main institute in Freimann was moved to Garching to become part of the new Max Planck Institute for Extraterrestrial Physics, the aim being to continue research from outer space, making this thus an early example of recycling internal expertise to launch a new research field. While the founder Reimar Lüst and his collaborators focused on space plasmas and cosmic ray particles in near-Earth space, the cosmic ray people from Munich instead joined forces with the second director at the institute, Klaus Pinkau, who had moved in late 1965 from Bagge's institute in Kiel to the Max Planck Institute for Extraterrestrial Physics. Pinkau, who had been educated at Bagge's institute in Hamburg first, and then in Kiel, had developed into a skilled cosmic ray physicist during his stay in Britain with Powell's group in Bristol in the second half of the 1950s, gaining his PhD in 1958 on electromagnetic cascades that originated in high-energy gamma rays.[213] He had later started balloon experiments in

212 See institute timeline with cosmic ray division: MPI für Kernphysik: *50 Jahre Max-Planck-Institut für Kernphysik. Von Kernphysik und Kosmochemie zu Quantendynamik und Astroteilchenphysik*. Heidelberg: MPIK 2008, 8–9.

213 On Pinkau's cosmic-ray activities in Kiel and his appointment at MPE, see Chapter 3. Work for his PhD dissertation was done in Bristol, based on research on electromagnetic air showers: Klaus Pinkau: Energy Determination of Electromagnetic Cascades in Nuclear Emulsions. *The Philosophical Magazine: A Journal of Theoretical Experimental and Applied Physics* 2/23 (1957), 1389–1392. doi:10.1080/14786435708243215.

Kiel,[214] before being involved by Lüst in starting gamma-ray astronomy and in launching space research at the newly founded Institute for Extraterrestrial Physics.[215] Through Bagge, Pinkau had indirectly benefited from the research tradition established by Heisenberg and he was now becoming part of that same tradition, contributing to its evolution toward the new fields being opened up by novel observational and technological windows. Throughout the 1960s, Pinkau transitioned from high-altitude cosmic ray physics to space-based gamma-ray astronomy.

Satellites

Cosmic gamma rays are mostly absorbed in the upper atmosphere and their signal has to be separated from the background produced by the interaction of cosmic rays with local matter. The turning point in interest in cosmic gamma rays had thus come with the opportunities afforded by the launch of Sputnik. In a seminal paper published just one month after the launch of the Soviet satellite, Philip Morrison, inspired and encouraged by the cosmic ray physicist Giuseppe Cocconi, his colleague at Cornell University,[216] listed the

---

214 Diederich Köhn, Klaus Pinkau, and Gerd Wibberrens: Gamma-Ray Spectrometer for Balloon Flights. *Composition, Origin, and Prehistory, Proceedings from the 8th International Cosmic Ray Conference.* Jaipur 1963, 203–205. http://adsabs.harvard.edu/abs/1963ICRC....3..203K. Last accessed 2/7/2020. Klaus Pinkau et al.: Balloon Experiment Using Spark Chambers and an Ionization Spectrometer. *Proceedings of the 9th International Cosmic Ray Conference. London, UK.* 1965, 821–823. http://adsabs.harvard.edu/abs/1965ICRC....2..821P. Last accessed 10/30/2018. Instead of using simple nuclear emulsions, the study of high-energy cascades both from charged particles and gamma rays was improved, combining spark chambers with a ionization spectrometer. Diederich Köhn, Klaus Pinkau, and Gerd Wibberenz: Messung des sekundären Gammaspektrums von 0,2 bis 2 GeV in der oberen Atmosphäre. *Zeitschrift für Physik* 193/3 (1966), 443–458. doi:10.1007/BF01326442. See also, Klaus Pinkau: The Early Days of Gamma-Ray Astronomy. *Astronomy and Astrophysics Supplement* 120 (1996), 43–47. https://ui.adsabs.harvard.edu/#abs/1996A&AS..120C..43P. Last accessed 10/30/2018.

215 Reimar Lüst, and Klaus Pinkau: Theoretical Aspects of Celestial Gamma-Rays. In: J.C. Emming (ed.): *Electromagnetic Radiation in Space. Proceedings of the Third ESRO Summer School in Space Physics, Held in Alpbach, Austria, from 19 July to 13 August, 1965.* Dordrecht: Springer 1967, 231–248.

216 Philip Morrison: interview by Owen Gingerich, Session I, February 22, 2003, AIP, https://www.aip.org/history-programs/niels-bohr-library/oral-histories/30591-1. Last accessed 10/3/2019. Morrison acknowledged discussions with Cocconi—who in the meantime had moved to high-energy physics with accelerators, and would soon after become Director of CERN's Proton Synchrotron—and with the well-known cosmic-ray physicist Kenneth Greisen, an expert in cascade theory, who had earned his PhD at Cornell University in 1942 under Bruno Rossi.

possible physical mechanisms for the production of gamma rays in astrophysical sources, and calculated that outside the atmosphere it should be possible to detect a very high flux of gamma rays leading to a potential new 'window' of astronomy, similar to what was occurring with radio astronomy.[217] His publication single-handedly triggered the start of gamma-ray astronomy as a means of obtaining fresh information about high-energy astrophysical processes and on cosmic structures in far regions of the Universe. Beyond Sputnik, the link between the rising interest in gamma rays, and the spectacular successes of radio astronomy since the postwar era is quite direct. Until the 1940s, almost all the information on the cosmos was obtained via optical channels, and the appearance and increasing use of two new channels of astrophysical information, the 'radio channel' and the 'cosmic ray channel,' introduced a fundamental new feature into the development of astronomy during the 1950s.

Establishing a connection between cosmic radio emission and cosmic ray particles—mainly relativistic electrons, gyrating in galactic and intergalactic magnetic fields, and radiating electromagnetic waves in such circular, accelerated motion—led to a renewed interest in the old problem of the origin of cosmic rays, which, broadly speaking, is the problem of their acceleration and propagation under various conditions. It was recognized that in all regions of the Universe there are sources of acceleration mechanisms leading to the emission of *synchrotron radiation*, which is a *non-thermal* radiation determined by processes different from thermal random motion of particles in matter with temperature above absolute zero, which are instead at the origin of the electromagnetic radiation generated in hot objects such as stars, hot gases, and galaxies.[218]

These brilliant successes combining radio astronomy with optical astronomy emphasized that much could be gained from exploring new wavelengths. But radio emissions appeared rather complex in origin, while gamma ray is a form of radiation which is more directly related to high-energy and nuclear processes in astronomical objects of various classes than is optical or radio emission.

---

217 Philip Morrison: On Gamma-Ray Astronomy. *Il Nuovo Cimento* 7/6 (1958), 858–865. doi:10.1007/BF02745590.

218 Cosmic rays are the best-known example of non-thermal particle population, with a spectral energy distribution showing no characteristic temperature scale and with energies (up to 10^{20} eV and above), which are well beyond what can be produced by any conceivable thermal emission mechanism.

With the prediction by Morrison in 1958, and the first observation of solar gamma-ray lines in 1959,[219] particle physicists around the world sought to deploy their instruments for the new space-based challenge. As we have just mentioned, this was notably the case with Klaus Pinkau, who had started his scientific career in Hamburg and Kiel under the direction of Erich Bagge, who was one of the world's pioneers in a new kind of particle detector called the spark chamber, a descendant of particle counters based on wartime advances in electronic timing circuits, spawned from work on cosmic rays.[220] This device was the first attempt to use electronic equipment to reconstruct the trajectories of particles inside a chamber; it could operate up to a thousand times faster than the traditional methods of cloud chambers, bubble chambers, and photographic emulsion stacks.[221]

The spark chamber, which emerged during the 1960s as one of the primary instruments of particle physics research (notably used for the discovery of muon neutrino in 1962),[222] was also a ready-made gamma-ray telescope. It needed only to be made robust and compact enough to be carried by high-altitude balloons and satellites. Through the early 1960s, Pinkau, like many others around the world, initiated a program of high-altitude balloons, first in Kiel and later at MPE, which sought to detect the gamma rays at an altitude

219 Laurence E. Peterson, and J. R. Winckler: Short Gamma-Ray Burst from a Solar Flare. *Physical Review Letters* 1/6 (1958), 205–206. doi:10.1103/PhysRevLett.1.205. Laurence E. Peterson, and John R. Winckler: Gamma-Ray Burst from a Solar Flare. *Journal of Geophysical Research* 64/7 (1959), 697–707. doi:10.1029/JZ064i007p00697. The highly influential paper by the Burbidges, Fowler, and Hoyle on nucleosynthesis in stars also raised hopes to discover gamma rays from excited nuclei in supernova explosions. E. Margaret Burbidge et al.: Synthesis of the Elements in Stars. *Reviews of Modern Physics* 29/4 (1957), 547–650. doi:10.1103/RevModPhys.29.547.

220 Arthur Roberts: Development of the Spark Chamber. A Review. *Review of Scientific Instruments* 32/5 (1961), 482–485. doi:10.1063/1.1717420. On Bagge's group and the development of spark chambers, see Peter Galison: *Image and Logic. A Material Culture of Microphysics.* Chicago, IL: University of Chicago Press 1997, 467–470.

221 A spark chamber is an instrument created for detecting electrically charged particles. When a charged particle travels through a stack of metal plates inside a sealed box filled with a noble gas it ionizes the gas and sparks form along the trajectory, if a high enough voltage is applied between each adjacent pair of plates. The effect becomes visible as a line of sparks. In the case of gamma rays, their presence is made visible through the electron-positron pair production. In this interaction, the incident gamma ray is completely annihilated, with its energy transferred to the pair, which is strongly beamed forward; and thus, the trajectory of the gamma ray can be inferred from the pair's trajectory.

222 Danby et al., Observation of High-Energy Neutrino Reactions, 1962, 36–44.

around 10 km. These early observations were due in large part to the development of high-altitude ballooning and experimental techniques pioneered by the cosmic-ray physics community, many of which could be quickly adopted by the early gamma ray researchers.[223] It would turn out that the prediction was too optimistic, as the flux was too low to trigger the small detectors carried by balloons. Moreover, balloon experiments are complicated by the fact that the flux of primary gamma rays of extraterrestrial origin has to be separated from the background produced by secondary gamma rays generated by cosmic ray interactions in the atmosphere of the Earth. The mid-1960s, when this disappointment had sunk in—and when X-ray astronomy was beginning to be a consolidated field with a number of successful important detections made employing balloon or rocket launches—was exactly the moment when space-based platforms became possible, so allowing the true flux of primary gamma rays to be reliably measured by means of satellite instrumentation, which afforded the new reality of observations from space in several wavelengths, including the low-energy gamma-ray regions of the electromagnetic spectrum. Since then, the development of relativistic astrophysics has increasingly showed how high-energy astrophysical processes can produce relativistic particles and associated gamma radiation over an enormous range of energies. Gamma rays can traverse great distances in space without being absorbed by intergalactic dust and gas; however, as most gamma rays coming from space are absorbed by the earth's atmosphere, gamma-ray astronomy could not develop until it was possible to get detectors above the Earth's atmosphere, using balloon, spacecraft and, in particular, satellites. In 1961, the American satellite Explorer XI, launched by MIT physicists, provided the first view of the Universe at the shortest wavelength of the electromagnetic spectrum, by identifying gamma rays originated outside the Earth, with a detector consisting in a scintillation-Cherenkov device.[224] However, in the five months during which it remained operational, Explorer XI detected only 22 cosmic gamma rays, too few to establish where they came from. Firm detection of gamma-ray sources was finally achieved with the short-lived SAS-2 satellite launched by NASA in 1972, which was able to show that the galactic center and the Vela and the

223 For a historical review of gamma-ray astrophysics at MPE see Volker Schönfelder, and Jochen Greiner: Half-a-Century of Gamma-Ray Astrophysics at the Max-Planck Institute for Extraterrestrial Physics. 1. *The European Physical Journal H* 46/1 (2021), 27. doi:10.1140/epjh/s13129-021-00031-8.

224 William L. Kraushaar, and George W. Clark: Search for Primary Cosmic Gamma Rays with the Satellite Explorer XI. *Physical Review Letters* 8/3 (1962), 106–109. doi:10.1103/PhysRevLett.8.106.

Crab pulsars are strong gamma-ray emitters.[225] The next major high-energy gamma-ray mission was Pinkau's COS-B, the first ESA scientific satellite (after the ESRO-ELDO merger), with a high-energy gamma telescope as single payload, launched in 1975 and successfully operational until April 1982.[226] This satellite, by establishing the field of low-energy gamma-ray astronomy, was able to provide the first complete gamma-ray map of the disc of the Milky Way, recording galactic continuum emission (mainly from interactions of cosmic rays with gas and radiation in the interstellar medium), as well as a catalogue of point sources, and detailed studies of several of them, also thanks to its long-lived spark chamber, the result of metal-ceramic technology.[227]

During the 1970s and '80s, this first successful generation of gamma-ray satellites revealed the large-scale features of the gamma-ray sky and the diffuse emission from the galactic plane resulting from the interaction of cosmic rays with interstellar matter. The Crab Nebula and the Vela pulsars were identified as emitters of high-energy gamma rays and the periodicity of their light curves was mapped.[228] The observation of intense point sources of gamma-rays was somewhat surprising, because of the incredible amount of energy

225 On this phase of research in gamma-ray astronomy, see Klaus Pinkau: History of Gamma-Ray Telescopes and Astronomy. *Experimental Astronomy* 25/1–3 (2009), 157–171. doi:10.1007/s10686-009-9143-z.

226 Klaus Pinkau at the Institute for Extraterrestrial Physics in Garching had been the Principal Investigator of this European mission, using as detector a spark chamber descendant from the ones pioneered while he was still in Kiel, which was able to determine the direction and energy of incoming gamma rays. The COS-B gamma-ray instrument was designed and built by the Caravane collaboration, a consortium of five institutes from Germany, the Netherlands, France, and Italy. The premises for such collaboration dates back to the years the young Pinkau spent in Great Britain with Powell's group, which paved the way to his participation in European space activities.

227 COS-B long period of observation (1975–1982) created the premises for the preparation of new ESA missions like EXOSAT, XMM-Newton, and the INTEGRAL mission. Pinkau, The Early Days, 1996, 43–47. Pinkau, History of Gamma-Ray Telescopes and Astronomy, 2009, 157–171. About the history of COS-B satellite see John Krige, Arturo Russo, and Lorenza Sebesta: *A History of the European Space Agency 1958–1987. The Story of ESA, 1973 to 1987*. Vol. 2. European Space Agency 2000. Chapter 7. Volker Schönfelder: The History of Gamma-Ray Astronomy. *Astronomische Nachrichten* 323/6 (2002), 524–529. doi:10.1002/1521-3994(200212)323:6<524::AID-ASNA524>3.0.CO;2-Z. Volker Schönfelder (ed.): *The Universe in Gamma Rays*. Berlin: Springer 2001. H. A. Mayer-Hasselwander et al.: Large-Scale Distribution of Galactic Gamma Radiation Observed by COS-B. *Astronomy and Astrophysics* 105/1 (1982), 164–175. http://adsabs.harvard.edu/abs/1982A%26A...105..164M. Last accessed 5/15/2018.

228 Several point gamma-ray sources were observed, including four radio pulsars, a result considered particularly striking, since only one radio pulsar had been seen at either optical or X-ray frequencies. C. E. Fichtel: Gamma-Ray Astrophysics. *Space Science Reviews*

required to produce them. Surprising, too, was the fact that the two most intense sources, namely the Vela pulsar and Geminga, were relatively weak emitters in other wavebands. In parallel to all these developments, the completely unexpected phenomenon of gamma-ray bursts, the ultra-high energy transient emissions of gamma rays of cosmic origin, was likewise discovered by the Vela defense satellites.[229] In the 1970s, the interaction between X-ray, optical, and radio astronomy had already begun to unravel the physics and astrophysics of neutron stars and black holes. These results from gamma-ray spacecraft were now also showing that gamma-ray astronomy—as in the case of radio and X-ray astronomy—might well find new types of sources with the potential to disclose further aspects of the relativistic universe.

The Revival of Extensive Air Shower Arrays

However, at energies above 100 GeV, the fluxes of nearly all possible sources (see more details later) are in the order of one or less photons per year for square meter instruments, the maximum practical size of balloon- or satellite-borne gamma-detectors.[230] This problem of detector size is unsurmountable, as the collection areas would need to be in the order of square kilometers.

20/2 (1977), 191–234. doi:10.1007/BF02186864. For a summary of the discoveries of the 1970s, see section 4 in Pinkau, The Early Days, 1996, 43–47. For a general overview on space-based gamma-ray astronomy, see Pinkau, History of Gamma-Ray Telescopes and Astronomy, 2009, 157–171. For a general review, see also Giovanni F. Bignami, and W. Hermsen: Galactic Gamma-Ray Sources. *Annual Review of Astronomy and Astrophysics* 21 (1983), 67–108. doi:10.1146/annurev.aa.21.090183.000435.

229 The first of these gamma-ray bursts, lasting typically less than one minute and whose nature remained a mystery, was detected in 1967, but they were not reported in the scientific literature until 1973. Ray W. Klebesadel, Ian B. Strong, and Roy A. Olson: Observations of Gamma-Ray Bursts of Cosmic Origin. *Astrophysical Journal* 182 (1973), L85–L88. doi:10.1086/181225. For almost a quarter of a century, the origin of gamma ray bursts remained unknown and their extragalactic nature was finally established by the satellite Beppo-Sax only in 1997. van Paradijs, Jan et al.: Transient Optical Emission from the Error Box of the γ-Ray Burst of 28 February 1997. *Nature* 386/6626 (1997), 686–689. doi:10.1038/386686a0. Later observations confirmed that sources were extragalactic, sometimes located at cosmological distances.

230 From observations, we know that the flux of cosmic rays, especially gamma rays, drops as a function of energy following approximately a smooth power law $F(E) = constant \times E^{-\gamma}$, with a spectral index γ roughly equal to 3, apart from some features, the most important being the cosmic ray *knee*, at about 3×10^{15} eV, where the spectrum steepens up to 10^{16} eV, changing to a flatter slope around 10^{17} to 10^{18} eV, and the *ankle*, at about 3×10^{19} eV, where the spectrum again becomes flatter. At the extreme energies of 10^{20} eV (100 EeV), the flux is typically one particle per square kilometer per century. To enhance the probability of detecting such high energy particles, the Pierre Auger Observatory in Argentina has an active surface area of 3000 km^2, which allows exploration of the energy range above EeV

Finally, at even higher energies, one reaches the range above around 100 TeV. With gamma-ray primaries at this energy range—providing evidence of the existence of energetic accelerating mechanisms—the flux is extremely low, as in the case of ultra-high-energy-charged cosmic rays, whose estimated flux is so low that even large arrays of detectors spread across areas tens of square kilometers would catch only a few events in several years of operation.

What became pivotal, here, was the air shower detection method developed since the 1930s by investigating the cascades of ionized particles and electromagnetic radiation produced in the atmosphere, to a width of several kilometers, by primary cosmic rays.[231] The extensive showers of secondary particles generated by high-energy primaries and interacting with nuclei in the high atmosphere manage to reach the ground and can be detected, thus providing adequate information about the energy, direction, and type of primary particle or cosmic gamma ray originating the shower, based on the study of spatial and temporal properties of secondary cascade products. The deployment of shower arrays (appropriate particle detectors spaced at different distances from each other, depending on the energy range) over large surfaces (order of 10000–100000 m^2 and above) allows direct detection of the shower particles (electrons, muons, hadrons), so increasing the effective area of detection by several orders of magnitude greater than the compact experimental set-ups that can be flown to balloon altitudes or carried in satellites. This technique had long been crucial for the historical development of ground-based gamma-ray astronomy, during the 1980s, and it underwent a revival also at the

(10^{18} eV). The power-law spectrum confirms the *non-thermal* origin of such steady rain of high-energy particles, connected to relativistic processes and not to the standard thermal processes typical of our Sun and stars, representing the conversion of kinetic energy of random movements of atoms and molecules (thermal energy) into electromagnetic radiation. The latter is instead a temperature-dependent spectrum of light and follows the typical blackbody radiation curve, which also characterizes the cosmic microwave background, relic of the Big Bang.

231 For a detailed review of the history of studies on Extensive Air Showers, see Karl-Heinz Kampert, and Alan A. Watson: Extensive Air Showers and Ultra High-Energy Cosmic Rays. A Historical Review. *The European Physical Journal H* 37/3 (2012), 359–412. doi:10.1140/epjh/e2012-30013-x. Kampert and Watson duly emphasized the key role of the development of the coincidence approach (Bothe and Rossi, see Chapter 1), also for the discovery and study of Extensive Air Showers. Despite the early crucial contributions of Rossi, Bothe, Kolhörster, Regener, and his collaborators, the credit for the discovery of Extensive Air Showers has usually been given to Auger and his collaborators. The first of the giant shower arrays was constructed at Volcano Ranch, New Mexico, by members of Rossi's group at MIT at the end of the 1950s.

Max-Planck Institutes for Physics in Munich and for Nuclear Physics in Heidelberg.

In an extensive air shower, the energy of the primary cosmic ray interacting with a nucleus in the upper atmosphere is shared among the shower particles. As the energy of the primary increases, the number of particles produced in the first few collisions increases, and also the number of generations that contribute to the nuclear cascade.[232] But it took many decades to understand the intricacies and mechanisms of the showering process, so progress in disentangling the fine structure of the shower was slow. When the first powerful accelerators like the Cosmotron and the Bevatron went into operation in the early 1950s in the United States, elementary-particle and cosmic-ray physics began to drift apart and only nuclear interactions at exceptionally high energies remained within the province of cosmic rays. Particle detectors and large air shower arrays were developed because it was recognized that the phenomenon of Extensive Air Showers could offer a major tool for the study of the very-high-energy interactions provided by cosmic rays, still well beyond the realm of particle beams that could be obtained in terrestrial laboratories. Studying the showering process still offered the possibility of answering main questions regarding nuclear and particle physics, the arrival direction of the high-energy particles, the energy spectrum, and the mass composition of the primary cosmic rays.[233] In the late 1950s and early '60s, it was recognized that the very high energies can extend, in the rarest, highest-energy events, up to millions of TeV ($\sim 10^{20}$ eV).[234] To put such energy in context, it can be contrasted to those achieved by proton beams in the Large Hadron Collider (LHC),

---

232 The shower is a mixture of nucleonic and electronic cascades, but the nuclear active particles constitute the backbone of the shower, and though relatively few in number these particles are the most energetic and keep supplying secondaries around the core of the shower. The most obvious differentiating characteristic of hadronic versus purely electromagnetic cascades is the presence of pions, and subsequent production of muons in pion decays as well as electromagnetic sub-showers. The muons are very penetrating: they interact so weakly and have such a long lifetime that they can practically cross the rest of the atmosphere undisturbed. A part of the muon component decays as electrons and neutrinos. Low-energy muons can be detected with scintillation or tracking detectors. High-energy muons are captured with deep underground detectors.

233 Kampert, and Watson, Extensive Air Showers and Ultra High-Energy Cosmic Rays, 2012, 359–412, 360.

234 John Linsley: Evidence for a Primary Cosmic-Ray Particle with Energy 10^{20} eV. *Physical Review Letters* 10/4 (1963), 146–148. doi:10.1103/PhysRevLett.10.146. These events were recorded at the giant array of Volcano Ranch, New Mexico, constructed by Bruno Rossi's group at MIT. This array, consisting in 19 plastic large-area scintillator counters, each viewed with a photomultiplier enclosing an area of 8.1 km^2, set the trend for highly quantitative investigations at energies above 10^{17} eV, by determining the shower directions and

which reach only 7 orders of magnitude lower energies than the most energetic cosmic rays. But already at an energy beyond 10^{14} eV, the flux of primary particles is about one per square meter per hour. The most energetic particles have a rate less than 1 particle per km^2 per century, but such ultra-high-energy cosmic rays are at the energy frontier, and by studying them, one can aim to understand the nature and mechanisms of cosmic accelerators able to boost particles to energies far beyond what terrestrial laboratories will attain in the foreseeable future. The observation of particles with energies of 10^{18}–10^{20} eV provided evidence that a purely galactic origin of cosmic rays was untenable and that at least the most energetic particles cannot be accelerated and confined inside our Galaxy and accordingly must come from outside. However, the problem of their origin, arising from the presence of the tangled galactic magnetic fields, which causes the paths of the vast majority of the charged particle components to bear no relationship to the direction of their respective source, continued to be a very difficult question to answer, given the relatively primitive observations and the theoretical insights of that period.

By the early 1960s, the search for 'point sources' of charged particles had been virtually abandoned and thoughts had turned to looking for extraterrestrial gamma-rays, with considerations revolving around the mechanisms by which such gamma rays might be produced, and in which celestial objects—at the least, those gamma rays which could be identified as the products of cosmic ray interaction with gas nuclei and photons in the interstellar medium and elsewhere. Thus, if the target gas could be identified and the gamma rays could be measured, then the cosmic ray intensity could be inferred at places remote from the Earth. One of the problems with gamma-ray astronomy is that the rays are scarce relative to other entities that resemble them experimentally, and more so, as one looks for the more energetic ones. Gamma rays actually constitute only one or two out of every 100000 cosmic rays, which means a big signal-to-background challenge. The flux of cosmic rays at 'low' energies is great enough that they can be observed *directly* with

---

core locations with high accuracy; and it also established unambiguously the presence of primaries with energies at least up to 10^{20} eV. For a first-hand account of Rossi's group research activity at MIT, see George W. Clark: Air Shower Experiments at MIT. In: Yataro Sekido, and Harry Elliot (eds.): *Early History of Cosmic Ray Studies. Personal Reminiscences with Old Photographs.* Dordrecht: D. Reidel Publishing Company 1985, 239–246. The study of ultra-high-energy cosmic rays has continued ever since, with increasingly large detector arrays. Kampert, and Watson, Extensive Air Showers and Ultra High-Energy Cosmic Rays, 2012, 359–412. See also Giorgio Matthiae: The Cosmic Ray Energy Spectrum as Measured Using the Pierre Auger Observatory. *New Journal of Physics* 12/7 (2010), 075009. doi:10.1088/1367-2630/12/7/075009.

detectors above the atmosphere, which can be shielded from practically all incoming radiations *other than gamma rays*, by using the so-called anticoincidence technique to reject the majority of charged cosmic-ray particles; but already above a few GeV, one would need a very large detector area in orbit for years to increase the detection probability for such far rarer, *very-high-energy and ultrahigh-energy events*; moreover, these detectors are also very massive. Thus, it was beyond this natural limit for space gamma-astronomy that the appeal of ground-based gamma-ray astronomy grew.

As we have seen, gamma-ray astronomy in the 100 MeV region was boosted by the 1958 seminal paper of Philip Morrison,[235] while detectability of the higher-energy TeV gamma rays from the Crab Nebula—through the detection of gamma-induced air showers by ground-based arrays of detectors—was suggested the following year by Giuseppe Cocconi during the 6th International Cosmic Ray Conference held in Moscow.[236] Although, like Morrison, he overestimated the eventual detected flux, his article is considered to have "sowed the seeds for the first serious atmospheric Cherenkov experiments to detect very high energy gamma rays from cosmic sources."[237] It certainly stimulated further attempts to use angular anisotropy as indirect evidence for the presence of gamma-ray sources.[238] At the end of the 1950s, the idea that the science of gamma ray astronomy could be a tool for answering questions in high-energy astrophysics and cosmology gained currency. Ground-based gamma-ray astronomy, however, would need to rely on discriminating primary gamma rays of extraterrestrial origin from the overwhelming background of

235 Morrison, On Gamma-Ray Astronomy, 1958, 858–865.

236 His proposal was to search for gamma-ray sources as a narrow-angle anisotropy in the distribution of Extensive Air Showers at mountain altitudes, near 1/2 of atmospheric depth with characteristic energy of 1 TeV and angular resolution about 1°. In particular, he suggested that high-energy protons in the Crab Nebula could produce neutral pions and thus generate a considerable flux of gamma rays from their decay. Giuseppe Cocconi: An Air Shower Telescope and the Detection of 10^{12} eV Photon Sources. *Proceedings from the 6th International Cosmic Ray Conference, Held in Moscow, Russia, 6–11 July 1959. Extensive Air Showers and Cascades Process*. 1960, 309–311. https://ui.adsabs.harvard.edu/#abs/1960ICRC....2..309C. Last accessed 10/20/2018.

237 Trevor C. Weekes: The Atmospheric Cherenkov Technique in Very High Energy Gamma-Ray Astronomy. *Space Science Reviews* 75/1 (1996), 1–15, 1. doi:10.1007/BF00195020.

238 Aleksandr E. Chudakov: VHE and UHE Gamma Ray Astronomy. History and Problems. In: Maurice M. Shapiro, and John P. Wefel (eds.): *Cosmic Gamma Rays, Neutrinos, and Related Astrophysics. Proceedings of the NATO Advanced Study Institute on Cosmic Gamma Rays and Cosmic Neutrinos Erice, Italy 20–30 April, 1988*. Dordrecht: Springer Netherlands 1989, 163–182, 164. doi:10.1007/978-94-009-0921-2_12.

cosmic ray particles and, especially, of secondary gamma rays generated by cosmic rays in the atmosphere on the basis of air shower characteristics.

Above 10 TeV, in the *ultrahigh-energy* (UHE) region, where the number of cascade particles reaching the observation level is large enough to be recorded directly by air shower arrays, particle detectors are dispersed over a large area, generally at mountain altitudes, even if detection beyond a few hundred TeV is possible at sea level. Other detectors record the associated muon content, an important feature depending on the nature of the primary. The muon content in gamma-ray-induced showers is much reduced, compared to a proton shower; consequently, the careful analysis and selection of such extensive air showers appeared to be practically the only possibility to obtain experimental information about the existence of primary gamma rays with energies larger than 10^5 GeV.[239]

Early efforts to identify point-source anomalies in the arrival direction were not successful, beyond establishing meaningful upper limits for the flux of diffuse gamma rays, and interest waned for several years. Moreover, these conventional ground-based scintillator arrays are only practical for the highest energies, leaving a gap in the energy range between 10 GeV to around 10 TeV (the Very-High-Energy region, VHE), where the majority of interesting, high-energy cosmic processes take place. This is actually the energy range dominated by the Imaging Atmospheric Cherenkov Technique (IACT), which will be the main focus in the following pages. We will also see, as with the previous two examples (solar neutrinos and gravitational waves), how the Cherenkov technique had originated in several decades of developments, in this case primarily in the United States, Britain, and the Soviet Union.

239 The ratio of the total number of muons to the total number of electrons of a shower was a parameter widely used and applied since the end of the 1960s to distinguish gamma-ray initiated showers from hadronic background. The muon component arises almost exclusively from the decay of charged pions and kaons within hadron showers. A small fraction, however, originates from *photonuclear reactions* occurring when primary high-energy cosmic gamma rays are absorbed by nuclei or by individual protons or neutrons, resulting in the production of charged and neutral pions or other elementary particles, whose decay products also contain muons. But in this case, the muon content is much reduced compared to a proton shower, as a consequence of the extremely low probability of photonuclear interaction with air nuclei (small cross-section). The low production or muons could thus be a criterion for identifying gamma-ray induced showers, helping in reducing the nearly overwhelming background from the dominant hadron-induced cosmic-ray showers. K. Kamata et al.: Predominantly Electromagnetic Air Showers of Energy 10^{14} eV to 10^{16} eV. *Canadian Journal of Physics* 65/10 (1968), S72–S74. doi:10.1139/p68-177. Ronald Maze et al.: Origin of Muon-Poor Extensive Air Showers. *Physical Review Letters* 22/17 (1969), 899–901. doi:10.1103/PhysRevLett.22.899.

New Horizons for Ground-Based Gamma-Ray Astronomy: The Imaging Atmospheric Cherenkov Technique

In the energy range of 10 GeV to 10 TeV, the cascades die out in the upper atmosphere and the number of cascade particles reaching the observation level is too small. In this case, showers can be detected by means of a technique based on a phenomenon discovered already in the 1930s. The ultra-relativistic electrons and positrons generated by gamma rays in the upper atmosphere have velocities exceeding the speed of light in air and consequently emit *Cherenkov radiation*, the electromagnetic equivalent of a sonic boom generated by an aircraft flying at supersonic speed.[240] Thousands of relativistic charged particles in the shower emit light almost simultaneously, as a fast light flash that can be detected on the ground during clear, dark nights. The required collection areas to work in the VHE region, which are in the range of 10^4–10^6 m^2 (that is, up to a square kilometer), would thus be reached using the atmosphere itself as detector. Moreover, in the case of a cosmic gamma ray having sufficient energy to initiate a shower of many hundreds of charged particles, the emitted flash of typical blue light would largely travel in a cone within a very small angle around the original photon trajectory, pointing back to the gamma-ray source. This provided the most ambitious promise of the incipient technique: to conduct gamma-ray astronomy without needing to go to outer space. Moreover, the relatively narrow angular spread of Cherenkov light produced by Extensive Air Showers made the possibility of detecting point sources of the primary cosmic rays more attractive.

During the 1950s and '60s, the technique was pioneered by John V. Jelley and William Galbraith of the Atomic Energy Research Establishment (AERE) at Harwell, UK, and by Alexander Chudakov from the Lebedev Physical Institute of the Soviet Academy of Sciences in Moscow.[241] In the late 1950s, while

240 Alan A. Watson: The Discovery of Cherenkov Radiation and Its Use in the Detection of Extensive Air Showers. *Nuclear Physics B—Proceedings Supplements* 212–213 (2011), 13–19. doi:10.1016/j.nuclphysbps.2011.03.003.

241 Independently of Jelley and Galbraith, the Air Cherenkov Technique applied to cosmic ray showers was also experimented in 1953–57 by Chudakov, at the same Lebedev Physical Institute where Pavel A. Cherenkov, experimental discoverer of the phenomenon, worked. On the Pamir mountains, in a high altitude site traditionally used for cosmic ray research, Chudakov pioneered studies of Cherenkov radiation from Extensive Air Showers with an array of eight Cherenkov detectors. N. M. Nesterova, and A. E. Chudakov: On the Observation of Cerenkov Radiation Accompanying Broad Atmospheric Showers of Cosmic Rays. *Journal of Experimental and Theoretical Physics* 1/2 (1955), 388–389. http://www.jetp.ac.ru/cgi-bin/e/index/e/1/2/p388?a=list. Last accessed 11/9/2018. An account of early efforts in the USSR can be found in Aleksandr S. Lid-

Chudakov was deeply involved in cosmic-ray experiments in space with rockets and satellites, Georgii T. Zatsepin, his colleague at the Lebedev Physical Institute and a real expert in cosmic-ray physics, proposed that he discuss the possibility of using Cherenkov light as a tool *to search for local gamma-ray sources.*[242] Inspired by Cocconi's talk at the Moscow conference about the detectability of TeV gamma rays from the Crab Nebula and other sources with an "air shower telescope," they formulated by 1960 the plan and principles of the VHE Cherenkov technique and the world's first gamma-ray telescopes to focus the Cherenkov light onto photon detectors—typically, photomultiplier tubes—were mounted in Katsiveli, Crimea, on the shores of the Black Sea. The following year, the number of mirrors was increased up to 12.[243] They conducted the first systematic searches for gamma ray sources, observing ten target sources, among others, the Crab Nebula, supernova remnants, and radio galaxies recently identified by radio telescopes as sources emitting non-thermal, synchrotron radiation; but no statistically significant positive effect from any of these was found. And so, contrary to Cocconi's overly optimistic predictions, no point sources of TeV photons could be detected.

However, Chudakov and collaborators' pioneering experiments in Crimea were followed with interest by other groups. In the early 1960s, a collaboration between Jelley, at Harwell, and Neil A. Porter, formerly at Harwell but meanwhile at the University College in Dublin, led to a further experiment with two 90 cm mirrors at Glencullen, a dark site in the Wicklow Mountains, Ireland, to where the installations built at Harwell were transferred in 1963.[244] It was with this experiment that the Dublin-born Trevor C. Weekes, a young member of the Irish–UK collaboration, began his career and the long quest to refine the Atmospheric Cherenkov Technique. Weekes then joined Giovanni Fazio's

---

vansky: Air Cherenkov Methods in Cosmic Rays. Review and Some History. *Radiation Physics and Chemistry* 75/8 (2006), 891–898. doi:10.1016/j.radphyschem.2005.12.019. For the history of cosmic ray research at the Lebedev Institute, see Georgii T. Zatsepin, and Tat'yana M. Roganova: Cosmic Ray Investigations. *Physics-Uspekhi* 52/11 (2009), 1139–1146. doi:10.3367/UFNe.0179.200911f.1203.

242 Aleksandr S. Lidvansky: G T Zatsepin and the Birth of Gamma-Ray Astronomy. *Physics-Uspekhi* 61/9 (2018), 921–925, 922. doi:10.3367/UFNe.2017.05.038184.

243 For a historical excursion on these early experiments at the Lebedev Institute, see Chudakov, VHE and UHE Gamma Ray Astronomy, 1989, 163–182.

244 John V. Jelley, and N. A. Porter: Čerenkov Radiation from the Night Sky, and Its Application to γ-Ray Astronomy. *Quarterly Journal of the Royal Astronomical Society* 4 (1963), 275–293. http://adsabs.harvard.edu/abs/1963QJRAS...4..275J. Last accessed 12/23/2018. J. H. Fruin et al.: Flux Limits for High—Energy γ—Rays from Quasi—Stellar and Other Radio Sources. *Physics Letters* 10/2 (1964), 176–177. doi:10.1016/0031-9163(64)90159-3.

group at the Smithsonian Astrophysical Observatory and the Harvard College Observatory, Cambridge, Massachusetts, which had been interested in detecting primary gamma rays with balloons and satellites since the early 1960s.

It turned out that neither the shower arrays nor the atmospheric Cherenkov experiments conducted for several years in the early 1960s could detect gamma-ray sources. Upper limits could be set only on the flux of high energy photons from supernova remnants, even if they also provided evidence that electrons are directly accelerated in the Crab Nebula.[245] Technically, it was a painstaking problem with 1960s electronics to detect the faint and very brief flashes of light, separate them from other light sources, and analyze them to obtain physically meaningful information about the originating gamma rays. The search for point sources thus proved frustrating for a long time.[246]

In 1968, one of the first pulsars was discovered at the center of the Crab Nebula, as a pulsating radio source,[247] and was identified as the remnant star of the supernova explosion.[248] The spinning neutron star hypothesis for the origin of the signal provided the high-energy scenario in which particles could be accelerated to high energies in several possible ways.[249] The possibility that pulsars could emit X-rays and even gravitational waves—and the observation of the star's pulsation also in the MeV gamma-ray energy region by satellites—led to a parallel revival of interest in further observations of the Crab Nebula in the TeV and PeV gamma-ray energy domains, inaccessible from small outer space satellites. In that same year, 1968, Giovanni Fazio and his group, now

245 Chudakov's group set an upper limit in the Katsiveli experiment, showing the gamma-ray flux above 5 TeV to be two orders of magnitude less than had been anticipated by Cocconi. Aleksandr E. Chudakov et al.: On the High Energy Photons from Local Sources. *Extensive Air Showers, Proceedings from the 8th International Cosmic Ray Conference*. 1963, 199–204. http://adsabs.harvard.edu/abs/1963ICRC....4..199C. Last accessed 11/8/2018.

246 For a survey of early atmospheric Cherenkov radiation studies see Chapter 2 in David Fegan: *Cherenkov Reflections. Gamma-Ray Imaging and the Evolution of TeV Astronomy*. Singapore: World Scientific 2019.

247 Antony Hewish et al.: Observation of a Rapidly Pulsating Radio Source. *Nature* 217/5130 (1968), 709–713. doi:10.1038/217709a0.

248 Thomas Gold: Rotating Neutron Stars as the Origin of the Pulsating Radio Sources. *Nature* 218/5143 (1968), 731–732. doi:10.1038/218731a0.

249 Franco Pacini: Energy Emission from a Neutron Star. *Nature* 216/5115 (1967), 567–568. doi:10.1038/216567a0. Pacini's article suggesting that strongly magnetized neutron stars could release their rotational energy and produce a large flow of relativistic particles was actually published before the actual discovery of the pulsar, and its subsequent identification as a neutron star. He pointed out that a spinning neutron star with a large magnetic field would emit electromagnetic waves, and might even be a source of gravitational waves.

including Trevor C. Weekes from the University College of Dublin, began to build a 10 m optical reflector at the Whipple Observatory on Mount Hopkins in Arizona, aiming in particular to detect gamma rays from the Crab Nebula.[250] Construction of this reflector had been boosted by the prediction that if the radiation from radio to X-rays from the Crab Nebula was due to synchrotron radiation by relativistic electrons, then these same electrons colliding with the low-energy synchrotron-radiated photons would boost them to gamma ray energies through the Compton interaction process. The resultant gamma-ray spectrum would be most easily detectable at 100–1000 GeV energies, according to the Compton-synchrotron model of the Crab Nebula developed by Robert J. Gould in 1965, following Morrison's early suggestions about the possibility of such an effect.[251]

Around the early 1970s, another newcomer to the field, Arnold Stepanian's group at the Crimean Astrophysical Observatory in USSR, also searched for point sources of high-energy gamma rays using extensive air shower Cherenkov flashes detection, and reported what was considered a controversial observation, namely a very-high-energy gamma ray outburst from Cygnus X-3, a well-known compact X-ray binary source discovered by a rocket flight as early as 1966.[252] During the 1970s, the search for gamma-rays from the direction of the X-ray source Cygnus X-3 was carried out by several groups applying different observation modes and experimental techniques and covering a wide range of energies. But the experimental results were somewhat contradictory, ranging from claims of a "clear excess" of gamma-rays from Cygnus X-3 to "no effect," as for example, in the data of the COS-B satellite mission.

---

250 G. G. Fazio et al.: A Search for Discrete Sources of Cosmic Gamma Rays of Energies Near 2×10^{12} eV. *The Astrophysical Journal* 154 (1968), L83. doi:10.1086/180275. G. G. Fazio et al.: An Experiment to Search for Discrete Sources of Cosmic Gamma Rays in the 10^{11} to 10^{12} eV Region. *Canadian Journal of Physics* 46/10 (1968), S451–S455. doi:10.1139/p68-268.

251 Robert J. Gould: High-Energy Photons from the Compton-Synchrotron Process in the Crab Nebula. *Physical Review Letters* 15/14 (1965), 577–579. doi:10.1103/PhysRevLett.15.577.

252 B. M. Vladimirsky, A. A. Stepanian, and V. P. Fomin: High-Energy Gamma-Ray Outburst in the Direction of the X-Ray Source CYG X-3. *Proceedings of the 13th International Cosmic Ray Conference, University of Denver, Denver, Colorado, August 17–30, 1973* 1 (1973), 456–460. https://ui.adsabs.harvard.edu/#abs/1973ICRC....1..456V. Last accessed 12/6/2018. The installation consisted of four 1.5 m parabolic mirrors with fast photomultipliers in the focal planes. The mirrors were combined to form a system of two completely independent pairs of detectors. During the 1950s, Arnold Stepanian and his group at the Crimean Astrophysical Observatory had studied cosmic rays also with balloon probes and from the mid-1960s had started to develop simple gamma-ray telescopes with the aim of detecting ultrahigh energy gamma rays.

By the mid-1970s, while observations from space were becoming a reality, ground-based gamma-ray detectors proved unsuccessful in the search for sources in the TeV range. More sensitive and sophisticated methods and detectors were needed to deal with an overwhelming background of charged cosmic rays, in order to study the cosmic gamma radiation by means of ground-based instruments.

The Cosmic-Ray Group in Kiel Moves Toward High-Energy Gamma-Ray Astronomy

Pinkau's trajectory from cosmic-ray showers, first in Hamburg then, from 1957, at the Pure and Applied Nuclear Physics Institute in Kiel, to space-based gamma-ray astronomy at the Max Planck Institute for Extraterrestrial Physics in Munich, brings us back to Bagge's cosmic-ray group, of which also Otto Claus Allkofer and Joachim Trümper were members.[253] This incubator for talented physicists, while also having minority participation in German and international satellite projects, could maintain its national leadership solely with cosmic-ray studies from the ground, by further specializing in its expertise in detectors and cosmic ray showers, and by following a more articulate path, eventually leading to high-energy ground-based gamma ray astronomy.

With the lifting of restrictions on nuclear research in 1955, Bagge benefited from the promises of the nuclear age and had a main role in the production of Germany's first nuclear-powered vessel.[254] In parallel to his activity as Technical Director of the nuclear facility, he had moved from Hamburg to Kiel in 1957, as Director of the brand-new Institute for Pure and Applied Nuclear Physics at the Christian-Albrechts-University. While he largely dedicated himself to the new applied nuclear enterprise, Bagge brought along the researchers he had trained in Hamburg, and continued to favor and promote research in cosmic ray physics at his new institute. As Trümper recalled:

253 In the early 1960s, Allkofer and Trümper collaborated in an experiment using spark chambers built in Kiel to measure the energy distribution of cosmic-ray muons at 3,000 m (an unprecedented altitude for this kind of experiment) in the cosmic-ray laboratory belonging to Heisenberg's Institute in Munich, on top of the Zugspitze, Germany's highest peak. Otto Claus Allkofer, and Joachim Trümper: Das Muonen-Spektrum in 3000 m Höhe. *Zeitschrift für Naturforschung A* 19/11 (1964), 1304–1309. doi:10.1515/zna-1964-1108. Joachim Trümper: interview by Luisa Bonolis and Juan-Andres Leon, Berlin, 6–7 May 2019.

254 See, for example, Hans-Willy Hohn, and Uwe Schimank: *Konflikte und Gleichgewichte im Forschungssystem. Akteurkonstellationen und Entwicklungspfade in der staatlich finanzierten außeruniversitären Forschung*. Frankfurt am Main: Campus Verlag 1990.

> At the Kiel Institute we were quite free in our choice of research topics. After the PhD (1959), I started a big experiment in cosmic rays, a so-called air shower experiment to study cosmic radiation at very high energies from 10^{14} to 10^{17} eV [...] One question was: What are the sources of cosmic radiation and how are the particles accelerated?[255]

And so, in the early 1960s, they redirected some of the activities carried out with traditional techniques and detectors—as well as with balloon flights—to a more ambitious project: the construction of a multipurpose shower array, including different detection systems and electronic equipment.[256] The extensive air shower experiment went into operation in Kiel in June 1965 and, together with Allkofer, until 1970–71, Joachim Trümper was a leading member of the Kiel cosmic-ray group, before moving to Tübingen University, where he was appointed the Chair of Astronomy, as successor to Heinrich Siedentopf. There, he began to develop his program in the promising field of X-ray astronomy. Sometime in 1969 or '70, Trümper visited the Max Planck Insti-

255 Joachim Trümper: interview by Helmuth Trischler and Matthias Knopp, March 18, 2010. Transcript, HAEU, https://archives.eui.eu/en/oral_history/INT076. Last accessed 1/4/2019.

256 The main aim was to study the electromagnetic core structure of showers, which could provide information about the chemical composition and mass of the primary particles at the highest energies and the nuclear interactions involved in the cascade process. The central laboratory building consisted of a former air raid shelter from a marine base in Kiel, hosting the muon spectrometers and a liquid Cherenkov counter, on top of which a light laboratory building was constructed. It contained most of the detectors: an array of 16 scintillation counters, each of 1 m^2 area, and a neon hodoscope of 32 m^2 area, consisting of 36 compact units comprising about 180000 neon tubes to track the charged particles. Erich Bagge et al.: The Extensive Air Shower Experiment at Kiel. *Proceedings of the 9th International Cosmic Ray Conference, London, UK*. London New York Paris Los Angeles: Institute of Physics and the Physical Society 1965, 738–741. https://ui.adsabs.harvard.edu/#abs/1965ICRC....2..738B. Last accessed 11/23/2018. See fig. 5, showing a cross section of the laboratory building. In order to use the necessary sufficiently broad class of models for nuclear interactions, the group used Monte Carlo calculations, but at that time it was not yet possible to do Monte Carlo simulations of the cascade. They were able to measure the hadron distribution in the air shower cores at energies above 800 GeV. This was about two orders of magnitude higher than the energy of the CERN Proton Synchrotron, or of the Brookhaven Alternating Gradient Synchrotron, two powerful accelerators put into operation between 1959–60. The work was supported by the Land Schleswig-Holstein and the Bundesministerium für Wissenschaftliche Forschung.

tute for Extraterrestrial Physics in Garching,[257] to where his colleague and friend Klaus Pinkau had moved from Kiel. The foundations for more ambitious X-ray astronomy projects were laid on this occasion, and led to Trümper's later appointment as Director of the Max Planck Institute for Extraterrestrial Physics (in 1975)—as well as to the planning and construction of ROSAT, the successful German X-ray satellite—although he still continued his strong collaboration with the Tübingen Institute, which became the MPE's partner in all X-ray astronomy activities. Trümper had long since felt that such a long-term enterprise could be tackled only within the Max Planck Society.[258]

The fate of these two researchers exemplifies quite well what happened in the field of cosmic ray research after the 1950s. On the one hand, particle accelerators replaced cosmic rays as the source of collisions for experimental particle physics. On the other, the advent of the space age heralded the availability of satellites for research formerly done with balloons. Many prominent researchers with astrophysical interests, and their disciples, too, left cosmic rays behind from the late 1950s onwards, and turned instead to the development of satellite-based gamma-ray astronomy, as described in Chapter 3.

On the other hand, their colleague Otto Claus Allkofer, who remained in Kiel, continued to work on cosmic-ray research,[259] and from 1975 to '76, considerably extended and modified their air shower experiment. After ten years of operation, further scintillation counters and larger detector areas were added; this improved the reliability of detector response and allowed more detailed and more accurate data to be compiled, also thanks to a now completely automatic scan of each neon hodoscope photograph by a computer-controlled device that stored data on a magnetic tape for further analysis.[260]

257 Ludwig Biermann, and Reimar Lüst: Max-Planck-Institut für Physik und Astrophysik. Institut für Astrophysik und Institut für extraterrestrische Physik. *Mitteilungen der Astronomischen Gesellschaft* 29 (1971), 86–112, 86. http://adsabs.harvard.edu/abs/1971MitAG..29...86B. Last accessed 10/30/2018.

258 Trümper, Joachim: interview by Luisa Bonolis and Juan-Andres Leon, Munich, August, 7–8, 2017. DA GMPG, BC 601036.

259 As evident from his publications, Allkofer's interests continued to be especially focused on muons, also later, during the 1980s, in connection with high-energy physics at CERN accelerators. See, for example, Otto Claus Allkofer et al.: The Kiel Cosmic Ray Muon Spectograph. *Nuclear Instruments and Methods* 83/2 (1970), 317–325. doi:10.1016/0029-554X(70)90479-9. He also wrote an introductory volume to cosmic-ray physics and a book on spark chambers. Otto Claus Allkofer: *Introduction to Cosmic Radiation*. München: Karl Thiemig 1975. Otto Claus Allkofer: *Spark Chambers*. München: Karl Thiemig 1969.

260 Erich Bagge, Manfred Samorski, and Wilhelm Stamm: A New Air Shower Experiment at Kiel. *15th International Cosmic Ray Conference, Plovdiv, Bulgaria, August 13–26, 1977*,

In the early 1980s, interest in continuing the search with the shower array technique arose in the wake of various successful studies of gamma ray astronomy: by satellites, up to the GeV energy region; cosmic gamma ray bursts observed by satellites; and interesting results obtained from Extensive Air Shower (EAS) measurements at ground level (threshold energies above 10^{11} eV). In particular, much attention continued to be devoted to Cygnus X-3, but controversial results were reported up to the early 1980s by satellite experiments, balloon flights, and by groups pioneering the Imaging Atmospheric Cherenkov Technique. As emphasized by Trevor Weekes,

> Before 1980 [...] despite considerable effort, there were few results. The general feeling amongst astrophysicists was disinterest, if not disbelief. 'Gamma-Ray Astronomy' was interpreted as a branch of space science and presumed to terminate where satellites ceased to be useful, i.e., at 1 GeV.[261]

The field was languishing and then, in 1983, Manfred Samorsky and Wilhelm Stamm from Kiel surprised the community by reporting that, after four years of operation (from March 18, 1976 to January 7, 1980), measurements from their air shower array showed "a significant excess of extensive air showers"—presumably from gamma rays—emanating from the direction of Cygnus X-3. The signal was claimed to have a significance of 4.4 standard deviations.[262] At the energies involved, it could be estimated that there must be charged particles with energies up to about 10^{17} eV in the emitting source.

Conference Papers. Budapest: Dept. of Cosmic Rays, Central Research Institute for Physics of the Hungarian Academy of Sciences 1977, 24–29. https://ui.adsabs.harvard.edu/#abs/1977ICRC....12...24B. Last accessed 11/13/2018. The new EAS experiment was designed to investigate the structure of extensive showers initiated in the atmosphere by primary cosmic rays having an energy range from 10^{15} to 10^{17} eV. Details of the experimental arrangement were also published in Erich Bagge, Manfred Samorski, and Wilhelm Stamm: Lateral Distribution of Electrons in the Core Region of Extensive Air Showers of 10^{15}–10^{17} eV. *Proceedings of the 16th International Cosmic Ray Conference (Kyoto)*. 1979, 260–265. https://ui.adsabs.harvard.edu/#abs/1979ICRC....13..260B. Last accessed 8/14/2018. The work was again supported by the Deutsche Forschungsgemeinschaft (DFG).

261 Trevor C. Weekes: Very High Energy Gamma-Ray Astronomy. *Physics Reports* 160 (1988), 1–121, 3. doi:10.1016/0370-1573(88)90177-9.

262 Manfred Samorski, and Wilhelm Stamm: Detection of 2×10^{15} to 2×10^{16} eV Gamma-Rays from Cygnus X-3. *The Astrophysical Journal* 268 (1983), L17–L21. doi:10.1086/184021. Previous negative results were also summarized in this publication. Samorski had been Trümper's doctoral student (personal communication to the author L.B., 01/10/2021).

This result seemed to point toward a solution to the mystery of high-energy cosmic ray sources: "Our thoughts on the 70-year-old question of the origin of cosmic rays have been radically changed…," remarked Alan A. Watson from the University of Leeds, who later became instrumental in the creation of the giant Pierre Auger cosmic-ray observatory in Argentina.[263] Results from the Kiel array were one of the highlights of the 18th International Cosmic Ray Conference held in Bangalore, India, from August 22 to September 3, 1983. The University of Leeds (UK) group also presented data from a small sub-array which was part of the giant Haverah Park air shower experiment, using Cherenkov water detectors, which appeared to confirm the results from Kiel.[264]

Part of the excitement aroused by Samorsky and Stamm's puzzling claim came from events observed in parallel at massive underground proton decay detectors, built after the advent of the Grand Unified Theory of the electroweak and strong interactions, and predicting a likely instability of the proton, with a lifetime of less than 10^{32} years.[265] The reported signals from these experiments, apparently related to Cygnus X-3, could not be understood in terms of known particles or interaction processes, i.e., in the framework of conventional physics.[266] Especially perplexing was the muon component of these showers and, as we will see, this problem was instrumental in attracting the attention of theoreticians and high-energy physicists to ultra-high-energy gamma rays from cosmic sources. There were two stark alternatives: either the interpretation of the underground evidence was somehow wrong or *a quite new physics* was involved, the structure and importance of which could not yet

263 Alan A. Watson: Ultra High Energy Cosmic Rays and Ultra High Energy γ-Rays. *Advances in Space Research* 4/2 (1984), 35–44, 41. doi:10.1016/0273-1177(84)90290-4.

264 J. Lloyd-Evans et al.: Observation of γ Rays >10^{15} eV from Cygnus X-3. *Nature* 305/5937 (1983), 784–787. doi:10.1038/305784a0. The authors mentioned observation of gamma rays from Cygnus X-3 by other groups and emphasized that Kiel's result could have considerable implications both for the understanding of the source and for the solution of the longstanding problem of the origin of cosmic rays.

265 Donald H. Perkins: Proton Decay Experiments. *Annual Review of Nuclear and Particle Science* 34 (1984), 1–52. doi:10.1146/annurev.ns.34.120184.000245.

266 Hypothetical 'exotic particles' might be responsible for the observed showers. A. Michael Hillas: Why Is Cygnus X-3 (with Related Sources) a Highlight of Cosmic-Ray Astrophysics? *Proceedings from the 19th International Cosmic Ray Conference, La Jolla, USA, August 11–12, 1985*. 1985, 407–414. http://adsabs.harvard.edu/abs/1985ICRC....9..407H. Last accessed 11/16/2018.

even be guessed at.[267] As Trevor Weekes of the Whipple Observatory remarked a few years later:

> No subject has aroused such interest (and controversy) as the apparent detection of Cygnus X-3 in underground nucleon-decay experiments [...] The existence of a new particle that would fit within the rather narrow constraints imposed by the underground experiments has come as a challenge to theoretical particle physicists *at a time when there is not too much excitement in the field* [our emphasis].[268]

Conventional physics could not explain those puzzling signals from Cygnus X-3, which was becoming a very topical object in conferences, and "in the flood of exotic theoretical predictions for an energy range inaccessible to HEP accelerator experiments,"[269] an obvious name—*cygnet*—was coined to denote such hypothetical exotic primaries, unobserved particles with unique characteristics, which were especially intriguing for theoretical particle physicists.[270] In

267 Aleksandr E. Chudakov: Is the Signal from CYG X-3, as Recorded in Some Underground Experiments, Real? *Proceedings from the 19th International Cosmic Ray Conference, La Jolla, USA, August 11–12, 1985* 9 (1985), 441–444. https://ui.adsabs.harvard.edu/#abs/1985ICRC....9..441C/abstract. Last accessed 3/23/2019.

268 Weekes, Very High Energy Gamma-Ray Astronomy, 1988, 1–121, Appendix B. For a review of controversies and claims about the unusual signals from Cygnus X-3, see also Trevor C. Weekes: TeV Radiation from Galactic Sources. *Space Science Reviews* 59/3 (1992), 315–364. doi:10.1007/BF00242089.

269 Eckart Lorenz, and Robert Wagner: Very-High Energy Gamma-Ray Astronomy. A 23-Year Success Story in High-Energy Astroparticle Physics. *European Physical Journal H* 37/3 (2012), 459–513, 470. doi:10.1140/epjh/e2012-30016-x.

270 M. L. Marshak et al.: Evidence for Muon Production by Particles from Cygnus X-3. *Physical Review Letters* 54/19 (1985), 2079–2082. doi:10.1103/PhysRevLett.54.2079. Edward W. Kolb: Searching for Cygnets. In: G. Lazarides, and Q. Shafi (eds.): *Proceedings of the International Symposium on Particles and the Universe, Thessaloniki, Greece, 24–29 June 1985*. Amsterdam: North Holland 1986, 247–255. http://inspirehep.net/record/218657?ln=it. Last accessed 11/27/2018. See also a discussion on the underground experiments connecting the *cygnet* production mechanisms to the existence of strange quark matter as a basic component of the compact member of the Cygnus X-3 binary system, either a neutron star or a black hole. Gordon Baym et al.: Is Cygnus X-3 Strange? *Physics Letters B* 160/1 (1985), 181–187. doi:10.1016/0370-2693(85)91489-3. See also Gordon Baym in the special issue of *Los Alamos Science* dedicated to Cygnus X-3: V. J. Stenger: Teravolt Astronomy. *Advances in Space Research* 3/10 (1984), 139–145. doi:10.1016/0273-1177(84)90079-6. Alan A. Watson: High-Energy Astrophysics. Is Cygnus X-3 a Source of Gamma Rays or of New Particles? *Nature* 315/6019 (1985), 454–455. doi:10.1038/315454b0. Alan A. Watson: Cygnus X-3. In: K. E. Turver (ed.): *Very High Energy Gamma Ray Astronomy*. Dordrecht:

this climate of unexpected results in the ultra-high-energy regime, in particular the one related to the striking observation of muon-rich air showers from the direction of Cygnus X-3, the name CYGNUS was given even to a new air-shower array located at Los Alamos National Laboratory in the US.[271]

In that same year, 1983, the experimental confirmation at CERN of the existence of the heavy vector bosons $W^{\pm}$ and Z, one of the main consequences of the Glashow-Weinberg-Salam unification of the weak and electromagnetic interactions, was achieved due to a major advance in high-energy physics: the creation of a proton-antiproton collider providing the necessary collision energy, which had been far beyond the reach of existing accelerators and detectors. Such detection showed that the unified electroweak theory had made a very good start, and supported the theoretical expectation that unification of strong, weak, and electromagnetic forces would reveal itself at the extremely high energies and particle densities available in the first instants of our Universe, so reinforcing the establishment of the new deep connection between particle physics and cosmology. Within the framework of the hot Big Bang model (based on Einstein's theory of general relativity and the hypothesis that the Universe is isotropic and homogeneous when viewed over sufficiently large distances), the laws of particle physics could be applied in an attempt to trace the evolution of the cosmos in very early times.[272]

The impact originating from Kiel's tantalizing observations was actually part of the growing symbiotic relationship between particle physics, astrophysics, and cosmology, which in those days of its very earliest appearances was being given a label of its own: the title of a talk given at the fourth Marcel

---

Springer Netherlands 1987, 53–61. doi:10.1007/978-94-009-3831-1_6. A special session at the Durham NATO Workshop on Ultra High Energy gamma-ray astronomy held during August 11–15, 1986, was dedicated to Cygnus X-3. For the state of art of the field at that time, see K. E. Turver: *Very High Energy Gamma Ray Astronomy*. Dordrecht: Springer Netherlands 1987. Some authors of these papers were already deeply involved in discussions about the cosmological and astrophysical implications of the existence of new particles predicted by Grand Unified Theories, also in connection to early life of the Universe following the Big Bang.

271 The experiment, aiming at the search for point sources, consisted of an array of scintillation detectors and an associated muon detector, which underwent expansion in several stages after going into operation in 1986. Todd J. Haines et al.: The Status of the CYGNUS Experiment: Past, Present, and Future. *Nuclear Physics B—Proceedings Supplements* 14/1 (1990), 244–249. doi:10.1016/0920-5632(90)90428-W.

272 For a book providing an excellent orientation in the field of astroparticle physics, bridging the gap between a presentation of the field at a simple level and a textbook for more expert readers, see Claus Grupen: *Astroparticle Physics*. 2nd ed. Cham: Springer 2020. doi:10.1007/978-3-030-27339-2.

Grossmann Meeting on General Relativity in 1985 by Abdus Salam, one of the protagonists of the unified electroweak theory, was *Astro-Particle-Physics*.[273]

In the wake of the excitement aroused by their 1983 announcement, the Kiel group was stimulated to continue the work in this new field of research with a far more ambitious experiment dedicated to the detection of gamma-ray sources in the energy region 10^{14}–10^{17} eV. This experiment was to be run at an altitude of 2200 m, much better than the sea-level of their existing facility, and at a site far south of Kiel. On La Palma, one of the Canary Islands, the summit of Roque de los Muchachos, already home to several major astronomical observatories, would be a suitable site for installation of the new Kiel High Energy Gamma Ray Array (HEGRA).[274] This choice was a brilliant stroke of luck—or perhaps thanks to the prescience of Claus Allkofer; for even though,

273 Abdus Salam: Astro-Particle-Physics. In: Remo Ruffini (ed.): *Proceedings of the Fourth Marcel Grossmann Meeting on General Relativity, Held at the University of Rome La Sapienza, 17–21 June, 1985*. Elsevier Science 1986, 3–7. Abdus Salam had recently been awarded the Nobel Prize in Physics 1979, jointly with Sheldon Lee Glashow and Steven Weinberg, for his contribution to the theory of the unified weak and electromagnetic interactions, and was still working on the extension of unification of fundamental forces, including articles on magnetic monopoles and supersymmetry. It appears that the first to use the term in a printed source was Gary Steigman, as early as 1984, mentioning the 'astro-particle physics community' in a review article on a book on the very early Universe. Gary Steigman: Inflationary Cosmology. *Nature* 309/5967 (1984), 473–474. doi:10.1038/309473a0. Interestingly, especially the Russian tradition uses the term 'cosmoparticle physics,' which is related to the tradition of studies originating in the works of Andrei Sakharov, Ya. B. Zeldovich, Moisey A. Markov, and their scientific schools, whose approach naturally embedded since the 1960s the study of fundamental links between cosmology and particle physics, as presented in Maxim Yu Khlopov: *Cosmoparticle Physics*. Singapur: World Scientific Publishing 1999. In this regard, see, for example, Zeldovich, The Universe as a Hot Laboratory, 1970, 12–17. Zeldovich compared the early Universe—a superhot, superdense state of matter—to "the poor man's accelerator." In "extrapolating physical laws to energies 10^{15} times larger than those achieved in the most powerful accelerators," astronomical knowledge could be applied "to find (or at least to constrain) the fundamental laws of physics in regions inaccessible to direct experiments." Yakov Borisovich Zeldovich: Cosmology from Robertson to Today. *Physics Today* 41/3 (1988), 27–29, 29. doi:10.1063/1.881146.

274 At this latitude, Cygnus X-3 and Hercules X-1 could be detected and, in addition, three further candidate sources were in the observation field: the Crab Nebula, Geminga (whose nature was still quite unknown), and the pulsar PSR 1937, all three known as gamma-ray emitters around 10^{12} eV. Information about the first stage of the Kiel array HEGRA, by that time under construction, was given in Otto Claus Allkofer, M. Samorski, and W. Stamm: The Kiel EAS Experiment for Detecting UHE Gamma Rays. In: Turver, K.E. (ed.): *Very High Energy Gamma Ray Astronomy. Proceedings of the NATO Advanced Research Workshop, Durham, England, Aug. 11–15, 1986*. Dordrecht: Springer 1987, 281–284. doi:10.1007/978-94-009-3831-1_44.

at the time, the proposed cosmic ray experiments had nothing to do with the optical observational quality of the site, it was precisely this, five years later, that made the Roque de los Muchachos an obvious location for the Cherenkov telescopes, too. The preparation for the new air shower experiment started in Kiel in 1986, and the construction on La Palma in April 1988. The initial HEGRA array consisted of 37 scintillator detectors which went into operation in July 1988.[275]

The Max Planck Society's Return to Air Shower Arrays: Munich and Heidelberg at HEGRA

The Kiel group's claims that very energetic gamma-rays from Cygnus X-3 produced copious hadronic showers attracted the attention of several elementary particle physicists. Eckart Lorenz, then at the Max Planck Institute for Physics, after a successful career at CERN and other laboratories,[276] was fascinated by the Kiel announcements. Lorenz later recalled his scientific 'leap,' from the underground tunnels hosting CERN accelerators to the highest mountaintops:

> I remember it perfectly: I was standing in the dark, in the CERN grounds, and a colleague approached me and said, "Somewhere up there is this Cygnus X-3. And there you have it: new physics." Then I said: "I'm interested in what is going on up there…" [our translation].[277]

Particle physicists, accustomed since many years to working on immense experiments in huge collaborations, were bringing their skills and strategies—and of course novel technologies—to astrophysical projects rapidly growing in size and complexity, hopefully rejuvenating the field of particle physics itself and expanding its boundaries. The Max Planck Institute for Physics thus

275 As in the old Kiel experiment, the detectors were scintillation counters of 1 m^2 area each, with two photomultipliers, one for particle density and the other for fast-timing measurements.

276 Razmik Mirzoyan, and Christian Spiering: Nachruf auf Eckart Lorenz. *Physik Journal* 13/12 (2014), 50–50. www.pro-physik.de/details/articlePdf/7074441/issue.html. Last accessed 8/14/2018. Razmik Mirzoyan: Eckart Lorenz 1938–2014. *CERN Courier* 55/1 (2015), 40–40. https://cds.cern.ch/record/1983144. Last accessed 12/18/2018.

277 Dirk H. Lorenzen: Stimmen zum Gammateleskop MAGIC. *Physik der Welt*, 1/14/2004. https://www.weltderphysik.de/gebiet/universum/kosmische-strahlung/detektoren/magic/stimmen-zu-magic/. Last accessed 8/14/2018.

joined the HEGRA project with the initial intention of increasing the scintillator array by a significant factor.[278]

The 37 detectors of the initial HEGRA array planned by the Kiel group were twice read out for control purposes, from July 1989 to November 1990, by two independent electronic systems. The collected data were analyzed, including potential objects for which claims for very-high-energy or ultra-high-energy gamma-ray emission existed; but there was not

> the slightest indication for an excess from any of the 9 sources. Especially for Cygnus X-3 (and some other sources), even less showers were detected than due to the expected average background.[279]

The search for ultrahigh-energy gamma-ray emission from the direction of sources observed from the satellite COS-B were likewise unsuccessful. Moreover, by April 1991, the data gathered by Jim Cronin's array CASA-MIA (the Chicago Air Shower Array—Michigan muon Array), at the time the EAS experiment with the highest sensitivity, put a stringent upper limit to the signal

---

278 See Annual Report for 1988 of the Institute for Physics and Astrophysics, where the HEGRA experiment is mentioned for the first time ("Suche nach kosmischen gamma-Quellen oberhalb 10^{14} eV" [Search for cosmic gamma sources above 10^{14} eV]), and the planned scintillation counters as well as the scheme for the enlarged array are shown (AMPG, IX. Abt. Rep. 5, No. 630, pp. 69–81). The official beginning of the collaboration was announced in the Annual Report for the year 1989. Together with the MPIP, it included the Universities of Kiel and Madrid, but it was also announced that new groups would join from the Universities of Hamburg, Wuppertal, and Yerevan, contributing several improvements which would enhance the sensitivity of the HEGRA detecting system. The starting group at MPIP was formed by M. Bott-Bodenhausen, I. Holl, H. Fischer, A. Karle, E. Lorenz, C. Sesena. In the preliminary phase, they would build a detector formed by an array of 169 scintillation counters with a surface each of 1 m^2, distributed over a square surface of 135 m x 120 m (Max-Planck-Institut für Physik und Astrophysik. Werner-Heisenberg-Institut für Physik. Jahresbericht 1989, 55–59. AMPG, IX Abt. Rep. 5, No. 631). From June 1989, a second array of 66 counters (similar in type to the Kiel counters), from the Munich Institute and the University of Madrid began to take data, also reading out the signals of the Kiel detectors.

279 Otto Claus Allkofer et al.: Results of the HEGRA Experiment. In: M. Nagano, and F. Takahara (eds.): *Astrophysical Aspects of the Most Energetic Cosmic Rays. Proceedings of the ICRR International Symposium, Kofu, Japan, 26–29 November 1990*. Singapore: World Scientific 1991, 200–211, 203. See in particular table 1 and table 2. For data collected from summer 1989 to late spring 1991, see Victoria Fonseca: The HEGRA Experiment (High Energy Gamma Ray Array). *Nuclear Physics B—Proceedings Supplements* 28/1 (1992), 409–412. doi:10.1016/0920-5632(92)90205-7.

from Cygnus X-3, which ruled out this source excluding earlier observations.[280] Cronin, a former accelerator physicist, had switched to the study of cosmic rays shortly after being awarded, together with Val L. Fitch, the Nobel Prize in Physics 1980, for the discovery of a slight asymmetry between matter and antimatter known as CP violation. Cronin, too, had become intrigued by the Kiel report and, leading a team from the Universities of Chicago and Michigan, had proposed that the large air shower array CASA-MIA search for high-energy gamma-ray sources, in particular for signals from Cygnus X-3.[281] His array, which went into operation in early 1990, had pushed to the limit the possibility of a point source of cosmic rays, like Cygnus X-3, almost a factor of 100 lower than the original reports. So, it appeared that small experiments like the HEGRA/Kiel array, employing electron detectors only, would have no chance of finding sources, even if run for many years. The only alternative to the situation was to make HEGRA much larger and more sophisticated, increasing the detection area and, at the same time, the angular resolution, both by using many more electron detectors with fast-timing facilities, and adding large-area muon and Cherenkov light detectors to suppress the background showers. And this is the way the HEGRA experiment eventually evolved, becoming the most comprehensive instrument for ground-based gamma-ray astronomy, with a unique combination of detector capabilities. By early 1990, a proposal for extension of the installation at La Palma was in progress under the name of 'HEGRA Collaboration,' now including seven groups of scientists from German and Spanish institutions, and supported by funds from the Deutsche Forschungsgemeinschaft (DFG, German Research Association) and the Land Schleswig-Holstein.[282]

280 A. Borione et al.: High Statistics Search for Ultrahigh Energy Gamma-Ray Emission from Cygnus X-3 and Hercules X-1. *Physical Review D* 55/4 (1997), 1714–1731. doi:10.1103/PhysRevD.55.1714. The stimulus for CASA-MIA, an array on a different scale, in terms of numbers of detectors, from anything built previously, had come from the Kiel claim. However, as Cronin recalled, "The astronomy division of the National Science Foundation just couldn't conceive of this, and as for particle physics, it's a stretch of imagination saying we were doing particle physics. Gary Taubes: Astronomers Turn New Eyes On the Cosmic Ray Sky. *Science* 259/5092 (1993), 177–179, 178. doi:10.1126/science.259.5092.177.

281 Cronin's decision to build the shower array CASA-MIA was *directly* related to Samorski and Stamm's claim, as confirmed by Joachim Trümper to the author L.B., 01/10/2021).

282 The collaboration included the Department of Atomic Physics and Astrophysics of the Universidad Complutense of Madrid, the University of Hamburg II (Institute for Experimental Physics), Institute for Pure and Applied Nuclear Physics of Kiel University, the University of Wuppertal, *and* the Max Planck Institute for Physics and Astrophysics in Munich. See the HEGRA detector arrangement with Kiel's detectors already in operation

Meanwhile, the particle physicist Werner Hofmann, recently appointed Director at the Max Planck Institute for Nuclear Physics in Heidelberg,[283] had witnessed the enormous interest triggered by the observation of the Kiel group of an excess of high-energy extensive air showers from the direction of Cygnus X-3—in particular its puzzling muon content—and subsequent revival of this field of cosmic ray physics.[284] Starting from October 1988, in parallel with his intense high-energy physics research at accelerators, Hofmann began to set up his own cosmic ray project (Cosmic Ray Tracking), in collaboration with the Physics Institute of Heidelberg University.[285] And so, besides improving the

---

since July 1988, and the Hamburg-Munich-Madrid set-up under construction in Fig. 1, on p. 346 of Otto Claus Allkofer et al.: The HEGRA Project for UHE Gamma Ray Astronomy. *Nuclear Physics B—Proceedings Supplements* 14/1 (1990), 345–347. doi:10.1016/0920-5632(90)90443-X. There were plans to extend the array to more than 200 detectors within 1991. The next important step would be to supplement the array with large-area muon track detectors also improving the angular resolution of the experiment.

283 Hofmann's call was related in part to Peter Brix's retirement (CPTS meeting minutes of 27.06.1984, 18.09.1984, AMPG, II. Abt., Rep. 62, No. 1802, 1803), but also to plans for the future developments at the Max Planck Institute for Nuclear Physics, which were to tackle new fundamental questions in the realm between nuclear and elementary particle physics (CPTS meeting minutes of 04.02.1987, AMPG, II. Abt., Rep. 62, No. 1810). Hofmann, who had been working since 1982 at Lawrence Berkeley Laboratory in the US, with a focus on heavy quark phenomena, appeared to be the right person for reorienting experimental research, definitely shifting the focus from the more 'classical' nuclear physics, moving up in energy towards the boundary between nuclear physics and high-energy physics, and to the next stage, the quark level.

284 J. Heintze et al.: Measuring the Chemical Composition of Cosmic Rays at ~10^{13}-10^{15} eV by Utilizing the Solar and Geomagnetic Fields. *Experimental Astronomy* 1/1 (1989), 21–34. doi:10.1007/BF00414793. Hofmann himself emphasized how "At that time, high-energy astrophysics was dealing essentially with the same processes, the same physics [...] It was evident that there were interesting physics questions and with the particle physics knowledge and technology, you could move things on a financial scale which was visibly modest." For such physicists, building such detectors for cosmic ray-physics—having the in-house capacity to do so—was an interesting challenge ("but not a big deal") and they also liked to experience smaller-scale research groups. Werner Hofmann: interview by Luisa Bonolis and Juan-Andres Leon, Heidelberg, November 10–11, 2017. DA GMPG, BC 601010.

285 The proposal for an Extensive Shower Array appeared as Heidelberg Report HD-PY 88/05 (related to the parallel proposal to the Bundesministerium fur Forschung und Technologie "Cosmic Ray Tracking—ein neuer Weg für die gamma-Astronomie bei höhen Energien"). See also, Annual Report of the Max-Planck-Institut für Physik und Astrophysik. Werner-Heisenberg-Institut für Physik. Jahresbericht 1989, 65–69 (AMPG, IX Abt. Rep. 5, No. 632). The Cosmic Ray Tracking project was based on the measurement of individual cosmic ray tracks and muon identification using large-area drift chambers, applying an electronic system suitable for the read-out of several thousand channels,

sensitivity for point sources, the aim of the proposal was to fill the gap left by the Air Cherenkov Telescopes and the Extensive Air Showers detectors with a new type of EAS array based on the measurement of the direction of *individual* shower particles.[286] As we will see, Hofmann's step into the realm of particle astrophysics would be a premise for the later involvement of Heidelberg's Institute for Nuclear Physics in the HEGRA project.

By the beginning of the 1990s, it was definitely clear that cosmological and astrophysical observations were a valuable complement to accelerator experiments. High-energy gamma-ray astrophysics could explore energy regions beyond the reach of accelerators. In October 1993, due to budget problems, the US Congress officially canceled the Superconducting Super Collider (SSC) project, after about ten years of planning and some 2 billion dollars already spent.[287] It is interesting to recall here Jim Cronin's comment about the different scales of estimated costs for building a big accelerator, in comparison with those for a large cosmic ray project like CASA-MIA, which he had successfully constructed and operated: 50–60 million dollars would be "only 10 percent of an SSC detector and 1 percent of the cost of the SSC itself." Taken as "dollars per electron volt," it might "sound like a bargain."[288]

Not long after the cancellation of SSC, a "small and simple" space-borne cosmic-ray detector was proposed by the high-energy physicist Samuel Ting, who had been awarded the Nobel Prize in Physics 1976, jointly with Burton Richter, for the discovery of a new heavy particle known as J/Ψ, which confirmed the existence of the charmed quark. The new particle physics experiment in space, coming out of high-energy physics, was meant as an alternative project to the large coalition for a detector at SSC, put together by Ting, which

---

techniques similar to those they had successfully used in track-detectors at electron-positron storage rings (JADE at PETRA/DESY, TPC at PEP/SLAC, OPAL at LEP/CERN).

286 J. Heintze et al.: Cosmic Ray Tracking—A New Approach to High-Energy γ-Astronomy. *Nuclear Instruments and Methods in Physics Research Section A: Accelerators, Spectrometers, Detectors and Associated Equipment* 277/1 (1989), 29–41. doi:10.1016/0168-9002(89)90532-9. The performance of the system proposed had been simulated through an extensive Monte Carlo study. In total, 385 detector modules would form the EAS array at mountain altitude. M. Feuerstack et al.: Cosmic Ray Tracking. A New Approach to High-Energy γ-Astronomy. *Nuclear Instruments and Methods in Physics Research A* 310/1–2 (1991), 287–291. doi:10.1016/0168-9002(91)91045-W. In the planned cosmic-ray experiment, the electronic system would be able to record up to 50 tracks in one chamber. All the information would be transferred to a central electronic station for calculation of shower parameters.

287 Michael Riordan, Lillian Hoddeson, and Adrienne W. Kolb: *Tunnel Visions: The Rise and Fall of the Superconducting Super Collider*. Chicago, IL: University of Chicago Press 2015.

288 Taubes, Astronomers, 1993, 177–179, 179.

had been rejected in 1991. Similarly, CERN had rejected Ting's proposal for an experiment at the future Large Hadron Collider, an upgraded version of the successful L3 experiment he had led at the Large Electron-Positron Collider. The Alpha Magnetic Spectrometer (AMS) experiment, to be hosted at the International Space Station (ISS), was considered controversial,[289] but the context in which it was born, and its main scientific goals, both in the domain of astrophysics (cosmic-ray origin, age, and propagation, as well as the exploration of the most energetic gamma-ray sources) and in the domain of particle astrophysics (the search for cosmic antimatter and dark matter), are a further relevant example of the migration of high-energy physicists—Ting, like Jim Cronin, was actually a high-profile refugee from the world of particle accelerators—now taking advantage of the opportunities offered by the Universe as a 'great cosmic accelerator.'

Similarly, particle physicists awaiting the planned yet still remote future Large Hadron Collider were hoping to get answers from Imaging Atmospheric Cherenkov telescopes (IACT), the new type of Cherenkov detectors working on high-energy physics at the TeV scale, which could not be addressed by accelerators. In the months following the first ever observation, in February 1987, of a burst of neutrinos from the explosion of the Supernova 1987A, the possibility of detecting very-high-energy gamma rays from this source was likewise examined. However, within a year, it was the Crab Nebula, the result of a supernova explosion first recorded by Chinese astronomers in 1054, which came once again to the fore. The group working at the Whipple Observatory at Mount Hopkins, Arizona, submitted to the *Astrophysical Journal* a paper announcing the observation of a steady flux of gamma rays above 0.7 TeV from this source, at a very high level of statistical significance.[290] Such a 9-sigma signal had been recorded using a refined version of the Imaging Atmospheric Cherenkov Technique, the Imaging Air Cherenkov Telescope (IACT), in which a camera containing an array of photo multipliers is placed in the focal plane of a large mirror.[291] After so many statistically suspicious and controversial

289 Eric Hand: Particle Physics: Sam Ting's Last Fling. 7215. *Nature* 455/7215 (2008), 854–857. doi:10.1038/455854a.

290 Trevor C. Weekes et al.: Observation of TeV Gamma Rays from the Crab Nebula Using the Atmospheric Cerenkov Imaging Technique. *Astrophysical Journal* 342 (1989), 379–395. doi:10.1086/167599.

291 Gamma-ray showers can be discriminated from the overwhelming background due to cosmic ray-initiated air showers based on the image shape and orientation. Gamma-ray images from purely electromagnetic cascades appear as narrow, elongated ellipses in the camera plane. The long axis of the ellipse corresponds to the vertical extension of the air shower and points back towards the source position in the field of view.

'discoveries' of gamma ray sources, the Whipple Telescope result represented the first uncontroversial detection of a source of TeV gamma rays, completely changing the future research perspective of high-energy gamma-ray astronomy.[292]

In order to fully grasp the significance of this breakthrough, one has to go back to the 1960s, when, on the ground, resources for detecting gamma rays with the Cherenkov Technique were practically nonexistent, and it took many years with the detection techniques then available to solve several fundamental problems, step by step. This was "largely due to inadequate instruments, slowly developing theories about particle interaction, slowly oncoming additional information from accelerator experiments and the lack of powerful computers."[293] Astrophysics depends on theory and modeling to a greater degree than most other physical sciences, because observations can be done only remotely. The ability of astrophysicists to extract physical insight from observational data necessarily relies on more powerful computers and computer programs that incorporate realistic physics. The rapid advance of computers, and the concurrent development of analytical techniques mentioned earlier, revitalized the community. But the 1970s, too, saw little progress and very few gamma-ray observatories were in operation. Only in the 1980s, when also computing power increased enormously and several major programs were developed for the interpretation of data and modeling, did the field of ground-based gamma-ray astronomy start to look promising to outsiders.

After decades of slow improvements, the American team led by Trevor Weekes, who had initiated gamma-ray studies with a reflector in 1968, started developing the 'imaging' technique in 1981.[294] By 1985, Michael Hillas's Monte Carlo simulations indicated that it should be possible to discriminate between the gamma-ray and proton-induced showers, rejecting up to 97 percent of the

292 In this regard, as recalled by Lorenz and Wagner later, "The community followed a suggestion of Trevor Weekes that observed sources were accepted as discoveries only if their significance exceeded 5 sigma and all sources on the sky map were at least confirmed by one other experiment." Lorenz, and Wagner, Very-High Energy Gamma-Ray Astronomy, 2012, 459–513, 492.

293 Lorenz, and Wagner, Very-High Energy Gamma-Ray Astronomy, 2012, 459–513, 462.

294 Trevor C. Weekes: A Fast Large Aperture Camera for Very High Energy Gamma-Ray Astronomy. *17th International Cosmic Ray Conference (ICRC 1981), 13–25 Jul 1981. Paris, France.* Gif-sur-Yvette, France: Centre d'Etudes Nucleaires, Saclay 1981, 34–37. https://ui.adsabs.harvard.edu/#abs/1981ICRC....8...34W. Last accessed 12/2/2018. For the coming of age of the IACT technique, after many years of investigation and instrumentation development at Whipple Observatory, see Fegan, *Cherenkov Reflections*, 2019.

background events.[295] This was indeed the really effective strategy for tackling the overwhelming and unwanted background of charged cosmic rays: by comparing the real-data parameter distributions with those of the gamma-ray and hadronic shower simulations, it became possible to enhance the gamma-ray content of any set of observations. Promising results were finally reached in 1989, when the first robust detection of TeV gamma rays from an astrophysical object was announced, a steady 9-sigma gamma-ray signal from the Crab Nebula, obtained with the 10 m optical reflector of the Whipple Observatory, equipped with a fast camera for imaging Cherenkov light from Extensive Air Showers.[296]

As we already emphasized, the opening of this new observational window onto the cosmos occurred in concomitance with the gradual appearance, over the 1980s, of astroparticle physics, as well as with a crisis in accelerator-based physics related to the end of the Cold War

Stereoscopic Cherenkov Imaging: An Armenian Tradition with International Reach

But the end of the Cold War also gave a key boost to cosmic ray research in Europe, crucially aided by the involvement of Soviet researchers. In the mid-1980s, physicists at the prestigious Yerevan Physics Institute in Armenia also decided to move into the field, which was already being pursued in the Soviet Union by Arnold Stepanian in Crimea. Stepanian had been using a four-mirror

295 A. Michael Hillas: Cerenkov Light Images of EAS Produced by Primary Gamma Rays and by Nuclei. *Proceedings from the 19th International Cosmic Ray Conference, La Jolla, USA, August 11–12, 1985*. 1985, 445–448. https://ui.adsabs.harvard.edu/#abs/1985ICRC....3..445H. Last accessed 11/28/2018. Michael Hillas had started to work on simulations in the mid-1970s, when advances in the performance of computers had facilitated the Monte Carlo simulation of Extensive Air Showers on a new scale, improving significantly the signal-to-noise ratio.

296 Weekes et al., Observation of TeV Gamma Rays from the Crab, 1989, 379–395. Weekes, TeV Radiation, 1992, 315–364. As emphasized by Lorenz and Wagner, a main ingredient of the 'discovery' was the use, for the first time, of a camera allowing an efficient gamma/hadron separation of the data. The "third and most important achievement" was "the introduction of a refined gamma/hadron separation method based on the calculation of image moments," an analysis developed by the Whipple collaboration in the mid-1980s, based on the combination of the measurement of the shower image orientation proposed by Trevor Weekes in 1981, with an analysis to evaluate the difference in images between gamma-ray showers and hadron showers originally proposed by Stepanian and his group in 1983. Lorenz, and Wagner, Very-High Energy Gamma-Ray Astronomy, 2012, 459–513, 474.

system since the end of the 1960s,[297] but the more recent arrivals in the field felt that he, as an astronomer, had little particle physics insight and lacked credibility among the global community.

As mentioned earlier, the Yerevan Physics Institute in Armenia had begun developing the concept of stereoscopic approach in the mid-1980s, using the novel technique of multiple Imaging Atmospheric Telescopes.[298] These were also part of larger efforts in the field of high-energy cosmic rays, in the course of which a complex shower array site similar to HEGRA was to be installed on Mount Aragats.[299]

---

297 B. M. Vladimirskii et al.: Some Results of a Search for Point Sources of High-Energy Gamma Rays. *Soviet Astronomy* 16/1 (1972), 1–5. https://ui.adsabs.harvard.edu/#abs/1972SvA....16....1V. Last accessed 12/11/2018. Vladimirsky, Stepanian, and Fomin, High-Energy Gamma-Ray Outburst, 1973, 456–460. A. A. Stepanian et al.: A Search for Discrete Gamma-Ray Sources of Energy Greater than 2×10^{12} eV. *Astrophysics and Space Science* 38/2 (1975), 267–282. doi:10.1007/BF00647127. V. P. Fomin et al.: Angular Distribution of Cerenkov Light in EAS with Energy > 10^{13} eV. *18th International Cosmic Ray Conference, Bangalore, India, 22 August–3 September, 1983*. 1983, 223. https://ui.adsabs.harvard.edu/#abs/1983ICRC....6..223F. Last accessed 12/25/2018. Fomin et al., Angular Distribution, 1983, 223.

298 With two or more images of the same shower, a 3-dimensional reconstruction of the shower axis becomes possible. The first page of the proposal presented by the Yerevan Physics Institute in February 1985 and the scheme for the five Imaging Cherenkov Telescopes are reproduced in Razmik Mirzoyan: Early Days of Cherenkov Emission and Milestones. Presented at the SPSAS School, San Paulo, 5/23/2017, 44. http://www.astro.iag.usp.br/~highenastro/Talks/Lecture_III_Razmik_Mirzoyan_1.pdf. Last accessed 12/9/2018.

299 Their original idea had been to build a system of five imaging Cherenkov telescopes surrounded by an array of Cherenkov detectors. The planned array comprised about 50 scintillation detectors of 1 m^2 area each (to which further 150 detectors should be added) dislocated on an area of about 10^5 m^2, a second larger system of scintillators to register muons, and a system of about 30 spherical mirrors to collect the Cherenkov radiation which were to substitute a 10 m^2 surface area mirror, much more difficult to construct. The Cherenkov light collected by the mirror system would be detected by means of a set of 100 photomultipliers. As announced in 1987, the system of 5 Imaging Cherenkov telescopes (diameter 3 m), each consisting of 19 spherical mirrors (diameter 60 cm), was under construction on Mount Aragats, at Nor Amberd cosmic ray station, where cosmic ray research had begun already in 1934. Ashot A. Chilingarian, R. G. Mirzoyan, and M. Z. Zazyan: Cosmic Ray Research in Armenia. *Journal of Contemporary Physics (Armenian Academy of Sciences)* 44/5 (2009), 219–230. doi:10.3103/S106833720905003X. The Crimean Astrophysical Observatory installation for Cherenkov light consisted in two independent detectors with four 1.5 m parabolic mirrors, with photomultipliers placed at the focus of each mirror, and which were connected pairwise for coincidence. From 1972, they also carried out regular observations from Cygnus X-3. S. A. Agadjanyan et al.: Complex Installation for Investigation of Primary Gamma-Rays in Energy Range 10^{12}-10^{16} eV. *Proceedings of the 20th International Cosmic Ray Conference, 2–15 August 1987, Moscow*. 1987,

However, these plans were laid in the final years of the Soviet Union, and economic collapse paralyzed such efforts. Previous contact with Claus Allkofer at the Nuclear Physics Institute in Kiel led to Razmik Mirzoyan and electronic engineers from the Yerevan Institute being invited in 1990 to work with Samorsky and Stamm.[300] Allkofer unfortunately passed away in January 1990. Their initial bargaining chip was not the Cherenkov telescopes, but rather their access to military-grade scintillation detectors with which they offered to extend the collection area of the shower array. Before 1992, German interest in the Armenian group was largely due to such detectors, to be used to enlarge the HEGRA array. The Cherenkov array, alien to the tradition of the other participating groups, was a rather independent addition to the site, whose worth remained unproven during its first years of existence.

The final goal of the HEGRA Collaboration was now to build a large-area, multi-detector experiment that would enable the simultaneous measurement of extensive air showers and many shower parameters.[301] Plans were made by scientists from the Yerevan Physics Institute, the Crimean Astrophysical

332–335. http://adsabs.harvard.edu/abs/1987ICRC....2..332A. Last accessed 12/3/2018. See a reproduction of the original scheme of the five planned Imaging Cherenkov Telescopes in the proposal dated February 1985 in Razmik Mirzoyan: Brief History of Ground-Based Very High Energy Gamma-Ray Astrophysics with Atmospheric Air Cherenkov Telescopes. *Astroparticle Physics* 53 (2014), 91–99, 44. doi:10.1016/j.astropartphys.2013.11.004.

300 "We had special photomultipliers, at that time they were produced only a few hundred pieces per year, most of them disappeared for military purposes. We prepared very high-quality mirrors by ourselves... We prepared everything for building five telescopes. While we were commissioning the first telescope, the country decayed, Soviet Union decayed. At some moment, the situation became obsolete. I was working in a powerful institute and you ordered this and that and then suddenly there was nothing, we had a cut of electricity, and then there was confusion everywhere, supermarkets became empty. Everything was decaying, we could not continue like that. We remembered we had met Prof. Allkofer from Kiel, at the Institute for Nuclear Physics; we got in contact with him. He told us they were building an array at La Palmas, HEGRA: 'Why don't you join us?'" Razmik Mirzoyan: interview by Juan-Andres Leon, Munich, August 13–14, 2018. DA GMPG, BC 601021. In 1992, the level of state support decreased by two times that of the previous years and by 1994, the level of financing of Russian science was almost six times lower than in developed countries of the West. On the breakup of the Soviet Union and crisis in Russian science, see Loren Graham, and Irina Dezhina: *Science in the New Russia. Crisis, Aid, Reform*. Bloomington, IN: Indiana University Press 2008.

301 See a brief summary of the physics program in Eckart Lorenz: The HEGRA Experiment. In: P.C. Bosetti (ed.): *Trends in Astroparticle-Physics*. Wiesbaden: Vieweg+Teubner Verlag 1994, 139–151, 139. A substantial completion would occur between 1991–92, with 49 muon detector stations, allowing an effective suppression of the background of hadron-induced showers, thus helping to clarify whether the gamma-induced showers really had a much lower muon content than hadron-induced showers, as predicted by theory.

Observatory, and the Munich Max Planck Institute for Physics and Astrophysics to complement the HEGRA array by a system of five Imaging Air Cherenkov Telescopes, each consisting of a multi-mirror reflector of 5 m^2 collection area and a fast, 37-pixel camera in its focus. For each event, the light level in each pixel would be digitized and recorded. The darkness of La Palma and the large number of clear nights made the area one of the best sites for the atmospheric Cherenkov technique. The simultaneous operation of the ultra-high-energy air shower detector and the system of atmospheric Cherenkov light receivers would cover the energy spectrum from 10^{12} eV to 10^{17} eV (1–1000 TeV). Such a unique combination of detector systems was expected to help answer the open questions at the interface of high-energy gamma-ray astronomy and particle physics. The Armenian group was officially invited to participate in the HEGRA Collaboration in October 1990,[302] and on March 7, 1991, a first official agreement of cooperation between the Kiel and Yerevan institutes was signed by the two directors and by Wilhelm Stamm and Razmik Mirzoyan.[303] Armenia became independent in September that same year, 1991.

While Kiel had first made contact with the Armenians, and scientists were invited to Kiel for short periods in 1990, the center of gravity was starting to shift to Munich. In June 1990, Eckart Lorenz, the most senior researcher from the Max Planck Institute for Physics in the HEGRA Collaboration, invited Razmik Mirzoyan to Munich, and it was decided that the mechanical mountings of the detector systems would be redesigned and built at the Institute for Physics, while the imaging camera and electronics would be made in Kiel.[304]

302 An official letter of invitation was sent from the Institute for Nuclear Physics in Kiel, dated October 24, 1990, and signed by Manfred Samorski (Spokesman of the HEGRA Collaboration). As specified in the document, participants in the collaboration at the time were the Max Planck Institute in Munich and the Universities of Hamburg, Kiel, Madrid, Nottingham, Wuppertal. Courtesy of Razmik Mirzoyan.

303 The agreement implied the construction of a system of 5 imaging Cherenkov light receivers on La Palma, which might work as a standalone system, with the potential to be operated simultaneously with the HEGRA particle array: "By this unique combination of an extended particle array with detectors for atmospheric Cherenkov light, the observation of cosmic gamma-ray sources will be possible over an extended energy range from 10^{11} to 10^{17} eV." The document specified that it had originally been proposed that Yerevan should bring the mounts for the telescopes to La Palma. But due to the strong winds there, the mechanical construction was to be fortified in Germany, with the help of the Max Planck Institute for Physics in Munich. A copy of this document was kindly shown to the author J-A. L. by Razmik Mirzoyan.

304 A proposal for "Imaging Air Cherenkov Telescopes in the HEGRA Particle Array," dated May 31, 1991, was signed by F. A. Aharonian, A. G. Akhperjanian, A. S. Kankanian, R. G. Mirzoyan (Yerevan Physics Institute), A. A. Stepanian (Crimean Astrophysical Obser-

After many years of work in particle physics, Lorenz was now changing field, entering into cosmic rays and astroparticle physics: Lorenz "was someone with a vision," had an "immense energy," and "was the driving force in Munich," Mirzoyan recalls.[305] Eckart Lorenz was then setting up his shower-array-based contribution called AIROBICC,[306] the first version of which was completed in fall 1992, but he was already seeing the potential in the Cherenkov telescopes. The five telescopes for atmospheric Cherenkov light—of 5 m diameter, with 19 mirrors to be operated in the brand-new Imaging Atmospheric Cherenkov Technique—would actually extend the energy range of the experiment down to about 10^{11} eV. The Yerevan Institute of Physics, together with the Max Planck Institute for Physics in Munich and the Kiel cosmic ray group, were to take care of this last extension, fully transforming HEGRA in a very special multi-component EAS detector.[307]

HEGRA's first telescope was designed as a somewhat modified version of the first prototype of the five-telescope array the Armenians had planned to build

---

vatory), M. Samorski, W. Stamm (Institute for Nuclear Physics, University of Kiel), M. Bott-Bodenhausen, E. Lorenz, P. Sawalisch (Max Planck Institute for Physics and Astrophysics, Munich). This document was kindly shown by Mirzoyan to one of us (J.-A. L). On the presentation of the new set-up at the HEGRA site, see Felix A. Aharonian et al.: Status and Extensions of the HEGRA Detector on La Palma. *Proceedings of the 22nd International Cosmic Ray Conference. 11–23 August, 1991. Dublin, Ireland* 4 (1991), 452–455. https://ui.adsabs.harvard.edu/#abs/1991ICRC....4..452A. Last accessed 5/18/2018. Felix A. Aharonian et al.: A System of Air Cherenkov Telescopes in the HEGRA Array. *Proceedings of the 22nd International Cosmic Ray Conference. 11–23 August, 1991. Dublin, Ireland* 2 (1991), 615–617. https://ui.adsabs.harvard.edu/#abs/1991ICRC....2..615A. Last accessed 5/18/2018. Each telescope was planned to have a 3 m diameter tessellated mirror of 5 m^2 area, to be equipped with a 37-pixel imaging camera in the focal plane at 5 m.

305 Razmik Mirzoyan: interview by Juan-Andres Leon, Munich, August 13–14, 2018. DA GMPG, BC 601021. The MPIP Annual Report for 1990 mentioned that five Cherenkov Telescopes would be added at the HEGRA site, the first of which was under construction (Annual Report of the Max-Planck-Institut für Physik und Astrophysik. Werner-Heisenberg-Institut für Physik. Jahresbericht 1990, 80. AMPG, IX Abt. Rep. 5, No. 632). Unfortunately, Eckart Lorenz passed away on June 20, 2014. Mirzoyan, and Spiering, Nachruf auf Eckart Lorenz, 2014, 50–50.

306 M. Bott-Bodenhausen et al.: Airobicc–a New Array of Angle Integrating Cerenkov Counters for Improved γ/Hadron Separation in Extended Air Showers. *AIP Conference Proceedings* 220/1 (1991), 305–309. doi:10.1063/1.40314. The AIROBICC array, including 169 large diameter photomultipliers, directly viewing the night sky, was installed in 1994 inside the HEGRA scintillator array. A. Karle et al.: Design and Performance of the Angle Integrating Čerenkov Array AIROBICC. *Astroparticle Physics* 3/4 (1995), 321–347. doi:10.1016/0927-6505(95)00009-6. The array was eventually decommissioned after having been nearly destroyed in 2000 by a forest fire.

307 See status and planned extensions of the HEGRA detector array, showing the different detectors, in Fig. 1 of Allkofer et al., Results of the HEGRA Experiment, 1991, 200–211, 402.

at Nor Amberd in 1989.[308] It was installed in spring 1992. Until that moment, HEGRA had been an overlap of several types of detectors occupying the same space. The Cherenkov telescopes were just one more component of the mix, with the promise of extending the lower end of the detection energy range; and they comprised the only system based on direct observation of the interaction of gamma rays and the showers they cause with the atmosphere above, rather than on detection of the 'tail' end of the showers on the ground.[309]

The first Imaging Atmospheric Cherenkov Telescope at HEGRA, commissioned in the summer of 1992, confirmed the Crab Nebula as a source of very high-energy gamma rays, only two months after installation of the electronics and the imaging camera.[310] This detection put HEGRA on at least an equal footing with the Americans of Whipple Observatory, and boosted worldwide confidence in the new technique. But more significantly, the plans underway for an array of five telescopes promised to quickly make HEGRA the most advanced system in the world.

---

308 Felix A. Aharonian et al.: The System of Imaging Atmospheric Cherenkov Telescopes. The New Prospects for VHE Gamma Ray Astronomy. *Experimental Astronomy* 2/6 (1992), 331–344. doi:10.1007/BF00395984.

309 M. Bott-Bodenhausen et al.: A New Air Cherenkov Counter Concept for the Observation of Extended Air Showers. *Nuclear Instruments and Methods in Physics Research Section A: Accelerators, Spectrometers, Detectors and Associated Equipment* 315/1–3 (1992), 236–251. doi:10.1016/0168-9002(92)90709-D. See also results of Monte Carlo studies on the performance of the five telescope system. Aharonian et al., The System of Imaging Atmospheric Cherenkov Telescopes, 1992, 331–344. A further addition was the so-called 'muon towers,' or 'Geiger towers,' built with the objective to measure the local energy deposition and to reconstruct and identify muon tracks. Each tower could also serve as an independent high resolution muon flux monitor. Wolfgang Rhode et al.: Design and Performance of the Lead-Concrete Geiger Tower Array within the HEGRA Experiment. *Nuclear Instruments and Methods in Physics Research Section A: Accelerators, Spectrometers, Detectors and Associated Equipment* 378/3 (1996), 399–409. doi:10.1016/0168-9002(96)00488-3.

310 General results were based on data recorded from September 1992 to February 1993. F. Krennrich et al.: Observation of VHE γ-Emission from the Crab Nebula with the Prototype of the HEGRA Air Cerenkov Telescope Array. In: D. A. Leahy, R. B. Hicks, and D. Venkatesan (eds.): *23rd International Cosmic Ray Conference, Vol. 1, Held 19–30 July, 1993 at University of Calgary, Alberta, Canada*. 1993, 251–254. https://ui.adsabs.harvard.edu/#abs/1993ICRC....1..251K. Last accessed 12/11/2018. A preliminary estimate of the flux was in agreement with extrapolations from the Whipple data. Razmik Mirzoyan et al.: The First Telescope of the HEGRA Air Cherenkov Imaging Telescope Array. *Nuclear Instruments and Methods in Physics Research Section A: Accelerators, Spectrometers, Detectors and Associated Equipment* 351/2 (1994), 513–526. doi:10.1016/0168-9002(94)91381-1. For a description of the HEGRA project, see also the section "Hochenergetische Strahlungsquellen im Universum" in Max-Planck-Gesellschaft (ed.): *Max-Planck-Gesellschaft. Berichte und Mitteilungen 1/93. Max-Planck-Institut für Physik München*. München 1993.

A Second Entry Point for Heidelberg

The detection by the Whipple Observatory of the Crab as a steady source of TeV gamma rays—which, like the replication of this detection at HEGRA, had put the field of Very-High-Energy gamma-ray astronomy on a firm observational basis—had also triggered the interest of Heinrich Völk at the Institute for Nuclear Physics in Heidelberg (see Chapters 3 and 4 for more on his trajectory in the Max Planck Society). Throughout his career in the United States, Munich, and Heidelberg, Völk had investigated the question of acceleration mechanisms of cosmic rays from a theoretical perspective,[311] which led him in turn to consider this problem in the light both of the connection to gamma-ray sources[312] and the possible role, in the production of gamma rays, of cosmic rays penetrating a dense interstellar medium (the so-called molecular clouds), a research question raised by COS-B observations.[313] As Völk himself has emphasized,

> One knows that essentially all the known universe is filled with this non-thermal component… That's what I called "the non-thermal Universe." And so, our idea was to study the non-thermal Universe, which in other terms you could call cosmic rays… but which is much, much more than just what people call cosmic rays…

---

311 The roots of this interest go back to the 1960s, when Völk was still based at the Institute for Extraterrestrial Physics, before moving in 1975, as Director, to the Heidelberg Institute for Nuclear Physics, and he applied his plasma physics background to cosmic-ray physics to understand the transport of cosmic rays in turbulent interstellar medium or in solar wind. In 1982, the 'Kosmochemie' section was renamed 'Kosmophysik' in the Annual Report. The origin of cosmic rays as well as the sources of gamma rays were widely discussed: Generalverwaltung der Max-Planck Gesellschaft (ed.): *Max-Planck-Gesellschaft Jahrbuch 1982*. Göttingen: Vandenhoeck & Ruprecht 1982, 534. All these developments were going on in parallel with the development of the GALLEX project for the detection of solar neutrinos. At the same time, Völk's interest in gamma-ray astronomy had been aroused because of Pinkau's involvement at the Institute for Extraterrestrial Physics in the satellite COS-B. Völk, Heinrich: interview by Luisa Bonolis and Juan-Andres Leon, Heidelberg, October 9–10, 2017. DA GMPG, BC 601037.

312 Heinrich J. Völk, and M. Forman: Cosmic Rays and Gamma-Rays from OB Stars. *The Astrophysical Journal* 253 (1982), 188–198. doi:10.1086/159623. Heinrich J. Völk: Cosmic-Ray Acceleration and Transport, and Diffuse Galactic γ-Ray Emission. *Space Science Reviews* 36/1 (1983), 3–25. doi:10.1007/BF00171897.

313 Catherine Cesarsky, and Heinrich Völk: Cosmic Ray Penetration into Molecular Clouds. *Astronomy and Astrophysics* 70 (1978), 367–377. https://ui.adsabs.harvard.edu/#abs/1978A&A....70..367C. Last accessed 5/16/2018.

In this perspective, *"the combination of plasma astrophysics and particle physics is basically gamma-ray astronomy"* [our emphasis].[314] Considering the high energies that the gamma rays achieve, thermal mechanism cannot be responsible for their production and one needs to evoke non-thermal processes to explain their origin. Such non-thermal radiation, typical of all gamma-ray sources, may originate from different processes, in particular from the interaction of non-thermal particle populations with photons and matter. In this sense, as Völk emphasized, gamma rays provide a window onto the non-thermal physics in our Universe, a view on a variety of objects—such as neutron stars, black holes, stellar explosions and the remnants thereof—which emit a significant fraction of their energy through non-thermal processes. This *non-thermal* astrophysics-oriented research became a basic hallmark of the Heidelberg Institute.

During a trip to Chicago, Völk met Felix Aharonian, the theoretician and director of the Armenian team, who shared with him very similar theoretical interests.[315] Völk was very impressed and thus invited Aharonian to Heidelberg in 1993, where they started collaborating.[316] Aharonian in turn brought with him the team's leading engineer, Ruben Kankanian. In July 1993, at the 21st International Cosmic Ray Conference, the official announcement of the observation of a very-high-energy gamma emission from the Crab with the first of the HEGRA Air Cherenkov Telescope Array was cosigned by physicists from Heidelberg.[317]

Since the late 1980s, as we have already seen, Werner Hofmann's group was proposing at MPIK—in collaboration with the Institute for Physics of the University of Heidelberg—the Cosmic Ray Tracking (CRT) project, a new type of extensive air shower array, the basic concept and construction and performance details of which were presented at various conferences.[318] Their

---

314 Heinrich Völk: interview by Luisa Bonolis and Juan-Andres Leon, Heidelberg, August 9–10, 2017. DA GMPG, BC 601037.

315 Felix A. Aharonian: Very High Energy Gamma-Ray Astronomy and the Origin of Cosmic Rays. *Nuclear Physics B—Proceedings Supplements* 39/1 (1995), 193–206. doi:10.1016/0920-5632(95)00022-2.

316 See, for example, Felix A. Aharonian, and H. J. Völk: Very High Energy Gamma-Ray Astronomy with Ground-Based Imaging Cherenkov Telescopes. In: W. Wamsteker, M.S. Longair, and Y. Kondo (eds.): *Frontiers of Space And Ground-Based Astronomy*. Dordrecht: Springer Science & Business Media 1994, 705–706. doi:10.1007/978-94-011-0794-5_118.

317 Krennrich et al., Observation of VHE γ-Emission from the Crab Nebula, 1993, 251–254.

318 J. Heintze et al.: The Heidelberg Cosmic Ray Project—Aims and Status. *Nuclear Physics B—Proceedings Supplements* 14A/1 (1990), 148–152. doi:10.1016/0920-5632(90)90411-M. J. Heintze et al.: The Heidelberg Cosmic Ray Tracking Project—A New Approach to High

Cosmic Ray Tracking system, based on the high-energy physics detector technology of a time projection chamber, promised to improve the sensitivity for the detection of point sources by about a factor about 100 greater than existing conventional EAS arrays or Atmospheric Cherenkov Telescopes, which extended the energy range to a few TeV, while there were indications for gamma point sources at higher energies. The proposed project was expected to bridge the observational gap for cosmic rays in the energy range from TeV to PeV, opening a new energy window in astronomical observation and the potential to discover new phenomena in high-energy physics. Initially, the proposed location was the astronomical observatory at Llano del Hato, close to Merida in the Venezuelan Andes, at 3600 m above sea level. But in May 1992, when presenting the design and construction of the first full-size detector module, which was already running in coincidence with a small scintillator array in Heidelberg, it was announced that a series of ten full-size prototypes was under construction to form a "realistic small array," and would be later moved to La Palma, to be *tested* in the HEGRA EAS array.[319] In January–February 1993, the first two detectors of the CRT project were delivered and

---

Energy γ—Astronomy. In: Protheroe, R. J. (ed.): *21st International Cosmic Ray Conference (ICRC 1990), 6–19 January 1990. Adelaide, Australia.* 1990, 266–269. http://adsabs.harvard.edu/abs/1990ICRC....4..266H. Last accessed 12/1/2018. J. Heintze: Cosmic Ray Tracking for Gamma-Ray Astronomy. In: O. Fackler, and J. Tran Thanh Van (eds.): *New and Exotic Phenomena '90. Proceedings, 25th Rencontres de Moriond, 10th Moriond Workshop, Les Arcs, France, January 20–27, 1990.* Gif-sur-Yvette, France: Editions Frontières, 405–410. https://inspirehep.net/record/309842. Last accessed 11/30/2018. Interestingly, the Time Projection Chamber itself can be considered a successor of systems developed in Kiel in the 1950s. See Peter Galison: *Image and Logic. A Material Culture of Microphysics*. Chicago: The University of Chicago Press 1997, 469–470. The planned CRT array of about 400 large-area drift-chambers, to be installed at an altitude of more than 3000 meters, would be sensitive in the energy range from the domain of the Atmospheric Cherenkov Telescopes to that of conventional Extensive Air Shower Arrays.

319 Successful operation of this prototype would lead to a detailed proposal for the funding agencies by mid-1992. M. Feuerstack et al.: Cosmic Ray Tracking—a New Approach to High-Energy γ-Astronomy. *Nuclear Instruments and Methods in Physics Research Section A: Accelerators, Spectrometers, Detectors and Associated Equipment* 315/1 (1992), 257–259. doi:10.1016/0168-9002(92)90711-C. In December 1992, the Merida Observatory in Venezuela was still indicated as a suitable site for the full-scale array in D. Brecht et al.: Cosmic Ray Tracking—a New Approach to High-Energy γ-Astronomy. *Nuclear Instruments and Methods in Physics Research Section A: Accelerators, Spectrometers, Detectors and Associated Equipment* 323/1 (1992), 60–64. doi:10.1016/0168-9002(92)90269-A. By spring 1995, ten detectors had been operating for a couple of years on the HEGRA site. The detector performance had been evaluated using data from the detectors and from the HEGRA array, as well as by Monte Carlo simulations. Konrad Bernlöhr et al.: The Cosmic Ray Tracking (CRT) Detector System. *Nuclear Instruments and Methods in Physics*

installed in La Palma. They could be operated standalone, as well as together with the HEGRA array.[320] At the 23rd International Cosmic Ray Conference (ICRC), held in July 1993 in Canada, the aforementioned observation of the Crab with the first operative HEGRA Imaging Atmospheric Cherenkov Telescope was announced by a collaboration now also including the Heidelberg scientists.[321] By 1995, both the Max Planck Institutes for Physics and Nuclear Physics were fully participating in the multi-detector experiment HEGRA, and in August that year, Heinrich Völk and Werner Hofmann were among the signatories of the HEGRA Collaboration report on the results of observation of gamma rays from the Crab Nebula by the *second* HEGRA Imaging Atmospheric Cherenkov Telescope installed and taking data since February 1994.[322] The analysis of the observations during the period October 1994–March 1995 revealed a positive signal at 10-sigma confidence level, a remarkable result that was submitted to *Astroparticle Physics*, the journal founded in 1992 as a dedicated publication channel for this nascent field. In this way, two Max Planck directors, from Heidelberg, strongly supported the scientific operations of the HEGRA project, marking the beginning of a new era in which major experiments in the field were to be designed and run from the start by larger groups. This was a definite change of scale in Max Planck involvement, compared to the situation in Munich. Once again, as at several times in the history of the Cherenkov technique, newcomers with a significantly higher scale of resources and prestige seemed to be gradually taking over from a more modest,

Research Section A: Accelerators, Spectrometers, Detectors and Associated Equipment 369/1 (1996), 284–292. doi:10.1016/0168-9002(95)00772-5. Konrad Bernlöhr et al.: Operation and Performance of the Cosmic Ray Tracking (CRT) Detector System. *Nuclear Instruments and Methods in Physics Research Section A: Accelerators, Spectrometers, Detectors and Associated Equipment* 369/1 (1996), 293–305. doi:10.1016/0168-9002(95)00771-7.

320 A preliminary analysis of data obtained during the first weeks of operation was reported at the 23rd International Cosmic Ray Conference held in July 1993 in Canada. Konrad Bernlöhr et al.: Tracking Detectors for High Energy γ-Astronomy. In: D. A. Leahy, R. B. Hicks, and D. Venkatesan (eds.): *23rd International Cosmic Ray Conference, Vol. 4, Held 19–30 July, 1993 at University of Calgary, Alberta, Canada*. 1993, 199–202. https://ui.adsabs.harvard.edu/#abs/1993ICRC....4..199B. Last accessed 12/11/2018. A 10-detectors CRT test array at La Palma, although being too small for a search for cosmic gamma-ray sources, would enable them to test shower reconstruction algorithms under realistic conditions and to evaluate the performance that could be expected from the planned full-sized CRT array consisting of almost 400 modules.

321 Krennrich et al., Observation of VHE γ-Emission from the Crab Nebula, 1993, 251–254.

322 A. Konopelko et al.: Detection of Gamma Rays above 1 TeV from the Crab Nebula by the Second HEGRA Imaging Atmospheric Cherenkov Telescope at La Palma. *Astroparticle Physics* 4/3 (1996), 199–215. doi:10.1016/0927-6505(95)00044-5.

previous generation. The Heidelberg Institute contributed to building up the HEGRA–IACT system, which from 1998 consisted of five telescopes and thus could convincingly demonstrate the power of stereoscopic observations and the great potential of the technique; and it also served as the prototype for the third-generation instruments, guiding the evolution of TeV astronomy from a branch of cosmic ray studies into a full-fledged astronomical discipline.[323]

By the mid-1990s, the two originally 'nuclear' Max Planck Institutes in Munich and Heidelberg, which had participated in cosmic ray research since the middle of the century, before abandoning it during the space age, had again clearly attained global leadership in the field, thanks to their involvement in HEGRA and the absorption of experts such as Mirzoyan, Aharonian, and Kankanian from the Armenian group, and, in particular, their combined capacity to mobilize their in-house workshops for the construction of Cherenkov telescopes, a task beyond the ability of the smaller partners in HEGRA.

At the same time, there were already significant differences in the approach of each institute. Munich had been first on the scene, but was limited in scope by a '*Mittelbau*' (non-director) researcher in a struggling experimental department. Heidelberg had arrived later, and with the full power of not one, but two Max Planck directors. At the experimental level, there was also a divergence in scientific style: Munich's approach was based on the collaboration between Eckart Lorenz and Razmik Mirzoyan, both experimental physicists with an interest in understanding and innovating their instruments. Heidelberg was a more heterogeneous mix, with Völk and Aharonian on the theoretical end, Hofmann, a newcomer to Cherenkov astronomy, as the most important experimental physicist, with a wide expertise in instrumentation for high-energy physics, and technicians like Ruben Kankanian, who had a firm preference for stable, reliable instruments.

---

323 A. Daum et al.: First Results on the Performance of the HEGRA IACT Array. *Astroparticle Physics* 8/1 (1997), 1–11. doi:10.1016/S0927-6505(97)00031-5. HEGRA Collaboration et al.: Performance of the Stereoscopic System of the HEGRA Imaging Air Čerenkov Telescopes. Monte Carlo Simulations and Observations. *Astroparticle Physics* 10/4 (1999), 275–289. doi:10.1016/S0927-6505(98)00062-0. Niels Götting, and HEGRA Collaboration: Recent Results from HEGRA. *The European Physical Journal C—Particles and Fields* 33/1 (2004), 932–934. doi:10.1140/epjcd/s2004-03-1628-3. After running in its final form from 1998 to 2002, the HEGRA ACT system was switched off and efforts were redeployed to the bigger projects MAGIC and H.E.S.S. See also an arXiv preprint encompassing thirteen individual contributions of the HEGRA Collaboration at the 28th ICRC, highlighting results up to 2003. HEGRA Collaboration et al.: HEGRA Contributions to the 28th International Cosmic Ray Conference. arXiv:Astro-Ph/0307334, 2003. http://arxiv.org/abs/astro-ph/0307334. Last accessed 12/19/2018.

Parallel Continuation of HEGRA *as* H.E.S.S. *and* MAGIC

By the mid-1990s, this divergence in approach was expressing itself in tensions between the factions gravitating towards the different Max Planck Institutes. As a consequence of such divergences and personal differences, Heidelberg and Munich ended up taking separate paths when proposing the next generation and scale of ground-based Cherenkov telescopes: H.E.S.S. (High Energy Stereoscopic System) in the south,[324] and MAGIC (Major Atmospheric Gamma Imaging Cherenkov Telescopes) in the north.[325]

In the case of Heidelberg, the preference was for a conservative approach that would be a gradual continuation of the systems proven successful at the Whipple Observatory and HEGRA. This was to be an array of multiple, middle-sized telescopes consisting of easily manufactured steel structures, for the study of sources in the energy range between 100 GeV and 100 TeV. Such an array favored a site away from La Palma, in which Heidelberg had less of a stake anyway. The end of apartheid in South Africa and the independence of Namibia spelled a new opportunity to build on the Gamsberg, where the Max Planck Institute for Astronomy had planned to set up its observatory in the early 1970s (see Chapters 3 and 4). This would be the first gamma-ray observatory in the southern hemisphere.[326] In choosing the southern location, Heidelberg also gained strong backing from the French groups, which were to become the main collaborators. Finally, the choice between northern or southern hemisphere was related also to which kind of sources the new

---

324 Heinrich J. Völk: Max-Planck-Institut für Kernphysik–Astrophysik. Jahresbericht für 1997. *Mitteilungen der Astronomischen Gesellschaft Hamburg* 81 (1998), 469–484, 469. https://ui.adsabs.harvard.edu/#abs/1998MitAG..81..469V. Last accessed 9/26/2018. Werner Hofmann, and HESS Collaboration: The High Energy Stereoscopic System (HESS) Project. *AIP Conference Proceedings* 515/1 (2000), 500–509. doi:10.1063/1.1291416.

325 See March 1998 version of the design study (supported in part by a contract of the Bundesministerium für Bildung und Forschung, the Spanish Inter-ministerial Commission for Science and Technology, and the European Union), and related early publications. MAGIC Collaboration: *The* MAGIC *Telescope. Design Study for the Construction of a 17 m Čerenkov Telescope for Gamma-Astronomy above 10 GeV*, 1998. Eckart Lorenz: The MAGIC Telescope Project for Gamma Ray Astronomy in the 15 to 300 GeV Energy Range. *Nuclear Physics B Proceedings Supplements* 48 (1996), 494–496. doi:10.1016/0920-5632(96)00302-7. Razmik Mirzoyan: 17m Diameter MAGIC Telescope Project For Sub-100 GeV Gamma Ray Astronomy. *Nuclear Physics B Proceedings Supplements* 54/3 (1997), 350–361. doi:10.1016/S0920-5632(97)00134-5.

326 In fact, there are reminiscences that one of the early pioneers of ground-based gamma astronomy, Arnold Stepanian, had been setting up a pioneering system in Chile in the early 1970s, but this was interrupted by the right-wing military coup, after which all Soviet astronomers had to swiftly leave the country. Razmik Mirzoyan: interview by Juan-Andres Leon, Munich, August 13–14, 2018. DA GMPG, BC 601031.

astronomy was expecting to focus on: the southern hemisphere makes available most of our own galaxy, the Milky Way—and in particular, the galactic center—so was the obvious choice for the detection of potentially smaller, closer sources. Conversely, a northern location would be more limited to extragalactic sources, which would need to be more powerful to be detected; plus, in terms of accessible skies, a northern site would be in direct competition, but also constructive overlap, with the Americans.

In proposing two separate projects, the institutes entered into competition with each other. While the Munich researchers favored a larger detection dish and finer detector technology to obtain higher sensitivity,[327] the Heidelberg team, following on the already proven tradition with stereoscopic systems, went instead for an array of multiple, medium-sized dishes.[328] But the key to the worldwide leadership of the Max Planck Institutes at this scale was that both Heidelberg and Munich had enough in-house technical competence in their respective workshops to produce the experimental systems internally, with relatively minor financial help from the Max Planck Society.[329] In the mid-1990s, both Heidelberg and Munich had the independent capacity to create the world's most important projects in the field, in competition with each other. This 'competition' was, therefore, more a case of massive potential rooted in complementarity, brought about by the different experimental

327 The ambition was to collect enough photoelectrons to lower the threshold energy to about 20 GeV, thus giving overlap with the satellite detectors that had discovered a great number of point sources, but with much higher sensitivity to faint sources. MAGIC started normal operations, with the first telescope in 2004, and stereo observations with both telescopes in 2009. The twin ultra-large MAGIC telescopes, with 17 m diameter mirrors, were built incorporating several novel features. In this regard, it has been emphasized that "As it was obvious that a stereo system of such telescopes with so many new features would never have been funded in the first round, a second telescope was built only after the new items of the first one proved to work." Lorenz, and Wagner, Very-High Energy Gamma-Ray Astronomy, 2012, 459–513, 494.

328 Felix A. Aharonian et al.: The Potential of Ground Based Arrays of Imaging Atmospheric Cherenkov Telescopes. I. Determination of Shower Parameters. *Astroparticle Physics* 6/3 (1997), 343–368. doi:10.1016/S0927-6505(96)00069-2. Felix A. Aharonian et al.: The Potential of the Ground Based Arrays of Imaging Atmospheric Cherenkov Telescopes. II. Gamma Ray Flux Sensitivities. *Astroparticle Physics* 6/3 (1997), 369–377. doi:10.1016/S0927-6505(96)00070-9. Felix A. Aharonian et al.: On the Performance of Ground Based Arrays of Imaging Atmospheric Cherenkov Telescopes. *International Cosmic Ray Conference* 5 (1997), 109–113. https://ui.adsabs.harvard.edu/#abs/1997ICRC....5..109A. Last accessed 5/17/2018.

329 Heinrich Völk: interview by Luisa Bonolis and Juan-Andres Leon, Heidelberg, August 9–10, 2017. DA GMPG, BC 601037; Werner Hofmann: interview by Luisa Bonolis and Juan-Andres Leon, Heidelberg, August 10–11, 2017. DA GMPG, BC 601010.

choices which had driven their divergence in the first place. Munich was betting on a few large telescopes built with novel materials, and an emphasis on detector systems, while being conservative on their siting. Heidelberg more closely followed the Soviet tradition and American third-generation proposals (the future VERITAS, the Very Energetic Radiation Imaging Telescope Array System), opting for a larger number of cheaper, smaller, but also sturdier telescopes. As previously mentioned, Heidelberg was also making the best of the necessity of finding a new site, following the break-up with the project in La Palma, siting projects in Namibia instead, thanks to contacts within the Max Planck Institute for Astronomy (see Chapter 3),[330] and the encouragement of their French partners.[331]

The HEGRA telescopes system ceased operations after six years of observations, in September 2002, in coincidence with the start-up of H.E.S.S., whose first telescope was inaugurated in late summer 2002.[332] The first-stage array of H.E.S.S. was quickly built between 2002 and 2004 with, thanks to Völk's requests, the help of external funding from the German Ministry of Research, which complemented the internal backing by the two participating departments in Heidelberg. The decision to opt for reliable, proven technologies paid off, for the system quickly yielded results, such as the more than one hundred new sources identified by the first survey of the galactic center, at a time when no other system was yet in stable operation.[333]

330 Heinrich Völk: interview by Luisa Bonolis and Juan-Andres Leon, Heidelberg, August 9–10, 2017. DA GMPG, BC 601037.

331 The first of the four telescopes of the first phase of the H.E.S.S. project went into operation in summer 2002 and started stereoscopic observations in 2003. These detectors were actually located next to the mountain that had been acquired by the Max Planck Society for the southern observatory of the Max Planck Institute for Astronomy in Heidelberg. When the Spanish in La Palma, loyal to Munich, obstructed the placement of H.E.S.S. on their island, Völk was advised by Hans Elsässer, Director of the Astronomy Institute, to use their site in Namibia instead. In the end, for cost-saving reasons, the detectors were placed next to the mountain, not on top of it as originally intended (Heinrich Völk: interview by Luisa Bonolis and Juan-Andres Leon, Heidelberg, August 9–10, 2017. DA GMPG, BC 601037).

332 The progress report 2001/2002, providing an overview of scientific work conducted at the Institute for Nuclear Physics, had a new general heading for fields such as high-energy, theoretical and infrared astrophysics, neutrino physics, and heavy flavor physics: "Crossroads of Particle Physics and Astrophysics," officially inaugurating one of the two new research directions formulated at the end of the 1990s (the second one being "Many Particle Dynamics of Atoms and Molecules") and mirrored in the new structure of the Annual Report (AMPG, IX Abt., Rep. 5, No. 413).

333 H.E.S.S. has surveyed the Milky Way in gamma-ray light for the last 15 years. To celebrate this anniversary, the H.E.S.S. collaboration has published its largest set of results

The situation in Munich was more haphazard, with success slower to mature. The funding of MAGIC had been more difficult in the first place, with only internal funds and resources from the institute, supplemented somewhat by an insurance payment following severe fire damage to the HEGRA site in 1997.[334] Still, Italian and Spanish partners helped maintain the project. Furthermore, when it started, the search for a new director was still on in Lorenz and Mirzoyan's department. In 1993, with the arrival in Munich of Masahiro Teshima,[335] MAGIC was finally on a par with Heidelberg, as far as directorial support went. Technically, MAGIC was more innovative and daring than H.E.S.S., but this created many delays. Also, since the funding for MAGIC had been restricted to a single telescope, in its first iteration it could not benefit from the advantages of the stereoscopic approach. The operation of MAGIC started to stabilize around 2005, by which stage Munich was ready to apply for the funds to expand the system to the intended scale.[336] While the earliest plans had suggested up to four dishes, the mid-2000s expansion stage foresaw only a second one; and it was autumn 2009 when the second telescope (MAGIC II) went into operation, so enabling stereoscopic observations. Mostly owing to input by Eckart Lorenz and the Munich MPI group, MAGIC was specifically designed to search for VHE emission from gamma-ray bursts (GRBs) and fast

---

in a series of papers, in a special issue of the journal *Astronomy & Astrophysics*: Thierry Forveille, Sergio Campana, and Steve Shore: H.E.S.S. Phase-I Observations of the Plane of the Milky Way. *Astronomy & Astrophysics* 612 (2018), E1. doi:10.1051/0004-6361/201833049. A burst, which Fermi and Swift discovered on July 20, 2018, dubbed GRB 180720B, was pointed by H.E.S.S. 10 hours after the alert. The data collected over two hours, from 10 to 12 hours after the gamma-ray burst, showed a new point-like gamma-ray source, a surprising detection of VHE gamma rays with energies up to 440 GeV many hours after the initial event, deep in the afterglow phase. H. Abdalla et al.: A Very-High-Energy Component Deep in the γ-Ray Burst Afterglow. *Nature* 575/7783 (2019), 464–467. doi:10.1038/s41586-019-1743-9.

334 Science New Staff: Fire Damages Gamma-Ray Observatory. *Science* | AAAS, 10/24/1997. https://www.sciencemag.org/news/1997/10/fire-damages-gamma-ray-observatory. Last accessed 3/27/2021.

335 Teshima had led the Japanese efforts in gamma-ray astronomy in the 1980s in Japan, and 1990s in Utah. The proposal for such a position had been specifically motivated by the need for a director who could lead research in experimental astroparticle physics, a field which was viewed as having strong potential for future developments at the Institute for Physics. In particular, it was stressed how the rising costs for instruments at accelerators favored the development of promising areas, such as astroparticle physics, which could be operated on a relatively small financial basis (CPTS meeting minutes of 18/19.10.2001, AMPG, II. Abt., Rep. 62, No. 1855, p. 19).

336 Eckart Lorenz, and Manel Martinez: The Magic Telescope. *Astronomy & Geophysics* 46/6 (2005), 6.21-6.25. doi:10.1111/j.1468-4004.2005.46621.x.

transient phenomena in general. This required a light mechanical structure, fast and precise movement, and the capability to quickly focus the mirrors after pointing to a source when alerted by satellites to do so.[337] In any case, the lower energy threshold of MAGIC meant an area of overlap with the Fermi satellite that turned out to benefit the system. In 2008, the MAGIC collaboration detected for the first time a pulsed gamma emission from the pulsar in the Crab Nebula at energies above 25 GeV, creating again a bridge between satellite detectors and Imaging Atmospheric Cherenkov Telescopes.[338] MAGIC became the first facility to report unambiguous VHE emission, with energies up to 1 TeV.[339] MAGIC repointing procedure and new VHE data opened a novel pathway for understanding GRBs that will be further extended by current instruments, a new generation of ground-based gamma-ray telescopes.

Finally, as mentioned earlier, the situation in the northern hemisphere necessitated specialization in extragalactic sources, which, even though fewer in number than the abundant 'harvest' of the galactic plane, were very different in kind: extreme, faraway objects with more potential for answering fundamental questions, including the 'holy grail' of the field: establishing the

337 The author L. B. is grateful to Alessandro de Angelis for this specific remark and for emphasizing the relevance of this major innovation that increasingly results in a successful specialty for MAGIC (personal communication, November 27, 2019).

338 The MAGIC Collaboration: Observation of Pulsed γ-Rays Above 25 GeV from the Crab Pulsar with MAGIC. *Science* 322/5905 (2008), 1221–1224. doi:10.1126/science.1164718. In August 2019, MAGIC detected afterglow photons at TeV energy coming from a gamma ray burst recorded by NASA's Neil Gehrels Swift satellite. The ultra-high energy photons were at least 10 times more energetic than the highest energy photons detected previously from any burst and the authors show that observed gamma radiation must have originated in a relativistic jet moving at 0.9999 the speed of light toward us. Combination of data with observations of lower energy (X-ray) photons carried by Swift have enabled, for the first time, discrimination between different emission models as well as discovery of the exact conditions in the explosion, still one of the most puzzling questions involving these cosmic phenomena. Evgeny Derishev, and Tsvi Piran: The Physical Conditions of the Afterglow Implied by MAGIC's Sub-TeV Observations of GRB 190114C. *The Astrophysical Journal* 880/2 (2019), L27. doi:10.3847/2041-8213/ab2d8a.

339 On January 14, 2019, both the Fermi and Swift satellites detected a spike of gamma rays from the constellation Fornax. The missions transmitted to the astronomical community the location of the burst, dubbed GRB 190114C, 30 seconds after the event. The two MAGIC 64-ton telescopes automatically turned to the direction of the burst. They began observing it just 50 seconds after the explosion, and captured the most energetic gamma rays yet seen from these events. MAGIC Collaboration: Teraelectronvolt Emission from the γ-Ray Burst GRB 190114C. *Nature* 575/7783 (2019), 455–458. doi:10.1038/s41586-019-1750-x. P. Veres et al.: Observation of Inverse Compton Emission from a Long γ-Ray Burst. *Nature* 575/7783 (2019), 459–463. doi:10.1038/s41586-019-1754-6.

origin of high-energy cosmic rays.[340] Today, the study of cosmic rays is in fact increasingly based on complementary approaches: on the one hand, the measurement of the energy spectrum, chemical composition, and anisotropy of charged ultra-high-energy cosmic; and on the other, the search for their sources through the observation of neutral radiation like photons and neutrinos, which point back to the emitting sources and are tracers of acceleration sites of charged cosmic rays. The physical connection between high-energy cosmic ray interactions and the resulting very-high-energy neutrinos and gamma-rays can in fact provide clues about their unknown astrophysical sources. Hopes in this direction were, for example, confirmed in the period 2017–18 by the first detection of a high-energy cosmic ray source, identified via the sequential, multi-messenger detection, firstly, of energetic neutrinos with the IceCube telescope in the South Pole; then by the Fermi gamma-ray observatory in space determining the approximate location of the associated gamma rays; and, thirdly, by MAGIC identifying the particular source of the highest energy photons, a so-called Blazar—a particular class of Active Galactic Nuclei (AGN), whose relativistic jet of ionized matter points toward the observer—that had already been registered in previous surveys.[341] Previous detections of individual astrophysical sources of neutrinos had been limited to the Sun and the supernova 1987A, and thus this event was a giant leap in the growing field of *multi-messenger astrophysics*, whose fundamental goal is to understand the properties of the high-energy astrophysical sources by means of new observational strategies, integrating the study of high-energy charged cosmic rays, neutrinos, and electromagnetic radiation across a broad range of wavelengths.[342]

---

340 In 2018, for the first time, dedicated observations of the microquasar SS 433, taken from 2006 to 2011, from both MAGIC and H.E.S.S., were combined, accounting for a total effective observation time of 16.5 hours. Such data were used to place constraints on the particle acceleration fraction at the inner jet regions and on the physics of the jet/medium interactions, so providing hints on the behavior of relativistic particles in the source. MAGIC Collaboration et al.: Constraints on Particle Acceleration in SS433/W50 from MAGIC and H.E.S.S. Observations. *Astronomy & Astrophysics* 612 (2018), A14. doi:10.1051/0004-6361/201731169.

341 IceCube Collaboration: Neutrino Emission from the Direction of the Blazar TXS 0506+056 Prior to the IceCube-170922A Alert. *Science* 361 (2018), 147–151. doi:10.1126/science.aat2890. The IceCube Collaboration et al.: Multimessenger Observations of a Flaring Blazar Coincident with High-Energy Neutrino IceCube-170922A. *Science* 361/6398 (2018). doi:10.1126/science.aat1378.

342 Felix A. Aharonian et al.: 5@5—a 5 GeV Energy Threshold Array of Imaging Atmospheric Cherenkov Telescopes at 5 Km Altitude. *Astroparticle Physics* 15/4 (2001), 335–356. doi:10.1016/S0927-6505(00)00164-X.

Over the past few years, detecting gravitational waves and neutrinos has become almost routine. Using information carried by photons, cosmic rays, neutrinos, *and* gravitational waves to investigate violent astrophysical phenomena, multi-messenger astrophysics has definitely been established as a "new kind of big science" (involving big collaborations, big instruments, and big data), thus, not only deeply affecting and expanding our scientific understanding of astrophysical processes, but also reshaping "the very way science is carried out."[343]

Evolution towards a Single Global Collaboration

In the year 2005, when both MAGIC and H.E.S.S. were applying to the Max Planck Society and external sources for funding, competition between the two projects peaked. Based on their initial successes, each system was planning to expand into its final configuration as a stereoscopic array. By this point, there were deep concerns within the Max Planck Society regarding the projects' competition, potential duplication of efforts, and, especially, the perceptibly growing animosities.[344] In a rare case of direct intervention, the MPG President and Vice-President brought together the main players in Heidelberg and Munich and indicated that it expected them to eventually reconcile and join forces; and, furthermore, that the current stage of expansion, H.E.S.S. II and MAGIC II, was contingent on their promise to do so. This ultimatum would be the origin of the next-generation project, later named the Cherenkov Telescope Array (CTA). The immediate effect of this high-level meeting, which did indeed result in an agreement on future cooperation, was the release of funds allowing both extant systems to expand and fulfill their intended potential; and hence, over the next decade, while still separate and in competition, they cemented their global dominance. And this dominance assured them enormous leverage when negotiating the next-generation project with their international partners. Moreover, this approved second stage permitted the competing projects, while still separate, to converge in terms of their technical capabilities, their respective upgrades moving each in the direction of its rival:

---

343 A special collection of review and commentary articles on multi-messenger astrophysics published in *Nature Review Physics* was highlighted on September 2, 2020 (https://astronomycommunity.nature.com/posts/a-collection-on-multi-messenger-astrophysics. Last accessed 7/19/2021). See, in particular, Mészáros et al., Multi-Messenger, 2019, 585–599.

344 Heinrich Völk: interview by Luisa Bonolis and Juan-Andres Leon, Heidelberg, August 9–10, 2017. DA GMPG, BC 601037. Razmik Mirzoyan: interview by Juan-Andres Leon, Munich, August 13–14, 2018. DA GMPG, BC 601031. Juan-Andres Leon, conversation with Masahiro Teshima, Munich, August 14, 2018.

MAGIC II, by adding a second large telescope, finally obtained stereoscopic capabilities for the Palma site, while H.E.S.S. II brought to Namibia the largest Cherenkov telescope ever built, with a 28 m-diameter mirror.[345]

By the first decade of the 21st century, it was firmly established that the Max Planck Society should not own or administer large scientific infrastructures (Chapter 4). So, since its inception, CTA included one additional German partner in the form of the Helmholtz Institute of DESY-Zeuthen, successor to the Institute for High-Energy Physics, the main East German particle physics institute, which after reunification had been merged with the much larger and famous Hamburg-based research center for particle physics, founded in 1959 around a powerful electron-synchrotron project.[346] The DESY branch location at Zeuthen had been traditionally involved since the 1980s in neutrino astrophysics and at the time was a main collaborator in the large-scale IceCube experiment, the neutrino telescope at the South Pole, now also extending its astroparticle physics program to gamma rays, through participation in CTA.

The two Max Planck Institutes, their international partners in H.E.S.S. and MAGIC, and the Helmholtz, represented by DESY-Zeuthen, constituted the core of CTA, which was being bolstered at the time of its constitution by the good performance of the existing competing projects in Namibia and La Palma. The plans reflected what has become the standard logic of expansion for each new generation of large scientific projects: a change in scale and sensitivity of an order of magnitude. That is, the total amount of telescopes in CTA would reach beyond one hundred, and the expected costs were in the order of half a billion dollars, thus, were starting to resemble the scale of upcoming astronomical projects such as the ALMA in radio astronomy and the European Extremely Large Telescope (ELT) in optical astronomy. Moreover, the resemblance with ALMA is striking, in that the CTA project was proposing a practical monopoly in the field of ground-based gamma astronomy. The crucial step in this direction was the absorption of the American competitors into CTA. This was the first ever such case in a scientific field, of Americans joining an exist-

---

345 In 2015–16, the cameras of the four H.E.S.S. I telescopes were fully refurbished using state-of-the-art electronics and, in particular, the newly developed NECTAr readout chip especially designed for the next big experimental Cherenkov Telescope Array. S. Vorobiov et al.: NECTAr: New Electronics for the Cherenkov Telescope Array. *Nuclear Instruments and Methods in Physics Research Section A* 639/1 (2011), 62–64. doi:10.1016/j.nima.2010.08.112.

346 See a short account of this merging in Frank Grotelüschen: *Insight Starts Here. 50 Years of DESY*. Hamburg: Deutsches Elektronen-Synchrotron DESY 2009.

ing collaboration as 'minority partners.'[347] This was, however, a very special kind of monopoly, and could more aptly be named an agglomeration of heterogeneous partners within a single organization.[348] The geographical choices reflect this heterogeneity, maintaining a foot in each hemisphere, with each site showcasing a path-dependent continuity with either MAGIC or H.E.S.S, while adding features that are the specialty of the other global partners. Once the US, Brazilian, and Indian groups had joined, along with the strong Japanese participation, CTA represented a worldwide effort, extending well beyond its European roots.[349]

On the northern site of La Palma, a traditional stronghold of the Max Planck Institute for Physics in Munich, now the Heisenberg Institute, the new CTA telescopes are already under construction, on the site that previously hosted HEGRA and still hosts MAGIC. The southern site already selected will be near the ESO Paranal Observatory in Chile (see Chapter 4), and is to maintain a stronger connection to H.E.S.S. and the Max Planck Institute for Nuclear Physics in Heidelberg. The deployment of the project's German telescopes, a responsibility of DESY, still reflects the convergence of two separate paths (large telescopes and middle-sized arrays), and each of these ultimately derives heavily from Armenian expertise and a vast tradition in high-energy astrophysics extending back over half a century into the Soviet era.

CTA, in building on the technology of current ground-based detectors and utilizing three classes of telescopes to cover the full CTA's energy range, and in improving the performance of the current IACTs, is expected to facilitate the expansion of our knowledge of several scientific subjects, including the study of the origin of cosmic rays and the exploration of extreme particle acceleration (investigating in detail processes happening close to black holes, and within relativistic jets or winds); and, last but not least, also to shed light on the nature of the still mysterious dark matter and its distribution in the Universe.[350]

---

347 Americans are notably reluctant to be minority partners in international collaborations. Even in ALMA, their current participation (2018) is 37.5 percent, exactly the same as in ESO, and above the remaining 25 percent of "East Asia."

348 This is again very similar to ALMA. See Chapter 4.

349 The CTA Consortium: Design Concepts for the Cherenkov Telescope Array CTA. An Advanced Facility for Ground-Based High-Energy Gamma-Ray Astronomy. *Experimental Astronomy* 32/3 (2011), 193–316. doi:10.1007/s10686-011-9247-0.

350 A whole issue of the journal *Astroparticle Physics* has been dedicated to the science explored with the CTA: Jim Hinton et al. (eds.): A New Era in Gamma-Ray Astronomy with the Cherenkov Telescope Array. *Astroparticle Physics* 43 (2013), 1–2. doi:10.1016/j.astropartphys.2012.12.002.

From January to February 2020, the prototype Large-Sized Telescope (LST), LST-1, while still in the commissioning phase, detected very high-energy emission from the Crab Pulsar: "This milestone shows us that the LST-1 is already performing at an extraordinary level, detecting a challenging source in record time," said Masahiro Teshima, Director of the Max Planck Institute for Physics in Munich and Principal Investigator of LST.[351]

In joining "the field of telescopes capable of detecting gamma-ray pulsars"—particularly challenging sources, because of their weak signals and the dominance of the foreground gamma-ray signal from the surrounding nebulae—this last-generation telescope, covering the low-energy sensitivity range, has inaugurated the CTA Observatory era for the worldwide astronomical and particle physics communities.

351 Cherenkov Telescope Array Observatory: CTA Prototype LST-1 Detects Very High-Energy Emission from the Crab Pulsar, 6/22/2020. https://www.cta-observatory.org/lst1-detects-vhe-emission-from-crab-pulsar/. Last accessed 3/27/2021.

APPENDIX

The History of Cosmic Research in the Max Planck Society through Its Finances

1 A Complementary Analysis

Quantitative financial data was examined early in our research and continued to inform the narrative that is presented in the book. It was decided, however, not to detail finances within the chapters themselves, where the interaction of sociopolitical dynamics and scientific developments was to remain the primary focus. Instead, at any mention of a particularly relevant financial event, we refer the reader to this Financial Appendix. The Appendix constitutes a complementary analysis of the cosmic research cluster, based on financial planning and money flow that, as the reader will see, significantly mirrored or even propelled the events described thoughout the book; and on occasion, this offers additional insight that no other historical source could have afforded.

The graphs used in this Appendix trace the overall expenditure of the Max Planck Society (MPG) from 1955, when it was first systematically recorded in all the institutes, to 1997, the last year for which figures in Deutschmarks are available.

The major source of data is the Budget Plans of the Max Planck Society (AMPG, Haushaltspläne, II. Abt, Rep. 69) of *prospective* data namely, inasmuch as the Plans' main purpose was to declare the respective financial needs of each institute in the coming year.

While this *prospective* data on expenditure ('outgoings') is highly systematic, the same unfortunately cannot be said of the sections and tables in the Plans that show income. To reveal sources of income was patently not a priority. Indeed, the *intransparency of funding sources* across different institutes, as well as between the MPG itself and the state and Länder, has itself become a vital strand of contemporary research into the history of the Max Planck Society, and is evidently related to the Society's particular role in the postwar West German corporatist system. The final *Synthesis Volume* of our Research Program on the History of the Max Planck Society (GMPG) will dedicate a section to this complex topic.

The actual costs incurred by each institute over the previous two years were recorded nonetheless, as a basis for the calculation and justification of future needs. In combination, these past outgoings and prospective needs allow us

DOI:10.1163/9789004529137_008

to reconstruct the financial evolution of the MPIs and the MPG. It must be emphasized, however, that the outgoings are not necessarily more definitive than the future figures. In practice, outgoings were presented, discussed, and revised at the Annual General Meeting (AGM) of the Society, and only the so-called revision data was approved and recorded. And recorded rather haphazardly, as it happens: it is not nearly as well organized or easily accessible as the Budget Plans, which for their part used the same categories across all the institutes for a given year and are accordingly ideal systematic aggregates for the intra-MPG comparison found in this Appendix.

Yet the expenditure designations did change considerably over the decades and we were therefore obliged to settle on the most stable categories possible, namely the total sum of personnel costs, the total sum of regular costs, (in Figure 1 only), and the total costs overall; and it is the last of these that allows us to distinguish the 'actual size' of a working institute (as reflected in this sum of the first two categories) from those large, one-off investments that characteristically involve construction, equipment, instrumentation, and project-specific research and development.

The reader is warned that we are interested not in exact absolute numbers but rather in the *qualitative* trends that these Budget Plans help us identify. This is one more reason to focus, (except in Figure 1), not on actual sums of money but on what proportion of the MPG's overall expenditure such sums represent (recorded here as a percentage of the total). As the reader will doubtless note, this proportionality affected the distribution of resources within the Society to a striking—and highly political—degree: most institutes gravitated towards a fixed proportion of the overall MPG budget and remained in this stable orbit, so to speak, for decades. Divergences from this trend, especially in personnel costs, can be explained by the specific developments at each institute.

The sequence of figures presented in this Appendix is as follows.

The initial step, in order to build the large-scale narrative in this book, was to identify a periodization scheme. We opted to divide the history of cosmic research in the Max Planck Society into three phases, which are analyzed in Figure 1 by means of contrast with trends within the MPG overall. The periodization of the MPG itself is informed by that which our Research Program proposes, which will be published in its *Synthesis Volume*.

As we see in detail below, cosmic research represents a crucial exception within this global periodization, as, unlike the MPG overall, the institutes benefiting from the space age experienced no stagnation in the 1970s and '80s.

In this Appendix, we illustrate the evolution of the cluster of cosmic research institutes in the Max Planck Society by first comparing its behavior within the Society overall, which helps us define the most useful metric for subsequent comparisons: the relative size of an institute relative to the Society. This metric filters out the effects of the generalized economic growth of the MPG in Germany (to which the reader can always refer in Figure 1), and it also preempts the need for inflation-adjusted sums such as were needed in Figure 1.

In a second analysis, in Figure 2, we describe the problem of defining what constitutes cosmic as opposed to 'nuclear' research. Some institutes (those we here call the 'core') are dedicated exclusively to cosmic research, so they function best as proxy to the actual development of cosmic research proper, which is why they are used in Figure 1. However, they constitute the strictest 'lower bounds' and to ignore institutes only partially dedicated to cosmic research would be to hide the crucial impact, on research as well as on political gravitas, of those institutes with a significant presence in other clusters. This is most notably the case with institutes participating in 'nuclear' research, and is a central theme throughout the book. But there are others, too, such as the MPI for Aeronomy, which suffered a continuous identity crisis over several decades, due to some of its research lines being better described as geophysical than as cosmic; and nor did it ever make a significant transition to the Earth System Science, unlike another institute in the cluster, the MPI for Chemistry in Mainz.

After analyzing the effect of using this 'expanded' view of the cluster, which features heavily in the book, the collective behavior of the cluster and its uncertain boundaries are described. To follow up on the crucial effects of 'nuclear' research on the cosmic cluster, we then zoom in for the first time to the individual institute level, the subject of Figure 3. In it, we look at the two 'nuclear age' institutes in Munich and Heidelberg, (whose competitivity is a central topic throughout the book), for it is they that best illustrate the shifting balance of the nuclear and cosmic research conducted in both places. Still, the evolution of each of them differed remarkably, the first being at the root of the Munich-area 'cell division,' the second remaining monolithic to this day, yet while holding within it many different lines of research spanning several clusters.

Subsequently, we examine the financial evolution of the other institutes in the cluster, likewise in pairs, to underpin many significant facets featured in Chapters 1 to 5; for this sometimes yields novel insights into episodes and general patterns that so far had been treated only qualitatively.

Figure 4 compares the two 'crisis' institutes in the cluster, the MPI for Aeronomy and the MPI for Chemistry, whose activities fall to a large degree outside the bounds of cosmic research proper yet are significantly connected to it, in both epistemic and social terms.

Figure 5 compares the 'twin' institutes in the Munich 'family,' namely the Institute for Astrophysics and the Institute for Extraterrestrial Physics, which clearly epitomize the differences between a theoretically, respectively an experimentally focused institute.

Figure 6, the last in the Appendix, compares the two Max Planck Institutes dedicated principally to ground-based, observational astronomy. The MPI for Radio Astronomy, in Bonn, and the MPI for (Optical) Astronomy, in Heidelberg. This pair serves to illustrate how such institutes related to the large, and often, even colossal 'national' infrastructures that were under their administration yet predominantly not a part of their regular budget.

We chart the financial development solely of those institutes with a significant timeline and hence have omitted two entities that emerged as independent financial entities only in the 1990s: the gravitational wave research outpost of the MPI for Quantum Optics in Hanover, and the MPI for Gravitational Physics in Potsdam. Figures for these are included, however, in the aggregates used in Figures 1 and 2. The main budget of the MPI for Quantum Optics itself, which, as we saw in Chapter 5, was host to the (smaller-scale) gravitational wave experiments prior to creation of the Hanover site, is omitted entirely. It is too difficult to tease out the financial impact of these experiments from that of the main research conducted at the MPQ, which in any case only emerged as an independent institute in the 1980s, out of the IPP. Lastly, the IPP itself is not analyzed at all in this Financial Appendix; this, because the standard practice of the MPG itself is to handle the IPP budget as an entirely separate entity; unlike other institutes' budgets, therefore, it is not included in the 'total' expenditure of the MPG overall. As we have described throughout the book, the role of the IPP in cosmic research and the clustering of its institutes is remarkable; but part of the 'magic' of its relationship to the cluster lies precisely in this ambivalence.

2 Financial Periodization of the Cosmic Sciences in the Max Planck Society

One crucial analytical angle of this book is to explain, firstly, to what degree the trajectory of cosmic research corresponds to larger trends in the Max Planck Society overall; and, secondly, how, (if at all), these Society-wide trends are manifestations of broader developments in West Germany. Of course, such relationships are often bi-directional, as the 'Sputnik shock' and other examples have shown.

Our point of departure is, necessarily, a comparison of the financial development of the cosmic research cluster with that of the Max Planck Society overall.

The different phases that we enumerate below manifest themselves also in a multitude of quantitative but non-financial metrics, such as employment statistics, and the launch or cessation of institutes or scientific publications. While other GMPG historians are currently analyzing such metrics in detail, at the time of our researching and writing this book, the most readily available source of quantitative data was financial, and, as we will see below, it is striking how much examining decades of money flow can reveal about cosmic research.

In the cumulative graph in the top half of Figure 1 we divide the evolution of expenditure in the Max Planck Society overall into four distinct phases using vertical lines, creating four periods which roughly correspond to the general economic development of West Germany:[1]

1. Postwar Reconstruction (1940s–1950s)
2. 'Economic Miracle' (late 1950s–early 1970s)
3. Slowdown Cycles (1970s–1980s)
4. Pre- and Post-Reunification Expansion (late 1980s–1990s)

Now take a look at the bottom half of Figure 1, which highlights the evolution of the cosmic sciences cluster as a proportion of the Max Planck Society's expenditure overall. By comparing this with the evolution of the total MPG budget, one can highlight the following links between general trends and 'cosmic' developments.

Postwar Reconstruction: The 'Nuclear Age'

During the first postwar decade, which corresponds roughly to Chapter 1 of this book, the size and budget of the Max Planck Society overall was extremely modest in comparison to the decades that followed. This was the era of refounding and reconstructing the institutes in the original Kaiser Wilhelm Society, and sometimes also of incorporating other 'orphan' research groups dating from before 1945. Research during the first postwar decade was constrained not only by the shortage of resources but also by restrictions (Law 25, see Chapter 1) on the kinds of scientific research permitted, specifically

1 Further analysis of the financial history of the Max Planck Society is being conducted for the final *Synthesis Volume* by Jaromir Balcar, and Jürgen Kocka. In this book we present just a basic outline, to draw comparisons with the specific path taken by the cosmic sciences.

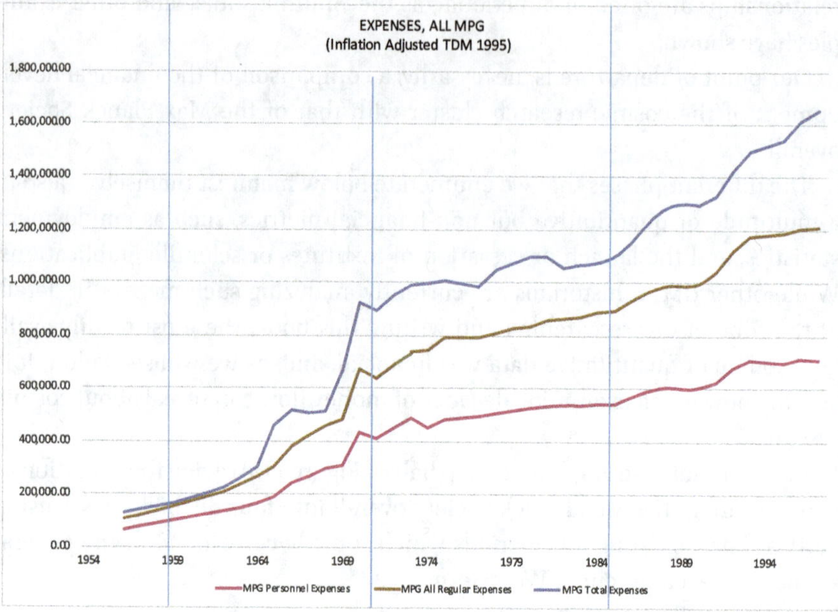

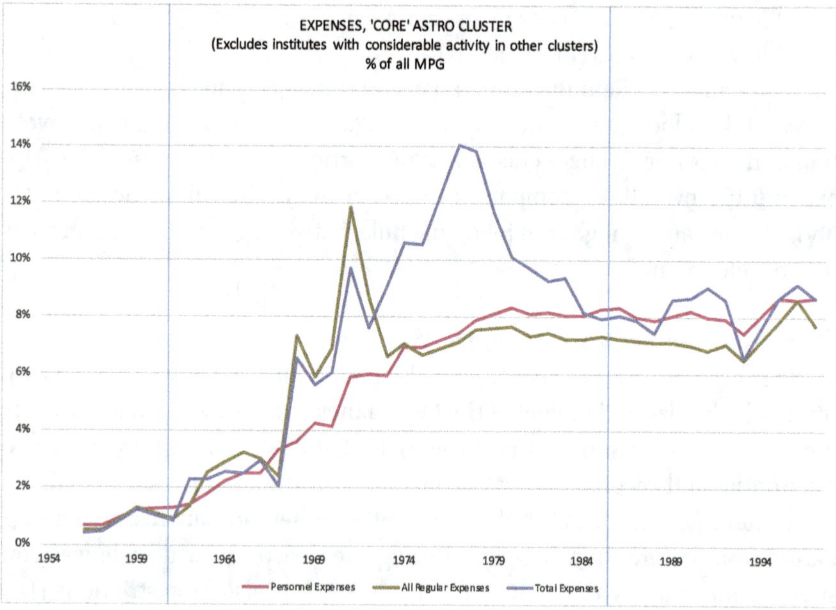

on nuclear weapons, aircraft, rocketry, radar, and other fields closely related to military technologies. At the same time, this was the era of maximum optimism regarding the promises of nuclear energy, when the most dynamic research fields were connected in some way with nuclear physics. Given the outstanding role and status of nuclear physics—a research field principally propelled by wartime interests and comprising at the time, (under the then still broadly interpreted term 'nuclear'), all research related to the fundamental structure of matter and the laws behind it—researchers in other fields often framed their own work in reference to 'nuclear' topics. In the cosmic sciences, this was evident in the nascent traditions of theoretical plasma astrophysics, in Göttingen, and cosmochemistry, in the southwest of the Federal Republic.

Economic Miracle': The Space Age

In the case of the cosmic sciences, the crucial transformation came with the launch of Sputnik in 1957, which was further magnified by the fact that West Germany (like the rest of Western Europe) then embarked on what was to be its period of greatest economic growth. During the German 'economic miracle,' the cosmic sciences' proportion of the Max Planck Society's overall budget grew spectacularly, and as we can see in the graph, this 'leap' was almost immediate. In the Max Planck Society overall, by contrast, the effects of the 'Sputnik

---

FIGURE 1 If represented as financial quantites, the lines used in our figures would be cumulative, such as in the one above, still in the original unit of thousands of Deutschmarks: the personnel costs are part of the regular costs, which in turn are part of the total costs. The difference between regular and total costs indicates investment in one-off items such as large infrastructural projects. In all subsequent graphs, we use instead percentages of total MPG expenditure, on the same concept so they no longer appear as cumulative. Rather, their relative position illustrates the balance at a given institute (or group of institutes) between personnel and one-off investments. Since the total regular costs did not turn out to provide significant new information for our analyses, they are not shown in subsequent figures in this Appendix.
The vertical lines demarcate the different periods which can be identified in the shape of the curves, and which correspond to distinct historical epochs, as described in the periodization. The vertical line of the early 1970s does not extend to the bottom graph as the cosmic sciences did not experience the general MPG stagnation of the period 1970s–80s, and instead had their largest infrastructural costs at this time. The bottom graph shows expenditure in institutes in the cosmic sciences cluster as a proportion of the total MPG budget. For the bottom graph we used only data from the Institutes for Astrophysics (Göttingen–Munich-Garching), Extraterrestrial Physics (Garching), Radio Astronomy (Bonn), Astronomy (Heidelberg), and Gravitational Physics (Hanover–Potsdam), which constitute the 'core' of the cluster. We omitted institutes with a significant presence in other clusters: Aeronomy (Lindau), Physics (Göttingen–Munich), Nuclear Physics (Heidelberg), and Chemistry (Mainz). These follow a different trajectory due to the effects of nuclear and environmental research, as described further below.

shock' were more gradual; but even so, between 1964 and the end of the 'boom' in the early 1970s, the Society's total expenditure increased almost five-fold!

Meanwhile, between the launch of Sputnik and the end of the bonanza, even if taking into account solely the effects on the exclusively cosmic institutes, the proportion of personnel costs (which best illustrate the long-term trends of a cluster) grew about eight-fold, from ca. 1 percent to over 8 percent of the total. Given the overall growth of the MPG, the level at which purely cosmic research stabilized in the 1980s was about 30 times its pre-Sputnik participation.

'Missing Stagnation'

Most striking, in the entire periodization of the cosmic sciences in the MPG, is that this field's 'space age' growth did not suffer the stagnation faced by other fields in the Society during the economic slowdown of the 1970s and early '80s. Indeed, the proportion of cosmic research within the MPG continued to grow without interruption, before stabilizing in the mid-1980s. The book so far has shown that this was a global trend in the cosmic research of this era, in part owing to the field's connection with Cold War technologies; but the trend was even more marked in the Max Planck Society, since the cosmic research cluster had established a strong position there even before the launch of Sputnik, and eventually had a virtual monopoly in Germany. This predominance was further magnified by its 'protectors,' namely pivotal (and 'nuclear') figures of the Max Planck Society, such as Werner Heisenberg, Wolfgang Gentner, and in the next generation, Reimar Lüst. Given Lüst's deep roots in the cosmic cluster, and the fact that he presided over the MPG precisely during the era of stagnation, it is difficult to dismiss the informal claims advanced by many of his generation, that the austerity measures he instituted in the Society overall, which included the closure of many institutes and even the virtual disappearance of whole research lines, were not quite so deeply felt in his own cluster.

This was not just because of his personal protection, however, since the significant momentum of infrastructural growth likewise played a substantial role: from the 1960s to the '80s, one can see the very great impact of individual large infrastructural projects, which show up as distinctive peaks in the total expenditure. Many such peaks corresponded to the execution of projects that predated the austerity period.

However, we must point out that such large-scale expenditure is of use only qualitatively, as it is recorded only when publicly funded. For example, the large telescopes of the Max Planck Institute for Radio Astronomy and other donations made in those years did not feature in the Budget Plans and accordingly cannot be traced in the following graphs.

In contrast with these, the costs related to the Calar Alto observatory of the Max Planck Institute for Astronomy are clearly visible in the volcano-shaped feature that shows the degree to which the expenditures of the cosmic cluster reached their peak in the late 1970s, and continued to have a significant impact until the mid-1980s. In the following graphs, we see that the space age bonanza marked by the Max Planck Institutes' control over large national projects concluded around the mid-1980s, even before the end of the Cold War or German reunification.

Post-Reunification Expansion: The Globalization / Collaboration Era

Two characteristics mark the cosmic science cluster's transition to a new era in the second half of the 1980s. Firstly, its proportion of expenditure within the Max Planck Society overall reached a roughly constant plateau of around 8 percent (for the exclusively cosmic institutes in the cluster). And secondly, the large infrastructural 'peaks' of previous eras did not reoccur. The book has shown how this transition corresponded to the maturation and even crisis of the model prevailing in the 1960s–1980s, when the Max Planck Society was in charge of West Germany's large infrastructural projects in the cosmic sciences. However, as our detailed treatment here of individual institutes shows, striking differences persisted, depending on the types of research conducted: ground-based astronomy, space-based astronomy, theoretical astrophysics, planetary science, and research in multi-messenger astroparticle physics.

While the transition to the space age is clearly identifiable as Sputnik-induced, these financial graphs are a quite useful means to emphasize that the subsequent transition to the globalization-era regime was not triggered solely by the fall of the Iron Curtain and German reunification, but was already underway in the 1980s, as the wave of infrastructural investment trickled out and was not renewed. The 1987 *Denkschrift* (memorandum), a central feature of Chapter 4, marked the initial move towards ending the MPG's cosmic monopoly, immediately after the large telescope in Calar Alto opened, and precisely when an upswing in the Society's (and West Germany's) finances signaled the end of the Society's era of financial stagnation.

After the mid-1980s, we can see the effects of 'defensive' growth in the cosmic research cluster, as its financial backing was 'locked in' at a steady level; yet although this rise in resources allowed growth to continue, it was nothing like the post-Sputnik explosion but rather simply enough to ensure that the cluster would keep up with the other institutes in the Society.

German reunification had surprisingly little impact and certainly did not lead to an 'expansion'; rather, what probably would otherwise have manifested as growth in West Germany was now shifted to the new federal state in a 'neutral' way that did not alter the scale of the cluster overall.

3 Shifting Balances of 'Nuclear,' 'Cosmic,' and 'Earth-System' Research

As this book has described, a significant proportion of cosmic research was conducted within institutes primarily dedicated to 'nuclear' topics, that is, research on the atomic nucleus proper, as well as on subnuclear, elementary particle, and high-energy physics. Of these institutes, the ones in Munich and Heidelberg were significantly larger than the typical Max Planck Institute. Unfortunately, the available financial data for those institutes does not show exactly what 'significant proportion of cosmic research' was conducted there; and in fact, as we have described throughout this book, these institutes often thrived on this ambiguity.

However, if not just the purely cosmic institutes are included in the graph but also those that conducted research in other fields, too, the shifting power balance between cosmic and non-cosmic research is very clear from the financial data. This is what we see in Figure 2, in aggregate, and in Figure 3, where the focus is on the two institutes primarily dedicated to 'nuclear' research.

We witness how, in this case, the growth of the cluster relative to the MPG overall starts long before Sputnik and carries over into the early years of the space age with remarkable momentum, before the institutes completely dedicated to cosmic research take off. The magnificent first peak corresponds to the foundational years of Gentner's Max Planck Institute for Nuclear Physics in Heidelberg, and the simultaneous move to Munich of Heisenberg's Max Planck Institute for Physics and Astrophysics. We see how it is around the middle of the 1960s that the new space age institutes seriously blast off. The 'extended' cosmic institutes, however, continue to provide a magnificent 'baseline'; and if they are counted in, their participation in the Society is almost triple that of the core group: whereas the core moved from about 5 percent to 15 percent in the mid-1960s, and 8 percent to 22 percent in the mid-1980s. And this is without counting the Max Planck Institute for Plasma Physics (IPP), which is on so large a scale as to distort any meaningful financial analysis and has therefore always been recorded separately in MPG Budget Plans and Annual Reports. Even without the IPP, institutes with some form of presence in the cosmic research cluster already amounted to about 25 percent of the entire budget of the Max Planck Society, so represented a formidable bastion in economic terms alone, besides their political and scientific clout.

From the 1970s onwards, however, instead of the steady plateau that we see in the 'core' group, one sees a gradual decline through the 1980s and '90s, if one counts in this larger set of institutes. These decades of 'decline' (with respect only to the rest of the MPG, while continuing to represent growth in absolute terms) are precisely the period when these institutes were shifting their weight from 'nuclear' to cosmic subjects; so that even if we go by the conservative estimate, namely that the scale of cosmic research remained constant in them (rather than grew, as was, in reality, most often the case), the decline in other kinds of research within those institutes was more dramatic than the graphs indicate here, and thus pushed down the figures overall.

Moreover, the figures even as early as the mid-1960s show that the scale of infrastructural investments is less pronounced, when these other institutes are counted in. Experimental physics saw a much earlier transition to intra-institutional collaborations and, for example, much particle research was conducted not at MPI-owned installations, but at external sites such as the federal nuclear facilities in Karlsruhe and Jülich, as well as at DESY, the Laue-Langevin Institute in Grenoble, and CERN.

This nuclear 'parabola' was evident also at an institute whose cosmic research is known to have actually declined in this period, in parallel with a considerable decline in its 'nuclear' activities: this is the case of Mainz, which is treated further below. Although it shifted its weight increasingly towards Earth-system research (in which it came to excel), this did not completely cancel out the overall downward trend; that the trend here looks only slightly downward is due to the significant rise in Earth-system research, which replaced activities in both 'nuclear' physics and cosmochemistry (Figure 4).

This is not to say that some aspects of cosmic research avoided decline altogether in the last decades of the century. The best case to illustrate this is Max Planck Institute for Aeronomy in Lindau, also shown in Figure 4. It is amply discussed elsewhere in the book how this institute was the problem child ('Sorgenkind') of the cluster, due to the early death of its most influential patriarch, Erich Regener. Nonetheless, we see that this institute experienced spectacular growth during the space age, particularly in the 1960s. As we described in the book, this institute was in deep crisis in the early 1970s. In view of the fragile political standing of this 'orphan' institute within the cosmic research cluster, it is no surprise that it evinced the same trend to stagnation as the MPG overall, in this era. We even see a significant and sustained reduction in its personnel costs from the late 1970s onwards—rare indeed, in the cosmic cluster, and a result, here, of the drastic cuts in staff numbers introduced after Ian Axford took charge, even though his regime is generally regarded as the

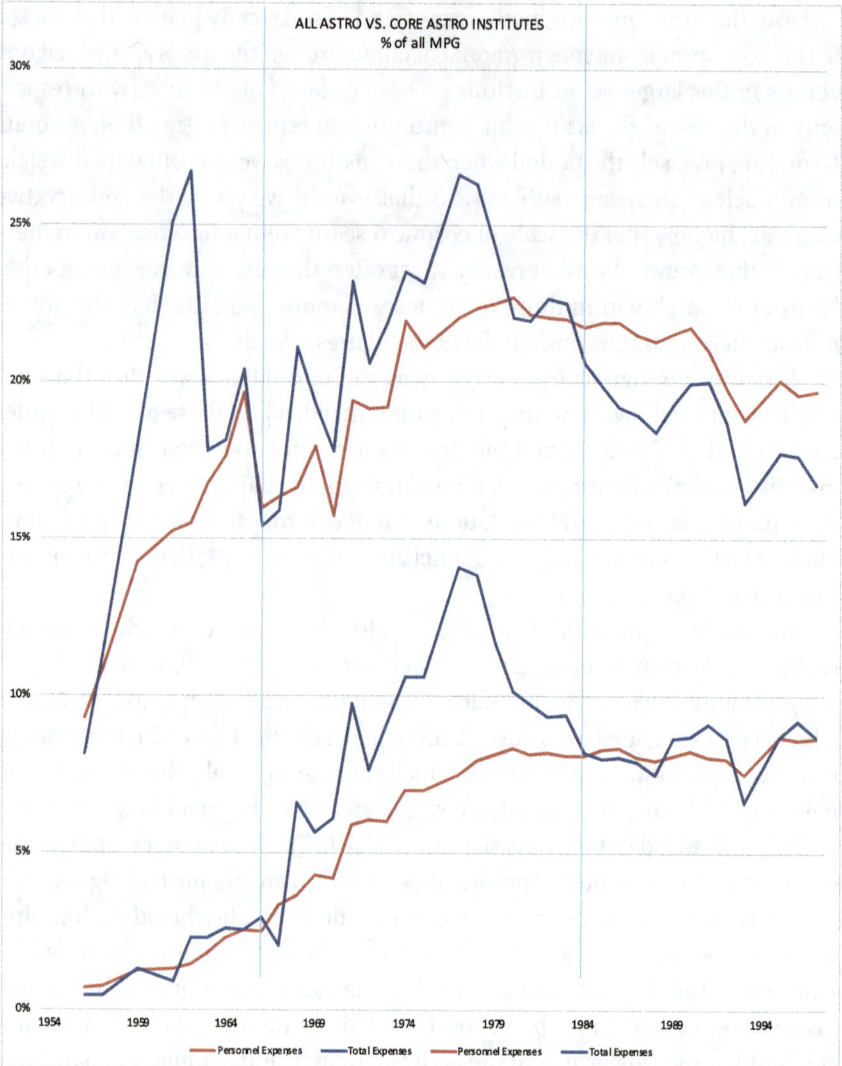

institute's heyday, owing to its participation in quite successful and expensive space exploration missions and ionosphere research projects.

The multiple, middle-sized peaks in total costs characterize Lindau as a typical site of space-based projects, as we will see with even more clarity at the Max Planck Institute for Extraterrestrial Physics in Figure 5. This contrasts significantly with the Mainz institute, which, after the arrival of Christian Junge and its transition to Earth System Science outsourced many of its large investments to national and international collaborations.

4 Financial Lock-Ins and the Complementarity of Theoretical and Experimental Research

Looking at the graphs for single institutes presented so far, the reader may have begun to notice a remarkable pattern, by which institutes, after a period of initial growth, seem to settle at a relatively stable proportion of the Max Planck Society expenditure overall. Personnel costs, which best manifest the long-term commitments of an institute, diverge only slightly from these stable percentages, and such divergences are well explained by the narrative that we have presented in this book, most notably by the Max Planck Institute for Aeronomy.

That these percentages are an instance of MPG-wide and cluster-wide coordination is most evident when looking at the first institute fully dedicated to cosmic research in the Society, the Sub-Institute for Astropysics of the Max Planck Institute for Physics and Astrophysics, that would later become the MPI for Astrophysics (MPA) (Figure 5). After the founding of the Sub-Institute for Extraterrestrial Physics (also in Figure 5), the MPA became an eminently theoretical institute whose expenditure was largely determined by human activities, leading to an almost parallel behavior between its personnel costs and total costs (which are well below average costs in the cluster and the MPG). The MPA always benefited from 'tapping into' resources from other members of the Munich 'family,' firstly, the original Institute for Physics in Munich-Freimann, and then the Institute for Extraterrestrial Physics in Garching. All along, it also benefited from the computational resources of the Institute for Plasma Physics (IPP).

FIGURE 2 This graph compares the financial scale of the cosmic cluster, depending on whether one includes in the count institutes that are only partially dedicated to cosmic research. In addition to the 'core' institutes used in Figure 1, reproduced at the bottom, the aggregate graph that results further up includes the Institutes for Aeronomy (Lindau), Physics (Göttingen–Munich), Nuclear Physics (Heidelberg), and Chemistry (Mainz). The two key differences in the upper graph are the significant early peaks owing to participation in 'nuclear' research, and a prominent decline in expenditures from the 1980s onwards, which disappears when one counts only those institutes dedicated exclusively to cosmic research. Taking into consideration that, in the institutes in the top graph, cosmic research did generally not diminish throughout the century, the decline in 'nuclear' activities would be even more striking, if these could be seen in isolation. It is, however, important to emphasize that the 'nuclear' institutes of the MPG thrived particularly owing to the persistent ambiguity regarding the identification of research as 'nuclear' or 'cosmic,' throughout the Society's history; an ambiguity that is reflected in the current hybrid identity of Astroparticle Physics.

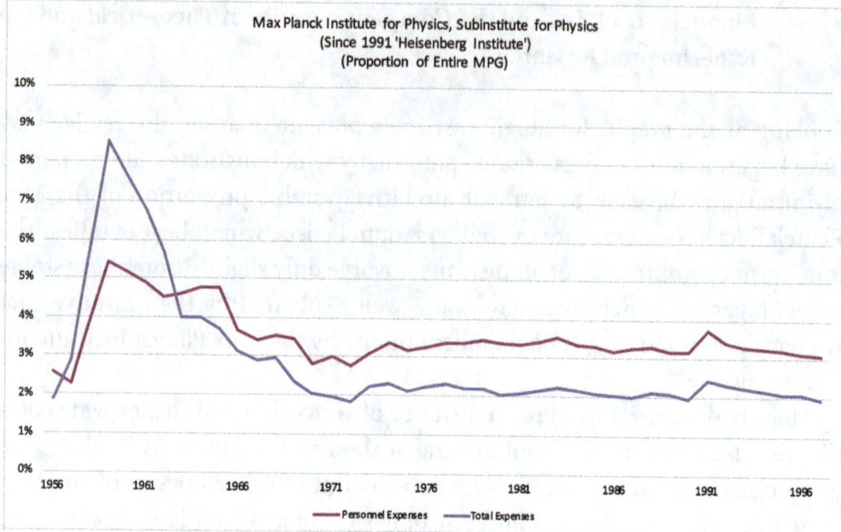

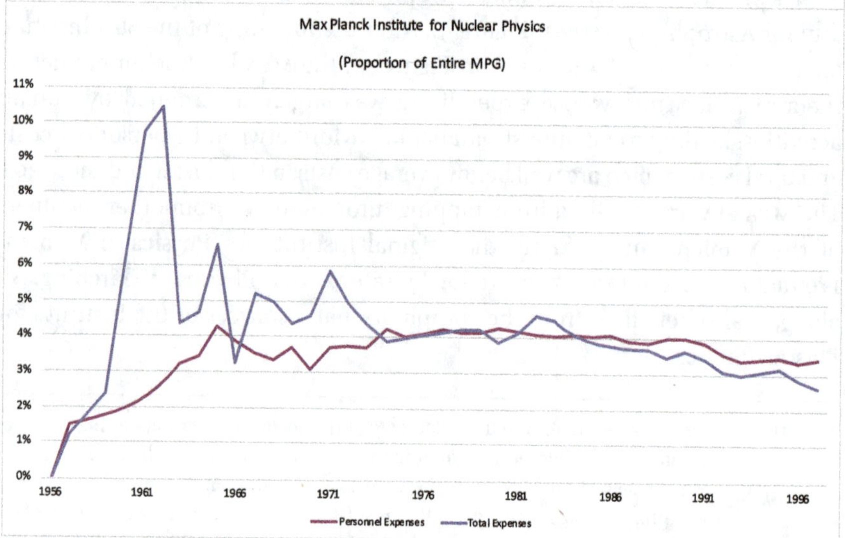

This is in sharp contrast with the heavily project-based Institute for Extraterrestrial Physics (also in Figure 5), which was dedicated first to space plasma experiments launched on rockets, then moving to space-based astronomy in the high-energy realms (x-rays and gamma rays), as well as space- and ground-based infrared astronomy. The MPE is the archetypical space-age institution, with multiple sharp peaks throughout its existence, while the underlying employment conditions of its staff were stable, contrasting markedly with the MPE for Aeronomy in Lindau treated earlier. The MPE is recognized to this day as the most successful institute of the entire cluster, so it is particularly remarkable that even under these circumstances, its long-term size, as expressed in the personnel costs, is still almost horizontal with respect to the MPG development overall, yet another example of the financial 'lock-in' described earlier.

The financial evolution of the MPE is also a useful means to remark on the diminishing scale of MPG-led projects. Peaks in the 1960s captured a significantly larger proportion of the total expenditure of the Society. The smaller peaks in later decades are due not to the projects having become 'cheaper' over time, but rather, to the fact that the growth in the scale of individual projects was slightly outstripped by that of other MPG expenditure over the decades. The fact that single Max Planck Institutes no long host larger projects is due to their growing tendency to outsource them to international organizations and collaborations, to which, often, the federal government and other supporters contribute a significant portion of the funding through channels other than the MPG (such as the DLR), so that the MPG-adminstered budget relates only to the direct 'home' costs of these projects. Finally, the more institutes there

---

FIGURE 3 A comparison between the main two 'nuclear age' institutes of the MPG illustrates their generally similar evolution from spectacular beginnings to a period of stabilization and even decline. Still, the two institutes had a very different function within their 'families': Heisenberg's Institute for Physics quickly underwent a process of 'cell division,' from the late 1950s onwards, and its experimental nuclear research activities were generally based at external sites like CERN and DESY. For this reason, the total costs do not show the characteristic peaks of project-based research. Its graph also clearly shows the considerable decline through the 1960s, which contributed to the succession crisis described in Chapter 3. One more detail to note: all the institutes of the Munich 'family' which belonged at some point to the Max Planck Institute for Physics and Astrophysics show a 'kink' in the early 1990s, which can be attributed to the change in accounting standards and institutional organization following the breakup of this bastion into several fully independent Max Planck Institutes.
The Max Planck Institute for Nuclear Physics, stronghold of Gentner's southwestern 'family,' did not separate into sub-institutes but remained a single unit throughout the 20th century, being one of the largest Max Planck Institutes overall, and the largest single financial unit in the cosmic research cluster. As this institute integrates significant experimental 'nuclear' activities and space missions, its non-personnel costs are proportionally larger than those in Munich above. The MPIK also shows peaks characteristic of 'space' institutes like the MPE (treated further below). Finally, this institute, while very successful, did considerably scale down its activities in classical nuclear physics (as a proportion of total MPG expenditure), and while the activities in cosmic research grew, they did not quite manage to keep up with the levels of investment and personnel reached in 1965 at the crossroads of the nuclear and space ages.

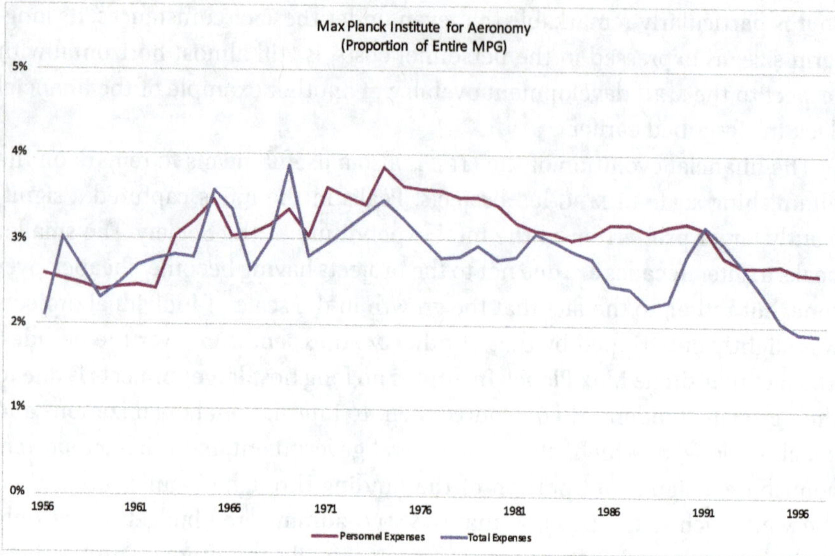

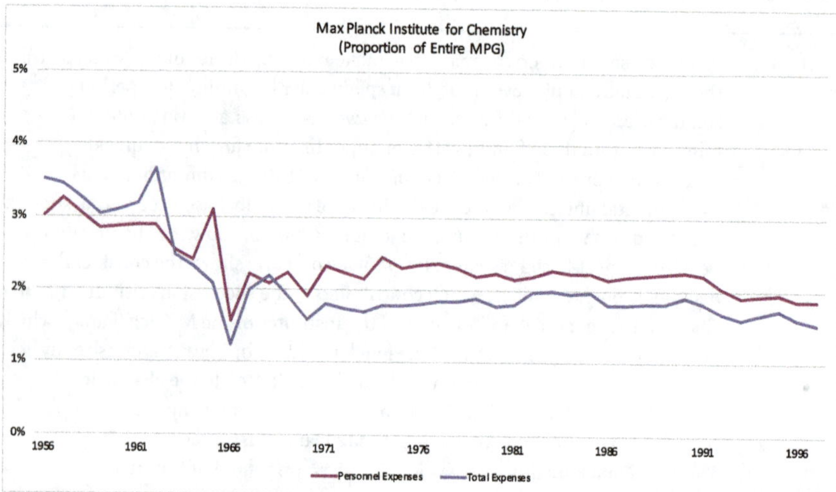

are in the MPG, overall, the more the expenditure on large projects averages out, globally, so eliminating the peak magnification that occurs when there are relatively few projects of similar scale underway in the Society in any one year.

5 Astronomical Institutes, Their Infrastructures, and the End of an Era for the MPG

There is a significant contrast between the financial behavior of typically space-based and project-based institutes, such as the Institute for Extraterrestrial Physics (Figure 5), and the practices of the predominantly ground-based astronomical institutes (Figure 6). As we have seen already with both the Institutes of Aeronomy (Figure 4) and Extraterrestrial Physics (Figure 5), 'space' institutes show a seesaw pattern of expenditure that corresponds to multiple ongoing 'medium-size' projects. In fact, this project-influenced seesaw pattern is evident also in the Max Planck Institute for Nuclear Physics (Figure 3), which, as we saw earlier, encompasses nuclear research, space exploration, and astrophysical activities at one site.

In contrast, the ground-based astronomical Max Planck Institutes (Figure 6) depended on a smaller number of projects requiring proportionally much larger funding than their regular costs. These institutes were created also to administer astronomical infrastructures that, since regarded as West German 'national' projects, were not funded directly by the Max Planck Society.

One 'well-behaved' example of such an institute was Radio Astronomy in Bonn (Figure 6). As was described in the book, this institute has tradition-

---

FIGURE 4 Here we take a look at the two 'crisis' institutes in the cosmic research cluster. The Max Planck Institute for Aeronomy shows the typical pattern of a space research institute's total expenditure: a seesaw of peaks related to specific projects; but in contrast to successful ones like the Institute for Extraterrestrial Physics (treated later), the personnel costs are some of the most erratic in the MPG, and after the mid-1970s evince a significant gradual decline. Interestingly, however, this financial 'decline' actually coincides with the directorship of Ian Axford, the 'savior' of this institute. It is evident that great sacrifices were made, unparalleled by any other institute in the cluster.
Meanwhile, the Max Planck Institute for Chemistry is the oddball institute, insofar as it had promising beginnings that never quite panned out. Its founding figure Otto Hahn never conducted research there, as he became President of the Max Planck Society. Under his presidency, this institute (unlike Heisenberg's and Gentner's) did not even grow. Its senior figures Mattauch and Paneth were quickly approaching retirement, and younger generations were not considered as permanent successors with scientific authority enough to keep the institute open. In the 1960s, this institute went into a period of precipitous decline, coincident with its succession crisis, even though the younger generation was by then conducting highly influential cosmochemical research there. The transition to the directorship of Christian Junge and Earth-system research is very evident in the graph, for it created a radically new regime of stability and growth that was 'locked in' with that of the MPG overall.

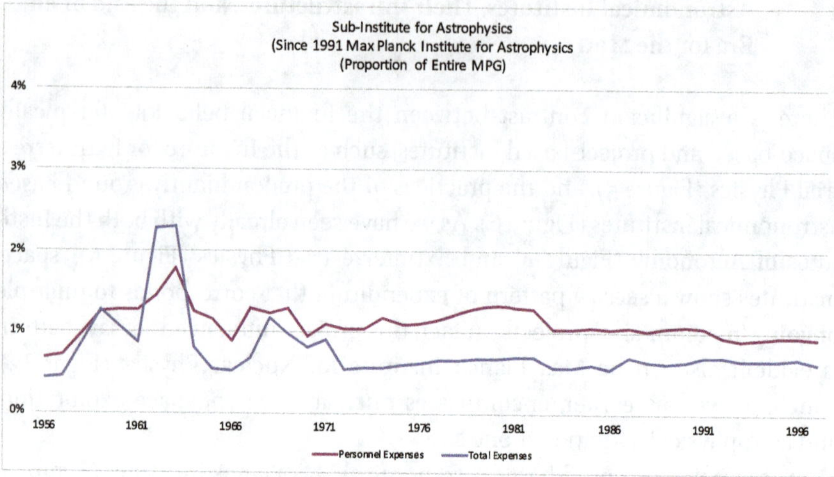

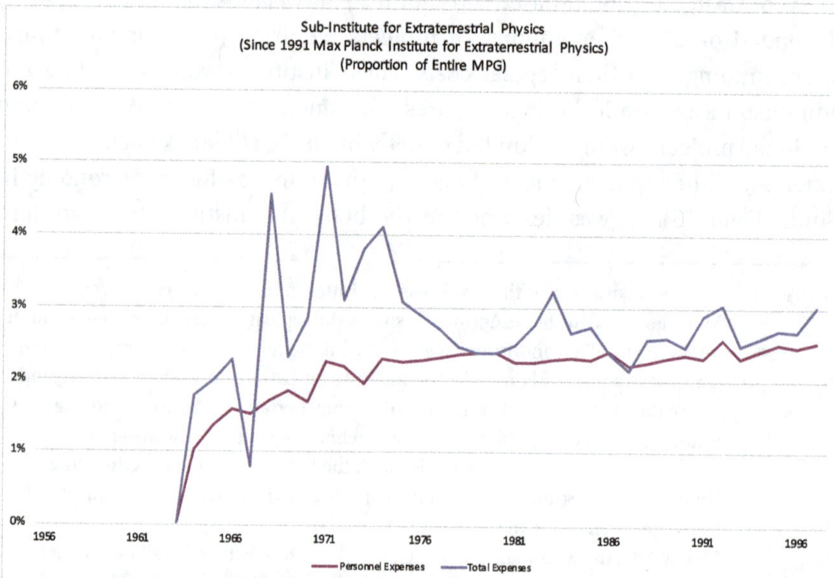

ally been one of the most fiercely independent in the cosmic research cluster, thanks to its close political and industrial links in northwestern Germany. From the financial perspective, this meant that the MPG had oversight only over the 'home' institute in Bonn, so, based on the graph of its financial evolution, one might well confound it with a theoretical institute such as the Institute for Astrophysics analyzed earlier (Figure 5). The reality was markedly different: on the one hand, its three major observatory projects (Effelsberg, Pico Veleta–IRAM, Mount Graham–Heinrich Hertz Telescope) were the result of considerable financial and in-kind donations by its allied industrial partners. While in the Effelsberg case, the Bonn institute directly administered the telescope (so incurring a portion of its regular costs), the second telescope was built as a contribution to IRAM, the pioneering international collaboration, initially with France, later also with the direct participation of the Spanish hosts. The German contribution to IRAM was administered via the Max Planck Society (giving it a voice in its governance) and accounted for in its global budget, within the budget of the Bonn institute, which nonetheless for several decades was considered to 'own' this collaboration, as the telescope contributed was the masterpiece of Peter Mezger. Only since the late 1990s has the IRAM collaboration been increasingly used by the other astronomical institutes in Heidelberg and Garching, and increasingly also as a means of access to the Plateau de Bure interferometric array, rather than just to the German-built antenna in Spain. Figure 6 includes the MPG's financial contribution to IRAM as a dotted line, as it is separate from any institute's budget.

The Heinrich Hertz Telescope on Mount Graham in Arizona was a fifty-fifty binational collaboration with the University of Arizona, the Germans contributing the equipment, and the Americans, the site and the buildings. Given the considerable difficulties experienced in Mount Graham throughout the 1980s and '90s (see chapter 4), it is remarkable that none of the drama manifests in the Bonn institute expenditure, which, as we see below, could have been quite different.

FIGURE 5 Shown here are the highly complementary 'twins,' the Institutes for Astrophysics (MPA) and Extraterrestrial Physics (MPE). The first one was already dedicated significantly to theoretical research, because that is what was possible in the immediate postwar decades, and after its separation from the MPE, it actually specialized further in this direction, while the MPE was an eminently experimental facility, initially in the same field of research, plasma astrophysics. This differentiation led the MPA to display the most strikingly 'locked in' budget of any institute in the cluster, which is evident here in the horizontal parallel lines almost unchanging from the early 1970s to the end of the century, untouched even by the succession crisis of the 1990s. Typical of a theoretical institute is its relatively low expenditure on non-personnel items, in comparison with the MPG in general.

The Institute for Extraterrestrial Physics founded by Reimar Lüst is arguably the most successful institute in the cosmic cluster. The first striking qualitative feature of its graphs is the significant weight of project-based research since the beginning, as indicated by the serrated profile of its total costs. At the same time, even in this successful institute, the personnel costs reach a plateau by the mid-1970s, which illustrates the strong influence of MPG mandates on the medium-sized institutes.

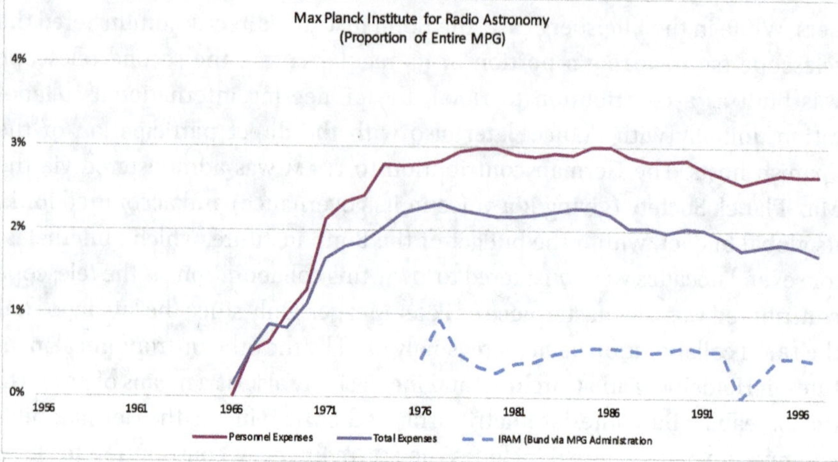

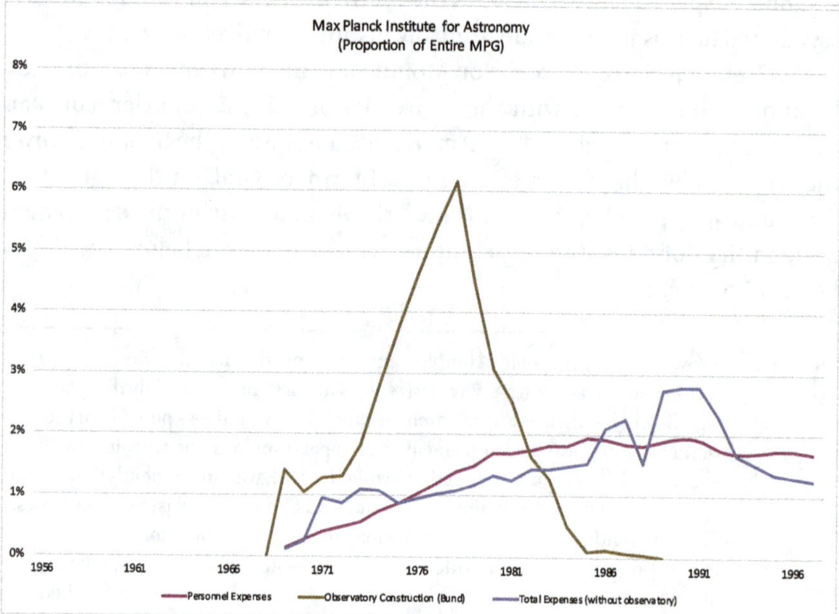

The final graph in this appendix (at the bottom of Figure 6) is also probably the most telling with regard to how finances themselves impacted the evolution of cosmic research in the Max Planck Society and in Germany in general. In principle, the Max Planck Institute for Astronomy could have functioned similarly to the MPIfR in Bonn, which it was modelled on. The main initial difference was that instead of a large private donation, such as the Volkswagen Foundation's for Effelsberg, the Institute for Astronomy received funding from the federal government, initially to build two observatories, one in each hemisphere. From the Bonn graphs, it is evident that the finalization of a flagship observatory is a good indicator of when an astronomical institute becomes fully operational; and indeed, by the early 1970s the Bonn institute had already attained its target size, at about 2 percent of annual MPG personnel costs and 3 percent total annual costs.

In contrast, it took the institute in Heidelberg two decades to reach its peak staffing, at around 2 percent of the MPG's personnel costs. This slow growth was due to the long-delayed construction process in Calar Alto, Spain. The expenditure directly related to this project was by far the largest ever seen within the cosmic cluster, constituting at its peak close to 6 percent of the total outgoings of the MPGs.

FIGURE 6 In this final graph we compare the two predominantly ground-based observatory institutes of the Max Planck Society. The creation of the Max Planck Institute for Radio Astronomy in Bonn was the prototype of an MPI that would be the operator of 'national' astronomical infrastructures. Its three main telescopes do not feature in its own budgets, as they were the result of external funding and in-kind donations. After Effelsberg opened in the early 1970s, the institute reached the characteristic 'lock-in' plateau that we have witnessed in many institutes in this Appendix. In this case, the lower proportion of non-personnel costs signals not predominantly theoretical acivities, such as we saw in the MPA in Figure 5, but quite the opposite: the one-off investments underpinning it are so large that they are not channeled through the institute, and many not even through the MPG. The MPIfR was an early pioneer of multinational collaborations, contributing its second telescope to the French–German partnership IRAM. The portion of this organization's budget paid through the MPG administration is displayed as a dotted line for comparison. Were one to include the sticker-price costs of Effelsberg, Pico Veleta, Mount Graham, and APEX, the peaks would be sharper even than in a project-based institute like the MPE, although never as massive as Calar Alto, immediately below.

The final graph in our analysis is also the most striking of all, corresponding to the Max Planck Institute for Astronomy in Heidelberg. This institute could have followed the model of the Bonn institute, above, with the difference that its observatory investments were funded federally rather than by donations, and hence appear in the general MPG budgets (but not in the institute's own). As described in the book, these observatory projects turned into one of the most significant failures of the MPG in the second half of the century. One of the two originally foreseen observatory sites could not be built for political reasons, while the other experienced long delays and overbudgets. Upon completion, 10 years off schedule, its large telescope was already technologically way behind the state of the art. After the inauguration of the large telescope in 1986, the financial consequences for the Max Planck Society and the institute proper were more direct, as seen in the 'humps' of the late 1980s. These running overcosts significantly increased the pressure on the MPG to avoid operating large infrastructures in the future. Besides these non-financial issues, the location itself turned out less than ideal, climatically.

Meanwhile, much of the research activity in these two decades occurred rather in the field of space-based infrared astronomy, a field granted resources independently, by the Federal Ministry of Research; and this made the Heidelberg institute the second most important space astronomy institute, after the MPE.

While the construction stage peaked in 1979 and tapered off with the inauguration of the 3.5 m telescope in 1986, the aftermath had no less dramatic an impact: while significantly less costly than the federally funded construction, the operational running costs were now a factor in the institute's own expenditure; which is also to say, the money flowed directly out of the MPG budget. It was these overcosts that threw the institute into deep crisis; and they became the prime reminder of why the MPG should never again commit to operating large research infrastructures.

Regardless of all the radical historical shifts in the late 1980s, and the changed nature of astronomical research, it is impossible to overstate the direct impact Calar Alto had on the way the Max Planck Society operated in Astronomy after 1986.

Bibliography

Abazov, A. I., D. N. Abdurashitov, O. L. Anosov, L. A. Eroshkina, E. L. Faizov, V. N. Gavrin, A. V. Kalikhov, et al.: First Results from the Soviet-American Gallium Experiment. *Nuclear Physics B Proceedings Supplements* 19 (1991), 84–93. doi:10.1016/0920-5632(91)90191-G.

Abazov, A. I., O. L. Anosov, E. L. Faizov, V. N. Gavrin, A. V. Kalikhov, T. V. Knodel, I. I. Knyshenko, et al.: Search for Neutrinos from the Sun Using the Reaction $^{71}Ga(\nu_e,e^-)^{71}Ge$. *Physical Review Letters* 67/24 (1991), 3332–3335. doi:10.1103/PhysRevLett.67.3332.

Abbott, Alison: Battery Fault Ends X-Ray Satellite Mission. 6732. *Nature* 399/6732 (1999), 93–93. doi:10.1038/20029.

Abbott, Alison: German Astronomers Fight Rumoured Closing of Institute. *Nature* 358 (1992), 267–267. doi:10.1038/358267a0.

Abbott, Alison: Max Planck Institute Gets New Life. *Nature* 360 (1992), 6. doi:10.1038/360006b0.

Abbott, B. P., R. Abbott, T. D. Abbott, F. Acernese, K. Ackley, C. Adams, T. Adams, et al.: Gravitational Waves and Gamma-Rays from a Binary Neutron Star Merger. GW170817 and GRB 170817A. *The Astrophysical Journal Letters* 848/2 (2017), L13. doi:10.3847/2041-8213/aa920c.

Abbott, B. P., R. Abbott, T. D. Abbott, F. Acernese, K. Ackley, C. Adams, T. Adams, et al.: Multi-Messenger Observations of a Binary Neutron Star Merger. *The Astrophysical Journal Letters* 848/2 (2017), L12. doi:10.3847/2041-8213/aa91c9.

Abbott, B. P., R. Abbott, R. Adhikari, A. Ageev, B. Allen, R. Amin, S. B. Anderson, et al.: Detector Description and Performance for the First Coincidence Observations between LIGO and GEO. *Nuclear Instruments and Methods in Physics Research Section A. Accelerators, Spectrometers, Detectors and Associated Equipment* 517/1 (2004), 154–179. doi:10.1016/j.nima.2003.11.124.

Abdalla, H., R. Adam, F. Aharonian, F. Ait Benkhali, E. O. Angüner, M. Arakawa, C. Arcaro, et al.: A Very-High-Energy Component Deep in the γ-Ray Burst Afterglow. *Nature* 575/7783 (2019), 464–467. doi:10.1038/s41586-019-1743-9.

Abdurashitov, J. N., E. L. Faizov, V. N. Gavrin, A. O. Gusev, A. V. Kalikhov, T. V. Knodel, I. I. Knyshenko, et al.: Results from SAGE (The Russian-American Gallium Solar Neutrino Experiment). *Physics Letters B* 328/1 (1994), 234–248. doi:10.1016/0370-2693(94)90454-5.

Abdurashitov, J. N., V. N. Gavrin, S. V. Girin, V. V. Gorbatchev, A. O. Gusev, A. V. Kalikhov, T. V. Knodel, et al.: Results from SAGE. *Nuclear Physics B—Proceedings Supplements* 48/1 (1996), 299–303. doi:10.1016/0920-5632(96)00264-2.

Abdurashitov, J. N., E. P. Veretenkin, V. M. Vermul, V. N. Gavrin, and S. V. Girin: Solar Neutrino Flux Measurements by the Soviet-American Gallium Experiment (SAGE) for Half the 22-Year Solar Cycle. *Journal of Experimental and Theoretical Physics* 95/2 (2002), 181–193. doi:10.1134/1.1506424.

About, Jean: *Les débuts de la recherche spatiale française. Au temps des fusées-sondes.* Paris: Institut français d'histoire de l'espace 2007.

Abuter, R., A. Amorim, N. Anugu, M. Bauböck, M. Benisty, J. P. Berger, N. Blind, et al.: Detection of the Gravitational Redshift in the Orbit of the Star S2 near the Galactic Centre Massive Black Hole. *Astronomy and Astrophysics* 615 (2018), L15. doi:10.1051/0004-6361/201833718.

Abuter, R., A. Amorim, M. Bauböck, J. P. Berger, H. Bonnet, W. Brandner, V. Cardoso, et al.: Detection of the Schwarzschild Precession in the Orbit of the Star S2 near the Galactic Centre Massive Black Hole. *Astronomy & Astrophysics* 636 (2020), L5. doi:10.1051/0004-6361/202037813.

Accelerator Tradition is Thriving at Brookhaven. *CERN Courier* 42/2 (2002), 24–27. https://cds.cern.ch/record/1733305. Last accessed 3/17/2021.

Adams, John Bertram: *European Fusion Research. Report of the CERN Study Group on Fusion Problems.* CERN Yellow Reports. Monographs, CERN-59-16. Geneva: CERN 1959. http://cds.cern.ch/record/214328. Last accessed 10/30/2018.

Adams, John Bertram: Wolfgang Gentner and CERN. In: Dieter Hoffmann, and Ulrich Schmidt-Rohr (eds.): *Wolfgang Gentner. Festschrift zum 100. Geburtstag.* Berlin: Springer 2006, 139–145.

Aerodynamische Versuchsanstalt Göttingen: *Die Aerodynamische Versuchsanstalt Göttingen von 1945 bis 1969.* Göttingen: Max-Planck-Gesellschaft 1969.

Agadjanyan, S. A., F. A. Aharonian, A. Ts. Amatuni, A. S. Ambartsumyan, V. A. Ambartsumyan, A. G. Arutunov, V. V. Avakyan, et al.: Complex Installation for Investigation of Primary Gamma-Rays in Energy Range 10^{12}-10^{16} eV. *Proceedings of the 20th International Cosmic Ray Conference, 2–15 August 1987, Moscow.* 1987, 332–335. http://adsabs.harvard.edu/abs/1987ICRC....2..332A. Last accessed 12/3/2018.

Ageorges, N., L.E. Tacconi-Garman, and C. Lidman: One Year of NACO Operations. In: Wolfgang Brandner, and Markus E. Kasper (eds.): *Science with Adaptive Optics. Proceedings of the ESO Workshop Held at Garching, Germany, 16–19 September 2003.* Berlin: Springer 2005, 53–61.

Agostini, M., K. Altenmüller, S. Appel, V. Atroshchenko, Z. Bagdasarian, D. Basilico, G. Bellini, et al.: Experimental Evidence of Neutrinos Produced in the CNO Fusion Cycle in the Sun. 7835. *Nature* 587/7835 (2020), 577–582. doi:10.1038/s41586-020-2934-0.

Aharonian, Felix A.: Very High Energy Gamma-Ray Astronomy and the Origin of Cosmic Rays. *Nuclear Physics B—Proceedings Supplements* 39/1 (1995), 193–206. doi:10.1016/0920-5632(95)00022-2.

Aharonian, Felix A., A. G. Akhperjanian, A. S. Kankanian, R. G. Mirzoyan, A. A. Stepanian, N. Müller, M. Samorski, et al.: A System of Air Cherenkov Telescopes in the HEGRA Array. *Proceedings of the 22nd International Cosmic Ray Conference. 11–23 August, 1991. Dublin, Ireland* 2 (1991), 615–617. https://ui.adsabs.harvard.edu/#abs/1991ICRC....2..615A. Last accessed 5/18/2018.

Aharonian, Felix A., F. Arqueros, H. Badran, K. H. Becker, De La Cruz, R. M, M. Bott-Bodenhausen, et al.: Status and Extensions of the HEGRA Detector on La Palma. *Proceedings of the 22nd International Cosmic Ray Conference. 11–23 August, 1991. Dublin, Ireland* 4 (1991), 452–455. https://ui.adsabs.harvard.edu/#abs/1991ICRC....4..452A. Last accessed 5/18/2018.

Aharonian, Felix A., A. A. Chilingarian, R. G. Mirzoyan, A. K. Konopelko, and A. V. Plyasheshnikov: The System of Imaging Atmospheric Cherenkov Telescopes. The New Prospects for VHE Gamma Ray Astronomy. *Experimental Astronomy* 2/6 (1992), 331–344. doi:10.1007/BF00395984.

Aharonian, Felix A., W. Hofmann, A. K. Konopelko, and H. J. Völk: On the Performance of Ground Based Arrays of Imaging Atmospheric Cherenkov Telescopes. *International Cosmic Ray Conference* 5 (1997), 109–113. https://ui.adsabs.harvard.edu/#abs/1997ICRC....5..109A. Last accessed 5/17/2018.

Aharonian, Felix A., W. Hofmann, A. K. Konopelko, and H. J. Völk: The Potential of Ground Based Arrays of Imaging Atmospheric Cherenkov Telescopes. I. Determination of Shower Parameters. *Astroparticle Physics* 6/3 (1997), 343–368. doi:10.1016/S0927-6505(96)00069-2.

Aharonian, Felix A., W. Hofmann, A. K. Konopelko, and H. J. Völk: The Potential of the Ground Based Arrays of Imaging Atmospheric Cherenkov Telescopes. II. Gamma Ray Flux Sensitivities. *Astroparticle Physics* 6/3 (1997), 369–377. doi:10.1016/S0927-6505(96)00070-9.

Aharonian, Felix A., A. K. Konopelko, H. J. Völk, and H. Quintana: 5@5—a 5 GeV Energy Threshold Array of Imaging Atmospheric Cherenkov Telescopes at 5 km Altitude. *Astroparticle Physics* 15/4 (2001), 335–356. doi:10.1016/S0927-6505(00)00164-X.

Aharonian, Felix A., and H. J. Völk: Very High Energy Gamma-Ray Astronomy with Ground-Based Imaging Cherenkov Telescopes. In: W. Wamsteker, M.S. Longair, and Y. Kondo (eds.): *Frontiers of Space And Ground-Based Astronomy*. Dordrecht: Springer Science & Business Media 1994, 705–706. doi:10.1007/978-94-011-0794-5_118.

Akasofu, Syun: Chapman and Alfvén. A Rigorous Mathematical Physicist Versus an Inspirational Experimental Physicist. *EOS Transactions* 84 (2003), 269–274. doi:10.1029/2003EO290002.

Albrecht, Helmuth, and Armin Hermann: Die Kaiser-Wilhelm-Gesellschaft im Dritten Reich (1933–1945). In: Rudolf Vierhaus, and Bernhard vom Brocke (eds.): *Forschung im Spannungsfeld von Politik und Gesellschaft. Geschichte und Struktur der Kaiser-Wilhelm-/Max-Planck-Gesellschaft*. Stuttgart: Deutsche Verlags-Anstalt 1990, 356–406.

Aldrich, L.T., W.S. Broecker, S. Epstein, R. R. Marshall, and G. R. Tilton (eds.): *Cosmological and Geological Implications of Isotope Ratio Variations. Proceedings of an Informal Conference, Massachusetts Institute of Technology, June 13–15, 1957*. Washington, D.C.: National Academy of Sciences, National Research Council 1958.

Alexander, Tal: Stellar Processes near the Massive Black Hole in the Galactic Center. *Physics Reports* 419/2–3 (2005), 65–142. doi:10.1016/j.physrep.2005.08.002.

Alfvén, Hannes: Plasma Universe. *Physica Scripta* T18 (1987), 20–28. doi:10.1088/0031-8949/1987/T18/002.

Alfvén, Hannes: The Plasma Universe. *Physics Today* 39/9 (1986), 22. doi:10.1063/1.881039.

Alfvén, Hannes: Paradigm Transition in Cosmic Plasma Physics. *Geophysical Research Letters* 10/6 (1983), 487–488. doi:10.1029/GL010i006p00487.

Alfvén, Hannes: *Cosmical Electrodynamics*. Oxford: Oxford University Press 1950.

Alfvén, Hannes, and Nicolai Herlofson: Cosmic Radiation and Radio Stars. *Physical Review* 78/5 (1950), 616–616. doi:10.1103/PhysRev.78.616.

Alimonti, G., C. Absmaier, H. Back, M. Balata, T. Beau, G. Bellini, J. Benziger, et al.: Science and Technology of Borexino. A Real-Time Detector for Low Energy Solar Neutrinos. *Astroparticle Physics* 16/3 (2002), 205–234. doi:10.1016/S0927-6505(01)00110-4.

Alimonti, G., C. Arpesella, G. Bacchiocchi, M. Balata, G. Bellini, J. Benziger, S. Bonetti, et al.: A Large-Scale Low-Background Liquid Scintillation Detector. The Counting Test Facility at Gran Sasso. *Nuclear Instruments and Methods in Physics Research Section A. Accelerators, Spectrometers, Detectors and Associated Equipment* 406/3 (1998), 411–426. doi:10.1016/S0168-9002(98)00018-7.

Allan, William: Sir William Ian Axford. 2 January 1933–13 March 2010. *Biographical Memoirs of Fellows of the Royal Society* 59 (2013), 5–31. doi:10.1098/rsbm.2013.0007.

Allen, Bruce, Karsten Danzmann, Gerhard Huisken, Hermann Nicolai, and Bernard Frederick Schutz: Jürgen Ehlers. 29.12.1929-20.05.2008. *Jahresbericht der Max-Planck-Gesellschaft. Annual Report 2008*. Max-Planck-Gesellschaft zur Förderung der Wissenschaften e.V. 2009, 24–26.

Allkofer, Otto Claus: *Introduction to Cosmic Radiation*. München: Karl Thiemig 1975.

Allkofer, Otto Claus: *Spark Chambers*. München: Karl Thiemig 1969.

Allkofer, Otto Claus, H. Badran, T. Behrmann, G. Börger, M. Bruhn, P. Fernandez, H.-J. Gaube, et al.: The HEGRA Project for UHE Gamma Ray Astronomy. *Nuclear Physics B—Proceedings Supplements* 14/1 (1990), 345–347. doi:10.1016/0920-5632(90)90443-X.

Allkofer, Otto Claus, H. Badran, T. Behrmann, G. Börger, P. Fernandez, H.-J. Gaube, C. Gruber, et al.: Results of the HEGRA Experiment. In: M. Nagano, and F. Takahara (eds.): *Astrophysical Aspects of the Most Energetic Cosmic Rays. Proceedings of the ICRR International Symposium, Kofu, Japan, 26–29 November 1990*. Singapore: World Scientific 1991, 200–211.

Allkofer, Otto Claus, K. Carstensen, W. D. Dau, W. Heinrich, E. Kraft, and M. Weinert: The Kiel Cosmic Ray Muon Spectograph. *Nuclear Instruments and Methods* 83/2 (1970), 317–325. doi:10.1016/0029-554X(70)90479-9.

Allkofer, Otto Claus, M. Samorski, and W. Stamm: The Kiel EAS Experiment for Detecting UHE Gamma Rays. In: Turver, K.E. (ed.): *Very High Energy Gamma Ray Astronomy. Proceedings of the NATO Advanced Research Workshop, Durham, England, Aug. 11–15, 1986*. Dordrecht: Springer 1987, 281–284. doi:10.1007/978-94-009-3831-1_44.

Allkofer, Otto Claus, and Joachim Trümper: Das Muonen-Spektrum in 3000 m Höhe. *Zeitschrift für Naturforschung A* 19/11 (1964), 1304–1309. doi:10.1515/zna-1964-1108.

Altenmüller, Hartmut: BMBW und BMFT. Fusionen und Teilungen. *Spektrum der Wissenschaft* 12 (1994), 127. https://www.spektrum.de/magazin/bmbw-und-bmft-fusionen-und-teilungen/821997. Last accessed 5/15/2019.

Alter, Peter: Die Kaiser-Wilhelm-Gesellschaft in den deutsch-britischen Wissenschaftsbeziehungen. In: Rudolf Vierhaus, and Bernhard vom Brocke (eds.): *Forschung im Spannungsfeld von Politik und Gesellschaft. Geschichte und Struktur der Kaiser-Wilhelm-/Max-Planck-Gesellschaft*. Stuttgart: Deutsche Verlags-Anstalt 1990, 726–746.

Amaldi, Edoardo: 1939, Perché scelsi l'America. *La Repubblica* (12/6/1996). https://ricerca.repubblica.it/repubblica/archivio/repubblica/1996/12/06/1939-perche-scelsi-america.html?ref=search. Last accessed 1/20/2022.

Amaldi, Edoardo: *Da Via Panisperna all'America. I fisici italiani e la seconda guerra mondiale*. Edited by Giovanni Battimelli, and Michelangelo De Maria. Roma: Editori Riuniti 1997.

Amaldi, Edoardo: From the Discovery of the Neutron to the Discovery of Nuclear Fission. *Physics Reports* 111/1 (1984), 1–331. doi:10.1016/0370-1573(84)90214-X.

Amaldi, Edoardo: *The Adventurous Life of Friedrich Georg Houtermans, Physicist (1903–1966)*. Edited by Saverio Braccini, Antonio Ereditato, and Paola Scampoli. Berlin: Springer-Verlag 2012.

Amaldi, Edoardo: The Years of Reconstruction. In: Giovanni Battimelli, and Giovanni Paoloni (eds.): *20th Century Physics. Essays and Recollections. A Selection of Historical Writings by Edoardo Amaldi*. Singapore: World Scientific 1998, 263–294.

Amorim, A., M. Bauböck, J. P. Berger, H. Bonnet, W. Brandner, Y. Clénet, V. Coudé du Foresto, et al.: Detection of Orbital Motions near the Last Stable Circular Orbit of the Massive Black Hole SgrA*. *Astronomy & Astrophysics* 618/L10 (2018), 1–15. doi:10.1051/0004-6361/201834294.

Amorim, A., M. Bauböck, J. P. Berger, W. Brandner, Y. Clénet, V. Coudé Du Foresto, P. T. de Zeeuw, et al.: Test of the Einstein Equivalence Principle near the Galactic Center

Supermassive Black Hole. *Physical Review Letters* 122/10 (2019), 101102. doi:10.1103/PhysRevLett.122.101102.

Anderl, Sybille: Pforte zum Kosmos. *FAZ.NET* (11/3/2017). https://www.faz.net/-gwz-938nm. Last accessed 4/10/2018.

Anderson, Herbert L., Enrico Fermi, and Leo Szilard: Neutron Production and Absorption in Uranium. *Physical Review* 56/3 (1939), 284–286. doi:10.1103/PhysRev.56.284.

Anselmann, P., W. Hampel, G. Heusser, J. Kiko, T. Kirsten, M. Laubenstein, E. Pernicka, et al.: GALLEX Results from the First 30 Solar Neutrino Runs. *Physics Letters B* 327/3 (1994), 377–385. doi:10.1016/0370-2693(94)90744-7.

Anselmann, P., W. Hampel, G. Heusser, J. Kiko, T. Kirsten, E. Pernicka, R. Plaga, et al.: Implications of the GALLEX Determination of the Solar Neutrino Flux. *Physics Letters B* 285/4 (1992), 390–397. doi:10.1016/0370-2693(92)91522-B.

Anselmann, P., W. Hampel, G. Heusser, J. Kiko, T. Kirsten, E. Pernicka, R. Plaga, et al.: Solar Neutrinos Observed by GALLEX at Gran Sasso. *Physics Letters B* 285/4 (1992), 376–389. doi:10.1016/0370-2693(92)91521-A.

Antrack, D., Ludwig Biermann, and Reimar Lüst: Some Statistical Properties of Comets with Plasma Tails. *Annual Review of Astronomy and Astrophysics* 2/1 (1964), 327–340. doi:10.1146/annurev.aa.02.090164.001551.

Arnulf Schlüter und Rudolf Wienecke. Pioniere der Fusionsforschung. IPP-Report, 16/22. Garching: Max-Planck-Institut für Plasmaphysik 2012.

Aschenbach, Bernd, Hermann-Michael Hahn, and Joachim Trümper: *The Invisible Sky. Rosat and the Age of X-Ray Astronomy*. New York, NY: Springer 1998.

Ash, Mitchell G.: *Die Max-Planck-Gesellschaft im Kontext der Vereinigung 1989–1995*. Berlin: GMPG-Preprint 2020.

Ashton, Nigel J.: Harold Macmillan and the "Golden Days" of Anglo-American Relations Revisited, 1957–63. *Diplomatic History* 29/4 (2005), 691–723. https://www.jstor.org/stable/24915066. Last accessed 5/21/2019.

Astone, P., and G. Pizzella: Upper Limits in the Case That Zero Events Are Observed: An Intuitive Solution to the Background Dependence Puzzle. hep-ex/0002028. *1st Workshop on Confidence Limits, CERN, Geneva, Switzerland, 17–18 Jan 2000*. Geneva: CERN 2000, 199–205. doi:10.5170/CERN-2000-005.199.

Astronomy. Politics and Science. *Nature* 236/5346 (1972), 320–321. doi:10.1038/236320b0.

Baade, Walter, and Fritz Zwicky: Cosmic Rays from Super-Novae. *Proceedings of the National Academy of Sciences* 20/5 (1934), 254–259. doi:10.1073/pnas.20.5.259.

Baars, Jacob W.M.: *International Radio Telescope Projects. A life among their designers, builders and users*. Rheinbach: Createspace Independent Publishing Platform 2013.

Babul, Arif, and Simon D. M. White: Quasar-Modulated Galaxy Clustering in a Cold Dark Matter Universe. *Monthly Notices of the Royal Astronomical Society* 253/1 (1991), 31P–34P. doi:10.1093/mnras/253.1.31P.

Bagge, Erich: Die Sonne und die Fixsterne als Quellen kosmischer Strahlung. *Il Nuovo Cimento* 6/3 (1949), 327–329. doi:10.1007/BF02822006.

Bagge, Erich: Ursprung und Eigenschaften der kosmischen Strahlung. In: S. Flügge, and F. Trendelenburg (eds.): *Ergebnisse der Exakten Naturwissenschaften*. Berlin: Springer-Verlag 1949, 202–262. doi:10.1007/978-3-662-25834-7_6.

Bagge, Erich: Kernzertrümmerungen und schwere Teilchen in der kosmischen Strahlung. *Naturwissenschaften* 29/21 (1941), 318–318. doi:10.1007/BF01479547.

Bagge, Erich: Beiträge zur Theorie der schweren Atomkerne. I. Zur Frage des Neutronenüberschusses in den schweren Atomkernen. *Annalen der Physik* 425/4 (1938), 359–388. doi:10.1002/andp.19384250406.

Bagge, Erich, and Ludwig Biermann: Die Erzeugung von Ultrastrahlung auf der Sonne. *Die Naturwissenschaften* 35/4 (1948), 120–121. doi:10.1007/BF00626776.

Bagge, Erich, and Ludwig Biermann: Über die Entstehung der solaren Komponente der Ultrastrahlung. *Zeitschrift für Naturforschung A* 4/4 (1949), 303–315. doi:10.1515/zna-1949-0410.

Bagge, Erich, E. Böhm, R. Fritze, U. J. Roose, M. Samorski, C. Schnier, R. Staubert, et al.: The Extensive Air Shower Experiment at Kiel. *Proceedings of the 9th International Cosmic Ray Conference, London, UK*. London New York Paris Los Angeles: Institute of Physics and the Physical Society 1965, 738–741. https://ui.adsabs.harvard.edu/#abs/1965ICRC....2..738B. Last accessed 11/23/2018.

Bagge, Erich, Kurt Diebner, and Kenneth Jay: *Von der Uranspaltung bis Calder Hall*. Edited by Ernesto Grassi. Hamburg: Rowohlt 1957.

Bagge, Erich, Manfred Samorski, and Wilhelm Stamm: Lateral Distribution of Electrons in the Core Region of Extensive Air Showers of 10^{15}-10^{17} eV. *Proceedings of the 16th International Cosmic Ray Conference (Kyoto)*. 1979, 260–265. https://ui.adsabs.harvard.edu/#abs/1979ICRC...13..260B. Last accessed 8/14/2018.

Bagge, Erich, Manfred Samorski, and Wilhelm Stamm: A New Air Shower Experiment at Kiel. *15th International Cosmic Ray Conference, Plovdiv, Bulgaria, August 13–26, 1977, Conference Papers*. Budapest: Dept. of Cosmic Rays, Central Research Institute for Physics of the Hungarian Academy of Sciences 1977, 24–29. https://ui.adsabs.harvard.edu/#abs/1977ICRC...12...24B. Last accessed 11/13/2018.

Bahcall, John N.: Two Solar Neutrino Problems. *Nuclear Physics B—Proceedings Supplements* 43/1–3 (1995), 41–46. doi:10.1016/0920-5632(95)00447-H.

Bahcall, John N.: Neutrinos from the Sun. *Scientific American* 221/1 (1969), 28–37.

Bahcall, John N.: Solar Neutrinos. Theoretical. *Physical Review Letters* 12/11 (1964), 300–302. doi:10.1103/PhysRevLett.12.300.

Bahcall, John N., Neta A. Bahcall, and Giora Shaviv: Present Status of the Theoretical Predictions for the 37Cl Solar-Neutrino Experiment. *Physical Review Letters* 20/21 (1968), 1209–1212. doi:10.1103/PhysRevLett.20.1209.

Bahcall, John N., B. T. Cleveland, R. Davis, I. Dostrovsky, J. C. Evans, W. Frati, G. Friedlander, et al.: Proposed Solar-Neutrino Experiment Using ^{71}Ga. *Physical Review Letters* 40/20 (1978), 1351–1354. doi:10.1103/PhysRevLett.40.1351.

Bahcall, John N., and Raymond Jr. Davis: The Evolution of Neutrino Astronomy. *Publications of the Astronomical Society of the Pacific* 112/770 (2000), 429–433. doi:10.1086/316545.

Bahcall, John N., and Raymond Jr. Davis: Solar Neutrinos. A Scientific Puzzle. *Science* 191/4224 (1976), 264–267. doi:10.1126/science.191.4224.264.

Bahcall, John N., and R. L. Sears: Solar Neutrinos. *Annual Review of Astronomy and Astrophysics* 10 (1972), 25–44. doi:10.1146/annurev.aa.10.090172.000325.

Balcar, Jaromír: *Wandel durch Wachstum in »dynamischen Zeiten«. Die Max-Planck-Gesellschaft 1955/57 bis 1972*. Berlin: GMPG-Preprint 2020.

Balcar, Jaromír: *Die Ursprünge der Max-Planck-Gesellschaft. Wiedergründung–Umgründung–Neugründung*. Berlin: GMPG-Preprint 2019.

Balcar, Jaromír: *Instrumentenbau–Patentvermarktung–Ausgründungen. Die Geschichte der Garching Instrumente GmbH*. Berlin: GMPG-Preprint 2018.

Balcar, Jaromír: *Politik auf dem Land. Studien zur bayerischen Provinz 1945 bis 1972*. München: Oldenbourg 2004.

Balick, Bruce, and Robert L. Brown: Intense Sub-Arcsecond Structure in the Galactic Center. *The Astrophysical Journal* 194 (1974), 265–270. doi:10.1086/153242.

Balsiger, Hans, Hugo Fechtig, and Johannes Geiss: A Close Look at Halley's Comet. *Scientific American* 259/3 (1988), 96–103. http://www.jstor.org/stable/24989232. Last accessed 10/30/2018.

Baneke, David: Let's Not Talk About Science: The Normalization of Big Science and the Moral Economy of Modern Astronomy. *Science, Technology, & Human Values* 45/1 (2020), 164–194. doi:10.1177/0162243919846600.

Bartels, Julius: Weltraumforschung. Methoden und Ergebnisse. In: Generalverwaltung der Max-Planck Gesellschaft zur Förderung der Wissenschaften e.V. (ed.): *Jahrbuch 1962 der Max-Planck-Gesellschaft zur Förderung der Wissenschaften e.V.* Göttingen 1962, 19–50.

Bartels, Julius: Erdmagnetismus II. Zeitliche Variationen, Beziehungen zur Sonnenphysik, zum Polarlicht, zur Ionosphäre. *Geophysics Part I.* Wiesbaden: Dieterich'sche Verlagsbuchhandlung 1948, 39–91.

Bartels, Julius: *Geophysics. Part I.* Wiesbaden: Dieterich'sche Verlagsbuchhandlung 1948.

Bartels, Julius: *Geophysics. Part II.* Wiesbaden: Dieterich'sche Verlagsbuchhandlung 1948.

Bartels, Julius: Zur Vorgeschichte der Weltraumforschung. *Die Naturwissenschaften* 49/14 (1962), 313–323. doi:10.1007/BF00602195.

Bartels, Julius, Walter Dieminger, and Alfred Ehmert: Max-Planck-Institut für Aeronomie in Lindau. In: Generalverwaltung der Max-Planck-Gesellschaft zur Förderung der Wissenschaften e.V. (ed.): *Jahrbuch 1961 der Max-Planck-Gesellschaft zur Förderung der Wissenschaften e.V. Teil II*. Göttingen 1962, 16–45.

Bartz, Olaf: *Wissenschaftsrat und Hochschulplanung. Leitbildwandel und Planungsprozesse in der Bundesrepublik Deutschland zwischen 1957 und 1975*. Dissertation/PhD Thesis. Köln: Universität zu Köln 2006. http://kups.ub.uni-koeln.de/volltexte/2006/1879/. Last accessed 7/31/2015.

Barucci, M.A., and M. Fulchignoni: Major Achievements of the Rosetta Mission in Connection with the Origin of the Solar System. *The Astronomy and Astrophysics Review* 25/1 (2017), 1–52. doi:10.1007/s00159-017-0103-8.

Baschek, B.: Albrecht Unsöld (20. April 1905–23. September 1995). *Mitteilungen der Astronomischen Gesellschaft* 79 (1996), 11–15. https://ui.adsabs.harvard.edu/#abs/1996PhyBl..52..890B. Last accessed 10/30/2018.

Baym, Gordon, Edward W. Kolb, Larry McLerran, T. P. Walker, and R. L. Jaffe: Is Cygnus X-3 Strange? *Physics Letters B* 160/1 (1985), 181–187. doi:10.1016/0370-2693(85)91489-3.

Becker, Astrid: Zurück in die Zukunft. *Süddeutsche Zeitung* (5/5/2018). https://www.sueddeutsche.de/muenchen/starnberg/25-jahre-spacelab-d2-zurueck-in-die-zukunft-1.3967737. Last accessed 6/3/2019.

Becker, Thomas, and Philip Rosin: *Geschichte der Universität Bonn. Die Natur- und Lebenswissenschaften*. Vol. 4. Göttingen: Vandenhoeck & Ruprecht 2018.

Begemann, Friedrich: Edelgase in Meteoriten. In: Generalverwaltung der Max-Planck-Gesellschaft (ed.): *Jahrbuch der Max-Planck-Gesellschaft zur Förderung der Wissenschaften e.V. 1972*. Göttingen 1972, 59–82.

Begemann, Friedrich: Noble Gases and Meteorites. *Meteoritics & Planetary Science* 31/2 (1996), 171–176. doi:10.1111/j.1945-5100.1996.tb02012.x.

Behrens, Ulrike, Ulrich Gensch, Thomas Naumann, and Michael Walter: Zeitreise. Vom Institut X zum DESY—eine deutsche Geschichte. Edited by Deutsches Elektronen-Synchrotron (DESY). DESY 2012. https://pr.desy.de/sites/sites_desygroups/sites_extern/site_pr/content/e104098/e104108/Broschuere_DESY_Zeuthen_Zeitreise_Web_ger.pdf. Last accessed 8/15/2020.

Bellini, Gianpaolo: The Impact of Borexino on the Solar and Neutrino Physics. *Nuclear Physics B* 908 (2016), 178–198. doi:10.1016/j.nuclphysb.2016.04.011.

Bellini, Gianpaolo, J. Benziger, D. Bick, S. Bonetti, G. Bonfini, D. Bravo, M. Buizza Avanzini, et al.: First Evidence of pep Solar Neutrinos by Direct Detection in Borexino. *Physical Review Letters* 108/5 (2012), 051302. doi:10.1103/PhysRevLett.108.051302.

Bellini, Gianpaolo, and J. A. Villar: The Borexino Experiment and the Results of the Counting Test Facility. *Nuclear Physics B—Proceedings Supplements*. Edited by A. Morales, and J. Morales 48/1 (1996), 363–369. doi:10.1016/0920-5632(96)00273-3.

Bennett, K., J. J. Burger, G. G. Lichti, B. G. Taylor, R. D. Wills, G. F. Bignami, G. Boella, et al.: Preliminary Results from the European Space Agency's COS-B Satellite for Gamma-Ray Astronomy. *NASA Conference Publication* 2 (1977), 27. https://ui.adsabs.harvard.edu/?#abs/1977NASCP...2...27B. Last accessed 11/20/2018.

Berezinsky, Vadim L., G. Fiorentini, and M. Lissia: LAST HOPE for an Astrophysical Solution to the Solar Neutrino Problem. *Physics Letters B* 365/1 (1996), 185–192. doi:10.1016/0370-2693(95)01241-9.

Bergman, Ronen: *Rise and Kill First. The Secret History of Israel's Targeted Assassinations*. New York, NY: Random House 2018.

Bernardini, Carlo, and Luisa Bonolis (eds.): *Enrico Fermi. His Work and Legacy*. Berlin: Springer-Verlag 2004. doi:10.1007/978-3-662-01160-7.

Bernhart, Simone, Alessandra Bertarini, Arno Muskens, and Walter Alef: The Bonn Astro/Geo Mark IV Correlator. *Annual Report*, 2008, 201–204.

Bernlöhr, Konrad, R. Eckmann, M. Feuerstack, S. Gamp, J. Heintze, G. Hermann, W. Hofmann, et al.: Tracking Detectors for High Energy γ-Astronomy. In: D. A. Leahy, R. B. Hicks, and D. Venkatesan (eds.): *23rd International Cosmic Ray Conference, Vol. 4, Held 19–30 July, 1993 at University of Calgary, Alberta, Canada*. 1993, 199–202. https://ui.adsabs.harvard.edu/#abs/1993ICRC....4..199B. Last accessed 12/11/2018.

Bernlöhr, Konrad, S. Gamp, G. Hermann, W. Hofmann, T. Kihm, J. Knöppler, G. Leffers, et al.: The Cosmic Ray Tracking (CRT) Detector System. *Nuclear Instruments and Methods in Physics Research Section A: Accelerators, Spectrometers, Detectors and Associated Equipment* 369/1 (1996), 284–292. doi:10.1016/0168-9002(95)00772-5.

Bernlöhr, Konrad, W. Hofmann, G. Leffers, V. Matheis, M. Panter, U. Trunk, M. Ulrich, et al.: Operation and Performance of the Cosmic Ray Tracking (CRT) Detector System. *Nuclear Instruments and Methods in Physics Research Section A: Accelerators, Spectrometers, Detectors and Associated Equipment* 369/1 (1996), 293–305. doi:10.1016/0168-9002(95)00771-7.

Bernstein, Jeremy: *Hitler's Uranium Club. The Secret Recordings at Farm Hall*. Woodbury, NY: American Institute of Physics 1996.

Besser, Bruno Philipp: *Austria's History in Space*. Noordwijk: ESA Publications Division 2004.

Bethe, Hans: Energy Production in Stars. *Physical Review* 55/1 (1939), 103–103. doi:10.1103/PhysRev.55.103.

Bethe, Hans: Energy Production in Stars. *Physical Review* 55/5 (1939), 434–456. doi:10.1103/PhysRev.55.434.

Bettini, Alessandro: *The Gran Sasso Laboratory, 1979–1999. A Vision Becomes Reality*. Assergi: INFN 1999.

Betz, Albert: Aus der Geschichte der Aerodynamischen Versuchsanstalt Göttingen. In: Generalverwaltung der Max-Planck-Gesellschaft (ed.): *Jahrbuch 1957 der Max-Planck-Gesellschaft zur Förderung der Wissenschaften e.V.* Göttingen 1957, 40–59.

Bibring, J.-P., H. Rosenbauer, H. Boehnhardt, S. Ulamec, J. Biele, S. Espinasse, B. Feuerbacher, et al.: The Rosetta Lander ("Philae") Investigations. *Space Science Reviews* 128/1 (2007), 205–220. doi:10.1007/s11214-006-9138-2.

Biermann, Ludwig: Plasmaphysik im Kosmos und für die Fusion. In: Max-Planck Gesellschaft (ed.): *Ludwig Biermann. 1907–1986.* München 1988, 63–76.

Biermann, Ludwig: On the History of the Solar Wind Concept. In: Wilfried Schröder, and International Association of Geomagnetism and Aeronomy (eds.): *Historical Events and People in Geosciences. Selected Papers from the Symposia of the Interdivisional Commission on History of IAGA during the IUGG General Assembly, Held in Hamburg, 1983.* Frankfurt am Main: Peter Lang 1985, 39–47.

Biermann, Ludwig: *The Plasma in Interplanetary Space.* Technical Report NASA-TN-D-1901. United States: NASA 1963. https://ntrs.nasa.gov/search.jsp?R=19630012233&hterms=Biermann&qs=N%3D0%26Ntk%3DAuthor-Name%26Ntt%3DBiermann%2C%2520L.%26Ntx%3Dmode%2520matchall. Last accessed 11/2/2017.

Biermann, Ludwig: Jahresberichte deutscher astronomischer Institute für 1961 München, Max-Planck-Institut für Physik und Astrophysik, Institut für Astrophysik. *Mitteilungen der Astronomischen Gesellschaft* 15 (1962), 68–74. http://adsabs.harvard.edu/abs/1962MitAG..15...68. Last accessed 10/30/2018.

Biermann, Ludwig: 50 Jahre physikalische Grundlagenforschung in der Kaiser-Wilhelm und der Max-Planck-Gesellschaft. *Die Naturwissenschaften* 48/1 (1961), 2–10. doi:10.1007/BF00600935.

Biermann, Ludwig: Relations between Plasma Physics and Astrophysics. *Reviews of Modern Physics* 32/4 (1960), 1008–1011. doi:10.1103/RevModPhys.32.1008.

Biermann, Ludwig: Recent Work on Controlled Thermonuclear Fusion in Germany (Federal Republic). *Proceedings of the Second United Nations International Conference on the Peaceful Uses of Atomic Energy. Held in Geneva 1 September–13 September 1958. Vol. 31. Theoretical and Experimental Aspects of Controlled Nuclear Fusion.* Geneva: United Nations 1958, 21–26. http://www-naweb.iaea.org/napc/physics/2ndgenconf/data/Proceedings%201958/NG900088.pdf. Last accessed 10/30/2018.

Biermann, Ludwig: Solar Corpuscular Radiation and the Interplanetary Gas. *The Observatory* 77 (1957), 109–110. https://ui.adsabs.harvard.edu/abs/1957Obs....77.109B/abstract. Last accessed 8/14/2020.

Biermann, Ludwig: Elektronische Rechenmaschinen und physikalische Forschung. In: Generalverwaltung der Max-Planck-Gesellschaft (ed.): *Jahrbuch 1956 der Max-Planck-Gesellschaft zur Förderung der Wissenschaften e.V.* Göttingen 1956, 19–33.

Biermann, Ludwig: Physical Processes in Comet Tails and Their Relation to Solar Activity. In: P. Swings (ed.): *La Physique Des Comètes. Communications Présentées Au Quatrième Colloque International d'Astrophysique, Tenu à Liège Les 19, 20 et 21 Septembre 1952. Mémoires de La Société Royale Des Sciences de Liège. Quatrième Série.* Liège: Institut d'Astrophysique de l'Université de Liège 1953, 251–262.

Biermann, Ludwig: Origin and Propagation of Cosmic Rays. *Annual Review of Nuclear Science* 2 (1953), 235–364. doi:10.1146/annurev.ns.02.120153.002003.

Biermann, Ludwig: Entstehung von Magnetfeldern in bewegten Plasmen. *Annalen der Physik* 445/8 (1952), 413–417. doi:10.1002/andp.19524450802.

Biermann, Ludwig: Über den Schweif des Kometen Halley im Jahre 1910. *Zeitschrift für Naturforschung A* 7/1 (1952), 127–136. doi:10.1515/zna-1952-0122.

Biermann, Ludwig: Kometenschweife und solare Korpuskularstrahlung. *Zeitschrift für Astrophysik* 29 (1951), 274–286. https://ui.adsabs.harvard.edu/#abs/1951ZA.....29..274B. Last accessed 10/30/2018.

Biermann, Ludwig: Über den Ursprung der Magnetfelder auf Sternen und im interstellaren Raum. *Zeitschrift für Naturforschung A* 5/2 (1950), 65–71. doi:10.1515/zna-1950-0201.

Biermann, Ludwig: Der innere Aufbau der Sterne. In: Paul Bruggencate (ed.): *Astronomie, Astrophysik und Kosmogonie.* Wiesbaden: Dieterich'sche Verlagsbuchhandlung 1948, 161–179.

Biermann, Ludwig: Über die Ursache der chromosphärischen Turbulenz und des UV-Exzesses der Sonnenstrahlung. *Zeitschrift fur Astrophysik* 25 (1948), 161–177. https://ui.adsabs.harvard.edu/#abs/1948ZA.....25..161B. Last accessed 2/19/2019.

Biermann, Ludwig: Normierte Wellenfunktionen verschiedener Zustände des Leuchtelektrons und Oszillatorenstärken der Übergänge zwischen ihnen für Na I, K I, Mg II, Si II und Al I. *Veröffentlichungen der Universitäts-Sternwarte zu Göttingen* 5 (1947), 245–248. http://adsabs.harvard.edu/abs/1947VeGoe...5..245B. Last accessed 10/30/2018.

Biermann, Ludwig: Neuere Fortschritte der Theorie des inneren Aufbaues und der Entwicklung der Sterne. In: Ferdinand Trendelenburg (ed.): *Ergebnisse der Exakten Naturwissenschaften.* Berlin: Springer-Verlag 1945, 1–49.

Biermann, Ludwig: Die Oszillatorenstärken einiger Linien in den Spektren des Na I, K I und Mg II. *Zeitschrift für Astrophysik* 22 (1943), 157–164. http://adsabs.harvard.edu/abs/1943ZA.....22..157B. Last accessed 10/30/2018.

Biermann, Ludwig: Untersuchungen über den inneren Aufbau der Sterne, III. Über Sternmodelle mit entartetem Kern. *Veroeffentlichungen der Universitaets-Sternwarte zu Goettingen* 2 (1931), 163–171.

Biermann, Ludwig, and Heinz Billing: Der Stand der Entwicklung und Ausnutzung der elektronischen Rechenanlagen in Göttingen. *Mitteilungen aus der Max-Planck-Gesellschaft zur Förderung der Wissenschaften* Heft 1 (1954), 35–38.

Biermann, Ludwig, and Paul ten Bruggencate: Über die Ursachen der hohen Temperatur der Sonnenkorona nebst einer Bemerkung über das Nachthimmellicht. *Veröffentlichungen der Universitäts-Sternwarte zu Göttingen* 5 (1947), 223–228.

Biermann, Ludwig, Otto Haxel, and Arnulf Schlüter: Neutrale Ultrastrahlung von der Sonne. *Zeitschrift für Naturforschung A* 6/1 (1951), 47–48. doi:10.1515/zna-1951-0107.

Biermann, Ludwig, and Reimar Lüst: Max-Planck-Institut für Physik und Astrophysik. Institut für Astrophysik und Institut für extraterrestrische Physik. *Mitteilungen der Astronomischen Gesellschaft* 31 (1972), 323–350. https://ui.adsabs.harvard.edu/#abs/1972MitAG..31..323B. Last accessed 4/29/2018.

Biermann, Ludwig, and Reimar Lüst: Max-Planck-Institut für Physik und Astrophysik. Institut für Astrophysik und Institut für extraterrestrische Physik. *Mitteilungen der Astronomischen Gesellschaft* 29 (1971), 86–112. http://adsabs.harvard.edu/abs/1971MitAG..29...86B. Last accessed 10/30/2018.

Biermann, Ludwig, and Reimar Lüst: Max-Planck-Institut für Physik und Astrophysik. Institut für Astrophysik und Institut für extraterrestrische Physik. *Mitteilungen der Astronomischen Gesellschaft* 28 (1970), 79–105. http://adsabs.harvard.edu/abs/1970MitAG..28...79B. Last accessed 10/30/2018.

Biermann, Ludwig, and Reimar Lüst: Jahresberichte astronomischer Institute für 1965, München Max-Planck-Institut für Physik und Astrophysik, Institute für Astrophysik und extraterrestrische Physik. *Mitteilungen der Astronomischen Gesellschaft* 20 (1966), 67–79. http://adsabs.harvard.edu/abs/1966MitAG..20...66. Last accessed 10/30/2018.

Biermann, Ludwig, and Reimar Lüst: Jahresberichte astronomischer Institute für 1964, München Max-Planck-Institut für Physik und Astrophysik, Institute für Astrophysik und extraterrestrische Physik. *Mitteilungen der Astronomischen Gesellschaft* 18 (1965), 57–66. https://ui.adsabs.harvard.edu/abs/1965MitAG..18...57. Last accessed 10/30/2018.

Biermann, Ludwig, and Reimar Lüst: Jahresberichte astronomischer Institute 1963. München Max-Planck-Institut für Physik und Astrophysik. Institut für Astrophysik. *Mitteilungen der Astronomischen Gesellschaft Hamburg* 17 (1964), 180–186. http://adsabs.harvard.edu/abs/1964MitAG..17..180. Last accessed 4/13/2020.

Biermann, Ludwig, and Reimar Lüst: Remarks on the Energy of the Non-Thermal Radio-Frequency Emission. In: Hendrik Christoffel Van de Hulst (ed.): *Radio Astronomy, Proceedings from 4th IAU Symposium. International Astronomical Union.* Radio astronomy. Cambridge: Cambridge University Press 1957, 354–355. http://adsabs.harvard.edu/abs/1957IAUS....4..354B. Last accessed 6/25/2020.

Biermann, Ludwig, Reimar Lüst, Rhea Lüst, and Heinrich U. Schmidt: Zur Untersuchung des interplanetarischen Mediums mit Hilfe künstlich eingebrachter Ionenwolken. *Zeitschrift für Astrophysik* 53 (1961), 226–236. http://articles.adsabs.harvard.edu/pdf/1961ZA.....53..226B. Last accessed 10/30/2018.

Biermann, Ludwig, Reimar Lüst, Rhea Lüst, and H.U. Schmidt: Zur Untersuchung des interplanetaren Mediums mit Hilfe künstlich eingebrachter Ionenwolken. *Zeitschrift für Astrophysik* 53 (1961), 163–176.

Biermann, Ludwig, and Arnulf Schlüter: Cosmic Radiation and Cosmic Magnetic Fields. II. Origin of Cosmic Magnetic Fields. *Physical Review* 82/6 (1951), 863–868. doi:10.1103/PhysRev.82.863.

Biermann, Ludwig, and Arnulf Schlüter: Interstellare Magnetfelder. *Zeitschrift für Naturforschung A* 5/5 (1950), 237–251. doi:10.1515/zna-1950-0501.

Biermann, Ludwig, and Eleonore Trefftz: Wellenfunktionen und Übergangswahrscheinlichkeiten der Leuchtelektronen des Atoms Mg I. I. Teil. *Zeitschrift für Astrophysik* 26 (1949), 213–239. http://adsabs.harvard.edu/abs/1949ZA.....26..213B. Last accessed 11/2/2017.

Biermann, Ludwig, and Carl F. Weizsäcker: Jahresberichte deutscher Sternwarten und Institute für 1955, Göttingen, Max-Planck-Institut für Physik. *Mitteilungen der Astronomischen Gesellschaft* 7 (1956), 106–108. http://articles.adsabs.harvard.edu/pdf/1956MitAG...7..106. Last accessed 10/30/2018.

Biermann, Ludwig, and Carl Friedrich von Weizsäcker: Jahresberichte deutscher Sternwarten für 1950. Göttingen. Max-Planck-Institut für Physik (Abteilungen für Astrophysik und für theoretische Physik). *Mitteilungen der Astronomischen Gesellschaft* 2 (1950), 41–43. http://adsabs.harvard.edu/abs/1950MitAG...2...30. Last accessed 10/30/2018.

Biermann, Ludwig, and Peter Wellmann: Physik der Sternatmosphären. In: Paul ten Bruggencate (ed.): *Astronomie, Astrophysik und Kosmogonie*. Wiesbaden: Dieterich'sche Verlagsbuchhandlung 1948, 119–159.

Bignami, Giovanni F., and W. Hermsen: Galactic Gamma-Ray Sources. *Annual Review of Astronomy and Astrophysics* 21 (1983), 67–108. doi:10.1146/annurev.aa.21.090183.000435.

Bilenky, Samoil M.: Bruno Pontecorvo and the Neutrino. *Physics-Uspekhi* 57/5, 489–496. doi:10.3367/UFNe.0184.201405g.0531.

Billing, Heinz: Ludwig Biermann und die Rechenmaschinen. In: Max-Planck Gesellschaft (ed.): *Ludwig Biermann. 1907–1986*. München 1988, 51–62.

Billing, Heinz: *Meine Lebenserinnerungen*. Garching: Selbstverlag 1994.

Billing, Heinz: Schnelle Rechenmaschinenspeicher und ihre Geschwindigkeits- und Kapazitätsgrenzen. In: Generalverwaltung der Max-Planck Gesellschaft zur Förderung der Wissenschaften e.V. (ed.): *Jahrbuch 1962 der Max-Planck-Gesellschaft zur Förderung der Wissenschaften e.V.* Göttingen 1962, 52–79.

Billing, Heinz, Peter Kafka, Karl Maischberger, Friedrich Meyer, and Walther Winkler: Results of the Munich-Frascati Gravitational-Wave Experiment. *Nuovo Cimento, Lettere* 12/4 (1975), 111–116. doi:10.1007/BF02790471.

Billing, Heinz, Walter Winkler, Roland Schilling, Albrecht Rüdiger, Karl Maischberger, and Lise Schnupp: The Munich Gravitational Wave Detector Using Laser Interferometry. In: Pierre Meystre, and M.O. Scully (eds.): *Quantum Optics, Experimental Gravity, and Measurement Theory*. Conference "Quantum Optics and Experimental General Relativity", held August 16–29, 1981 in Bad Windsheim, Federal Republic of Germany. Boston, MA: Springer 1983, 525–566. doi:10.1007/978-1-4613-3712-6_23.

Billings, Lee: Space Science: The Telescope That Ate Astronomy. *Nature* 467/7319 (2010), 1028–1030. doi:10.1038/4671028a.

Biswas, N. N., L. Ceccarelli-Fabbrichesi, M. Ceccarelli, K. Gottstein, N. C. Varshneya, and P. Waloschek: Decay Modes and Mean Life of Scattered K^+-mesons. *Il Nuovo Cimento (1955-1965)* 4/3 (1956), 631–636. doi:10.1007/BF02745387.

Blaauw, Adriaan: *ESO Historical Archives (EHA). Inventory per December 1992*. Garching: European Southern Observatory 1992.

Blaauw, Adriaan: *ESO's Early History. The European Southern Observatory from Concept to Reality*. ESO 1991.

Blackett, P. M. S.: Prof. E. Regener. 4469. *Nature* 175/4469 (1955), 1107–1108. doi:10.1038/1751107a0.

Blamont, Jacques: The Beginning of Space Experiments in Munich. In: Gerhard Haerendel, and Bruce Battrick (eds.): *Topics in Plasma-, Astro- and Space Physics. A Volume Dedicated to Reimar Lüst on the Occasion of His 60th Birthday*. Garching: Max-Planck-Institut für Physik und Astrophysik 1983, 161–164.

Blamont, Jacques E.: Alkali Metal Cloud Experiments in the Upper Atmosphere. In: Johan A. M. Bleeker, Johannes Geiss, and Martin C. E. Huber (eds.): *The Century of Space Science*. Dordrecht: Kluwer Academic Publishers 2001, 189–202.

Bleeker, Johan A. M., Johannes Geiss, and Martin C. E. Huber (eds.): *The Century of Space Science*. Dordrecht: Kluwer Academic Publishers 2001.

Bleuler, Konrad: Werner Heisenberg's Ideas on Particle Physics in the Light of Recent Achievements. *Zeitschrift Für Naturforschung A* 45/9 und 10 (2014), 1051–1058. doi:10.1515/zna-1990-9-1001.

Blum, Alexander S.: *Heisenberg's 1958 Weltformel and the Roots of Post-Empirical Physics*. Cham, Switzerland: Springer Nature 2019. doi:10.1007/978-3-030-20645-1.

Blum, Alexander S.: The State Is Not Abolished, It Withers Away. How Quantum Field Theory Became a Theory of Scattering. *Studies in History and Philosophy of Modern Physics*, 2017, 33, 33. doi:10.1016/j.shpsb.2017.01.004.

Blum, Alexander, Domenico Giulini, Roberto Lalli, and Jürgen Renn: Editorial Introduction to the Special Issue "The Renaissance of Einstein's Theory of Gravitation". *The European Physical Journal H* 42/2 (2017), 45–105. doi:10.1140/epjh/e2017-80023-3.

Blum, Alexander, Roberto Lalli, and Jürgen Renn: Gravitational Waves and the Long Relativity Revolution. *Nature Astronomy* 2 (2018), 534–543. doi:10.1038/s41550-018-0472-6.

Blum, Alexander, Roberto Lalli, and Jürgen Renn: The Reinvention of General Relativity. A Historiographical Framework for Assessing One Hundred Years of Curved Space-Time. *Isis* 106/3 (2015), 598–620. doi:10.1086/683425.

Blum, Alexander, Roberto Lalli, and Jürgen Renn: The Renaissance of General Relativity. How and Why It Happened. *Annalen Der Physik* 528/5 (2016), 344–349. doi:10.1002/andp.201600105.

Blum, Alexander S., Roberto Lalli, and Jürgen Renn (eds.): *The Renaissance of General Relativity in Context*. Vol. 16. Basel: Birkhäuser 2020. doi:10.1007/978-3-030-50754-1.

Bodenheimer, Peter H.: Louis George Henyey. February 3, 1910-February 18, 1970. In: National Academy of Sciences (ed.): *Biographical Memoirs*. Washington, DC: The National Academies Press 1995, 169–189.

Bodenschatz, Eberhard, and Michael Eckert: Prandtl and the Göttingen School. In: Peter A. Davidson, Yukio Kaneda, Keith Moffatt, and Katepalli R. Sreenivasan (eds.): *A Voyage Through Turbulence*. Cambridge: Cambridge University Press 2011, 40–100.

Boenke, Susan: *Entstehung und Entwicklung des Max-Planck-Instituts für Plasmaphysik 1955–1971*. Frankfurt am Main: Campus Verlag 1991.

Bohr, Niels: Nuclear Photo-Effects (Letter to the Editor). *Nature* 141/3564 (1938), 326–327. doi:10.1038/141326a0.

Boller, Th., M.J. Freyberg, J. Trümper, F. Haberl, W. Voges, and K. Nandra: Second ROSAT All-Sky Survey (2RXS) Source Catalogue. *Astronomy & Astrophysics* 588/A103 (2016), 1–26. doi:10.1051/0004-6361/201525648.

Bonnell, J. T., and R. W. Klebesadel: A Brief History of the Discovery of Cosmic Gamma-Ray Bursts. *AIP Conference Proceedings*. Gamma-ray bursts. 3rd Huntsville symposium. Huntsville, Alabama (USA): AIP 1996, 977–980. doi:10.1063/1.51630.

Bonnet, Roger, and Gerhard Haerendel: Reimar Lüst (1923–2020). *Space Research Today* 208 (2020), 4–6. doi:10.1016/j.srt.2020.07.006.

Bonolis, Luisa: Emilio Gino Segrè. Research Profile. *Lindau Nobel Mediatheque*, 4/11/2018. http://www.mediatheque.lindau-nobel.org/research-profile/laureate-segr. Last accessed 4/11/2018.

Bonolis, Luisa: Hannes Olof Gösta Alfvén. *Lindau Nobel Mediatheque*. http://www.mediatheque.lindau-nobel.org/research-profile/laureate-alfveen#page=all. Last accessed 2/28/2018.

Bonolis, Luisa: Stellar Structure and Compact Objects before 1940. Towards Relativistic Astrophysics. *The European Physical Journal H* 42/2 (2017), 311–393. doi:10.1140/epjh/e2017-80014-4.

Bonolis, Luisa: J. Hans D. Jensen. Research Profile. *Lindau Nobel Laureate Meetings*, 2014. https://www.mediatheque.lindau-nobel.org/research-profile/laureate-jensen. Last accessed 9/9/2018.

Bonolis, Luisa: Stellar Structure and Compact Objects before 1940. Towards Relativistic Astrophysics. *The European Physical Journal H* 42/2 (2017), 311–393. doi:10.1140/epjh/e2017-80014-4.

Bonolis, Luisa: International Scientific Cooperation During the 1930s. Bruno Rossi and the Development of the Status of Cosmic Rays into a Branch of Physics. *Annals of Science* 71/3 (2014), 355–409. doi:10.1080/00033790.2013.827074.

Bonolis, Luisa: From Cosmic Ray Physics to Cosmic Ray Astronomy. Bruno Rossi and the Opening of New Windows on the Universe. *Astroparticle Physics* 53 (2014), 67–85. doi:10.1016/j.astropartphys.2013.05.008.

Bonolis, Luisa: *Il sogno di Prometeo. Dagli anni della ricostruzione alla nascita delle ricerche sulla fusione nucleare in Italia*. Labirinti 2022.

Bonolis, Luisa: Walther Bothe and Bruno Rossi. The Birth and Development of Coincidence Methods in Cosmic-Ray Physics. *American Journal of Physics* 79/11 (2011), 1133–1150. doi:10.1119/1.3619808.

Bonolis, Luisa: Bruno Pontecorvo. From Slow Neutrons to Oscillating Neutrinos. *American Journal of Physics* 73/6 (2005), 487–499. doi:10.1119/1.1852540.

Bonolis, Luisa, and Juan-Andres Leon: Gravitational-Wave Research as an Emerging Field in the Max Planck Society: The Long Roots of GEO600 and of the Albert Einstein Institute. In: Alexander S. Blum, Roberto Lalli, and Jürgen Renn (eds.): *The Renaissance of General Relativity in Context*. Basel: Birkhäuser 2020, 285–361. doi:10.1007/978-3-030-50754-1_9.

Borione, A., M. C. Chantell, C. E. Covault, J. W. Cronin, B. E. Fick, J. W. Fowler, L. F. Fortson, et al.: High Statistics Search for Ultrahigh Energy Gamma-Ray Emission from Cygnus X-3 and Hercules X-1. *Physical Review D* 55/4 (1997), 1714–1731. doi:10.1103/PhysRevD.55.1714.

Bothe, Walther: Coincidence Method. *Science* 122/3175 (1955), 861–863. http://www.jstor.org/stable/1749457. Last accessed 10/30/2018.

Bothe, Walther: Erich Regener 70 Jahre. *Zeitschrift für Naturforschung A* 6/11 (1951), 564–567. doi:10.1515/zna-1951-1101.

Bothe, Walther, and Siegfried Flügge (eds.): *Nuclear Physics, Cosmic Rays*. Vol. 1. Wiesbaden: Dieterich'sche Verlagsbuchhandlung 1948.

Bothe, Walther, and Wolfgang Gentner: Atomumwandlungen durch γ-Strahlen. *Zeitschrift für Physik* 106/3–4 (1937), 236–248. doi:10.1007/BF01340320.

Bothe, Walther, and Werner Kolhörster: Die Natur der Höhenstrahlung. *Naturwissenschaften* 17/17 (1929), 271–273. doi:10.1007/BF01507590.

Bott-Bodenhausen, M., I. Holl, A. Kabelschacht, A. Karle, E. Lorenz, R. Maier, M. Merck, et al.: A New Air Cherenkov Counter Concept for the Observation of Extended Air Showers. *Nuclear Instruments and Methods in Physics Research Section A: Accelerators, Spectrometers, Detectors and Associated Equipment* 315/1–3 (1992), 236–251. doi:10.1016/0168-9002(92)90709-D.

Bott-Bodenhausen, M., I. Holl, A. Kabelschact, A. Karle, E. Lorenz, R. Maier, M. Merck, et al.: Airobicc–a New Array of Angle Integrating Cerenkov Counters for Improved γ/Hadron Separation in Extended Air Showers. *AIP Conference Proceedings* 220/1 (1991), 305–309. doi:10.1063/1.40314.

Bower, Tom: *The Paperclip Conspiracy. The Hunt for the Nazi Scientists.* Boston: Little, Brown 1987.

Bradaschia, Carlo: VIRGO 20th Anniversary. *H—The Gravitational Voice. Special Anniversary Edition* (2009), 2–12. http://www.ego-gw.it/public/hletter/doc/h_Special_Edition.pdf. Last accessed 1/29/2019.

Bradaschia, Carlo, R. Del Fabbro, A. Di Virgilio, A. Giazotto, H. Kautzky, V. Montelatici, D. Passuello, et al.: The VIRGO Project: A Wide Band Antenna for Gravitational Wave Detection. *Nuclear Instruments and Methods in Physics Research Section A: Accelerators, Spectrometers, Detectors and Associated Equipment* 289/3 (1990), 518–525. doi:10.1016/0168-9002(90)91525-G.

Bradt, Hale, Takaya Ohashi, and Kenneth A. Pounds: X-Ray Astronomy Missions. *Annual Review of Astronomy and Astrophysics* 30 (1992), 391–427. doi:10.1146/annurev.aa.30.090192.002135.

Braginskii, Vladimir B.: Gravitational Radiation and the Prospect of Its Experimental Discovery. *Soviet Physics Uspekhi* 8/4 (1966), 513–521. http://stacks.iop.org/0038-5670/8/i=4/a=R02. Last accessed 7/24/2018.

Braginskii, Vladimir B., A. B. Manukin, E. I. Popov, V. N. Rudenko, and A. A. Khorev: Search for Gravitational Radiation of Extraterrestrial Origin. *Journal of Experimental and Theoretical Physics Letters* 16/3 (1972), 108–112. http://www.jetpletters.ac.ru/ps/1759/article_26757.shtml. Last accessed 6/12/2018.

Braginskii, Vladimir B., Ya B. Zel'Dovich, and V. N. Rudenko: Reception of Gravitational Radiation of Extraterrestrial Origin. *Soviet Journal of Experimental and Theoretical Physics Letters* 10 (1969), 280–283. https://ui.adsabs.harvard.edu/#abs/1969JETPL..10..280B. Last accessed 10/25/2018.

Bramanti, Donato, and Karl Maischberger: Construction and Operation of a Weber-Type Gravitational-Wave Detector and of a Divided-Bar Prototype. *Nuovo Cimento, Lettere* 4/17 (1972), 1007–1013. doi:10.1007/BF02757124.

Bramanti, Donato, Karl Maischberger, and Donald Parkinson: Optimization and Data Analysis of the Frascati Gravitational-Wave Detector. *Lettere al Nuovo Cimento* 7/14 (1973), 665–670. doi:10.1007/BF02728048.

Brandt, Leo: *Forschen und Gestalten. Reden und Aufsätze 1930–1962*. Köln: Westdeutscher Verlag 1962.

Brecht, D., S. Gamp, W. Hofmann, T. Kihm, J. Knöppler, V. Matheis, M. Panter, et al.: Cosmic Ray Tracking—a New Approach to High-Energy γ-Astronomy. *Nuclear Instruments and Methods in Physics Research Section A: Accelerators, Spectrometers, Detectors and Associated Equipment* 323/1 (1992), 60–64. doi:10.1016/0168-9002(92)90269-A.

Breuer, Reinhard, and Uwe Schumacher: *Max-Planck-Institut für Plasmaphysik. Garching bei München*. Edited by Max-Planck-Gesellschaft. München: Max-Planck-Gesellschaft 1982.

Bridge, H.S., C. Dilworth, A.J. Lazarus, E.F. Lyon, B. Rossi, and F. Scherb: Direct Observations of the Interplanetary Plasma. *Journal of the Physical Society of Japan* 17/Supplement A-II (1962), 553–559.

Briel, U. G., E. Pfeffermann, B. Aschenbach, H. Brauninger, J. Trumper, E. P. Ney, T. F. Flowers, et al.: Supernova 1987A in the Large Magellanic Cloud. *International Astronomical Union Circular* 4452 (1987), 1. https://ui.adsabs.harvard.edu/abs/1987IAUC.4452....1B/abstract. Last accessed 6/5/2019.

Brocke, Bernhard vom: Die Kaiser-Wilhelm-Gesellschaft im Kaiserreich. Vorgeschichte, Gründung und Entwicklung bis zum Ausbruch des Ersten Weltkriegs. In: Rudolf Vierhaus, and Bernhard vom Brocke (eds.): *Forschung im Spannungsfeld von Politik und Gesellschaft. Geschichte und Struktur der Kaiser-Wilhelm-/Max-Planck-Gesellschaft*. Stuttgart: Deutsche Verlags-Anstalt 1990, 17–162.

Brocke, Bernhard vom: Die Kaiser-Wilhelm-Gesellschaft in der Weimarer Republik. Ausbau zu einer gesamtdeutschen Forschungsorganisation (1918–1933). In: Rudolf Vierhaus, and Bernhard vom Brocke (eds.): *Forschung im Spannungsfeld von Politik und Gesellschaft. Geschichte und Struktur der Kaiser-Wilhelm-/Max-Planck-Gesellschaft*. Stuttgart: Deutsche Verlags-Anstalt 1990, 197–355.

Brown, Laurie M., Max Dresden, and Lillian Hoddeson (eds.): *Pions to Quarks. Particle Physics in the 1950s*. Cambridge: Cambridge University Press 1989.

Brown, M. Laurie, and Lillian Hoddeson (eds.): *The Birth of Particle Physics. Based on a Fermilab Symposium*. Cambridge: Cambridge University Press 1983.

Brüche, Ernst: Physiker-Tagung in Göttingen. Vorträge von der Göttinger Tagung. *Physikalische Blätter* 3/9 (1947), 317–325. doi:10.1002/phbl.19470030909.

Bruggencate, Paul ten: *Astronomie, Astrophysik und Kosmogonie*. Wiesbaden: Dieterich'sche Verlagsbuchhandlung 1948.

Büchner, Jörg (ed.): *Geschichte des Fachverbands Extraterrestrische Physik und der Arbeitsgemeinschaft Extraterrestrische Forschung*. Katlenburg-Lindau: Arbeitsgemeinschaft Extraterrestrische Forschung E.V. 2009.

Buras, Andrzej J., John Ellis, Mary K. Gaillard, and Dimitri V. Nanopoulos: Aspects of the Grand Unification of Strong, Weak and Electromagnetic Interactions. *Nuclear Physics B* 135/1 (1978), 66–92. doi:10.1016/0550-3213(78)90214-6.

Burbidge, E. Margaret, Geoffrey R. Burbidge, William A. Fowler, and Fred Hoyle: Synthesis of the Elements in Stars. *Reviews of Modern Physics* 29/4 (1957), 547–650. doi:10.1103/RevModPhys.29.547.

Burbidge, Geoffrey R.: Estimates of the Total Energy in Particles and Magnetic Field in the Non-Thermal Radio Sources. *Astrophysical Journal* 129 (1959), 849–852. doi:10.1086/146680.

Butenandt, Adolf: *Das Werk eines Lebens. Wissenschaftspolitische Aufsätze, Ansprachen und Reden*. Vol. 2. Göttingen: Vandenhoeck & Ruprecht 1981.

Cameron, Alastair G.W.: Nuclear Astrophysics. *Annual Review of Nuclear Science* 8/1 (1958), 299–326. doi:10.1146/annurev.ns.08.120158.001503.

Carlson, Per, and Alessandro De Angelis: Nationalism and Internationalism in Science. The Case of the Discovery of Cosmic Rays. *The European Physical Journal H* 35/4 (2011), 309. doi:10.1140/epjh/e2011-10033-6.

Carlson, Per, and Alan A. Watson: Erich Regener and the Ionisation Maximum of the Atmosphere. *History of Geo- and Space Sciences* 5 (2014), 175–182. doi:10.5194/hgss-5-175-2014.

Carpenter, John: Introduction to Radio Facilities: Millimeter and Submillimeter Interferometers. Academia Sinica, 2016. https://events.asiaa.sinica.edu.tw/school/20160815/talk/jcarpenter0817.pdf. Last accessed 10/8/2020.

Carson, Cathryn: *Heisenberg in the Atomic Age. Science and the Public Sphere*. Cambridge: Cambridge University Press 2010.

Carson, Cathryn: Beyond Reconstruction. CERN's Second-Generation Accelerator Program as an Indicator of Shifts in West-German Science. In: Helmuth Trischler, and Mark Walker (eds.): *Physics and Politics. Research and Research Support in Twentieth Century Germany in International Perspective*. Stuttgart: Steiner 2010, 107–130.

Carson, Cathryn: Heisenberg als Wissenschaftsorganisator. In: Christian Kleint, Helmut Rechenberg, and Gerald Wiemers (eds.): *Werner Heisenberg, 1901–1976. Beiträge, Berichte, Briefe. Festschrift zu seinem 100. Geburtstag*. Leipzig: Verlag der Sächsischen Akademie der Wissenschaften zu Leipzig 2005, 214–222.

Carson, Cathryn: Nuclear Energy Development in Postwar West Germany. Struggles over Cooperation in the Federal Republic's First Reactor Station. *History and Technology* 18/3 (2002), 233–270. doi:10.1080/07341510220000020166.

Carson, Cathryn: *Particle Physics and Cultural Politics. Werner Heisenberg and the Shaping of a Role for the Physicist in Postwar West Germany*. Dissertation. Cambridge, MA: Harvard University 1995. ProQuest.

Cassidy, David: Controlling German Science, I: U.S. and Allied Forces in Germany, 1945–1947. *Historical Studies in the Physical and Biological Sciences* 24/2 (1994), 197–235. doi:10.2307/27757723.

Cassidy, David: Controlling German Science, II: Bizonal Occupation and the Struggle over West German Science Policy, 1946–1949. *Historical Studies in the Physical and Biological Sciences* 26/2 (1996), 197–239. doi:10.2307/27757762.

Cassidy, David: *Farm Hall and the German Atomic Project of World War II. A Dramatic History*. Cham, Switzerland: Springer 2017.

Cassidy, David C.: *Uncertainty. The Life and Science of Werner Heisenberg*. New York: W. H. Freeman 1992.

Cesarsky, Catherine, and Alberto Salama (eds.): *ISO Science Legacy. A Compact Review of ISO Major Achievements*. Dordrecht: Springer 2005.

Cesarsky, Catherine, and Heinrich Völk: Cosmic Ray Penetration into Molecular Clouds. *Astronomy and Astrophysics* 70 (1978), 367–377. https://ui.adsabs.harvard.edu/#abs/1978A&A....70..367C. Last accessed 5/16/2018.

Chadwick, James: Possible Existence of a Neutron. *Nature* 129/3252 (1932), 312. doi:10.1038/129312a0.

Chaisson, Eric: *The Hubble Wars. Astrophysics Meets Astropolitics in the Two-Billion-Dollar Struggle over the Hubble Space Telescope*. New York, NY: HarperCollins Publishers 1994.

Chamberlain, Owen, W. W. Chupp, A. G. Ekspong, G. Goldhaber, S. Goldhaber, E. J. Lofgren, E. Segrè, et al.: Example of an Antiproton-Nucleon Annihilation. *Physical Review* 102/3 (1956), 921–923. doi:10.1103/PhysRev.102.921.

Chapman, Sydney: Julius Bartels. *Quarterly Journal of the Royal Astronomical Society* 6 (1965), 235–245. https://ui.adsabs.harvard.edu/abs/1965QJRAS...6..235. Last accessed 5/3/2020.

Chapman, Sidney: A Theory of Upper-Atmospheric Ozone. *Memoirs of the Royal Meterological Society* 3/26 (1930), 103–125.

Chapman, Sidney: On Ozone and Atomic Oxygen in the Upper Atmosphere. *The London, Edinburgh, and Dublin Philosophical Magazine and Journal of Science* 10/64 (1930), 369–383. https://doi.org/10.1080/14786443009461588. Last accessed 3/1/2018.

Chapman, Sydney, and Harold Zirin: Notes on the Solar Corona and the Terrestrial Ionosphere. *Smithsonian Contributions to Astrophysics* 2 (1957), 1–14. http://adsabs.harvard.edu/abs/1957SCoA....2....1C. Last accessed 2/2/2020.

Cherenkov Telescope Array Observatory: CTA Prototype LST-1 Detects Very High-Energy Emission from the Crab Pulsar, 6/22/2020. https://www.cta-observatory.org/lst1-detects-vhe-emission-from-crab-pulsar/. Last accessed 3/27/2021.

Chilingarian, Ashot A., R. G. Mirzoyan, and M. Z. Zazyan: Cosmic Ray Research in Armenia. *Journal of Contemporary Physics (Armenian Academy of Sciences)* 44/5 (2009), 219–230. doi:10.3103/S106833720905003X.

Chiu, Hong-Yee: Gravitational Collapse. *Physics Today* 17/5 (1964), 21. doi:10.1063/1.3051610.

Chudakov, Aleksandr E.: Is the Signal from CYG X-3, as Recorded in Some Underground Experiments, Real? *Proceedings from the 19th International Cosmic Ray Conference, La Jolla, USA, August 11–12, 1985* 9 (1985), 441–444. https://ui.adsabs.harvard.edu/#abs/1985ICRC....9..441C/abstract. Last accessed 3/23/2019.

Chudakov, Aleksandr E.: VHE and UHE Gamma Ray Astronomy. History and Problems. In: Maurice M. Shapiro, and John P. Wefel (eds.): *Cosmic Gamma Rays, Neutrinos, and Related Astrophysics. Proceedings of the NATO Advanced Study Institute on Cosmic Gamma Rays and Cosmic Neutrinos Erice, Italy 20–30 April, 1988*. Dordrecht: Springer Netherlands 1989, 163–182. doi:10.1007/978-94-009-0921-2_12.

Chudakov, Aleksandr E., V. L. Dadykin, V. I. Zatsepin, and N. M. Nesterova: On the High Energy Photons from Local Sources. *Extensive Air Showers, Proceedings from the 8th International Cosmic Ray Conference*. 1963, 199–204. http://adsabs.harvard.edu/abs/1963ICRC....4..199C. Last accessed 11/8/2018.

Ciesla, Burghard: Das "Project Paperclip". Deutsche Naturwissenschaftler und Techniker in den USA (1946 bis 1952). In: Jürgen Kocka (ed.): *Historische DDR-Forschung. Aufsätze und Studien*. Berlin: Akademie Verlag 1993, 287–301.

Ciesla, Burghard, and Helmuth Trischler: Legitimation through Use. Rockets and Aeronautical Research in the Third Reich and the USA. In: Mark Walker (ed.): *Science and Ideology. A Comparative History*. London: Routledge 2003, 156–185.

Cirkel-Bartelt, Vanessa: History of Astroparticle Physics and Its Components. *Living Reviews in Relativity* 11 (2008), 2–58. doi:10.12942/lrr-2008-2.

Clark, George W.: Air Shower Experiments at MIT. In: Yataro Sekido, and Harry Elliot (eds.): *Early History of Cosmic Ray Studies. Personal Reminiscences with Old Photographs*. Dordrecht: D. Reidel Publishing Company 1985, 239–246.

Clerico, Mona (ed.): *MPA 50 Years. 50th Anniversary Brochure of the Max Planck Institute for Astrophysics*. Garching: Max-Planck-Institut für Astrophysik 2008. https://www.mpa-garching.mpg.de/69639/MPA_englisch.pdf. Last accessed 10/31/2018.

Clery, Daniel: The Bizarre Reactor That Might Save Nuclear Fusion. *Science*, 10/21/2015. doi:10.1126/science.aad4746.

Close, Frank: *Half-Life. The Divided Live of Bruno Pontecorvo, Physicist or Spy*. New York, NY: Basic Books 2015.

Cocconi, Giuseppe: An Air Shower Telescope and the Detection of 10^{12} eV Photon Sources. *Proceedings from the 6th International Cosmic Ray Conference, Held in Moscow, Russia, 6–11 July 1959. Extensive Air Showers and Cascades Process*. 1960,

309–311. https://ui.adsabs.harvard.edu/#abs/1960ICRC....2..309C. Last accessed 10/20/2018.

Cockcroft, Sir John: British Research in Controlled Thermonuclear Fusion. Kurchatov Memorial Article for Atomnaya Energiya. *Journal of Nuclear Energy. Part C, Plasma Physics, Accelerators, Thermonuclear Research* 5/6 (1963), 388–391. doi:10.1088/0368-3281/5/6/311.

Colgate, Stirling A., and Richard H. White: The Hydrodynamic Behavior of Supernovae Explosions. *Astrophysical Journal* 143 (1966), 626–681. doi:10.1086/148549.

Collins, Harry: *Gravity's Shadow. The Search for Gravitational Waves*. Chicago, IL: University of Chicago Press 2004.

Comella, J. M., H. D. Craft, R. V. E. Lovelace, J. M. Sutton, and G. Leonard Tyler: Crab Nebula Pulsar NP 0532. *Nature* 221/5179 (1969), 453–454. doi:10.1038/221453a0.

Control Council and Coordinating Committee of the Allied Control Authority: Enactments and Approved Papers of the Control Council and Coordinating Committee. Allied Control Authority, Germany (1945–1948). 9 Volumes. *Military Legal Resources. Federal Research Division. Library of Congress*. https://www.loc.gov/rr/frd/Military_Law/enactments-home.html. Last accessed 10/30/2018.

Conversi, M., E. Pancini, and O. Piccioni: On the Disintegration of Negative Mesons. *Physical Review* 71/3 (1947), 209–210. doi:10.1103/PhysRev.71.209.

Cowan, Clyde L., F. Reines, F. B. Harrison, H. W. Kruse, and A. D. McGuire: Detection of the Free Neutrino. A Confirmation. *Science* 124/3212 (1956), 103–104. doi:10.1126/science.124.3212.103.

Cowling, Thomas G., and Leon Mestel: Obituary—Biermann, L.F.B. *Quarterly Journal of the Royal Astronomical Society* 27/4 (1986), 698–700. http://adsabs.harvard.edu/abs/1986QJRAS..27..698C. Last accessed 1/13/2020.

Cramer, Katharina C., and Olof Hallonsten (eds.): *Big Science and Research Infrastructures in Europe*. Cheltenham: Edward Elgar Publishing 2020.

Crawford, M. K., R. Genzel, A. I. Harris, D. T. Jaffe, J. H. Lacy, J. B. Lugten, E. Serabyn, et al.: Mass Distribution in the Galactic Centre. *Nature* 315 (1985), 467. doi:10.1038/315467a0.

Cronin, James W.: The 1953 Cosmic Ray Conference at Bagnères de Bigorre. The Birth of Sub Atomic Physics. *The European Physical Journal H* 36/2 (2011), 183–201. doi:10.1140/epjh/e2011-20014-4.

Crutzen, Paul J.: Christian Junge. 2.7.1912-18.6.1996. *Jahresbericht der Max-Planck-Gesellschaft*, 1996, 196–199, 196–199.

Czechowsky, Peter, and Rüdiger Rüster (eds.): *60 Jahre Forschung in Lindau. 1946–2006. Vom Fraunhofer-Institut zum Max-Planck-Institut für Sonnensystemforschung. Eine Sammlung von Erinnerungen*. Katlenburg-Lindau: Copernicus 2007.

Danby, G., J. M. Gaillard, K. Goulianos, L. M. Lederman, N. Mistry, M. Schwartz, and J. Steinberger: Observation of High-Energy Neutrino Reactions and the Existence of Two Kinds of Neutrinos. *Physical Review Letters* 9/1 (1962), 36–44. doi:10.1103/PhysRevLett.9.36.

Danzmann, Karsten: Laser Interferometric Gravitational Wave Detectors. In: R.J. Gleiser, C.N. Kozameh, and O.M. Moreschi (eds.): *Proceedings. 13th International Conference on General Relativity and Gravitation. Cordoba, Argentina, June 28-July 4, 1992*. Bristol: Institute of Physics 1993, 3–19. https://inspirehep.net/record/348400. Last accessed 1/24/2019.

Danzmann, Karsten, J. Chen, P. G. Nelson, T. M. Niebauer, A. Rüdiger, Roland Schilling, L. Schnupp, et al.: The GEO—Project a Long-Baseline Laser Interferometer for the Detection of Gravitational Waves. In: J. Ehlers, and G. Schäfer (eds.): *Relativistic Gravity Research With Emphasis on Experiments and Observations. Proceedings of the 81 WE-Heraeus-Seminar Held at the Physikzentrum, Bad Honnef, Germany 2–6 September 1991*. Berlin: Springer 1992, 184–209. doi:10.1007/3-540-56180-3_9.

Danzmann, Karsten, H. Lück, A. Rüdiger, Roland Schilling, M. Schrempel, W. Winkler, J. Hough, et al.: GEO 600—A 600m Laser Interferometric Gravitational Wave Antenna. In: Eugenio Coccia, Guido Pizzella, and Francesco Ronga (eds.). First Edoardo Amaldi Conference on Gravitational Wave Experiments. World Scientific 1995, 100–111. doi:10.1142/9789814533652.

Danzmann, Karsten, A. Rüdiger, Roland Schilling, M. Schrempel, W. Winkler, J. Hough, G.P. Newton, et al.: *GEO 600. Proposal for a 600 m Laser-Interferometric Gravitational Wave Antenna*. MPQ-190. Garching: Max-Planck-Institut für Quantenoptik 1994.

Daum, A., G. Hermann, M. Heß, W. Hofmann, H. Lampeitl, G. Pühlhofer, F. Aharonian, et al.: First Results on the Performance of the HEGRA IACT Array. *Astroparticle Physics* 8/1 (1997), 1–11. doi:10.1016/S0927-6505(97)00031-5.

Davies, Richard I., Wolfgang K. P. Hackenberg, Thomas Ott, Andreas Eckart, Hans-Christoph Holstenberg, Sebastian Rabien, Andreas Quirrenbach, et al.: ALFA: First Operational Experience of the MPE/MPIA Laser Guide Star System for Adaptive Optics. *Adaptive Optical System Technologies*. Astronomical Telescopes and Instrumentation. Kona, HI: International Society for Optics and Photonics 1998, 116–124. doi:10.1117/12.321747.

Davis, Raymond: A Review of the Homestake Solar Neutrino Experiment. *Progress in Particle and Nuclear Physics* 32 (1994), 13–32. doi:10.1016/0146-6410(94)90004-3.

Davis, Raymond Jr.: Attempt to Detect the Antineutrinos from a Nuclear Reactor by the Cl^{37}(Anti ν, e^-)A^{37} Reaction. *Physical Review* 97/3 (1955), 766–769. doi:10.1103/PhysRev.97.766.

Davis, Raymond Jr., Don S. Harmer, and Kenneth C Hoffman: Search for Neutrinos from the Sun. *Physical Review Letters* 20/21 (1968), 1205–1209. doi:10.1103/PhysRevLett.20.1205.

Derishev, Evgeny, and Tsvi Piran: The Physical Conditions of the Afterglow Implied by MAGIC's Sub-TeV Observations of GRB 190114C. *The Astrophysical Journal* 880/2 (2019), L27. doi:10.3847/2041-8213/ab2d8a.

Deutsche Physikalische Gesellschaft: Stellungnahmen zur Bemannten Raumfahrt. Homepage, 2018. https://www.dpg-physik.de/veroeffentlichungen/publikationen/stellungnahmen-der-dpg/bemannte-raumfahrt. Last accessed 6/6/2019.

Deutsches Zentrum für Luft- und Raumfahrt e.V.: *50 Jahre DLR Lampoldshausen. 1959–2009*. Köln: Bernd Rölle 2009.

Deutschmann, Martin: Untersuchung Der Kosmischen Strahlenschauer Mit Hilfe Einer Großen Wilson-Kammer. *Zeitschrift Für Naturforschung A* 2/2 (1947), 61–69. doi:10.1515/zna-1947-0201.

DeVorkin, David H.: The Space Age and Disciplinary Change in Astronomy. In: Steven J. Dick (ed.): *NASA's First 50 Years. Historical Perspectives*. Washington, D.C.: NASA 2010, 389–426.

DeVorkin, David H.: Who Speaks for Astronomy? How Astronomers Responded to Government Funding After World War II. *Historical Studies in the Physical and Biological Sciences* 31/1 (2000), 55–92. doi:10.2307/27757846.

DeVorkin, David H.: *Science With A Vengeance. How the Military Created the US Space Sciences After World War II*. New York, NY: Springer-Verlag 1992.

De Maria, Michelangelo: *Europe in Space. Edoardo Amaldi and the Inception of ESRO*. ESA-HSR-5. Noordwijk, the Netherlands: ESA Publications Division 1993.

De Maria, Michelangelo, Mario Grilli, and Fabio Sebastiani (eds.): *The Restructuring of Physical Sciences in Europe and the United States, 1945–1960. Proceedings of the International Conference*. Singapore: World Scientific 1989.

De Maria, Michelangelo, Lucia Orlando, and Filippo Pigliacelli: *Italy in Space—1946–1988*. Noordwijk: ESA Publications Division 2003.

Dick, Wolfgang R. (ed.): *300 Jahre Astronomie in Berlin und Potsdam. Eine Sammlung von Aufsätzen aus Anlaß des Gründungsjubiläums der Berliner Sternwarte*. Thun: Deutsch 2000.

Dieminger, Walter: Radio Communication and Solar-Terrestrial Physics. In: Wilfried Schröder, and International Association of Geomagnetism and Aeronomy (eds.): *Historical Events and People in Geosciences. Selected Papers from the Symposia of the Interdivisional Commission on History of IAGA during the IUGG General Assembly, Held in Hamburg, 1983*. Frankfurt am Main: Peter Lang 1985, 11–27.

Dieminger, Walter: Die Ionosphäre als Grenzschicht zwischen Erdatmosphäre und extraterrestrischem Raum. In: Generalverwaltung der Max-Planck-Gesellschaft der Max-Planck-Gesellschaft zur Förderung der Wissenschaft e.V. (ed.): *Jahrbuch 1954 der Max-Planck-Gesellschaft zur Förderung der Wissenschaften e.V.* Göttingen 1955, 42–80.

Dieminger, Walter: Ionosphäre. *Geophysics. Part I.* Wiesbaden: Dieterich'sche Verlagsbuchhandlung 1948, 93–163.

Dieminger, Walther, Alfred Ehmert, and Georg Pfotzer: *Sonderheft Professor em. Dr. Walter Dieminger zum 70. Geburtstag am 7.7.1977. Ansprachen und Vorträge anlässlich seiner feierlichen Verabschiedung aus seinem Amt als Direktor des Max-Planck-Instituts für Aeronomie am 9. und 10.7.1975.* Edited by Julius Bartels. Berlin: Springer 1977.

Doeleman, Sheperd S., Eric Agol, Don Backer, Fred Baganoff, Geoffrey C Bower, Avery Broderick, Andrew Fabian, et al.: *Imaging an Event Horizon: Submm-VLBI of a Super Massive Black Hole. A Science White Paper to the Decadal Review Committee.* https://arxiv.org/abs/0906.3899. Last accessed 3/23/2021.

Doeleman, Sheperd S., and Thomas P. Krichbaum: Status of VLBI Observations at 1 MM Wavelength and Future Prospects. In: A. Greve, and T. P. Krichbaum (eds.): *Proceedings of the Second Millimeter VLBI Science Workshop, Held at IRAM in Granada (Spain) on May 27–29, 1999.* St. Martin d'Heres: IRAM 1999, 73.

Doeleman, Sheperd S., Jonathan Weintroub, Alan E. E. Rogers, Richard Plambeck, Robert Freund, Remo P. J. Tilanus, Per Friberg, et al.: Event-Horizon-Scale Structure in the Supermassive Black Hole Candidate at the Galactic Centre. 7209. *Nature* 455/7209 (2008), 78–80. doi:10.1038/nature07245.

Dongen, Jeroen van (ed.): *Cold War Science and the Transatlantic Circulation of Knowledge.* Vol. 1. Leiden: Brill 2015.

Dooley, Katherine L., J. R. Leong, Teneisha Adams, C. Affeldt, Aparna Bisht, Christina Bogan, J. Degallaix, et al.: GEO 600 and the GEO-HF Upgrade Program: Successes and Challenges. *Classical and Quantum Gravity* 33/7 (2016), 075009. doi:10.1088/0264-9381/33/7/075009.

Dorado, Jose M., Manuel Bautista, and Pedro Sanz-Aránguez: Spain in Space. A Short History of Spanish Activity in the Space Sector. *History Study Reports.* Edited by European Space Agency 26 (2002). http://www.esa.int/esapub/hsr/HSR_26.pdf. Last accessed 5/2/2021.

Doroshkevich, Andrey Georgievich, Malcolm S. Longair, and Yakov Borisovich Zeldovich: The Evolution of Radio Sources at Large Redshifts. *Monthly Notices of the Royal Astronomical Society* 147/2 (1970), 139–148. doi:10.1093/mnras/147.2.139.

Douglass, D. H., R. Q. Gram, J. A. Tyson, and R. W. Lee: Two-Detector-Coincidence Search for Bursts of Gravitational Radiation. *Physical Review Letters* 35/8 (1975), 480–483. doi:10.1103/PhysRevLett.35.480.

Dreisigacker, E.: Außeruniversitäre Forschung in den östlichen Bundesländern vor Neuanfang. *Physikalische Blätter* 47/8 (1991), 763–767. doi:10.1002/phbl.19910470808.

Dreisigacker, E.: Grundlagenforschung—quo vadis? Großmann-Kommission legt Empfehlungen vor. *Physikalische Blätter* 48/5 (1992), 372–374. doi:10.1002/phbl.19920480514.

Drever, Ronald William P., J. Hough, A. J. Munley, S.-A. Lee, R. Spero, S. E. Whitcomb, H. Ward, et al.: Gravitational Wave Detectors Using Laser Interferometers and Optical Cavities. Ideas, Principles and Prospects. In: Pierre Meystre, and Marlan O. Scully (eds.): *Quantum Optics, Experimental Gravity, and Measurement Theory*. Conference "Quantum Optics and Experimental General Relativity", held August 16–29, 1981 in Bad Windsheim, Federal Republic of Germany. Boston, MA: Springer US 1983, 503–514. doi:10.1007/978-1-4613-3712-6_21.

Drukier, Andrzej K., and Leo Stodolsky: Principles and Applications of a Neutral-Current Detector for Neutrino Physics and Astronomy. *Physical Review D* 30/11 (1984), 2295–2309. doi:10.1103/PhysRevD.30.2295.

Dunar, Andrew J, and Stephen P Waring: *Power to Explore. A History of Marshall Space Flight Center, 1960–1990*. Washington, DC: National Aeronautics and Space Administration, NASA History Office, Office of Policy and Plans 1999.

Dunar, Andrew J., and Stephen P. Waring: The Hubble Space Telescope. *Power to Explore. A History of Marshall Space Flight Center, 1960–1990*. Washington, DC: National Aeronautics and Space Administration, NASA History Office, Office of Policy and Plans 1999, 473–525.

Dyson, Freeman J.: *Maker of Patterns. An Autobiography through Letters*. New York, NY: Liveright 2018.

Dyson, Freeman J.: *Selected Papers of Freeman Dyson. With Commentary*. Cambridge, MA: American Mathematical Society 1996.

Dyson, F. J.: Photon Noise and Atmospheric Noise in Active Optical Systems. *Journal of the Optical Society of America* 65 (1975), 551–558. http://adsabs.harvard.edu/abs/1975JOSA...65..551D. Last accessed 8/13/2020.

Dyson, Freeman J.: Gravitational Machines. In: Alastair G.W. Cameron (ed.): *Interstellar Communication*. New York, NY: Benjamin Press 1963, 115–120.

e.V., Deutsches Zentrum für Luft- und Raumfahrt, and Institut für Raumfahrtantriebe (eds.): *50 Jahre DLR Lampholdshausen*, 2009.

Eckart, A., and Reinhard Genzel: Observations of the Galactic Center with SHARP: First Stellar Proper Motions. In: Roland Gredel (ed.): *The Galactic Center, Astronomical Society of the Pacific Conference Series*. 4th international meeting jointly organized by the European Southern Observatory (ESO) and Cerro Tololo Inter-American Observatory (CTIO), held March 10–15, 1996 in La Serena, Chile, San Francisco. Astronomical Society of the Pacific. 1996, 196–202. http://adsabs.harvard.edu/abs/1996ASPC..102..196E. Last accessed 5/9/2020.

Eckart, Andreas, and Reinhard Genzel: First Conclusive Evidence for a Massive Black Hole in the Center of the Milky Way. In: Friedrich W. Hehl, Claus Kiefer, and Ralph J.K. Metzler (eds.): *Black Holes: Theory and Observation.* Berlin: Springer-Verlag 1998, 60–68. doi:10.1007/978-3-540-49535-2_3.

Eckart, Wolfgang U.: Max-Planck-Institut für medizinische Forschung Heidelberg. In: Reinhard Rürup, and Peter Gruss (eds.): *Denkorte. Max-Planck-Gesellschaft und Kaiser-Wilhelm-Gesellschaft. Brüche und Kontinuitäten 1911–2011.* Dresden: Sandstein 2010, 174–183.

Eckert, Michael: *Ludwig Prandtl. Strömungsforscher und Wissenschaftsmanager. Ein unverstellter Blick auf sein Leben.* Berlin: Springer 2017.

Eckert, Michael: Heisenberg and the Beginnings of Nuclear Energy in the FRG. In: Michelangelo De Maria, Mario Grilli, and Fabio Sebastiani (eds.): *Proceedings of the International Conference on The Restructuring of Physical Sciences in Europe and the United States 1945–1960. Università "La Sapienza", Rome, Italy, 19–23 September 1988.* Singapore: World Scientific 1989, 247–256.

Eckert, Michael, and Maria Osietzki: *Wissenschaft für Macht und Markt. Kernforschung und Mikroelektronik in der Bundesrepublik Deutschland.* München: Beck 1989.

Eddington, Arthur S.: *The Internal Constitution of the Stars.* Cambridge: Cambridge University Press 1926.

Edgerton, David: *The Rise and Fall of the British Nation. A Twentieth-Century History.* London: Allen Lane 2018.

Edgerton, David: *Warfare State. Britain, 1920–1970.* Cambridge: Cambridge University Press 2006.

Edingshaus, Anna-Lydia: *Heinz Maier-Leibnitz. Ein halbes Jahrhundert experimentelle Physik.* München: Piper 1986.

Edingshaus, Anne-Lydia: *Heinz Maier-Leibnitz. Eine halbes Jahrhundert experimentelle Physik.* München: Piper & Co. Verlag 1986.

Edwards, Paul N.: Entangled Histories. Climate Science and Nuclear Weapons Research. *Bulletin of the Atomic Scientists* 68/4 (2012), 28–40. doi:10.1177/0096340212451574.

Ehlers, Jürgen: Probleme und Ergebnisse der modernen Kosmologie. *Mitteilungen der Astronomischen Gesellschaft* 27 (1969), 73–86. http://adsabs.harvard.edu/abs/1969MitAG..27...73E. Last accessed 4/25/2018.

Ehlers, Jürgen, J. J. Perry, and M. Walker (eds.): *9th Texas Symposium on Relativistic Astrophysics. 14–19 December 1978, Munich.* New York, NY: New York Academy of Sciences 1980.

Ehmert, A.: In memoriam Julius Bartels. *Space Science Reviews* 3/1 (1964), 2–4. doi:10.1007/BF00226642.

Ehmert, A. (ed.): *The Space Environment. Le Milieu Spatial. Report on the Survey Meeting of Information on the Space Environment Paris, 27 September 1963*. Wien: Springer 1964.

Ehmert, Alfred: Ultrastrahlung von der Sonne. *Zeitschrift für Naturforschung A* 3/5 (1948), 264–285. doi:10.1515/zna-1948-0504.

Ehmert, Alfred, and Erwin Schopper: In Memoriam Erich Regener. *Die Naturwissenschaften* 43/4 (1956), 69–71. doi:10.1007/BF00631846.

Eisenhauer, F., Reinhard Genzel, T. Alexander, R. Abuter, T. Paumard, T. Ott, A. Gilbert, et al.: SINFONI in the Galactic Center: Young Stars and Infrared Flares in the Central Light-Month. *The Astrophysical Journal* 628/1 (2005), 246–259. doi:10.1086/430667.

Elbers, Astrid: *The Rise of Radio Astronomy in the Netherlands. The People and the Politics*. Cham: Springer 2016.

Elliot, S. R., A. A. Hahn, and M. K. Moe: Search for Double Beta Decay in ^{82}Se. In: T. Kitagaki, and H. Yuta (eds.): *Proceedings of the 12th International Conference on Neutrino Physics and Astrophysics, Sendai, Japan, June 3–8, 1986*. Singapore: World Scientific 1986, 93–100.

Elliott, S. R., A. A. Hahn, and M. K. Moe: Direct Evidence for Two-Neutrino Double-Beta Decay in ^{82}Se. *Physical Review Letters* 59/18 (1987), 2020–2023. doi:10.1103/PhysRevLett.59.2020.

Elliott, Samuel R., A. A. Hahn, and M. K. Moe: *A Direct Laboratory Measurement of Two-Neutrino Double Beta Decay in ^{82}Se*. Neutrino Physics. Proceedings of an International Workshop Held in Heidelberg, October 20–22, 1987. Berlin: Springer 1988.

Elsässer, Hans: Nachrufe. H. Siedentopf. *Mitteilungen der Astronomischen Gesellschaft* 17 (1964), 33–41. https://ui.adsabs.harvard.edu/#abs/1964MitAG..17...33. Last accessed 10/30/2018.

Elsässer, Hans: *Weltall im Wandel. Die neue Astronomie*. Reinbek bei Hamburg: Rowohlt 1989.

Encrenaz, Pierre, Jesus Gómez-Gónzalez, James Lequeux, and Wayne Orchiston: Highlighting the History of French Radio Astronomy. 7. The Genesis of the Institute of Radioastronomy at Millimeter Wavelengths (IRAM). *Journal of Astronomical History and Heritage* 14/2 (2011), 83–92. http://www.narit.or.th/en/files/2011JAHHvol14/2011JAHH...14...83E.pdf. Last accessed 10/30/2018.

European Space Agency: Professor Reimar Lüst (1923–2020). *European Space Agency*, 4/2/2020. https://www.esa.int/About_Us/ESA_history/Professor_Reimar_Luest_1923-2020. Last accessed 4/3/2020.

European Strategy Forum on Research Infrastructures: *Roadmap 2018. Strategy Report on Research Infrastructures*. Milan: ESFRI 2018. http://roadmap2018.esfri.eu/media/1066/esfri-roadmap-2018.pdf. Last accessed 8/15/2020.

Event Horizon Telescope. https://eventhorizontelescope.org/home. Last accessed 3/23/2021.

Falcke, Heino: *Light in the Darkness. Black Holes, The Universe and Us*. San Francisco, CA: HarperOne 2021.

Falcke, Heino, Fulvio Melia, and Eric Agol: Viewing the Shadow of the Black Hole at the Galactic Center. *The Astrophysical Journal Letters* 528 (2000), L13–L16. doi:10.1086/312423.

Falkenburg, Brigitte, and Wolfgang Rhode: *From Ultra Rays to Astroparticles. A Historical Introduction to Astroparticle Physics*. Dordrecht: Springer 2012.

Fälthammar, Carl-Gunne, and Alexander J. Dessler: Hannes Alfvén. 30 May 1908–2 April 1995. *Proceedings of the American Philosophical Society* 150/4 (2006), 649–662. https://www.jstor.org/stable/i412827. Last accessed 3/27/2018.

Fazio, G. G., H. F. Helmken, G. H. Rieke, and T. C. Weekes: A Search for Discrete Sources of Cosmic Gamma Rays of Energies Near 2x10^{12} eV. *The Astrophysical Journal* 154 (1968), L83. doi:10.1086/180275.

Fazio, G. G., H. F. Helmken, G. H. Rieke, and T. C. Weekes: An Experiment to Search for Discrete Sources of Cosmic Gamma Rays in the 10^{11} to 10^{12} eV Region. *Canadian Journal of Physics* 46/10 (1968), S451–S455. doi:10.1139/p68-268.

Fechtig, Hugo, Eberhard Grün, Bo Gustafson, and Stanley Dermott (eds.): *Interplanetary Dust*. Berlin: Springer 2001.

Feder, Toni: Max Planck Institute for Aeronomy Averts Closure. *Physics Today* 50/5 (1997), 51–52. doi:10.1063/1.881823.

Fegan, David: *Cherenkov Reflections. Gamma-Ray Imaging and the Evolution of TeV Astronomy*. Singapore: World Scientific 2019.

Fermi, Enrico: Galactic Magnetic Fields and the Origin of Cosmic Radiation. *The Astrophysical Journal* 119/1 (1954), 1–6. doi:10.1086/145789.

Fermi, Enrico: On the Origin of the Cosmic Radiation. *Physical Review* 75/8 (1949), 1169–1174. doi:10.1103/PhysRev.75.1169.

Fermi, Enrico: Versuch einer Theorie der Beta-Strahlen. *Zeitschrift für Physik* 88 (1934), 161–171. doi:10.1007/BF01351864.

Fernandez, Bernard, and Georges Ripka: *Unravelling the Mystery of the Atomic Nucleus. A Sixty Year Journey 1896–1956*. New York: Springer 2013.

Ferry, Georgina: *EMBO in Perspective. A Half-Century in the Life Sciences*. EMBO 2014.

Feuerstack, M., J. Heintze, P. Lennert, B. Müller, S. Polenz, B. Schmidt, J. Spitzer, et al.: Cosmic Ray Tracking. A New Approach to High-Energy γ-Astronomy. *Nuclear Instruments and Methods in Physics Research A* 310/1–2 (1991), 287–291. doi:10.1016/0168-9002(91)91045-W.

Feuerstack, M., J. Heintze, P. Lennert, S. Polenz, B. Schmidt, J. Spitzer, B. Stadler, et al.: Cosmic Ray Tracking—a New Approach to High-Energy γ-Astronomy. *Nuclear Instruments and Methods in Physics Research Section A: Accelerators, Spectrometers,*

Detectors and Associated Equipment 315/1 (1992), 257–259. doi:10.1016/0168-9002(92)90711-C.

Fichtel, C. E.: Gamma-Ray Astrophysics. *Space Science Reviews* 20/2 (1977), 191–234. doi:10.1007/BF02186864.

Fick, Dieter, and Horst Kant: Walther Bothe's Contributions to the Understanding of the Wave-Particle Duality of Light. *Studies in History and Philosophy of Science Part B: Studies in History and Philosophy of Modern Physics* 40/4 (2009), 395–405. doi:10.1016/j.shpsb.2009.08.005.

Finkbeiner, Ann: JASON Past, Present, and Future. The World's Most Independent Defence Science Advisers. *Nature* 477 (2011), 397–399. doi:10.1038/477397a.

Finkbeiner, Ann K.: *The Jasons. The Secret History of Science's Postwar Elite*. New York, NY: Viking Press 2006.

Fischer, Peter: *The Origins of the Federal Republic of Germany's Space Policy 1959–1965. European and National Dimensions*. Noordwijk: European Space Agency Publications 1994.

Fisher, David E.: *Much Ado About (Practically) Nothing. A History of the Noble Gases*. New York, NY: Oxford University Press 2010.

Flachowsky, Sören: Von der Wagenburg der Autarkie zu transnationaler Zusammenarbeit. Der Verein Deutscher Eisenhüttenleute und das KWI/MPI für Eisenforschung 1917–2009. In: Helmut Maier (ed.): *150 Jahre Stahlinstitut VDEh 1860–2010*. Essen: Klartext 2010, 671–708.

Fleischmann, Rudolf von: Walter Bothe und sein Beitrag zur Atomkernforschung. *Die Naturwissenschaften* 44/17 (1957), 457–460. doi:10.1007/BF00640879.

Fletcher, Seth: *Einstein's Shadow. A Black Hole, a Band of Astronomers, and the Quest to See the Unseeable*. New York, NY: Ecco 2018.

Fomin, V. P., Y. I. Neshpor, V. G. Shitov, A. A. Stepanian, B. M. Vladimirsky, and Y. L. Zyskin: Angular Distribution of Cerenkov Light in EAS with Energy > 10^{13} eV. *18th International Cosmic Ray Conference, Bangalore, India, 22 August–3 September, 1983*. 1983, 223. https://ui.adsabs.harvard.edu/#abs/1983ICRC....6..223F. Last accessed 12/25/2018.

Fonseca, Victoria: The HEGRA Experiment (High Energy Gamma Ray Array). *Nuclear Physics B—Proceedings Supplements* 28/1 (1992), 409–412. doi:10.1016/0920-5632(92)90205-7.

Forman, W., C. Jones, L. Cominsky, P. Julien, S. Murray, G. Peters, H. Tananbaum, et al.: The Fourth Uhuru Catalog of X-Ray Sources. *The Astrophysical Journal Supplement Series* 38 (1978), 357–412. doi:10.1086/190561.

FORS—Das Arbeitspferd am Very Large Telescope. In: Bundesministerium für Bildung und Forschung (BMBF) (ed.): *Stark im Verbund: Naturwissenschaftliche Grundlagenforschung an Großgeräten*. Bonn 2009, 49–50.

Forveille, Thierry, Sergio Campana, and Steve Shore: H.E.S.S. Phase-I Observations of the Plane of the Milky Way. *Astronomy & Astrophysics* 612 (2018), E1. doi:10.1051/0004-6361/201833049.

Fowler, William A.: Completion of the Proton-Proton Reaction Chain and the Possibility of Energetic Neutrino Emission by Hot Stars. *The Astrophysical Journal* 127 (1958), 551–556. doi:10.1086/146487.

Fowler, William A., and Fred Hoyle: Nature of Strong Radio Sources. *Nature* 197/4867 (1963), 533–535. doi:10.1038/197533a0.

Frank, Charles (ed.): *Operation Epsilon. The Farm Hall Transcripts*. London: Institute of Physics Publishing 1993.

Freier, Phyllis, Edward J. Lofgren, Edward P. Ney, Frank Oppenheimer, Helmut L. Bradt, and Benno Peters: Evidence for Heavy Nuclei in the Primary Cosmic Radiation. *Physical Review* 74/2 (1948), 213–217. doi:10.1103/PhysRev.74.213.

Freise, Andreas, Gerhard Heinzel, Kenneth A. Strain, Jun Mizuno, Kenneth David Skeldon, Harald Lück, Benno Willke, et al.: Demonstration of Detuned Dual Recycling at the Garching 30 m Laser Interferometer. *Physics Letters A* 277/3 (2000), 135–142. doi:10.1016/S0375-9601(00)00704-0.

Frenkel, Viktor J.: *Professor Friedrich Houtermans. Arbeit, Leben, Schicksal. Biographie eines Physikers des zwanzigsten Jahrhunderts*. Preprint 414. Edited by Dieter Hoffmann. Max-Planck-Institut für Wissenschaftsgeschichte 2011. https://www.mpiwg-berlin.mpg.de/Preprints/P414.PDF. Last accessed 10/30/2018.

Freytag, Carl: »Bürogenerale« und »Frontsoldaten« der Wissenschaft. Atmosphärenforschung in der Kaiser-Wilhelm-Gesellschaft während des Nationalsozialismus. In: Helmut Maier (ed.): *Gemeinschaftsforschung, Bevollmächtigte und der Wissenstransfer. Die Rolle der Kaiser-Wilhelm-Gesellschaft im System kriegsrelevanter Forschung des Nationalsozialismus*. Göttingen: Wallstein Verlag 2007, 215–267.

Friedensinitiative Garchinger Naturwissenschaftler: *30 Jahre Göttinger Erklärung. Nachdenken über die Rolle des Wissenschaftlers in der Gesellschaft*. Marburg: Bund Demokratischer Wissenschaftlerinnnen und Wissenschaftler 1987.

Friedlander, G.: Report on an Informal Conference on the Status and Future of Solar-Neutrino Research. *Comments on Astrophysics* 8 (1978), 47–54. http://adsabs.harvard.edu/abs/1978ComAp...8...47F. Last accessed 1/5/2019.

Friedman, Herbert: From the Ionosphere to High Energy Astronomy. A Personal Experience. In: Johan A. M. Bleeker, Johannes Geiss, and Martin C. E. Huber (eds.): *The Century of Space Science*. Dordrecht: Kluwer Academic Publishers 2001, 277–286.

Fruin, J. H., J. V. Jelley, C. D. Long, N. A. Porter, and T. C. Weekes: Flux Limits for High—Energy γ—Rays from Quasi—Stellar and Other Radio Sources. *Physics Letters* 10/2 (1964), 176–177. doi:10.1016/0031-9163(64)90159-3.

Fukuda, Y., T. Hayakawa, E. Ichihara, K. Inoue, K. Ishihara, H. Ishino, Y. Itow, et al.: Evidence for Oscillation of Atmospheric Neutrinos. *Physical Review Letters* 81/8 (1998), 1562–1567. doi:10.1103/PhysRevLett.81.1562.

Galison, Peter: *Image and Logic. A Material Culture of Microphysics*. Chicago: The University of Chicago Press 1997.

GALLEX Collaboration: Results of the Whole GALLEX Experiment. *Nuclear Physics B. Proceedings Supplements* 70/1–3 (1999), 284–291. doi:10.1016/S0920-5632(98)00438-1.

Gambke, Gotthard, Rudolf Kerscher, and Walter Kertz: *Denkschrift zur Lage der Weltraumforschung*. Wiesbaden: Franz Steiner Verlag 1961.

Gamow, George: Expanding Universe and the Origin of Elements. *Physical Review* 70/7–8 (1946), 572–573. doi:10.1103/PhysRev.70.572.2.

Garcia, Alexandre, Sidney Servulo Cunha Yamanaka, Alexandre Nogueira Barbosa, Francisco Carlos Parquet Bizarria, Wolfgang Jung, and Frank Scheuerpflug: VSB-30 Sounding Rocket. History of Flight Performance. *Journal of Aerospace Technology and Management* 3/3 (2011), 325–330. doi:10.5028/jatm.2011.03032211.

Garrett, M. A.: Ground Based VLBI Facilities—the European and Global VLBI Network. In: Franco Mantovani, and Andrzej Kus (eds.): *The Role of VLBI in Astrophysics, Astrometry and Geodesy*. Dordrecht: Springer 2005, 403–413.

Garrett, M. A., H. Rampadarath, E. Lenc, and Olaf Wucknitz: LOFAR, E-LOFAR and Low-Frequency VLBI. arXiv:0902.2534 [*Astro-Ph*]. European VLBI Network IX Symposium. Bologna 2008. http://arxiv.org/abs/0902.2534. Last accessed 8/15/2020.

Gast, Robert: Bild des Schwarzen Lochs: Das Monster zeigt seine Zähne. *Spektrum der Wissenschaft (online)*. https://www.spektrum.de/news/das-bild-des-schwarzen-lochs/1638154. Last accessed 8/19/2019.

Gavrin, Vladimir N.: The Russian-American Gallium Experiment SAGE. *Physics-Uspekhi* 54/9 (2011), 941–948. doi:10.3367/UFNe.0181.201109g.0975.

Geiger, Hans, and Walther Müller: Das Elektronenzählrohr. *Physikalische Zeitschrift* 29 (1928), 839–841.

Geiger, Roger L.: What Happened after Sputnik? Shaping University Research in the United States. *Minerva* 35/4 (1997), 349–367. https://www.jstor.org/stable/41821079. Last accessed 5/24/2019.

Geiger, Tim: *Atlantiker gegen Gaullisten. Außenpolitischer Konflikt und innerparteilicher Machtkampf in der CDU/CSU 1958–1969*. München: Oldenbourg 2008.

Gelmini, G. B., L. J. Hall, and M. J. Lin: What Is the Cosmion? *Nuclear Physics B* 281/3–4 (1987), 726–735. doi:10.1016/0550-3213(87)90424-X.

Generalverwaltung der Max-Planck-Gesellschaft (ed.): *Max-Planck-Gesellschaft Jahrbuch 1993*. Göttingen: Vandenhoeck & Ruprecht 1993.

Generalverwaltung der Max-Planck-Gesellschaft (ed.): *Max-Planck-Gesellschaft Jahrbuch 1982*. Göttingen: Vandenhoeck & Ruprecht 1982.

Generalverwaltung der Max-Planck-Gesellschaft (ed.): *Gedenkfeier Wolfgang Gentner*. München 1981.

Generalverwaltung der Max-Planck-Gesellschaft (ed.): *Mitteilungen aus der Max-Planck-Gesellschaft zur Förderung der Wissenschaften. Heft 6/1960*. Göttingen 1960.

Generalverwaltung der Max-Planck-Gesellschaft (ed.): *Mitteilungen aus der Max-Planck-Gesellschaft zur Förderung der Wissenschaften. Heft 6/1958*. Göttingen 1958.

Generalverwaltung der Max-Planck-Gesellschaft (ed.): *Mitteilungen aus der Max-Planck-Gesellschaft zur Förderung der Wissenschaften. Heft 2/1957*. Göttingen 1957.

Generalverwaltung der Max-Planck-Gesellschaft (ed.): *Mitteilungen aus der Max-Planck-Gesellschaft zur Förderung der Wissenschaften. Heft 3/1957*. Göttingen 1957.

Generalverwaltung der Max-Planck-Gesellschaft (ed.): *Mitteilungen aus der Max-Planck-Gesellschaft. Heft 2*. Göttingen 1952.

Generalverwaltung der Max-Planck-Gesellschaft (ed.): *Mitteilungen aus der Max-Planck-Gesellschaft. Heft 3*. Göttingen 1952.

Generalverwaltung der Max-Planck-Gesellschaft zur Förderung der Wissenschaften (ed.): *Max-Planck-Gesellschaft Jahrbuch 2001*. Göttingen: Vandenhoeck & Ruprecht 2001.

Generalverwaltung der Max-Planck-Gesellschaft zur Förderung der Wissenschaften (ed.): *Max-Planck-Gesellschaft Jahrbuch 1981*. Göttingen: Vandenhoeck & Ruprecht 1981.

Generalverwaltung der Max-Planck-Gesellschaft zur Förderung der Wissenschaften (ed.): *Max-Planck-Gesellschaft Jahrbuch 1977*. Göttingen: Vandenhoeck & Ruprecht 1977.

Gentner, Wolfgang: Einige Rückblicke auf die Anfänge der 50 jährigen Forschung über die kosmische Strahlung. *Die Naturwissenschaften* 50/8 (1963), 317–318. doi:10.1007/BF00645924.

Gentner, Wolfgang: Max-Planck-Institut für Kernphysik in Heidelberg. In: Generalverwaltung der Max-Planck-Gesellschaft zur Förderung der Wissenschaften e.V. (ed.): *Jahrbuch 1961 der Max-Planck-Gesellschaft zur Förderung der Wissenschaften e.V. Teil II*. Göttingen 1962, 486–491.

Gentner, Wolfgang: Nachruf für Walter Bothe. *Zeitschrift für Naturforschung A* 12/2 (1957), 175–176. doi:10.1515/zna-1957-0213.

Gentner, Wolfgang: Das Heidelberger Zyklotron. *Nuclear Physics and Cosmic Rays*. Vol. 2, 1948.

Gentner, Wolfgang: Die Erzeugung schneller Ionenstrahlen für Kernreaktionen. In: Ferdinand Trendelenburg (ed.): *Ergebnisse der Exakten Naturwissenschaften*. Berlin: Verlag von Julius Springer 1940, 107–169.

Gentner, Wolfgang: Mitteilungen aus der Kernphysik. Das neue 1,5 Meter-Zyklotron in Berkeley (Calif.). *Die Naturwissenschaften* 28/25 (1940), 394–396. doi:10.1007/BF01479460.

Gentner, Wolfgang, Hugo Fechtig, and G. Kistner: Edelgase und ihre Isotopenverschiebung im Eisenmeteorit Treysa. *Zeitschrift Naturforschung Teil A* 13/7 (1958), 569–570b. doi:10.1515/zna-1958-0719.

Gentner, Wolfgang, and Joseph Zähringer: Kalium-Argon-Alter einiger Tektite. *Zeitschrift Naturforschung Teil A* 14/7 (1959), 686–687. doi:10.1515/zna-1959-0716.

Gentner, Wolfgang, and Josef Zähringer: Argon und Helium als Kernreaktionsprodukte in Meteoriten. *Geochimica et Cosmochimica Acta* 11/1–2 (1957), 60–71.

Gentner, Wolfgang, and Josef Zähringer: Argon- und Heliumbestimmungen in Eisenmeteoriten. *Zeitschrift für Naturforschung A* 10/6 (1955), 498–499. doi:10.1515/zna-1955-0610.

Genzel, Reinhard: Autobiography of Reinhard Genzel. *The Shaw Prize*, 9/9/2008. https://www.shawprize.org/prizes-and-laureates/astronomy/2008/autobiography-of-reinhard-genzel. Last accessed 8/15/2020.

Genzel, Reinhard, F. Eisenhauer, and S. Gillessen: The Galactic Center Massive Black Hole and Nuclear Star Cluster. *Reviews of Modern Physics* 82/4 (2010), 3121–3195. doi:10.1103/RevModPhys.82.3121.

Genzel, Reinhard, and C. H. Townes: Physical Conditions, Dynamics, and Mass Distribution in the Center of the Galaxy. *Annual Review of Astronomy and Astrophysics* 25 (1987), 377–423. doi:10.1146/annurev.aa.25.090187.002113.

Genzel, Reinhard, D. M. Watson, C. H. Townes, H. L. Dinerstein, D. Hollenbach, D. F. Lester, M. Werner, et al.: Far-Infrared Spectroscopy of the Galactic Center—Neutral and Ionized Gas in the Central 10 Parsecs of the Galaxy. *The Astrophysical Journal* 276 (1984), 551–559. doi:10.1086/161644.

Genzel, Reinhard, D. Watson, C. Townes, D. Lester, H. Dinerstein, M. Werner, and J. Storey: OI and OIII in SGR A. Neutral and Ionized Gas at the Galactic Center. *AIP Conference Proceedings* 83 (1982), 72–76. doi:10.1063/1.33504.

German Astronomical Society: BMBF Project Funding Review Board of the BMBF (Bundesministerium Für Bildung und Forschung) Project Funding Scheme Astrophysics and Astroparticle Physics. *German Astronomical Society*. http://www.astronomische-gesellschaft.org/en/rds/bodies/bmbf?set_language=en. Last accessed 4/10/2018.

German Space Projects Described. *Flight International* (1963), 892–893.

Gershtein, Semen Solomonovich, and Yakov Borisovich Zeldovich: Rest Mass of Muonic Neutrino and Cosmology. *Soviet Journal of Experimental and Theoretical Physics Letters* 4 (1966), 120–122. http://adsabs.harvard.edu/abs/1966JETPL...4..120G. Last accessed 8/8/2020.

Giacconi, Riccardo, E. Kellogg, P. Gorenstein, H. Gursky, and H. Tananbaum: An X-Ray Scan of the Galactic Plane from UHURU. *The Astrophysical Journal Letters* 165 (1971), L27–L35. doi:10.1086/180711.

Giacconi, Riccardo: History of X-Ray Telescopes and Astronomy. *Experimental Astronomy* 25/1–3 (2009), 143–156. doi:10.1007/s10686-009-9139-8.

Giacconi, Riccardo, Herbert Gursky, Frank R. Paolini, and Bruno B. Rossi: Evidence for x Rays From Sources Outside the Solar System. *Physical Review Letters* 9/11 (1962), 439–443. doi:10.1103/PhysRevLett.9.439.

Giacconi, Riccardo, and Harvey D. Tananbaum: The High Energy X-Ray Universe. *Proceedings of the National Academy of Sciences of the United States of America* 107/16 (2010), 7202–7207. https://www.jstor.org/stable/pdf/25665331.pdf. Last accessed 10/30/2018.

Gillessen, S., F. Eisenhauer, G. Perrin, W. Brandner, C. Straubmeier, K. Perraut, A. Amorim, et al.: GRAVITY: A Four Telescope Beam Combiner Instrument for the VLTI. *Optical and Infrared Interferometry II*. SPIE Astronomical Telescopes + Instrumentation. San Diego, CA: International Society for Optics and Photonics 2010, 77340Y. doi:10.1117/12.856689.

Gillessen, S., F. Eisenhauer, S. Trippe, T. Alexander, R. Genzel, F. Martins, and T. Ott: Monitoring Stellar Orbits around the Massive Black Hole in the Galactic Center. *The Astrophysical Journal* 692/2 (2009), 1075–1109. doi:10.1088/0004-637X/692/2/1075.

Gilman, Nils: *Mandarins of the Future. Modernization Theory in Cold War America*. Baltimore, MD: Johns Hopkins University Press 2003.

Ginzburg, Vitalii L.: Cosmic Ray Astrophysics. History and General Review. *Physics-Uspekhi* 39/2 (1996), 155–168. doi:10.1070/PU1996v039n02ABEH000132.

Ginzburg, Vitaly L.: On the Birth and Development of Cosmic Ray Astrophysics. In: Yataro Sekido, and Harry Elliot (eds.): *Early History of Cosmic Ray Studies. Personal Reminiscences with Old Photographs*. Dordrecht: Springer Netherlands 1985, 411–426. doi:10.1007/978-94-009-5434-2_37.

Ginzburg, Vitaly L.: The Nature of Cosmic Radio Emission and the Origin of Cosmic Rays. *Il Nuovo Cimento* 3/1 (1956), 38–48. doi:10.1007/BF02745509.

Glaßmeier, Karl-Heinz, Manfred Siebert, and Emilio Herrero-Bervera: Bartels, Julius (1899–1964). *Encyclopedia of Geomagnetism and Paleomagnetism*. Edited by David Gubbins. Dordrecht: Springer 2007. https://doi.org/10.1007/978-1-4020-4423-6_15. Last accessed 11/3/2018.

Globig, Michael: Mit der Tonne in die Atmosphäre. *Max Planck Forschung* 4 (2006), 56–57. https://www.mpg.de/971205/S003_Rueckblende_056_057.pdf. Last accessed 10/30/2018.

Globig, Michael: Zur Person. Richard Wielebinski. *Max Planck Forschung* 4 (2001), 98–103.

GMT Scientific Advisory Committee: *GMT Scientific Advisory Committee. Operations Concept White Paper*, 2012. http://www.gmto.org/Resources/GMT_SAC_Operations_White_Paper.pdf. Last accessed 10/8/2010.

Godwin, Matthew: *The Skylark Rocket. British Space Science and the European Space Research Organisation. 1957–1972*. Paris: Beauchesne 2007.

Goebel, K., P. Schmidlin, and J. Zähringer: Das Tritium-Helium- und das Kalium-Argon-Alter des Meteoriten "Ramsdorf." *Zeitschrift für Naturforschung A* 14/11 (1959), 996–998. doi:10.1515/zna-1959-1112.

Goenner, Hubert: General Relativity and the Growth of a Sub-Discipline "Gravitation" in Germany. *The European Physical Journal H* 42/3 (2017), 395–430. doi:10.1140/epjh/e2017-70057-4.

Goenner, Hubert: Some Remarks on the Early History of the Albert Einstein Institute. arXiv:1612.01338 [physics.hist-ph] 2016. doi:10.48550/arXiv.1612.01338.

Goenner, Hubert, and F.W. Hehl: Zur Gründung des Albert-Einstein-Instituts für Gravitationsphysik. *Physikalische Blätter* 47/10 (1991), 936–936. doi:10.1002/phbl.19910471015.

Gold, Thomas: Rotating Neutron Stars as the Origin of the Pulsating Radio Sources. *Nature* 218/5143 (1968), 731–732. doi:10.1038/218731a0.

Goss, W. M., and R. X. McGee: The Discovery of the Radio Source Sagittarius A (Sgr A). In: Roland Gredel (ed.): *The Galactic Center, Astronomical Society of the Pacific Conference Series*. The Galactic Center. Astronomical Society of the Pacific 4th international meeting, March 10–15, 1996. La Serena 1996, 369.

Götting, Niels, and HEGRA Collaboration: Recent Results from HEGRA. *The European Physical Journal C—Particles and Fields* 33/1 (2004), 932–934. doi:10.1140/epjcd/s2004-03-1628-3.

Gottstein, Klaus: Die Blasenkammer und ihre Anwendung in der Physik der Elementarteilchen. *Die Naturwissenschaften* 46/3 (1959), 97–102. doi:10.1007/BF00638309.

Gottstein, Klaus: On the Fragmentations of Heavy Cosmic Ray Nuclei. *Il Nuovo Cimento (1943-1954)* 11/2 (1954), 377–380. doi:10.1007/BF02781100.

Gottstein, Klaus: Zur Aufspaltung der schweren Kerne in der kosmischen Strahlung. *Naturwissenschaften* 40/3 (1953), 104–105. doi:10.1007/BF00597050.

Gottstein, Klaus, B. Boederer, J. Boederer, N. Varshneya, and P. Waloschek: Heavy Unstable Particles in Nuclear Emulsions. *Il Nuovo Cimento (1955–1965)* 4/2 (1956), 440–444. doi:10.1007/BF02747915.

Gottstein, Klaus, and Martin Teucher: Zur Mesonenerzeugung beim Zusammenstoß energiereicher Nukleonen. *Zeitschrift für Naturforschung A* 8/2–3 (1953), 120–126. doi:10.1515/zna-1953-2-303.

Goudsmit, Samuel A.: *Alsos*. Woodbury, NY: AIP Press 1996.

Gould, Robert J.: High-Energy Photons from the Compton-Synchrotron Process in the Crab Nebula. *Physical Review Letters* 15/14 (1965), 577–579. doi:10.1103/PhysRevLett.15.577.

Graham, Loren, and Irina Dezhina: *Science in the New Russia. Crisis, Aid, Reform.* Bloomington, IN: Indiana University Press 2008.

Gravity Collaboration, R. Abuter, M. Accardo, A. Amorim, N. Anugu, G. Ávila, N. Azouaoui, et al.: First Light for GRAVITY. Phase Referencing Optical Interferometry for the Very Large Telescope Interferometer. *Astronomy and Astrophysics* 602 (2017), A94. doi:10.1051/0004-6361/201730838.

Gredel, Roland: Guido Münch (1921–2020). *Max-Planck-Institut für Astronomie*,. 5/4/2020. http://www.mpia.de/aktuelles/mpia-news/2020-05-04-muench-en. Last accessed 8/16/2020.

Greenstein, J.L., and T.A. Matthews: Red-Shift of the Unusual Radio Source. 3C 48. *Nature* 197/4872 (1963), 1041–1042. doi:10.1038/1971041a0.

Gribov, Vladimir N., and Bruno Pontecorvo: Neutrino Astronomy and Lepton Charge. *Physics Letters B* 8/7 (1969), 493–496. doi:10.1016/0370-2693(69)90525-5.

Grieger, G., C. Beidler, E. Harmeyer, W. Lotz, J. KißLinger, P. Merkel, J. Nührenberg, et al.: Modular Stellarator Reactors and Plans for Wendelstein 7-X. *Fusion Technology* 21/3P2B (1992), 1767–1778. doi:10.13182/FST92-A29977.

Gringauz, K.I., V.G. Kurt, V.I. Moroz, and I.S. Shklovskii: Results of Observations of Charged Particles Observed Out to R = 100,000 Km, with the Aid of Charged-Particle Traps on Soviet Space Rockets. *Soviet Astronomy* 4 (1961), 680–695. https://ui.adsabs.harvard.edu/#abs/1961SvA.....4..680G. Last accessed 10/30/2018.

Grote, Hartmut, Karsten Danzmann, Kathrine L. Dooley, R. Schnabel, J. Slutsky, and H. Vahlbruch: First Long-Term Application of Squeezed States of Light in a Gravitational-Wave Observatory. *Physical Review Letters* 110/18 (2013), 181101. doi:10.1103/PhysRevLett.110.181101.

Grote, Hartmut, Gerhard Heinzel, Andreas Freise, Stefan Goßler, Benno Willke, Harald Lück, Harry Ward, et al.: Automatic Beam Alignment for the Mode-Cleaner Cavities of GEO 600. *Applied Optics* 43/9 (2004), 1938–1945. doi:10.1364/AO.43.001938.

Grotelüschen, Frank: *Insight Starts Here. 50 Years of DESY.* Hamburg: Deutsches Elektronen-Synchrotron DESY 2009.

Grotz, K., H. V. Klapdor, and J. Metzinger: Microscopic Calculation of Neutrino Capture Rates in ^{69,71}Ga and the Detection of Solar and Galactic Neutrinos. *Physical Review C* 33/4 (1986), 1263–1269. doi:10.1103/PhysRevC.33.1263.

Grotz, Klaus, and Hans Volker Klapdor: *The Weak Interaction in Nuclear, Particle and Astrophysics*. Bristol: Adam Hilger 1990.

Ground-Based Astronomy. A Ten-Year Program. A Report Prepared by the Panel on Astronomical Facilities for the Committee on Science and Public Policy of the National Academy of Sciences. *Science* 146/3652 (1964), 1641–1648. doi:10.1126/science.146.3652.1641.

Grupen, Claus: *Astroparticle Physics*. 2nd ed. Cham: Springer 2020. doi:10.1007/978-3-030-27339-2.

Gruss, Peter, Gunnar Klack, and Matthias Seidel (eds.): *Fehling+Gogel. Die Max-Planck-Gesellschaft als Bauherr der Architekten Hermann Fehling und Daniel Gogel*. Berlin: Jovis 2009.

Guerra, Francesco, Matteo Leone, and Nadia Robotti: The Discovery of Artificial Radioactivity. *Physics in Perspective* 14/1 (2012), 33–58. doi:10.1007/s00016-011-0064-7.

Guerra, Francesco, and Nadia Robotti: *The Lost Notebook of Enrico Fermi. The True Story of the Discovery of Neutron-Induced Radioactivity*. Cham: Springer Verlag 2018. doi:10.1007/978-3-319-69254-8.

Gursky, Herbert: Herbert Friedman. *Physics Today* 54/3 (2001), 94. doi:10.1063/1.1366078.

Gursky, Herbert: Technology and the Emergence of X-Ray Astronomy. *Journal of Astronomical History and Heritage* 3/1 (2000), 1–12. http://www.narit.or.th/en/files/2000JAHHvol03/2000JAHH....3....1G.pdf. Last accessed 10/30/2018.

Güsten, Rolf, Helmut Wiesemeyer, David Neufeld, Karl M. Menten, Urs U. Graf, Karl Jacobs, Bernd Klein, et al.: Astrophysical Detection of the Helium Hydride Ion HeH^+. *Nature* 568/7752 (2019), 357–359. doi:10.1038/s41586-019-1090-x.

Güsten, Rolf, Helmut Wiesemeyer, David Neufeld, Karl M. Menten, Urs U. Graf, Karl Jacobs, Bernd Klein, et al.: Astrophysical Detection of the Helium Hydride Ion HeH+. 7752. *Nature* 568/7752 (2019), 357–359. doi:10.1038/s41586-019-1090-x.

Habersack, Michael: *Friedrich Dessauer (1881–1963). Eine politische Biographie des Frankfurter Biophysikers und Reichstagsabgeordneten*. Paderborn: Ferdinand Schöningh 2011.

Habfast, Claus: *Großforschung mit kleinen Teilchen. Das Deutsche Elektronen-Synchrotron DESY 1956–1970*. Berlin: Springer 1989.

Haerendel, Gerhard: History of EISCAT. Part 4. On the German Contribution to the Early Years of EISCAT. *History of Geo- and Space Sciences* 7/2 (2016), 67–72. doi:10.5194/hgss-7-67-2016.

Haerendel, Gerhard: Towards an Artificial Comet. In: Gerhard Haerendel, and Bruce Battrick (eds.): *Topics in Plasma-, Astro- and Space Physics. A Volume Dedicated to Reimar Lüst on the Occasion of His 60th Birthday*. Garching: Max-Planck-Institut für Physik und Astrophysik, Institut Extraterrestrische Physik 1983, 165–177.

Haerendel, Gerhard, S. Grzedzielski, G. Cavallo, and Bruce Battrick (eds.): *40 Years of Cospar*. Noordwijk: ESA Publications Division 1998.

Haerendel, Gerhard: Stand und Ergebnisse des deutschen Höhenforschungsprogramms. *Raumfahrtforschung* 1 (1976), 34.

Haerendel, Gerhard, and Reimar Lüst: Artificial Plasma Clouds in Space. *Scientific American* 219/5 (1968), 80–95. https://www.jstor.org/stable/24927565. Last accessed 5/16/2021.

Haerendel, Gerhard: *My Life in Space Exploration*. Cham: Springer 2022. doi:10.1007/978-3-031-10286-8.

Hahn, Otto, A. Flammersfeld, and W. Kroebel: Persönliche Erinnerungen an Frédéric Joliot. *Physikalische Blätter* 14/11 (1958), 510–514. doi:10.1002/phbl.19580141106.

Hahn, Otto: Friedrich A. Paneth. *Zeitschrift für Elektrochemie. Berichte der Bunsengesellschaft für physikalische Chemie* 61/9 (1957), 1121–1122. doi:10.1002/bbpc.19570610902.

Hahn, Otto: Erich Regener und die Max-Planck-Gesellschaft. *Zeitschrift für Naturforschung A* 6/11 (1951), 568–569. doi:10.1515/zna-1951-1102.

Hahn, Otto, and Fritz Straßmann: Über den Nachweis und das Verhalten der bei der Bestrahlung des Urans mittels Neutronen entstehenden Erdalkalimetalle. *Die Naturwissenschaften* 27/1 (1939), 11–15. doi:10.1007/BF01488241.

Haines, Todd J., D. Berley, C. Y. Chang, B. L. Dingus, J. A. Goodman, T. J. Haines, M. Stark, et al.: The Status of the CYGNUS Experiment: Past, Present, and Future. *Nuclear Physics B—Proceedings Supplements* 14/1 (1990), 244–249. doi:10.1016/0920-5632(90)90428-W.

Halban, Hans von, Frédéric Joliot, and Lew Kowarski: Liberation of Neutrons in the Nuclear Explosion of Uranium. *Nature* 143/3620 (1939), 470–471. doi:10.1038/143470a0.

Halban, Hans von, Frédéric Joliot, and Lew Kowarski: Number of Neutrons Liberated in the Nuclear Fission of Uranium. *Nature* 143/3625 (1939), 680–680. doi:10.1038/143680a0.

Hammersley, P. L., and J. M. Rodríguez Espinosa: *Guaranteed Time for PI Instruments on the GTC*. 1, 2005. http://www.gtc.iac.es/instruments/media/GTpolicy.pdf. Last accessed 10/8/2020.

Hampel, W.: The Gallium Solar Neutrino Detector. *AIP Conference Proceedings*. American Institute of Physics 1983, 88–96. doi:10.1063/1.33928.

Hampel, Wolfgang: Particle Physics. Can the Gallium Detector Solve the Solar Neutrino Problem? *Nature* 308/5957 (1984), 312. doi:10.1038/308312a0.

Hampel, Wolfgang: Low-Energy Neutrinos in Astrophysics. In: Ettore Fiorini (ed.): *Neutrino Physics and Astrophysics*. Boston MA: Springer 1982, 61–79.

Hampel, Wolfgang: The Gallium Solar Neutrino Detector. In: M.L. Cherry, K. Lande, and W.A. Fowler (eds.). Solar Neutrinos and Neutrino Astronomy. 23–25 August 1984, Lead, SD, USA. American Institute of Physics 1985, 162–174. doi:10.1063/1.35172.

Hampel, Wolfgang, G. Heusser, M. Hübner, J. Kiko, T. Kirsten, K. Schneider, and R. Schlotz: Results of Ultra-Low Level ^{71}Ge Counting for Application in the Neutrino Experiment at the Gran Sasso Underground Physics Laboratory. In: F.C. Jones, J. Adams, and G.M. Mason (eds.): *Proceedings from the 19th International Cosmic Ray Conference, La Jolla, USA, August 11–23, 1985*. NASA. Goddard Space Flight Center 1985, 422–425. https://ntrs.nasa.gov/api/citations/19850026678/downloads/19850026678.pdf. Last accessed 10/31/2018.

Hampel, Wolfgang, G. Heusser, J. Kiko, T. Kirsten, M. Laubenstein, E. Pernicka, W. Rau, et al.: Final Results of the ^{51}Cr Neutrino Source Experiments in GALLEX. *Physics Letters B* 420/1–2 (1998), 114–126. doi:10.1016/S0370-2693(97)01562-1.

Hand, Eric: Particle Physics: Sam Ting's Last Fling. 7215. *Nature* 455/7215 (2008), 854–857. doi:10.1038/455854a.

Harwit, Martin: *Cosmic Messengers. The Limits of Astronomy in an Unruly Universe*. Cambridge: Cambridge University Press 2021.

Harwit, Martin: *Cosmic Discovery. The Search, Scope, and Heritage of Astronomy*. Cambridge, MA: Cambridge University Press 2019.

Harwit, Martin: Physicists and Astronomy—Will You Join the Dance? *Physics Today* 34/11 (1981), 172–187. doi:10.1063/1.2914355.

Hashagen, Ulf, and Raul Rojas (eds.): *The First Computers. History and Architectures*. Cambridge, MA: MIT Press 2000.

Hawking, S. W.: Gravitational Radiation from Colliding Black Holes. *Physical Review Letters* 26/21 (1971), 1344–1346. doi:10.1103/PhysRevLett.26.1344.

Hazard, C., M.B. Mackey, and A.J. Shimmins: Investigation of the Radio Source 3C 273 By The Method of Lunar Occultations. *Nature* 197/4872 (1963), 1037–1039. doi:10.1038/1971037a0.

Heckmann, Otto: Nachrufe. Hans Kienle. *Mitteilungen der Astronomischen Gesellschaft* 38 (1976), 9–10. https://ui.adsabs.harvard.edu/#abs/1976MitAG..38....9H. Last accessed 10/30/2018.

HEGRA Collaboration, G. Heinzelmann, W. Hofmann, D. Horns, D. Kranich, G. Puehlhofer, G. Rowell, et al.: HEGRA Contributions to the 28th International Cosmic Ray Conference. arXiv:Astro-Ph/0307334, 2003. http://arxiv.org/abs/astro-ph/0307334. Last accessed 12/19/2018.

HEGRA Collaboration, A. Konopelko, M. Hemberger, F. Aharonian, A. Daum, W. Hofmann, C. Köhler, et al.: Performance of the Stereoscopic System of the HEGRA Imaging Air Čerenkov Telescopes. Monte Carlo Simulations and Observations. *Astroparticle Physics* 10/4 (1999), 275–289. doi:10.1016/S0927-6505(98)00062-0.

Heinemann, Manfred: Der Wiederaufbau der Kaiser-Wilhelm-Gesellschaft und die Neugründungen der Max-Planck-Gesellschaft (1945–1949). In: Rudolf Vierhaus, and Bernhard vom Brocke (eds.): *Forschung im Spannungsfeld von Politik und*

Gesellschaft. Geschichte und Struktur der Kaiser-Wilhelm-/Max-Planck-Gesellschaft. Stuttgart: Deutsche Verlags-Anstalt 1990, 407–470.

Heinemann, Manfred: Überwachung und »Inventur« der deutschen Forschung. Das Kontrollratsgesetz Nr. 25 und die alliierte Forschungskontrolle im Bereich der Kaiser-Wilhelm-/Max-Planck-Gesellschaft (KWG/MPG) 1945–1955. In: Lothar Mertens (ed.): *Politischer Systemumbruch als irreversibler Faktor von Modernisierung in der Wissenschaft?* Berlin: Duncker & Humblot 2001, 167–199.

Heintze, J.: Cosmic Ray Tracking for Gamma-Ray Astronomy. In: O. Fackler, and J. Tran Thanh Van (eds.): *New and Exotic Phenomena '90. Proceedings, 25th Rencontres de Moriond, 10th Moriond Workshop, Les Arcs, France, January 20–27, 1990.* Gif-sur-Yvette, France: Editions Frontières, 405–410. https://inspirehep.net/record/309842. Last accessed 11/30/2018.

Heintze, J., P. Lennert, S. Polenz, B. Schmidt, W. Brückner, B. Povh, and J. Spitzer: Cosmic Ray Tracking—A New Approach to High-Energy γ-Astronomy. *Nuclear Instruments and Methods in Physics Research Section A: Accelerators, Spectrometers, Detectors and Associated Equipment* 277/1 (1989), 29–41. doi:10.1016/0168-9002(89)90532-9.

Heintze, J., P. Lennert, S. Polenz, B. Schmidt, J. Spitzer, W. Brückner, W. Hofmann, et al.: Measuring the Chemical Composition of Cosmic Rays at ~10^{13}-10^{15} eV by Utilizing the Solar and Geomagnetic Fields. *Experimental Astronomy* 1/1 (1989), 21–34. doi:10.1007/BF00414793.

Heintze, J., P. Lennert, S. Polenz, B. Schmidt, J. Spitzer, B. Stadler, W. Hofmann, et al.: The Heidelberg Cosmic Ray Project—Aims and Status. *Nuclear Physics B—Proceedings Supplements* 14A/1 (1990), 148–152. doi:10.1016/0920-5632(90)90411-M.

Heintze, J., P. Lennert, S. Polenz, B. Schmidt, J. Spitzer, B. Stadler, W. Hofmann, et al.: The Heidelberg Cosmic Ray Tracking Project—A New Approach to High Energy γ—Astronomy. In: Protheroe, R. J. (ed.): *21st International Cosmic Ray Conference (ICRC 1990), 6–19 January 1990. Adelaide, Australia.* 1990, 266–269. http://adsabs.harvard.edu/abs/1990ICRC....4..266H. Last accessed 12/1/2018.

Heinze, Thomas, Olof Hallonsten, and Steffi Heinecke: From Periphery to Center. Synchrotron Radiation at DESY, Part I. 1962–1977. *Historical Studies in the Natural Sciences* 45/3 (2015), 447–492. doi:10.1525/hsns.2015.45.3.447.

Heinze, Thomas, Olof Hallonsten, and Steffi Heinecke: From Periphery to Center. Synchrotron Radiation at DESY, Part II 1977–1993. *Historical Studies in the Natural Sciences* 45/4 (2015), 513–548. doi:10.1525/hsns.2015.45.4.513.

Heinzel, Gerhard, Andreas Freise, Hartmut Grote, Kenneth Strain, and Karsten Danzmann: Dual Recycling for GEO 600. *Classical and Quantum Gravity* 19/7 (2002), 1547–1553. doi:10.1088/0264-9381/19/7/343.

Heinzel, Gerhard, Jun Mizuno, Roland F. Schilling, Walter Winkler, Albrecht Rüdiger, and Karsten Danzmann: An Experimental Demonstration of Resonant Sideband

Extraction for Laser-Interferometric Gravitational Wave Detectors. *Physics Letters A* 217/6 (1996), 305–314. doi:10.1016/0375-9601(96)00361-1.

Heinzel, Gerhard, Albrecht Rüdiger, Roland F. Schilling, Kenneth A. Strain, Walter Winkler, Jun Mizuno, and Karsten Danzmann: Automatic Beam Alignment in the Garching 30-m Prototype of a Laser-Interferometric Gravitational Wave Detector. *Optics Communications* 160/4 (1999), 321–334. doi:10.1016/S0030-4018(98)00654-3.

Heinzel, Gerhard, Kenneth A. Strain, Jun Mizuno, Kenneth David Skeldon, Benno Willke, Walter Winkler, Roland F. Schilling, et al.: Experimental Demonstration of a Suspended Dual Recycling Interferometer for Gravitational Wave Detection. *Physical Review Letters* 81/25 (1998), 5493–5496. doi:10.1103/PhysRevLett.81.5493.

Heisenberg, Werner: *Encounters with Einstein and Other Essays on People, Places, and Particles*. Princeton, NJ: Princeton University Press 1989.

Heisenberg, Werner: *Collected Works. Series A/1: Original Scientific Papers*. Edited by Walter Blum, Hans-Peter Dürr, and Helmut Rechenberg. Vol. 1. Berlin: Springer-Verlag 1985. doi:10.1007/978-3-642-61659-4.

Heisenberg, Werner: *Collected Works. Series A/2: Original Scientific Papers*. Edited by Walter Blum, Hans-Peter Dürr, and Helmut Rechenberg. Vol. 2. Berlin: Springer-Verlag 1985. doi:10.1007/978-3-642-70078-1.

Heisenberg, Werner: *Collected Works. Series A/3: Original Scientific Papers*. Edited by Walter Blum, Hans-Peter Dürr, and Helmut Rechenberg. Vol. 3. Berlin: Springer-Verlag 1985. doi:10.1007/978-3-642-70078-1.

Heisenberg, Werner: Das Kaiser-Wilhelm-Institut für Physik. Geschichte eines Instituts. In: Generalverwaltung der Max-Planck Gesellschaft zur Förderung der Wissenschaften e.V. (ed.): *Jahrbuch 1971 der Max-Planck-Gesellschaft zur Förderung der Wissenschaften e.V.* Göttingen 1971, 45–89.

Heisenberg, Werner: *Physics and Beyond. Encounters and Conversations*. New York: Harper & Row Publishers 1971.

Heisenberg, Werner: Max-Planck-Institut für Physik und Astrophysik in München. In: Generalverwaltung der Max-Planck-Gesellschaft (ed.): *Jahrbuch 1961 der Max-Planck-Gesellschaft zur Förderung der Wissenschaften e.V. Teil II*. Göttingen 1962, 632–643.

Heisenberg, Werner: The Production of Mesons in Very High Energy Collisions. *Il Nuovo Cimento Series 10* 2/1 (1955), 96–103. doi:10.1007/BF02746079.

Heisenberg, Werner (ed.): *Kosmische Strahlung. Vorträge gehalten im Max-Planck-Institut für Physik Göttingen*. 2nd ed. Berlin: Springer 1953.

Heisenberg, Werner: Bemerkungen zur Theorie der Vielfacherzeugung von Mesonen. *Naturwissenschaften* 39 (1952), 69–69. doi:10.1007/BF00596818.

Heisenberg, Werner: Mesonenerzeugung als Stoßwellenproblem. *Zeitschrift für Physik A Hadrons and nuclei* 133/1 (1952), 65–79. doi:10.1007/BF01948683.

Heisenberg, Werner: Kosmische Strahlungen und Atomphysik. *Jahrbuch 1951 der Max-Planck-Gesellschaft zur Förderung der Wissenschaften e.V.* Göttingen: Max-Planck-Gesellschaft zur Förderung der Wissenschaften 1951, 229–263.

Heisenberg, Werner: Die Erzeugung von Mesonen in Vielfachprozessen. *Il Nuovo Cimento (1943-1954)* 6/3 (1949), 493–497. doi:10.1007/BF02822044.

Heisenberg, Werner: Production of Meson Showers. *Nature* 164/4158 (1949), 65–66. doi:10.1038/164065c0.

Heisenberg, Werner: Über die Entstehung von Mesonen in Vielfachprozessen. *Zeitschrift für Physik* 126/6 (1949), 569–582. doi:10.1007/BF01330108.

Heisenberg, Werner: On the Theory of Statistical and Isotropic Turbulence. *Proceedings of the Royal Society of London* 195/1042 (1948), 402–406. www.jstor.org/stable/98337. Last accessed 10/30/2018.

Heisenberg, Werner: Zur statistischen Theorie der Turbulenz. *Zeitschrift für Physik* 124/7 (1948), 628–657. doi:10.1007/BF01668899.

Heisenberg, Werner: Über die Arbeiten zur technischen Ausnutzung der Atomkernenergie in Deutschland. *Die Naturwissenschaften* 33/11 (1946), 325–329. doi:10.1007/BF00842932.

Heisenberg, Werner (ed.): *Vorträge über Kosmische Strahlung.* Berlin: Springer-Verlag 1943.

Heisenberg, Werner: Theoretische Überlegungen zur Höhenstrahlung. *Annalen der Physik* 405/4 (1932), 430–452. doi:10.1002/andp.19324050404.

Heisenberg, Werner: Über die durch Ultrastrahlung hervorgerufenen Zertrümmerungsprozesse. *Naturwissenschaften* 20/21 (1932), 365–366. doi:10.1007/BF01504936.

Heisenberg, Werner, and Elisabeth Heisenberg: *My Dear Li. Correspondence, 1937–1946.* Edited by Anna Maria Hirsch-Heisenberg. New Haven, CT: Yale University Press 2016.

Heisenberg, Werner, and Carl F. von Weizsäcker: Die Gestalt der Spiralnebel. *Zeitschrift für Physik* 125/4–6 (1948), 290–292.

Henning, Eckart, and Marion Kazemi: *Chronik der Kaiser-Wilhelm-/Max-Planck-Gesellschaft zur Förderung der Wissenschaften 1911–2011. Daten und Quellen.* Berlin: Duncker & Humblot 2011.

Henning, Eckart, and Marion Kazemi: *Handbuch zur Institutsgeschichte der Kaiser-Wilhelm-/Max-Planck-Gesellschaft zur Förderung der Wissenschaften 1911–2011. Daten und Quellen.* Vol. 1. Berlin: Archiv der Max-Planck-Gesellschaft 2016.

Henning, Eckart, and Marion Kazemi: *Handbuch zur Institutsgeschichte der Kaiser-Wilhelm-/Max-Planck-Gesellschaft zur Förderung der Wissenschaften 1911–2011. Daten und Quellen.* Vol. 2. Berlin: Archiv der Max-Planck-Gesellschaft 2016.

Hentschel, Klaus: *The Einstein Tower. An Intertexture of Dynamic Construction, Relativity Theory, and Astronomy.* California: Stanford University Press 1997.

Henyey, Louis G.: The Evolution of Stars Near the Main Sequence. *Publications of the Astronomical Society of the Pacific* 68/405 (1956), 503–504. doi:10.1086/126986.

Henyey, Louis G., L. Wilets, K.H. Böhm, R. Lelevier, and R.D. Levee: A Method for Automatic Computation of Stellar Evolution. *Astrophysical Journal* 129 (1959), 628–636. doi:10.1086/146661.

Herbert, Ulrich, and Axel Schildt (eds.): *Kriegsende in Europa. Vom Beginn des deutschen Machtzerfalls bis zur Stabilisierung der Nachkriegsordnung 1944–1948*. 1st ed. Essen: Klartext 1998.

Hermann, Armin: *Und trotzdem Brüder. Die deutsch-deutsche Geschichte der Firma Carl Zeiss*. München: Piper 2002.

Hermann, Armin: Germany's Part in the Setting-up of CERN. In: Armin Hermann, John Krige, Ulrike Merseits, and Dominique Pestre (eds.): *History of CERN. Launching the European Organization for Nuclear Research*. Amsterdam: North-Holland 1987, 383–429.

Hermann, Armin, John Krige, Ulrike Mersits, Dominique Pestre, and Lanfranco Belloni: *History of CERN. Launching the European Organization for Nuclear Research*. Vol. 1. Amsterdam: North-Holland 1987.

Hermann, Armin, John Krige, Ulrike Mersits, Dominique Pestre, and Laura Weiss: *History of CERN. Building and Running the Laboratory, 1954–1965*. Vol. 2. Amsterdam: North-Holland 1990.

Hermsen, Wim: COS-B Views on the Diffuse Galactic Gamma-Ray Emission and Some Point Sources. *Space Science Reviews* 49/1 (1989), 17–39. doi:10.1007/BF00173739.

Hess, Victor: Über Beobachtungen der durchdringenden Strahlung bei sieben Freiballonfahrten. *Physikalische Zeitschrift* 13 (1913), 1084–1091.

Hess, W.N.: Energetic Particles in the Inner Van Allen Belt. *Space Science Reviews* 1/2 (1962), 278–312. doi:10.1007/BF00240580.

Hetherington, Norris S.: Walter Baade: A Life in Astrophysics. *Physics Today* 55/11 (2002), 69–69. doi:10.1063/1.1535009.

Hewish, Antony, S.J. Bell, J.D.H. Pilkington, P.F. Scott, and R.A. Collins: Observation of a Rapidly Pulsating Radio Source. *Nature* 217/5130 (1968), 709–713. doi:10.1038/217709a0.

Hild, Stefan, Hartmut Grote, J. Degallaix, Szczepan Chelkowski, Karsten Danzmann, Andreas Freise, M. Hewitson, et al.: DC-Readout of a Signal-Recycled Gravitational Wave Detector. *Classical and Quantum Gravity* 26/5 (2009), 055012. doi:10.1088/0264-9381/26/5/055012.

Hill, Charles N.: *A Vertical Empire. The History of the UK Rocket and Space Programme, 1950–1971*. London: Imperial College Press 2001.

Hillas, A. Michael: Why Is Cygnus X-3 (with Related Sources) a Highlight of Cosmic-Ray Astrophysics? *Proceedings from the 19th International Cosmic Ray Confer-*

ence, La Jolla, USA, August 11–12, 1985. 1985, 407–414. http://adsabs.harvard.edu/abs/1985ICRC....9..407H. Last accessed 11/16/2018.

Hillas, A. Michael: Cerenkov Light Images of EAS Produced by Primary Gamma Rays and by Nuclei. *Proceedings from the 19th International Cosmic Ray Conference, La Jolla, USA, August 11–12, 1985*. 1985, 445–448. https://ui.adsabs.harvard.edu/#abs/1985ICRC....3..445H. Last accessed 11/28/2018.

Hillebrandt, Wolfgang: Kraft schöpfen aus dem Wandel: Die deutsche Astronomie in der Zeit der Wiedervereingung. In: Dietrich Lemke (ed.): *Die Astronomische Gesellschaft 1863–2013*. Heidelberg: Astronomische Gesellschaft 2013, 127–141.

Hillebrandt, Wolfgang: Supernova 1987A. Core Collapse Models and Neutrinos. *ESO Workshop on the SN 1987A, Garching, Federal Republic of Germany, July 6–8, 1987, Proceedings (A88-35301 14–90)*. Garching (Germany): Federal Republic of Germany, European Southern Observatory 1987, 301–312. http://adsabs.harvard.edu/abs/1987ESOC...26..301H. Last accessed 10/31/2018.

Hintenberger, Heinrich: Josef Mattauch. 21.11.1895–10.8.1976. *Berichte und Mitteilungen Max-Planck-Gesellschaft* Nachrufe, Jahresbericht 1976, Jahresrechnung 1975 (1977), 19–21.

Hinton, Jim, Subir Sarkar, Diego Torres, and Johannes Knapp (eds.): A New Era in Gamma-Ray Astronomy with the Cherenkov Telescope Array. *Astroparticle Physics* 43 (2013), 1–2. doi:10.1016/j.astropartphys.2012.12.002.

Hirakawa, Hiromasa, and Kazumichi Narihara: Search for Gravitational Radiation at 145 Hz. *Physical Review Letters* 35/6 (1975), 330–334. doi:10.1103/PhysRevLett.35.330.

Hirata, K., T. Kajita, M. Koshiba, M. Nakahata, Y. Oyama, N. Sato, A. Suzuki, et al.: Observation of a Neutrino Burst from the Supernova SN1987A. *Physical Review Letters* 58/14 (1987), 1490–1493. doi:10.1103/PhysRevLett.58.1490.

Hoare, M.R.: Max-Planck-Gesellschaft: A Model for "Small Science"? *Nature* 237/5352 (1972), 206–209. doi:10.1038/237206a0.

Hoddeson, Lillian, Paul W. Henriksen, Roger A. Meade, and Catherine Westfall (eds.): *Critical Assembly. A Technical History of Los Alamos during the Oppenheimer Years, 1943–1945*. Cambridge: Cambridge University Press 1993.

Hoerner, Sebastian von: Design of Large Steerable Antennas. *The Astronomical Journal* 72/1 (1967), 35–47. doi:10.1086/110198.

Hoerner, Sebastian von: Die numerische Integration des n-Körper-Problems für Sternhaufen. II. *Zeitschrift für Physik* 57 (1963), 47–82. http://adsabs.harvard.edu/abs/1963ZA.....57...47V. Last accessed 10/30/2018.

Hoerner, Sebastian von: Die numerische Integration des n-Körper-Problemes für Sternhaufen. I. *Zeitschrift für Astrophysik* 50 (1960), 184–214. http://adsabs.harvard.edu/abs/1960ZA.....50..184V. Last accessed 10/30/2018.

Hoerner, Sebastian von: Herstellung von Zufallszahlen auf Rechenautomaten. *Zeitschrift für angewandte Mathematik und Physik* 8/1 (1957), 26–52. doi:10.1007/BF01601153.

Hofferbert, Ralph, Harald Baumeister, Thomas Bertram, Jürgen Berwein, Peter Bizenberger, Armin Böhm, Michael Böhm, et al.: LINC-NIRVANA for the Large Binocular Telescope: Setting up the World's Largest near Infrared Binoculars for Astronomy. *Optical Engineering* 52/8 (2013), 081602. doi:10.1117/1.OE.52.8.081602.

Hoffmann, Dieter (ed.): *Physik im Nachkriegsdeutschland.* Frankfurt am Main: Harri Deutsch 2002.

Hoffmann, Dieter, and Ulrich Schmidt-Rohr: Wolfgang Gentner. Ein Physiker als Naturalist. In: Dieter Hoffmann, and Ulrich Schmidt-Rohr (eds.): *Wolfgang Gentner. Festschrift zum 100. Geburtstag.* Berlin: Springer 2006, 1–60.

Hoffmann, Dieter, and Ulrich Schmidt-Rohr (eds.): *Wolfgang Gentner. Festschrift zum 100. Geburtstag.* Berlin: Springer 2006.

Hofmann, Werner, and HESS Collaboration: The High Energy Stereoscopic System (HESS) Project. *AIP Conference Proceedings* 515/1 (2000), 500–509. https://aip.scitation.org/doi/abs/10.1063/1.1291416. Last accessed 12/12/2018.

Hofmeister, Emmi: Sternentwicklung VII. Zur Entwicklung von Sternen mit 5 und 9 Sonnenmassen. *Zeitschrift fur Astrophysik* 65 (1967), 164–184. https://ui.adsabs.harvard.edu/#abs/1967ZA.....65..164H. Last accessed 9/20/2018.

Hofmeister, Emmi, Rudolf Kippenhahn, and Alfred Weigert: Sternentwicklung I. Ein Programm zur Lösung der zeitabhängigen Aufbaugleichungen. *Zeitschrift für Astrophysik* 59 (1964), 215–241. https://ui.adsabs.harvard.edu/#abs/1964ZA.....59..215H. Last accessed 10/31/2018.

Hofmeister, Emmi, Rudolf Kippenhahn, and Alfred Weigert: Sternentwicklung II. Die Wasserstoff-brennende Phase eines Sterns von 7.0 Sonnenmassen. *Zeitschrift fur Astrophysik* 59 (1964), 242–260. https://ui.adsabs.harvard.edu/#abs/1964ZA.....59..242H. Last accessed 9/19/2018.

Hofmeister, Emmi, Rudolf Kippenhahn, and Alfred Weigert: Sternentwicklung III. Die Helium-brennende Phase und die Cepheidenstadien eines Sterns von 7.0 Sonnenmassen. Mit 6 Textabbildungen. *Zeitschrift fur Astrophysik* 60 (1964), 57–75. https://ui.adsabs.harvard.edu/#abs/1964ZA.....60...57H. Last accessed 9/19/2018.

Hohn, Hans-Willy, and Uwe Schimank: *Konflikte und Gleichgewichte im Forschungssystem. Akteurkonstellationen und Entwicklungspfade in der staatlich finanzierten außeruniversitären Forschung.* Frankfurt am Main: Campus Verlag 1990.

Holton, Gerald: Striking Gold in Science: Fermi's Group and the Recapture of Italy's Place in Physics. *Minerva* 12/2 (1974), 159–198. https://www.jstor.org/stable/41820198. Last accessed 1/8/2020.

Horeis, Heinz: Hans-Walter Rix: Der Jung-Star. *Bild Der Wissenschaft* 11 (2004), 46.

Hough, James, S. Hoggan, G. A. Kerr, J. B. Mangan, B. J. Meers, G. P. Newton, N. A. Robertson, et al.: The Development of Long Baseline Gravitational Radiation Detectors at Glasgow University. *Gravitation, Geometry and Relativistic Physics. Proceedings of the "Journées Relativistes" Held at Aussois, France, May 2–5, 1984*. Berlin: Springer 1984, 204–212. doi:10.1007/BFb0012592.

Hough, James, B. J. Meers, G. P. Newton, N. A. Robertson, H. Ward, B. F. Schutz, R. W. P. Drever, et al.: *A British Long Baseline Gravitational Wave Observatory*. Design study report GWD/RAL/86-001. Glasgow: Rutherford Appleton Laboratory, Glasgow University 1986.

Hough, James, B.J. Meers, G.P. Newton, N.A. Robertson, H. Ward, G. Leuchs, T.M. Niebauer, et al.: *Proposal for a Joint German-British Interferometric Gravitational Wave Detector*. MPQ 147. Garching: Max-Planck-Institut für Quantenoptik 1989. http://eprints.gla.ac.uk/114852/7/114852.pdf. Last accessed 4/18/2018.

Hough, James, J. R. Pugh, R. Bland, and R. W. P. Drever: Search for Continuous Gravitational Radiation. *Nature* 254/5500 (1975), 498–501. doi:10.1038/254498a0.

Hoyle, Fred: The Nature of Cosmic Radio Sources. *New Scientist* 17/332 (1963), 681–683.

Hoyle, Fred, and William A. Fowler: On the Nature of Strong Radio Sources. *Monthly Notices of the Royal Astronomical Society* 125/2 (1962), 169–176. doi:10.1093/mnras/125.2.169.

Hoyle, F., and M. Schwarzschild: On the Evolution of Type II Stars. *The Astrophysical Journal Supplement* 2 (1955), 1–40. doi:10.1086/190015.

Huebner, Walter F.: Kometen, Sonnenwind und Wasserstoffkoma. In: Max-Planck Gesellschaft (ed.): *Ludwig Biermann. 1907–1986*. München 1988, 35–50.

Hüfner, Klaus, and Jens Naumann: *Konjunkturen der Bildungspolitik in der Bundesrepublik Deutschland. Der Aufschwung (1960–1967)*. Vol. 1. Stuttgart: Klett 1977.

Hughes, Thomas Parke: *American Genesis. A Century of Invention and Technological Enthusiasm, 1870–1970*. New York: Viking Press 1989.

Hulse, Russel A., and Joseph H. Taylor: Discovery of a Pulsar in a Binary System. *Astrophysical Journal* 195 (1975), L51–L53. doi:10.1086/181708.

Hunt, Linda: *Secret Agenda. The United States Government, Nazi Scientists, and Project Paperclip, 1945–1990*. New York: St. Martin's Press 1991.

IceCube Collaboration: Neutrino Emission from the Direction of the Blazar TXS 0506+056 Prior to the IceCube-170922A Alert. *Science* 361 (2018), 147–151. doi:10.1126/science.aat2890.

Israel, Werner: Black Hole 2000: The Astrophysical Era. *Publications of the Astronomical Society of the Pacific* 112/771 (2000), 583–585. doi:10.1086/316557.

Jacobsen, Annie: *Operation Paperclip. The Secret Intelligence Program to Bring Nazi Scientists to America*. New York, NY: Little, Brown and Company 2014.

Jacrot, Bernard: *Des neutrons pour la science. Histoire de l'Institut Laue-Langevin, une coopération internationale particulièrement réussie*. Les Ulis: EDP sciences 2006.

Jaenicke, Ruprecht: Christian Junge. The Pioneer of Aerosol Chemistry. In: Sem J. Gilmore, Denis Boulaud, Peter Brimblecombe, David S. Ensor, James W. Gentry, Johannes C.M Marijnissen, and Othmar Preining (eds.): *History & Reviews of Aerosol Science. Proceedings of the Second Symposium on the History of Aerosol Science, 13–14 October 2001, Portland, Oregon, USA*. Reston: American Association for Aerosol Research 2005, 37–47.

Jaenicke, Ruprecht: Die Erfindung der Luftchemie. Christian Junge. In: Horst Kant, and Carsten Reinhardt (eds.): *100 Jahre Kaiser-Wilhelm-, Max-Planck-Institut für Chemie (Otto-Hahn-Institut). Facetten seiner Geschichte*. Berlin: Archiv der Max-Planck-Gesellschaft 2012, 187–202.

Jager, Cornelis de: Early Solar Space Research. In: Johan A. M. Bleeker, Johannes Geiss, and Martin C. E. Huber (eds.): *The Century of Space Science*. Dordrecht: Kluwer Academic Publishers 2001, 203–223.

Jäger, F.W.: Nachruf auf P. ten Bruggencate. *Mitteilungen der Astronomischen Gesellschaft* 15 (1962), 21–23. http://adsabs.harvard.edu/abs/1962MitAG..15...21J. Last accessed 10/30/2018.

Jäger, Klaus: Schwarzschild-Medaille der Astronomischen Gesellschaft für Richard Wielebinski. *idw-Informationsdienst Wissenschaft*, 9/5/2017. https://idw-online.de/de/news680475. Last accessed 4/12/2018.

Janka, H.-T., and W. Hillebrandt: Neutrino Emission from Type II Supernovae. An Analysis of the Spectra. *Astronomy and Astrophysics* 224/1–2 (1989), 49–56. http://adsabs.harvard.edu/abs/1989A%26A...224...49J. Last accessed 10/31/2018.

Jelley, John V., and N. A. Porter: Čerenkov Radiation from the Night Sky, and Its Application to γ-Ray Astronomy. *Quarterly Journal of the Royal Astronomical Society* 4 (1963), 275–293. http://adsabs.harvard.edu/abs/1963QJRAS...4..275J. Last accessed 12/23/2018.

Jerke, G.: Welche Zukunft hat das MPI für Astrophysik? *Physikalische Blätter* 48/7–8 (1992), 514, 605. doi:10.1002/phbl.19920480704.

Johnson, Karen E.: From Natural History to the Nuclear Shell Model: Chemical Thinking in the Work of Mayer, Haxel, Jensen, and Suess. *Physics in Perspective* 6/3 (2004), 295–309. doi:10.1007/s00016-003-0203-x.

Judt, Matthias, and Burghard Ciesla (eds.): *Technology Transfer out of Germany after 1945*. Amsterdam: Harwood Academic Publishers 1996.

Junge, Christian: Die Entstehung der Erdatmosphäre und ihre Beeinflussung durch den Menschen. In: Generalverwaltung der Max-Planck Gesellschaft zur Förderung der Wissenschaften e.V. (ed.): *Max-Planck-Gesellschaft Jahrbuch 1975*. Göttingen: Vandenhoeck & Ruprecht 1975, 36–48.

Jüngling, Simone: *Röntgenastronomie in Deutschland. Entstehungsgeschichte, Institutionalisierung und instrumentelle Entwicklungen*. Hamburg: Verlag Dr. Kovač 2007.

Kafka, Peter: On the Evaluation of the Munich-Frascati Weber-Type Experiment. In: Y. Choquet-Bruhat (ed.): *Ondes et Radiations Gravitationelles.* International Conference, Paris, France, June 18–22, 1973. Paris: Centre National de la Recherche Scientifique 1974, 181–200.

Kafka, Peter: Discussion of Possible Sources of Gravitational Radiation. *Mitteilungen Der Astronomischen Gesellschaft* 27 (1969), 134–138. https://ui.adsabs.harvard.edu/#abs/1969MitAG..27..134K. Last accessed 11/3/2017.

Kafka, Peter: How to Count Quasars. *Nature* 213/5074 (1967), 346–350. doi:10.1038/213346a0.

Kafka, Peter, and Lise Schnupp: Final Result of the Munich-Frascati Gravitational Radiation Experiment. *Astronomy and Astrophysics* 70 (1978), 97–103. http://adsabs.harvard.edu/abs/1978A&A....70...97K. Last accessed 11/18/2017.

Kaiser, David: Bringing the Human Actors Back on Stage. The Personal Context of the Einstein-Bohr Debate. *The British Journal for the History of Science* 27/2 (1994), 129–152. http://www.jstor.org/stable/4027432. Last accessed 10/30/2018.

Kaiser, David: *Drawing Theories Apart. The Dispersion of Feynman Diagrams in Postwar Physics*. Chicago, IL: University of Chicago Press 2005.

Kaiser, David, and Benjamin Wilson: Calculating Times. Radar, Ballistic Missiles, and Einstein's Relativity. In: Naomi Oreskes, and John Krige (eds.): *Science and Technology in the Global Cold War.* Cambridge, MA: MIT Press 2014, 273–316.

Kajita, Takaaki, and Superkamiokande and Kamiokande collaborations: Atmospheric Neutrino Results from Super-Kamiokande and Kamiokande—Evidence for ν_μ Oscillations. *Nuclear Physics B—Proceedings Supplements* 77/1 (1999), 123–132. doi:10.1016/S0920-5632(99)00407-7.

Kalvius, Georg Michael, and Paul Kienle (eds.): *The Rudolf Mössbauer Story. His Scientific Work and Its Impact on Science and History.* Berlin: Springer 2012.

Kamata, K., S. Shibata, O. Saavedra, V. Domingo, K. Suga, K. Murakami, Y. Toyoda, et al.: Predominantly Electromagnetic Air Showers of Energy 10^{14} eV to 10^{16} eV. *Canadian Journal of Physics* 65/10 (1968), S72–S74. doi:10.1139/p68-177.

Kampert, Karl-Heinz, and Alan A. Watson: Extensive Air Showers and Ultra High-Energy Cosmic Rays. A Historical Review. *The European Physical Journal H* 37/3 (2012), 359–412. doi:10.1140/epjh/e2012-30013-x.

Kant, Horst: Albert Einstein, Max von Laue, Peter Debye und das Kaiser-Wilhelm-Institut für Physik in Berlin (1917–1939). *Die Kaiser-Wilhelm-/Max-Planck-Gesellschaft und ihre Institute: Studien zu ihrer Geschichte: Das Harnack-Prinzip,* 1996, 227–243, 227–243. https://pure.mpg.de/pubman/faces/ViewItemOverviewPage.jsp?itemId=item_2276913_1. Last accessed 1/17/2020.

Kant, Horst: Das Max-Planck-Institut für Physik. Berlin–München. In: Peter Gruss, and Reinhard Rürup (eds.): *Denkorte. Max-Planck-Gesellschaft und Kaiser-Wilhelm-*

Gesellschaft. Brüche und Kontinuitäten 1911–2011. Dresden: Sandstein Verlag 2010, 316–323.

Kant, Horst: Von der Lichttherapie zum Zyklotron. Das Institut für Physik im Heidelberger Kaiser-Wilhelm-Institut für medizinische Forschung bis 1945. *Dahlemer Archivgespräche* 13 (2008), 49–92. http://opac.ifz-muenchen.de/webOPACClient.ifzsis/start.do?Login=woifz&Query=10="BV040926399". Last accessed 10/30/2018.

Kant, Horst, and Jürgen Renn: Eine utopische Episode. Carl Friedrich von Weizsäcker in den Netzwerken der Max-Planck-Gesellschaft. In: Klaus Hentschel, and Dieter Hoffmann (eds.): *Carl Friedrich von Weizsäcker. Physik–Philosophie–Friedensforschung*. Stuttgart: Wissenschaftliche Verlagsgesellschaft 2014, 213–242.

Karle, A., M. Merck, R. Plaga, F. Arqueros, V. Haustein, G. Heinzelmann, I. Holl, et al.: Design and Performance of the Angle Integrating Čerenkov Array AIROBICC. *Astroparticle Physics* 3/4 (1995), 321–347. doi:10.1016/0927-6505(95)00009-6.

Kasteren, Joost van: *An Overview of Space Activities in the Netherlands*. Noordwijk: ESA Publications Division 2002.

Kauffmann, Guinevere, Simon D.M. White, and Bruno Guiderdoni: The Formation and Evolution of Galaxies Within Merging Dark Matter Haloes. *Monthly Notices of the Royal Astronomical Society* 264/1 (1993), 201–218. doi:10.1093/mnras/264.1.201.

Kaufman, Scott: *Project Plowshare The Peaceful Use of Nuclear Explosives in Cold War America*. Ithaca, NY: Cornell University Press 2013.

Kay, W. D.: *Defining NASA. The Historical Debate over the Agency's Mission*. Albany, NY: State University of New York Press 2005.

Kellermann, Kenneth (ed.): *A Brief History of Radio Astronomy in the USSR. A Collection of Scientific Essays*. Translated by Denise C. Gabuzda. Dordrecht: Springer 2012.

Kellermann, Kenneth I., Ellen N. Bouton, and Sierra S. Brandt: VLBI and the Very Long Baseline Array. In: Kenneth I. Kellermann, Ellen N. Bouton, and Sierra S. Brandt (eds.): *Open Skies. The National Radio Astronomy Observatory and Its Impact on US Radio Astronomy*. Cham: Springer 2020, 391–459. doi:10.1007/978-3-030-32345-5_8.

Kellermann, Kenneth I., Ellen N. Bouton, and Sierra S. Brandt: The Largest Feasible Steerable Telescope. In: Kenneth I. Kellermann, Ellen N. Bouton, and Sierra S. Brandt (eds.): *Open Skies: The National Radio Astronomy Observatory and Its Impact on US Radio Astronomy*. Cham: Springer International Publishing 2020, 461–531. doi:10.1007/978-3-030-32345-5_9.

Kellermann, K. I., and J. M. Moran: The Development of High-Resolution Imaging in Radio Astronomy. *Annual Review of Astronomy and Astrophysics* 39/1 (2001), 457–509. doi:10.1146/annurev.astro.39.1.457.

Kellermann, Kenneth, and B. Sheets: *Serendipitous Discoveries in Radio Astronomy. Proceedings of a Workshop Held at the National Radio Astronomy Observatory Green Bank, West Virginia on May 4, 5, 6, 1983*. Green Bank, WV: National Radio Astronomy

Observatory, Associated Universities 1983. http://library.nrao.edu/public/collection/02000000000280.pdf. Last accessed 3/21/2021.

Keppler, Erhard: *Der Weg zum Max Planck Institut für Aeronomie. Von Regener bis Axford—eine persönliche Rückschau*. Katlenburg-Lindau: Copernicus 2003.

Kevles, Daniel J.: *The Physicists: The History of a Scientific Community in Modern America*. Harvard University Press 1995.

Khlopov, Maxim Yu: *Cosmoparticle Physics*. Singapur: World Scientific Publishing 1999.

Khriplovich, Iosif B.: The Eventful Life of Fritz Houtermans. *Physics Today* 45/7 (1992), 29–37. doi:10.1063/1.881313.

Kienle, Hans: Paul ten Bruggencate. 24.2.1901–14.9.1961. *Die Naturwissenschaften* 49/4 (1962), 73–74. doi:10.1007/BF00622019.

Kienle, Paul: Festveranstaltung anlässlich des 100. Geburtstags von Prof. Heinz Maier-Leibnitz. "Rückblick eines Zeitzeugens" (1952–2000). Meinem verehrten Lehrer, Kollegen und Freund zum 100ten Geburtstag gewidmet. Physics Department of the Technical University of Munich, 3/28/2011. https://www.frm2.tum.de/fileadmin/w00bnv/www/_migrated_content_uploads/Wie_Kernphysik_in_Muenchen_einzog.pdf. Last accessed 10/30/2018.

Kiepenheuer, Karl-Otto: Cosmic Rays as the Source of General Galactic Radio Emission. *Physical Review* 79/4 (1950), 738–739. doi:10.1103/PhysRev.79.738.

Kiepenheuer, Karl-Otto: Solar-terrestrische Erscheinungen. In: Paul ten Bruggencate (ed.): *Astronomie, Astrophysik und Kosmogonie*. Wiesbaden: Dieterich'sche Verlagsbuchhandlung 1948, 229–284.

Kiko, Jürgen, and Ulrich Schmidt-Rohr: *Max-Planck-Institut für Kernphysik. Heidelberg*. Edited by Generalverwaltung der Max-Planck-Gesellschaft. München 1975.

Kippenhahn, Rudolf: Als die Computer die Astronomie eroberten. In: Siegfried Röser (ed.): *Cosmic Matter*. Weinheim: Wiley-VCH 2008, 1–14.

Kippenhahn, Rudolf: Ludwig Biermann und die Theorie der Konvektionszonen in Sternen. In: Max-Planck Gesellschaft (ed.): *Ludwig Biermann. 1907–1986*. München 1988, 11–23.

Kippenhahn, Rudolf, H. C. Thomas, and A. Weigert: Sternentwicklung IV. Zentrales Wasserstoff- und Heliumbrennen bei einem Stern von 5 Sonnenmassen. Mit 4 Textabbildungen. *Zeitschrift fur Astrophysik* 61 (1965), 241. https://ui.adsabs.harvard.edu/#abs/1965ZA.....61..241K. Last accessed 9/19/2018.

Kippenhahn, Rudolf, H. C. Thomas, and A. Weigert: Sternentwicklung V. Der Kohlenstoff-Flash bei einem Stern von 5 Sonnenmassen. *Zeitschrift fur Astrophysik* 64 (1966), 373–394. https://ui.adsabs.harvard.edu/#abs/1966ZA.....64..373K. Last accessed 9/20/2018.

Kippenhahn, Rudolf, and Klaus Pinkau: Max-Planck-Institut für Physik und Astrophysik. Institut für Astrophysik und Institut für extraterrestrische Physik. I. Institut für Astrophysik; II. Institut für extraterrestrische Physik. Reports 1975. *Mitteilungen der Astronomischen Gesellschaft* 39 (1976), 112–134. http://adsabs.harvard.edu/abs/1976MitAG..39..112K. Last accessed 4/29/2018.

Kirsten, Till A.: Gentner und die Kosmochemie. Hobby oder Symbiose? In: Dieter Hoffmann, and Ulrich Schmidt-Rohr (eds.): *Wolfgang Gentner. Festschrift zum 100. Geburtstag*. Berlin: Springer 2006, 177–208.

Kirsten, Till A.: Radiochemical Solar Neutrino Experiments and Implications. *Physica Scripta* 2000/T85 (2000), 71–81. doi:10.1238/Physica.Topical.085a00071.

Kirsten, Till A.: Solar Neutrino Spectroscopy. *Nuclear Physics B—Proceedings Supplements* 87/1–3 (2000), 152–161. doi:10.1016/S0920-5632(00)00655-1.

Kirsten, Till A.: Solar Neutrino Experiments: Results and Implications. *Reviews of Modern Physics* 71/4 (1999), 1213–1232. doi:10.1103/RevModPhys.71.1213.

Kirsten, Till A.: Results from Solar-Neutrino Experiments. *Il Nuovo Cimento C* 19/6 (1996), 821–833. doi:10.1007/BF02508123.

Kirsten, Till A.: Das GALLEX-Sonnenneutrino-Experiment. *Mitteilungen der Astronomischen Gesellschaft* 68 (1987), 59–70. https://ui.adsabs.harvard.edu/#abs/1987MitAG..68...59K. Last accessed 4/30/2018.

Kirsten, Till A.: GALLEX/GNO. Context and Recollections. In: Mikko Meyer, and Kai Zuber (eds.): *Solar Neutrinos. Proceedings of the 5th International Solar Neutrino Conference, Dresden, Germany, 11–14 June 2018*. Singapore: World Scientific 2019, 47–68. doi:doi:10.1142/11384.

Kirsten, Till A.: Oliver Adam Schaeffer. 20.02.1919-11.11.1981. *Berichte und Mitteilungen Max-Planck-Gesellschaft. Jahresbericht 1981 und Jahresrechnung 1980, Nachrufe*. München 1982, 33–36.

Kirsten, Till A., W. Gentner, and O. Müller: Isotopenanalyse der Edelgase in einem Tellurerz von Boliden (Schweden). *Zeitschrift für Naturforschung A* 22/11 (1967), 1783–1792. doi:10.1515/zna-1967-1116.

Kirsten, Till A., W. Gentner, and O. A. Schaeffer: Massenspektrometrischer Nachweis von ββ-Zerfallsprodukten. *Zeitschrift für Physik* 202/1 (1967), 273–292. doi:10.1007/BF01331214.

Kirsten, Till A., O. A. Schaeffer, E. Norton, and R. W. Stoenner: Experimental Evidence for the Double-Beta Decay of Te^{130}. *Physical Review Letters* 20/23 (1968), 1300–1303. doi:10.1103/PhysRevLett.20.1300.

Kistiakowsky, George B.: *A Scientist at the White House. The Private Diary of President Eisenhower's Special Assistant for Science and Technology*. Cambridge, MA: Harvard University Press 1976.

Klapdor, Hans Volker: From Nuclear Physics to Fundamental Questions of Particle Physics, Cosmology and Reactor Physics. *Iadernaia Energiia* 25 (1987), 74–137. https://ui.adsabs.harvard.edu/#abs/1987IadEn..25...74K. Last accessed 10/14/2018.

Klapdor, Hans Volker (ed.): *Weak and Electromagnetic Interactions in Nuclei. Proceedings of the International Symposium, Heidelberg, July 1–5, 1986*. Berlin Heidelberg: Springer 1986.

Klebesadel, Ray W., Ian B. Strong, and Roy A. Olson: Observations of Gamma-Ray Bursts of Cosmic Origin. *Astrophysical Journal* 182 (1973), L85–L88. doi:10.1086/181225.

Klein, Thomas, Francisco M. Montenegro-Montes, Miroslaw Ciechanowicz, Claudio Agurto, Juan-Pablo Araneda, Oriel Arriagada, Francisco Azagra, et al.: APEX beyond 2016. The Evolution of an Experiment into an Efficient and Productive Submillimeter Wavelength Observatory. *Society of Photo-Optical Instrumentation Engineers (SPIE) Conference Series* 10704 (2018), 107041V. doi:10.1117/12.2312687.

Kohl, Wilfrid L.: *French Nuclear Diplomacy*. Princeton, NJ: Princeton University Press 1971.

Kohl, Wilfrid L.: The French Nuclear Deterrent. *Proceedings of the Academy of Political Science* 29/2 (1968), 80–94. doi:10.2307/1173251.

Kohlrausch, Martin, and Helmuth Trischler: *Building Europe on Expertise. Innovators, Organizers, Networkers*. Basingstoke: Palgrave Macmillan 2014.

Köhn, Diederich, Klaus Pinkau, and Gerd Wibberenz: Messung des sekundären Gammaspektrums von 0,2 bis 2 GeV in der oberen Atmosphäre. *Zeitschrift für Physik* 193/3 (1966), 443–458. doi:10.1007/BF01326442.

Köhn, Diederich, Klaus Pinkau, and Gerd Wibberrens: Gamma-Ray Spectrometer for Balloon Flights. *Composition, Origin, and Prehistory, Proceedings from the 8th International Cosmic Ray Conference*. Jaipur 1963, 203–205. http://adsabs.harvard.edu/abs/1963ICRC....3..203K. Last accessed 2/7/2020.

Kolb, Edward W., and Michael S. Turner: David N. Schramm (1945–97). 6666. *Nature* 391/6666 (1998), 444. doi:10.1038/35044.

Kolb, Edward W.: Searching for Cygnets. In: G. Lazarides, and Q. Shafi (eds.): *Proceedings of the International Symposium on Particles and the Universe, Thessaloniki, Greece, 24–29 June 1985*. Amsterdam: North Holland 1986, 247–255. http://inspirehep.net/record/218657?ln=it. Last accessed 11/27/2018.

Kolboske, Birgit: *Hierarchien. Das Unbehagen der Geschlechter mit dem Harnack-Prinzip. Arbeits- und Lebenswelten von Frauen in der Max-Planck-Gesellschaft, 1948–1998*. Dissertation. Leipzig 2021.

Kolboske, Birgit: Hierarchies. Lotta Support, Little Science? Scientists and Secretaries in the Max Planck Society. In: Ulla Weber (ed.): *Fundamental Questions. Gender Dimensions in Max Planck Research Projects*. Baden-Baden: Nomos 2021, 105–134. doi:10.5771/9783748924869.

Konopelko, A., F. Aharonian, A. Akhperjanian, F. Arqueros, G. Aslanian, A. Beglarian, S. Bradbury, et al.: Detection of Gamma Rays above 1 TeV from the Crab Nebula by the Second HEGRA Imaging Atmospheric Cherenkov Telescope at La Palma. *Astroparticle Physics* 4/3 (1996), 199–215. doi:10.1016/0927-6505(95)00044-5.

Kopfermann, Hans: Zur Geschichte der Heidelberger Physik seit 1945. In: Universitäts-Gesellschaft (ed.): *Heidelberger Jahrbücher*. Berlin: Springer 1960, 159–164. doi:10.1007/978-3-642-45950-4_9.

Kovalev, Y. Y., N. S. Kardashev, K. I. Kellermann, and P. G. Edwards: The RadioAstron Space VLBI Project. *2014 XXXIth URSI General Assembly and Scientific Symposium (URSI GASS)*. 2014 XXXIth URSI General Assembly and Scientific Symposium (URSI GASS). Beijing 2014. doi:10.1109/URSIGASS.2014.6929994.

Kragh, Helge: An Unlikely Connection: Geochemistry and Nuclear Structure. *Physics in Perspective* 2/4 (2000), 381–397. doi:10.1007/s000160050051.

Kragh, Helge: *Cosmology and Controversy. The Historical Development of Two Theories of the Universe*. Princeton, New Jersey: Princeton University Press 1996.

Kragh, Helge: Bohr—Kramers—Slater Theory. In: Daniel Greenberger, Klaus Hentschel, and Friedel Weinert (eds.): *Compendium of Quantum Physics*. Berlin: Springer 2009, 62–64. doi:10.1007/978-3-540-70626-7_19.

Kramer, Michael, Karl M. Menten, Gerd Weigelt, J. Anton Zensus, and Richard Wielebinski: Peter Georg Mezger. 19.11.1928–09.07.2014. *Personalien 2014. Beileger zum Jahresbericht der Max-Planck-Gesellschaft*, 2015, 37–38, 37–38.

Kraus, Elisabeth: *Von der Uranspaltung zur Göttinger Erklärung. Otto Hahn, Werner Heisenberg, Carl Friedrich Weizsäcker und die Verantwortung des Wissenschaftlers*. Würzburg: Königshausen & Neumann 2001.

Kraushaar, William L., and George W. Clark: Search for Primary Cosmic Gamma Rays with the Satellite Explorer XI. *Physical Review Letters* 8/3 (1962), 106–109. doi:10.1103/PhysRevLett.8.106.

Krennrich, F., R. Mirzoyan, N. Müller, H. Sander, F. Aharonian, A. G. Akhperjanian, F. Arqueros, et al.: Observation of VHE γ-Emission from the Crab Nebula with the Prototype of the HEGRA Air Cerenkov Telescope Array. In: D. A. Leahy, R. B. Hicks, and D. Venkatesan (eds.): *23rd International Cosmic Ray Conference, Vol. 1, Held 19–30 July, 1993 at University of Calgary, Alberta, Canada*. 1993, 251–254. https://ui.adsabs.harvard.edu/#abs/1993ICRC....1..251K. Last accessed 12/11/2018.

Krichbaum, Thomas P.: Millimeter-VLBI with a Large Millimeter-Array: Future Possibilities. In: Peter A. Shaver (ed.): *Science with Large Millimetre Arrays*. Berlin, Heidelberg: Springer Berlin Heidelberg 1996, 95–102. doi:10.1007/978-3-540-69999-6_12.

Krichbaum, Thomas P., D A Graham, A Witzel, A Greve, J E Wink, M Grewing, F Colomer, et al.: VLBI Observations of the Galactic Center Source Sgr A* at 86 GHz and 215 GHz. *Astronomy and Astrophysics* 335, L106–L110.

Krige, John: NASA's International Relations in Space. An Historical Overview. *NASA's First 50 Years. Historical Perspectives*. Washington, DC: NASA 2010, 109–150.

Krige, John: *American Hegemony and the Postwar Reconstruction of Science in Europe*. Cambridge, MA: MIT Press 2006.

Krige, John: Atoms for Peace, Scientific Internationalism, and Scientific Intelligence. *Osiris* 21/1 (2006), 161–181. doi:10.1086/507140.

Krige, John: NATO and the Strengthening of Western Science in the Post-Sputnik Era. *Minerva* 38/1 (2000), 81–108. https://www.jstor.org/stable/41821156. Last accessed 12/7/2018.

Krige, John: The History of the European Launcher—An Overview. In: R. A. Harris (ed.): *The History of the European Space Agency. Proceedings of an International Symposium 11–13 November 1998, the Science Museum, London*. Noordwijk: ESA Publication Division 1999, 69–78.

Krige, John (ed.): *History of CERN*. Vol. 3. Amsterdam: North Holland 1996.

Krige, John, Angelina Long Callahan, and Ashok Maharaj: *NASA in the World. Fifty Years of International Collaboration in Space*. New York: Palgrave MacMillan 2013.

Krige, John, and Arturo Russo: *A History of the European Space Agency 1958–1987. The Story of ESRO and ELDO, 1958–1973*. Vol. 1. Noordwijk: European Space Agency 2000.

Krige, John, Arturo Russo, and Lorenza Sebesta: *A History of the European Space Agency 1958–1987. The Story of ESA, 1973 to 1987*. Vol. 2. European Space Agency 2000.

Krige, John, Arturo Russo, and Lorenza Sebesta: *A History of the European Space Agency 1958–1987. The Story of ESA, 1973 to 1987*. Vol. 2. European Space Agency 2000.

Kudritzki, Rolf-Peter, and Reinhold Häfner: German Astronomy. *Encyclopedia of Astronomy and Astrophysics*. Edited by Paul Murdin, 2001, 1–3, 1–3. doi:10.1888/0333750888/2933.

Kuiper, Gerard P.: German Astronomy during the War. *Popular Astronomy* 54 (1946), 263–286. https://ui.adsabs.harvard.edu/abs/1946PA.....54..263K/abstract. Last accessed 9/17/2018.

Kuzmin, Vadim A.: Detection of Solar Neutrinos by Means of the $Ga^{71}(\nu,e^{-})Ge^{71}$ Reaction. *Journal of Experimental and Theoretical Physics* 22/5 (1966), 1051–1052. http://www.jetp.ac.ru/cgi-bin/e/index/e/22/5/p1051?a=list. Last accessed 4/30/2019.

Kuzmin, Vadim A., and Georgii T. Zatsepin: On the Neutrino Spectroscopy of the Sun. *Proceedings of the 8th International Cosmic Ray Conference, December 2–14, 1963, Jaipur, India*. Bombay, India: Commercial Printing Press 1965, 1023–1024. https://ui.adsabs.harvard.edu/#abs/1965ICRC....2.1023K. Last accessed 1/6/2019.

Kuzminov, V. V.: The Baksan Neutrino Observatory. *The European Physical Journal Plus* 127/9 (2012), 113. doi:10.1140/epjp/i2012-12113-0.

La Rana, Adele: EUROGRAV 1986–1989: The First Attempts for a European Interferometric Gravitational Wave Observatory. *European Physical Journal H* 47/1 (2022), 1–23. doi:10.1140/epjh/s13129-022-00036-x.

La Rana, Adele: The Origins of Virgo and the Emergence of the International Gravitational Wave Community. In: Alexander S. Blum, Roberto Lalli, and Jürgen Renn (eds.): *The Renaissance of General Relativity in Context*. Cham: Springer International Publishing 2020, 363–406. doi:10.1007/978-3-030-50754-1_10.

La Rana, Adele, and Leopoldo Milano: The Early History of Gravitational Wave Detection in Italy. From the First Resonant Bars to the Beginning of the Virgo Collaboration. In: Salvatore Esposito (ed.): *Società Italiana Degli Storici Della Fisica e Dell'Astronomia. Atti Del XXXVI Convegno Annuale / Proceedings of the 36th Annual Conference. Napoli 2016*. Pavia: Pavia University Press 2017, 185–196. doi:10.23739/9788869520709/c17.

Lalli, Roberto: Crafting Europe from CERN to Dubna: Physics as Diplomacy in the Foundation of the European Physical Society. *Centaurus* 63/1 (2021), 103–131. doi:https://doi.org/10.1111/1600-0498.12304.

Lalli, Roberto: *Building the General Relativity and Gravitation Community During the Cold War*. Cham: Springer International Publishing 2017. doi:10.1007/978-3-319-54654-4.

Lalli, Roberto, Riaz Howey, and Dirk Wintergrün: The Dynamics of Collaboration Networks and the History of General Relativity, 1925–1970. *Scientometrics* 122 (2020), 1129–1170. doi:10.1007/s11192-019-03327-1.

Lambrecht, Wolfgang: Deutsch-Deutsche Reformdebatten vor "Bologna". Die "Bildungskatastrophe" der 1960er Jahre. *Zeithistorische Forschung* 4/3 (2007), 472–477.

Lande, K.: Experimental Neutrino Astrophysics. *Annual Review of Nuclear and Particle Science* 29 (1979), 395–410. doi:10.1146/annurev.ns.29.120179.002143.

Landrock, Konrad: Friedrich Georg Houtermans (1903–1966)—Ein bedeutender Physiker des 20. Jahrhunderts. *Naturwissenschaftliche Rundschau* 56/4 (2003), 187–199. http://www.naturwissenschaftliche-rundschau.de/navigation/dokumente/Beitrag-Landrock-4-2003.pdf. Last accessed 10/30/2018.

Laney, Monique: *German Rocketeers in the Heart of Dixie. Making Sense of the Nazi Past During the Civil Rights Era*. New Haven, CT: Yale University Press 2015.

Lattes, Cesare M. G., H. Muirhead, G. P. S. Occhialini, and C. F. Powell: Processes Involving Charged Mesons. *Nature* 159/4047 (1947), 694–697. doi:10.1038/159694a0.

Laursen, Lucas: Cutbacks Kick off Kerfuffle over Spanish-German Observatory. *Nature News Blog*, 6/17/2013. http://blogs.nature.com/news/2013/06/calar-alto-cutback-kerfuffle.html. Last accessed 4/12/2018.

Lax, Gregor: *From Atmospheric Chemistry to Earth System Science. Contributions to the Recent History of the Max Planck Institute for Chemistry (Otto Hahn Institute), 1959–2000*. Diepholz: GNT-Verlag 2018.

Lax, Gregor: *Wissenschaft zwischen Planung, Aufgabenteilung und Kooperation. Zum Aufstieg der Erdsystemforschung in der MPG, 1968–2000*. Berlin: GMPG-Preprint 2020.

Lax, Gregor: *Von der Atmosphärenchemie zur Erforschung des Erdsystems. Beiträge zur jüngeren Geschichte des Max-Planck-Instituts für Chemie (Otto-Hahn-Institut), 1959–2000*. Berlin: GMPG-Preprint 2018.

Lax, Gregor: Zum Aufbau der Atmosphärenwissenschaften in der BRD seit 1968. *NTM Zeitschrift für Geschichte der Wissenschaften, Technik und Medizin* 24/1 (2016), 81–107. http://nbn-resolving.de/urn:nbn:de:1111-2016040114869. Last accessed 8/5/2016.

Lee, J.A.N.: Heinz Billing. *Computer Pioneers*. Edited by J.A.N. Lee, and IEEE Computer Society. IEEE Computer Society 2012. http://history.computer.org/pioneers/billing.html. Last accessed 10/30/2018.

Lehnert, B. (ed.): *Electromagnetic Phenomena in Cosmical Physics*. Cambridge, MA: Cambridge University Press 1958.

Leinert, Ch., U. Graser, F. Przygodda, L.B.F.M. Waters, G. Perrin, W. Jaffe, B. Lopez, et al.: MIDI—the 10 Mm Instrument on the VLTI. *Astrophysics and Space Science* 286/1 (2003), 73–83. doi:10.1023/A:1026158127732.

Lemke, Dietrich: *Im Himmel über Heidelberg. 40 Jahre Max-Planck-Institut für Astronomie in Heidelberg (1969–2009)*. Edited by Archiv der Max-Planck-Ges. Vol. 21. Berlin 2011.

Lemke, Dietrich: The Short History of Infrared Space Telescopes. *Astronomische Nachrichten* 330/6 (2009), 562–567. doi:10.1002/asna.200911217.

Lemke, Dietrich, and Astronomische Gesellschaft (eds.): *Die Astronomische Gesellschaft 1863–2013. Bilder und Geschichten aus 150 Jahren*. Heidelberg: Astronomische Gesellschaft 2013.

Lemke, Dietrich, and M. Kessler: The Infrared Space Observatory ISO. In: G. Klare (ed.): *Reviews in Modern Astronomy*. Berlin: Springer 1989, 53–71.

Leon, Juan-Andres: *Citizens of the Chemical Complex. Industrial Expertise and Science Philanthropy in Imperial and Weimar Germany*. Dissertation. Cambridge, MA: Harvard University 2013.

Leuchs, Gerd, Karl Maischberger, Albrecht Rüdiger, Roland Schilling, Lise Schnupp, and Walter Winkler: *Proposal for the Construction of a Large Laser Interferometer for the Measurement of Gravitational Waves*. MPQ 131. Garching: Max-Planck-Institut für Quantenoptik 1987.

Leverington, David: *Observatories and Telescopes of Modern Times. Ground-Based Optical and Radio Astronomy Facilities since 1945*. Cambridge: Cambridge University Press 2017.

Levine, James L., and Richard L. Garwin: Absence of Gravity-Wave Signals in a Bar at 1695 Hz. *Physical Review Letters* 31/3 (1973), 173–176. doi:10.1103/PhysRevLett.31.173.

Lewis, Jeffrey: Kalter Krieg in der Max-Planck-Gesellschaft. Göttingen und Tübingen—eine Vereinigung mit Hindernissen, 1948–1949. In: Wolfgang Schieder, and Achim

Trunk (eds.): *Adolf Butenandt und die Kaiser-Wilhelm-Gesellschaft. Wissenschaft, Industrie und Politik im »Dritten Reich«*. Göttingen: Wallstein Verlag 2004, 403–443.

Libby, Willard F.: Atmospheric Helium Three and Radiocarbon from Cosmic Radiation. *Physical Review* 69/11–12 (1946), 671–672. doi:10.1103/PhysRev.69.671.2.

Libby, Willard F.: Radiocarbon Dating. Nobel Lecture, 12/12/1960. https://www.nobelprize.org/nobel_prizes/chemistry/laureates/1960/libby-lecture.pdf. Last accessed 10/30/2018.

Lidvansky, Aleksandr S.: Air Cherenkov Methods in Cosmic Rays. Review and Some History. *Radiation Physics and Chemistry* 75/8 (2006), 891–898. doi:10.1016/j.radphyschem.2005.12.019.

Lidvansky, Aleksandr S.: G T Zatsepin and the Birth of Gamma-Ray Astronomy. *Physics-Uspekhi* 61/9 (2018), 921–925. doi:10.3367/UFNe.2017.05.038184.

Liftin, Gerd, and Jürgen Mlynek: Zum 80. Geburtstag von Herbert Welling. *Physik Journal* 8/10 (2009), 45. https://www.pro-physik.de/restricted-files/98106. Last accessed 4/14/2020.

Light, Jennifer S.: When Computers Were Women. *Technology and Culture* 40/3 (1999), 455–483. http://www.jstor.org/stable/25147356. Last accessed 7/3/2018.

LIGO Scientific Collaboration & Virgo Collaboration, B. P. Abbott, R. Abbott, T. D. Abbott, M. R. Abernathy, F. Acernese, K. Ackley, et al.: Observation of Gravitational Waves from a Binary Black Hole Merger. *Physical Review Letters* 116/6 (2016), 061102. doi:10.1103/PhysRevLett.116.061102.

Linsley, John: Evidence for a Primary Cosmic-Ray Particle with Energy 10^{20} eV. *Physical Review Letters* 10/4 (1963), 146–148. doi:10.1103/PhysRevLett.10.146.

Linsley, John, Livio Scarsi, and Bruno Rossi: Extremely Energetic Cosmic-Ray Event. *Physical Review Letters* 6/9 (1961), 485–487. doi:10.1103/PhysRevLett.6.485.

Lloyd-Evans, J., R. N. Coy, A. Lambert, J. Lapikens, M. Patel, R. J. O. Reid, and A. A. Watson: Observation of γ Rays $>10^{15}$ eV from Cygnus X-3. *Nature* 305/5937 (1983), 784–787. doi:10.1038/305784a0.

Logsdon, John M. (ed.): *Exploring the Unknown. Selected Documents in the History of the U.S. Civil Space Program. Space and Earth Science*. Vol. VI. Washington, D.C.: NASA 2004.

Lohrmann, Erich, and Paul Söding: *Von schnellen Teilchen und hellem Licht. 50 Jahre Deutsches Elektronen-Synchrotron DESY*. Weinheim: Wiley-VCH Verlag 2009.

Longair, Malcolm: 100 Years of Astronomy, Astrophysics and Cosmology. A Celebration of the Centenary of the IAU. In: C. Sterken, J. Hearnshaw, and D. Valls-Gabaud (eds.): *Under One Sky: The IAU Centenary Symposium Proceedings IAU Symposium*. 349. Cambridge: Cambridge University Press 2019, 3–24. doi:10.1017/S1743921319000097.

Longair, Malcolm S.: *The Cosmic Century. A History of Astrophysics and Cosmology*. Cambridge: Cambridge University Press 2006.

Looking Up. Europe's Quiet Revolution in Microgravity Research. New York City, NY: Scientific American Custom Publishing 2008.

Lorenz, Eckart, and Manel Martinez: The Magic Telescope. *Astronomy & Geophysics* 46/6 (2005), 6.21–6.25. doi:10.1111/j.1468-4004.2005.46621.x.

Lorenz, Eckart: The MAGIC Telescope Project for Gamma Ray Astronomy in the 15 to 300 GeV Energy Range. *Nuclear Physics B Proceedings Supplements* 48 (1996), 494–496. https://doi.org/10.1016/0920-5632(96)00302-7. Last accessed 5/17/2018.

Lorenz, Eckart: The HEGRA Experiment. In: P.C. Bosetti (ed.): *Trends in Astroparticle-Physics*. Wiesbaden: Vieweg+Teubner Verlag 1994, 139–151.

Lorenz, Eckart, and Robert Wagner: Very-High Energy Gamma-Ray Astronomy. A 23-Year Success Story in High-Energy Astroparticle Physics. *European Physical Journal H* 37/3 (2012), 459–513. doi:10.1140/epjh/e2012-30016-x.

Lorenz, Robert: *Protest der Physiker. Die »Göttinger Erklärung« von 1957*. Bielefeld: transcript Verlag 2011.

Lorenzen, Dirk H.: Stimmen zum Gammateleskop MAGIC. *Physik der Welt*, 1/14/2004. https://www.weltderphysik.de/gebiet/universum/kosmische-strahlung/detektoren/magic/stimmen-zu-magic/. Last accessed 8/14/2018.

Lovell, Bernard: The Effects of Defence Science on the Advance of Astronomy. *Journal for the History of Astronomy* 8 (1977), 151–173. https://journals.sagepub.com/doi/abs/10.1177/002182867700800301. Last accessed 5/24/2019.

Lovell, Bernard A. C.: Patrick Maynard Stuart Blackett, Baron Blackett, of Chelsea, 18 November 1897–13 July 1974. *Biographical Memoirs of Fellows of the Royal Society* 21 (1975), 1–115. doi:10.1098/rsbm.1975.0001.

Low, Frank J., G. H. Rieke, and R. D. Gehrz: The Beginning of Modern Infrared Astronomy. *Annual Review of Astronomy and Astrophysics* 45 (2007), 43–75. doi:10.1146/annurev.astro.44.051905.092505.

Lucha, Gerda Maria (ed.): *Dokumente zu Entstehung und Entwicklung des Max-Planck-Instituts für Plasmaphysik 1955–1971*. Garching: IPP 2005.

Lück, Harald, Andreas Freise, Stefan Goßler, S. Hild, K. Kawabe, and Karsten Danzmann: Thermal Correction of the Radii of Curvature of Mirrors for GEO 600. *Classical and Quantum Gravity* 21/5 (2004), S985–S989. doi:10.1088/0264-9381/21/5/090.

Lück, Harald, K. O. Müller, Peter Aufmuth, and Karsten Danzmann: Correction of Wavefront Distortions by Means of Thermally Adaptive Optics. *Optics Communications* 175/4 (2000), 275–287. doi:10.1016/S0030-4018(00)00468-5.

Lück, Harald, and the Geo600 Team: The GEO600 Project. *Classical and Quantum Gravity* 14/6 (1997), 1471–1476. doi:10.1088/0264-9381/14/6/012.

Lüst, Reimar, and Klaus Pinkau: Theoretical Aspects of Celestial Gamma-Rays. In: J.C. Emming (ed.): *Electromagnetic Radiation in Space. Proceedings of the Third ESRO Summer School in Space Physics, Held in Alpbach, Austria, from 19 July to 13 August, 1965*. Dordrecht: Springer 1967, 231–248.

Lüst, Reimar: Carl Friedrich Weizsäcker. Ein Doktorand erinnert sich. In: Klaus Hentschel, and Dieter Hoffmann (eds.): *Carl Friedrich von Weizsäcker. Physik, Philosophie, Friedensforschung*. Stuttgart: Wissenschaftliche Verlagsgesellschaft 2014, 263–270.

Lüst, Reimar: Heisenberg and the Scientist's Responsibility. In: Gerd Buschhorn, and Julius Wess (eds.): *Fundamental Physics—Heisenberg and Beyond. Werner Heisenberg Centennial Symposium "Developments in Modern Physics."* Berlin: Springer 2004, 15–24.

Lüst, Reimar: Barium Cloud Experiments in the Upper Atmosphere. In: Johan A. M. Bleeker, Johannes Geiss, and Martin C. E. Huber (eds.): *The Century of Space Science*. Dordrecht: Kluwer Academic Publishers 2001, 179–187.

Lüst, Reimar: Léon Van Hove and the Max-Planck-Institute for Physics. *Scientific Highlights in Memory of Léon Van Hove*. World Scientific 1993, 51–59.

Lüst, Reimar: Ludwig Biermann. 13.3.1907–12.1.1986. *Berichte und Mitteilungen der Max-Planck-Gesellschaft*. München 1986, 78–81.

Lüst, Reimar: Terrestrische und extraterrestrische Plasmen. *Die Naturwissenschaften* 62/6 (1975), 255–263. doi:10.1007/BF00608951.

Lüst, Reimar: Künstliche Wolken. Ein Mittel der Weltraumforschung. In: Generalverwaltung der Max-Planck-Gesellschaft (ed.): *Jahrbuch der Max-Planck-Gesellschaft zur Förderung der Wissenschaften e.V. 1968*. Göttingen 1968, 150–169.

Lüst, Reimar: Weltraumforschung in der Bundesrepublik und Europa. *Weltraumforschung in der Bundesrepublik und Europa. Sonnenforschung*. VS Verlag für Sozialwissenschaften 1966, 7–29.

Lüst, Reimar: *Aktuelle Probleme der Weltraumforschung. Festvortrag anläßlich der Jahresversammlung des Stifterverbandes in Wiesbaden am 11. Mai 1965*. Essen-Bredeney: Gemeinnützige Verwaltungsgesellschaft für Wissenschaftspflege 1965.

Lüst, Reimar: Extraterrestrische Physik. Über Arbeiten im Institut für Extraterrestrische Physik am Max-Planck-Institut für Physik und Astrophysik. *Die Naturwissenschaften* 52/19 (1965), 525–529. doi:10.1007/BF00645816.

Lüst, Reimar: The European Space Research Organization. *Science* 149/3682 (1965), 394–397. doi:10.1126/science.149.3682.394.

Lüst, Reimar: *Die Gegenwärtigen Probleme Der Weltraumforschung*. München: Oldenbourg; Düsseldorf: VDI-Verl 1964.

Lüst, Reimar: Eindrücke von einer Reise in die Sowjetunion. Gegenbesuch einer Delegation der Max-Planck-Gesellschaft bei der Akademie der Wissenschaften der UdSSR. In: Generalverwaltung der Max-Planck-Gesellschaft (ed.): *Mitteilungen aus der Max-Planck-Gesellschaft. Heft 1–2*. Max-Planck-Gesellschaft 1964, 55–63.

Lüst, Reimar: Extraterrestrische Forschung. *Mitteilungen der Astronomischen Gesellschaft* 16 (1963), 37–47. http://articles.adsabs.harvard.edu/pdf/1963MitAG..16..37L. Last accessed 8/28/2017.

Lüst, Reimar: Interplanetary Plasma. *Space Science Reviews* 1/3 (1963), 522–552. doi:10.1007/BF00225270.

Lüst, Reimar: Impact Zones for Solar Cosmic-Ray Particles. *Physical Review* 105/6 (1957), 1827–1839. doi:10.1103/PhysRev.105.1827.

Lüst, Reimar: Magneto-hydrodynamische Stoßwellen in einem Plasma unendlicher Leitfähigkeit. *Zeitschrift für Naturforschung A* 8/5 (1953), 277–284. http://zfn.mpdl.mpg.de/data/Reihe_A/8/ZNA-1953-8a-0277.pdf. Last accessed 11/2/2017.

Lüst, Reimar: Die Entwicklung einer um einen Zentralkörper rotierenden Gasmasse. I. Lösungen der hydrodynamischen Gleichungen mit turbulenter Reibung. *Zeitschrift für Naturforschung A* 7/1 (1952), 87–98. doi:10.1515/zna-1952-0118.

Lüst, Reimar, and Paul Nolte: *Der Wissenschaftsmacher. Reimar Lüst im Gespräch mit Paul Nolte*. München: C.H. Beck 2008.

Lüst, Reimar, and Klaus Pinkau: Theoretical Aspects of Celestial Gamma-Rays. In: J.C. Emming (ed.): *Electromagnetic Radiation in Space. Proceedings of the Third ESRO Summer School in Space Physics, Held in Alpbach, Austria, from 19 July to 13 August, 1965*. Dordrecht: Springer 1967, 231–248.

Lüst, Reimar, and M. Scholer: Kompressionswellen in Einer Isothermen Atmosphäre Mit Vertikalem Magnetfeld. *Zeitschrift Für Naturforschung A* 21/7 (1966), 1098–1106. doi:10.1515/zna-1966-0734.

Lüst, Reimar, and John A. Simpson: Initial Stages in the Propagation of Cosmic Rays Produced by Solar Flares. *Physical Review* 108/6 (1957), 1563–1576. doi:10.1103/PhysRev.105.1827.

Lüst, Rhea, and Rudolf Kippenhahn: Max-Planck-Institut für Physik und Astrophysik. Institut für Astrophysik München. *Max-Planck-Gesellschaft. Berichte und Mitteilungen* 1/77 (1977), 1–64.

Lutz, Harald: Die vergessenen Raketenexperimente von Cuxhaven. *Sterne und Weltraum* 44/3 (2005), 40–45.

Lynden-Bell, Donald: Galactic Nuclei as Collapsed Old Quasars. *Nature* 223/5207 (1969), 690–694. doi:10.1038/223690a0.

Lynden-Bell, Donald, and M. J. Rees: On Quasars, Dust and the Galactic Centre. *Monthly Notices of the Royal Astronomical Society* 152/4 (1971), 461–475. doi:10.1093/mnras/152.4.461.

MacGregor, Robert R.: Imagining an Aerospace Agency in the Atomic Age. *NASA's First 50 Years: Historical Perspectives*. Washington, DC: NASA 2010, 31–48.

Maddox, John: Science in West Germany. Discovery and Disappointment. *Nature* 297/5864 (1982), 261–280. doi:10.1038/297261a0.

Madsen, Claus: *The Jewel on the Mountaintop. The European Southern Observatory through Fifty Years*. ESO 2012.

MAGIC Collaboration: Teraelectronvolt Emission from the γ-Ray Burst GRB 190114C. *Nature* 575/7783 (2019), 455–458. doi:10.1038/s41586-019-1750-x.

MAGIC Collaboration: *The MAGIC Telescope. Design Study for the Construction of a 17 m Čerenkov Telescope for Gamma-Astronomy above 10 GeV*, 1998.

MAGIC Collaboration, M. L. Ahnen, S. Ansoldi, L. A. Antonelli, C. Arcaro, A. Babić, B. Banerjee, et al.: Constraints on Particle Acceleration in SS433/W50 from MAGIC and H.E.S.S. Observations. *Astronomy & Astrophysics* 612 (2018), A14. doi:10.1051/0004-6361/201731169.

Mahoney, Leo J.: *A History of the War Department Scientific Intelligence Mission (ALSOS) 1943–1945*. Dissertation/ PhD Thesis. Ann Arbor, MI: University of Michigan 1981.

Maier-Leibnitz, Heinz: W. Bothe, Experimental Nuclear Physicist. *Science* 126/3267 (1957), 246–247. doi:10.1126/science.126.3267.246.

Maischberger, Karl, Albrecht Rüdiger, Roland Schilling, Lise Schnupp, David Shoemaker, and Walter Winkler: *Vorschlag Zum Bau Eines Grossen Laser-Interferometers Zur Messung von Gravitationswellen*. MPQ 96. Max-Planck-Institut für Quantenoptik 1985.

Maischberger, Karl, Arnold Rüdiger, Roland Schilling, Lise Schnupp, Walter Winkler, and Gerd Leuchs: Status of the Garching 30-Meter Prototype for a Large Gravitational Wave Detector. In: Peter F. Michelson, Hu En-ke, and Guido Pizzella (eds.): *International Symposium on Experimental Gravitational Physics. 3–8 August 1987. Guangzhou, China*. Singapore: World Scientific 1988, 316–321. https://ui.adsabs.harvard.edu/1988egp..conf..316M/abstract. Last accessed 10/29/2019.

Mandel, H., I. Appenzeller, W. Seifert, W. Xu, T. Herbst, R. Lenzen, T. Thatte, et al.: LUCIFER-LBT NIR Spectroscopic Utility with Camera and Integral Field Unit for Extragalactic Research. *Optical and Infrared Spectroscopy of Circumstellar Matter, ASP Conference Series, Vol. 188*. 1999, 321.

Marshak, M. L., J. Bartelt, H. Courant, K. Heller, T. Joyce, E. A. Peterson, K. Ruddick, et al.: Evidence for Muon Production by Particles from Cygnus X-3. *Physical Review Letters* 54/19 (1985), 2079–2082. doi:10.1103/PhysRevLett.54.2079.

Marshall, Eliot: Astrophysics Institute at Risk. *Science* 257/5070 (1992), 606–606. doi:10.1126/science.257.5070.606-a.

Martin, B. R.: Radio Astronomy Revisited. A Reassessment of the Role of Competition and Conflict in the Development of Radio Astronomy. *The Sociological Review* 26/1 (1978), 27–55. doi:10.1111/j.1467-954X.1978.tb00123.x.

Marvin, Ursula B.: Oral Histories in Meteoritics and Planetary Science. VIII. Friedrich Begemann. *Meteoritics & Planetary Science* 37/S12 (2002), B69–B77. doi:10.1111/j.1945-5100.2002.tb00905.x.

Marvin, Ursula B.: Oral Histories in Meteoritics and Planetary Science. IX. Heinrich Wänke. *Meteoritics & Planetary Science* 37 (2002), B79–B88. doi:10.1111/j.1945-5100.2002.tb00906.x.

Massey, Harrie Stewart Wilson, and M. O. Robins: *History of British Space Science*. Cambridge: Cambridge University Press 1986.

Mather, J. C., E. S. Cheng, D. A. Cottingham, R. E. Eplee, D. J. Fixsen, T. Hewagama, R. B. Isaacman, et al.: Measurement of the Cosmic Microwave Background Spectrum by the COBE FIRAS Instrument. *The Astrophysical Journal* 420 (1994), 439–444. doi:10.1086/173574.

Mattauch, Josef: Max-Planck-Institut für Chemie (Otto-Hahn-Institut) in Mainz. In: Generalverwaltung der Max-Planck-Gesellschaft zur Förderung der Wissenschaften e.V. (ed.): *Jahrbuch 1961 der Max-Planck-Gesellschaft zur Förderung der Wissenschaften e.V. Teil II*. Göttingen 1962, 215–224.

Matthiae, Giorgio: The Cosmic Ray Energy Spectrum as Measured Using the Pierre Auger Observatory. *New Journal of Physics* 12/7 (2010), 075009. doi:10.1088/1367-2630/12/7/075009.

Mattig, Wolfgang: Nachrufe. Karl-Otto Kiepenheuer. *Mitteilungen der Astronomischen Gesellschaft* 38 (1976), 11–13. https://ui.adsabs.harvard.edu/#abs/1976MitAG.38...11M. Last accessed 10/30/2018.

Matveyenko, L.: Early VLBI in the USSR. *Astronomische Nachrichten* 328/5 (2007), 411–419. doi:10.1002/asna.200710763.

Mauersberger, Konrad: Measurement of Heavy Ozone in the Stratosphere. *Geophysical Research Letters* 8/8 (1981), 935–937. doi:10.1029/GL008i008p00935.

Max Planck Institute for Astrophysics: Successful Launch of the Spectrum-RG Mission, 7/13/2019. https://www.mpa-garching.mpg.de/720407/news20190713. Last accessed 3/23/2021.

Max-Planck-Gesellschaft (ed.): *Max-Planck-Gesellschaft. Berichte und Mitteilungen 1/93. Max-Planck-Institut für Physik München*. München 1993.

Max-Planck-Gesellschaft (ed.): *Max-Planck-Institut für Radioastronomie. Bonn*. München 1992.

Max-Planck-Gesellschaft (ed.): *Max-Planck-Gesellschaft. Berichte und Mitteilungen 4/81. Max-Planck-Institut für Aeronomie Katlenburg-Lindau*. München 1981.

Max-Planck-Gesellschaft: Max-Planck-Institut für Quantenoptik Garching b. München. *Max-Planck-Gesellschaft. Berichte und Mitteilungen* 6/86 (1986), 1–129.

Mayer-Hasselwander, H. A., K. Bennett, G. F. Bignami, R. Buccheri, P. A. Caraveo, W. Hermsen, G. Kanbach, et al.: Large-Scale Distribution of Galactic Gamma Radiation Observed by COS-B. *Astronomy and Astrophysics* 105/1 (1982), 164–175. http://adsabs.harvard.edu/abs/1982A%26A...105..164M. Last accessed 5/15/2018.

Maze, Ronald, J. Wdowczyk, A. W. Wolfendale, and A. Zawadzki: Origin of Muon-Poor Extensive Air Showers. *Physical Review Letters* 22/17 (1969), 899–901. doi:10.1103/PhysRevLett.22.899.

McDougall, Walter A.: *The Heavens and the Earth. A Political History of the Space Age*. 2nd ed. Baltimore, MD: Johns Hopkins University Press 1997.

McPartland, Mary A.: *The Farm Hall Scientists. The United States, Britain, and Germany in the New Atomic Age, 1945–46*. Dissertation. Washington, DC: The George Washington University 2013. https://search.proquest.com/pqdtglobal/docview/1436978271/abstract/DDC811FAFC464047PQ/2. Last accessed 4/1/2019.

Mehra, Jagdish, and Helmut Rechenberg: *The Historical Development of Quantum Theory*. Vol. 4. New York, NY: Springer Verlag 1982.

Meitner, Lise: Prof. Walter Bothe. *Nature* 179/4561 (1957), 654–655. doi:10.1038/179654a0.

Meitner, Lise, and Otto Frisch: Disintegration of Uranium by Neutrons. A New Type of Nuclear Reaction. *Nature* 143/3615 (1939), 239–240. doi:10.1038/143239a0.

Menten, Karl M.: Leo Brandt. Pionier der Funkmesstechnik und Initiator der Radioastronomie in Deutschland. In: Bernhard Mittermaier, and Bernd-A. Rusinek (eds.): *Leo Brandt (1908–1971). Ingenieur—Wissenschaftsförderer—Visionär*. Jülich: Forschungszentrum Jülich, Zentralbibliothek 2009, 41–53.

Merkel, Angela: Rede von Bundeskanzlerin Dr. Angela Merkel beim Besuch des Max-Planck-Instituts für Plasmaphysik (IPP) am 3. Februar 2016 in Greifswald. *Bulletin der Bundesregierung 14–1*, 2/5/2016. https://www.bundesregierung.de/Content/DE/Bulletin/2016/02/14-1-bk-besuch-max-planck-institut.html. Last accessed 4/12/2018.

Merloni, A., P. Predehl, W. Becker, H. Böhringer, T. Boller, H. Brunner, M. Brusa, et al.: eROSITA Science Book. Mapping the Structure of the Energetic Universe. arXiv:1209.3114 [*Astro-Ph.HE*], 2012. http://arxiv.org/abs/1209.3114. Last accessed 11/21/2018.

Merloni, Andrea, Kirpal Nandra, and Peter Predehl: eROSITA's X-Ray Eyes on the Universe. *Nature Astronomy*, 2020, 1–3, 1–3. doi:10.1038/s41550-020-1133-0.

Mersits, Ulrike: From Cosmic-Ray and Nuclear Physics to High-Energy Physics. In: Armin Hermann, John Krige, Ulrike Mersits, and Lanfranco Belloni (eds.): *History of CERN. Launching the European Organization for Nuclear Research*. Amsterdam: North-Holland 1987, 3–52.

Mészáros, Péter, Derek B. Fox, Chad Hanna, and Kohta Murase: Multi-Messenger Astrophysics. *Nature Reviews Physics* 1/10 (2019), 585–599. doi:10.1038/s42254-019-0101-z.

Metropolis, Nicholas: The Beginning of the Monte Carlo Method. *Los Alamos Science* 15/Special Issue (1987), 125–130. https://library.lanl.gov/cgi-bin/getfile?00326866.pdf. Last accessed 7/19/2020.

Metzler, Gabriele: Kernphysik und Politik. Werner Heisenberg in der Wissenschafts- und Zeitgeschichte. Ein Forschungsbericht. *Historisches Jahrbuch* 115 (1995), 208–

222. http://opac.ifz-muenchen.de/webOPACClient.ifzsis/start.do?Login=woifz &Query=10="BV011522186". Last accessed 10/30/2018.

Metzler, Gabriele: Wissenschaft im Krieg. Frédéric Joliot-Curie und die deutschen Besatzer am Collège de France. In: Stefan Martens, and Maurice Vaïsse (eds.): *Frankreich und Deutschland im Krieg (November 1942–Herbst 1944). Okkupation, Kollaboration, Résistance*. Bonn: Bouvier 2000, 685–700.

Meyer, Peter, and John A. Simpson: Reminiscences of Solar Flares and the Chicago Years of Reimar Lüst. *Topics in Plasma-, Astro- and Space Physics. A Volume Dedicated to Reimar Lüst on the Occasion of His 60th Birthday*. München: Max-Planck-Institut für Physik und Astrophysik, Institut Extraterrestrische Physik 1983, 117–134.

Mezger, Peter G.: The Center of the Galaxy. In: A. Reiz (ed.): *Research Programmes for the New Large Telescopes, Proceedings of the ESO/SRC/CERN Conference Held 27–31 May, 1974 in Geneva, Switzerland*. Research Programmes for the New Large Telescopes. 1974, 79–107. http://adsabs.harvard.edu/abs/1974rpnl.conf...79M. Last accessed 5/8/2020.

Mezger, Peter G., Wolfgang J. Duschl, and Robert Zylka: The Galactic Center: A Laboratory for AGN? *The Astronomy and Astrophysics Review* 7/4 (1996), 289–388. doi:10.1007/s001590050007.

Michel, Karl-Wolfgang, Ludwig Biermann, Hans-Jürgen Bolle, Erich Kirste, Till Kirsten, Horst Bücker, Hugo Fechtig, et al.: *Denkschrift Planetenforschung*. Edited by Deutsche Forschungsgemeinschaft. Boppard: Harald Boldt Verlag 1977.

Milch, Isabella: Persönliches. Zum Tode von Ewald Fünfer. *Physik Journal* 51/10 (1995), 965–965. doi:10.1002/phbl.19950511014.

Millard, Douglas: *An Overview of United Kingdom Space Activity 1957–1987*. Noordwijk: ESA Publications Division 2005.

Millikan, Max F., Universities National Bureau Committee for Economic Research, and Universities–National Bureau Committee for Economic Research: *National Economic Planning. A Conference of the Universities-National Bureau Committee for Economic Research*. New York City, NY: National Bureau of Economic Research 1967.

Mirzoyan, Razmik: 17m Diameter MAGIC Telescope Project For Sub-100 GeV Gamma Ray Astronomy. *Nuclear Physics B Proceedings Supplements* 54/3 (1997), 350–361. doi:10.1016/S0920-5632(97)00134-5.

Mirzoyan, Razmik: Brief History of Ground-Based Very High Energy Gamma-Ray Astrophysics with Atmospheric Air Cherenkov Telescopes. *Astroparticle Physics* 53 (2014), 91–99. doi:10.1016/j.astropartphys.2013.11.004.

Mirzoyan, Razmik: Early Days of Cherenkov Emission and Milestones. Presented at the SPSAS School, San Paulo, 5/23/2017. http://www.astro.iag.usp.br/~highenastro/Talks/Lecture_III_Razmik_Mirzoyan_1.pdf. Last accessed 12/9/2018.

Mirzoyan, Razmik: Eckart Lorenz 1938–2014. *CERN Courier* 55/1 (2015), 40–40. https://cds.cern.ch/record/1983144. Last accessed 12/18/2018.

Mirzoyan, Razmik, R. Kankanian, F. Krennrich, N. Müller, H. Sander, P. Sawallisch, F. Aharonian, et al.: The First Telescope of the HEGRA Air Cherenkov Imaging Telescope Array. *Nuclear Instruments and Methods in Physics Research Section A: Accelerators, Spectrometers, Detectors and Associated Equipment* 351/2 (1994), 513–526. doi:10.1016/0168-9002(94)91381-1.

Mirzoyan, Razmik, and Christian Spiering: Nachruf auf Eckart Lorenz. *Physik Journal* 13/12 (2014), 50–50. www.pro-physik.de/details/articlePdf/7074441/issue.html. Last accessed 8/14/2018.

Mittermaier, Bernhard, and Bernd-A. Rusinek (eds.): *Leo Brandt (1908–1971): Ingenieur—Wissenschaftsförderer—Visionär*. Jülich: Forschungszentrum, Zentralbibliothek 2009.

Mittermaier, Bernhard, and Bernd-A. Rusinek: *Leo Brandt (1908–1971). Ingenieur—Wissenschaftsförderer—Visionär. Wissenschaftliche Konferenz zum 100. Geburtstag des nordrhein-westfälischen Forschungspolitikers und Gründers des Forschungszentrums Jülich*. Jülich: Forschungszentrum Jülich 2009.

Mitzner, Veera: *European Union Research Policy. Contested Origins*. London: Palgrave Macmillan 2020. doi:10.1007/978-3-030-41395-8.

Miyoshi, M., S. Kameno, and H. Falcke: A Proposal for Constructing a New VLBI Array, Horizon Telescope 289 (2003), 33–36.

Mizuno, Jun, Kenneth A. Strain, Peter G. Nelson, J. M. Chen, Roland F. Schilling, Albrecht Rüdiger, Walter Winkler, et al.: Resonant Sideband Extraction: A New Configuration for Interferometric Gravitational Wave Detectors. *Physics Letters A* 175/5 (1993), 273–276. doi:10.1016/0375-9601(93)90620-F.

Moll, Eberhard: The Franco-German High Flux Reactor and Its Facilities for Nuclear Research. In: János Erő, and J. Szűcs (eds.): *Nuclear Structure Study with Neutrons. Proceedings of the International Conference on Nuclear Structure Study with Neutrons. Budapest, 31 July–5 August 1972*. Boston, MA: Springer 1974, 313–326. doi:10.1007/978-1-4613-4499-5_14.

Möller, Horst: *Franz Josef Strauß. Herrscher und Rebell*. München: Piper 2015.

Moran, J. M.: Thirty Years of VLBI: Early Days, Successes, and Future. *International Astronomical Union Colloquium* 164 (1998), 1–10. doi:10.1017/S0252921100044353.

Morrison, Philip: On Gamma-Ray Astronomy. *Il Nuovo Cimento* 7/6 (1958), 858–865. doi:10.1007/BF02745590.

Moss, G. E., L. R. Miller, and R. L. Forward: Photon-Noise-Limited Laser Transducer for Gravitational Antenna. *Applied Optics* 10/11 (1971), 2495–2498. doi:10.1364/AO.10.002495.

Mossavi, Kasem, Martin R. Hewitson, Stefan Hild, Frank Seifert, U. Weiland, Joshua R. Smith, Harald Lück, et al.: A Photon Pressure Calibrator for the GEO 600 Gravitational Wave Detector. *Physics Letters A* 353/1 (2006), 1–3. doi:10.1016/j.physleta.2005.12.053.

Mössbauer, Rudolf L.: Kernresonanzabsorption von Gammastrahlung in Ir191. *Die Naturwissenschaften* 45/22 (1958), 538–539. doi:10.1007/BF00632050.

Motley, Robert: *Q Machines*. Oxford: Elsevier Science 1975.

Moulin, Hervé: *La France dans l'Espace 1959–1979. Contribution à l'effort spatial européen*. Vol. 37. Nordwijk: ESA Publications Division 2006.

MPI für Kernphysik: *50 Jahre Max-Planck-Institut für Kernphysik. Von Kernphysik und Kosmochemie zu Quantendynamik und Astroteilchenphysik*. Heidelberg: MPIK 2008.

Muir-Harmony, Teasel: American Foreign Policy and the Space Race. *Oxford Research Encyclopedia of American History*, 2017. doi:10.1093/acrefore/9780199329175.013.274.

Mulcahy, Kevin V.: Walt Rostow as National Security Adviser, 1966–69. *Presidential Studies Quarterly* 25/2 (1995), 223–236. https://www.jstor.org/stable/27551419. Last accessed 5/29/2019.

Müller, Ewald: Computational Problems in Supernova Simulations. *Computer Physics Communications* 44/3 (1987), 271–277. doi:10.1016/0010-4655(87)90082-8.

National Academy of Sciences: *Ground-Based Astronomy. A Ten-Year Program. A Report Prepared by the Panel on Astronomical Facilities for the Committee on Science and Public Policy of the National Academy of Sciences*. Washington, D.C.: The National Academy Press 1964.

National Security Archive: Preoccupations with West Germany's Nuclear Weapons Potential Shaped Kennedy-Era Diplomacy, 2/2/2018. https://nsarchive.gwu.edu/briefing-book/nuclear-vault/2018-02-02/german-nuclear-question-nonproliferation-treaty. Last accessed 6/10/2019.

Nesterova, N. M., and A. E. Chudakov: On the Observation of Cerenkov Radiation Accompanying Broad Atmospheric Showers of Cosmic Rays. *Journal of Experimental and Theoretical Physics* 1/2 (1955), 388–389. http://www.jetp.ac.ru/cgi-bin/e/index/e/1/2/p388?a=list. Last accessed 11/9/2018.

Nesvizhevsky, Valery, and Jacques Villain: The Discovery of the Neutron and Its Consequences (1930–1940). *Comptes Rendus Physique* 18/9 (2017), 592–600. doi:10.1016/j.crhy.2017.11.001.

Neufeld, Michael J.: The Nazi Aerospace Exodus. Towards a Global, Transnational History. *History and Technology* 28/1 (2012), 49–67.

Newell, Catherine L.: The Strange Case of Dr. von Braun and Mr. Disney. Frontierland, Tomorrowland, and America's Final Frontier. *Journal of Religion and Popular Culture* 25/3, 416–429.

Newell, Homer Edward: *Beyond the Atmosphere. Early Years of Space Science*. Vol. NASA SP-4211. Washington, DC: NASA 1980.

Nickel, Dietmar K.: *Brückenpfeiler. Vierzig Jahre wissenschaftliche Zusammenarbeit zwischen Deutschland und Israel*. München: Minerva Stiftung 1998.

Nickel, Dietmar K., and Ulrich Schmidt-Rohr: Wolfgang Gentner und die Begründung der deutsch-israelischen Wissenschaftsbeziehungen. In: Dieter Hoffmann, and Ulrich Schmidt-Rohr (eds.): *Wolfgang Gentner. Festschrift zum 100. Geburtstag.* Berlin: Springer 2006, 147–170.

Nobelprize.org. Nobel Media AB 2014.: The Nobel Prize in Physics 2002. https://www.nobelprize.org/nobel_prizes/physics/laureates/2002/. Last accessed 3/31/2018.

Nothnagel, Axel, Wolfgang Schlüter, and Hermann Seeger: Die geodätische VLBI in Deutschland. *Zeitschrift für Geodäsie* 129/4 (2004), 219–226.

Nye, Mary Jo: *Blackett. Physics, War, and Politics in the Twentieth Century.* Cambridge, MA: Harvard University Press 2004.

Oexle, Otto Gerhard: *The British Roots of the Max-Planck-Gesellschaft.* Translated by Jane Rafferty. London: German Historical Institute 1995.

Oexle, Otto Gerhard: Wie in Göttingen die Max-Planck-Gesellschaft entstand. In: Max-Planck-Gesellschaft zur Förderung der Wissenschaften e.V. (ed.): *Max-Planck-Gesellschaft Jahrbuch 1994.* Göttingen: Vandenhoeck & Ruprecht 1994, 43–60.

OHB: *Faszination Raumfahrt. Geschäftsbericht 2018.* Bremen 2016. https://www.ohb-system.de/files/images/mediathek/downloads/OHB_GB_2016_dt_s.pdf. Last accessed 10/15/2020.

Oke, J.B.: Absolute Energy Distribution in the Optical Spectrum of 3C 273. *Nature* 197/4872 (1963), 1040–1041. doi:10.1038/1971040b0.

Olivotto, Cristina: The G-Stack Collaboration (1954): An Experiment of Transition. *Historical Studies in the Natural Sciences* 39/1 (2009), 63–103. doi:10.1525/hsns.2009.39.1.63.

Orchiston, Wayne (ed.): *The New Astronomy. Opening the Electromagnetic Window and Expanding Our View of Planet Earth. A Meeting to Honor Woody Sullivan on His 60th Birthday.* Dordrecht: Springer 2005.

Oreskes, Naomi, and John Krige (eds.): *Science and Technology in the Global Cold War.* Cambridge, MA: MIT Press 2014.

Ormes, Jonathan F. (ed.): *Centenary Symposium 2012: Discovery of Cosmic Rays, 26–28 June 2012, Denver, Colorado, USA. AIP Conference Proceedings.* Vol. 1516. New York, NY: American Institute of Physics 2013.

Osietzki, Maria: The Ideology of Early Particle Accelerators: An Association between Knowledge and Power. In: Monika Renneberg, and Mark Walker (eds.): *Science, Technology and National Socialism.* Cambridge: Cambridge University Press 1994, 255–270.

Osietzki, Maria: *Wissenschaftsorganisation und Restauration. Der Aufbau ausseruniversitärer Forschungseinrichtungen und die Gründung des westdeutschen Staates.* Köln: Böhlau 1984.

Osterbrock, Donald E.: *Walter Baade. A Life in Astrophysics.* Princeton, NJ: Princeton University Press 2001.

Oswatitsch, K., and K. Wieghardt: Ludwig Prandtl and His Kaiser-Wilhelm-Institut. *Annual Review of Fluid Mechanics* 19/1 (1987), 1–25. doi:10.1146/annurev.fl.19.010187.000245.

Pacini, Franco: Energy Emission from a Neutron Star. *Nature* 216/5115 (1967), 567–568. doi:10.1038/216567a0.

Paetzold, Hans-Karl: Erich Regener. A Pioneer of Geophysical Research. In: Wilfried Schröder, and International Association of Geomagnetism and Aeronomy (eds.): *Historical Events and People in Geosciences. Selected Papers from the Symposia of the Interdivisional Commission on History of IAGA during the IUGG General Assembly, Held in Hamburg, 1983*. Frankfurt am Main: Peter Lang 1985, 59–63.

Paetzold, Hans-Karl, Georg Pfotzer, and Erwin Schopper: Erich Regener als Wegbereiter der extraterrestrischen Physik. *Zur Geschichte der Geophysik. Festschrift zur 50jährigen Wiederkehr der Gründung der Deutschen Geophysikalischen Gesellschaft*. Berlin Heidelberg: Springer 1974, 167–188.

Palme, Herbert: Heinrich Wänke und die Erforschung des Mondes und der terrestrischen Planeten. In: Horst Kant, and Carsten Reinhard (eds.): *100 Jahre Kaiser-Wilhelm-/ Max-Planck-Institut für Chemie (Otto-Hahn-Institut). Facetten seiner Geschichte*. Berlin: Archiv der Max-Planck-Gesellschaft 2012, 203–239.

Palme, Herbert: Cosmochemistry along the Rhine. *Geochemical Perspectives* 7/1 (2018), 1–116. doi:10.7185/geochempersp.7.1.

Paneth, Friedrich A.: Scientific Research in the British Zone of Germany. *Nature* 161/4084 (1948), 191–192. doi:10.1038/161191a0.

Paneth, Friedrich A., P. Reasbeck, and K.I. Mayne: Production by Cosmic Rays of Helium-3 in Meteorites. *Nature* 172/4370 (1953), 200–201. doi:10.1038/172200a0.

Paneth, Friedrich A., P. Reasbeck, and K.I. Mayne: Helium 3 Content and Age of Meteorites. *Geochimica et Cosmochimica Acta* 2/5 (1952), 300–303. doi:10.1016/0016-7037(52)90013-6.

Paradijs, Jan van, Paul J. Groot, Titus Galama, Chryssa Kouveliotou, R.G. Strom, John H. Telting, R.G.M. Rutten, et al.: Transient Optical Emission from the Error Box of the γ-Ray Burst of 28 February 1997. *Nature* 386/6626 (1997), 686–689. doi:10.1038/386686a0.

Parker, Eugene N.: John Alexander Simpson. November 3, 1916–August 31, 2000. *Biographical Memoirs. Volume 81*. Washington, DC: The National Academies Press 2002, 318–339.

Parker, Eugene N.: A History of the Solar Wind Concept. In: Johan A. M. Bleeker, Johannes Geiss, and Martin C. E. Huber (eds.): *The Century of Space Science*. Dordrecht: Kluwer Academic Publishers 2001, 225–255.

Parker, Eugene N.: John Alexander Simpson. *Physics Today* 53/12 (2000), 83–84. doi:10.1063/1.4808481.

Parker, Eugene N.: Coronal Expansion and Solar Corpuscular Radiation. In: C. C. Chang, and S. S. Huang (eds.): *Proceedings of the Plasma Space Science Symposium. Held at the Catholic University of America Washington, D.C., June 11–14, 1963*. New York, NY: Springer 1965, 99–114.

Parker, Eugene: Dynamics of the Interplanetary Gas and Magnetic Fields. *Astrophysical Journal* 128 (1958), 664–676. doi:10.1086/146579.

Pash, Boris T.: *The Alsos Mission*. New York, NY: Award House 1969.

Paul, Ewald: 50 Years of Experimental Particle Physics in Bonn. A Personal Recollection. *European Physical Journal H* 38/4 (2013), 471–506. doi:10.1140/epjh/e2013-30028-7.

Pauli, Wolfgang: *Wissenschaftlicher Briefwechsel mit Bohr, Einstein, Heisenberg u.a. / Scientific Correspondence with Bohr, Einstein, Heisenberg a.o. Teil/Part IV: 1957–1958*. Edited by Karl von Meyenn. Vol. 4. Berlin: Springer 2005.

Pauli, Wolfgang: On the Earlier and More Recent History of the Neutrino. In: K. Winter (ed.): *Neutrino Physics*. Cambridge: Cambridge University Press 1991, 1–25.

Peacock, Vita S.: *We, the Max Planck Society. A Study of Hierarchy in Germany*. London: Doctoral thesis, University College London 2014.

Pearce, Kimber Charles: Narrative Reason and Cold War Economic Diplomacy in W. W. Rostow's "Stages of Economic Growth." *Rhetoric and Public Affairs* 2/3 (1999), 395–414. https://www.jstor.org/stable/41940179. Last accessed 5/29/2019.

Pease, R. S., and S. Lindqvist: Hannes Olof Gösta Alfvén. 30 May 1908–2 April 1995. *Biographical Memoirs of Fellows of the Royal Society* 44 (1998), 3–19. doi:10.1098/rsbm.1998.0001.

Penzias, Arno A., and Robert W. Wilson: A Measurement of Excess Antenna Temperature at 4080 Mc/s. *Astrophysical Journal* 142 (1965), 419–421. doi:10.1086/148307.

Perkins, Donald H.: Proton Decay Experiments. *Annual Review of Nuclear and Particle Science* 34 (1984), 1–52. doi:10.1146/annurev.ns.34.120184.000245.

Perkins, Donald H.: The Remarkable History of the Discovery of Neutrino Oscillations. *The European Physical Journal H* 39/5 (2014), 505–515. doi:10.1140/epjh/e2014-50037-4.

Perl, Martin L., G. S. Abrams, A. M. Boyarski, M. Breidenbach, D. D. Briggs, F. Bulos, W. Chinowsky, et al.: Evidence for Anomalous Lepton Production in e^+-e^- Annihilation. *Physical Review Letters* 35/22 (1975), 1489–1492. doi:10.1103/PhysRevLett.35.1489.

Persson, Lars: Pierre Auger—A Life in the Service of Science. *Acta Oncologica* 35/7 (1996), 785–787. doi:10.3109/02841869609104027.

Peruzzi, Giulio, and Sofia Talas: The Italian Contributions to Cosmic-Ray Physics from Bruno Rossi to the G-Stack. A New Window into the Inexhaustible Wealth of Nature. *La Rivista Del Nuovo Cimento* 30/5 (2007), 197–257. doi:10.1393/ncr/i2007-10020-0.

Peterson, Laurence E., and J. R. Winckler: Short Gamma-Ray Burst from a Solar Flare. *Physical Review Letters* 1/6 (1958), 205–206. doi:10.1103/PhysRevLett.1.205.

Peterson, Laurence E., and John R. Winckler: Gamma-Ray Burst from a Solar Flare. *Journal of Geophysical Research* 64/7 (1959), 697–707. doi:10.1029/JZ064i007p00697.

Pfalzner, S., M.B. Davies, M. Gounelle, A. Johansen, C. Münker, P. Lacerda, S. Portegies Zwart, et al.: The Formation of the Solar System. *Physica Scripta* 90/6 (2015), 068001–068019. doi:10.1088/0031-8949/90/6/068001.

Pfotzer, Georg: Early Evolution of Coincidence Counting a Fundamental Method in Cosmic Ray Physics. In: Yataro Sekido, and Harry Elliot (eds.): *Early History of Cosmic Ray Studies. Personal Reminiscences with Old Photographs.* Dordrecht: Springer Netherlands 1985, 39–44. doi:10.1007/978-94-009-5434-2_5.

Pfotzer, Georg: On Erich Regener's Cosmic Ray Work in Stuttgart and Related Subjects. In: Yataro Sekido, and Harry Elliot (eds.): *Early History of Cosmic Ray Studies: Personal Reminiscences with Old Photographs.* Dordrecht: Springer Netherlands 1985, 75–89. doi:10.1007/978-94-009-5434-2_8.

Pfotzer, Georg: History of the Use of Balloons in Scientific Experiments. *Space Science Reviews* 13/2 (1972), 199–242. doi:10.1007/BF00175313.

Pfotzer, Georg: Kosmische Strahlung als Informationsquelle für Zustände und Vorgänge im Weltraum. In: Generalverwaltung der Max-Planck Gesellschaft zur Förderung der Wissenschaften e.V. (ed.): *Jahrbuch 1960 der Max-Planck-Gesellschaft zur Förderung der Wissenschaften e.V.* Göttingen 1960, 126–160.

Phillips, Thomas G., and Reinhard Genzel: An International Program for Submillimeter and Far-Infrared Astronomy from Space. In: B. H. Kaldeich (ed.): *From Ground-Based to Space-Borne Sub-Mm Astronomy. Proceedings of the 29th Liège International Astrophysical Colloquium Held at the Institut D'Astrophysique, Liège, Belgium, 3–5 July 1990.* Paris: European Space Agency 1990, 407–414.

Picht, Georg: *Die deutsche Bildungskatastrophe. Analyse und Dokumentation.* Olten: Walter Verlag 1964.

Pinault, Michel: *Frédéric Joliot-Curie.* Paris: Editions Odile Jacob 2000.

Pinch, Trevor: *Confronting Nature: The Sociology of Solar-Neutrino Detection.* Dordrecht: Springer Science & Business Media 2013.

Pinkau, Klaus: History of Gamma-Ray Telescopes and Astronomy. *Experimental Astronomy* 25/1–3 (2009), 157–171. doi:10.1007/s10686-009-9143-z.

Pinkau, Klaus: The Early Days of Gamma-Ray Astronomy. *Astronomy and Astrophysics Supplement* 120 (1996), 43–47. https://ui.adsabs.harvard.edu/#abs/1996A&AS..120C..43P. Last accessed 10/30/2018.

Pinkau, Klaus: Energy Determination of Electromagnetic Cascades in Nuclear Emulsions. *The Philosophical Magazine: A Journal of Theoretical Experimental and Applied Physics* 2/23 (1957), 1389–1392. doi:10.1080/14786435708243215.

Pinkau, Klaus, U. Pollvogt, W. Schmidt, and R.W. Huggett: Balloon Experiment Using Spark Chambers and an Ionization Spectrometer. *Proceedings of the 9th International Cosmic Ray Conference. London, UK.* 1965, 821–823. http://adsabs.harvard.edu/abs/1965ICRC....2..821P. Last accessed 10/30/2018.

Pontecorvo, Bruno: Neutrino Experiments and the Problem of Conservation of Leptonic Charge. *Pis'ma v Zhurnal Èksperimental'noi i Teoreticheskoi Fiziki* 53 (1967), 984–988. http://inspirehep.net/record/51319?ln=en. Last accessed 11/10/2017.

Porsche, Herbert: HELIOS-Mission. Mission Objectives, Mission Verification, Selected Results. In: Burke, W.~R. (ed.): *The Solar System and Its Exploration.* ESA 1981, 43–50.

Präsidialbüro der Max-Planck-Gesellschaft (ed.): *Mitteilungen aus der Max-Planck-Gesellschaft zur Förderung der Wissenschaften. Heft 6/1970.* München 1970.

Präsidialbüro der Max-Planck-Gesellschaft (ed.): *Mitteilungen aus der Max-Planck-Gesellschaft zur Förderung der Wissenschaften. Heft 2/1966.* München 1966.

Präsidialbüro der Max-Planck-Gesellschaft (ed.): *Mitteilungen aus der Max-Planck-Gesellschaft zur Förderung der Wissenschaften. Heft 4/1965.* München 1965.

Präsidialbüro der Max-Planck-Gesellschaft (ed.): *Mitteilungen aus der Max-Planck-Gesellschaft zur Förderung der Wissenschaften. Heft 1–2/1963.* München 1963.

Presas i Puig, Albert: Science on the Periphery. The Spanish Reception of Nuclear Energy. An Attempt at Modernity? *Minerva* 43/2 (2005), 197–218. doi:10.1007/s11024-005-2332-7.

Press, W. H., and D. N. Spergel: Capture by the Sun of a Galactic Population of Weakly Interacting, Massive Particles. *The Astrophysical Journal* 296 (1985), 679–684. doi:10.1086/163485.

Preuss, E.: The Beginnings of VLBI at the 100-m Radio Telescope. In: E. Ros, R.W. Parcas, A.P. Lobanov, and J.A. Zensus (eds.): *Proceedings of the 6th European VLBI Network Symposium.* Bonn 2002.

Price, S. D.: *History of Space-Based Infrared Astronomy and the Air Force Infrared Celestial Backgrounds Program.* AFRL-RV-HA-TR-2008-1039. Fort Belvoir, VA: Air Force Research Laboratory, Space Vehicles Directorate, Hanscom Air Force Base 2008. doi:10.21236/ADA513643.

Quirrenbach, A, and W Hackenberg: The ALFA Dye Laser System. In: N. Hubin, and H. Friedmann (eds.): *Laser Technology for Laser Guide Star Adaptive Optics Astronomy.* ESO 1997, 126–131.

Radkau, Joachim: *Geschichte der Zukunft. Prognosen, Visionen, Irrungen in Deutschland von 1945 bis heute.* München: Carl Hanser Verlag 2017.

Radkau, Joachim, and Lothar Hahn: *Aufstieg Und Fall Der Deutschen Atomwirtschaft.* München: Oekom 2013.

Radkau, Joachim: *Aufstieg und Krise der deutschen Atomwirtschaft 1945–1975. Verdrängte Alternativen in der Kerntechnik und der Ursprung der nuklearen Kontroverse.* Reinbek: Rowohlt 1983.

Raffelt, Georg, and Leo Stodolsky: New Particles from Nuclear Reactions in the Sun. *Physics Letters B* 119/4 (1982), 323–327. doi:10.1016/0370-2693(82)90680-3.

Ranucci, G. et al.: Overview and Accomplishments of the Borexino Experiment. *Journal of Physics: Conference Series* 675/1 (2016), 012036. doi:10.1088/1742-6596/675/1/012036.

Rasch, Manfred: *Geschichte des Kaiser-Wilhelm-Instituts für Kohlenforschung 1913–1943.* Weinheim: VCH 1989.

Rauchhaupt, Ulf von: Coping with a New Age. The Max Planck Society and the Challenge of Space Science in the Early 1960s. In: Max-Planck-Gesellschaft zur Förderung der Wissenschaften (ed.): *Innovative Structures in Basic Research. Ringberg-Sympsium, 4–7 October 2000.* München 2002, 197–205.

Rauchhaupt, Ulf von: Colorful Clouds and Unruly Rockets: Early Research Programs at the Max Planck Institute for Extraterrestrial Physics. *Historical Studies of the Physical and Biological Sciences* 32/1 (2001), 115–124. doi:10.1525/hsps.2001.32.1.115.

Rauchhaupt, Ulf von: *To Venture Beyond the Atmosphere. Aspects of the Foundation of the Max Planck Institute for Extraterrestrial Physics.* Berlin: Max-Planck-Institut für Wissenschaftsgeschichte 2000.

Rawer, Karl: *Meine Kinder umkreisen die Erde. Der Bericht eines Satellitenforschers.* Freiburg im Breisgau: Herder 1986.

Rechenberg, Helmut: Gentner und Heisenberg. Partner bei der Erneuerung der Kernphysik- und Elementarteilchenforschung im Nachkriegsdeutschland (1946–1958). In: Dieter Hoffmann, and Ulrich Schmidt-Rohr (eds.): *Wolfgang Gentner. Festschrift zum 100. Geburtstag.* Berlin: Springer 2006, 63–94.

Rechenberg, Helmut: The Early S-Matrix Theory and Its Propagation (1942–1952). In: Laurie M. Brown, Max Dresden, and Lillian Hoddeson (eds.): *Pions to Quarks. Particle Physics in the 1950s.* Cambridge, MA: Cambridge University Press 1989, 551–578.

Rechenberg, Helmut: Vom Atomkern Zum Kosmischen Wirbel. *Physik Journal* 1/6 (2002), 59–61. https://www.pro-physik.de/restricted-files/114961. Last accessed 4/29/2020.

Rechenberg, Helmut: *Werner Heisenberg—Die Sprache der Atome. Leben und Wirken—Eine wissenschaftliche Biographie. Die "Fröhliche Wissenschaft" (Jugend bis Nobelpreis).* Berlin: Springer-Verlag 2009.

Rechenberg, Helmut: Werner Heisenberg und das Forschungsprogramm des Kaiser-Wilhelm-Instituts für Physik (1940–1948). In: Bernhard Vom Brocke, and Hubert Laitko (eds.): *Die Kaiser-Wilhelm-/Max-Planck-Gesellschaft und ihre Institute. Studien zu ihrer Geschichte. Das Harnack-Prinzip.* Berlin: Walter de Gruyter 1996, 245–262.

Rechenberg, Helmut: Werner Heisenberg und das Kaiser-Wilhelm- (Max-Planck-) Institut für Physik. *Physikalische Blätter* 37/12 (1981), 357–364. doi:https://doi.org/10.1002/phbl.19810371206.

Reed, Bruce Cameron: *The History and Science of the Manhattan Project*. Berlin: Springer 2014. doi:10.1007/978-3-642-40297-5.

Regener, Erich: Das atmosphärische Ozon. In: Field Information Agencies Technical, and Office of Military Government for Germany (eds.): *Geophysics. Part II*. Wiesbaden: Dieterich'sche Verlagsbuchhandlung 1948, 297–307.

Regener, Erich: Intensity of Cosmic Radiation in the High Atmosphere. *Nature* 130/3279 (1932), 364–364. doi:10.1038/130364a0.

Regener, Erich: Messung der Ultrastrahlung in der Stratosphäre. *Naturwissenschaften* 20/38 (1932), 695–699. doi:10.1007/BF01494465.

Regener, Erich: Spectrum of Cosmic Rays. *Nature* 127/3198 (1931), 233–234. doi:10.1038/127233b0.

Regener, Erich: Über die chemische Wirkung kurzwelliger Strahlung auf gasförmige Körper. *Annalen der Physik* 325/10 (1906), 1033–1046. doi:10.1002/andp.19063251008.

Regener, Erich: Über die Temperatur der höchsten Atmosphärenschichten. *Die Naturwissenschaften* 29/32–33 (1941), 479–484. doi:10.1007/BF01485940.

Regener, Erich, and Alfred Ehmert: Über die Schauer der kosmischen Ultrastrahlung in der Stratosphäre. *Zeitschrift für Physik* 111/7–8 (1939), 501–507. doi:10.1007/BF01329511.

Regener, Erich, and Victor H. Regener: Ultra-Violet Solar Spectrum and Ozone in the Stratosphere. *Nature* 134 (1934), 380. doi:10.1038/134380a0.

Reinhard, Carsten: Max-Planck-Institut für Chemie–Mainz. In: Peter Gruss, and Reinhard Rürup (eds.): *Denkorte. Max-Planck-Gesellschaft und Kaiser-Wilhelm-Gesellschaft. Brüche und Kontinuitäten 1911–2011*. Dresden: Sandstein 2010, 256–265.

Reinhard, Carsten, and Horst Kant: *100 Jahre Kaiser-Wilhelm-/Max-Planck-Institut für Chemie (Otto-Hahn-Institut). Facetten seiner Geschichte*. Vol. 22. Berlin 2012.

Reinke, Niklas: *The History of German Space Policy. Ideas, Influences, and Interdependence 1923–2002*. Translated by Barry Smerin, and Barbara Wilson. Paris: Beauchesne 2007.

Renn, Jürgen, Horst Kant, and Birgit Kolboske: Stationen der Kaiser-Wilhelm-/Max-Planck-Gesellschaft. In: Jürgen Renn, Birgit Kolboske, and Dieter Hoffmann (eds.): *»Dem Anwenden muss das Erkennen vorausgehen«. Auf dem Weg zu einer Geschichte der Kaiser-Wilhelm-/Max-Planck-Gesellschaft*. 2nd ed. Berlin: epubli 2015, 5–120.

Renneberg, Monika, and Mark Walker: *Science, Technology, and National Socialism*. Cambridge: Cambridge University Press 2003.

Rese, Alexandra: *Wirkung politischer Stellungnahmen von Wissenschaftlern am Beispiel der Göttinger Erklärung zur atomaren Bewaffnung*. Vol. 835. Frankfurt am Main: Lang 1999.

Rhode, Wolfgang, K. H Becker, K Daum, F. Falkenberg, V. Fonseca, B. Funk, W. Greve, et al.: Design and Performance of the Lead-Concrete Geiger Tower Array within the HEGRA Experiment. *Nuclear Instruments and Methods in Physics Research Section A: Accelerators, Spectrometers, Detectors and Associated Equipment* 378/3 (1996), 399–409. doi:10.1016/0168-9002(96)00488-3.

Rhodes, Richard: *Dark Sun. The Making of the Hydrogen Bomb*. New York, NY: Simon & Schuster 2005.

Rhodes, Richard: *The Making of the Atomic Bomb*. 2nd ed. Simon and Schuster: London 1988.

Rieke, G. H.: History of Infrared Telescopes and Astronomy. *Experimental Astronomy* 25/1–3 (2009), 125–141. doi:10.1007/s10686-009-9148-7.

Riordan, Michael, Lillian Hoddeson, and Adrienne W. Kolb: *Tunnel Visions: The Rise and Fall of the Superconducting Super Collider*. Chicago, IL: University of Chicago Press 2015.

Roberts, Arthur: Development of the Spark Chamber. A Review. *Review of Scientific Instruments* 32/5 (1961), 482–485. doi:10.1063/1.1717420.

Robertson, N. A., R. W. P. Drever, I. Kerr, and J. Hough: Passive and Active Seismic Isolation for Gravitational Radiation Detectors and Other Instruments. *Journal of Physics E: Scientific Instruments* 15/10 (1982), 1101–1105. doi:10.1088/0022-3735/15/10/032.

Robinson, Ivor, Alfred Schild, and Engelbert Schücking (eds.): *Quasi-Stellar Sources and Gravitational Collapse, Proceedings of the 1st Texas Symposium on Relativistic Astrophysics*. Chicago, IL: Chicago University Press 1965.

Rogalski, A.: History of Infrared Detectors. *Opto-Electronics Review* 20/3 (2012), 279–308. doi:10.2478/s11772-012-0037-7.

Roka, E.G. von: Sonnenaktivität und kosmische Strahlung. *Zeitschrift für Naturforschung A* 6/3 (1950), 117–122. doi:10.1515/zna-1951-0301.

Rompe, Robert, and Max Steenbeck: Der Plasmazustand der Gase. In: Ferdinand Trendelenburg (ed.): *Ergebnisse der Exakten Naturwissenschaften*. Berlin: Springer 1939, 257–376. doi:10.1007/BFb0112028.

Rosen, Stephen (ed.): *Selected Papers on Cosmic Ray Origin Theories*. New York: Dover 1969.

Rossi, Bruno: *Cosmic Rays*. New York, NY: McGraw-Hill 1964.

Rostow, Walt Whitmann: *The Stages of Economic Growth, a Non-Communist Manifesto*. Cambridge: Cambridge University Press 1960.

Röthlein, Brigitte, and Uwe Schumacher: *Max-Planck-Institut für Plasmaphysik. Garching bei München*. Edited by Generalverwaltung der Max-Planck-Gesellschaft. München: Generalverwaltung der Max-Planck Gesellschaft zur Förderung der Wissenschaften e.V. 1977.

Rubbia, Carlo: *Edoardo Amaldi: Scientific Statesman.* Vol. 91–09. Geneva: CERN 1991. doi:10.5170/CERN-1991-009.

Rüdiger, Albrecht, Roland Schilling, Lise Schnupp, David Shoemaker, Walter Winkler, Gerd Leuchs, and Karl Maischberger: The Garching 30-Meter Prototype and Plans for a Large Gravitational Wave Detector. In: Melville P. Ulmer (ed.): *13th Texas Symposium on Relativistic Astrophysics. Chicago, December 14–19, 1986.* Singapore: World Scientific 1987, 20–22.

Rüdiger, Albrecht, Roland Schilling, Lise Schnupp, Walter Winkler, Heinz Billing, and Karl Maischberger: Gravitational Wave Detection by Laser Interferometry. In: I. Ursu, and A.M. Prokhorov (eds.): *Lasers and Applications.* Bucharest: CIPPress 1983, 155–179.

Ruffini, Remo, and John Archibald Wheeler: Introducing the Black Hole. *Physics Today* 24/1 (1971), 30–41. doi:10.1063/1.3022513.

Runcorn, S. K.: Prof. J. Bartels. Obituary. *Nature* 203/4947 (1964), 814–815. doi:10.1038/203814a0.

Rusinek, Bernd-A.: Europas 300-GeV-Maschine: Der grösste Teilchenbeschleuniger der Welt an einem westfälischen Standort? *Geschichte im Westen* 11/2 (1996), 135–153.

Russell, Alexander J. B.: 75th Anniversary of 'Existence of Electromagnetic–Hydrodynamic Waves'. *Solar Physics* 293/5 (2018), 83. doi:10.1007/s11207-018-1296-3.

Rydbeck, Olof E. H: *Femtio år som rymdforskare och ingenjörsutbildare. Från skånska horisonter till fjärran galaxer (Vol 2: 1951–1989).* Vol. 2. Göteborg: Chalmers tekniska högsk 1991.

Sabel, Bernhard A.: Science Reunification in Germany: A Crash Program. *Science* 260/5115 (1993), 1753–1758. www.jstor.org/stable/2881355. Last accessed 11/17/2019.

Sachse, Carola: *Die Max-Planck-Gesellschaft und die Pugwash Conference on Science and World Affairs (1955–1984).* Preprint 479. Berlin: Max Planck Institute for the History of Science 2016.

Sachse, Carola: Wissenschaft und Diplomatie. *Die Max-Planck-Gesellschaft im Feld der internationalen Politik (1945–2000).* Göttingen: Vandenhoeck & Ruprecht 2023.

Salam, Abdus: Astro-Particle-Physics. In: Remo Ruffini (ed.): *Proceedings of the Fourth Marcel Grossmann Meeting on General Relativity, Held at the University of Rome La Sapienza, 17–21 June, 1985.* Elsevier Science 1986, 3–7.

Samorski, Manfred, and Wilhelm Stamm: Detection of $2x10^{15}$ to $2x10^{16}$ eV Gamma-Rays from Cygnus X-3. *The Astrophysical Journal* 268 (1983), L17–L21. doi:10.1086/184021.

Sänger, Eugen: *Raumfahrt. Heute–Morgen–Übermorgen.* Düsseldorf: Econ-Verlag 1963.

Sänger, Eugen, and Heinz Gartmann: *Raketenantriebe. Ihre Entwicklung, Anwendung und Zukunft. Eine Einführung in das Wesen des Raketenantriebes, sowie Raketen- und Weltraumfluges.* Zürich: Schweizer Druck- und Verlagshaus 1952.

Schaeffer, Oliver, and Josef Zähringer: Helium- und Argon-Erzeugung in Eisentargets durch energiereiche Protonen. *Zeitschrift für Naturforschung A* 13/4 (1958), 346–347. doi:10.1515/zna-1958-0413.

Schaeffer, Oliver, and Josef Zähringer: High-Sensitivity Mass Spectrometric Measurement of Stable Helium and Argon Isotopes Produced by High-Energy Protons in Iron. *Physical Review* 113/2 (1959), 674–678. doi:10.1103/PhysRev.113.674.

Schein, Marcel, William P. Jesse, and E. O. Wollan: The Nature of the Primary Cosmic Radiation and the Origin of the Mesotron. *Physical Review* 59/7 (1941), 615–615. doi:10.1103/PhysRev.59.615.

Schlüter, Arnulf: Besprechungen. Cosmical Electrodynamics von H. Alfvén. Oxford. Clarendon Press, 1950. *Zeitschrift für Naturforschung A* 6/1 (1951), 55–56. doi:10.1515/zna-1951-0111.

Schlüter, Arnulf: Biermanns Göttinger Schule der Kosmischen Elektrodynamik. In: Max-Planck-Gesellschaft zur Förderung der Wissenschaften (ed.): *Ludwig Biermann. 1907–1986*. München 1988, 24–34.

Schlüter, Arnulf: Dynamik des Plasmas I. Grundgleichungen, Plasma in gekreuzten Feldern. *Zeitschrift für Naturforschung A* 5/2 (1950), 72–78. doi:10.1515/zna-1950-0202.

Schlüter, Arnulf: Dynamik des Plasmas II. Plasma mit Neutralgas. *Zeitschrift für Naturforschung A* 6/2 (1951), 73–78. doi:10.1515/zna-1951-0202.

Schlüter, Arnulf: Fusion at Venice, Remembered 32 Years Later. *Plasma Physics and Controlled Fusion* 31/10 (1989), 1725–1726. http://stacks.iop.org/0741-3335/31/i=10/a=318. Last accessed 10/30/2018.

Schlüter, Arnulf: Solare Ultrastrahlung und Erdmagnetfeld. *Zeitschrift für Naturforschung A* 6/11 (1951), 613–618. doi:10.1515/zna-1951-1108.

Schlüter, Arnulf: Wozu Plasmaphysik? In: Generalverwaltung der Max-Planck-Gesellschaft (ed.): *Jahrbuch der Max-Planck-Gesellschaft zur Förderung der Wissenschaften e.V. 1970*. Göttingen 1970, 45–61.

Schlüter, Arnulf: Zur Theorie der Kurzwellenstrahlung der Sonne. *Die Naturwissenschaften* 35/5 (1948), 154–155. doi:10.1007/BF00631599.

Schlüter, Arnulf, and Gerd Burkhardt: Ausbreitung und Ausstrahlung radiofrequenter Wellen in der Sonnenkorona. *Zeitschrift für Astrophysik* 26 (1949), 295–304. http://adsabs.harvard.edu/abs/1949ZA.....26..295B. Last accessed 10/30/2018.

Schmaltz, Florian: Aeronautical Research Under Nazi Occupation in Paris. The Aerodynamische Versuchsanstalt Göttingen and the Mobilisation of Resources, French Scientists and Engineers, 1940–1944. In: Claudine Fontanon, and Irina Gouzévich (eds.): *Les Ingénieurs Civils et La Circulation Des Savoirs En XIXe–XXe Siècles*. Paris: Garnier im Erscheinen.

Schmaltz, Florian: Luftfahrtforschung auf Expansionskurs. Die Aerodynamische Versuchsanstalt in den besetzten Gebieten. In: Sören Flachowsky, Rüdiger Hachtmann, and Florian Schmaltz (eds.): *Ressourcenmobilisierung. Wissenschaftspolitik und Forschungspraxis im NS-Herrschaftssystem.* Göttingen: Wallstein Verlag 2016, 326–382.

Schmaltz, Florian, Jürgen Renn, Carsten Reinhardt, and Jürgen Kocka (eds.): *Research Program History of the Max Planck Society. Report 2014–2017*. 2014–2017. Berlin 2017.

Schmidt, Alexander, Andreas Stamminger, and Peter Turner: DLR's Mobile Rocket Base. 47 Years of Microgravity and Technical Experiments on Suborbital Flights. *65th International Astronautical Congress (IAC 2014)*. Toronto 2014. https://elib.dlr.de/93678/. Last accessed 5/9/2019.

Schmidt, M.: 3C 273. A Star-Like Object with Large Red-Shift. *Nature* 197/4872 (1963), 1040–1040. doi:10.1038/1971040a0.

Schmidt-Rohr, Ulrich: *Die Aufbaujahre des Max-Planck-Instituts für Kernphysik*. Heidelberg: Max-Planck-Institut für Kernphysik 1998.

Schmidt-Rohr, Ulrich: *Die deutschen kernphysikalischen Laboratorium*. Heidelberg: Max-Planck-Institut für Kernphysik Heidelberg 2005.

Schmidt-Rohr, Ulrich: *Die deutschen Teilchenbeschleuniger von den 30er Jahren bis zum Ende des Jahrhunderts*. Heidelberg: Max-Planck-Institut für Kernphysik Heidelberg 2001.

Schmidt-Rohr, Ulrich: *Erinnerungen an die Vorgeschichte und die Gründerjahre des Max-Planck-Instituts für Kernphysik*. Heidelberg: Selbstverlag 1996.

Schmitz, Norbert: Léon van Hove. 10.2.1924–2.9.1990. *Berichte und Mitteilungen Max-Planck-Gesellschaft*, 1991, 99–102.

Schmucker, Georg: *Jonathan Zenneck 1871–1959. Eine technisch-wissenschaftliche Biographie*. Universität Stuttgart 1999.

Schneider, Christoph (ed.): *Forschung in der Bundesrepublik Deutschland. Beispiele, Kritik, Vorschläge*. Weinheim: Chemie 1983.

Schnier, D., Jun Mizuno, Gerhard Heinzel, Harald Lück, Albrecht Rüdiger, Roland Schilling, Mathias Schrempel, et al.: Power Recycling in the Garching 30 m Prototype Interferometer for Gravitational-Wave Detection. *Physics Letters A* 225/4 (1997), 210–216. doi:10.1016/S0375-9601(96)00893-6.

Scholz, Juliane: *Partizipation und Mitbestimmung in der Forschung. Das Beispiel Max-Planck-Gesellschaft (1945–1980)*. Berlin: GMPG-Preprint 2019.

Schönfelder, Volker: The History of Gamma-Ray Astronomy. *Astronomische Nachrichten* 323/6 (2002), 524–529. doi:10.1002/1521-3994(200212)323:6<524::AID-ASNA524>3.0.CO;2-Z.

Schönfelder, Volker (ed.): *The Universe in Gamma Rays*. Berlin: Springer 2001.

Schönfelder, Volker, and Jochen Greiner: Half-a-Century of Gamma-Ray Astrophysics at the Max-Planck Institute for Extraterrestrial Physics. 1. *The European Physical Journal H* 46/1 (2021), 27. doi:10.1140/epjh/s13129-021-00031-8.

Schopper, Erwin: Early History of Shock Waves in Heavy Ion Collisions (The Frankfurt Group). In: Walter Greiner, and Horst Stöcker (eds.): *The Nuclear Equation of State: Part A: Discovery of Nuclear Shock Waves and the EOS*. Boston, MA: Springer US 1989, 427–446. doi:10.1007/978-1-4613-0583-5_33.

Schopper, Erwin: Janossy: Cosmic Rays and Nuclear Physics/Powell und Occhialini: Nuclear Physics in Photographs. *Physikalische Blätter* 4/10 (1948), 449–450. doi:10.1002/phbl.19480041012.

Schramm, David N.: Neutrinos and Cosmology. *Nuclear Physics B—Proceedings Supplements* 38/1–3 (1995), 349–362. doi:10.1016/0920-5632(94)00769-R.

Schramm, David N., and Xiangdong Shi: Solar Neutrinos: Solar Physics and Neutrino Physics. *Nuclear Physics B—Proceedings Supplements* 35 (1994), 321–333. doi:10.1016/0920-5632(94)90271-2.

Schröder, Reinald: Die "schöne deutsche Physik" von Gustav Hertz und der "weiße Jude" Heisenberg. Johannes Starks ideologischer Antisemitismus. In: Helmuth Albrecht (ed.): *Naturwissenschaft und Technik in der Geschichte. 25 Jahre Lehrstuhl für Geschichte der Naturwissenschaften und Technik am Historischen Institut der Universität Stuttgart*. Stuttgart: Verlag für Geschichte der Naturwissenschaften und Technik 1993.

Schubert, G., C. Covey, A. Del Genio, L. S. Elson, G. Keating, A. Seiff, R. E. Young, et al.: Structure and Circulation of the Venus Atmosphere. *Journal of Geophysical Research* 85/A13 (1980), 8007–8025. doi:10.1029/JA085iA13p08007.

Schumacher, Uwe: Arnulf Schlüter. 22. August 1922–24. Juni 2011. *Jahresbericht der Max-Planck-Gesellschaft. Beilage Personalien*, 2012, 29–30, 29–30.

Schüring, Michael: *Minervas verstoßene Kinder. Vertriebene Wissenschaftler und die Vergangenheitspolitik der Max-Planck-Gesellschaft*. Göttingen: Wallstein Verlag 2006.

Schüßler, Kersten: Showdown am Nil. Der Mossad, Die Nazis, und die Raketen. Dokumentation. 45 Min. ARD 2018.

Schutz, Bernard S.: Max-Planck-Institut Für Gravitationsphysik (Albert-Einstein-Institut). *Mitteilungen Der Astronomischen Gesellschaft* 82 (1999), 649–656. https://ui.adsabs.harvard.edu/abs/1999MitAG..82..649S/abstract. Last accessed 12/1/2019.

Schwartz, Rolf: *Chronik des Max-Planck-Instituts für Radioastronomie*. Bonn: Siering 2010.

Schwartz, Rolf: Otto Hachenberg. 1.7.1911–24.3.2001. In: Max-Planck-Gesellschaft zur Förderung der Wissenschaften (ed.): *Max-Planck-Gesellschaft Jahrbuch 2002*. Göttingen: Vandenhoeck & Ruprecht 2002, 863.

Schwehm, Oliver: Fly Rocket Fly—Mit Macheten zu den Sternen. Spielfilm. 90 Min. Lunabeach TV und Media GmbH 2018.

Science New Staff: Fire Damages Gamma-Ray Observatory. *Science | AAAS*, 10/24/1997. https://www.sciencemag.org/news/1997/10/fire-damages-gamma-ray-observatory. Last accessed 3/27/2021.

Scribner, Richard A., Theodore J. Ralston, and William D. Metz: *The Verification Challenge. Problems and Promise of Strategic Nuclear Arms Control Verification*. Boston, MA: Birkhäuser 1985.

Sebesta, Lorenza: *US-European Cooperation in Space during the 1960s*. Noordwijk: ESA Publications Division 1994.

Seibert, Günther: *The History of Sounding Rockets and Their Contribution to European Space Research*. Noordwijk: ESA Publications Division 2006.

Seifert, Frank, Patrick Kwee, Michèle Heurs, Benno Willke, and Karsten Danzmann: Laser Power Stabilization for Second-Generation Gravitational Wave Detectors. *Optics Letters* 31/13 (2006), 2000–2002. doi:10.1364/OL.31.002000.

Seiler, Michael P.: *Kommandosache "Sonnengott". Geschichte der deutschen Sonnenforschung im Dritten Reich und unter alliierter Besatzung*. Frankfurt am Main: Verlag Harri Deutsch 2007.

Sengbusch, K. von: Sternentwicklung IX. Die erste hydrostatische Kontrakionsphase für einen Stern von 1 solar mass. *Zeitschrift fur Astrophysik* 69 (1968), 79–111. https://ui.adsabs.harvard.edu/#abs/1968ZA.....69...79V. Last accessed 9/20/2018.

Setti, Giancarlo, and Léon van Hove (eds.): *Cosmology, Astronomy and Fundamental Physics. Second ESO-CERN Symposium, ESO, Garching Bei München, 17–21 March 1986. Proceedings*. Garching: European Southern Observatory 1986.

Setti, Giancarlo, and Léon van Hove (eds.): *Large-Scale Structure of the Universe, Cosmology and Fundamental Physics. First ESO-CERN Symposium, CERN, Geneva, 21–25 November 1983. Proceedings*. Garching: European Southern Observatory 1984.

Shakura, Nicolai Ivanovich, and Rashid Alievich Sunyaev: Black Holes in Binary Systems. Observational Appearance. *Astronomy and Astrophysics* 24 (1973), 337–355. http://adsabs.harvard.edu/abs/1973A%26A....24..337S. Last accessed 10/31/2018.

Siedentopf, Heinrich: *Methoden und Ergebnisse der Radioastronomie*. Vol. 6, 1954.

Siegfried, Tom: 50 Years Later, It's Hard to Say Who Named Black Holes. *Science News* (2013). https://www.sciencenews.org/blog/context/50-years-later-it's-hard-say-who-named-black-holes. Last accessed 7/19/2018.

Sime, Ruth Lewin: *From Exceptional Prominence to Prominent Exception. Lise Meitner at the Kaiser Wilhelm Institute for Chemistry*. Vol. 24. Berlin: Forschungsprogramm »Geschichte der Kaiser-Wilhelm-Gesellschaft im Nationalsozialismus« 2005.

Sime, Ruth Lewin: *Otto Hahn und die Max-Planck-Gesellschaft. Zwischen Vergangenheit und Erinnerung*. Edited by Carola Sachse. Vol. 14. Berlin: Max-Planck-Gesellschaft zur Förderung der Wissenschaft 2004.

Sime, Ruth Lewin: The Politics of Forgetting. Otto Hahn and the German Nuclear-Fission Project in World War II. *Physics in Perspective* 14/1 (2012), 59–94. doi:10.1007/s00016-011-0065-6.

Simpson, John: The Cosmic Radiation. In: Johan A. M. Bleeker, Johannes Geiss, and Martin C. E. Huber (eds.): *The Century of Space Science*. Dordrecht: Kluwer Academic Publishers 2001, 117–151.

Simsarian, James: Outer Space Co-Operation in the United Nations. *American Journal of International Law* 57 (1963), 854–867.

SNO Collaboration: Direct Evidence for Neutrino Flavor Transformation from Neutral-Current Interactions in the Sudbury Neutrino Observatory. *Physical Review Letters* 89/1 (2002), 011301. doi:10.1103/PhysRevLett.89.011301.

Snyder, C.W., and M. Neugebauder: Interplanetary Solar-Wind Measurements by Mariner II BT. Proceedings of the Plasma Space Science Symposium. Held at the Catholic University of America Washington, D.C., June 11–14, 1963. Dordrecht: Springer Netherlands 1965, 67–90.

Snyder, C.W., M. Neugebauer, and U.R. Rao: The Solar Wind Velocity and Its Correlation with Cosmic-Ray Variations and with Solar and Geomagnetic Activity. *Journal of Geophysical Research* 68/24 (1963), 6361–6370. doi:10.1029/JZ068i024p06361.

Soergel, V., H. Joliot-Langevin, A. Citron, J. Adams, and O. Schaeffer: *Wolfgang Gentner 1906–1980*. Geneva: CERN 1982.

Spergel, D. N., and W. H. Press: Effect of Hypothetical, Weakly Interacting, Massive Particles on Energy Transport in the Solar Interior. *The Astrophysical Journal* 294 (1985), 663–673. doi:10.1086/163336.

Spiering, Christian: Towards High-Energy Neutrino Astronomy. A Historical Review. *The European Physical Journal H* 37/3 (2012), 515–565. doi:10.1140/epjh/e2012-30014-2.

Staab, Heinz A.: 50 Jahre Kaiser-Wilhelm/Max-Planck-Institut für Medizinische Forschung Heidelberg. In: Universitäts–Gesellschaft (ed.): *Heidelberger Jahrbücher*. Berlin: Springer 1980, 47–70.

Stamm, Thomas: *Zwischen Staat und Selbstverwaltung. Die deutsche Forschung im Wiederaufbau 1945–1965*. Köln: Wissenschaft und Politik 1981.

Stamm-Kuhlmann, Thomas: Deutsche Forschung und internationale Integration 1945–1955. In: Rudolf Vierhaus, and Bernhard vom Brocke (eds.): *Forschung im Spannungsfeld von Politik und Gesellschaft. Geschichte und Struktur der Kaiser-Wilhelm-/Max-Planck-Gesellschaft*. Stuttgart: Deutsche Verlags-Anstalt 1990, 886–909.

Staude, Jakob: Hans Elsässer. 29.3.1929–10.6.2003. In: Generalverwaltung der Max-Planck-Gesellschaft zur Förderung der Wissenschaften (ed.): *Max-Planck-Gesellschaft Jahrbuch 2004*. München: Max-Planck-Gesellschaft zur Förderung der Wissenschaften 2004, 111–112.

Stecker, F.W.: Gamma Ray Astrophysics. In: J.L. Osborne, and A.W. Wolfendale (eds.): *Origin of Cosmic Rays. Proceedings of the NATO Advanced Study Institute Held in Durham, England, August 26-September 6, 1974*. Dordrecht: Springer 1975, 267–334.

Steigman, Gary: Inflationary Cosmology. *Nature* 309/5967 (1984), 473–474. doi:10.1038/309473a0.

Steinhauser, Thomas, Hanoch Gutfreund, and Jürgen Renn: *A Special Relationship. Turning Points in the History of German-Israeli Scientific Cooperation*. Berlin: GMPG-Preprint 2017.

Stenger, V. J.: Teravolt Astronomy. *Advances in Space Research* 3/10 (1984), 139–145. doi:10.1016/0273-1177(84)90079-6.

Stepanian, A. A., B. M. Vladimirsky, Yu. I. Neshpor, and V. P. Fomin: A Search for Discrete Gamma-Ray Sources of Energy Greater than 2×10^{12} eV. *Astrophysics and Space Science* 38/2 (1975), 267–282. doi:10.1007/BF00647127.

Sterken, Veerle J.: Sir Hermann Bondi: A Journey through His Life and the Early Endeavours of Europe into Space. *Acta Astronautica* 61/1 (2007), 514–525. doi:10.1016/j.actaastro.2007.01.059.

Stern, David P.: A Brief History of Magnetospheric Physics During the Space Age. *Reviews of Geophysics* 34/1 (1996), 1–31. doi:10.1029/95RG03508.

Sullivan III, Woodruff Turner: *Cosmic Noise. A History of Early Radio Astronomy*. Cambridge: Cambridge University Press 2009.

Sullivan III, Woodruff T.: The History of Radio Telescopes, 1945–1990. *Experimental Astronomy* 25/1 (2009), 107–124. doi:10.1007/s10686-009-9140-2.

Sullivan III, Woodruff T.: *The Early Years of Radio Astronomy. Reflections Fifty Years after Jansky's Discovery*. Cambridge: Cambridge University Press 1984.

Sunyaev, R. A., V. Efremov, A. Kaniovsky, D. Stepanov, S. Unin, A. Kuznetsov, V. Loznikov, et al.: Hard X-Rays from Supernova 1987A. *Advances in Space Research* 10/2 (1990), 47–53. doi:10.1016/0273-1177(90)90117-I.

Sunyaev, Rashid Alievich, A. Kaniovskii, V. Efremov, M. Gilfanov, E. Dhurazov, S.A. Grebenev, A.V. Kuznetsov, et al.: Detection of Hard X-Rays from Supernova 1987A. Preliminary Mir-Kvant Results. *Soviet Astronomy Letters* 13/6 (1987), 431. http://adsabs.harvard.edu/abs/1987SvAL...13..431S. Last accessed 12/13/2017.

Sunyaev, Rashid Alievich, and Yakov Borisovich Zel'dovich: The Spectrum of Primordial Radiation, Its Distortions and Their Significance. *Comments on Astrophysics and Space Physics* 29 (1970), 66–73. http://articles.adsabs.harvard.edu/cgi-bin/get_file?pdfs/ComAp/0002/1970ComAp...2...66S.pdf. Last accessed 10/31/2018.

Super-Kamiokande Collaboration, S. Fukuda, Y. Fukuda, M. Ishitsuka, Y. Itow, T. Kajita, J. Kameda, et al.: Solar 8B and Hep Neutrino Measurements from 1258 Days of Super-Kamiokande Data. *Physical Review Letters* 86/25 (2001), 5651–5655. doi:10.1103/PhysRevLett.86.5651.

Sutton, Christine: ESO and CERN: A Tale of Two Organizations. *CERN Courier* 52/8 (2012), 26–30. https://cds.cern.ch/record/1734856. Last accessed 5/4/2020.

Svertilov, S. I., V. G. Stolpovskii, V. V. Bogomolov, M. I. Kudryavtsev, I. G. Mitrofanov, and H. V. Klapdor-Kleingrothaus: Hard X-Ray and Gamma-Ray Spectrometer of High

Resolution and Sensitivity on Board the International Space Station (ISS). *Advances in Space Research* 25/3–4 (2000), 901–904. doi:10.1016/S0273-1177(99)00864-9.

Swanner, Leandra A.: *Mountains of Controversy. Narrative and the Making of Contested Landscapes in Postwar American Astronomy*. Dissertation/ PhD Thesis. Cambridge, MA: Harvard University 2013.

Tananbaum, Harvey, Ethan J. Schreier, and Wallace Tucker: Riccardo Giacconi (1931–2018). *Proceedings of the National Academy of Sciences* 116/26 (2019), 12587–12589. doi:10.1073/pnas.1902399116.

Tarter, C. Bruce: *The American Lab. An Insider's History of the Lawrence Livermore National Laboratory*. Baltimore, MD: Johns Hopkins University Press 2018.

Tätigkeitsbericht der Kaiser-Wilhelm-Gesellschaft und der Max-Planck-Gesellschaft Für die Zeit vom 1. 1. 1946 bis 31. 3. 1951. Chemisch-Physikalisch-Technische Sektion. Berichte aus den Einzelnen Instituten. *Die Naturwissenschaften* 38/16 (1951), 365–372. doi:10.1007/BF00637817.

Taubes, Gary: Astronomers Turn New Eyes On the Cosmic Ray Sky. *Science* 259/5092 (1993), 177–179. doi:10.1126/science.259.5092.177.

Taylor, Joseph H., L. A. Fowler, and P. M. McCulloch: Measurements of General Relativistic Effects in the Binary Pulsar PSR1913 + 16. *Nature* 277/5696 (1979), 437–440. doi:10.1038/277437a0.

Taylor, Joseph H., A. Wolszzan, T. Damour, and J. M. Weisberg: Experimental Constraints on Strong-Field Relativistic Gravity. *Nature* 355/6356 (1992), 132–136. doi:10.1038/355132a0.

Temesváry, Stefan: Vorträge gehalten im Max-Planck-Institut für Physik. *Mitteilungen der Astronomischen Gesellschaft* 5 (1954), 99–101. http://adsbit.harvard.edu/full/seri/MitAG/0005//0000100.000.html. Last accessed 5/24/2020.

Tenorth, Heinz-Elmar: *Geschichte der Erziehung. Einführung in die Grundzüge ihrer neuzeitlichen Entwicklung*. 5. Weinheim: Juventa 2010.

Teucher, M.: Die Absorption der Nukleonenkomponente der kosmischen Strahlung in Luft zwischen Seehöhe und 4000 m. *Zeitschrift für Naturforschung A* 7/1 (1952), 61–63. doi:10.1515/zna-1952-0111.

The Borexino Collaboration: Comprehensive Measurement of pp-Chain Solar Neutrinos. *Nature* 562/7728 (2018), 505–510. doi:10.1038/s41586-018-0624-y.

The Coming of Age of Adaptive Optics. *ESO Press Release* (10/23/1995). https://www.eso.org/public/news/eso9527/. Last accessed 8/13/2020.

The CTA Consortium: Design Concepts for the Cherenkov Telescope Array CTA. An Advanced Facility for Ground-Based High-Energy Gamma-Ray Astronomy. *Experimental Astronomy* 32/3 (2011), 193–316. https://doi.org/10.1007/s10686-011-9247-0.

The Event Horizon Telescope Collaboration: First M87 Event Horizon Telescope Results. I. The Shadow of the Supermassive Black Hole. *The Astrophysical Journal Letters* 875/1 (2019), L1–L17. doi:10.3847/2041-8213/ab0ec7.

The IceCube Collaboration, Fermi-LAT, Magic, Agile, Asas-Sn, Hawc, H.e.s.s, et al.: Multimessenger Observations of a Flaring Blazar Coincident with High-Energy Neutrino IceCube-170922A. *Science* 361/6398 (2018). doi:10.1126/science.aat1378.

The MACRO Collaboration: MACRO, a Large-Area Detector at the Gran Sasso Laboratory. *Il Nuovo Cimento C* 9/2 (1986), 281–292. doi:10.1007/BF02514848.

The MAGIC Collaboration: Observation of Pulsed γ-Rays Above 25 GeV from the Crab Pulsar with MAGIC. *Science* 322/5905 (2008), 1221–1224. doi:10.1126/science.1164718.

Thomas, H.-C.: Sternentwicklung VIII. Der Helium-Flash bei einem Stern von 1. 3 Sonnenmassen. *Zeitschrift für Astrophysik* 67 (1967), 420–455. https://ui.adsabs.harvard.edu/#abs/1967ZA.....67..420T. Last accessed 10/31/2018.

Thompson, Laird A.: Adaptive Optics in Astronomy. *Physics Today* 47/12 (2008), 24. doi:10.1063/1.881406.

Thorne, Kip S.: *Black Holes & Time Warps. Einstein's Outrageous Legacy*. New York, NY: Norton 1994.

Toptica Photonics: Laser Guide Star. Improving the Resolution of Telescopes with Artificial Stars. https://www.toptica.com/applications/astronomy-geology/laser-guide-star/. Last accessed 8/13/2020.

Tousey, Richard, C. V. Strain, F. S. Johnson, and J. J. Oberly: The Solar Ultraviolet Spectrum from a V-2 Rocket. *The Astronomical Journal* 52/1162 (1947), 158–159. doi:10.1086/106028.

Trefftz, Eleonore: In memoriam. Nachruf Professor Ludwig Biermann. *Journal of Geophysics* 60/3 (1986), 204–206. http://resolver.sub.uni-goettingen.de/purl?PPN1015067948_0060. Last accessed 2/9/2021.

Trefftz, Eleonore: Obituary—Temesvary, S. *Quarterly Journal of the Royal Astronomical Society* 27/1 (1986), 129–130. http://adsabs.harvard.edu/abs/1986QJRAS..27..129T. Last accessed 1/13/2020.

Trefftz, Eleonore: Zur Berechnung der Eigenschaften von Atomen und Molekülen. *Mitteilungen aus der Max-Planck-Gesellschaft* 5/73 (1973), 311–320.

Trefftz, Eleonore: Zur Entwicklung einer rotierenden Gasmasse. *Zeitschrift für Naturforschung A* 7/1 (1952), 99–103. doi:10.1515/zna-1952-0119.

Trefftz, Eleonore: Zur Statistik der Mischkristalle und Ferromagnetica. *Zeitschrift für Physik* 127/4 (1950), 371–380. doi:10.1007/BF01329834.

Trimble, Virginia: Dark Matter in the Universe. Where, What, and Why? *Contemporary Physics* 29/4 (1988), 373–392. doi:10.1080/00107518808213765.

Trimble, Virginia, and Paul Zaich: Productivity and Impact of Radio Telescopes. *Publications of the Astronomical Society of the Pacific* 118/844 (2006), 933–938. doi:10.1086/505182.

Trischler, Helmuth: A Talkative Artefact: Germany and the Development of a European Launcher in the 1960s. In: Martin Collins, and Doug Millard (eds.): *Showcasing Space*. London: Science Museum 2005, 7–28.

Trischler, Helmuth: Nationales Innovationssystem und regionale Innovationspolitik. Forschung in Bayern im westdeutschen Vergleich 1945 bis 1980. In: Thomas Schlemmer, and Hans Woller (eds.): *Bayern im Bund. Politik und Kultur im föderativen Staat 1949 bis 1973*. München: Oldenbourg 2004, 117–194.

Trischler, Helmuth: *The "Triple Helix" of Space. German Space Activities in a European Perspective*. Noordwijk: ESA Publications Division 2002.

Trischler, Helmuth: *Dokumente zur Geschichte der Luft- und Raumfahrtforschung in Deutschland: 1900–1970*. Köln: DLR, Deutsche Forschungsanstalt für Luft- und Raumfahrt 1993.

Trischler, Helmuth: *Luft-und Raumfahrtforschung in Deutschland 1900–1970. Politische Geschichte einer Wissenschaft*. Frankfurt am Main: Campus 1992.

Trischler, Helmuth, and Dieter Hoffmann: Wolfgang Gentner und die Großforschung im bundesdeutschen und europäischen Raum. In: Dieter Hoffmann, and Ulrich Schmidt-Rohr (eds.): *Wolfgang Gentner. Festschrift zum 100. Geburtstag*. Berlin: Springer 2006, 95–120.

Trümmer einer Vision. *Der Spiegel* 32 (1992), 180–182.

Trümper, Joachim: The History of X-Ray Astronomy in Germany. *Memorie Della Società Astronomica Italiana* 84 (2013), 493–500. http://adsabs.harvard.edu/abs/2013MmSAI..84..493T. Last accessed 5/11/2018.

Trümper, Joachim: Astronomy, Astrophysics and Cosmology in the Max Planck Society. In: André Heck (ed.): *Organizations and Strategies in Astronomy*. Dordrecht: Springer Netherlands 2004, 169–187.

Trümper, Joachim: X-Ray Astronomy in Europe. In: T.D. Guyenne, and B. Battrick (eds.): *Twenty Years of the ESA Convention. Proceedings of an International Symposium, Held at Deutsches Museum, Munich, Germany, 4–6 September 1995*. Paris: European Space Agency 1995, 85–88. https://ui.adsabs.harvard.edu/abs/1995ESASP.387...85T. Last accessed 4/23/2019.

Trümper, Joachim: Bizarre Röntgenquellen im Kosmos. Erste Ergebnisse von Rosat. In: Generalverwaltung der Max-Planck Gesellschaft zur Förderung der Wissenschaften e.V. (ed.): *Max-Planck-Gesellschaft Jahrbuch 1991*. Göttingen: Vandenhoeck & Ruprecht 1991, 75–89.

Trümper, Joachim: Kosmische Röntgenquellen. In: Generalverwaltung der Max-Planck Gesellschaft zur Förderung der Wissenschaften e.V. (ed.): *Max-Planck-Gesellschaft Jahrbuch 1983*. Göttingen: Vandenhoeck & Ruprecht 1983, 81–91.

Trümper, Joachim, Bernd Aschenbach, and Heinrich Brauninger: Development Of Imaging X-Ray Telescopes At Max-Planck-Institut Garching. *Proc. SPIE 0184, Space Optics Imaging X-Ray Optics Workshop*, (*9 August 1979*). Space Optics—Imaging X—Ray Optics Workshop. 1979, 12–19. doi:10.1117/12.957429.

Trümper, J., W. Pietsch, C. Reppin, W. Voges, R. Staubert, and E. Kendziorra: Evidence for Strong Cyclotron Line Emission in the Hard X-Ray Spectrum of Hercules X-1. *The Astrophysical Journal Letters* 219 (1978), L105–L110. doi:10.1086/182617.

Turner, Michael S.: Neutrinos and Cosmology. *AIP Conference Proceedings* 72/1 (1981), 335–355. doi:10.1063/1.33010.

Turver, K. E.: *Very High Energy Gamma Ray Astronomy*. Dordrecht: Springer Netherlands 1987.

Tyson, J. A.: Null Search for Bursts of Gravitational Radiation. *Physical Review Letters* 31/5 (1973), 326–329. doi:10.1103/PhysRevLett.31.326.

Tyson, Neil deGrasse, and Avis Lang: *Accessory to War. The Unspoken Alliance between Astrophysics and the Military*. New York, NY: W. W. Norton & Company 2018.

United Nations (ed.): *Proceedings of the Second United Nations International Conference on the Peaceful Uses of Atomic Energy. Held in Geneva 1 September–13 September 1958*. Geneva: United Nations Publication 1958.

United Nations Treaties: Treaty Banning Nuclear Weapon Tests in the Atmosphere, in Outer Space and under Water. Submitters. United States of America United Kingdom of Great Britain and Northern Ireland Union of Soviet Socialist Republics. Moscow 05/08/1963. https://treaties.un.org/pages/showDetails.aspx?objid=08000002801313d9. Last accessed 12/7/2017.

United States. Office of Scientific Research and Development: *Science, the Endless Frontier*. Washington, DC: United States Government Printing Office 1945.

Valenzuela, A., G. Haerendel, H. Föppl, F. Melzner, H. Neuss, E. Rieger, J. Stöcker, et al.: The AMPTE Artificial Comet Experiments. *Nature* 320 (1986), 700–703. doi:10.1038/320700a0.

Van Allen, James A.: What Is a Space Scientist? An Autobiographical Example. *Annual Review of Earth and Planetary Sciences* 18/1 (1997), 1–27. doi:10.1146/annurev.ea.18.050190.000245.

Van Allen, James A. Van, and Louis A. Frank: Radiation Around the Earth to a Radial Distance of 107,400 Km. 4659. *Nature* 183/4659 (1959), 430–434. doi:10.1038/183430a0.

Vasyliunas, Vytenis M.: Ian Axford. 07.01.1933–13.03.2010. *Jahresbericht Der Max-Planck-Gesellschaft*. Edited by Max-Planck Gesellschaft Beileger Personalien (2010), 15–17.

Vasyliunas, Vytenis M.: Sir Ian Axford FRS 1933–2010. *Astronomy & Geophysics* 51/3 (2010), 3.37–3.38. doi:10.1111/j.1468-4004.2010.51336_1.x.

Veres, P., P. N. Bhat, M. S. Briggs, W. H. Cleveland, R. Hamburg, B. Mailyan, C. A. Wilson-Hodge, et al.: Observation of Inverse Compton Emission from a Long γ-Ray Burst. *Nature* 575/7783 (2019), 459–463. doi:10.1038/s41586-019-1754-6.

Vernov, Sergei N., Naum L. Grigorov, George B. Khristiansen, Agasi N. Charakhchan, and Alexander E. Chudakov: From Balloons to Space Stations. In: Yataro Sekido, and Harry Elliot (eds.): *Early History of Cosmic Ray Studies. Personal Reminiscences with Old Photographs*. Dordrecht: Springer 1985, 357–374. doi:10.1007/978-94-009-5434-2_34.

Verschuur, Gerrit L.: *The Invisible Universe. The Story of Radio Astronomy*. 2nd ed. New York, NY: Springer 2007.

Vignaud, D.: The GALLEX Solar Neutrino Experiment. *Nuclear Physics B—Proceedings Supplements* 60/3 (1998), 20–29. doi:10.1016/S0920-5632(97)00498-2.

Vladimirskii, B. M., I. V. Pavlov, A. A. Stepanyan, and V. P. Fomin: Some Results of a Search for Point Sources of High-Energy Gamma Rays. *Soviet Astronomy* 16/1 (1972), 1–5. https://ui.adsabs.harvard.edu/#abs/1972SvA....16....1V. Last accessed 12/11/2018.

Vladimirsky, B. M., A. A. Stepanian, and V. P. Fomin: High-Energy Gamma-Ray Outburst in the Direction of the X-Ray Source CYG X-3. *Proceedings of the 13th International Cosmic Ray Conference, University of Denver, Denver, Colorado, August 17–30, 1973* 1 (1973), 456–460. https://ui.adsabs.harvard.edu/#abs/1973ICRC....1..456V. Last accessed 12/6/2018.

Voelter, Markus: *Once You Start Asking. Insights, Stories and Experiences from Ten Years of Reporting on Science and Engineering*, 2020.

Voges, W., B. Aschenbach, Th. Boller, H. Bräuninger, U. Briel, W. Burkert, K. Dennerl, et al.: The ROSAT All-Sky Survey Bright Source Catalogue. *Astronomy and Astrophysics* 349/2 (1999), 389–405. http://cdsads.u-strasbg.fr/abs/1999A%26A...349..389V. Last accessed 10/21/2018.

Vogt, Annette: Astronominnen in Berlin und Potsdam. In: Wolfgang R. Dick, and Klaus Fritze (eds.): *300 Jahre Astronomie in Berlin und Potsdam. Eine Sammlung von Aufsätzen aus Anlaß des Gründungsjubiläums der Berliner Sternwarte*. Frankfurt am Main: Thun 2000, 121–141.

Voigt, Hans-Heinrich: Nachruf auf Otto Heckmann. *Mitteilungen der Astronomischen Gesellschaft* 60 (1983), 9–12. https://ui.adsabs.harvard.edu/#abs/1983MitAG..60....9V. Last accessed 10/30/2018.

Voigt, Hans-Heinrich, Fr. Becker, H. Elsässer, W. Fricke, and E. Kirste: *Denkschrift zur Lage der Astronomie*. Wiesbaden: Steiner 1962.

Völk, Heinrich J.: Max-Planck-Institut für Kernphysik—Astrophysik. Jahresbericht für 1997. *Mitteilungen der Astronomischen Gesellschaft Hamburg* 81 (1998), 469–484. https://ui.adsabs.harvard.edu/#abs/1998MitAG..81..469V. Last accessed 9/26/2018.

Völk, Heinrich J. (ed.): *Facetten der Astronomie*. Leipzig: Barth 1993.

Völk, Heinrich J.: Cosmic-Ray Acceleration and Transport, and Diffuse Galactic γ-Ray Emission. *Space Science Reviews* 36/1 (1983), 3–25. doi:10.1007/BF00171897.

Völk, Heinrich J., Peter Biermann, Hugo Fechtig, Franz Kneer, Fritz Manfred Neubauer, Dieter Reimers, Manfred Schüssler, et al.: *Denkschrift Astronomie*. Weinheim: VCH 1987.

Völk, Heinrich J., and M. Forman: Cosmic Rays and Gamma-Rays from OB Stars. *The Astrophysical Journal* 253 (1982), 188–198. doi:10.1086/159623.

Völter, Markus: Alfred Krabbe, Thomas Keilig, Christian Fischer, Dörte Mehlert, and Zaheer Ali. 189—SOFIA Part 1, Basics. Podcast. 2015. Omega Tau. http://omegataupodcast.net/189-sofia-part-1-basics/. Last accessed 10/31/2018.

Vorobiov, S., J. Bolmont, P. Corona, E. Delagnes, F. Feinstein, D. Gascón, J.-F. Glicenstein, et al.: NECTAr: New Electronics for the Cherenkov Telescope Array. *Nuclear Instruments and Methods in Physics Research Section A* 639/1 (2011), 62–64. doi:10.1016/j.nima.2010.08.112.

Votano, Lucia: Origin and Status of the Gran Sasso INFN Laboratory. *Modern Physics Letters A* 29/36 (2014), 1430040. doi:10.1142/S0217732314300407.

Wäffler, Hermann: Die kosmische Strahlung. Ergebnisse und Probleme. In: Generalverwaltung der Max-Planck-Gesellschaft (ed.): *Jahrbuch 1959 der Max-Planck-Gesellschaft zur Förderung der Wissenschaften e.V.* Göttingen 1959, 202–222.

Walker, Mark: Physics, History, and the German Atomic Bomb. *Berichte Zur Wissenschaftsgeschichte* 40/3 (2017), 1–18. doi:10.1002/bewi.201701817.

Walker, Mark: "Mit der Bombe leben"—Carl Friedrich von Weizsäckers Weg von der Physik zur Bombe. In: Klaus Hentschel, and Dieter Hoffmann (eds.): *Carl Friedrich von Weizsäcker. Physik–Philosophie–Friedensforschung. Leopoldina-Symposium vom 20. bis 22. Juni 2012 in Halle (Saale)*. Stuttgart: Wissenschaftliche Verlagsgesellschaft 2014, 343–356.

Walker, Mark: *Eine Waffenschmiede? Kernwaffen- und Reaktorforschung am Kaiser-Wilhelm-Institut für Physik*. Ergebnisse. Vorabdrucke aus dem Forschungsprogramm »Geschichte der Kaiser-Wilhelm-Gesellschaft im Nationalsozialismus«, 26. Berlin: Forschungsprogramm »Geschichte der Kaiser-Wilhelm-Gesellschaft im Nationalsozialismus«. Max-Planck-Institut für Wissenschaftsgeschichte 2005.

Walker, Mark: *Nazi Science. Myth, Truth, and the German Atomic Bomb*. New York, NY: Plenum Press 1995.

Walker, Mark: Physics and Propaganda: Werner Heisenberg's Foreign Lectures under National Socialism. *Historical Studies in the Physical and Biological Sciences* 22/2 (1992), 339–389. doi:10.2307/27757685.

Walker, Mark: *Die Uranmaschine. Mythos und Wirklichkeit der deutschen Atombombe*. Berlin: Siedler Verlag 1990.

Walker, Mark: *German National Socialism and the Quest for Nuclear Power 1939–1949*. Cambridge: Cambridge University Press 1989.

Walker, Mark: National Socialism and German Physics. *Journal of Contemporary History* 24/1 (1989), 63–89. https://www.jstor.org/stable/260700. Last accessed 1/27/2020.

Wang, Zuoyue: *In Sputnik's Shadow. The President's Science Advisory Committee and Cold War America*. New Brunswick, NJ: Rutgers University Press 2008.

Wänke, Heinrich: 100 Gramm Mond nach Mainz. Interview mit Professor Heinrich Wänke vom Max-Planck-Institut für Chemie in Mainz. *Der Spiegel* 29 (1969), 105. http://www.spiegel.de/spiegel/print/d-45549231.html. Last accessed 3/1/2018.

Wänke, H.: Meteoritenalter und verwandte Probleme der Kosmochemie. In: E. Heilbronner, U. Hofmann, K.L. Schäfer, and G. Witting (eds.): *Kosmochemie. Fortschritte der Chemischen Forschung*. Berlin: Springer 1966, 322–408.

Wänke, H.: Methoden und Probleme der kosmochemischen Forschung. In: Generalverwaltung der Max-Planck-Gesellschaft (ed.): *Jahrbuch der Max-Planck-Gesellschaft zur Förderung der Wissenschaften e.V. 1966*. Göttingen 1966, 168–197.

Wänke, Heinrich, and Heinrich Hintenberger: Notizen. Helium und Neon als Reaktionsprodukte der Höhenstrahlung in Eisenmeteoriten. *Zeitschrift für Naturforschung A* 13/10 (1958), 895–897. doi:10.1515/zna-1958-1017.

Watson, Alan A.: The Discovery of Cherenkov Radiation and Its Use in the Detection of Extensive Air Showers. *Nuclear Physics B—Proceedings Supplements* 212–213 (2011), 13–19. doi:10.1016/j.nuclphysbps.2011.03.003.

Watson, Alan A.: Cygnus X-3. In: K. E. Turver (ed.): *Very High Energy Gamma Ray Astronomy*. Dordrecht: Springer Netherlands 1987, 53–61. doi:10.1007/978-94-009-3831-1_6.

Watson, Alan A.: High-Energy Astrophysics. Is Cygnus X-3 a Source of Gamma Rays or of New Particles? *Nature* 315/6019 (1985), 454–455. doi:10.1038/315454b0.

Watson, Alan A.: Ultra High Energy Cosmic Rays and Ultra High Energy γ-Rays. *Advances in Space Research* 4/2 (1984), 35–44. doi:10.1016/0273-1177(84)90290-4.

Watson, M. G.: ROSAT's View of the X-Ray Sky. Commentary on: Voges W., Aschenbach B., Boller Th., et al., 1999, A&A, 349, 389. *Astronomy & Astrophysics* 500/1 (2009), 581–582. doi:10.1051/0004-6361/200912206.

Weart, Spencer R.: Scientists in Power: France and the Origins of Nuclear Energy, 1900–1950. *Bulletin of the Atomic Scientists* 35/3 (1979), 41–50. doi:10.1080/00963402.1979.11458599.

Weber, Hermann: Max-Planck-Institut für medizinische Forschung in Heidelberg. In: Generalverwaltung der Max-Planck-Gesellschaft zur Förderung der Wissenschaften e.V. (ed.): *Jahrbuch 1961 der Max-Planck-Gesellschaft zur Förderung der Wissenschaften e. V. Teil II*. Göttingen 1962, 535–556.

Weber, Joseph: Evidence for Discovery of Gravitational Radiation. *Physical Review Letters* 22/24 (1969), 1320–1324. doi:10.1103/PhysRevLett.22.1320.

Weber, Joseph: Gravitational Radiation from the Pulsars. *Physical Review Letters* 21/6 (1968), 395–396. doi:10.1103/PhysRevLett.21.395.

Weber, Joseph: Remarks on Gravitational Experiments. *Il Nuovo Cimento* 29/4 (1963), 930–934. doi:10.1007/BF02827954.

Weekes, Trevor C.: The Atmospheric Cherenkov Technique in Very High Energy Gamma-Ray Astronomy. *Space Science Reviews* 75/1 (1996), 1–15. doi:10.1007/BF0019 5020.

Weekes, Trevor C.: TeV Radiation from Galactic Sources. *Space Science Reviews* 59/3 (1992), 315–364. doi:10.1007/BF00242089.

Weekes, Trevor C.: Very High Energy Gamma-Ray Astronomy. *Physics Reports* 160 (1988), 1–121. doi:10.1016/0370-1573(88)90177-9.

Weekes, Trevor C.: A Fast Large Aperture Camera for Very High Energy Gamma-Ray Astronomy. *17th International Cosmic Ray Conference (ICRC 1981), 13–25 Jul 1981. Paris, France*. Gif-sur-Yvette, France: Centre d'Etudes Nucleaires, Saclay 1981, 34–37. https://ui.adsabs.harvard.edu/#abs/1981ICRC....8...34W. Last accessed 12/2/2018.

Weekes, Trevor C., M.F. Cawley, D.J. Fegan, K.G. Gibbs, A.M. Hillas, P.W. Kowk, R.C. Lamb, et al.: Observation of TeV Gamma Rays from the Crab Nebula Using the Atmospheric Cerenkov Imaging Technique. *Astrophysical Journal* 342 (1989), 379–395. doi:10.1086/167599.

Weidenmüller, Hans-Arwed: Nuclear Physics in Heidelberg in the Years 1950 to 1980. Personal Recollections. *European Physical Journal H* 40/3 (2015), 279–299. doi:10 .1140/epjh/e2015-60019-4.

Weidenmüller, Hans-Arwed: Peter Brix 25.10.1918–21.01.2007. *Jahresbericht der Max-Planck-Gesellschaft* Beilage Personalien (2008), 14–15.

Weigert, A.: Sternentwicklung VI. Entwicklung mit Neutrinoverlusten und thermische Pulse der Helium-Schalenquelle bei einem Stern von 5 Sonnenmassen. *Zeitschrift fur Astrophysik* 64 (1966), 395–425. https://ui.adsabs.harvard.edu/#abs/1966ZA.... .64..395W. Last accessed 9/20/2018.

Weisel, Gary J.: Properties and Phenomena. Basic Plasma Physics and Fusion Research in Postwar America. *Physics in Perspective* 10/4 (2008), 396–437. doi:10.1007/s00016 -007-0371-1.

Weisel, Gary J.: The Plasma Archipelago. Plasma Physics in the 1960s. *Physics in Perspective* 19/3 (2017), 183–226. doi:10.1007/s00016-017-0205-8.

Weiss, Burghard: The "Minerva" Project. The Accelerator Laboratory at the Kaiser Wilhelm Institute / Max Planck Institute for Chemistry. Continuity in Fundamental Research. In: Monika Renneberg, and Mark Walker (eds.): *Science, Technology and National Socialism*. Cambridge: Cambridge University Press 2003, 271–290.

Weisskopf, Victor F.: Wolfgang Gentner—ein Forscherleben in unserer Zeit. In: Generalverwaltung der Max-Planck-Gesellschaft (ed.): *Max-Planck-Gesellschaft. Berichte und Mitteilungen. Gedenkfeier Wolfgang Gentner*. München 1981, 23–27.

Weisskopf, Victor F., and Willibald Jentschke: *Die Zukunft der Elementarteilchenforschung. Das Deutsche Elektronen-Synchrotron (DESY) Eigenschaften und Forschungsmöglichkeiten.* Vol. 153. Wiesbaden: VS Verlag für Sozialwissenschaften 1965.

Weizsäcker, Carl F. von: Werner Heisenberg 5.12.1901–1.2.1976. In: Generalverwaltung der Max-Planck-Gesellschaft (ed.): *Berichte und Mitteilungen Max-Planck-Gesellschaft. Nachrichten, Personalien, Würdigungen 1.4.1975–1.4.1976.* München: Generalverwaltung der Max-Planck Gesellschaft zur Förderung der Wissenschaften e.V. 1976, 5–11.

Weizsäcker, Carl F. von: Die Rotation kosmischer Gasmassen. *Zeitschrift für Naturforschung A* 3/8–11 (1948), 524–539. https://ui.adsabs.harvard.edu/#abs/1948ZNatA..3..524W. Last accessed 11/2/2017.

Weizsäcker, Carl F. von: Kosmogonie. In: Paul Bruggencate (ed.): *Astronomie, Astrophysik und Kosmogonie.* Wiesbaden: Dieterich'sche Verlagsbuchhandlung 1948, 413–426.

Weizsäcker, Carl F. von: Zur Kosmogonie. *Zeitschrift für Astrophysik* 24/1/2 (1948), 181–206. http://adsabs.harvard.edu/abs/1948ZA.....24..181V. Last accessed 10/30/2018.

Weizsäcker, Carl F. von: Über die Entstehung des Planetensystems. *Zeitschrift für Astrophysik* 22 (1943), 319–355. http://adsabs.harvard.edu/abs/1943ZA.....22..319W. Last accessed 10/30/2018.

Weizsäcker, Carl F. von: Über Elementumwandlungen im Inneren der Sterne. II. *Physikalische Zeitschrift* 39/17/18 (1938), 633–646.

Weizsäcker, Carl F. von: Über Elementumwandlungen im Inneren der Sterne. I. *Physikalische Zeitschrift* 38/6 (1937), 176–191.

Welch, Wm. J.: Millimeter and Submillimeter Interferometry. In: Graeme D. Watt, and Adrian S. Webster (eds.): *Submillimetre Astronomy. Proceedings of the Kona Symposium on Millimetre and Submilimetre Astronomy, Held at Kona, Hawaii, October 3–6, 1988.* Dordrecht: Springer 1990, 81–86.

Werth, Karsten: A Surrogate for War—The U.S. Space Program in the 1960s. *Amerikastudien / American Studies* 49/4 (2004), 563–587.

Weyer, Johannes: *Akteurstrategien und strukturelle Eigendynamiken. Raumfahrt in Westdeutschland 1945–1965.* Göttingen: Schwartz 1993.

Weyer, Johannes: Die Raumfahrtpolitik des Bundesforschungsministeriums. In: Peter Weingart, and Niels C. Taubert (eds.): *Das Wissensministerium. Ein halbes Jahrhundert Forschungs- und Bildungspolitik in Deutschland.* Weilerswist: Velbrück Wissenschaft 2006, 64–91.

Weyer, Johannes (ed.): *Technische Visionen—politische Kompromisse. Geschichte und Perspektiven der deutschen Raumfahrt*. Berlin: Edition Sigma 1993.

Whalen, David J.: *The Rise and Fall of COMSAT. Technology, Business, and Government in Satellite Communications*. London: Palgrave MacMillan 2014.

Wheeler, John Archibald: Our Universe. The Known and the Unknown. *American Scientist* 56/1 (1968), 1–20.

Wheeler, John Archibald, J. Barclay Adams, B. Kent Harrison, Louis T. Klauder Jr., Raymond Mjolsness, Masami Wakano, and Raymond Willey: Some Implications of General Relativity for the Structure and Evolution of the Universe. *Proceedings, 11ème Conseil de Physique de l'Institut International de Physique Solvay. La Structure et l'évolution de l'univers. Rapports et Discussions. Brussels, Belgium, June 9–13, 1958*. Bruxelles: R. Stoops 1958, 96–148.

White, Simon D. M.: Fundamentalist Physics: Why Dark Energy Is Bad for Astronomy. *Reports on Progress in Physics* 70/6 (2007), 883–897. doi:10.1088/0034-4885/70/6/R01.

White, Simon D. M.: The Early ISM and Galaxy Formation. In: Harley A. Thronson Jr., and J. Michael Shull (eds.): *The Interstellar Medium in Galaxies*. Dordrecht: Springer 1990, 371–386.

White, Simon D. M., and Luiz A. N. da Costa: Real and Imaginary Clusters. In: David W. Latham (ed.): *Large-Scale Structures and Peculiar Motions in the Universe*. 1991, 285–298. https://ui.adsabs.harvard.edu/#abs/1991ASPC...15..285W. Last accessed 9/19/2018.

White, Simon D. M., R. Pallavicini, and M. R. Kundu: Radio Flares and Magnetic Fields on Weak-Line T Tauri Stars. *Astronomy and Astrophysics* 259 (1992), 149–154. https://ui.adsabs.harvard.edu/#abs/1992A&A...259..149W/abstract. Last accessed 9/19/2018.

White, Simon D. M., and Volker Springel: Fitting the Universe on a Supercomputer. *Computing in Science & Engineering* 1/2 (1999), 36–45. http://ieeexplore.ieee.org/document/753045/. Last accessed 10/31/2018.

White, Simon D. M., and Dennis Zaritsky: Models for Galaxy Halos in an Open Universe. *The Astrophysical Journal* 394/1 (1992), 1–6. doi:10.1086/171552.

Wielebinski, Richard: Reminiscences of a Radio Astronomer. *Journal of Astronomical History and Heritage* 24/4 (2021), 1103–1122. https://ui.adsabs.harvard.edu/abs/2021JAHH...24.1103W. Last accessed 1/13/2022.

Wielebinski, Richard: Ludwig Franz Benedikt Biermann: The Doyen of German Post-War Astrophysics. *Journal of Astronomical History and Heritage* 18/3 (2015), 277–284. http://www.narit.or.th/en/files/2015JAHHvol18/2015JAHH...18..277W.pdf. Last accessed 10/30/2018.

Wielebinski, Richard: Albrecht Unsöld. His Role in the Interpretation of the Origin of Cosmic Radio Emission and in the Beginning of Radio Astronomy in Germany.

Journal of the Astronomical History and Heritage 16/1 (2013), 66–80. http://adsabs.harvard.edu/abs/2013JAHH...16...67W. Last accessed 10/30/2018.

Wielebinski, Richard: Fifty Years of the Stockert Radio Telescope and What Came Afterwards. *Astronomische Nachrichten* 328/5 (2007), 388–394. doi:10.1002/asna.200710766.

Wielebinski, Richard: Sebastian von Hoerner. *Mitteilungen der Astronomischen Gesellschaft* 86 (2003), 9–10. http://adsabs.harvard.edu/abs/2003MitAG..86....9. Last accessed 10/30/2018.

Wielebinski, Richard: The New Era of Large Paraboloid Antennas. The Life of Prof. Dr. Otto Hachenberg. *Advances in Radio Science* 1 (2003), 321–324. doi:10.5194/ars-1-321-2003.

Wielebinski, Richard: The Development of Radio Astronomy from Metre to Sub-mm Wavelengths. *Acta Cosmologica* 23/2 (1997), 53–58. http://www.narit.or.th/files/JAHH/2011JAHHvol14/2011JAHH...14....3W.pdf. Last accessed 10/30/2018.

Wielebinski, Richard: The Effelsberg 100-m Radio Telescope. *Naturwissenschaften* 58/3 (1971), 109–116. doi:10.1007/BF00593099.

Wielebinski, Richard: Coordination of VLBI Observations. In: C. Jaschek, and C. Sterken (eds.): *Coordination of Observational Projects in Astronomy. Proceedings of an International Conference Held in Strasbourg, November 23–26 1987*. Cambridge: Cambridge University Press 1988, 91–96.

Wielebinski, Richard, Norbert Junkes, and Berndt H. Grahl: The Effelsberg 100-m Radio Telescope. Construction and Forty Years of Radio Astronomy. *Journal of Astronomical History and Heritage* 14/1 (2011), 3–21. http://adsabs.harvard.edu/abs/2011JAHH...14....3W. Last accessed 10/30/2018.

Will, Clifford M.: The Renaissance of General Relativity. In: P.C.W. Davies (ed.): *The New Physics*. London: Cambridge University Press 1989, 7–33.

Winkler, Walter: History of Physics (19). Fundamental Research for the Development of Gravitational Wave Detectors in Germany. *SPG Mitteilungen* 54 (2018), 14–18. https://www.sps.ch/fileadmin/doc/Mitteilungen/Mitteilungen.54.pdf. Last accessed 5/30/2018.

Winston H. Bostick: Stockholm, August 1956, Revisited (Plasma Astrophysics). *IEEE Transactions on Plasma Science* 17/2 (1989), 69–75. doi:10.1109/27.24610.

Wittel, H., Harald Lück, C. Affeldt, K. L. Dooley, Hartmut Grote, J. R. Leong, M. Prijatelj, et al.: Thermal Correction of Astigmatism in the Gravitational Wave Observatory GEO\hspace0.167em600. *Classical and Quantum Gravity* 31/6 (2014), 065008. doi:10.1088/0264-9381/31/6/065008.

Witze, Alexandra: Delays Mount for NASA's $8-Billion Hubble Successor. 7712. *Nature* 559/7712 (2018), 16–17. doi:10.1038/d41586-018-05567-2.

Witze, Alexandra: Teething Troubles at Huge Telescope. *Nature News* 499/7457 (2013), 133. doi:10.1038/499133a.

Wolfe, Audra J.: *Competing with the Soviets. Science, Technology, and the State in Cold War America.* Baltimore, MD: John Hopkins University Press 2013.

Woltjer, Lodewijk: *Europe's Quest for the Universe.* ESO *and the* VLT, ESA *and Other Projects.* Les Ulis: EDP Sciences 2006.

Woosley, S.E, and M.M Phillips: Supernova 1987A! *Science* 240/4853 (1988), 750–759. doi:10.1126/science.240.4853.750.

Worcester, David L., Antonio Faraone, and Giuseppe Zaccai: The Summer of 1954 and Paths to the Institut Laue-Langevin. *Neutron News* 28/3 (2017), 15–19. doi:10.1080/10448632.2017.1342480.

Wright, C.M., E.F. van Dishoeck, J.H. Black, H. Feuchtgruber, J. Cernicharo, E. Gonzales-Alfonso, and Th. de Graauw: ISO–SWS Observations of Pure Rotational H2O Absorption Lines Toward Orion–IRc2. *Astronomy and Astrophysics* 358 (2000), 689–700. http://adsabs.harvard.edu/abs/2000A&A...358..689W. Last accessed 10/31/2018.

X-Raying the Universe. 6. *Nature Astronomy* 4/6 (2020), 549–549. doi:10.1038/s41550-020-1137-9.

Zacher, Hans F.: Die Max-Planck-Gesellschaft im Prozeß der deutschen Einigung. In: Max-Planck-Gesellschaft zur Förderung der Wissenschaften e.V. (ed.): *Max-Planck-Gesellschaft Jahrbuch 1991.* Göttingen: Vandenhoeck & Ruprecht 1991, 11–23.

Zahn, U. von, K. H. Fricke, D. M. Hunten, D. Krankowsky, K. Mauersberger, and A. O. Nier: The Upper Atmosphere of Venus during Morning Conditions. *Journal of Geophysical Research* 85/A13 (1980), 7829–7840. doi:10.1029/JA085iA13p07829.

Zähringer, Josef: Rätselhafte Mondproben. In: Generalverwaltung der Max-Planck-Gesellschaft (ed.): *Jahrbuch der Max-Planck-Gesellschaft zur Förderung der Wissenschaften e.V. 1970.* Göttingen 1970, 169–199.

Zähringer, Josef: Isotope Chronology of Meteorites. *Annual Review of Astronomy and Astrophysics* 2 (1964), 121–148. doi:10.1146/annurev.aa.02.090164.001005.

Zank, Wolfgang: Adenauers Griff nach der Atombombe. *Die Zeit* (7/26/1996). https://www.zeit.de/1996/31/Adenauers_Griff_nach_der_Atombombe/komplettansicht. Last accessed 6/3/2019.

Zatsepin, Georgii T., and Tat'yana M. Roganova: Cosmic Ray Investigations. *Physics-Uspekhi* 52/11 (2009), 1139–1146. doi:10.3367/UFNe.0179.200911f.1203.

Zeldovich, Yakov Borisovich: Cosmology from Robertson to Today. *Physics Today* 41/3 (1988), 27–29. doi:10.1063/1.881146.

Zeldovich, Yakov Borisovich: The Universe as a Hot Laboratory for the Nuclear and Particle Physicist. *Comments on Astrophysics and Space Physics* 2 (1970), 12–17. http://adsabs.harvard.edu/abs/1970CoASP...2...12Z. Last accessed 9/19/2020.

Zensus, Anton: Radio Astronomy / VLBI. Research. *Max Planck Institute for Radio Astronomy.* https://www.mpifr-bonn.mpg.de/research/vlbi. Last accessed 3/23/2021.

Zensus, J. Anton, and Eduardo Ros: European VLBI Network: Present and Future. *Proceedings of Science*. 12th European VLBI Network Symposium and Users Meeting—7–10 October 2014. Cagliari 5, 001. doi:10.22323/1.230.0001.

Zerilli, Frank J.: Gravitational Field of a Particle Falling in a Schwarzschild Geometry Analyzed in Tensor Harmonics. *Physical Review D* 2/10 (1970), 2141–2160. doi:10.1103/PhysRevD.2.2141.

Zichichi, Antonino: The Gran Sasso Project. *AIP Conference Proceedings* 96/1 (1983), 52–64. doi:10.1063/1.33925.

Index

www.ingramcontent.com/pod-product-compliance
Lightning Source LLC
LaVergne TN
LVHW010636300926
849161LV00005B/962